Cholesterol Binding and Cholesterol Transport Proteins

SUBCELLULAR BIOCHEMISTRY

SERIES EDITOR

J. ROBIN HARRIS, University of Mainz, Mainz, Germany

ASSISTANT EDITOR

P.J. QUINN, King's College London, London, U.K.

Recent Volumes in this Series

Volume 33 **Bacterial Invasion into Eukaryotic Cells**
Tobias A. Oelschlaeger and Jorg Hacker
Volume 34 **Fusion of Biological Membranes and Related Problems**
Edited by Herwig Hilderson and Stefan Fuller
Volume 35 **Enzyme-Catalyzed Electron and Radical Transfer**
Andreas Holzenburg and Nigel S. Scrutton
Volume 36 **Phospholipid Metabolism in Apoptosis**
Edited by Peter J. Quinn and Valerian E. Kagan
Volume 37 **Membrane Dynamics and Domains**
Edited by P.J. Quinn
Volume 38 **Alzheimer's Disease: Cellular and Molecular Aspects of Amyloid beta**
Edited by R. Harris and F. Fahrenholz
Volume 39 **Biology of Inositols and Phosphoinositides**
Edited by A. Lahiri Majumder and B.B. Biswas
Volume 40 **Reviews and Protocols in DT40 Research**
Edited by Jean-Marie Buerstedde and Shunichi Takeda
Volume 41 **Chromatin and Disease**
Edited by Tapas K. Kundu and Dipak Dasgupta
Volume 42 **Inflammation in the Pathogenesis of Chronic Diseases**
Edited by Randall E. Harris
Volume 43 **Subcellular Proteomics**
Edited by Eric Bertrand and Michel Faupel
Volume 44 **Peroxiredoxin Systems**
Edited by Leopold Flohd J. Robin Harris
Volume 45 **Calcium Signalling and Disease**
Edited by Ernesto Carafoli and Marisa Brini
Volume 46 **Creatine and Creatine Kinase in Health and Disease**
Edited by Gajja S. Salomons and Markus Wyss
Volume 47 **Molecular Mechanisms of Parasite Invasion**
Edited by Barbara A. Burleigh and Dominique Soldati-Favre
Volume 48 **The Coronin Family of Proteins**
Edited by Christoph S. Clemen, Ludwig Eichinger and Vasily Rybakin
Volume 49 **Lipids in Health and Disease**
Edited by Peter J. Quinn and Xiaoyuan Wang
Volume 50 **Genome Stability and Human Diseases**
Edited by Heinz Peter Nasheuer

J. Robin Harris
Editor

Cholesterol Binding and Cholesterol Transport Proteins

Structure and Function in Health and Disease

Editor
Prof. J. Robin Harris
11 Hackwood Park
Hexham, Northumberland
United Kingdom NE46 1AX
Tel. (00)44 (0)1434 606981

ISBN 978-90-481-8621-1   e-ISBN 978-90-481-8622-8
DOI 10.1007/978-90-481-8622-8
Springer Dordrecht Heidelberg London New York

Library of Congress Control Number: 2010921984

© Springer Science+Business Media B.V. 2010
No part of this work may be reproduced, stored in a retrieval system, or transmitted in any form or by any means, electronic, mechanical, photocopying, microfilming, recording or otherwise, without written permission from the Publisher, with the exception of any material supplied specifically for the purpose of being entered and executed on a computer system, for exclusive use by the purchaser of the work.

Printed on acid-free paper

Springer is part of Springer Science+Business Media (www.springer.com)

INTERNATIONAL ADVISORY EDITORIAL BOARD

R. BITTMAN, Queens College, City University of New York, New York, USA
D. DASGUPTA, Saha Institute of Nuclear Physics, Calcutta, India
L. FLOHE, MOLISA Gmbh, Magdeburg, Germany
H. HERRMANN, German Cancer Research Center, Heidelberg, Germany
A. HOLZENBURG, Texas A&M University, Texas, USA
H-P. NASHEUER, National University of Ireland, Galway, Ireland
S. ROTTEM, The Hebrew University, Jerusalem, Israel
M. WYSS, DSM Nutritional Products Ltd., Basel, Switzerland

Frontispiece

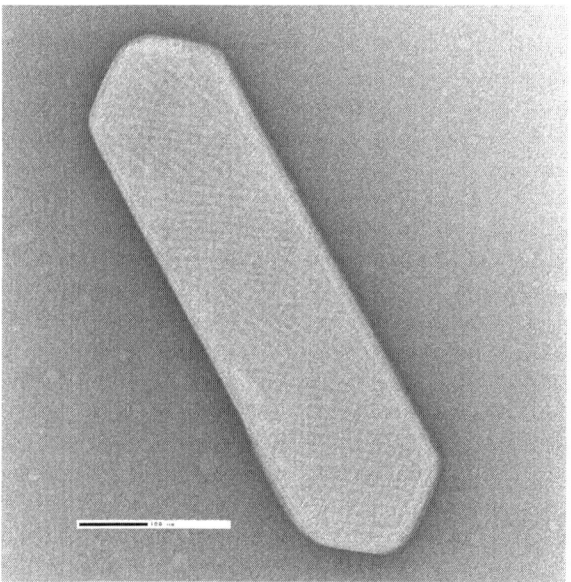

An electron micrograph showing a negatively stained cholesterol microcrystal, surface-decorated with Pyolysin domain 4 fragment (the cholesterol-binding domain), The underlying crystalline cholesterol imposes the *quasi* 2-dimensional crystal lattice of the Pyolysin domain 4 molecules (J. Robin Harris and Michael Palmer, previously unpublished data).

Preface

This book was conceived as a result of the long-standing interest of the Editor in structural and functional aspects of cholesterol, most particularly in relation to the formation and experimental use of aqueous suspensions of cholesterol microcrystals, and the role of cholesterol in Alzheimer's disease and for the study of the cholesterol-dependent cytolysins (*see* Frontispiece and Chapters 2 and 20 to 22). It also serves as an extension from the earlier volume in the *Subcellular Biochemistry* series, dealing with "Cholesterol: Its Functions and Metabolism in Biology and Medicine" (Vol. 28, ed. Robert Bittman, 1997). Although the theme of cellular membranes and cholesterol-rich plasma membrane "raft" domains appears several times throughout the book, the chapters within the book fall loosely into two sections; the opening group concentrate primarily upon soluble proteins that bind cholesterol and the later chapters place emphasis on membrane-bound proteins and membrane-active toxins that have an affinity for cholesterol. Throughout there is a strong emphasis on fundamental cellular and biochemical aspects, in particular detailed considerations of membrane "raft" domains, as well as clinical/pathological conditions involving deviant cholesterol metabolism and homeostasis.

The opening chapter is by *Gerald Gimple*, who presents a detailed account on cholesterol reporter molecules for the study of cholesterol-protein interactions. This chapter serves well as an Introduction for the book and provides a thorough technical survey, whilst at the same time introducing the recurring theme of cholesterol-containing membrane "raft" domains. Thus, this opening chapter links well with topics presented in several of the subsequent chapters. Then, my colleague *Nathaniel Milton* and I present a survey of the role of cholesterol in Alzheimer's disease and other amyloidogenic disorders. This is a rapidly advancing field, currently not without some controversy, yet may ultimately prove to be of considerable significance, particularly if clinical benefit can be proven following cholesterol lowering by statins. Several viral proteins have the ability to bind cholesterol. This topic is covered by *Cornelia Schroeder*, who considers HIV and influenza virus proteins from a strong molecular stance. *Emma de Fabiani* and colleagues then present a detailed survey on sterol-protein interactions in cholesterol and bile acid biosynthesis. The role of bile acids in metabolic signalling is give due emphasis. Although the cholesterol oxidases have been most intensively studies in bacterial systems, their likely importance in animals can be predicted in view of the bioreactivity of oxidized cholesterol products. *Alice Vrierlink* provides a detailed account of the structure and enzymic mechanism of the bacterial cholesterol oxidases. Appropriately, the oxysterol-binding proteins are then dealt with by *Neale D. Ridgeway,* who places emphasis upon the eukaryotic oxysterol binding protein (OSBP) family. The role of high density lipoprotein (HDL) in reverse cholesterol transport is given a thorough handling by *Sissel Lund-Katz and Michael C. Phillips,* who deal with the apolipoprotein components in depth. The topic of lipoprotein modification and uptake by macrophages, within the context of the pathologic role of cholesterol in atherosclerosis, is reviewed in detail by *Yury I. Miller* and colleagues. The next chapter, by *Richard M. Epand* and colleagues provides a thorough overview on the involvement of cholesterol in membrane-related phenomena. Under the title of: Cholesterol Interaction with Proteins that Partition into Membrane Domains, these authors provide a useful link between all the following chapters that have an emphasis on cell membrane cholesterol. The topic of intracellular cholesterol transport is handled by *Fiedhelm Schroeder* and colleagues, who place emphasis on caveolin and sterol carrier protein-2 in relation to cholesterol-rich microdomains. Niemann-Pick type C (NPC) disease is a cholesterol storage disease due to a defined genetic lesion. *Xiaoning Bi and Guanghong Liao* review recent achievements in the investigation of the disruption of cholesterol homeostasis-induced neurodegeneration in NPC disease, and provide new insight into the development of a potential therapeutic strategy. Sterol transport across the intestinal brush border membrane is mediated by a number of proteins. *J. Mark Brown and Liqing Yu* survey this complex situation, the first line of cholesterol entry into and passage across the intestinal enterocytes and the animal body, which can have a major impact upon atherosclerotic cardiovascular disease. The role of cholesterol in the endoplasmic reticulum (ER) is presented by *Teruo Hayashi and Tsung-Ping Su,* who place emphasis upon the sigma-1 receptor chaperones and other ER proteins in relation to the diverse ER functions, such as protein folding, compartmentalization and segregation of ER proteins, and sphingolipid biosynthesis. *Denis Corbeil* and colleagues then give a thorough account on Prominin-1, a distinct cholesterol-binding protein of the apical plasma membrane of epithelial cells. The StAR-related lipid transfer (START) domain is an evolutionary conserved protein. This topic is presented by *Pierre Lavigne* and colleagues, who review the understanding of the structure and reversible cholesterol binding mechanism of START domains. The role of membrane cholesterol in the function of G-protein coupled receptors (GPCRs) is handled by *Yamuna Devi Paila and Amitabha Chattopadhyay.* These authors consider that deciphering molecular details of the GPCR-cholesterol interaction in the membrane should lead to better insight into the overall understanding of GPCR function in health and disease. *Francisco J Barrantes* then discusses cellular aspects of the role of cholesterol in the nicotinic acetylcholine receptor (AchR). Indeed, the cholesterol content of the plasmalemma may homeostatically modulate AchR dynamics, cell-surface organization and the lifetime of receptor nanodomains, in turn exerting control over the ion permeation process. The brain contains the highest content of cholesterol of all organs of the animal body, largely contained in myelinated nerves. In their chapter on cholesterol and myelin biogeneis, *Gesine Saher and Mikael Simons* consider the role of cholesterol in both the central and peripheral nervous system. The diversity of plasma membrane ion channels is almost overwhelming! *Irena Levitan* and colleagues present a thorough account of the role of cholesterol in regulation of the major types of ion channels and discuss this in the context of the current models for channel function. As indicated above, the cholesterol-dependent cytolysins (CDCs) are toxin molecules of great personal interest. This topic is covered with a strong molecular slant by *Alejandro P. Heuck* and colleagues, in their chapter on the cholesterol-dependent cytolysin family of Gram-positive bacterila toxins. Evidence for the role of cholesterol in the activity and pore-formation by a range of other β-barrel pore-forming Gram-negative bacterial toxins is covered by my colleague *Michael Palmer* and myself. The final chapter, by *Yoshiko Ohno-Iwashita* and colleagues expands upon themes introduced earlier, in relation to the value of having specific probes for cholesterol localization. These authors describe in detail studies using non-cytolytic molecular fragments of the CDC perfringolysin, together with anti-cholesterol antibodies, as tools for membrane cholesterol localization.

Overall, when compiling the contents of this book I have attempted to include almost all topics of significance, and have been greatly encouraged by the positive responses I have received from the chapter authors, from the early contacts thorough to the preparation of the chapter manuscripts for publication.

Hexham, UK   J. Robin Harris

Contents

1. **Cholesterol–Protein Interaction: Methods and Cholesterol Reporter Molecules** 1
   Gerald Gimpl

2. **Cholesterol in Alzheimer's Disease and other Amyloidogenic Disorders** 47
   J. Robin Harris and Nathaniel G.N. Milton

3. **Cholesterol-Binding Viral Proteins in Virus Entry and Morphogenesis** 77
   Cornelia Schroeder

4. **Sterol–Protein Interactions in Cholesterol and Bile Acid Synthesis** 109
   Emma De Fabiani, Nico Mitro, Federica Gilardi, and Maurizio Crestani

5. **Cholesterol Oxidase: Structure and Function** 137
   Alice Vrielink

6. **Oxysterol-Binding Proteins** 159
   Neale D. Ridgway

7. **High Density Lipoprotein Structure–Function and Role in Reverse Cholesterol Transport** 183
   Sissel Lund-Katz and Michael C. Phillips

8. **Lipoprotein Modification and Macrophage Uptake: Role of Pathologic Cholesterol Transport in Atherogenesis** 229
   Yury I. Miller, Soo-Ho Choi, Longhou Fang, and Sotirios Tsimikas

9. **Cholesterol Interaction with Proteins That Partition into Membrane Domains: An Overview** 253
   Richard M. Epand, Annick Thomas, Robert Brasseur, and Raquel F. Epand

10	Caveolin, Sterol Carrier Protein-2, Membrane Cholesterol-Rich Microdomains and Intracellular Cholesterol Trafficking	279
	Friedhelm Schroeder, Huan Huang, Avery L. McIntosh, Barbara P. Atshaves, Gregory G. Martin, and Ann B. Kier	
11	Cholesterol in Niemann–Pick Type C disease	319
	Xiaoning Bi and Guanghong Liao	
12	Protein Mediators of Sterol Transport Across Intestinal Brush Border Membrane	337
	J. Mark Brown and Liqing Yu	
13	Cholesterol at the Endoplasmic Reticulum: Roles of the Sigma-1 Receptor Chaperone and Implications thereof in Human Diseases	381
	Teruo Hayashi and Tsung-Ping Su	
14	Prominin-1: A Distinct Cholesterol-Binding Membrane Protein and the Organisation of the Apical Plasma Membrane of Epithelial Cells	399
	Denis Corbeil, Anne-Marie Marzesco, Christine A. Fargeas, and Wieland B. Huttner	
15	Mammalian StAR-Related Lipid Transfer (START) Domains with Specificity for Cholesterol: Structural Conservation and Mechanism of Reversible Binding	425
	Pierre Lavigne, Rafael Najmanivich, and Jean-Guy LeHoux	
16	Membrane Cholesterol in the Function and Organization of G-Protein Coupled Receptors	439
	Yamuna Devi Paila and Amitabha Chattopadhyay	
17	Cholesterol Effects on Nicotinic Acetylcholine Receptor: Cellular Aspects	467
	Francisco J. Barrantes	
18	Cholesterol and Myelin Biogenesis	489
	Gesine Saher and Mikael Simons	
19	Cholesterol and Ion Channels	509
	Irena Levitan, Yun Fang, Avia Rosenhouse-Dantsker, and Victor Romanenko	
20	The Cholesterol-Dependent Cytolysin Family of Gram-Positive Bacterial Toxins	551
	Alejandro P. Heuck, Paul C. Moe, and Benjamin B. Johnson	
21	Cholesterol Specificity of Some Heptameric β-Barrel Pore-Forming Bacterial Toxins: Structural and Functional Aspects	579
	J. Robin Harris and Michael Palmer	

22	**Cholesterol-Binding Toxins and Anti-cholesterol Antibodies as Structural Probes for Cholesterol Localization**	597
	Yoshiko Ohno-Iwashita, Yukiko Shimada, Masami Hayashi, Machiko Iwamoto, Shintaro Iwashita, and Mitsushi Inomata	

Index 623

Contributors

Barbara P. Atshaves Department of Physiology and Pharmacology, Texas A&M University, TVMC College Station, TX 77843-4, USA

Francisco J. Barrantes UNESCO Chair of Biophysics and Molecular Neurobiology and Instituto de Investigaciones Bioquímicas de Bahía Blanca, C.C. 857, B8000FWB Bahía Blanca, Argentina

Xiaoning Bi Department of Basic Medical Sciences, COMP, Western University of Health Sciences, Pomona, CA 91766, USA, xbi@westernu.edu

Robert Brasseur Centre de Biophysique Moléculaire Numérique, AgroBiotech of Gembloux, ULg, Passage des déportés, 2, 5030 Gembloux, Belgium

J. Mark Brown Department of Pathology Section on Lipid Sciences, Wake Forest University School of Medicine, Winston-Salem, NC, USA

Amitabha Chattopadhyay Centre for Cellular and Molecular Biology, Council of Scientific and Industrial Research, Uppal Road, Hyderabad 500 007, India

Soo-Ho Choi Department of Medicine, University of California, San Diego, La Jolla, CA 92037-0682, USA

Denis Corbeil Tissue Engineering Laboratories, BIOTEC, Technische Universität Dresden, Tatzberg 47-49, 01307 Dresden, Germany

Maurizio Crestani "Giovanni Galli" Laboratory of Biochemistry and Molecular Biology of Lipids, Department of Pharmacological Sciences, Via Balzaretti, 9, 20133 Milan, Italy

Raquel F. Epand Department of Biochemistry and Biomedical Sciences, McMaster University, Hamilton, Ontario L8N 3Z5, Canada

Richard M. Epand Department of Biochemistry and Biomedical Sciences, McMaster University, Hamilton, Ontario L8N 3Z5, Canada, epand@mcmaster.ca

Emma De Fabiani "Giovanni Galli" Laboratory of Biochemistry and Molecular Biology of Lipids, Department of Pharmacological Sciences, Via Balzaretti, 9, 20133 Milan, Italy, emma.defabiani@unimi.it

Longhou Fang Department of Medicine, University of California, San Diego, La Jolla, CA 92037-0682, USA

Yun Fang Institute for Medicine and Engineering, University of Pennsylvania, 3340 Smith Walk, Philadelphia, PA, USA

Christine A. Fargeas Tissue Engineering Laboratories, BIOTEC, Technische Universität Dresden, Tatzberg 47-49, 01307 Dresden, Germany

Federica Gilardi "Giovanni Galli" Laboratory of Biochemistry and Molecular Biology of Lipids, Department of Pharmacological Sciences, Via Balzaretti, 9, 20133 Milan, Italy

Gerald Gimpl Institut für Biochemie, Johannes Gutenberg-University of Mainz, Johann-Joachim-Becherweg 30, D-55128 Mainz, Germany, Gimpl@uni-mainz.de

J. Robin Harris Institute of Zoology, University of Mainz, D-55099 Mainz, Germany and Institute of Cell and Molecular Biosciences, University of Newcastle, Newcastle-upon-Tyne, NE2 4HH, UK, rharris@uni-mainz.de

Masami Hayashi Research Team for Functional Genomics, Tokyo Metropolitan Institute of Gerontology, 35-2 Sakae-cho, Itabashi-ku, Tokyo 173-0015, Japan

Teruo Hayashi Cellular Pathobiology Section, Cellular Neurobiology Research Branch, Intramural Research Program, National Institute on Drug Abuse, Department of Health and Human Services, National Institutes of Health, 333 Cassell Drive, Baltimore, Maryland 21224, USA

Alejandro P. Heuck Department of Biochemistry and Molecular Biology, University of Massachusetts, Amherst, MA 01003, USA

Huan Huang Department of Physiology and Pharmacology, Texas A&M University, TVMC College Station, TX 77843-4, USA

Wieland B. Huttner Max-Planck-Institute of Molecular Cell Biology and Genetics, Pfotenhauerstrasse 108, 01307 Dresden, Germany

Mitsushi Inomata Research Team for Functional Genomics, Tokyo Metropolitan Institute of Gerontology, 35-2 Sakae-cho, Itabashi-ku, Tokyo 173-0015, Japan

Machiko Iwamoto Research Team for Functional Genomics, Tokyo Metropolitan Institute of Gerontology, 35-2 Sakae-cho, Itabashi-ku, Tokyo 173-0015, Japan

Shintaro Iwashita Faculty of Pharmacy, Iwaki Meisei University, 5-5-1 Chuodai Iino, Iwaki City, Fukushima 970-8551, Japan

Benjamin B. Johnson Department of Biochemistry and Molecular Biology, University of Massachusetts, Amherst, MA 01003, USA

Ann B. Kier Department of Pathobiology, Texas A&M University, TVMC, College Station, TX 77843-4467, USA

Contributors

Pierre Lavigne Département de Pharmacologie, Institut de Pharmacologie, Faculté de médecine et des sciences de la santé, Université de Sherbrooke 3001 12e Avenue Nord, Sherbrooke, QC, Canada J1H 5N4

Jean-Guy LeHoux Département de Biochimie, Faculté de médecine et des sciences de la santé, Université de Sherbrooke, 3001 12e Avenue Nord, Sherbrooke, QC, Canada J1H 5N4

Irena Levitan Department of Medicine, Sections of Pulmonary, Critical Care and Sleep Medicine, University of Illinois at Chicago, 840 S Wood St, Chicago 60612, USA

Guanghong Liao Department of Basic Medical Sciences, COMP, Western University of Health Sciences, Pomona, CA 91766, USA

Sissel Lund-Katz Division of Gastroenterology/Hepatology/Nutrition, Children's Hospital of Philadelphia, University of Pennsylvania School of Medicine, 3615 Civic Center Blvd, Suite 1102, Philadelphia, PA 19104-4318, USA

Gregory G. Martin Department of Physiology and Pharmacology, Texas A&M University, TVMC College Station, TX 77843-4, USA

Anne-Marie Marzesco Max-Planck-Institute of Molecular Cell Biology and Genetics, Pfotenhauerstrasse 108, 01307 Dresden, Germany

Avery L. McIntosh Department of Physiology and Pharmacology, Texas A&M University, TVMC College Station, TX 77843-4, USA

Yury I. Miller Department of Medicine, University of California, San Diego, La Jolla, CA 92037-0682, USA, yumiller@ucsd.edu

Nathaniel G. N. Milton Department of Human and Health Sciences, School of Life Sciences, University of Westminster, London W1W 6UW, UK

Nico Mitro "Giovanni Galli" Laboratory of Biochemistry and Molecular Biology of Lipids and "The Giovanni Armenise – Harvard Foundation" Laboratory, Department of Pharmacological Sciences, Via Balzaretti, 9, 20133 Milan, Italy

Paul C. Moe Department of Biochemistry and Molecular Biology, University of Massachusetts, Amherst, MA 01003, USA

Rafael Najmanivich Département de Biochimie. Faculté de médecine et des sciences de la santé Université de Sherbrooke, 3001 12e Avenue Nord, Sherbrooke, QC, Canada J1H 5N4

Yoshiko Ohno-Iwashita Research Team for Functional Genomics, Tokyo Metropolitan Institute of Gerontology, 35-2 Sakae-cho, Itabashi-ku, Tokyo 173-0015, Japan and Faculty of Pharmacy, Iwaki Meisei University, 5-5-1 Chuodai Iino, Iwaki City, Fukushima 970-8551, Japan

Yamuna Devi Paila Centre for Cellular and Molecular Biology, Council of Scientific and Industrial Research, Uppal Road, Hyderabad 500 007, India

Michael Palmer Department of Chemistry, University of Waterloo, 200 University Ave. W., Waterloo, Ontario, N2L 3G1, Canada

Michael C. Phillips Division of Gastroenterology/Hepatology/Nutrition, Children's Hospital of Philadelphia, University of Pennsylvania School of Medicine, 3615 Civic Center Blvd, Suite 1102, Philadelphia, PA 19104-4318, USA, phillipsmi@email.chop.edu

Neale D. Ridgway Departments of Pediatrics and Biochemistry & Molecular Biology, Atlantic Research Centre, Dalhousie University, 5849 University Av. Halifax, Nova Scotia, Canada B3H 4H7, nridgway@dal.ca

Victor Romanenko Department of Pharmacology and Physiology, University of Rochester Medical Center, Box 711, 601 Elmwood Ave, Rochester, NY 14642, USA

Avia Rosenhouse-Dantsker Department of Medicine, Sections of Pulmonary, Critical Care and Sleep Medicine, University of Illinois at Chicago, 840 S Wood St, Chicago 60612, USA

Gesine Saher Max-Planck-Institute for Experimental Medicine, Department of Neurogenetics, Hermann-Rein-Str. 3, Göttingen, Germany

Cornelia Schroeder Max Planck Institute for Molecular Cell Biology and Genetics, Pfotenhauerstr. 108, D-01307 Dresden, Germany, cornelia.schroeder@mpi-cbg.de

Friedhelm Schroeder Department of Physiology and Pharmacology, Texas A&M University, TVMC College Station, TX 77843-4466, USA fschroeder@cvm.tamu.edu

Mikael Simons Max-Planck-Institute for Experimental Medicine, Hermann-Rein-Str. 3, Göttingen, Germany and Department of Neurology, University of Göttingen, Robert-Koch-Str. 40, Göttingen, Germany

Yukiko Shimada Research Team for Functional Genomics, Tokyo Metropolitan Institute of Gerontology, 35-2 Sakae-cho, Itabashi-ku, Tokyo 173-0015, Japan

Tsung-Ping Su Cellular Pathobiology Section, Cellular Neurobiology Research Branch, Intramural Research Program, National Institute on Drug Abuse, Department of Health and Human Services, National Institutes of Health, 333 Cassell Drive, Baltimore, Maryland 21224, USA

Annick Thomas Centre de Biophysique Moléculaire Numérique, AgroBiotech of Gembloux, ULg, Passage des déportés, 2, 5030 Gembloux, Belgium

Sotirios Tsimikas Department of Medicine, University of California, San Diego, La Jolla, CA 92037-0682, USA

Alice Vrielink School of Biomedical, Biomolecular and Chemical Sciences, University of Western Australia, 35 Stirling Highway, Crawley, WA 6009, Australia, alice.vrielink@uwa.edu.au

Liqing Yu Department of Pathology Section on Lipid Sciences, Wake Forest University School of Medicine, Winston-Salem, Medical Center Blvd., Winston-Salem, NC 27157-1040, USA, lyu@wfubmc.edu

Chapter 1
Cholesterol–Protein Interaction: Methods and Cholesterol Reporter Molecules

Gerald Gimpl

Abstract Cholesterol is a major constituent of the plasma membrane in eukaryotic cells. It regulates the physical state of the phospholipid bilayer and is crucially involved in the formation of membrane microdomains. Cholesterol also affects the activity of several membrane proteins, and is the precursor for steroid hormones and bile acids. Here, methods are described that are used to explore the binding and/or interaction of proteins to cholesterol. For this purpose, a variety of cholesterol probes bearing radio-, spin-, photoaffinity- or fluorescent labels are currently available. Examples of proven cholesterol binding molecules are polyene compounds, cholesterol-dependent cytolysins, enzymes accepting cholesterol as substrate, and proteins with cholesterol binding motifs. Main topics of this report are the localization of candidate membrane proteins in cholesterol-rich microdomains, the issue of specificity of cholesterol– protein interactions, and applications of the various cholesterol probes for these studies.

Keywords Cholesterol binding proteins · Cyclodextrins · Fluorescent and photoreactive sterols · Polyenes · Cytolysins

Abbreviations

ACAT	acyl-coenzyme A:cholesterol acyltransferase
BCθ-toxin	a biotinylated and carlsberg protease-nicked derivative of perfringolysin O
Benzophenone-cholesterol	22-(p-benzoylphenoxy)-23,24-bisnorcholan-5-en-3β-ol
Bodipy	boron dipyrromethene(4,4-difluoro-5,7-dimethyl-4-bora-3a,4a-diazara-s-indacene)
CCKBR	cholecystokinin receptor type B
CCM	cholesterol consensus motif

G. Gimpl (✉)
Institut für Biochemie, Johannes Gutenberg-Universität, Johann-Joachim-Becherweg 30, Mainz, Germany
e-mail: Gimpl@uni-mainz.de

CDCs	cholesterol-dependent cytolysins
CRAC	cholesterol recognition/interaction amino acid consensus
DIG, (= DRM)	detergent-insoluble (= detergent resistant) glycosphingolipid-enriched membrane domains
GPCR	G protein coupled receptor
HDL	high-density lipoprotein
HPβCD	2-Hydroxypropyl)-β-cyclodextrin
5HT1A	5-hydroxytryptamine 1A
LDL	low-density lipoprotein
LDM	low-density microdomains
MβCD	methyl-β-cyclodextrin
22-NBD Cholesterol	22-(N-(7-nitrobenz-2-oxa-1,3-diazol-4-yl)amino)-23,24-bisnor-5-cholen-3β-ol
25-NBD Cholesterol	25-[N-[(7-nitro-2-1,3-benzoxadiazol-4-yl)methyl]amino]-27-norcholesterol
NMR	nuclear magnetic resonance
NPC	Niemann-Pick C
OTR	oxytocin receptor
PBR	peripheral benzodiazepine receptor (= TSPO)
SCAP	SREBP cleavage activating protein
SREBP	sterol regulatory element binding protein
SSD	sterol sensing domain
StAR	steroid acute regulatory protein
START	StAR related lipid transfer
TSPO	mitochondrial translocator protein (18 kDa) (= PBR)

1.1 Introduction

Cholesterol is a major constituent of the plasma membrane in most eukaryotic cells where it fulfils several functions. It regulates membrane fluidity, increases membrane thickness, establishes the permeability barrier of the membrane, modulates the activity of various membrane proteins, and is the precursor for steroid hormones and bile acids (Burger et al., 2000; Pucadyil and Chattopadhyay, 2006). Cholesterol is non-randomly distributed in cells and membranes (Yeagle, 1985) and plays an essential role in the formation of lateral membrane domains, often designated as 'lipid rafts' (Simons and Ikonen, 1997; Simons and Toomre, 2000).

Here, methods are described that are used to explore the binding/interaction of proteins to cholesterol. In case of membrane proteins such as receptors, transporters or ion channels, researchers often like to know whether their candidate protein is associated with cholesterol-rich microdomains or not. To address this issue, different subcellular fractionation protocols are usually performed. These approaches will be briefly compared herein. Even if a given membrane protein is shown to be excluded from cholesterol-rich microdomains, it is unlikely for any membrane protein to avoid molecular contacts with cholesterol. This is due to the fact that cholesterol is in large excess over any membrane protein within the plasma membrane. Some membrane proteins show functional changes dependent on the amounts of cholesterol. In this case, one has to discriminate between direct cholesterol–protein interactions and indirect effects caused by the influence of cholesterol on the biophysical state of the membrane. Approaches are described that help to prove the specificity of putative cholesterol–protein interactions. Various cholesterol-binding molecules are currently known. Among these are polyene compounds, cholesterol-dependent cytolysins, enzymes accepting cholesterol as substrate, and proteins with cholesterol binding motifs, many of which are covered in detail within the chapters of this book. In Table 1.1, proteins with defined or putative cholesterol binding motifs are listed. For cholesterol modifying enzymes cholesterol binding is obvious as they use cholesterol as substrate. For other proteins, cholesterol binding/interaction has been shown by various techniques such as binding studies, affinity labelling with photoreactive cholesterol analogues, or crystallography. Convincing proof for cholesterol binding has only been demonstrated for a handful of proteins, e.g. for NPC2 using various approaches such as binding studies with [^{3}H]cholesterol (Okamura et al., 1999; Ko et al., 2003; Infante et al., 2008c), spectroscopical measurements (Friedland et al., 2003; Liou et al., 2006; Cheruku et al., 2006), and crystallography (Friedland et al., 2003). The following criteria and techniques could support evidence for cholesterol–protein interaction: (i) presence in cholesterol-rich microdomains; (ii) alterations in protein function induced by changes of the cholesterol content in membranes/cells; (iii) alterations in protein function induced by substitution of cholesterol by sterol analogues; (iv) influence of cholesterol binding molecules (e.g. polyenes) as functional cholesterol 'competitors'; (v) binding studies with [^{3}H]cholesterol; (vi) spectroscopic binding assays using fluorescent sterol analogues (e.g. dehydroergosterol); (vii) affinity labelling of the protein with photoreactive cholesterol analogues; (viii) identification of cholesterol binding domains. The above mentioned criteria and topics will be critically discussed below. Finally, I will focus on currently available cholesterol probes bearing radio-, spin-, photoaffinity- or fluorescent labels and describe their utility for cholesterol research.

Table 1.1 Proteins with defined or putative binding sites for cholesterol

Protein	Cholesterol binding site/motif	Method	Reference[a]
ACAT1, ACAT2	Unknown	Enzyme assay	1, 2
Caveolin-1	Unknown	Photoaffinity labeling, liposome incorporation	3, 4
Cholesterol 24-hydroxylase CYP46A1	Banana-shaped hydrophobic cavity	Spectral binding, enzyme assay, crystal structure	5
Cholesterol dehydrogenases	Unknown	Enzyme assay	6, 7

Table 1.1 (continued)

Protein	Cholesterol binding site/motif	Method	Reference[a]
Cholesterol-dependent cytolysins (>20, e.g. perfringolysin O)	D4 domain	Pore formation, spectral binding	8, 9
Cholesterol oxidases	Hydrophobic tunnel	Enzyme assay, crystal structure	10, 11
Cholesterol sulfo-transferase (SULT2B1b)	Hydrophobic binding pocket	Enzyme assay, crystal structure	12
β-Cryptogein	Non-specific hydrophobic cavity	Crystal structure	13
HIV-1 env gp41	CRAC motif	Adsorption to cholesteryl beads	14
NPC1	Sterol sensing domain, luminal loop-1 NPC1(25-264)	Photoaffinity labeling, radioligand binding	15–18
NPC2	Loosely packed hydrophobic core between β-sheets	Radioligand binding, crystal structure, fluorescence spectroscopy	19–21
Prominin-1 and 2	Unknown	Photoaffinity labeling	22–24
Retinoic acid-related orphan receptor (RORα)	Binding pocket	Crystal structure of ligand binding domain	25
SCAP	Sterol sensing domain SCAP(TM1-8)	Radioligand binding, trypsin protection, photoaffinity labeling	26–28
Sigma-1 receptor	Two CRAC motifs	Adsorption to cholesteryl beads	29
StARD1 (=StAR)	START domain, cavity with StAR93-212 as cholesterol docking site	Homology modeling, photoaffinity labeling, fluorescence spectroscopy	30–32
StARD3 (=MLN64)	START and N-terminal domain, hydrophobic cavity	Crystal structure, photoaffinity labeling	33, 34
StARD4	START domain, tunnel	Radioligand binding, lipid protein overlay assay, crystal structure	35, 36
StARD5	START domain	Radioligand binding	37
TSPO translocator protein (=PBR)	CRAC motif	Radioligand binding, photoaffinity labeling	38
G protein coupled receptors			
β$_2$-Adrenergic	Unknown: cholesterol consensus motif (CCM)[b]	Crystal structure	39

Table 1.1 (continued)

Protein	Cholesterol binding site/motif	Method	Reference[a]
Cannabinoid (CB1)	Unknown	Radioligand binding, capacity increased	40
Galanin (Gal$_2$)	Unknown	Radioligand binding, affinity modulator?	41
5-Hydroxytryptamine (5HT1A)	unknown: CCM?	Radioligand binding, affinity and capacity decreased	42
5-Hydroxytryptamine (5-HT7)	Unknown: CCM?	Radioligand binding, affinity decreased	43
Opioid δ	Unknown	Radioligand binding, affinity decreased	44
Oxytocin	Unknown: CCM?	Radioligand binding affinity increased	45, 46
Ion channels			
Nicotinic acetylcholine receptor	15 putative cholesterol binding sites	Modeling based on structure	47
Metabotropic glutamate mGlu1	Unknown	Radioligand binding, photoaffinity labeling, affinity increased	48
K$^+$-channel Kir2	CD loop is cholesterol sensitive	Mutant analysis	49

[a]References for Table 1.1: 1. (Chang et al., 1998); 2. (Das et al., 2008); 3. (Thiele et al., 2000); 4. (Murata et al., 1995); 5. (Mast et al., 2008); 6. (Kishi et al., 2000); 7. (Chiang et al., 2008); 8. (Chiang et al., 2008); 9. (Rossjohn et al., 2007); 10. (Yue et al., 1999); 11. (MacLachlan et al., 2000); 12. (Lee et al., 2003); 13. (Lascombe et al., 2002); 14. (Vincent et al., 2002); 15. (Ohgami et al., 2004); 16. (Infante et al., 2008a); 17. (Infante et al., 2008b); 18. (Liu et al., 2009b); 19. (Friedland et al., 2003); 20. (Xu et al., 2007); 21. (Infante et al., 2008c); 22. (Roper et al., 2000); 23. (Florek et al., 2007); 24. (Marzesco et al., 2009); 25. (Kallen et al., 2002); 26. (Radhakrishnan et al., 2004); 27. (Brown et al., 2002); 28. (Adams et al., 2004); 29. (Palmer et al., 2007); 30. (Petrescu et al., 2001); 31. (Murcia et al., 2006); 32. (Reitz et al., 2008); 33. (Tsujishita and Hurley, 2000); 34. (Alpy et al., 2005); 35. (Romanowski et al., 2002); 36. (Rodriguez-Agudo et al., 2008); 37. (Rodriguez-Agudo et al., 2005); 38. (Li et al., 2001); 39. (Hanson et al., 2008); 40. (Bari et al., 2005); 41. (Pang et al., 1999); 42. (Pucadyil and Chattopadhyay, 2004); 43. (Sjogren et al., 2006); 44. (Huang et al., 2007); 45. (Klein et al., 1995); 46. (Gimpl et al., 1997); 47. (Brannigan et al., 2008); 48. (Eroglu et al., 2003); 49. (Epshtein et al., 2009).
[b]The strict CCM motif is found in about 40 GPCRs, details in Hanson et al. (2008).

1.2 Cholesterol-Rich Microdomains

According to our current understanding, biomembranes are much more ordered than postulated in the *fluid mosaic model* proposed by Singer and Nicolson in 1972. Flippases are involved to generate and maintain an asymmetric distribution of lipids across the bilayer of the plasma membrane. Moreover, within the plane of the membrane, lipids and proteins are unevenly distributed. The type of lateral membrane organization that exists in vivo is, however, still controversial. The concept of lipid 'rafts' is a widespread microdomain model that was originally developed by Simons and van Meer to explain the sorting of proteins to the apical membrane in polarized epithelial cells (Simons and van Meer, 1988; Simons and Ikonen, 1997; Simons and Toomre, 2000). This concept has been modified over the years, a process that is still going on. In a 2006 Keystone Symposium, membrane rafts were defined as 'small (10–200 nm), heterogeneous, highly dynamic, sterol- and sphingolipid-enriched domains that compartmentalize cellular processes. Small rafts can sometimes coalesce to form larger platforms through protein–protein and protein–lipid interactions.'

The flask-shaped caveolae that are seen in electron micrographs are also enriched in glycolipids and cholesterol and are regarded as subdomains of lipid rafts. The scaffolding protein caveolin-1 is necessary for the formation of the typical caveolae structures (Kurzchalia and Parton, 1999). Raft and caveolar microdomains may participate in signal transduction, cholesterol trafficking, and vesicular sorting. A variety of hormone-receptor complexes, toxins, viruses, and bacteria are internalized into cells by a caveolar/raft-dependent endocytosis pathway that is clathrin-independent, but requires dynamin and cholesterol (Nabi and Le, 2003). Biochemically, rafts are primarily based on the criterion of cholesterol enrichment. They are assumed to be composed of lipids that exist in a cholesterol-enriched liquid-ordered (L_o) state, separated from and coexisting with cholesterol-poor liquid-disordered (L_d) domains. Detergent resistance is often used as an experimental hallmark of L_o structures. However, there is no definitive evidence to identify detergent-resistant biomembranes with raft and L_o domains (Lichtenberg et al., 2005). Recently, the entire raft model has raised some criticism, and alternatives to this model have been developed (Munro, 2003; Shaw, 2006; Kenworthy, 2008). In these alternative models, lipid ordering plays a minor role. Instead, formation of microdomains (submicrometer-sized clusters) is mainly driven by protein–protein interactions, either through diffusional trapping or self-organization (Douglass and Vale, 2005; Sieber et al., 2007).

The following detergent-based criteria are used to verify the localization of a candidate membrane protein in lipid rafts: (i) Insolubility of the protein in non-ionic detergents (Triton X-100) at cold (and vice versa, solubility in Triton X-100 at 37°C). (ii) Flotation of the detergent-insoluble proteins to the upper low-density fractions following sucrose (or OptiPrep) gradient centrifugation. (iii) Decrease or disappearance of detergent-insolubility after removal of cholesterol (e.g. by cyclodextrins). Different subcellular fractionation protocols are employed to isolate membrane microdomains and to verify or exclude the raft association of a certain membrane protein. Basically, we can distinguish detergent-based and detergent-free fractionation methods.

1.2.1 Detergent-Based Methods

Detergent-based methods utilize the property of insolubility of raft proteins in cold non-ionic detergents, typically Triton X-100. In contrast, at 37°C, solubilization of raft proteins occurs in Triton X-100 (Brown and Rose, 1992). After incubation of the cells with detergent, the extract is fractionated in a density gradient (e.g. sucrose, OptiPrep) and the cholesterol-enriched low-density gradient fractions are harvested. Marker proteins such as caveolin or flotillin are used to identify these low-density fractions. Rafts prepared accordingly are also designated as 'detergent-insoluble (or detergent-resistant) glycosphingolipid-enriched' membrane domains (DIGs or DRMs). Different types of raft domains may be isolated when using detergents other than Triton X-100 (Roper et al., 2000; Chamberlain, 2004). Detergents that have been employed for this purpose were Triton X-114, Lubrol PX, Lubrol WX, Brij58, Brij96, Brij98, CHAPS, Nonident P40, and octylglucoside (Brown and Rose, 1992; Madore et al., 1999; Roper et al., 2000; Drevot et al., 2002; Slimane et al., 2003; Schuck et al., 2003). Recently, we introduced a novel cholesterol-based detergent (termed Chapsterol) to improve the isolation of cholesterol dependent raft proteins (Gehrig-Burger et al., 2005). Unfortunately, this detergent is not yet available commercially. The pentaspan protein prominin, for example, was soluble in Triton X-100, but insoluble in Lubrol WX. Nevertheless, several other properties of prominin classified it as a *bona fide* raft protein (Roper et al., 2000). Since rafts become solubilized at high detergent:lipid ratios (Chamberlain and Gould, 2002), it is necessary to use the lowest amount of detergent that maintains insolubility for raft proteins but completely solubilizes non-raft proteins (e.g. the transferrin receptor) (Chamberlain, 2004). Moreover, the level of detergent insolubility can change for some raft proteins (e.g. for H-ras at GTP-loading) (Prior et al., 2001). Some proteins shown to be raft-associated by other criteria (e.g. the insulin receptor residing in caveolae) could be solubilized by detergent (Gustavsson et al., 1999). Thus, proteins excluded from DRM fractions can still be associated with raft domains. In addition, detergents such as Triton X-100, can themselves promote domain formation in lipid mixtures (Heerklotz, 2002). In each case, the name of the employed detergent should be included to specify the type of microdomains that has been isolated, i.e. designate them as Triton X-100 rafts, Lubrol WX-rafts, etc. To exclude potential artifacts associated with the use of detergents, various detergent-free fractionation protocols have been developed.

1.2.2 Detergent-Free Methods

It is clear that the composition of proteins and lipids from detergent and non-detergent-based preparations significantly differ from each other. To avoid confusion, we have designated the low-buoyant density gradient fractions obtained by detergent-free preparations as low-density microdomains (LDM) (Gimpl and Fahrenholz, 2000). Two widely applied protocols use sonication steps to disrupt the cellular membranes (Smart et al., 1995; Song et al., 1996). In one approach, the cells were sonicated in sodium carbonate buffer (pH 11) prior to centrifugation in a discontinuous 5–45% sucrose gradient (Song et al., 1996). Smart et al. (1995) prepared LDM rafts according to the following subsequent steps: lysis of the cells in an isotonic buffer, purification of plasma membranes therefrom on a Percoll gradient, sonication of these membranes, and isolation of LDM rafts by flotation through a 10–20% OptiPrep gradient. Each of these protocols has subsequently been modified. The time-consuming OptiPrep protocol has been simplified (Macdonald and Pike, 2005). For the isolation of rafts from brain tissues, the usage of sucrose was preferred before OptiPrep in density gradients (Persaud-Sawin et al., 2009). Antibody-based immunoisolation approaches allow isolation of subpopulations of rafts enriched for different markers, as caveolin-1 or flotillin (Shah and Sehgal, 2007).

1.2.3 Receptors in Cholesterol-Rich Microdomains (Lipid Rafts)

Membrane proteins of different families such as G protein coupled receptors (GPCRs), transporters, or ion channels have been shown to be localized or enriched in lipid rafts. For GPCRs, Chini and Parenti (2004) have recently summarized their signaling, coupling efficacy, and trafficking in dependence on their distribution in lipid rafts and caveolae. We have studied in more detail the oxytocin receptor, a typical GPCR, that requires cholesterol to maintain its high-affinity state for oxytocin (Klein et al., 1995; Gimpl et al., 1995, 1997). Thus, a partial localization of this receptor in lipid rafts was expected. For this purpose, HEK293 cells expressing the human oxytocin receptor (HEKOTR) were fractionated by detergent-free and detergent-based methods (Gimpl and Fahrenholz, 2000). Only a minor fraction (~1%) of the receptor was found in Triton X-100 rafts, whereas substantially more oxytocin receptors (10–15%) were found in LDM rafts produced according

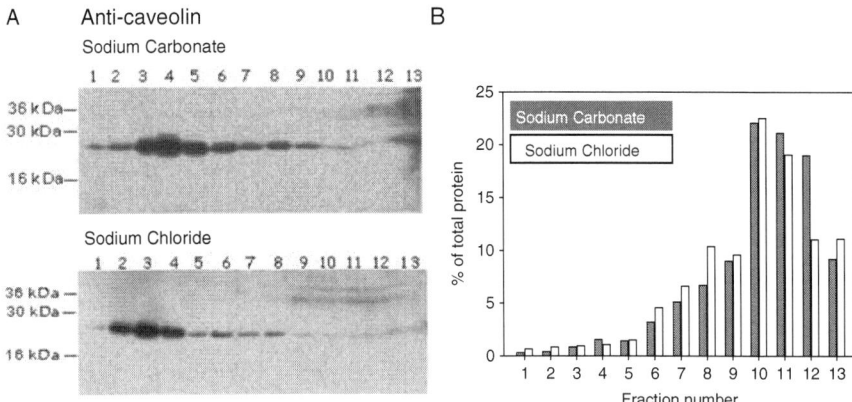

Fig. 1.1 Detergent-free subcellular fractionation of HEK293 cells expressing the human oxytocin receptor (HEKOTR). The cells were fractionated in a sucrose flotation gradient using either a sodium carbonate (Song et al., 1996) or sodium chloride (Gimpl and Fahrenholz, 2000) protocol. Aliquots of each of the 13 fractions were analyzed with respect to caveolin immunodetection (**A**) and determination of total protein contents (**B**). For methodical details see Gimpl and Fahrenholz (2000)

to a detergent-free protocol. To analyze the amounts of functional receptors, we modified the fractionation method based on sodium carbonate buffer (pH 11) (Song et al., 1996), because ligand-receptor binding is normally inhibited at such an alkaline pH. For this purpose, a subcellular fractionation protocol was developed in which the sodium carbonate buffer was substituted by sodium chloride (1 M) in 20 mM Hepes (pH 7.4)/EDTA (Gimpl and Fahrenholz, 2000). The distribution profile of total proteins and of the raft marker caveolin was shown to be similar for sodium carbonate and sodium chloride based fractionation (Fig. 1.1). In addition, the majority of cholesterol was found in the low-density fractions for both methods. The cholesterol enrichment of low-density fractions is one of the most important properties for lipid rafts. Vice versa, rafts and caveolae are disrupted when cholesterol is extracted from these microdomains, e.g. via cyclodextrins.

1.3 Manipulation of the Membrane Cholesterol Content by Cyclodextrins

Cyclodextrins (CDs) are torus-shaped cyclic oligosaccharides linked by α-1,4 glycosidic bonds (Fig. 1.2A). They are produced from the enzymatic degradation of starch. Cyclodextrins comprised of 6, 7 and 8 D-glucopyranosyl residues units (termed α-, β- and γ-forms, respectively) were used to alter the lipid composition of cells (Ohtani et al., 1989). They possess a hydrophilic outer surface and a hydrophobic inner cavity. In aqueous solution, this hydrophobic cavity contains low entropy and easily displaceable water molecules. Cyclodextrins enhance the solubility of non-polar substances (e.g. cholesterol) by incorporating them (at least partly) into their hydrophobic cavity and forming non-covalent water-soluble inclusion complexes. Particularly, β-cyclodextrins (βCDs) and its derivatives, such as methyl-β-cyclodextrin (MβCD) or 2-hydroxypropyl-β-cyclodextrin (HPβCD), were found to selectively extract cholesterol from the plasma membrane, in preference to other membrane lipids (Irie et al., 1992; Klein et al., 1995; Gimpl et al., 1995, 1997; Kilsdonk et al., 1995). However, the extraction of cholesterol does not seem to be selective for lipid rafts (Mahammad and Parmryd, 2008). Figure 1.2 shows that the size of the cholesterol molecule is too large to be fully incorporated into the cavity of βCDs. The exact structure of the soluble cholesterol–βCD inclusion complex is still unknown. An excellent review to the use of cyclodextrins for manipulation of the cholesterol content in membranes has been published (Zidovetzki and Levitan, 2007).

The kinetics of cyclodextrin-mediated cholesterol efflux provided information about cholesterol pools in cells (Yancey et al., 1996). While 'empty' β-cyclodextrins function as rather selective cholesterol acceptors, cholesterol–cyclodextrin complexes serve as very efficient sterol donors in vitro and in vivo (Gimpl et al., 1995, 1997; Kilsdonk et al., 1995). For example, up to 80% of the cholesterol can be extracted from living cells via MβCD within 10–30 min (Fig. 1.3). Vice versa, using cholesterol-MβCD as donor, cholesterol-depleted cells can be reloaded with cholesterol within the same time scale. Thereby, substantial cholesterol

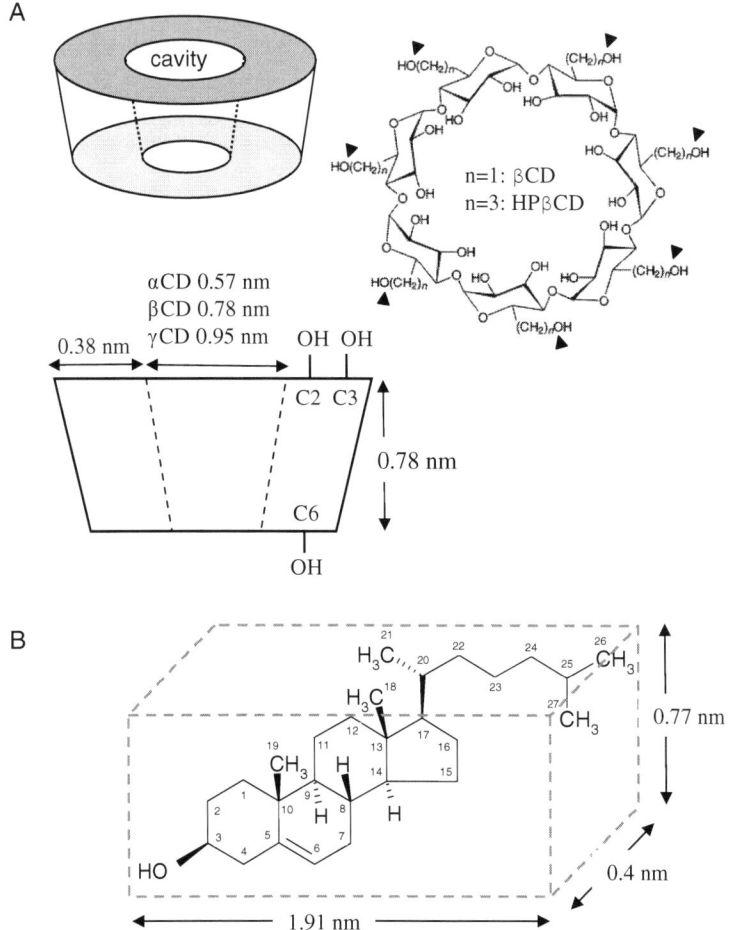

Fig. 1.2 Structure and dimensions of cyclodextrins (**A**) and cholesterol (**B**). Cyclodextrins (CDs) typically possess 6, 7 or 8 D-glucopyranosyl residues (α-, β-, and γ-cyclodextrin respectively). Their overall shape is like a truncated cone that accommodates guest molecules into a hydrophobic cavity. The hydrophilic OH groups are on the outside of the cavity with the C2- and C3-hydroxyls located around the wider ring and the C6-OH groups aligned around the smaller opening. The hydroxyl groups may be derivatized to modify the physical and chemical properties of the cyclodextrins. The 6-OH groups (black arrows) are most easily derivatized. Methyl-β-cyclodextrin (MβCD) and (2-Hydroxypropyl)-β-cyclodextrin (HPβCD) are often used for cholesterol depletion experiments. They normally contain about 0.5-2 mol methyl- or hydroxymethyl groups per unit anhydroglucose. Cholesterol is too large to be completely encapsulated within the cavity of a single βCD. So, cholesterol may be either partly incorporated into the cavity, or two stacked βCDs may be required for the complete complexation of cholesterol

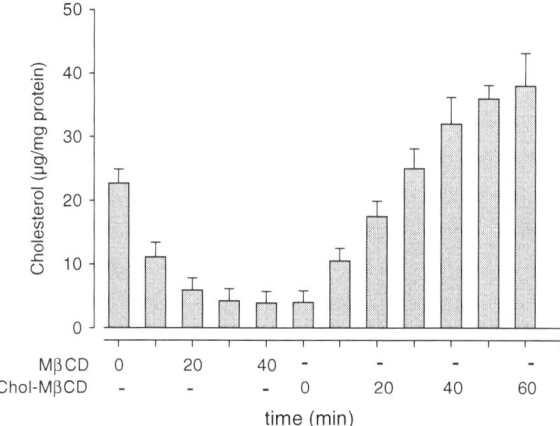

Fig. 1.3 Depletion and reloading of cholesterol in living HEKOTR cells. To extract cholesterol, the cells were incubated with 10 mM MβCD (stock 200 mM) for 0–40 min at 37°C in serum-free culture medium. The cells were then washed twice with medium. Cholesterol enrichment of the cholesterol-depleted cells was started using 0.3 mM Chol-MβCD (stock 10 mM) for 0–60 min in serum-free culture medium at 37°C. Cholesterol levels were determined using a diagnostic kit (data are means ± SD, n=3). Methods are described in detail (see Gimpl et al., 1997)

overloading of cells easily occurs, as shown in Fig. 1.3. The efficiency of cholesterol extraction and reloading varies with incubation time, temperature, cell type, and concentration of the cholesterol acceptor. It is also possible to stabilize membranes or cells at a certain cholesterol concentration by varying the molar ratio between βCD and cholesterol in the complex. The experimental conditions to achieve this 'cholesterol equilibrium' have to be determined for each cell system (Zidovetzki and Levitan, 2007). However, one should be aware that even when the total cholesterol levels are held constant, the distribution of intracellular cholesterol pools (e.g. between plasma membrane and endoplasmic reticulum) may change.

Membrane cholesterol can also be rapidly substituted with sterol analogues by adding the corresponding sterol–MβCD complexes to the cholesterol-depleted membranes or cells. Thus, cyclodextrin-based exchanges enable the researcher to explore the cholesterol specificity of a candidate membrane protein in a precise structure-activity analysis. We have performed such studies for two GPCRs: the oxytocin receptor (OTR) and cholecystokinin B receptor (CCKBR) (Gimpl et al., 1997). A similar study has been performed for SCAP, a protein with a sterol sensing domain (Table 1.1) (Brown et al., 2002). As shown in Table 1.2, the sterol requirements reveal some similarities between the GPCRs and SCAP, although these are unrelated proteins. Sterols supporting low membrane fluidity (equal to a high level of anisotropy of diphenylhexatriene, Table 1.2) (e.g. desmosterol, dihydrocholesterol, β-sitosterol, 5α-cholest-7-en-3βol) were most effective to maintain the binding function of the GPCRs and a certain conformation of SCAP. The structure-activity analysis also permits discrimination between cholesterol effects that are due

Table 1.2 Structure–activity relationship of cholesterol analogues with respect to membrane fluidity and protein activity

Sterol[a]	Anisotropy (DPH)[b]	OTR binding (%)[c]	CCKR binding (%)[c]	SCAP conformation[d]
Desmosterol	0.271	91	107	++
Dihydrocholesterol	0.264	86	105	++
Cholesterol	0.263	100	100	++
7-Dehydrocholesterol	0.258	91	88	n.d.
Stigmastanol	0.241	53	97	n.d.
20α-Hydroxycholesterol	0.233	74	86	n.d.
β-Sitosterol	0.231	92	67	+
Epicholesterol	0.224	10	63	−
5α-Cholest-7-en-3β-ol	0.223	67	82	+
Fucosterol	0.217	77	70	n.d.
4-Cholesten-3-one	0.213	6	88	−
Stigmasterol	0.213	56	62	n.d.
6-Ketocholestanol	0.213	23	86	n.d.
Ergosterol	0.211	68	69	n.d.
19-Hydroxycholesterol	0.209	70	58	−
22-Ketocholesterol	0.209	76	72	n.d.
Cholesteryl ethyl ether	0.208	11	67	n.d.
Campesterol	0.206	97	61	n.d.
Lanosterol	0.197	7	71	+
Coprostanol	0.196	57	69	n.d.
Epicoprostanol	0.194	6	68	n.d.
7β-Hydroxycholesterol	0.189	12	61	−
25-Hydroxycholesterol	0.187	13	54	−
22(R)-Hydroxycholesterol	0.183	16	47	n.d.
Allocholesterol	n.d.	n.d.	n.d.	++
7-Ketocholesterol	?	?	?	++
27-Hydroxycholesterol	n.d.	n.d.	n.d.	−
Dehydroergosterol	n.d.	85	67	n.d.
None (MβCD-depleted)	0.174	5	39	−

[a]Following cholesterol depletion with MβCD, cells/membranes were incubated with a variety of sterol-MβCD complexes to substitute cholesterol with the indicated sterol analogues.
[b]Membrane fluidity was measured by fluorescence anisotropy of diphenylhexatriene (DPH) (details from Gimpl et al., 1997). A high value indicates low fluidity. The data are sorted according to the anisotropy level in decreasing order. n.d., not determined.
[c]Radioligand binding (in %, 'cholesterol' set to 100%) of [^{3}H]oxytocin and [^{3}H]CCKS to the oxytocin receptor (OTR) and the cholecystokinin (type B) receptor (CCKR), respectively (details from Gimpl et al., 1997). n.d., not determined.
[d]SCAP conformation: the capability of sterols to support the accessibility of certain arginine residues to trypsin is indicated: '++', high; '+', moderate; '−', none; n.d., not determined (data from Brown et al., 2002).

to specific sterol–protein interactions or due to changes in the physical state of the membrane bilayer. In case of the CCKBR, the effects of sterols correlated with changes in membrane fluidity. For the oxytocin receptor and for SCAP, a unique requirement for cholesterol was observed suggesting that these proteins are regulated by specific cholesterol–protein interactions. For example, epicholesterol that maintains membrane fluidity moderately-good was completely inactive for both the OTR and SCAP. The data for lanosterol and 19-hydroxycholesterol indicate a distinct cholesterol specificity of SCAP and the OTR, respectively (Table 1.2). The cholesterol analogue 4-cholesten-3-one supports the membrane fluidity moderately and maintains the ligand binding of the CCKBR. However, 4-cholesten-3-one was found to be inactive for all cholesterol-dependent membrane proteins reported so far. Examples include the oxytocin receptor, SCAP (Table 1.2), the hippocampal 5HT1A receptor (Pucadyil et al., 2005), the galanin receptor (Pang et al., 1999), or ecto-nucleotidase CD39 (Papanikolaou et al., 2005).

Overall, the administration of β-cyclodextrins and sterol-β-cyclodextrins as cholesterol acceptor and sterol donor complexes, respectively, allows one to alter and exchange the cholesterol content in membranes and cells. This is now a standard methodology in the research of 'lipid rafts' (Simons and Toomre, 2000). Moreover, cyclodextrins have found a wide range of applications in food, pharmaceutical and textile industry, cosmetics, environmental engineering, and agrochemistry. For example, cyclodextrins are employed for the preparation of cholesterol-free products or for the delivery of drugs (Challa et al., 2005). Finally, cyclodextrins offer great therapeutic potential as recently been demonstrated for HPβCD. The cholesterol acceptor was able to reverse the defective lysosomal transport in a mouse model of Niemann Pick C disease (Liu et al., 2009a). However, one should be aware that application of β-cyclodextrins induce several ill-defined side effects that are associated with their non-specific action. They can extract a wide range of hydrophobic compounds including phospholipids, sphingomyelin, even GPI-anchored proteins and some protein kinases (Ilangumaran and Hoessli, 1998; Ottico et al., 2003; Monnaert et al., 2004). Some of these undesirable effects of β-cyclodextrins may be caused by inhibition of tyrosine kinases and/or vesicle shedding (Sheets et al., 1999). Thus, careful control experiments and the avoidance of overly high concentrations of cyclodextrins are recommended (Zidovetzki and Levitan, 2007).

1.4 Cholesterol-Binding Molecules

1.4.1 Polyenes

Among the family of polyenes, filipin is certainly the most important tool to visualize the localization of free cholesterol in cells. Filipin is an antibiotic with antifungal properties and a mixture of four macrolides with minor differences in their structure, the fraction known as filipin III being the major component (Bolard, 1986) (Fig. 1.4A). Filipin performs its antibiotic action by inducing a structural disorder in sterol containing membranes. The disintegration of the membranes then leads to the leakage of cellular components. Filipin requires a sterol partner with a free $3'$-OH group. So it does not recognize esterified cholesterol. Further details concerning the filipin–sterol interaction are unclear. Different models have been generated to explain the organization of the filipin-sterol complexes within the membrane

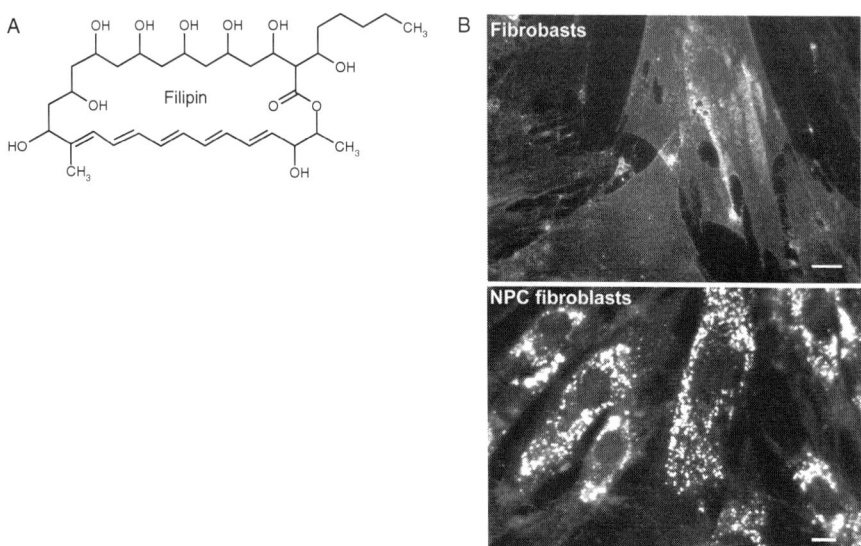

Fig. 1.4 Filipin (structure in panel A) staining is the standard method to visualize free cholesterol in cells and tissues. Human fibroblasts from a control person (B, 'fibroblasts') and from a patient with Niemann Pick C disease (B, 'NPC fibroblasts') were fixed by paraformaldehyde (3.7%), incubated with the filipin III fluorophore (0.05% in PBS/10% FCS) for 60 min, washed twice with PBS, and were mounted in Moviol. NPC fibroblasts show intense filipin accumulation in lysosomes/late endosomes whereas in control fibroblasts the filipin (and thus free cholesterol) is concentrated on the cell surface. [bar, 5 μm]

bilayer (Elias et al., 1979; Lopes et al., 2004). Filipin has been used for decades to localize the distribution of free cholesterol in cells and tissues (Kinsky et al., 1967; Elias et al., 1979; Orci et al., 1983; Butler et al., 1992). Filipin staining has been and still is a prominent diagnostic tool for the identification of cholesterol mislocalization in lysosomes of the Niemann-Pick C (NPC) phenotype (Butler et al., 1987, 1992) (Fig. 1.4B). However, filipin is a cytotoxic compound disrupting the integrity of sterol-containing membranes (Behnke et al., 1984). So, staining with filipin can only be used in prefixed cells or tissues. Moreover, it possesses unfavorable spectroscopic properties, e.g. high photobleaching and excitation within the UV range. Other polyene antibiotics such as nystatin and amphotericin B share the cholesterol binding property of filipin. Both substances form pores, unlike filipin. Electrophysiologists use nystatin for the so-called 'perforated patch clamping' technique. Nystatin forms complexes with cholesterol that lead to ion-selective 'perforations' in the bilayer inside the patch pipette. Polyenes, particularly filipin, can also be employed as cholesterol competitors in binding assays or functional interaction studies in order to verify or falsify a putative cholesterol interaction of candidate proteins. The oxytocin and galanin receptors, for example, showed a dose-dependent decrease in ligand binding in the presence of increasing concentrations of filipin (Gimpl et al., 1997; Pang et al., 1999).

Overall, although filipin is routinely used as to visualize free cholesterol, it is not entirely clear whether it's staining reflects the correct distribution of cholesterol, particularly at intracellular sites that are not easily accessible and/or prone to artefacts from fixation techniques. It has also been reported that some sterol-containing membranes are not labelled by filipin (Pelletier and Vitale, 1994; Steer et al., 1984; Severs and Simons, 1983).

1.4.2 Cholesterol-Dependent Cytolysins

Cholesterol-dependent cytolysins (CDCs) (*see also* Chap. 20) are a family of protein toxins produced by a variety of pathogenic Gram-positive bacteria including *Streptococcus pyogenes* (streptolysin O), *Streptococcus pneumoniae* (pneumolysin), *Listeria monocytogenes* (listeriolysin O), *Clostridium perfringens* (perfringolysin O), and *Bacillus anthracis* (anthrolysin) (Rossjohn et al., 1997; Palmer, 2001; Tweten et al., 2001). When the water-soluble cytolysin monomers bind to cholesterol-containing membranes, they self-associate to form large oligomeric complexes and aqueous pores in the bilayer. Structural features of the cholesterol molecule required for interaction with the toxins include the 3β-OH group, the stereochemistry of the sterol ring system, and the isooctyl side chain (Watson and Kerr, 1974; Prigent and Alouf, 1976; Nelson et al., 2008). This suggests that CDCs possess a specific cholesterol binding site.

Among the CDCs, perfringolysin O is one of the best studied cholesterol binding molecules. It is comprised of four domains. The C-terminal portion of perfringolysin O (D4 domain) folds into a separate β-sandwich domain composed of two four-stranded β-sheets at one end of the elongated molecule (Rossjohn et al., 1997). This D4 domain is involved in cholesterol recognition and binding (*see also* Chap. 21). Specifically, only the short hydrophobic loops at the tip of the D4 β-sandwich are responsible for mediating the interaction of the CDCs with cholesterol-rich membranes, whereas the remainder of the structure remains close to the membrane surface (Ramachandran et al., 2002; Soltani et al., 2007). Interestingly, the mammalian immune defense system, the complement membrane attack complex perforin, was found to be structurally homologous to the bacterial CDCs (Hadders et al., 2007; Anderluh and Lakey, 2008). A protease-nicked and biotinylated derivative of perfringolysin O (termed BCθ-toxin) was shown to retain specific binding to cholesterol without cytolytic activity. BCθ-toxin combined with fluorophore-labelled avidin has been introduced as a cholesterol reporter system (Fujimoto et al., 1997). This probe was used for the localization of membrane cholesterol in various cells by fluorescence microscopy and by electron microscopy in cryosections (Iwamoto et al., 1997; Waheed et al., 2001; Mobius et al., 2002; Sugii et al., 2003; Reid et al., 2004; Tashiro et al., 2004). Visualization of the D4 probe has been achieved both by conjugation of a fluorophore to D4 and by N-terminal fusion of green fluorescent protein (EGFP variant) to the protein (Shimada et al., 2002). Perfringolysin O derivatives detected cholesterol primarily in cholesterol-rich membrane microdomains (e.g. caveolae or 'lipid rafts') or liposomes with high (>20–25 mol%) cholesterol (Ohno-Iwashita et al., 1992; Waheed et al., 2001; Shimada et al., 2002; Sugii et al., 2003). Studying the pathophysiological cholesterol accumulation in the Niemann-Pick C mouse brain, staining with BCθ-toxin was found to be superior to that achieved by filipin. In brain regions known to be affected by the neurodegenerative NPC disease, cholesterol accumulations were observed both at a better signal-to-noise ratio and at earlier time points with BCθ-toxin as compared with filipin (Reid et al., 2004). In contrast, in a hippocampal culture system, cholesterol was detectable by BCθ only at the cell surface of fully matured neurons, whereas filipin stained intracellular and cell surface cholesterol in neurons at all developmental stages. Additionally, the two cholesterol reporters showed different labelling patterns in cultured hippocampal neurons. While BCθ staining was observed mainly on axons, filipin labeled axons, dendrites and somata (Tashiro et al., 2004). These authors reported also that neurons that were induced to the NPC phenotype by administration of certain reagents, lose cell surface BCθ staining on axons. Obviously, the distribution of cholesterol at the axonal surface is critical for recognition by BCθ. In addition, when membranes or cells were depleted of cholesterol by β-cyclodextrins, the binding of θ-toxin was completely abolished whereas significant filipin staining was retained (Waheed et al., 2001; Shimada et al., 2002). Thus, the toxin might recognize a certain arrangement of cholesterol at the outer leaflet of the membrane bilayer (Mobius et al., 2002).

Taken together, perfringolysin O derivatives might be good and selective probes for cholesterol-rich domains such as caveolae or rafts, but are neither suitable to label cholesterol-poor organelles nor for quantitative in situ determination of membrane cholesterol. In addition to their potency as cholesterol probes, cytolysins are useful tools in cell biology due to their pore-forming capacity. For example, application of a low concentration of streptolysin O facilitates the entry of macromolecules into cells of interest (Lafont et al., 1995).

1.4.3 Enzymes with Cholesterol as Substrate

Of course, all enzymes that accept cholesterol as substrate possess a specific cholesterol binding site and are potential candidate proteins to measure cholesterol amounts or to detect cholesterol in membranes and cells. These comprise cholesterol oxidases, the cholesterol esterifying ACAT enzymes, cholesterol sulfotransferase SULT2B1b (Lee et al., 2003), the cytochrome P450 family proteins (e.g. cholesterol hydroxylases like CYP46A1) (Mast et al., 2008), and other less well characterized enzymes (e.g. cholesterol transferase linking cholesterol to sonic hedgehog). However, most of these enzymes are very hydrophobic, require detergents for solubilization or are inconvenient for the development of highly sensitive enzymatic assays. So far, primarily the cholesterol oxidases (*see also* Chap. 5) have been used to measure cholesterol concentrations or to gain information about cellular and membrane cholesterol distribution.

The flavoenzyme cholesterol oxidase converts cholesterol and oxygen into the products 4-cholesten-3-one and hydrogen peroxide that can be quantitated by spectrophotometry (or fluorometry) via an oxidative coupling reaction in the presence of peroxidase to form a chromogen (or fluorophore). Currently, cholesterol oxidase is immobilized onto different surfaces for the fabrication of cholesterol biosensors. The properties of different cholesterol oxidases have been reviewed (MacLachlan et al., 2000). Cholesterol oxidase is produced by several microorganisms, e.g. *Nocardia erythropolis*, *Brevibacterium sterolicum*, *Streptomyces hygroscopicus*. Structures of cholesterol oxidase from *Brevibacterium sterolicum* and *Streptomyces hygroscopicus* in its free and substrate-bound states have been determined at atomic resolution (Vrielink et al., 1991; Li et al., 1993; Yue et al., 1999). The water-soluble enzyme associates peripherally with the surface of the membrane. Probably, it forms a complex with the lipid bilayer that allows cholesterol to move directly from the membrane into the active site (Bar et al., 1989; Ahn and Sampson, 2004). In fact, it is known that the properties of the membrane strongly influence the accessibility of the enzyme to its substrate. Thus, cholesterol oxidase is a valuable probe for studying membrane organization and a sensor of the bilayer lipid phase, with a preferential binding to the solid phase (Patzer and Wagner, 1978; Ahn and Sampson, 2004). Cellular cholesterol can be tracked by using its susceptibility to cholesterol oxidase (Lange, 1992). In living intact cells, cholesterol is only a poor substrate for cholesterol oxidase. This changes markedly, when certain substances or enzymes were added to the cells. Among the agents stimulating the enzymatic turnover are cholesterol, glutaraldehyde, low ionic strength buffer, phospholipase C, sphingomyelinase, detergents, and membrane intercalators such as decane or octanol (Lange et al., 1984; Slotte et al., 1989). Lysophosphatides are shown to inhibit the activity of cholesterol oxidase (Lange et al., 1984; Lange et al., 2005). It has been proposed that in the unperturbed plasma membrane, cholesterol is kept at a low chemical potential by its association with bilayer phospholipids (Radhakrishnan and McConnell, 2000). Thus, stimulators of enzyme activity might act by increasing the chemical activity of cholesterol leading to a better accessibility of cholesterol to the enzyme, whereas inhibitors such as lysophosphatidylcholine might associate with excess cholesterol and thereby lower its chemical activity (Lange et al., 2004). Variations of cholesterol oxidase accessibility have been explained by sterol superlattices in membranes (Wang et al., 2004). According to this model, cholesterol within sterol superlattices is tightly packed and more accessible to the aqueous phase (i.e. to cholesterol oxidase) as compared with cholesterol localized in irregularly distributed lipid areas (Wang et al., 2004). The cholesterol oxidase-accessible plasma membrane pool may be the same pool of cholesterol removed by high density lipoproteins (Vaughan and Oram, 2005). The susceptibility to cholesterol oxidase has been exploited to gain information about the localization, transfer kinetics, and transbilayer distribution of cholesterol (Lange, 1992). In human fibroblasts, 90% of the cholesterol in fixed (e.g. glutaraldehyde treated) cells was oxidized by the enzyme within ~1 min. The residual 10% of cholesterol resistant to cholesterol oxidase coincided with markers of endocytic membranes that are also in large part derived from the plasma membrane (Lange et al., 1989; Lange, 1991). This would indicate that in fibroblasts almost all of the cellular cholesterol is localized in the plasma membrane pool, whereas only minor cholesterol amounts (1% or less) are distributed to other organelles (e.g. the endoplasmic reticulum). Application of cholesterol oxidase in human erythrocytes suggested that cholesterol flips very rapidly (half time < 3 s at 37°C) across the plasma membrane (Lange et al., 1981). In contrast, another group reported a half-time of 1–2 h for the transmembrane movement of cholesterol using susceptibility to cholesterol oxidase as reporter (Brasaemle et al., 1988). The transfer of newly synthesized cholesterol to the cell surface occurred with a half-time of 10–60 min as measured by the cholesterol oxidase approach in fibroblasts (Lange and Matthies, 1984; Lange, 1991).

Overall, the application of cholesterol oxidase as a cholesterol probe requires careful selection of reaction conditions and rigorous control experiments. When the enzyme is used on living cells, alterations in protein localization or receptor signaling are possible (Smart et al., 1994; Gimpl et al., 1997; Okamoto et al., 2000). Inherent difficulties in this approach are related with the fact that the enzyme converts cholesterol to a steroid with substantially altered properties. For example, 4-cholesten-3-one does not condense a phospholipid monolayer to the same extent as cholesterol (Gronberg and Slotte, 1990). In addition, 4-cholestene-3-one is a raft-dissolving steroid that, unlike cholesterol, favours the liquid disordered phase (Xu and London, 2000). Its action promotes a certain rate of leakage of the plasma membrane (Ghoshroy et al., 1997). Mutant enzymes with unimpaired membrane binding and no catalytic activity may overcome this limitation (Yin et al., 2002). Also, other enzymes with cholesterol as substrate, e.g. cholesterol dehydrogenases or cholesterol sulfatases, should be evaluated for their applicability as cellular cholesterol reporters.

1.4.4 Other Cholesterol-binding Proteins

Enzymes using cholesterol as substrate usually possess binding cavities to accommodate large parts of the cholesterol molecule (Mast et al., 2008; Chen et al., 2008). In these and other cholesterol binding proteins, a couple of binding motifs have been described and may be classified as follows (Table 1.1):

(i) Cholesterol binding tunnels/cavities: typically these structures are hydrophobic pockets, sometimes closable by a lid. They accommodate a single cholesterol molecule with medium- to high-affinity and are found in enzymes and other unrelated proteins (e.g. NPC2, the retinoic acid-related orphan receptor (RORα)) (Kallen et al., 2002). (ii) Sterol-sensing domain (SSD): this domain comprises a penta-helical region that is weakly conserved across different polytopic proteins, such as sterol regulatory element-binding protein cleavage-activating protein (SCAP), hydroxymethylglutaryl-CoA reductase (HMG-CoA reductase), the Niemann-Pick C (NPC) disease protein NPC1, and the hedgehog receptor patched (Kuwabara and Labouesse, 2002). Although all these proteins are implicated in some aspects of cholesterol metabolism, proof of cholesterol binding at this domain are rather weak. In the case of NPC1, photoaffinity approaches suggested that the cholesterol binding site is present within the SSD domain (Ohgami et al., 2004), whereas cholesterol binding was localized to the large Cys-rich luminal loop-1 according to radioligand binding assays (Infante et al., 2008b). (iii) START domain: START proteins represent a superfamily of hydrophobic ligand binding proteins. The name-giving member of proteins bearing this ~200 amino acids domain is 'Steroid Acute Regulatory' Protein (StAR=STARD1). STARD1 is essentially involved in the rate-limiting step of steroidogenesis, the transport of cholesterol to the mitochondria. The structure of the cholesterol binding cavity of the related StARD3 (=MLN64) has been resolved (Tsujishita and Hurley, 2000). The START domain is however not a specific cholesterol binding domain as in other START proteins other lipids are bound in the cavity (e.g. phosphatidylcholine in StARD2). (iv) 'Cholesterol recognition amino acids consensus' (CRAC) domain: the peptide 'AT$\underline{V}$LNY$\underline{Y}$VW$\underline{R}$DNS' (underlined amino acids are purported to interact with cholesterol) has first been identified as a high-affinity cholesterol binding motif in the C-terminus of the peripheral benzodiazepine receptor (PBR) (= TSPO, 'Translocator protein, 18 kDa') (Li et al., 2001). Together with STARD1, this receptor enables the translocation of cholesterol into mitochondria (see above). In its generalized form (– L/V – $(X)_{1-5}$ – Y – $(X)_{1-5}$ – R/K), this motif is found in many proteins such as cholesterol-dependent GPCRs (receptors for oxytocin and 5HT1A possess this motif in their fifth transmembrane helix), the sigma-1 receptor (with two consecutive CRAC motifs), HIV-1 env gp41 protein, and others (Table 1.1). (v) 'Cholesterol Consensus Motif' (CCM): this motif has been described by the crystal structure of the β_2-adrenergic receptor (Hanson et al., 2008). Two cholesterol molecules were found in a receptor cleft formed by the segments of transmembrane helices 1-4. Based on homology, the following CCM has been defined: [4.39-4.43(R,K)]—[4.50(W,Y)]—[4.46(I,V,L)]—[2.41(F,Y)] (according to Ballesteros-Weinstein nomenclature). This motif is found in more than 40 class A GPCRs, including the cholesterol dependent oxytocin and 5HT1A receptor (Table 1.1).

Presumably, many proteins exist that interact with cholesterol but do not possess the above-mentioned motifs. The knowledge about these cholesterol binding motifs can be exploited to develop improved cholesterol binding assays. One of the main obstacles in cholesterol binding assays results from the low water solubility of cholesterol. Solubility of cholesterol can for example be achieved through inclusion in cyclodextrin complexes or stabilization by low amounts of detergents such as Triton X-100. However, these additional compounds can disturb binding reactions. For this purpose, it would be beneficial to have water-soluble cholesterol binding modules available that can act both as 'solubilizers' and competitors for cholesterol binding. To function as cholesterol donor, the affinity of these molecules to cholesterol must be lower than that of the candidate cholesterol binding protein under investigation. Further applications of these cholesterol binding proteins concern their potency to act as specific cholesterol donors or acceptors, that once may substitute the non-specific cholesterol donor and acceptor complexes MβCD/Chol-MβCD.

1.5 Binding Studies with Radiolabelled Cholesterol

In classical binding studies, radioligand and receptor protein are incubated to equilibrium, bound is separated from free radioligand by means of centrifugation or rapid filtration through presoaked glass fiber filters, and the pelleted membranes or the filters are washed and counted. However, when water-insoluble lipids such as cholesterol are the radioligands, direct binding assays are difficult and often not reproducible. Saturation and kinetic data have successfully been obtained when recombinant His-tagged proteins (e.g. NPC1 and SCAP) were used for cholesterol binding assays (Table 1.3). This allows the separation of bound and free

Table 1.3 Radioligand binding studies for some cholesterol binding proteins

Protein [tag], species	Ligand/Method[a]	Affinity	Reference[b]
NPC1 [His], human	[^{3}H]cholesterol, nickel-agarose	~100 nM	1
NPC1 luminal loop-1 [His], human	[^{3}H]cholesterol, nickel-agarose	50–130 nM	2, 3
NPC2 [His], mouse	[^{3}H]cholesterol, gel filtration	~30 nM	4
NPC2 [His], human	[^{3}H]cholesterol, nickel-agarose	90–150 nM	2, 3
NPC2 [none], human epididymis	[^{3}H]cholesterol, dextran-coated charcoal	2.3 μM	5
PBR [none], mouse	[^{3}H]cholesterol, centrifugation	10 nM	6
SCAP (TM1-8, solubilized) [His], hamster	[^{3}H]cholesterol, nickel-agarose	50–100 nM	7
StARD1-START [His], human	[^{14}H]cholesterol, nickel-agarose	>1 μM	8, 9
StARD3-START [His], human	[^{14}H]cholesterol, nickel-agarose	>1 μM	9
StARD4 [His], human	[^{14}H]cholesterol, nickel-agarose	<1 μM	10
StARD5 [His], human	[^{14}H]cholesterol, nickel-agarose	<1 μM	11

[a]The indicated method refers to the separation of free and bound radioligand.
[b]References for Table 1.3: 1. (Infante et al., 2008a); 2. (Infante et al., 2008b); 3. (Infante et al., 2008c); 4. (Ko et al., 2003); 5. (Okamura et al., 1999); 6. (Jamin et al., 2005); 7. (Radhakrishnan et al., 2004); 8. (Tsujishita and Hurley, 2000); 9. (Baker et al., 2007); 10. (Rodriguez-Agudo et al., 2008); 11. (Rodriguez-Agudo et al., 2005).

radioligand via affinity chromatography on nickel agarose columns: adsorption of the assay solution to the affinity matrix, washing off the free and elution of the bound radioligand by imidazol (>0.2 M). If the candidate protein is not His-tagged, separation in gel filtration columns may be possible. However, these assays are tedious because each data point requires a separate column. To characterize the cholesterol binding of CRAC domains in HIV-1 env gp 41 and sigma-1 receptor, cholesteryl-hemisuccinate coupled to agarose was used as the affinity matrix for these proteins (Vincent et al., 2002; Palmer et al., 2007). However, this adsorption method will not be generally applicable because most cholesterol binding proteins require the free hydroxyl group of cholesterol, which is esterified in the cholesteryl affinity matrix.

The K_D values obtained for cholesterol binding can vary markedly when different binding assays are used. This is well documented for NPC2 for which dissociation constants between 30 nM and 2.3 μM have been obtained (Table 1.3). The lowest affinity (2.3 μM) has been reported using a cholesterol binding assay where free [^{3}H]cholesterol was bound onto dextran-coated charcoal and was removed by centrifugation, a separation method often used in radioimmunoassays. Spectroscopic assays with NPC2 using dehydroergosterol resulted in K_D values of 120–660 nM (Friedland et al., 2003). When detergents are required, their usage can lead to dramatic alterations in cholesterol binding as documented for SCAP and NPC1 (Radhakrishnan et al., 2004; Infante et al., 2008a, 2008b). Detergent micelles may themselves sequester radiolabeled cholesterol and can thus disturb binding kinetics. The best detergents for SCAP and NPC1 were Fos-Choline 13 and Nonidet P-40. Recombinant NPC1 bound cholesterol, but binding was inhibited when the concentration of Nonidet P-40 exceeded the micellar threshold. In case of the more water soluble oxysterols (e.g. [^{3}H]25-hydroxycholesterol), traditional filter assays have been successfully performed in radioligand binding studies with NPC1 (Infante et al., 2008a). In these binding assays, NPC1 was actually identified as an oxysterol binding protein. The K_D for [^{3}H]25-hydroxycholesterol binding was ~10 nM in 0.004% Nonidet P-40, whereas at higher detergent concentrations (1%), the K_D increased several-fold to 80 nM (Infante et al., 2008a). Again, the binding results markedly depended on the concentrations of the employed detergent. Further studies showed that one (the putative second site within the SSD?) cholesterol binding site of NPC1 is localized in the N-terminal luminal loop-1 domain of the protein. NPC1 luminal loop-1 bound [^{3}H]cholesterol with a K_D of 130 nM, whereas binding of [^{3}H]25-hydroxycholesterol occurred with >10 fold higher affinity compared to [^{3}H]cholesterol (Infante et al., 2008b). More qualitative than quantitative cholesterol binding assays can be achieved by lipid protein overlay assays (Dowler et al., 2002; Rodriguez-Agudo et al., 2008). In this case, serial dilutions of cholesterol or other sterols are spotted onto a nitrocellulose membrane and are incubated with the candidate cholesterol binding protein possessing an epitope tag. After washing steps, the protein is detected with an anti-epitope antibody and a secondary fluorescence- or radiolabelled antibody. The method allows sensitive readouts and can be performed even if no radiolabelled sterol is available.

1.6 Fluorescent Cholesterol Analogues

The cholesterol molecule has achieved evolutionary perfection to fulfil its different functions in membrane organization (Yeagle, 1985). Features that have been found to be necessary for a biologically active cholesterol analogue are a free 3β-OH, a planar tetracyclic ring system with a $\Delta^{5(6)}$ double bond, angular methyl groups, and an isooctyl side chain at the 17β-position (Schroeder, 1984) (Fig. 1.2B). The aliphatic side chain may be necessary to allow flip-flops and/or tail-to-tail transbilayer interaction of cholesterol molecules, whereas the 3′-OH group could interact with the head group of phospholipids. The structural features of a biologically active cholesterol substitute supporting ordered lipid domains are very stringent (Yeagle, 1985; Schroeder et al., 1995; Vainio et al., 2006; Megha et al., 2006). None of the cholesterol probes designed so far can claim to mimic all properties of the multifunctional cholesterol molecule. Among the fluorescent sterols two classes of probes can be distinguished: (i) intrinsically fluorescent sterols (e.g. dehydroergosterol and cholestatrienol); (ii) cholesterol probes with chemically linked fluorophores (Fig. 1.5). Both classes of probes have their specific advantages and disadvantages. Sterols belonging to the first class are more cholesterol-like but possess unfavourable spectroscopic properties. To compensate for the low quantum yield and severe photobleaching of these fluorophores, cells must be loaded with a relatively high sterol concentration. It cannot be excluded that the high amounts required for these probes preferentially force them into pathways which are untypical for cholesterol. Sterol analogues of the second class bear bulky reporter groups. However, their fluorescence properties are much better so that these probes can be applied at lower concentrations. Several fluorescent cholesterol analogues have been employed to address fundamental issues of distribution and trafficking of cholesterol. In particular, they enable the researcher to design pulse-chase experiments and/or to image the sterol in living cells. Moreover, binding assays can be developed by exploiting specific fluorescence properties such as sensitized emission, polarization, lifetime, quenching behaviour, or resonance energy transfer (Table 1.4).

1.6.1 Dehydroergosterol

Dehydroergosterol (=ergosta-5,7,9(11),22-tetraene-3β-ol) is a cholesterol analogue with intrinsic fluorescence that naturally occurs in yeasts and certain sponges. Its structure differs from cholesterol only in possessing three additional double bonds

Fig. 1.5 Fluorescent cholesterol reporters used for cholesterol imaging and/or cholesterol binding assays: dehydroergosterol, cholestatrienol (=cholesta-5,7,9(11)-triene-3β-ol), 22-NBD-cholesterol (=22-(*N*-(7-nitrobenz-2-oxa-1,3-diazol-4-yl)amino)-23,24-bisnor-5-cholen-3-ol), 25-NBD-cholesterol (=25-[N-[(7-nitro-2-1,3-benzoxadiazol-4-yl)methyl]amino]-27-norcholesterol); 6-dansyl-cholestanol; Bodipy-cholesterol, 23-(dipyrrometheneboron difluoride)-24-norcholesterol

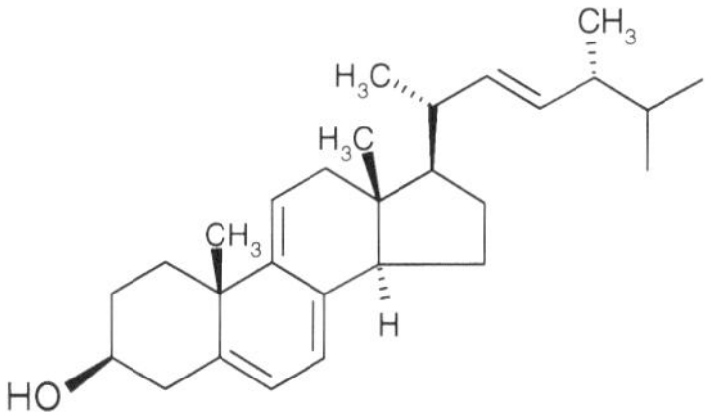

Dehydroergosterol

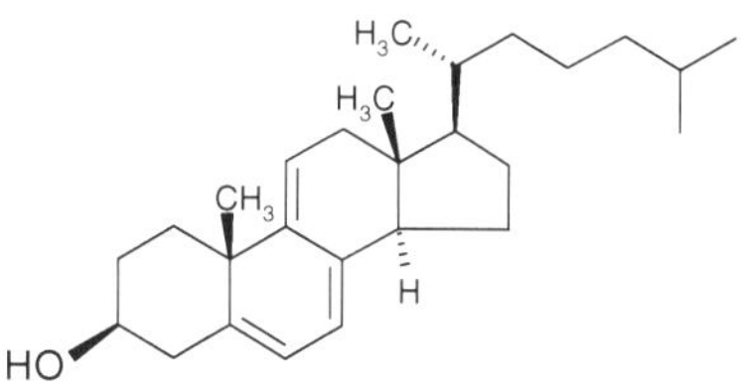

Cholestatrienol

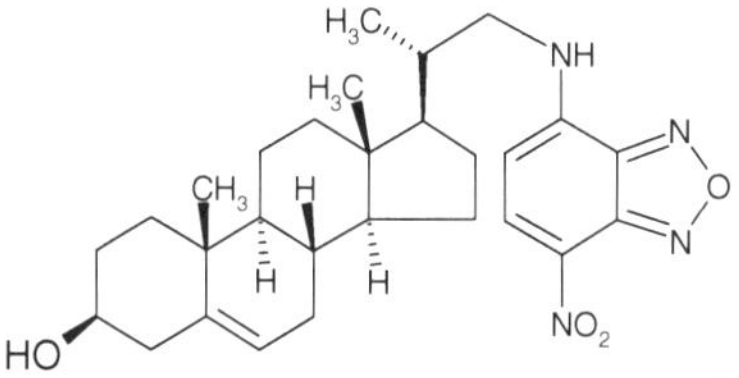

22-NBD-Cholesterol

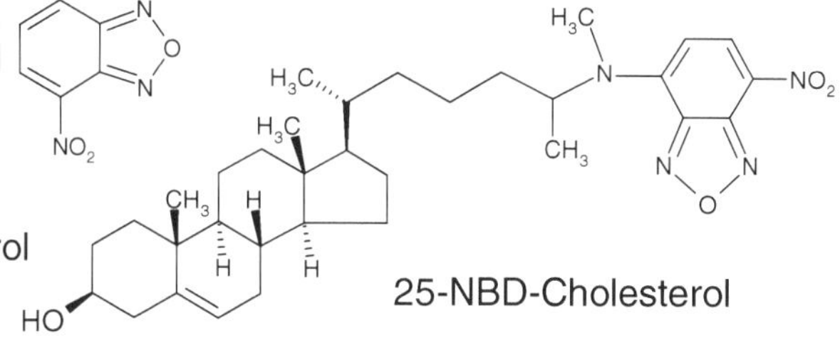

25-NBD-Cholesterol

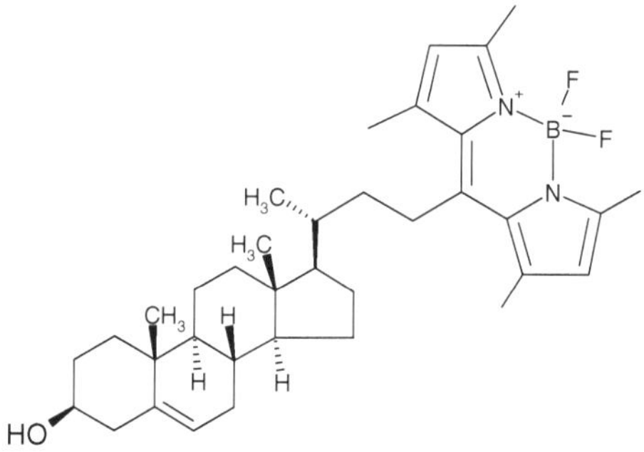

Bodipy-Cholesterol

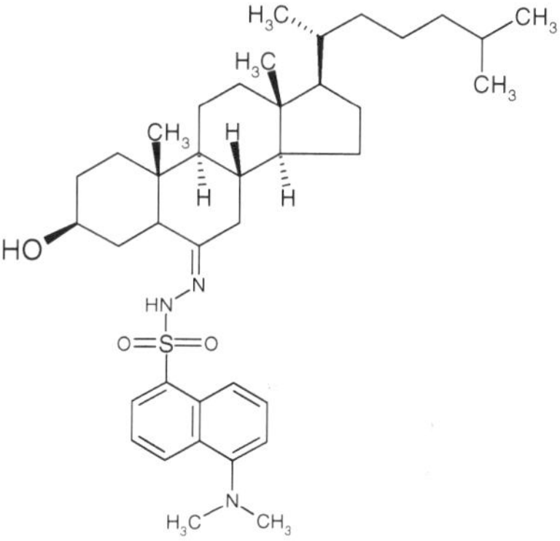

6-Dansyl-Cholestanol

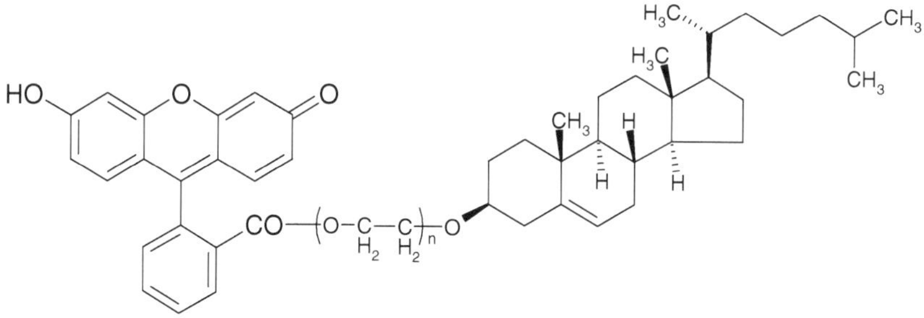

fPEG-Cholesterol (n=50 or 200)
Cholesterylpolyethyleneglycol resorcinophthalein ester

Fig. 1.5 (continued)

Table 1.4 Cholesterol binding assays using fluorescent cholesterol probes

Fluorescent probe	Protein/peptide	Method	Reference[a]
Bodipy-Cholesterol	NPC1	emission increase	1
Cholestatrienol	NPC1	emission increase	1
	Sterol carrier protein (rat liver)	polarization, lifetime energy transfer	2
	Rhodopsin	energy transfer	3
Dehydroergosterol	NPC1	emission increase	1
	NPC2	emission increase	4
	NPC2	emission increase	5
	Amyloidß-40	energy transfer, polarization	6
	Melittin	energy transfer	7
	Cryptogein	emission increase	8
	Fatty acid binding protein	emission increase polarization	9
22-NBD-cholesterol	NPC1	emission increase, collisional quenching Trp quenching	1
	STARD1	emission increase	10, 11
	Sterol carrier protein-2 (human)	emission increase	12

[a]References for Table 1.4: 1. (Liu et al., 2009b); 2. (Schroeder et al., 1985); 3. (Albert et al., 1996); 4. (Friedland et al., 2003); 5. (Liou et al., 2006); 6. (Qiu et al., 2009); 7. (Raghuraman and Chattopadhyay, 2004); 8. (Mikes et al., 1997); 9. (Nemecz and Schroeder, 1991); 10. (Baker et al., 2007); 11. (Petrescu et al., 2001); 12. (Avdulov et al., 1999).

and a methyl group at C-24 (Fig. 1.5). Dehydroergosterol is one of the best studied cholesterol probes concerning its physico-chemical properties (Schroeder, 1984; Schroeder et al., 1995). In many respects, it faithfully mimics cholesterol. For example, it co-distributes with cholesterol in both model and biological membranes, exhibits the same exchange kinetics as cholesterol in membranes, is nontoxic to cultured cells or animals, and is accepted by ACAT as a substrate for esterification (Schroeder, 1984; Smutzer et al., 1986; Schroeder et al., 1996; Frolov et al., 1996). It can also replace up to 85% of L-cell fibroblast cholesterol without causing significant detrimental effects (Schroeder, 1984). So, its employment as a cholesterol reporter should be a good choice. However, dehydroergosterol has unfavorable spectroscopic properties including a low quantum yield, excitation and emission in the UV region, and rapid photobleaching. Several imaging studies with dehydroergosterol have now been performed (Mukherjee et al., 1998; Frolov et al., 2000; Hao et al., 2002; Wustner et al., 2002, 2004; McIntosh et al., 2003; Zhang et al., 2005; Pipalia et al., 2007; Wustner, 2007; Mondal et al., 2009). Even short pulse-chase experiments were possible using dehydroergosterol complexed with methyl-β-cyclodextrin. In mouse L-fibroblasts, dehydroergosterol applied in the form of large unilamellar vesicles as donor was rapidly targeted from the plasma membrane to lipid droplets (Frolov et al., 2000; Zhang et al., 2005). In most studies,

UV-microscopy was performed using dehydroergosterol complexed with methyl-β-cyclodextrin as donor (Hao et al., 2002; Wustner et al., 2002, 2004). Hao et al. (2002) found that in a CHO cell line, dehydroergosterol was preferentially incorporated into the endocytic recycling compartment within minutes and remained there for hours. Likewise, in polarized HepG2 hepatocytes and in J774 macrophages, the influx of dehydroergosterol was reported to occur via a vesicular pathway with enrichment in recycling endosomes. In macrophages but not in hepatocytes, dehydroergosterol was also translocated to lipid droplets (Wustner et al., 2002, 2004). Since dehydroergosterol possesses a significantly higher esterification rate (>7-fold) as compared with [^{3}H]cholesterol, one may expect that after hours the sterol will be stored as ester into lipid droplets (Frolov et al., 2000). Because the enzymes responsible for esterification are localized in the endoplasmic reticulum, its substrate should also be localized there for some time. From unknown reasons dehydroergosterol has not been observed to be translocated into the endoplasmic reticulum. Dehydroergosterol has also been employed to address the question whether cholesterol-rich microdomains are present in vivo. The results of these studies provided some evidence in favour of the 'lipid rafts' concept (McIntosh et al., 2003; Zhang et al., 2005). Additionally, cholesterol binding assays for various cholesterol-binding proteins such as fatty acid binding protein (Nemecz and Schroeder, 1991), sterol carrier protein-2 (Schroeder et al., 1990), NPC1 (Liu et al., 2009b), and NPC2 (Friedland et al., 2003; Liou et al., 2006) have been established using the fluorescence properties of dehydroergosterol.

1.6.2 Cholestatrienol

Cholestatrienol (=cholesta-5,7,9(11)-triene-3β-ol) is a fluorescent cholesterol probe similar to dehydroergosterol. It differs from dehydroergosterol in the absence of both the double bond Δ^{22} and the methyl group at C-24 (Fig. 1.5). Thus, cholestatrienol possesses an isooctyl side chain like cholesterol and should therefore mimic cholesterol better than dehydroergosterol. This was indeed the case when their effects on phospholipid condensation by NMR spectroscopy were compared (Scheidt et al., 2003) Both fluorescent sterol analogues have been introduced at the same time as membrane and lipoprotein probes and can be used to measure the sterol exchange between membranes (Bergeron and Scott, 1982; Nemecz et al., 1988). In each case, cholestatrienol is regarded as a cholesterol analogue that mimics the membrane behaviour of cholesterol quite well (Fischer et al., 1984; Smutzer et al., 1986; Schroeder et al., 1988; Hyslop et al., 1990; Yeagle et al., 1990; Scheidt et al., 2003; Bjorkqvist et al., 2005).

Cholestatrienol associates with liquid ordered domains and its quenching by nitroxide-labelled lipids can report on the formation or separation of lipid domains (Bjorkqvist et al., 2005; Heczkova and Slotte, 2006). Cholestatrienol has also been evaluated as an appropriate reporter for sterol–protein interactions (Schroeder et al., 1985). A close interaction between cholesterol and rhodopsin has been demonstrated by fluorescence energy transfer from protein tryptophans to cholestatrienol in retinal rod outer segment disk membranes (Albert et al., 1996). Recently, even the imaging of cholestatrienol-specific fluorescence by confocal microscopy has been reported (Tserentsoodol et al., 2006). Low-density lipoproteins labelled with cholestatrienol crossed the blood-retina barrier and were taken up by the retina within 2 h of intravenous injection. The fluorescent sterol was observed to remain in photoreceptor outer segments for at least 24 h. Presumably, cholestatrienol became highly concentrated in retinal tissues, because imaging could not otherwise be expected under the described conditions (Tserentsoodol et al., 2006). In CHO cells, recent quenching studies performed with cholestatrienol and dehydroergosterol as cholesterol reporters and 2,4,6-trinitrobenzene sulfonic acid as membrane impermeant quencher, provided evidence that cholesterol is preferentially (~60–70%) localized in the cytoplasmic leaflet of the plasma membrane (Mondal et al., 2009).

1.6.3 NBD-Cholesterol

The 7-nitrobenz-2-oxa-1,3-diazol-4-yl (=NBD) fluorophore has been widely used as a reporter group for lipids (Chattopadhyay, 1990). The term NBD-cholesterol causes some confusion in the literature, because a couple of fluorescent cholesterol analogues are available with this name. The two predominantly applied analogues will be designated herein as 22-NBD-cholesterol and 25-NBD-cholesterol, respectively (Fig. 1.5). In further less appropriate NBD-cholesterol variants, the fluorophore including a spacer has been attached at the C-3 OH via an ester linkage. Both 22- and 25-NBD-cholesterol have been employed to study the distribution and dynamics of cholesterol in different systems. The behaviour of lateral phases in cholesterol and phosphatidylcholine monolayers has been visualized by fluorescence microscopy using 22-NBD-cholesterol (Slotte and Mattjus, 1995). Results with model membranes and 25-NBD-cholesterol as reporter indicated that cholesterol may form trans-bilayer, tail-to-tail dimers even at low sterol concentrations (Mukherjee and Chattopadhyay, 1996; Rukmini et al., 2001). McIntyre and Sleight (1991) introduced the fluorescence quenching of NBD by dithionite as an approach to measure the membrane lipid asymmetry. If the NBD group is localized at the outer leaflet of the membrane and is accessible from the aqueous phase, it can be chemically reduced to a non-fluorescent state by water-soluble dithionite. In lipid vesicles, which were selectively labelled with 22-NBD-cholesterol at the outer leaflet, dithionite reduced 95% of the NBD fluorescence (McIntyre and Sleight, 1991). Schroeder et al. (1991) explored the trans-bilayer cholesterol distribution of human erythrocytes using two approaches: photobleaching of 22-NBD-cholesterol and quenching of dehydroergosterol fluorescence. These results suggested an enrichment of cholesterol in the inner leaflet of the erythrocytes (Schroeder et al., 1991) similar to that observed in quenching studies with CHO cells (Mondal et al., 2009). In aggregates and micelles, taurocholic acid quenches the fluorescence of 22-NBD-cholesterol (Cai et al., 2002). Based on the dequenching of the NBD fluorescence, an in vitro assay has been developed to measure the exit of cholesterol from bile acid micelles (Cai et al., 2002). Concerning the quenching behavior of the NBD group, it was observed that the photostability of a NBD-labelled ceramide was strongly dependent on the cholesterol status of the Golgi apparatus where the ceramide accumulates. Cholesterol deprivation of the cells accelerated the photobleaching of the NBD-labelled ceramide several-fold, suggesting that this lipid may be used to monitor cholesterol at the Golgi compartment (Martin et al., 1993).

In recent years, 22-NBD-cholesterol has been mainly used, although the 25-NBD variant has some advantages as compared with 22-NBD-cholesterol. For example, 22-NBD-cholesterol revealed an anomalous distribution behaviour in phosphatidylcholine/cholesterol bilayers (Loura et al., 2001). 25-NBD-cholesterol contains the full isooctyl side chain like cholesterol. A chain length of at least 5 carbons at the 17β-position was necessary for sterols to form visible sterol/phospholipid domains in lipid monolayers (Mattjus et al., 1995). The spectroscopic properties of 25-NBD-cholesterol have been characterized in detail (Chattopadhyay and London, 1987, 1988). A critical evaluation of both NBD-cholesterols has been published by Scheidt et al. (2003), who observed that they adopt a reverse (up-side-down) orientation within a phospholipid bilayer. In contrast, Chattopadhyay and London (1987) found that the fluorophore of 25-NBD-cholesterol was deeply buried within the bilayer. Possibly, the high mobility of the sterol may explain these discrepant results. 22-NBD-cholesterol has successfully been employed to prove and characterize the cholesterol binding of the cholesterol-binding proteins 'steroidogenic acute regulatory protein' and 'sterol carrier protein-2' by spectroscopic techniques (Avdulov et al., 1999; Petrescu et al., 2001). When applied to CHO cells, a mistargeting to mitochondria has been observed for both 22- and 25-NBD-cholesterol (Mukherjee et al., 1998). However, in L-cell fibroblasts, 22-NBD-cholesterol distributed similarly as dehydroergosterol from the plasma membrane into lipid droplets (Frolov et al., 2000; Atshaves et al., 2000). In hamsters fed with a diet containing 22-NBD-cholesterol, the sterol was found to be absorbed (less efficiently than cholesterol) by intestinal epithelial cells and packaged into lipoproteins (Sparrow et al., 1999). Within the enterocytes most of the sterol was translocated into large apical droplets and was presumably stored there in esterified form. 22-NBD-cholesterol was verified as a good substrate for esterification in different cells (Sparrow et al., 1999; Frolov et al., 2000; Lada et al., 2004). HDL-associated 22-NBD-cholesterol was imaged in 3T3-L1 fibroblasts differentiating to adipocytes (Dagher et al., 2003). At early stages of differentiation, 22-NBD-cholesterol colocalized with scattered Golgi structures, while in developing adipocytes, the fluorescent sterol gradually concentrated in lipid droplets (Dagher et al., 2003).

1.6.4 Bodipy-Cholesterol

Several cholesterol analogs have been synthesized in which the Bodipy group was inserted into the aliphatic chain of cholesterol. The most promising reporter of this series is cholesterol linked to a Bodipy moiety at position C-24 (Fig. 1.5).

This compound preferentially partitioned into liquid-ordered domains in model membranes and giant unilamellar vesicles (Li et al., 2006; Shaw et al., 2006; Li and Bittman, 2007). Recently, this compound was used as promising tool to visualize sterol trafficking in living cells and organisms. When compared with [^{3}H]cholesterol, Bodipy-cholesterol has a higher tendency to be released from cells and its esterification rate was markedly lower (Holtta-Vuori et al., 2008).

1.6.5 Fluorescent PEG-Cholesterol

The fluorescein ester of poly(ethyleneglycol)cholesteryl ether (fPEG-Cholesterol) (Fig. 1.5) has been introduced as a special cholesterol probe (Ishiwata et al., 1997; Sato et al., 2004; Takahashi et al., 2007). Due to its water solubility and the absence of a hydroxyl group at C3, fPEG-Cholesterol is certainly not a good cholesterol mimic. However, this probe exclusively incorporates into the outer leaflet of the plasma membrane, co-localizes to some degree with rafts markers, and is thus useful to monitor the dynamics of cholesterol-rich membrane microdomains. The trafficking of fPEG-cholesterol was found to be different from that of dehydroergosterol. FPEG-cholesterol was not observed in the endocytic recycling compartment. Instead, it internalizes slowly via clathrin-independent pathways into endosomes and the Golgi region together with some raft markers. In fixed and permeabilized fibroblasts, the fluorescence pattern of fPEG-cholesterol was similar to that of filipin. The probe was also able to detect the mislocalization of cholesterol in NPC fibroblasts (Sato et al., 2004). However, one should be aware that the ester bond in fPEG-cholesterol could be easily cleaved by intracellular esterases.

1.6.6 Dansyl-Cholestanol

With the synthesis of 6-dansyl-cholestanol we have recently introduced a novel fluorescent cholesterol probe (Fig. 1.5) (Wiegand et al., 2003). The introduction of a photoreactive azo-group at the same position (6-azi-5α-cholestanol, see below) has proven to be a useful tool for cholesterol–protein interaction studies, as described below. Derivatization at position 6 did not change the biophysical parameters of the cholesterol analogue in model membranes (Mintzer et al., 2002). The 'dansyl'-group was chosen because it is one of the smallest fluorescent groups available. Using CHO cells we compared the behaviour of dansyl-cholestanol versus [^{3}H]cholesterol with respect to esterification rate, efflux kinetics, and distribution in detergent-insoluble lipid domains ('rafts'). Dansyl-cholestanol showed the same kinetics of esterification by ACAT as compared with [^{3}H]cholesterol. Also, the efflux kinetics and subcellular distribution profile were found to be same for both sterols (Wiegand et al., 2003). Further observations indicated the quality of dansyl-cholestanol as a probe for cholesterol: (i) The cellular influx of dansyl-cholestanol occurred rapidly by an energy-independent pathway via the endoplasmic reticulum. In previous biochemical studies with [^{3}H]cholesterol, it had been proposed that plasma membrane-derived cholesterol passed through the endoplasmic reticulum prior to its transfer to other intracellular sites (Lange et al., 1993; Liscum and Munn, 1999). (ii) Following inhibition of ACAT the unesterified dansyl-cholestanol accumulated in the endoplasmic reticulum in accordance with earlier predictions for cholesterol (Butler et al., 1992; Blanchette-Mackie, 2000). (iii) Dansyl-cholestanol was finally translocated to lipid droplets. This agrees well with the trafficking behavior of 22-NBD-cholesterol and dehydroergosterol as described above. (iv) In a recent imaging study, dansyl-cholestanol was also observed in cholesterol-rich microdomains and showed overall distribution patterns similar as dehydroergosterol (Petrescu et al., 2009). (v) Analysis of the membrane penetration depth revealed that the dansyl group of the probe is localized at the interfacial region of the membrane in agreement with the location of cholesterol in fluid-phase membranes (Shrivastava et al., 2009). Thus, 6-dansyl-cholestanol is certainly a promising cholesterol probe. One disadvantage concerns its substantial photobleaching that shortens the imaging time.

1.7 Spin-Labelled Cholesterol

Spin-labeled lipids provide information about the structure of biological membranes by using nuclear magnetic and electron spin resonance spectroscopy. Cholesterol analogues with a nitroxide spin-label (doxyl moiety) attached at the C-3 or C-25 position have been synthesized to analyze the orientation, distribution and transbilayer movements of the cholesterol probe in liposomes and biological membranes. Spin–spin interaction of 3β-doxyl-5α-cholestane in liposomes provided evidence for the formation of cholesterol-enriched domains (Tampe et al., 1991). Using this probe, it was also observed that cholesterol undergoes a rapid transbilayer movement (<1 min) in liposomes and human erythrocytes (Muller and Herrmann, 2002). An even faster flip-flop (<1 s) of cholesterol has been reported (Steck et al., 2002). Concerning its condensing effect on phospholipids the spin-labeled compound 25-doxyl-cholesterol (Fig. 1.6) was found to be an excellent cholesterol analogue. This sterol probe revealed a cholesterol-like orientation, with the doxyl group at C-25 facing the chain termini of the phospholipids (Scheidt et al., 2003). The localization of the doxyl group in the membrane interior was confirmed by the finding that the nitroxide label was inaccessible from the aqueous phase as it could not be reduced by ascorbate (Scheidt et al., 2003). Probes with the spin-label group accessible from the aqueous phase (e.g. at C-3 as in 3β-doxyl-5α-cholestane) allow to measure cholesterol flip-flop by chemical reduction of the nitroxide radical with ascorbate (Morrot et al., 1987). However, cholesterol analogues with modifications at C-3 are not regarded as faithful mimics of cholesterol. For example, 3β-doxyl-5α-cholestane was not able to exert a comparable condensing effect on phospholipids as cholesterol (Scheidt et al., 2003). Investigations on the transmembrane diffusion of lipids obtained with spin-labeled and fluorescent lipid probes have recently been summarized (Devaux et al., 2002).

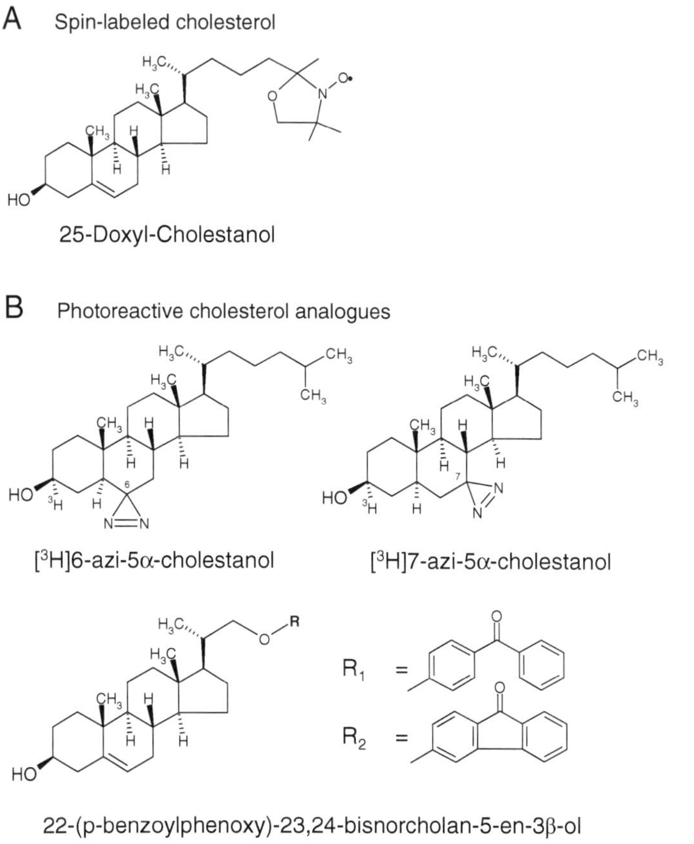

Fig. 1.6 Spin-labelled (**A**) and photoreactive (**B**) cholesterol analogues: (**A**) 25-doxyl-cholestanol; (**B**) [^{3}H]6-azi-5α-cholestanol, [^{3}H]7-azi-5α-cholestanol, 22-(p-benzoylphenoxy)-23,24-bisnorcholan-5-en-3β-ol (D, R=R$_1$), the fluorenone moiety (D, R=R$_2$)

1.8 Affinity Labelling with Photoreactive Cholesterol

Specific cholesterol binding proteins can be directly identified by the usage of photoreactive cholesterol analogues. In the first photoreactive cholesterol analogues that were synthesized, the photoreactive groups were incorporated either at the C-3 position (cholesteryl diazoacetate, 3α-azido-5-cholestene, or 3α-(4-azido-3-iodosalicylic)-cholest-5-ene) (Middlemas and Raftery, 1987; Corbin et al., 1998) or at the aliphatic side chain of cholesterol (25-azidonorcholesterol or sterols with diazoacetate, aryldiazirines or fluorodiazirine attached at C-22 or C-24) (Stoffel and Klotzbucher, 1978; Terasawa et al., 1986). Unfortunately, not many applications have been described for most of these compounds. The nicotinic acetylcholine receptor binds cholesterol but reveals very low structure–activity requirements for cholesterol. Even analogues derivatized at the C-3 positions with a broad range of substituents or bile acid derivatives support receptor activity (Fernandez et al., 1993; Corbin et al., 1998). Using 3α-(4-azido-3-iodosalicylic)-cholest-5-ene or the bile acid p-azidophenacyl 3α-hydroxy-5β-cholan-24-ate as photoreactive probes, all subunits of the nicotinic acetylcholine receptor could be labeled in membranes or proteoliposomes (Corbin et al., 1998). Photoreactive cholesteryl diazoacetate also labeled the nicotinic acetylcholine receptor (Middlemas and Raftery, 1987). Although this probe is modified at C-3, it immobilized in lipid bilayers like cholesterol and upon irradiation incorporated into the choline head group of phosphatidylcholine (Keilbaugh and Thornton, 1983). However, cholesteryl diazoacetate behaved differently to cholesterol concerning its exchange kinetics from unilamellar vesicles (Kan et al., 1992). The acetylcholine receptor may be an exception concerning its broad tolerance for cholesterol substitutes. To develop a probe with more general applicability, we synthesized [^{3}H]6-azi-5α-cholestanol in which both the C-3 and the isooctyl side chain left unattached (Fig. 1.6B) (Gimpl and Gehrig-Burger, 2007). The azi-group was introduced at position C-6 because this modification was functionally tolerated by the oxytocin receptor that we studied in detail with respect to its specific requirement for cholesterol (Klein et al., 1995; Gimpl et al., 1995, 1997; Burger, 2000). The first application of this compound (designated as photocholesterol) was published by Thiele et al. (2000). Up to now, several putative cholesterol binding proteins have been labelled with 6-azi-5α-cholestanol, among these are synaptophysin, caveolin (Thiele et al., 2000), vitellogenins (Matyash et al., 2001), proteolipid protein (Simons et al., 2000; Kramer-Albers et al., 2006), tetraspanins (Charrin et al., 2003), cholesterol absorption proteins in enterocytes (Kramer et al., 2003), STARD3 (Alpy et al., 2005; Reitz et al., 2008), and the E1 fusion protein from Semliki Forest virus (Umashankar et al., 2008). In all these studies, 6-azi-5α-cholestanol was primarily used to identify or confirm the cholesterol binding of a candidate protein. Recently, we demonstrated for STARD1 that this photoreactive probe could also be used to identify cholesterol binding sites within a protein (Reitz et al., 2008). Another related tritiated photoreactive cholesterol analogue, [^{3}H]7-azi-5α-cholestanol (Fig. 1.6B), has been synthesized by Cruz et al. (2002). A direct binding of this analogue with caveolin-1 and Niemann-Pick C1 (NPC1) protein has been demonstrated (Cruz et al., 2002; Ohgami et al., 2004; Liu et al., 2009b).

Spencer et al. (2004) synthesized a series of eight benzophenone-containing photoreactive cholesterol analogues. Due to the larger size of these photophores compared with the diazirines, these sterol analogues have the disadvantage of being less cholesterol-like. On the other hand, benzophenone derivatives show a high crosslinking yield and a preferential reaction with C-H bonds, which may be beneficial for the sterol labelling of some proteins. In one group of benzophenone-containing cholesterol probes, the photophore moiety extended, or replaced, most of the cholesterol isooctyl side chain. In another group of analogues, the photophore was attached at C-3 via an amide linkage. Surprisingly, for all of these analogues even those with modifications at C-3 were similarly effective as cholesterol when tested in an apolipoprotein A-I dependent sterol efflux assay. This indicates that at least in relation to certain transport pathways of cholesterol, biological membranes show an unexpected tolerance for cholesterol substitutes (Spencer et al., 2004). One of these analogues, tritiated 22-(p-benzoylphenoxy)-23,24-bisnorcholan-5-en-3β-ol (Fig. 1.6B, R=R_1), photolabelled caveolin effectively (Fielding et al., 2002). Fluorenone-containing cholesterol probes that are structurally similar to the corresponding benzophenone derivatives (Fig. 1.6B, R=R_2) represent a further interesting group of compounds since they are both photoreactive and fluorescent (Spencer et al., 2006). Two such analogues behaved similar to cholesterol in the above mentioned sterol efflux assay (Spencer et al., 2006).

In experiments with photoreactive cholesterol analogues, one has to consider that at least in the plasma membrane cholesterol is always present in large amounts, so that each integral membrane protein faces cholesterol in its direct environment. One expects that only membrane proteins are photolabelled that possess one ore more specific cholesterol docking site(s). Membrane proteins residing in cholesterol-enriched microdomains are good candidates. However, even if they are functionally dependent on cholesterol, their affinity for cholesterol could be low (~ millimolar range), since embedded in a cholesterol-rich environment they will always be in contact with cholesterol. In contrast, cholesterol-dependent membrane proteins residing in cholesterol-poor organelles such as the mitochondrion or the endoplasmic reticulum, should be evolutionary selected towards higher affinity for cholesterol. Therefore, it should be considered that the employment of high concentrations of photoreactive cholesterol could artefactually label proteins that do not possess a specific cholesterol binding site.

1.9 Concluding Remarks

A couple of membrane proteins such as GPCRs, receptor tyrosine kinases and ion channels, have been shown to reside in cholesterol-rich microdomains. Some of them may directly interact with cholesterol. However, due to the abundance of cholesterol in the plasma membrane, particularly in lipid rafts, the affinity of these proteins to cholesterol may be very low and therefore difficult to determine with traditional radioligand binding assays. Therefore, alternative binding protocols are required. To explore the binding or interaction of proteins to cholesterol, a variety of cholesterol probes bearing radio-, spin-, photoaffinity- or fluorescent labels are currently available. Examples of proven cholesterol binding molecules are polyene compounds, cholesterol-dependent cytolysins, enzymes accepting cholesterol as substrate, and proteins with cholesterol binding motifs. As far as we know to date, cholesterol binding domains are heterogenous structures existing either as hydrophobic cavities, an assembly of several transmembrane helices, or small stretches of amino acids. Possibly, water-soluble cholesterol binding modules could be applied in the future as 'solubilizers' and competitors for cholesterol binding. Of course, to be useful as potential cholesterol donors in binding assays, the affinity of these binding modules to cholesterol has to be lower than that of the candidate cholesterol binding protein. In contrast, high-affinity cholesterol binding domains might be useful as efficient cholesterol acceptors and could then substitute for the widely used β-cyclodextrins that currently function as non-specific cholesterol carrier molecules. Among the cholesterol modifying enzymes, the cholesterol oxidases have achieved a wide range of applications. They are used to determine the cholesterol concentration in all kinds of biological samples such as membranes, cells, serum, or food. Additionally, susceptibility to cholesterol oxidase provides information about the localization of cholesterol and the structure of cholesterol-containing membranes. Filipin, a member of the polyene compounds, is the standard reporter for the distribution of free cholesterol in fixed cells. Among the family of cholesterol-dependent cytolysin, perfringolysin O and fragments therefrom have been introduced to selectively label the localization of cholesterol-rich microdomains. Hopefully, other members of this family will be added as cholesterol probes in the future. Cholesterol research has been markedly stimulated by the development of different cholesterol derivatives such as photoreactive, spin-labelled, and fluorescent cholesterol analogues. The administration of photoreactive cholesterol probes offers a direct approach to identify and define cholesterol binding sites, whereas fluorescent sterols allow us to explore the trafficking and distribution of cholesterol in vivo. In addition, fluorescent cholesterol analogues have been established as important tools to analyze cholesterol-protein binding. Due to tremendous progress in microscopy/spectroscopy (e.g. FRET analysis, lifetime microscopy) in recent years, cholesterol research will focus more and more on living cells, thereby expanding our knowledge on all facets of cholesterol–protein interaction in the future.

Acknowledgements I thank Christa Wolpert for technical assistance and Falk Fahrenholz, Katja Gehrig-Burger, Volker Wiegand, Conny Trossen, and Julian Reitz for discussions and cooperations in cholesterol research over recent years.

References

Adams, C.M., Reitz, J., De Brabander, J.K., Feramisco, J.D., Li, L., Brown, M.S., Goldstein, J.L., 2004, Cholesterol and 25-hydroxycholesterol inhibit activation of SREBPs by different mechanisms, both involving SCAP and Insigs. *J. Biol. Chem.* **279**: 52772–52780.

Ahn, K.W., Sampson, N.S., 2004, Cholesterol oxidase senses subtle changes in lipid bilayer structure. *Biochemistry* **43**: 827–836.

Albert, A.D., Young, J.E., Yeagle, P.L., 1996, Rhodopsin-cholesterol interactions in bovine rod outer segment disk membranes. *Biochim. Biophys. Acta* **1285**: 47–55.

Alpy, F., Latchumanan, V.K., Kedinger, V., Janoshazi, A., Thiele, C., Wendling, C., Rio, M.C., Tomasetto, C., 2005, Functional characterization of the MENTAL domain. *J. Biol. Chem.* **280**: 17945–17952.

Anderluh, G., Lakey, J.H., 2008, Disparate proteins use similar architectures to damage membranes. *Trends Biochem. Sci.* **33**: 482–490.

Atshaves, B.P., Starodub, O., McIntosh, A., Petrescu, A., Roths, J.B., Kier, A.B., Schroeder, F., 2000, Sterol carrier protein-2 alters high density lipoprotein-mediated cholesterol efflux. *J. Biol. Chem.* **275**: 36852–36861.

Avdulov, N.A., Chochina, S.V., Igbavboa, U., Warden, C.S., Schroeder, F., Wood, W.G., 1999, Lipid binding to sterol carrier protein-2 is inhibited by ethanol. *Biochim. Biophys. Acta* **1437**: 37–45.

Baker, B.Y., Epand, R.F., Epand, R.M., Miller, W.L., 2007, Cholesterol binding does not predict activity of the steroidogenic acute regulatory protein, StAR. *J. Biol. Chem.* **282**: 10223–10232.

Bar, L.K., Chong, P.L., Barenholz, Y., Thompson, T.E., 1989, Spontaneous transfer between phospholipid bilayers of dehydroergosterol, a fluorescent cholesterol analog. *Biochim. Biophys. Acta* **983**: 109–112.

Bari, M., Battista, N., Fezza, F., Finazzi-Agro, A., Maccarrone, M., 2005, Lipid rafts control signaling of type-1 cannabinoid receptors in neuronal cells. Implications for anandamide-induced apoptosis. *J. Biol. Chem.* **280**: 12212–12220.

Behnke, O., Tranum-Jensen, J., van, D.B., 1984, Filipin as a cholesterol probe. II. Filipin-cholesterol interaction in red blood cell membranes. *Eur. J. Cell Biol.* **35**: 200–215.

Bergeron, R.J., Scott, J., 1982, Cholestatriene and ergostatetraene as in vivo and in vitro membrane and lipoprotein probes. *J. Lipid Res.* **23**: 391–404.

Bjorkqvist, Y.J., Nyholm, T.K., Slotte, J.P., Ramstedt, B., 2005, Domain formation and stability in complex lipid bilayers as reported by cholestatrienol. *Biophys. J.* **88**: 4054–4063.

Blanchette-Mackie, E.J., 2000, Intracellular cholesterol trafficking: role of the NPC1 protein. *Biochim. Biophys. Acta* **1486**: 171–183.

Bolard, J., 1986, How do the polyene macrolide antibiotics affect the cellular membrane properties? *Biochim. Biophys. Acta* **864**: 257–304.

Brannigan, G., Henin, J., Law, R., Eckenhoff, R., Klein, M.L., 2008, Embedded cholesterol in the nicotinic acetylcholine receptor. *Proc. Natl. Acad. Sci. U. S. A* **105**: 14418–14423.

Brasaemle, D.L., Robertson, A.D., Attie, A.D., 1988, Transbilayer movement of cholesterol in the human erythrocyte membrane. *J. Lipid Res.* **29**: 481–489.

Brown, A.J., Sun, L., Feramisco, J., Brown, M.S., Goldstein, J.L., 2002, Cholesterol addition to ER membranes alters conformation of SCAP, the SREBP escort protein that regulates cholesterol metabolism. *Mol. Cell* **10**: 237–245.

Brown, D.A., Rose, J.K., 1992, Sorting of GPI-anchored proteins to glycolipid-enriched membrane subdomains during transport to the apical cell surface. *Cell* **68**: 533–544.

Burger, K., 2000, Cholesterin und Progesteron – Modulatoren G-Protein gekoppelter Signaltransduktionswege. *Dissertation, Mainz*.

Burger, K., Gimpl, G., Fahrenholz, F., 2000, Regulation of receptor function by cholesterol. *Cell Mol. Life Sci.* **57**: 1577–1592.

Butler, J.D., Blanchette-Mackie, J., Goldin, E., O'Neill, R.R., Carstea, G., Roff, C.F., Patterson, M.C., Patel, S., Comly, M.E., Cooney, A., 1992, Progesterone blocks cholesterol translocation from lysosomes. *J. Biol. Chem.* **267**: 23797–23805.

Butler, J.D., Comly, M.E., Kruth, H.S., Vanier, M., Filling-Katz, M., Fink, J., Barton, N., Weintroub, H., Quirk, J.M., Tokoro, T., 1987, Niemann-Pick variant disorders: comparison of errors of cellular cholesterol homeostasis in group D and group C fibroblasts. *Proc. Natl. Acad. Sci. U. S. A* **84**: 556–560.

Cai, T.Q., Guo, Q., Wong, B., Milot, D., Zhang, L., Wright, S.D., 2002, Protein-disulfide isomerase is a component of an NBD-cholesterol monomerizing protein complex from hamster small intestine. *Biochim. Biophys. Acta* **1581**: 100–108.

Challa, R., Ahuja, A., Ali, J., Khar, R.K., 2005, Cyclodextrins in drug delivery: an updated review. *AAPS PharmSciTech.* **6**: E329–E357.

Chamberlain, L.H., 2004, Detergents as tools for the purification and classification of lipid rafts. *FEBS Lett.* **559**: 1–5.

Chamberlain, L.H., Gould, G.W., 2002, The vesicle- and target-SNARE proteins that mediate Glut4 vesicle fusion are localized in detergent-insoluble lipid rafts present on distinct intracellular membranes. *J. Biol. Chem.* **277**: 49750–49754.

Chang, C.C., Lee, C.Y., Chang, E.T., Cruz, J.C., Levesque, M.C., Chang, T.Y., 1998, Recombinant acyl-CoA:cholesterol acyltransferase-1 (ACAT-1) purified to essential homogeneity utilizes cholesterol in mixed micelles or in vesicles in a highly cooperative manner. *J. Biol. Chem.* **273**: 35132–35141.

Charrin, S., Manie, S., Thiele, C., Billard, M., Gerlier, D., Boucheix, C., Rubinstein, E., 2003, A physical and functional link between cholesterol and tetraspanins. *Eur. J. Immunol.* **33**: 2479–2489.

Chattopadhyay, A., 1990, Chemistry and biology of N-(7-nitrobenz-2-oxa-1,3-diazol-4-yl)-labeled lipids: fluorescent probes of biological and model membranes. *Chem. Phys. Lipids* **53**: 1–15.

Chattopadhyay, A., London, E., 1987, Parallax method for direct measurement of membrane penetration depth utilizing fluorescence quenching by spin-labeled phospholipids. *Biochemistry* **26**: 39–45.

Chattopadhyay, A., London, E., 1988, Spectroscopic and ionization properties of N-(7-nitrobenz-2-oxa-1,3-diazol-4-yl)-labeled lipids in model membranes. *Biochim. Biophys. Acta* **938**: 24–34.

Chen, L., Lyubimov, A.Y., Brammer, L., Vrielink, A., Sampson, N.S., 2008, The binding and release of oxygen and hydrogen peroxide are directed by a hydrophobic tunnel in cholesterol oxidase. *Biochemistry* **47**: 5368–5377.

Cheruku, S.R., Xu, Z., Dutia, R., Lobel, P., Storch, J., 2006, Mechanism of cholesterol transfer from the Niemann-Pick type C2 protein to model membranes supports a role in lysosomal cholesterol transport. *J. Biol. Chem.* **281**: 31594–31604.

Chiang, Y.R., Ismail, W., Heintz, D., Schaeffer, C., Van, D.A., Fuchs, G., 2008, Study of anoxic and oxic cholesterol metabolism by Sterolibacterium denitrificans. *J. Bacteriol.* **190**: 905–914.

Chini, B., Parenti, M., 2004, G-protein coupled receptors in lipid rafts and caveolae: how, when and why do they go there? *J. Mol. Endocrinol.* **32**: 325–338.

Corbin, J., Wang, H.H., Blanton, M.P., 1998, Identifying the cholesterol binding domain in the nicotinic acetylcholine receptor with [125I]azido-cholesterol. *Biochim. Biophys. Acta* **1414**: 65–74.

Cruz, J.C., Thomas, M., Wong, E., Ohgami, N., Sugii, S., Curphey, T., Chang, C.C., Chang, T.Y., 2002, Synthesis and biochemical properties of a new photoactivatable cholesterol analog 7,7-azocholestanol and its linoleate ester in Chinese hamster ovary cell lines. *J. Lipid Res.* **43**: 1341–1347.

Dagher, G., Donne, N., Klein, C., Ferre, P., Dugail, I., 2003, HDL-mediated cholesterol uptake and targeting to lipid droplets in adipocytes. *J. Lipid Res.* **44**: 1811–1820.

Das, A., Davis, M.A., Rudel, L.L., 2008, Identification of putative active site residues of ACAT enzymes. *J. Lipid Res.* **49**: 1770–1781.

Devaux, P.F., Fellmann, P., Herve, P., 2002, Investigation on lipid asymmetry using lipid probes: Comparison between spin-labeled lipids and fluorescent lipids. *Chem. Phys. Lipids* **116**: 115–134.

Douglass, A.D., Vale, R.D., 2005, Single-molecule microscopy reveals plasma membrane microdomains created by protein-protein networks that exclude or trap signaling molecules in T cells. *Cell* **121**: 937–950.

Dowler, S., Kular, G., Alessi, D.R., 2002, Protein lipid overlay assay. *Sci. STKE* **2002**: L6.

Drevot, P., Langlet, C., Guo, X.J., Bernard, A.M., Colard, O., Chauvin, J.P., Lasserre, R., He, H.T., 2002, TCR signal initiation machinery is pre-assembled and activated in a subset of membrane rafts. *EMBO J.* **21**: 1899–1908.

Elias, P.M., Friend, D.S., Goerke, J., 1979, Membrane sterol heterogeneity. Freeze-fracture detection with saponins and filipin. *J. Histochem. Cytochem.* **27**: 1247–1260.

Epshtein, Y., Chopra, A.P., Rosenhouse-Dantsker, A., Kowalsky, G.B., Logothetis, D.E., Levitan, I., 2009, Identification of a C-terminus domain critical for the sensitivity of Kir2.1 to cholesterol. *Proc. Natl. Acad. Sci. U. S. A* **106**: 8055–8060.

Eroglu, C., Brugger, B., Wieland, F., Sinning, I., 2003, Glutamate-binding affinity of Drosophila metabotropic glutamate receptor is modulated by association with lipid rafts. *Proc. Natl. Acad. Sci. U. S. A* **100**: 10219–10224.

Fernandez, A.M., Fernandez-Ballester, G., Ferragut, J.A., Gonzalez-Ros, J.M., 1993, Labeling of the nicotinic acetylcholine receptor by a photoactivatable steroid probe: effects of cholesterol and cholinergic ligands. *Biochim. Biophys. Acta* **1149**: 135–144.

Fielding, P.E., Russel, J.S., Spencer, T.A., Hakamata, H., Nagao, K., Fielding, C.J., 2002, Sterol efflux to apolipoprotein A-I originates from caveolin-rich microdomains and potentiates PDGF-dependent protein kinase activity. *Biochemistry* **41**: 4929–4937.

Fischer, R.T., Stephenson, F.A., Shafiee, A., Schroeder, F., 1984, delta 5,7,9(11)-Cholestatrien-3 beta-ol: a fluorescent cholesterol analogue. *Chem. Phys. Lipids* **36**: 1–14.

Florek, M., Bauer, N., Janich, P., Wilsch-Braeuninger, M., Fargeas, C.A., Marzesco, A.M., Ehninger, G., Thiele, C., Huttner, W.B., Corbeil, D., 2007, Prominin-2 is a cholesterol-binding protein associated with apical and basolateral plasmalemmal protrusions in polarized epithelial cells and released into urine. *Cell Tissue Res.* **328**: 31–47.

Friedland, N., Liou, H.L., Lobel, P., Stock, A.M., 2003, Structure of a cholesterol-binding protein deficient in Niemann-Pick type C2 disease. *Proc. Natl. Acad. Sci. U. S. A* **100**: 2512–2517.

Frolov, A., Petrescu, A., Atshaves, B.P., So, P.T., Gratton, E., Serrero, G., Schroeder, F., 2000, High density lipoprotein-mediated cholesterol uptake and targeting to lipid droplets in intact L-cell fibroblasts. A single- and multiphoton fluorescence approach. *J. Biol. Chem.* **275**: 12769–12780.

Frolov, A., Woodford, J.K., Murphy, E.J., Billheimer, J.T., Schroeder, F., 1996, Spontaneous and protein-mediated sterol transfer between intracellular membranes. *J. Biol. Chem.* **271**: 16075–16083.

Fujimoto, T., Hayashi, M., Iwamoto, M., Ohno-Iwashita, Y., 1997, Crosslinked plasmalemmal cholesterol is sequestered to caveolae: analysis with a new cytochemical probe. *J. Histochem. Cytochem.* **45**: 1197–1205.

Gehrig-Burger, K., Kohout, L., Gimpl, G., 2005, CHAPSTEROL. A novel cholesterol-based detergent. *FEBS J.* **272**: 800–812.

Ghoshroy, K.B., Zhu, W., Sampson, N.S., 1997, Investigation of membrane disruption in the reaction catalyzed by cholesterol oxidase. *Biochemistry* **36**: 6133–6140.

Gimpl, G., Burger, K., Fahrenholz, F., 1997, Cholesterol as modulator of receptor function. *Biochemistry* **36**: 10959–10974.

Gimpl, G., Fahrenholz, F., 2000, Human oxytocin receptors in cholesterol-rich vs. cholesterol-poor microdomains of the plasma membrane. *Eur. J. Biochem.* **267**: 2483–2497.

Gimpl, G., Gehrig-Burger, K., 2007, Cholesterol reporter molecules. *Biosci. Rep.* **27**: 335–358.

Gimpl, G., Klein, U., Reilander, H., Fahrenholz, F., 1995, Expression of the human oxytocin receptor in baculovirus-infected insect cells: high-affinity binding is induced by a cholesterol-cyclodextrin complex. *Biochemistry* **34**: 13794–13801.

Gronberg, L., Slotte, J.P., 1990, Cholesterol oxidase catalyzed oxidation of cholesterol in mixed lipid monolayers: effects of surface pressure and phospholipid composition on catalytic activity. *Biochemistry* **29**: 3173–3178.

Gustavsson, J., Parpal, S., Karlsson, M., Ramsing, C., Thorn, H., Borg, M., Lindroth, M., Peterson, K.H., Magnusson, K.E., Stralfors, P., 1999, Localization of the insulin receptor in caveolae of adipocyte plasma membrane. *FASEB J.* **13**: 1961–1971.

Hadders, M.A., Beringer, D.X., Gros, P., 2007, Structure of C8alpha-MACPF reveals mechanism of membrane attack in complement immune defense. *Science* **317**: 1552–1554.

Hanson, M.A., Cherezov, V., Griffith, M.T., Roth, C.B., Jaakola, V.P., Chien, E.Y., Velasquez, J., Kuhn, P., Stevens, R.C., 2008, A specific cholesterol binding site is established by the 2.8 A structure of the human beta2-adrenergic receptor. *Structure.* **16**: 897–905.

Hao, M., Lin, S.X., Karylowski, O.J., Wustner, D., McGraw, T.E., Maxfield, F.R., 2002, Vesicular and non-vesicular sterol transport in living cells. The endocytic recycling compartment is a major sterol storage organelle. *J. Biol. Chem.* **277**: 609–617.

Heczkova, B., Slotte, J.P., 2006, Effect of anti-tumor ether lipids on ordered domains in model membranes. *FEBS Lett.* **580**: 2471–2476.

Heerklotz, H., 2002, Triton promotes domain formation in lipid raft mixtures. *Biophys. J.* **83**: 2693–2701.

Holtta-Vuori, M., Uronen, R.L., Repakova, J., Salonen, E., Vattulainen, I., Panula, P., Li, Z., Bittman, R., Ikonen, E., 2008, BODIPY-cholesterol: a new tool to visualize sterol trafficking in living cells and organisms. *Traffic* **9**: 1839–1849.

Huang, P., Xu, W., Yoon, S.I., Chen, C., Chong, P.L., Liu-Chen, L.Y., 2007, Cholesterol reduction by methyl-beta-cyclodextrin attenuates the delta opioid receptor-mediated signaling in neuronal cells but enhances it in non-neuronal cells. *Biochem. Pharmacol.* **73**: 534–549.

Hyslop, P.A., Morel, B., Sauerheber, R.D., 1990, Organization and interaction of cholesterol and phosphatidylcholine in model bilayer membranes. *Biochemistry* **29**: 1025–1038.

Ilangumaran, S., Hoessli, D.C., 1998, Effects of cholesterol depletion by cyclodextrin on the sphingolipid microdomains of the plasma membrane. *Biochem. J.* **335 (Pt 2)**: 433–440.

Infante, R.E., bi-Mosleh, L., Radhakrishnan, A., Dale, J.D., Brown, M.S., Goldstein, J.L., 2008a, Purified NPC1 protein. I. Binding of cholesterol and oxysterols to a 1278-amino acid membrane protein. *J. Biol. Chem.* **283**: 1052–1063.

Infante, R.E., Radhakrishnan, A., bi-Mosleh, L., Kinch, L.N., Wang, M.L., Grishin, N.V., Goldstein, J.L., Brown, M.S., 2008b, Purified NPC1 protein: II. Localization of sterol binding to a 240-amino acid soluble luminal loop. *J. Biol. Chem.* **283**: 1064–1075.

Infante, R.E., Wang, M.L., Radhakrishnan, A., Kwon, H.J., Brown, M.S., Goldstein, J.L., 2008c, NPC2 facilitates bidirectional transfer of cholesterol between NPC1 and lipid bilayers, a step in cholesterol egress from lysosomes. *Proc. Natl. Acad. Sci. U. S. A* **105**: 15287–15292.

Irie, T., Fukunaga, K., Pitha, J., 1992, Hydroxypropylcyclodextrins in parenteral use. I: Lipid dissolution and effects on lipid transfers in vitro. *J. Pharm. Sci.* **81**: 521–523.

Ishiwata, H., Sato, S.B., Vertut-Doi, A., Hamashima, Y., Miyajima, K., 1997, Cholesterol derivative of poly(ethylene glycol) inhibits clathrin-independent, but not clathrin-dependent endocytosis. *Biochim. Biophys. Acta* **1359**: 123–135.

Iwamoto, M., Morita, I., Fukuda, M., Murota, S., Ando, S., Ohno-Iwashita, Y., 1997, A biotinylated perfringolysin O derivative: a new probe for detection of cell surface cholesterol. *Biochim. Biophys. Acta* **1327**: 222–230.

Jamin, N., Neumann, J.M., Ostuni, M.A., Vu, T.K., Yao, Z.X., Murail, S., Robert, J.C., Giatzakis, C., Papadopoulos, V., Lacapere, J.J., 2005, Characterization of the cholesterol recognition amino acid consensus sequence of the peripheral-type benzodiazepine receptor. *Mol. Endocrinol.* **19**: 588–594.

Kallen, J.A., Schlaeppi, J.M., Bitsch, F., Geisse, S., Geiser, M., Delhon, I., Fournier, B., 2002, X-ray structure of the hRORalpha LBD at 1.63 A: structural and functional data that cholesterol or a cholesterol derivative is the natural ligand of RORalpha. *Structure* **10**: 1697–1707.

Kan, C.C., Yan, J., Bittman, R., 1992, Rates of spontaneous exchange of synthetic radiolabeled sterols between lipid vesicles. *Biochemistry* **31**: 1866–1874.

Keilbaugh, S.A., Thornton, E.R., 1983, Synthesis and photoreactivity of cholesteryl diazoacetate: a novel photolabeling reagent. *J. Am. Chem. Soc.* **105**: 3283–3286.

Kenworthy, A.K., 2008, Have we become overly reliant on lipid rafts? Talking point on the involvement of lipid rafts in T-cell activation. *EMBO Rep.* **9**: 531–535.

Kilsdonk, E.P., Yancey, P.G., Stoudt, G.W., Bangerter, F.W., Johnson, W.J., Phillips, M.C., Rothblat, G.H., 1995, Cellular cholesterol efflux mediated by cyclodextrins. *J. Biol. Chem.* **270**: 17250–17256.

Kinsky, S.C., Luse, S.A., Zopf, D., van Deenen, L.L., Haxby, J., 1967, Interaction of filipin and derivatives with erythrocyte membranes and lipid dispersions: electron microscopic observations. *Biochim. Biophys. Acta* **135**: 844–861.

Kishi, K., Watazu, Y., Katayama, Y., Okabe, H., 2000, The characteristics and applications of recombinant cholesterol dehydrogenase. *Biosci. Biotechnol. Biochem.* **64**: 1352–1358.

Klein, U., Gimpl, G., Fahrenholz, F., 1995, Alteration of the myometrial plasma membrane cholesterol content with beta-cyclodextrin modulates the binding affinity of the oxytocin receptor. *Biochemistry* **34**: 13784–13793.

Ko, D.C., Binkley, J., Sidow, A., Scott, M.P., 2003, The integrity of a cholesterol-binding pocket in Niemann-Pick C2 protein is necessary to control lysosome cholesterol levels. *Proc. Natl. Acad. Sci. U. S. A.* **100**: 2518–2525.

Kramer, W., Girbig, F., Corsiero, D., Burger, K., Fahrenholz, F., Jung, C., Muller, G., 2003, Intestinal cholesterol absorption: identification of different binding proteins for cholesterol and cholesterol absorption inhibitors in the enterocyte brush border membrane. *Biochim. Biophys. Acta* **1633**: 13–26.

Kramer-Albers, E.M., Gehrig-Burger, K., Thiele, C., Trotter, J., Nave, K.A., 2006, Perturbed interactions of mutant proteolipid protein/DM20 with cholesterol and lipid rafts in oligodendroglia: implications for dysmyelination in spastic paraplegia. *J. Neurosci.* **26**: 11743–11752.

Kurzchalia, T.V., Parton, R.G., 1999, Membrane microdomains and caveolae. *Curr. Opin. Cell Biol.* **11**: 424–431.

Kuwabara, P.E., Labouesse, M., 2002, The sterol-sensing domain: multiple families, a unique role? *Trends Genet.* **18**: 193–201.

Lada, A.T., Davis, M., Kent, C., Chapman, J., Tomoda, H., Omura, S., Rudel, L.L., 2004, Identification of ACAT1- and ACAT2-specific inhibitors using a novel, cell based fluorescence assay: individual ACAT uniqueness. *J. Lipid Res.* **45**: 378–386.

Lafont, F., Simons, K., Ikonen, E., 1995, Dissecting the molecular mechanisms of polarized membrane traffic: reconstitution of three transport steps in epithelial cells using streptolysin-O permeabilization. *Cold Spring Harb. Symp. Quant. Biol.* **60**: 753–762.

Lange, Y., 1991, Disposition of intracellular cholesterol in human fibroblasts. *J. Lipid Res.* **32**: 329–339.

Lange, Y., 1992, Tracking cell cholesterol with cholesterol oxidase. *J. Lipid Res.* **33**: 315–321.

Lange, Y., Dolde, J., Steck, T.L., 1981, The rate of transmembrane movement of cholesterol in the human erythrocyte. *J. Biol. Chem.* **256**: 5321–5323.

Lange, Y., Matthies, H., Steck, T.L., 1984, Cholesterol oxidase susceptibility of the red cell membrane. *Biochim. Biophys. Acta* **769**: 551–562.

Lange, Y., Matthies, H.J., 1984, Transfer of cholesterol from its site of synthesis to the plasma membrane. *J. Biol. Chem.* **259**: 14624–14630.

Lange, Y., Strebel, F., Steck, T.L., 1993, Role of the plasma membrane in cholesterol esterification in rat hepatoma cells. *J. Biol. Chem.* **268**: 13838–13843.

Lange, Y., Swaisgood, M.H., Ramos, B.V., Steck, T.L., 1989, Plasma membranes contain half the phospholipid and 90% of the cholesterol and sphingomyelin in cultured human fibroblasts. *J. Biol. Chem.* **264**: 3786–3793.

Lange, Y., Ye, J., Steck, T.L., 2004, How cholesterol homeostasis is regulated by plasma membrane cholesterol in excess of phospholipids. *Proc. Natl. Acad. Sci. U. S. A* **101**: 11664–11667.

Lange, Y., Ye, J., Steck, T.L., 2005, Activation of membrane cholesterol by displacement from phospholipids. *J. Biol. Chem.* **280**: 36126–36131.

Lascombe, M.B., Ponchet, M., Venard, P., Milat, M.L., Blein, J.P., Prange, T., 2002, The 1.45 A resolution structure of the cryptogein-cholesterol complex: a close-up view of a sterol carrier protein (SCP) active site. *Acta Crystallogr. D. Biol. Crystallogr.* **58**: 1442–1447.

Lee, K.A., Fuda, H., Lee, Y.C., Negishi, M., Strott, C.A., Pedersen, L.C., 2003, Crystal structure of human cholesterol sulfotransferase (SULT2B1b) in the presence of pregnenolone and $3'$-phosphoadenosine $5'$-phosphate. Rationale for specificity differences between prototypical SULT2A1 and the SULT2BG1 isoforms. *J. Biol. Chem.* **278**: 44593–44599.

Li, H., Yao, Z., Degenhardt, B., Teper, G., Papadopoulos, V., 2001, Cholesterol binding at the cholesterol recognition/ interaction amino acid consensus (CRAC) of the peripheral-type benzodiazepine receptor and inhibition of steroidogenesis by an HIV TAT-CRAC peptide. *Proc. Natl. Acad. Sci. U. S. A* **98**: 1267–1272.

Li, J., Vrielink, A., Brick, P., Blow, D.M., 1993, Crystal structure of cholesterol oxidase complexed with a steroid substrate: implications for flavin adenine dinucleotide dependent alcohol oxidases. *Biochemistry* **32**: 11507–11515.

Li, Z., Bittman, R., 2007, Synthesis and spectral properties of cholesterol- and FTY720-containing boron dipyrromethene dyes. *J. Org. Chem.* **72**: 8376–8382.

Li, Z., Mintzer, E., Bittman, R., 2006, First synthesis of free cholesterol-BODIPY conjugates. *J. Org. Chem.* **71**: 1718–1721.

Lichtenberg, D., Goni, F.M., Heerklotz, H., 2005, Detergent-resistant membranes should not be identified with membrane rafts. *Trends Biochem. Sci.* **30**: 430–436.

Liou, H.L., Dixit, S.S., Xu, S., Tint, G.S., Stock, A.M., Lobel, P., 2006, NPC2, the protein deficient in Niemann-Pick C2 disease, consists of multiple glycoforms that bind a variety of sterols. *J. Biol. Chem.* **281**: 36710–36723.

Liscum, L., Munn, N.J., 1999, Intracellular cholesterol transport. *Biochim. Biophys. Acta* **1438**: 19–37.

Liu, B., Turley, S.D., Burns, D.K., Miller, A.M., Repa, J.J., Dietschy, J.M., 2009a, Reversal of defective lysosomal transport in NPC disease ameliorates liver dysfunction and neurodegeneration in the npc1-/- mouse. *Proc. Natl. Acad. Sci. U. S. A* **106**: 2377–2382.

Liu, R., Lu, P., Chu, J.W., Sharom, F.J., 2009b, Characterization of fluorescent sterol binding to purified human NPC1. *J. Biol. Chem.* **284**: 1840–1852.

Lopes, S.C., Goormaghtigh, E., Cabral, B.J., Castanho, M.A., 2004, Filipin orientation revealed by linear dichroism. Implication for a model of action. *J. Am. Chem. Soc.* **126**: 5396–5402.

Loura, L.M., Fedorov, A., Prieto, M., 2001, Exclusion of a cholesterol analog from the cholesterol-rich phase in model membranes. *Biochim. Biophys. Acta* **1511**: 236–243.

Macdonald, J.L., Pike, L.J., 2005, A simplified method for the preparation of detergent-free lipid rafts. *J. Lipid Res.* **46**: 1061–1067.

MacLachlan, J., Wotherspoon, A.T., Ansell, R.O., Brooks, C.J., 2000, Cholesterol oxidase: sources, physical properties and analytical applications. *J. Steroid Biochem. Mol. Biol.* **72**: 169–195.

Madore, N., Smith, K.L., Graham, C.H., Jen, A., Brady, K., Hall, S., Morris, R., 1999, Functionally different GPI proteins are organized in different domains on the neuronal surface. *EMBO J.* **18**: 6917–6926.

Mahammad, S., Parmryd, I., 2008, Cholesterol homeostasis in T cells. Methyl-beta-cyclodextrin treatment results in equal loss of cholesterol from Triton X-100 soluble and insoluble fractions. *Biochim. Biophys. Acta* **1778**: 1251–1258.

Martin, O.C., Comly, M.E., Blanchette-Mackie, E.J., Pentchev, P.G., Pagano, R.E., 1993, Cholesterol deprivation affects the fluorescence properties of a ceramide analog at the Golgi apparatus of living cells. *Proc. Natl. Acad. Sci. U. S. A* **90**: 2661–2665.

Marzesco, A.M., Wilsch-Brauninger, M., Dubreuil, V., Janich, P., Langenfeld, K., Thiele, C., Huttner, W.B., Corbeil, D., 2009, Release of extracellular membrane vesicles from microvilli of epithelial cells is enhanced by depleting membrane cholesterol. *FEBS Lett.* **583**: 897–902.

Mast, N., White, M.A., Bjorkhem, I., Johnson, E.F., Stout, C.D., Pikuleva, I.A., 2008, Crystal structures of substrate-bound and substrate-free cytochrome P450 46A1, the principal cholesterol hydroxylase in the brain. *Proc. Natl. Acad. Sci. U. S. A* **105**: 9546–9551.

Mattjus, P., Bittman, R., Vilcheze, C., Slotte, J.P., 1995, Lateral domain formation in cholesterol/phospholipid monolayers as affected by the sterol side chain conformation. *Biochim. Biophys. Acta* **1240**: 237–247.

Matyash, V., Geier, C., Henske, A., Mukherjee, S., Hirsh, D., Thiele, C., Grant, B., Maxfield, F.R., Kurzchalia, T.V., 2001, Distribution and transport of cholesterol in Caenorhabditis elegans. *Mol. Biol. Cell* **12**: 1725–1736.

McIntosh, A.L., Gallegos, A.M., Atshaves, B.P., Storey, S.M., Kannoju, D., Schroeder, F., 2003, Fluorescence and multiphoton imaging resolve unique structural forms of sterol in membranes of living cells. *J. Biol. Chem.* **278**: 6384–6403.

McIntyre, J.C., Sleight, R.G., 1991, Fluorescence assay for phospholipid membrane asymmetry. *Biochemistry* **30**: 11819–11827.

Megha, B.O., London, E., 2006, Cholesterol precursors stabilize ordinary and ceramide-rich ordered lipid domains (lipid rafts) to different degrees. Implications for the Bloch hypothesis and sterol biosynthesis disorders. *J. Biol. Chem.* **281**: 21903–21913.

Middlemas, D.S., Raftery, M.A., 1987, Identification of subunits of acetylcholine receptor that interact with a cholesterol photoaffinity probe. *Biochemistry* **26**: 1219–1223.

Mikes, V., Milat, M.L., Ponchet, M., Ricci, P., Blein, J.P., 1997, The fungal elicitor cryptogein is a sterol carrier protein. *FEBS Lett.* **416**: 190–192.

Mintzer, E.A., Waarts, B.L., Wilschut, J., Bittman, R., 2002, Behavior of a photoactivatable analog of cholesterol, 6- photocholesterol, in model membranes. *FEBS Lett.* **510**: 181–184.

Mobius, W., Ohno-Iwashita, Y., van Donselaar, E.G., Oorschot, V.M., Shimada, Y., Fujimoto, T., Heijnen, H.F., Geuze, H.J., Slot, J.W., 2002, Immunoelectron microscopic localization of cholesterol using biotinylated and non-cytolytic perfringolysin O. *J. Histochem. Cytochem.* **50**: 43–55.

Mondal, M., Mesmin, B., Mukherjee, S., Maxfield, F.R., 2009, Sterols are mainly in the cytoplasmic leaflet of the plasma membrane and the endocytic recycling compartment in CHO cells. *Mol. Biol. Cell* **20**: 581–588.

Monnaert, V., Tilloy, S., Bricout, H., Fenart, L., Cecchelli, R., Monflier, E., 2004, Behavior of alpha-, beta-, and gamma-cyclodextrins and their derivatives on an in vitro model of blood-brain barrier. *J. Pharmacol. Exp. Ther.* **310**: 745–751.

Morrot, G., Bureau, J.F., Roux, M., Maurin, L., Favre, E., Devaux, P.F., 1987, Orientation and vertical fluctuations of spin-labeled analogues of cholesterol and androstanol in phospholipid bilayers. *Biochim. Biophys. Acta* **897**: 341–345.

Mukherjee, S., Chattopadhyay, A., 1996, Membrane organization at low cholesterol concentrations: a study using 7-nitrobenz-2-oxa-1,3-diazol-4-yl-labeled cholesterol. *Biochemistry* **35**: 1311–1322.

Mukherjee, S., Zha, X., Tabas, I., Maxfield, F.R., 1998, Cholesterol distribution in living cells: fluorescence imaging using dehydroergosterol as a fluorescent cholesterol analog. *Biophys. J.* **75**: 1915–1925.

Muller, P., Herrmann, A., 2002, Rapid transbilayer movement of spin-labeled steroids in human erythrocytes and in liposomes. *Biophys. J.* **82**: 1418–1428.

Munro, S., 2003, Lipid rafts: elusive or illusive? *Cell* **115**: 377–388.

Murata, M., Peranen, J., Schreiner, R., Wieland, F., Kurzchalia, T.V., Simons, K., 1995, VIP21/caveolin is a cholesterol-binding protein. *Proc. Natl. Acad. Sci. U. S. A* **92**: 10339–10343.

Murcia, M., Faraldo-Gomez, J.D., Maxfield, F.R., Roux, B., 2006, Modeling the structure of the StART domains of MLN64 and StAR proteins in complex with cholesterol. *J. Lipid Res.* **47**: 2614–2630.

Nabi, I.R., Le, P.U., 2003, Caveolae/raft-dependent endocytosis. *J. Cell Biol.* **161**: 673–677.

Nelson, L.D., Johnson, A.E., London, E., 2008, How interaction of perfringolysin O with membranes is controlled by sterol structure, lipid structure, and physiological low pH: insights into the origin of perfringolysin O-lipid raft interaction. *J. Biol. Chem.* **283**: 4632–4642.

Nemecz, G., Fontaine, R.N., Schroeder, F., 1988, A fluorescence and radiolabel study of sterol exchange between membranes. *Biochim. Biophys. Acta* **943**: 511–521.

Nemecz, G., Schroeder, F., 1991, Selective binding of cholesterol by recombinant fatty acid binding proteins. *J. Biol. Chem.* **266**: 17180–17186.

Ohgami, N., Ko, D.C., Thomas, M., Scott, M.P., Chang, C.C., Chang, T.Y., 2004, Binding between the Niemann-Pick C1 protein and a photoactivatable cholesterol analog requires a functional sterol-sensing domain. *Proc. Natl. Acad. Sci. U. S. A* **101**: 12473–12478.

Ohno-Iwashita, Y., Iwamoto, M., Ando, S., Iwashita, S., 1992, Effect of lipidic factors on membrane cholesterol topology–mode of binding of theta-toxin to cholesterol in liposomes. *Biochim. Biophys. Acta* **1109**: 81–90.

Ohtani, Y., Irie, T., Uekama, K., Fukunaga, K., Pitha, J., 1989, Differential effects of alpha-, beta- and gamma-cyclodextrins on human erythrocytes. *Eur. J. Biochem.* **186**: 17–22.

Okamoto, Y., Ninomiya, H., Miwa, S., Masaki, T., 2000, Cholesterol oxidation switches the internalization pathway of endothelin receptor type A from caveolae to clathrin-coated pits in Chinese hamster ovary cells. *J. Biol. Chem.* **275**: 6439–6446.

Okamura, N., Kiuchi, S., Tamba, M., Kashima, T., Hiramoto, S., Baba, T., Dacheux, F., Dacheux, J.L., Sugita, Y., Jin, Y.Z., 1999, A porcine homolog of the major secretory protein of human epididymis, HE1, specifically binds cholesterol. *Biochim. Biophys. Acta* **1438**: 377–387.

Orci, L., Perrelet, A., Montesano, R., 1983, Differential filipin labeling of the luminal membranes lining the pancreatic acinus. *J. Histochem. Cytochem.* **31**: 952–955.

Ottico, E., Prinetti, A., Prioni, S., Giannotta, C., Basso, L., Chigorno, V., Sonnino, S., 2003, Dynamics of membrane lipid domains in neuronal cells differentiated in culture. *J. Lipid Res.* **44**: 2142–2151.

Palmer, C.P., Mahen, R., Schnell, E., Djamgoz, M.B., Aydar, E., 2007, Sigma-1 receptors bind cholesterol and remodel lipid rafts in breast cancer cell lines. *Cancer Res.* **67**: 11166–11175.

Palmer, M., 2001, The family of thiol-activated, cholesterol-binding cytolysins. *Toxicon* **39**: 1681–1689.

Pang, L., Graziano, M., Wang, S., 1999, Membrane cholesterol modulates galanin-GalR2 interaction. *Biochemistry* **38**: 12003–12011.

Papanikolaou, A., Papafotika, A., Murphy, C., Papamarcaki, T., Tsolas, O., Drab, M., Kurzchalia, T.V., Kasper, M., Christoforidis, S., 2005, Cholesterol-dependent lipid assemblies regulate the activity of the ecto-nucleotidase CD39. *J. Biol. Chem.* **280**: 26406–26414.

Patzer, E.J., Wagner, R.R., 1978, Cholesterol oxidase as a probe for studying membrane organisation. *Nature* **274**: 394–395.

Pelletier, R.M., Vitale, M.L., 1994, Filipin vs enzymatic localization of cholesterol in guinea pig, mink, and mallard duck testicular cells. *J. Histochem. Cytochem.* **42**: 1539–1554.

Persaud-Sawin, D.A., Lightcap, S., Harry, G.J., 2009, Isolation of rafts from mouse brain tissue by a detergent-free method. *J. Lipid Res.* **50**: 759–767.

Petrescu, A.D., Gallegos, A.M., Okamura, Y., Strauss, J.F., III, Schroeder, F., 2001, Steroidogenic acute regulatory protein binds cholesterol and modulates mitochondrial membrane sterol domain dynamics. *J. Biol. Chem.* **276**: 36970–36982.

Petrescu, A.D., Vespa, A., Huang, H., McIntosh, A.L., Schroeder, F., Kier, A.B., 2009, Fluorescent sterols monitor cell penetrating peptide Pep-1 mediated uptake and intracellular targeting of cargo protein in living cells. *Biochim. Biophys. Acta* **1788**: 425–441.

Pipalia, N.H., Hao, M., Mukherjee, S., Maxfield, F.R., 2007, Sterol, protein and lipid trafficking in Chinese hamster ovary cells with Niemann-Pick type C1 defect. *Traffic.* **8**: 130–141.

Prigent, D., Alouf, J.E., 1976, Interaction of steptolysin O with sterols. *Biochim. Biophys. Acta* **443**: 288–300.

Prior, I.A., Harding, A., Yan, J., Sluimer, J., Parton, R.G., Hancock, J.F., 2001, GTP-dependent segregation of H-ras from lipid rafts is required for biological activity. *Nat. Cell Biol.* **3**: 368–375.

Pucadyil, T.J., Chattopadhyay, A., 2004, Cholesterol modulates ligand binding and G-protein coupling to serotonin(1A) receptors from bovine hippocampus. *Biochim. Biophys. Acta* **1663**: 188–200.

Pucadyil, T.J., Chattopadhyay, A., 2006, Role of cholesterol in the function and organization of G-protein coupled receptors. *Prog. Lipid Res.* **45**: 295–333.

Pucadyil, T.J., Shrivastava, S., Chattopadhyay, A., 2005, Membrane cholesterol oxidation inhibits ligand binding function of hippocampal serotonin(1A) receptors. *Biochem. Biophys. Res. Commun.* **331**: 422–427.

Qiu, L., Lewis, A., Como, J., Vaughn, M.W., Huang, J., Somerharju, P., Virtanen, J., Cheng, K.H., 2009, Cholesterol modulates the interaction of beta-amyloid peptide with lipid bilayers. *Biophys. J.* **96**: 4299–4307.

Radhakrishnan, A., McConnell, H.M., 2000, Chemical activity of cholesterol in membranes. *Biochemistry* **39**: 8119–8124.

Radhakrishnan, A., Sun, L.P., Kwon, H.J., Brown, M.S., Goldstein, J.L., 2004, Direct binding of cholesterol to the purified membrane region of SCAP: mechanism for a sterol-sensing domain. *Mol. Cell* **15**: 259–268.

Raghuraman, H., Chattopadhyay, A., 2004, Interaction of melittin with membrane cholesterol: a fluorescence approach. *Biophys. J.* **87**: 2419–2432.

Ramachandran, R., Heuck, A.P., Tweten, R.K., Johnson, A.E., 2002, Structural insights into the membrane-anchoring mechanism of a cholesterol-dependent cytolysin. *Nat. Struct. Biol.* **9**: 823–827.

Reid, P.C., Sakashita, N., Sugii, S., Ohno-Iwashita, Y., Shimada, Y., Hickey, W.F., Chang, T.Y., 2004, A novel cholesterol stain reveals early neuronal cholesterol accumulation in the Niemann-Pick type C1 mouse brain. *J. Lipid Res.* **45**: 582–591.

Reitz, J., Gehrig-Burger, K., Strauss, J.F., III, Gimpl, G., 2008, Cholesterol interaction with the related steroidogenic acute regulatory lipid-transfer (START) domains of StAR (STARD1) and MLN64 (STARD3). *FEBS J.* **275**: 1790–1802.

Rodriguez-Agudo, D., Ren, S., Hylemon, P.B., Redford, K., Natarajan, R., Del, C.A., Gil, G., Pandak, W.M., 2005, Human StarD5, a cytosolic StAR-related lipid binding protein. *J. Lipid Res.* **46**: 1615–1623.

Rodriguez-Agudo, D., Ren, S., Wong, E., Marques, D., Redford, K., Gil, G., Hylemon, P., Pandak, W.M., 2008, Intracellular cholesterol transporter StarD4 binds free cholesterol and increases cholesteryl ester formation. *J. Lipid Res.* **49**: 1409–1419.

Romanowski, M.J., Soccio, R.E., Breslow, J.L., Burley, S.K., 2002, Crystal structure of the Mus musculus cholesterol-regulated START protein 4 (StarD4) containing a StAR-related lipid transfer domain. *Proc. Natl. Acad. Sci. U. S. A* **99**: 6949–6954.

Roper, K., Corbeil, D., Huttner, W.B., 2000, Retention of prominin in microvilli reveals distinct cholesterol-based lipid micro-domains in the apical plasma membrane. *Nat. Cell Biol.* **2**: 582–592.

Rossjohn, J., Feil, S.C., McKinstry, W.J., Tweten, R.K., Parker, M.W., 1997, Structure of a cholesterol-binding, thiol-activated cytolysin and a model of its membrane form. *Cell* **89**: 685–692.

Rossjohn, J., Polekhina, G., Feil, S.C., Morton, C.J., Tweten, R.K., Parker, M.W., 2007, Structures of perfringolysin O suggest a pathway for activation of cholesterol-dependent cytolysins. *J. Mol. Biol.* **367**: 1227–1236.

Rukmini, R., Rawat, S.S., Biswas, S.C., Chattopadhyay, A., 2001, Cholesterol organization in membranes at low concentrations: effects of curvature stress and membrane thickness. *Biophys. J.* **81**: 2122–2134.

Sato, S.B., Ishii, K., Makino, A., Iwabuchi, K., Yamaji-Hasegawa, A., Senoh, Y., Nagaoka, I., Sakuraba, H., Kobayashi, T., 2004, Distribution and transport of cholesterol-rich membrane domains monitored by a membrane-impermeant fluorescent polyethylene glycol-derivatized cholesterol. *J. Biol. Chem.* **279**: 23790–23796.

Scheidt, H.A., Muller, P., Herrmann, A., Huster, D., 2003, The potential of fluorescent and spin-labeled steroid analogs to mimic natural cholesterol. *J. Biol. Chem.* **278**: 45563–45569.

Schroeder, F., 1984, Fluorescent sterols: probe molecules of membrane structure and function. *Prog. Lipid Res.* **23**: 97–113.

Schroeder, F., Butko, P., Nemecz, G., Scallen, T.J., 1990, Interaction of fluorescent delta 5,7,9(11),22-ergostatetraen-3 beta-ol with sterol carrier protein-2. *J. Biol. Chem.* **265**: 151–157.

Schroeder, F., Dempsey, M.E., Fischer, R.T., 1985, Sterol and squalene carrier protein interactions with fluorescent delta 5,7,9(11)-cholestatrien-3 beta-ol. *J. Biol. Chem.* **260**: 2904–2911.

Schroeder, F., Frolov, A.A., Murphy, E.J., Atshaves, B.P., Jefferson, J.R., Pu, L., Wood, W.G., Foxworth, W.B., Kier, A.B., 1996, Recent advances in membrane cholesterol domain dynamics and intracellular cholesterol trafficking. *Proc. Soc. Exp. Biol. Med.* **213**: 150–177.

Schroeder, F., Nemecz, G., Gratton, E., Barenholz, Y., Thompson, T.E., 1988, Fluorescence properties of cholestatrienol in phosphatidylcholine bilayer vesicles. *Biophys. Chem.* **32**: 57–72.

Schroeder, F., Nemecz, G., Wood, W.G., Joiner, C., Morrot, G., yraut-Jarrier, M., Devaux, P.F., 1991, Transmembrane distribution of sterol in the human erythrocyte. *Biochim. Biophys. Acta* **1066**: 183–192.

Schroeder, F., Woodford, J.K., Kavecansky, J., Wood, W.G., Joiner, C., 1995, Cholesterol domains in biological membranes. *Mol. Membr. Biol.* **12**: 113–119.

Schuck, S., Honsho, M., Ekroos, K., Shevchenko, A., Simons, K., 2003, Resistance of cell membranes to different detergents. *Proc. Natl. Acad. Sci. U. S. A* **100**: 5795–5800.

Severs, N.J., Simons, H.L., 1983, Failure of filipin to detect cholesterol-rich domains in smooth muscle plasma membrane. *Nature* **303**: 637–638.

Shah, M.B., Sehgal, P.B., 2007, Nondetergent isolation of rafts. *Methods Mol. Biol.* **398**: 21–28.

Shaw, A.S., 2006, Lipid rafts: now you see them, now you don't. *Nat. Immunol.* **7**: 1139–1142.

Shaw, J.E., Epand, R.F., Epand, R.M., Li, Z., Bittman, R., Yip, C.M., 2006, Correlated fluorescence-atomic force microscopy of membrane domains: structure of fluorescence probes determines lipid localization. *Biophys. J.* **90**: 2170–2178.

Sheets, E.D., Holowka, D., Baird, B., 1999, Critical role for cholesterol in Lyn-mediated tyrosine phosphorylation of FcepsilonRI and their association with detergent-resistant membranes. *J. Cell Biol.* **145**: 877–887.

Shimada, Y., Maruya, M., Iwashita, S., Ohno-Iwashita, Y., 2002, The C-terminal domain of perfringolysin O is an essential cholesterol-binding unit targeting to cholesterol-rich microdomains. *Eur. J. Biochem.* **269**: 6195–6203.

Shrivastava, S., Haldar, S., Gimpl, G., Chattopadhyay, A., 2009, Orientation and dynamics of a novel fluorescent cholesterol analogue in membranes of varying phase. *J. Phys. Chem. B* **113**: 4475–4481.

Sieber, J.J., Willig, K.I., Kutzner, C., Gerding-Reimers, C., Harke, B., Donnert, G., Rammner, B., Eggeling, C., Hell, S.W., Grubmuller, H., Lang, T., 2007, Anatomy and dynamics of a supramolecular membrane protein cluster. *Science* **317**: 1072–1076.

Simons, K., Ikonen, E., 1997, Functional rafts in cell membranes. *Nature* **387**: 569–572.

Simons, K., Toomre, D., 2000, Lipid rafts and signal transduction. *Nat. Rev. Mol. Cell Biol.* **1**: 31–39.

Simons, K., van, M.G., 1988, Lipid sorting in epithelial cells. *Biochemistry* **27**: 6197–6202.

Simons, M., Kramer, E.M., Thiele, C., Stoffel, W., Trotter, J., 2000, Assembly of myelin by association of proteolipid protein with cholesterol- and galactosylceramide-rich membrane domains. *J. Cell Biol.* **151**: 143–154.

Sjogren, B., Hamblin, M.W., Svenningsson, P., 2006, Cholesterol depletion reduces serotonin binding and signaling via human 5-HT(7(a)) receptors. *Eur. J. Pharmacol.* **552**: 1–10.

Slimane, T.A., Trugnan, G., Van IJzendoorn, S.C., Hoekstra, D., 2003, Raft-mediated trafficking of apical resident proteins occurs in both direct and transcytotic pathways in polarized hepatic cells: role of distinct lipid microdomains. *Mol. Biol. Cell* **14**: 611–624.

Slotte, J.P., Hedstrom, G., Rannstrom, S., Ekman, S., 1989, Effects of sphingomyelin degradation on cell cholesterol oxidizability and steady-state distribution between the cell surface and the cell interior. *Biochim. Biophys. Acta* **985**: 90–96.

Slotte, J.P., Mattjus, P., 1995, Visualization of lateral phases in cholesterol and phosphatidylcholine monolayers at the air/water interface–a comparative study with two different reporter molecules. *Biochim. Biophys. Acta* **1254**: 22–29.

Smart, E.J., Ying, Y.S., Conrad, P.A., Anderson, R.G., 1994, Caveolin moves from caveolae to the Golgi apparatus in response to cholesterol oxidation. *J. Cell Biol.* **127**: 1185–1197.

Smart, E.J., Ying, Y.S., Mineo, C., Anderson, R.G., 1995, A detergent-free method for purifying caveolae membrane from tissue culture cells. *Proc. Natl. Acad. Sci. U. S. A* **92**: 10104–10108.

Smutzer, G., Crawford, B.F., Yeagle, P.L., 1986, Physical properties of the fluorescent sterol probe dehydroergosterol. *Biochim. Biophys. Acta* **862**: 361–371.

Soltani, C.E., Hotze, E.M., Johnson, A.E., Tweten, R.K., 2007, Structural elements of the cholesterol-dependent cytolysins that are responsible for their cholesterol-sensitive membrane interactions. *Proc. Natl. Acad. Sci. U. S. A* **104**: 20226–20231.

Song, K.S., Li, S., Okamoto, T., Quilliam, L.A., Sargiacomo, M., Lisanti, M.P., 1996, Co-purification and direct interaction of Ras with caveolin, an integral membrane protein of caveolae microdomains. Detergent-free purification of caveolae microdomains. *J. Biol. Chem.* **271**: 9690–9697.

Sparrow, C.P., Patel, S., Baffic, J., Chao, Y.S., Hernandez, M., Lam, M.H., Montenegro, J., Wright, S.D., Detmers, P.A., 1999, A fluorescent cholesterol analog traces cholesterol absorption in hamsters and is esterified in vivo and in vitro. *J. Lipid Res.* **40**: 1747–1757.

Spencer, T.A., Wang, P., Li, D., Russel, J.S., Blank, D.H., Huuskonen, J., Fielding, P.E., Fielding, C.J., 2004, Benzophenone-containing cholesterol surrogates: synthesis and biological evaluation. *J. Lipid Res.* **45**: 1510–1518.

Spencer, T.A., Wang, P., Popovici-Muller, J.V., Peltan, I.D., Fielding, P.E., Fielding, C.J., 2006, Preparation and biochemical evaluation of fluorenone-containing lipid analogs. *Bioorg. Med. Chem. Lett.* **16**: 3000–3004.

Steck, T.L., Ye, J., Lange, Y., 2002, Probing red cell membrane cholesterol movement with cyclodextrin. *Biophys. J.* **83**: 2118–2125.

Steer, C.J., Bisher, M., Blumenthal, R., Steven, A.C., 1984, Detection of membrane cholesterol by filipin in isolated rat liver coated vesicles is dependent upon removal of the clathrin coat. *J. Cell Biol.* **99**: 315–319.

Stoffel, W., Klotzbucher, R., 1978, Inhibition of cholesterol synthesis in cultured cells by 25-azidonorcholesterol. *Hoppe Seylers. Z. Physiol Chem.* **359**: 199–209.

Sugii, S., Reid, P.C., Ohgami, N., Shimada, Y., Maue, R.A., Ninomiya, H., Ohno-Iwashita, Y., Chang, T.Y., 2003, Biotinylated theta toxin derivative as a probe to examine intracellular cholesterol-rich domains in normal and Niemann-pick type C1 cells. *J. Lipid Res.* **44**: 1033–1041.

Takahashi, M., Murate, M., Fukuda, M., Sato, S.B., Ohta, A., Kobayashi, T., 2007, Cholesterol controls lipid endocytosis through Rab11. *Mol. Biol. Cell* **18**: 2667–2677.

Tampe, R., von, L.A., Galla, H.J., 1991, Glycophorin-induced cholesterol-phospholipid domains in dimyristoylphosphatidylcholine bilayer vesicles. *Biochemistry* **30**: 4909–4916.

Tashiro, Y., Yamazaki, T., Shimada, Y., Ohno-Iwashita, Y., Okamoto, K., 2004, Axon-dominant localization of cell-surface cholesterol in cultured hippocampal neurons and its disappearance in Niemann-Pick type C model cells. *Eur. J. Neurosci.* **20**: 2015–2021.

Terasawa, T., Ikekawa, N., Morisaki, M., 1986, Syntheses of cholesterol analogs with a carbene-generating substituent on the side chain. *Chem. Pharm. Bull.* **34**: 931–934.

Thiele, C., Hannah, M.J., Fahrenholz, F., Huttner, W.B., 2000, Cholesterol binds to synaptophysin and is required for biogenesis of synaptic vesicles. *Nat. Cell Biol.* **2**: 42–49.

Tserentsoodol, N., Sztein, J., Campos, M., Gordiyenko, N.V., Fariss, R.N., Lee, J.W., Fliesler, S.J., Rodriguez, I.R., 2006, Uptake of cholesterol by the retina occurs primarily via a low density lipoprotein receptor-mediated process. *Mol. Vis.* **12**: 1306–1318.

Tsujishita, Y., Hurley, J.H., 2000, Structure and lipid transport mechanism of a StAR-related domain. *Nat. Struct. Biol.* **7**: 408–414.

Tweten, R.K., Parker, M.W., Johnson, A.E., 2001, The cholesterol-dependent cytolysins. *Curr. Top. Microbiol. Immunol.* **257**: 15–33.

Umashankar, M., Sanchez-San, M.C., Liao, M., Reilly, B., Guo, A., Taylor, G., Kielian, M., 2008, Differential cholesterol binding by class II fusion proteins determines membrane fusion properties. *J. Virol.* **82**: 9245–9253.

Vainio, S., Jansen, M., Koivusalo, M., Rog, T., Karttunen, M., Vattulainen, I., Ikonen, E., 2006, Significance of sterol structural specificity. Desmosterol cannot replace cholesterol in lipid rafts. *J. Biol. Chem.* **281**: 348–355.

Vaughan, A.M., Oram, J.F., 2005, ABCG1 redistributes cell cholesterol to domains removable by high density lipoprotein but not by lipid-depleted apolipoproteins. *J. Biol. Chem.* **280**: 30150–30157.

Vincent, N., Genin, C., Malvoisin, E., 2002, Identification of a conserved domain of the HIV-1 transmembrane protein gp41 which interacts with cholesteryl groups. *Biochim. Biophys. Acta* **1567**: 157–164.

Vrielink, A., Lloyd, L.F., Blow, D.M., 1991, Crystal structure of cholesterol oxidase from Brevibacterium sterolicum refined at 1.8 A resolution. *J. Mol. Biol.* **219**: 533–554.

Waheed, A.A., Shimada, Y., Heijnen, H.F., Nakamura, M., Inomata, M., Hayashi, M., Iwashita, S., Slot, J.W., Ohno-Iwashita, Y., 2001, Selective binding of perfringolysin O derivative to cholesterol-rich membrane microdomains (rafts). *Proc. Natl. Acad. Sci. U. S. A* **98**: 4926–4931.

Wang, M.M., Olsher, M., Sugar, I.P., Chong, P.L., 2004, Cholesterol superlattice modulates the activity of cholesterol oxidase in lipid membranes. *Biochemistry* **43**: 2159–2166.

Watson, K.C., Kerr, E.J., 1974, Sterol structural requirements for inhibition of streptolysin O activity. *Biochem. J.* **140**: 95–98.

Wiegand, V., Chang, T.Y., Strauss, J.F., III, Fahrenholz, F., Gimpl, G., 2003, Transport of plasma membrane-derived cholesterol and the function of Niemann-Pick C1 Protein. *FASEB J.* **17**: 782–784.

Wustner, D., 2007, Plasma membrane sterol distribution resembles the surface topography of living cells. *Mol. Biol. Cell* **18**: 211–228.

Wustner, D., Herrmann, A., Hao, M., Maxfield, F.R., 2002, Rapid nonvesicular transport of sterol between the plasma membrane domains of polarized hepatic cells. *J. Biol. Chem.* **277**: 30325–30336.

Wustner, D., Mondal, M., Huang, A., Maxfield, F.R., 2004, Different transport routes for high density lipoprotein and its associated free sterol in polarized hepatic cells. *J. Lipid Res.* **45**: 427–437.

Xu, S., Benoff, B., Liou, H.L., Lobel, P., Stock, A.M., 2007, Structural basis of sterol binding by NPC2, a lysosomal protein deficient in Niemann-Pick type C2 disease. *J. Biol. Chem.* **282**: 23525–23531.

Xu, X., London, E., 2000, The effect of sterol structure on membrane lipid domains reveals how cholesterol can induce lipid domain formation. *Biochemistry* **39**: 843–849.

Yancey, P.G., Rodrigueza, W.V., Kilsdonk, E.P., Stoudt, G.W., Johnson, W.J., Phillips, M.C., Rothblat, G.H., 1996, Cellular cholesterol efflux mediated by cyclodextrins. Demonstration of kinetic pools and mechanism of efflux. *J. Biol. Chem.* **271**: 16026–16034.

Yeagle, P.L., 1985, Cholesterol and the cell membrane. *Biochim. Biophys. Acta* **822**: 267–287.

Yeagle, P.L., Albert, A.D., Boesze-Battaglia, K., Young, J., Frye, J., 1990, Cholesterol dynamics in membranes. *Biophys. J.* **57**: 413–424.

Yin, Y., Liu, P., Anderson, R.G., Sampson, N.S., 2002, Construction of a catalytically inactive cholesterol oxidase mutant: investigation of the interplay between active site-residues glutamate 361 and histidine 447. *Arch. Biochem. Biophys.* **402**: 235–242.

Yue, Q.K., Kass, I.J., Sampson, N.S., Vrielink, A., 1999, Crystal structure determination of cholesterol oxidase from Streptomyces and structural characterization of key active site mutants. *Biochemistry* **38**: 4277–4286.

Zhang, W., McIntosh, A.L., Xu, H., Wu, D., Gruninger, T., Atshaves, B., Liu, J.C., Schroeder, F., 2005, Structural analysis of sterol distributions in the plasma membrane of living cells. *Biochemistry* **44**: 2864–2884.

Zidovetzki, R., Levitan, I., 2007, Use of cyclodextrins to manipulate plasma membrane cholesterol content: evidence, misconceptions and control strategies. *Biochim. Biophys. Acta* **1768**: 1311–1324.

Chapter 2
Cholesterol in Alzheimer's Disease and other Amyloidogenic Disorders

J. Robin Harris and Nathaniel G.N. Milton

Abstract The complex association of cholesterol metabolism and Alzheimer's disease is presented in depth, including the possible benefits to be gained from cholesterol-lowering statin therapy. Then follows a survey of the role of neuronal membrane cholesterol in Aβ pore formation and Aβ fibrillogenesis, together with the link with membrane raft domains and gangliosides. The contribution of structural studies to Aβ fibrillogenesis, using TEM and AFM, is given some emphasis. The role of apolipoprotein E and its isoforms, in particular ApoE4, in cholesterol and Aβ binding is presented, in relation to genetic risk factors for Alzheimer's disease. Increasing evidence suggests that cholesterol oxidation products are of importance in generation of Alzheimer's disease, possibly induced by Aβ-produced hydrogen peroxide. The body of evidence for a link between cholesterol in atherosclerosis and Alzheimer's disease is increasing, along with an associated inflammatory response. The possible role of cholesterol in tau fibrillization, tauopathies and in some other non-Aβ amyloidogenic disorders is surveyed.

Keywords Cholesterol · Alzheimer's disease · Amyloid-β · Aβ · Oligomerization · Fibrillogenesis · Statin · HMG-CoA reductase inhibitor

Abbreviations

AD	Alzheimer's disease
Aβ	Amyloid-beta

2.1 Introduction

An understanding of the role of circulatory cholesterol in cardiovascular disease and cerebrovascular disease has long been at the forefront of medical science. Cholesterol is now also thought to impinge strongly upon the field of dementia and neurological disease, in particular its possible role in the development of

J.R. Harris (✉)
Institute of Zoology, University of Mainz, D-55099, Mainz, Germany
e-mail: rharris@uni-mainz.de

Alzheimer's disease (AD) and other neurological and peripheral amyloidogenic disorders. Because diverse medical, biomedical and basic scientific approaches are being used to pursue studies in relation to AD and the involvement of cholesterol, meaningful correlation of data is not always easy, but in the present state of rapidly expanding knowledge it is nevertheless thought to be appropriate to attempt this review.

Many studies on AD and cholesterol do not relate to the binding of cholesterol to the well-characterized predominantly extracellular amyloid-β (Aβ) protein fragments that occur in vivo, rather they impinge upon metabolic and biochemical considerations. For the sake of completeness it is necessary to include discussion of most of these; furthermore, it is likely that they will have a secondary impact of significance to the structural aspects. The brain synthesizes most of its own cholesterol, but dietary/circulatory cholesterol almost certainly impacts upon cerebrovascular amyloid in the condition termed cerebral amyloid angiopathy (CAA) or vascular dementia. It is not always clear whether one should place emphasis upon free cholesterol, esterified cholesterol or cholesterol oxidation products of both these forms. With the increasing interest in the deleterious effects of free radical and metal ion-induced oxidation within medical systems and the likely benefits to be gained from dietary and therapeutic antioxidants, cholesterol oxidation has emerged as a topic of considerable significance in neurological disease (*see also* Chapter 6).

This subject has been under discussion for some years and has been reviewed extensively, primarily from a metabolic stance (Hartmann, 2005; Koudinov and Koudinova, 2001a; Ledesna and Dotti, 2006; Lukiw et al., 2005; Raffaï and Weisgraber, 2003; Shobab et al., 2005), but less so from a more structural point of view (Yanagisawa, 2005); accordingly, this latter aspect will be given greater emphasis within the present chapter. Studies on the role of cholesterol in Alzheimer's disease range widely from in vivo human and animal experimentation, including dietary, immunological and pharmaceutical approaches and the use of transgenic knock-out and knock-in animals, through to cultures of neuronal and other cells and numerous in vitro biochemical and structural approaches. Overall, a remarkable strength comes from this technical diversity, despite the inevitable instances where data appears to be conflicting. Although the material below is presented as discrete topics, there is considerable overlap of subject matter between the sections.

2.2 Cholesterol Metabolism and Alzheimer's Disease

Implicit in this topic is the underlying concept that cholesterol participates in the control of the membrane-bound amyloid precursor protein (APP) expression, and the expression and activity of the proteases (β- and γ-secretases) involved in APP cleavage, to generate the extracellular soluble amyloid-β peptide fragments that participate in AD. Thus, cholesterolemia is widely considered as a major risk factor in AD (Canevari and Clark, 2007), although the multi-faceted cholesterol interaction in vivo remains far from being fully understood. The possibility that cholesterol

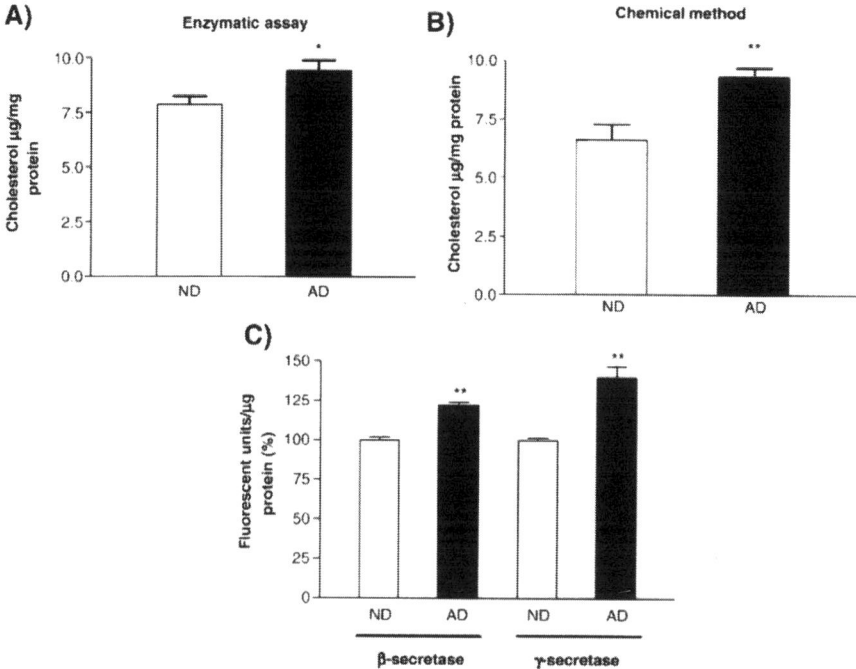

Fig. 2.1 The levels of cholesterol and the activities of β- and γ-secretases in normal and AD brain samples. *Panel* **A** shows that the enzymic determination of cholesterol in AD samples is significantly higher than in normal brain samples. *Panel* **B** shows the chemical determination of total brain cholesterol. AD brains have a significantly higher cholesterol content compared to non-AD brains. *Panel* **C** shows the analysis of brain lysates for β- and γ-secretase activity. The activities of both enzymes are significantly higher in the AD samples, compared to non-AD brain samples. From Xiong et al. (2008), with permission from Elsevier

retention in the Alzheimer's disease brain might be responsible for high β- and γ-secretase activities was advanced by Xiong et al. (2008), in an impressive study on the brains of AD patients, and cultured mouse cells stably transfected with the human APP gene (Fig. 2.1). These workers concluded that cholesterol homeostasis and transport was impaired, leading to increased retention in AD brains, due to altered levels or activities of nuclear receptors, and similar suggestions were made by Burns et al. (2003a). Brain cholesterol accumulation could be expressed by an increase in myelin membrane, neuronal plasma- and cyto-membrane membrane cholesterol content, by intra- or extra-cellular cholesterol inclusions, but the precise location(s) are yet to be defined. Indeed, that Aβ alters intracellular vesicle trafficking and cholesterol homeostasis, resulting in decreased cholesterol esterification and changes in neuronal free cholesterol distribution that are likely to be relevant to neurodegeneration, was advanced by Liu et al. (1998). Cholesterol accumulation in senile plaques of AD patients and transgenic APP (SW) mice was shown by Mori et al. (2001), using filipin fluorometric staining for cholesterol and an enzymatic technique (*see also* Section 1.4). It should, however, be mentioned that this work has recently been challenged by Lebouvier et al. (2009), who maintained that the purported presence of cholesterol in senile plaque was due to false positive results. Whilst it has to be accepted that cholesterol is naturally abundant in brain tissue, its presence bound within senile plaques could be at a low molecular level, rather than at the gross level of cholesterol crystalline or other lipid-rich deposits know to be present within vascular atherosclerotic plaques.

The presence of an increased amount of cholesterol in Aβ-positive presynaptic nerve terminals from AD brains led Gylys et al. (2007) to suggest that this might underlie neuronal dysfunction (synaptic loss), prior to or independent of subsequent extracellular Aβ deposition. A more general statement on the co-localization of cholesterol and raft lipids with extracellular disease-associated amyloid fibres extracted from tissues was advanced by Gellermann et al. (2005). Also at the neuronal level, earlier studies were performed on the role of cholesterol in synaptic plasticity and neuronal degeneration by Koudinov and Koudinova (2001b).

Independent support for the involvement of cholesterol in AD has come from dietary studies, where animals were subjected to high-fat or high-cholesterol feeding. Zatta et al. (2002) fed rabbits on a high cholesterol diet and found microglial activation and astrocytosis with over-expression of metallothionein-1 and -II, along with intraneuronal Aβ accumulation and occasional extracellular Aβ deposits. Intestinal epithelial cells of mice fed on a high fat diet were found by Galloway et al. (2007) to have an increased APP and Aβ concentration, leading to the interpretation that Aβ could serve as a chylomicron regulatory apolipoprotein, via its hydrophobic domain. A high cholesterol dietary-induced neuroinflammation and APP processing (Thirumangalakudi et al., 2008) was correlated with the loss of working memory in mice. Other supportive data has come from Hooijmans et al. (2009) who showed that cholesterol-containing diets influenced Alzheimer-like pathology, cognition and cerebral vasculature in transgenic mice. With a mouse genetic model for cholesterol loading, Fernández et al. (2009) showed that mitochondrial cholesterol loading enhanced Aβ-induced inflammation and neurotoxicity, modulated via mitochondrial glutathione. The work of Crameri et al. (2006) showed that deficiency in the cholesterol synthesising enzyme seladin-1 increases Aβ generation by increasing the β-secretase (β-site APP-cleaving enzyme: BACE) processing of the APP. Overexpression of seladin-1 had the reverse effect and increased the cholesterol in the membrane detergent resistant domains. Seladin-1 is also neuroprotective against Aβ and shows reduced expression in the AD brain (Greeve et al., 2000).

The muscle disorder termed sporadic inclusion body myositis (IBM) exhibits pathological similarities to AD, with respect to an increased skeletal muscle level of APP and Aβ. Rabbits fed a cholesterol-rich diet were found by Chen et al. (2008) to exhibit increased mRNA and proteins levels of APP and increased secretase activity favouring Aβ production, the pathological features of IBM.

Using the cholesterol-fed rabbit as a model for Alzheimer's disease, Sparks and his colleagues (Sparks and Schreurs, 2003; Sparks, 2004, 2007) have maintained that the presence of trace copper ions is necessary for Aβ to accumulate in the brain. In the absence of copper, rabbits with elevated cholesterol clear Aβ to the blood and liver. Supportive evidence has come from the studies of Opazo et al. (2002) and Puglielli et al. (2005), who claimed that copper-mediated oxidation of cholesterol might be responsible for AD pathogenesis and plaque formation. In a cholesterol-fed mouse model Lu et al. (2009) have shown that trace amounts of copper induced APP up-regulation, which activated the inflammatory pathway and exacerbated neurotoxicity. Other metal ions, such as those of zinc and iron, have also been implicated in AD (Ghribi et al., 2007; Gehman et al., 2008). Paradoxically, in rat brain tissue Bishop and Robinson (2004) claimed that the complex of Aβ with copper ions was not neurotoxic, whereas Aβ-iron and Aβ-zinc complexes were. The cholesterol-fed rabbit model has also been used by Prasanthi et al. (2008), who determined the extent to which brain hypercholesterolemia-induced Aβ levels were linked to a number of Aβ processing enzymes and receptors.

Even prior to the turn of the 21st century and more recently, cholesterol depletion induced by administration of statins to neuronal cultures and to humans was linked to a reduced risk of developing Alzheimer's disease (Simons et al., 1998, 2001; Fassbender et al., 2001; Wolozin et al., 2000; Wolozin, 2004; Zamrini et al., 2004) and the extended role of Aβ in lipid metabolism has been reviewed in depth by Zinser et al. (2007). Overall, the message that cholesterol-lowering strategies using statins (3-hydroxy-3-methylglutaryl coenzyme A reductase inhibitors) may provide a useful therapeutic approach to combat AD has emerged strongly, but extremely long-term epidemiological and clinical studies are required to provide the necessary proof. It is generally thought that the more lipophilic statins, such as lovastatin, are likely to carry greater protective potential (Ferrera et al., 2008). Atorvastatin administration to brain-injured rats has been found to be beneficial, in terms of reduced oedema and lipid peroxidation (Turkoglu et al., 2008), possibly mediated by a metabolite of atorvastatin that possesses antioxidant properties and inhibits membrane cholesterol-containing raft/domain formation (Mason et al., 2006).

Apart from the metabolic studies relating to statin reduction of APP production and processing, there is recent evidence that cholesterol depletion reduces Aβ aggregation (oligomer formation and fibrillogenesis) in hippocampal neurons (Schneider et al., 2006) and Aβ fibrillogenesis in macrophages (Gellermann et al., 2006), which provide support for the earlier concept that cholesterol also plays a key role in Aβ fibrillogenesis in vitro (Harris, 2002): *see below*.

2.3 Cholesterol Binding to Aβ and Aβ Fibrillogenesis

The interaction between Aβ and cholesterol has been assessed at the cellular and extracellular level. At the cell membrane level, cholesterol appears to be complexed with gangliosides and sphingolipids within neutral detergent-insoluble *raft domains* where it may influence APP cleavage by membrane-bound β-secretase (β-site APP-cleaving enzyme: BACE) followed by γ-secretase cleavage to release soluble Aβ (the amyloidogenic pathway) (*see* Zinser et al., 2007). In the cytoplasmic and most importantly extracellular compartments, cholesterol may promote the oligomerization and fibrillogenesis of the Aβ peptide. The extracellular deposition of brain senile plaques containing fibrillar Aβ in association with cholesterol, and a number of other proteins and glycoproteins, is of main concern within the present section, but it is necessary to briefly survey the relevant literature on other aspects.

2.3.1 Cholesterol and Membrane-Associated Aβ Pore Formation and Fibrillogenesis

The membrane disordering effect of Aβ has been largely investigated using cellular, liposomal and lipid monolayer systems, and related to the β-sheet conversion of Aβ and subsequent peptide aggregation/polymerization, in relation to membrane fluidity and neurotoxicity (reviewed by Eckert et al., 2005). That membrane cholesterol can act as a modulator of both membrane-associated Aβ fibrillogenesis and neurotoxicity has been advanced by McLaurin et al. (2002) and Yip et al. (2001). The requirement for cholesterol for the cytotoxic effects of Aβ on vascular smooth muscle cells and for Aβ binding to the muscle cell membrane was shown by Subashinghe et al. (2003), and clearly linked to beneficial drug-induced cholesterol lowering. Although implicit in the above studies, at a biophysical level Wood et al. (2002) have shown more specifically that it is the presence of cholesterol in the outer lipid monolayer of the neuronal cell membrane that is responsible for Aβ accumulation. Synaptic plasma membranes from cerebral cortex and hippocampus were shown by Chochina et al. (2001) to be enriched in cholesterol, compared to cerebellum. This neuronal membrane enrichment with cholesterol was linked to an increase in membrane fluidity, resulting in hydrophobic Aβ accumulation and fibrillogenesis.

The combined influence of metal ions and cholesterol on $A\beta_{1-42}$ interaction with model membranes (Lau et al., 2007; Gehman et al., 2008) has provided further evidence for the conversion of the peptide α-helix to β-sheet structure, the increase in hydrophobicity then enabling the peptide oligomer/pre-pore to penetrate the lipid bilayer as an ion channel. This concept has been advanced to account for the cytotoxicity of amyloidogenic proteins in general (Cheon et al., 2007; Rabzelj et al. 2008), together with the fact that these pore-forming proteins and peptides may share structural and functional homology (Yoshiike et al., 2007). That the dynamic formation of Aβ membrane-penetrating pores can act as calcium-selective ion channels responsible for neurotoxicity is gaining support (Jang et al., 2007b), and the role of cholesterol in this event was implied from studies on planar lipid bilayer membranes (Micelli et al., 2004). Molecular dynamics modelling of putative ion channels formed by neurotoxic Aβ ion channels (Jang et al., 2007a), formed by peptides of different length, has provided support for the overall hypothesis that an intermediate protein unfolding leads to exposure of hydrophobic β-sheet membrane-penetrating peptide hairpins (sometimes termed "U-shaped β-strand-turn-β-strands"; Jang et al., 2007b), in the form of oligomeric β-barrels. It is likely that there is a difference between $A\beta_{1-42}$ and $A\beta_{1-40}$ with respect to oligomer and ion channel formation (Kirkitadze and Kowalska, 2005) that may be responsible for the significantly higher $A\beta_{1-42}/A\beta_{1-40}$ ratio commonly found in familial AD. Using a liposomal model system Qui et al. (2009) have shown that the lateral organization of cholesterol, presumably in raft-like domains, controls the formation of oligomeric Aβ, which in turn could be linked to the toxicity of Aβ in neuronal membranes.

A strong parallel between the various amyloid cytotoxic cation channels and the cholesterol- and other lipid-dependent pore-forming toxins has been drawn by Lashuel and Lansbury (2006) and indeed suggested earlier by Gilbert et al. (1998), which will be expanded upon in Chapter 21. This concept is supported by the study of Srisailam et al. (2002) who investigated the transformation of the all β-barrel acidic fibroblast growth factor from *Notopthalmus viridescens* and found partially-structured intermediates leading to fibril formation. There are several documented instances where a low concentration of SDS or other alkyl sulfate induces or potentiates amyloid fibril formation, again indicating the importance of a critical level of protein unfolding with exposure of (paired) hydrophobic β-sheets that associate/polymerize in a linear manner as extremely stable (SDS resistant) crossed β-sheets.

2.3.2 Gangliosides and Aβ Fibrillogenesis

The role of monosialoganglioside (GM1) in AD, alone and in association with cholesterol, has been given considerable attention in recent years (reviewed by Yanagisawa, 2005, 2007). Both cellular and biochemical studies have been performed in relation to GM1-containing lipid rafts and Aβ production from APP (Ehehalt et al., 2003; Kalvodova et al., 2005). That the GM1-Aβ complex could act as a *seed* for the production of fibrillar Aβ on the neuronal surface has been proposed (Kakio et al., 2001, 2002), and this concept has been extended to intracellular release of Aβ via a deviant endocytic pathway into endosomes (Kimura and Yanagisawa, 2007; Yuyama et al., 2008).

Support for the involvement of both ganglioside and cholesterol in the formation of cell surface-bound fibrillar $A\beta_{1-42}$ has come from the study of Wakabayashi and Matsuzaki (2007), by showing degenerate neurites on NFG-differentiated PC12 neuron-like cells. Using the same neuronal-like cell line for Aβ cytotoxicity testing, Lin et al. (2008) concluded that both GM1 and cholesterol are essential for the formation of the GM1-Aβ complex on the cell surface and the modulation of the cytotoxicity of monomeric Aβ. In a liposomal model system, containing lipids similar to those in brain cortical membrane, Tashima et al. (2004) assessed Aβ release and fibril formation in the presence and absence of GM1 and cholesterol, and concluded that fibril formation required both these components. Further evidence that an age-dependent GM1 enrichment of neuronal presynaptic terminals forms zones where Aβ binds and promotes fibrillar amyloid assembly has recently been shown by Yamamoto et al. (2008).

2.3.3 Cholesterol and In Vitro Aβ Fibrillogenesis: Structural Studies

Surprisingly, of the numerous biochemical and structural publications dealing with in vitro oligomerization and fibrillogenesis of the amyloid-β peptides (the naturally occurring and chemically synthesised peptides of varying length), rather few have linked the influence of cholesterol to these events. Using fluorescently-labelled lipids, Avdulov et al. (1997) showed that Aβ aggregates, the nature of which was not fully defined, had a preferential binding for cholesterol rather than for phosphatidylcholine and fatty acids. Although most studies have been performed using the longer Aβ fragments, D'Errico et al. (2008) showed a cholesterol-dependent interaction between $A\beta_{25-35}$ and phospholipid bilayers, apparently due to an increased membrane fluidity. That native, and in particular oxidized, plasma lipoproteins can potentiate Aβ fibrillogenesis was shown by Stanyer et al. (2002), with the suggestion that this might be mediated via reactive aldehyde groups, as fibrillogenesis was inhibited in the presence of an aldehyde scavenger. More recent insights into lipid aldehyde-initiated fibrillogenesis of Aβ has come from Scheinost et al. (2008), who maintained that the ε-amino group of Aβ Lys16 adjacent to the central hydrophobic cluster (amino acids 17–21; Nelson and Alkon, 2007), could be a target for aldehyde adduction.

Native high density lipoprotein (HDL), with and without the three ApoE isoforms, was shown to inhibit Aβ fibrillogenesis (Olsen and Dragø, 2000), but this observation was not linked to cholesterol sequestration by HDL. Although relating more to Aβ production than fibrillogenesis, using a cholesterol-protein binding blot assay Yao and Papadopoulos (2002) showed that cholesterol binds to the hydrophobic amino acid sequence 10–20 of Aβ, thereby blocking the access of the α-secretase and cleavage of APP to $A\beta_{17-40}$, i.e. inhibiting the non-amyloidogenic pathway. Furthermore, the binding of cholesterol to LDL was inhibited by $A\beta_{1-40}$ and the binding of cholesterol to ApoE and LDL was completely abolished by $A\beta_{1-42}$.

The transmission electron microscope (TEM) and to a somewhat lesser extent the atomic force microscope (AFM) can provide detailed molecular information on the structure of Aβ oligomers through to protofibrils and fibrils, together with fibril polymorphism (Harris, 2002, 2008; Milton and Harris, 2009). X-ray fibre diffraction has also provided higher resolution data on the repeating crossed β-sheet structure that underlies fibril formation (Malinchik et al., 1998; Stromer and Serpell, 2005), which is broadly accepted as a structural feature of all known amyloid fibrils. As a slight variant Sinha et al. (2001) proposed a domain swapped interdigitating β-hairpin model for amyloid fibril elongation. Although several of the investigations relating to membrane-bound cholesterol in relation to Aβ fibrillogenesis have utilized AFM and TEM in a serious manner (Yip and McLaurin, 2001; Yip et al., 2001), other studies have only touched upon the possibilities that these microscopies, in particular TEM, can offer for the in vitro study of these interactions, particularly when placed alongside other more indirect biochemical and biophysical techniques (Castaño et al., 1995; Mizuno et al., 1999; Koppaka et al., 2003).

Other than for natural membrane, mixed-lipid liposomal and bilayer model systems, it is difficult to know how one should present cholesterol experimentally for interaction with Aβ in vitro. Cholesterol has a very low solubility in aqueous systems (Harberland and Reynolds, 1973) resulting in the formation of planar cholesterol microcrystals when an ethanolic solution of cholesterol is dispersed in an aqueous phase (Harris, 1988). Esterified cholesterol forms a suspension of globular particles in aqueous solution, similar in size to low density lipoproteins, when prepared in a similar manner (Harris, 2002). A commercially available "soluble" cholesterol (Sigma-Aldrich, termed cholesterol-PEG 600), is a waxy solid in which the 3β-OH group of cholesterol has been linked to a polyethylene glycol chain, resulting in an average mass of ~600 Da. This product more accurately creates a micellar solution, in all probability with the sterol at least partially buried beneath the surface hydrophilic polyethylene glycol chains. Nevertheless, this synthetic cholesterol derivative has been found to be a useful soluble cholesterol substitute for cellular studies (Ishiwata et al., 1997), as well as for studies on cholesterol binding to Aβ fibrils (Harris, 2008). Examples of cholesterol microcrystals, esterified cholesterol and LDL particles, and cholesterol PEG600 micelles are shown in Fig. 2.2. The clustering of Aβ$_{1-42}$ fibrils on and around cholesterol microcrystals is shown in Fig. 2.3. Detail of the Aβ$_{1-42}$ peptide binding to cholesterol microcrystals, with protofibrils visible on the crystal surface is shown in Fig. 2.4. Other Aβ peptide

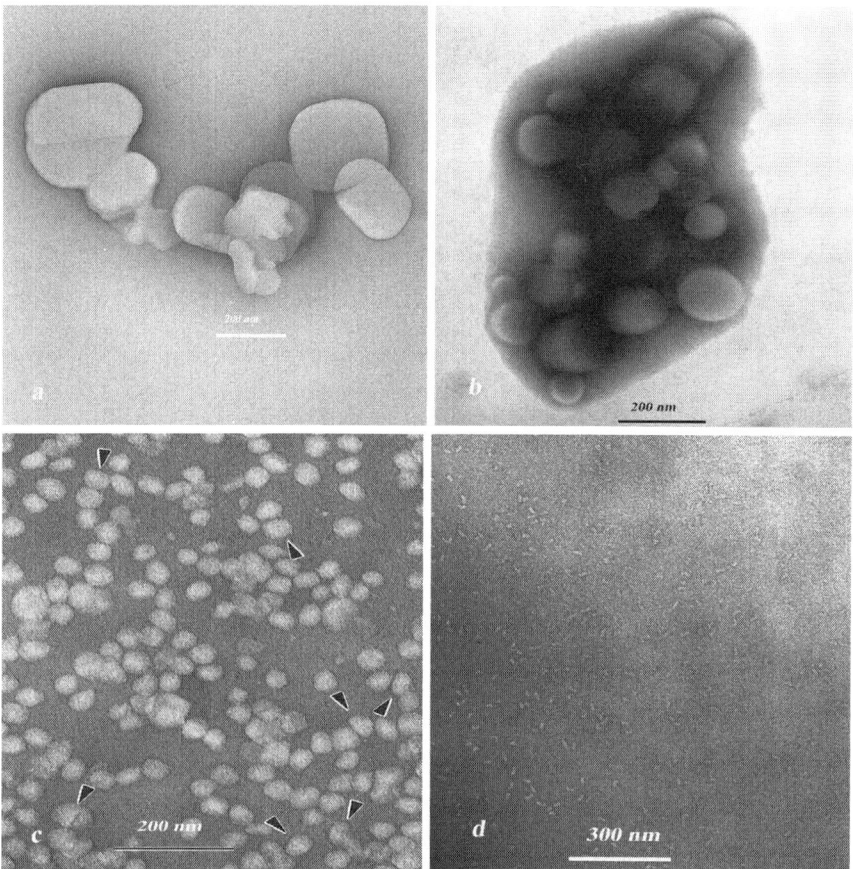

Fig. 2.2 Examples of four different experimental cholesterol substrates, shown in negatively stained TEM images, usable for Aβ fibrillogenesis studies. (**a**) A cluster of cholesterol microcrystals (Harris, 1988); (**b**) cholesterol acetate globular micelles; (**c**) human low density lipoprotein (LDL); (**d**) cholesterol-PEG600 micelles (soluble cholesterol)

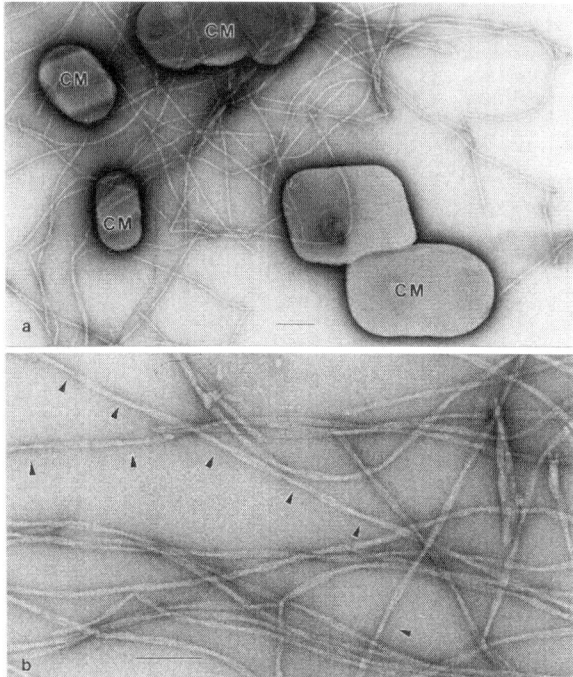

Fig. 2.3 (a) Cholesterol microcrystals (CM) surrounded by a cluster of $A\beta_{1-42}$ fibrils; (**b**) a higher magnification survey showing the double helical nature of the mature $A\beta_{1-42}$ fibrils formed in the presence of cholesterol microcrystals. The scale bars indicate 100 nm. From Harris (2002), with permission from Elsevier

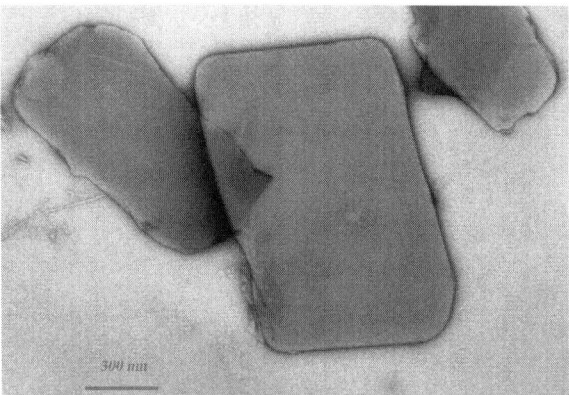

Fig. 2.4 Cholesterol microcrystals showing a thin surface coating of forming $A\beta_{1-42}$ protofibrils and fibrils, strongly indicative of the positive binding of the peptide to cholesterol and the subsequent promotion of fibril formation

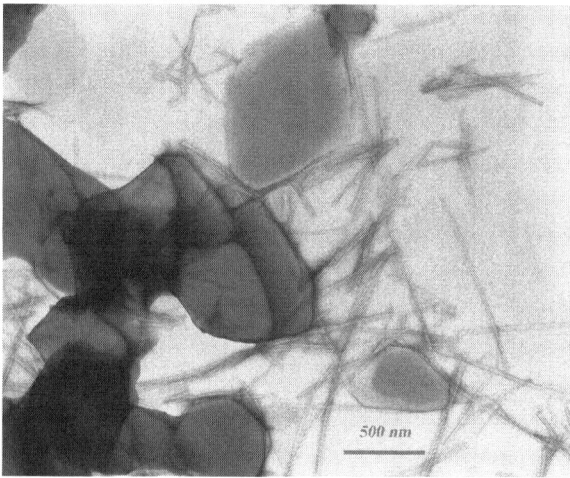

Fig. 2.5 Cholesterol microcrystals surrounded by and binding fibrils formed by the $A\beta_{22-35}$ fragment. These fibrils do not exhibit the characteristic double helical structure shown by mature $A\beta_{1-42}$ fibrils (Harris and Milton, previously unpublished data)

fragments, such as the $A\beta_{25-35}$ fragment, also show an affinity for cholesterol (Fig. 2.5), but it has yet to be demonstrated that $A\beta$ fibril-forming peptide fragments lacking the central hydrophobic domain have lost the capacity to bind to cholesterol. The binding of soluble cholesterol-PEG600 to preformed $A\beta_{1-42}$ fibrils, and fibrils formed in the presence of soluble cholesterol (Harris, 2002, 2008), is shown in Fig. 2.6. However, fibrils formed from the bacterial protease inhibitor Pepstatin A (an eight amino acid bacterial peptide), do not bind soluble cholesterol-PEG600 micelles and similarly, no evidence has been obtained by the authors to indicate that fibrils formed by the peptide amylin (islet amyloid peptide) have any cholesterol-binding potential. When $A\beta_{1-42}$ fibril formation is performed in the presence of both cholesterol and aspirin, fibril formation is prevented, but clusters of short rod-like $A\beta_{1-42}$ aggregates attach to the cholesterol microcrystals (Harris, 2002). This suggests that although aspirin inhibits fibril formation, it may not prevent oligomer formation by the $A\beta_{1-42}$ peptide.

2.4 Apolipoprotein E, Cholesterol and Alzheimer's Disease

There is an extensive literature to link the differing apolipoprotein E (ApoE) isoforms to $A\beta$ binding and late-onset AD (see: Carter, 2005; Crutcher, 2004; Hatters et al., 2006; Sullivan et al., 2008). However, until recently there has been relatively little evidence as to how the known cholesterol-binding of ApoE could modulate this $A\beta$ interaction (Hirsch-Reinshagen and Wellington, 2007; Reiss, 2005). The most significant fact to emerge from the early ApoE studies is that in individuals carrying the ApoE3 allele and most particularly the ApoE4 allele, and thus expressing these

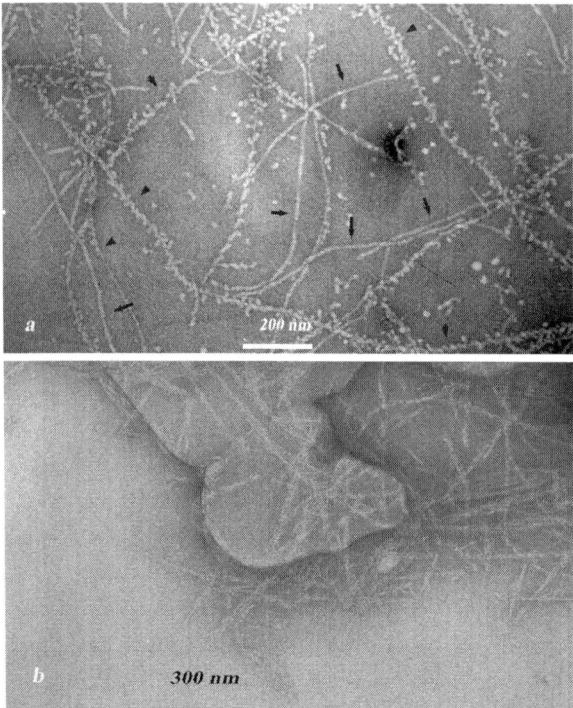

Fig. 2.6 (**a**) Amyloid $\beta_{1\text{-}42}$ fibrils formed in the presence of cholesterol PEG600. The negatively stained image shows that protofilaments have cholesterol-PEG600 micelles clustered obliquely along the length of the fibril (*arrowheads*), whereas the double helical mature filaments appear to have a smooth surface coating of the cholesterol derivative (*arrows*). From Harris (2002), with permission from Elsevier. (**b**) Amyloid $\beta_{1\text{-}42}$ fibrils prepared in the presence of 0.5 mg/ml cholesterol, followed by incubation with cholesterol-PEG600. The mature double helical fibrils, which are clustered around a stack of cholesterol microcrystals, are well-coated with cholesterol-PEG600 micelles, but there is no indication of the periodic binding shown in (**a**). From Harris (2008), with permission from Elsevier

isoforms in higher ratio, are susceptible to an increased AD risk. Those carrying the ApoE2 allele have the lowest AD risk. Although Castaño et al. (1995) showed that in vitro fibrillogenesis of Aβ was promoted by ApoE, this aspect has not be followed further in recent years in relation to the properties of the different ApoE isoforms.

Brain ApoE is synthesized primarily by glial cells and possibly also by neurons, and is present in the CSF within HLD-like lipoprotein particles. There is little or no evidence for transfer of peripheral ApoE, where it is present in plasma HDL and VLDL particles, across the blood-brain barrier to the CSF. The ApoE molecule has a molecular mass of 34.2 kDa (299 amino acids) and is present in the CSF as a major protein component, at the relatively high concentration of ~5 mg/ml. It is looked upon as a lipid transport protein and in the plasma is considered to have anti-atherogenic properties. The C-terminal domain of ApoE (amino acid residues

216–299) is thought to be responsible for both binding to Aβ and to lipids, whereas the N-terminal domain is responsible for binding of ApoE to the LDL receptor.

As already mentioned, ApoE exists as three main isoforms in man, ApoE2 (Cys^{112}, Cys^{258}), E3 (Cys^{112}, Arg^{158}) and E4 (Arg^{112}, Arg^{158}), with the gene for E3 being the most common allele. The blood plasma ApoE4 has a greater lipid-binding capacity, including cholesterol, and tends to locate to VLDL rather than HDL. A conformational change within the ApoE4 molecule exposing amphipathic α-helices results in a change in surface hydrophobicity that is thought to account for increased lipid binding, rather than the direct involvement of the arginine substitution at position 112, since this is out-with the N-terminal lipid-binding domain. The Apo E phenotype can also influence the effectiveness of lipid lowering therapies with more effects observed with fibrates compared to statins (Christidis et al., 2006).

An important finding using Aβ and ApoE immunostaining was the co-localization of Aβ and ApoE in AD-affected brain samples (Aizawa et al., 1997), in addition an antibody against the C-terminal of ApoE showed greater similarity of staining to the anti-ApoE mAb than did an anti-ApoE N-terminal antibody. Furthermore, in an AD transgenic mouse model Burns et al. (2003b) have shown the co-localization of extracellular cholesterol, ApoE and fibrillar Aβ in amyloid plaques. Figure 2.7 shows the co-localization of Aβ and cholesterol, in parallel with thioflavin S staining of fibrillar plaques (courtesy of Marc Burns). Also using transgenic mice, Fryer et al. (2005) investigated cerebral amyloid angiopathy (CAA), which is found in most AD patients, in relation to the ApoE3 and ApoE4 isoforms. They showed that the expression of human ApoE4 in mice led to substantial CAA plaques, but with few parenchymal amyloid plaques. Young ApoE4-expressing mice had an elevated ratio of Aβ 40:42 in the brain extracellular pool, but a lower ratio

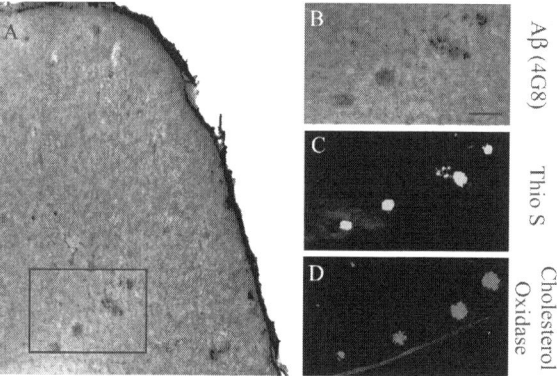

Fig. 2.7 Co-localization of fibrillar amyloid with cholesterol oxidase in the cortex of a 12-month-old PS/APP mouse. (**a**) A low-power overview of amyloid plaques stained with 4G8. (**b–d**) Plaques from within the boxed area double-labelled for Ab (**b**) and thioflavin S (c), and a consecutive section labeled for cholesterol oxidase (**d**). Scale bars 100 mM. From Burns et al. (2003b), with permission from Elsevier

in the CSF, suggestive of altered clearance and transport of Aβ. Although not implicating ApoE isoforms, support for the mediation of ApoE in cholesterol efflux from astrocytes came from the study of Abildayeva et al. (2006), showing that 24(S)-hydroxycholesterol induced ApoE-mediated efflux of cholesterol via a liver X receptor-controlled pathway, of likely relevance for neurological disease.

Cellular studies on ApoE isoforms in primary rat hippocampal neurons and astrocytes (Rapp et al., 2006) have indicated that the ApoE4 isoform is involved to a greater extent in neuronal cholesterol homeostasis than the other isoforms, and that this is more pronounced in neurons compared to astrocytes. However, Gong et al. (2002) showed that astrocytes from ApoE4 knock-in mice had a reduced cholesterol release into HDL-like particles compared to ApoE3 knock-in mice. Extending this study, Gong et al. (2007) have shown that in a neuronal culture system with ApoE bound to the surface of extracellular synthetic lipid particles, the ApoE4 isoform inhibited the release of cholesterol from neurons. However, in an attempt to link the ApoE isoform status to AD, Morishima-Kawashima et al. (2007) using ApoE4 knock-in mice were unable to show an ApoE4-specific effect on the increased association of Aβ with low-density brain membrane domains. In Down Syndrome (DS) the overexpression of APP is thought to contribute to the development of AD symptoms. The cholesterol levels in DS patients are not associated with AD symptoms, however, an Apo E4 allele was associated with susceptibility to hypercholesterolaemia (Prasher et al., 2008).

It has been recently suggested that Aβ binding to ApoE compromises physiological lipid binding and transport by ApoE, which in turn could have implications for amyloid plaque formation and cholesterol accumulation (Tamamizu-Kato et al., 2008). The convergence of risk factors, including the ApoE4 allele, in AD and cardiovascular disease (Martins et al., 2006) has emerged as a significant factor in cholesterol metabolism (*see also* Section 2.6). Other cholesterol-related genes such as those for hydroxy-methylglutaryl-coenzyme A reductase and the cholesterol transporter ABCA1 have also been claimed to modulate the risk of Alzheimer's disease (Rodríguez-Rodríguez et al., 2009).

2.5 Cholesterol Oxidation and Alzheimer's Disease

A link between oxidative stress and AD has been proposed for several years, but the precise mechanisms have yet to be fully defined (*reviewed by* Butterfield et al., 2002; Schöneich, 2002; Pappolla et al., 2002; Nelson, 2007). A broad survey on the role of oxysterols in neurodegenerative diseases has been recently presented by Björkhem et al. (2009). One has to consider the involvement of several reactive oxygen species when producing damage to membrane lipids, to Aβ and other proteins, together with the role of reactive metal ions, in particular copper and iron. Possible protection, particularly from the water- and lipid-soluble antioxidant vitamins, has been given considerable attention (Behl, 2005), yet it remains unclear whether this therapeutic approach really provides significant benefit, in the shorter or longer term. Remarkably, in a cellular study it has been claimed by Yao et al. (2002) that

22R-hydroxycholesterol, an intermediate in the production of pregnenolone from cholesterol, protected neuronal cells from Aβ-induced cytotoxicity by complexing with Aβ. Although implicit, it was not shown whether 22R-hydroxycholesterol bound to Aβ with a higher affinity than cholesterol.

Using lipid monolayers, oxidative damage to membrane lipids was claimed by Koppaka et al. (2003) to be linked synergically via Aβ$_{1-42}$ to the promotion of fibril formation by Aβ$_{1-40}$. That Aβ induces oxidation of membrane lipids emerged from a number of studies, exemplified clearly by Cutler et al. (2004), who claimed that in hippocampal neurones oxidative stress led to the accumulation of ceramides and cholesterol, preventable by inclusion of α-tocopherol. In several studies, the exact chemical nature of the oxysterols and cholesterol oxidation products are not defined, however Vaya and Schipper (2007) provided a detailed analysis of the range of oxysterol intermediates that can act as ligands for the liver X-activated receptor (LXR) nuclear receptors, regulators of genes involved in cholesterol homeostasis (*see* below). The principle cholesterol oxidation metabolites, derived from hydrogen peroxide and oxygen free radical interaction, are water soluble 24S-hydroxycholesterol and 7β-hydroxycholesterol; indeed, it has bee shown that hydrogen peroxide is produced catalytically by interaction of Aβ with cholesterol (Ferrera et al., 2008). Although 24S-hydroxycholesterol can cross the blood-brain-barrier (Björkhem et al., 2009) and is the primary cholesterol elimination product of the brain (with an increased level in the brains of AD patients) it has also been suggested that 24S-hydroxycholesterol has protective properties, by complexing with Aβ (Krištofiková et al., 2008). Differential expression and polymorphism of the gene encoding cholesterol 24S-hydroxylase, cytochrome P450 46 (*CYP450 46*), has been associated with AD, with a predisposition in certain genotypes (Kölsch et al., 2002; Papassotiropoulos et al., 2003; Borroni et al., 2004; Brown et al., 2004); this association has been challenged by others (Desai et al., 2002; Tedde et al., 2006). However, from a proteome analysis of cortical neurones Wang et al. (2008) concluded that 24S-hydroxycholesterol is a down-regulator of cholesterol synthesis and thereby important for brain cholesterol hemostasis. The crystal structure of the principal brain cholesterol hydroxylase, CYP450 46A1, has been determined by Mast et al. (2008) at 2.6 Å and at a slightly improved resolution, with and without bound substrate (White et al., 2008). This structural and biochemical data may ultimately lead to the development of therapeutically useful stimulatory and inhibitory agents. The alternative hydroxycholesterol, 27-hydroxycholesterol, synthesized by sterol 27-hydroxylase (CYP27A1), is known to facilitate the flow of cholesterol from the circulation across the blood brain barrier into the brain, and is likely to have a significant impact upon brain APP processing and Aβ production (Scott Kim et al., 2009). Further emphasis on the importance of 27-hydroxycholestrol has come from the studies of Prasanthi et al. (2009) and Ghibi et al. (2009).

At the genetic level, there has also been interest shown in a link between cholesterol and the enzyme heme oxygenase-1, which stimulates oxysterol production, but also activates the liver X receptor-β (Infante et al., 2008).

Other cholesterol oxidation products include 7-ketocholesterol and cholesterol epoxides (Ong et al., 2003), but their likely pathological roles have yet to be fully defined. Another reactive cholesterol oxidation metabolite that may be involved in inflammatory atherogenesis (Stewart et al., 2007) and Aβ aggregation (Zhang et al., 2004), is cholesterol *seco*-aldehyde (3β-hydroxy-5-oxo-5, *seco*cholestan-6-ol), produced by ozonolysis of cholesterol. Sathishkumar et al. (2007) found that although cholesterol *seco*-aldehyde-induced neurotoxicity could be prevented by *N*-acetyl-l-cysteine, this compound did not prevent Aβ aggregation. In their detailed in vitro study, Scheinost et al. (2008) found that fibrillogenesis of both $Aβ_{1-40}$ and $Aβ_{1-42}$ was accelerated by cholesterol *seco*-aldehyde, involving a site-specific adduction of the aldehyde to the ϵ-amino group of Lys16 pf Aβ. Cholesterol inhibited this cholesterol *seco*-aldehyde-induced fibrillogenesis of Aβ, perhaps unexpectedly.

That Aβ, and APP, possess an inherent copper-dependent enzyme-like activity in the presence of cholesterol and other substrates (Opazo et al., 2002; Yoshimoto et al., 2005), resulting in the catalytic production of hydrogen peroxide, with oxidation of the cholesterol at the C3β –OH group to produce 4-cholesten-3-one (Puglielli et al., 2005) or alternatively 7β-hydroxycholesterol (Nelson and Alkon, 2005), appears to be a highly significant observation. The correlation of this catalytic activity with the pathogenic mechanism leading to accumulation of cholesterol and cholesterol oxidation products remains to be clarified, but the fact that oxidative mechanisms are emerging strongly as mediators for both atherosclerotic and amyloid plaque formation indicates the likely importance of future research in this area (*see below* and Chapter 5).

2.6 Atherosclerosis and Alzheimer's Disease

That deviant cholesterol metabolism and deposition might be a link between peripheral vascular disease, cardiovascular disease and cerebrovascular disease, i.e. cerebral amyloid angiopathy, Alzheimer's disease and dementia, has been suggested for several years (Hofman et al., 1997; Li et al., 2003), and the topic has been widely reviewed (Sparks et al., 2000; Casserly and Topol, 2004; Kalman and Janka, 2005; Martins et al., 2006; Cechetto et al., 2008). That this link might also involve both peripheral and cerebral proinflammatory events is also apparent (Finch, 2005). In the periphery there is long-standing evidence that blood vessel macrophages accumulate free and esterified cholesterol (Klinknera et al., 1995), but despite the fact that brain astrocyte proliferation is associated with AD there is limited evidence for cholesterol accumulation by these cells or microglia. With circulatory macrophages, Aβ bound to modified LDL has been found to enhance cholesterol accumulation, foam cell formation and Aβ deposition in blood vessel walls (Schulz et al., 2007), but it is not clear whether this is due to the peptide monomer or to an oligomerized/fibrillar form of Aβ. On the other hand, microglial inflammatory activation has been found under conditions of cholesterol embolization (Rapp et al., 2008) and in dietary-induced hypercholesterolemia (Streit and Sparks, 1997; Xue et al., 2007). Modulation of this inflammatory response by liver X receptors has indicated that LXRs have the capacity to maintain phagocytosis in fibrillar Aβ-stimulated microglia (Zelcer et al., 2007). An overall improvement of cerebrovascular function, including reduced inflammation and soluble Aβ levels, following the administration of simvastatin to aged APP mice was shown by Tong et al. (2009). However, no reduction in the number of Aβ-containing plaques and memory improvement was detected.

Phagocytosis by activated microglia has also been implicated in demyelination disease (Smith, 2001), but this aspect has not received any emphasis in relation to the established presence of microglia within amyloid plaques. However, Stadler et al. (1999) concluded from their study on amyloid plaques in the brains of APP23 transgenic mice that microglial activation and phagocytosis might be associated with neuronal loss. In an attempt to relate the influence of Aβ on the cholesterol content of the Golgi complex in astrocytes, Igbavboa et al. (2003) concluded that extracellular $Aβ_{42,}$ in oligomeric rather than fibrillar state, disrupted cellular cholesterol homeostasis. Extending this study, Igbavboa et al. (2009) have shown that Aβ stimulates the trafficking of both cholesterol and caveolin-1 from the plasma membrane of primary astrocytes to the Golgi complex. The likely importance of the precise distribution of cholesterol, rather than just total brain cholesterol, was shown by Burns et al. (2006), who found that reduction of cholesterol level by statins also caused translocation of cholesterol from brain membrane cytofacial lipid monolayer to the exofacial monolayer. Cholesterol lowering by statins and the link between atherosclerosis and AD has been given due emphasis by Panza et al. (2005) and Orr (2008), and at the genetic level (Papassotiropoulos et al., 2005; Carter, 2007; Reiman et al., 2008) the likely association of multiple gene polymorphisms associated with cholesterol and lipoprotein metabolism in peripheral and cerebral vascular disease, and AD susceptibility, has been assessed.

2.7 Cholesterol and Tau Fibrillization in AD, the Tauopathies and Non-Aβ Amyloidogenic Disorders

The intracellular neuronal accumulation of paired helical filaments of the hyperphosphorylated tau protein represents a well-studied parallel aspect of AD and Niemann-Pick type C disease (NPC), in addition to Aβ studies, which in both diseases appears to be significantly influenced by cholesterol or cholesterol metabolism (*see also* Chapter 11). Despite the increasing interest in the similarities of AD and NPC, and the value of this comparison (Distl et al., 2003; Ohm et al., 2003; Burns and Duff, 2002; Michikawa, 2004; Adalbert et al., 2007), it is clear that the genetic lesion in NPC is clearly defined as a fatal autosomal recessive neurovisceral cholesterol storage disorder. In NPC there is intracellular tau fibrillization and secondary intracellular Aβ accumulation, which presents a significantly different feature to AD, where neuronal cholesterol accumulation although significant is less pronounced and where fibrillar Aβ formation is predominantly extracellular.

Dietary-induced cholesterol-dependent hyperphosphorylation of tau is common to both AD and NPC, and is ApoE isoform dependent (Saito et al., 2002; Rahman et al., 2005; Ghribi et al., 2006; Michikawa, 2006). Contrary to this Fan et al. (2001) had earlier claimed that inhibition of cholesterol synthesis in cultured neurons resulted in hyperphosphorylation of tau. The more widely accepted current point of view is that cells containing neurofibrillary tangles contain more free cholesterol than tangle-free NPC neurons (Distl et al., 2001). Contrary to AD, which is a neuronal disorder, in NCP the cholesterol storage defect is expressed by neurones, astrocytes and glial cells.

Early neuronal cholesterol accumulation is present within Purkinje neuronal dendritic trees (Reid et al., 2004), but subsequent age-dependent cellular changes in NPC are more clearly linked to the endosomal/lysosomal pathway, with respect to both cholesterol accumulation, and APP processing and Aβ aggregation (Yamazaki et al., 2001; Nixon, 2004). In an elegant cellular and subcellular study Liao et al. (2007) presented convincing evidence in Npc$^{-/-}$ mouse brain that autophagic dysfunction is linked to cholesterol accumulation, with the presence of neuronal vacuole-like structures and multivesicle bodies (*see also* Chapter 11). Also, in a *Drosophila* model for NPC1, Phillips et al. (2008) provided evidence for progressive filipin-staining of cholesterol aggregates in brain and retinal cells during ageing.

Over-expression of the protein α-synuclein is associated with Parkinson's disease, Lewy body formation, and other neurodegenerative α-synucleinopathies.

That normal membrane localization of cholesterol-containing lipid rafts is modified in the Parkinson's-associated A30P mutation, due to raft disruption and redistribution away from synapses, has been claimed by Fortin et al. (2004) to underlie the pathogenesis of Parkinson's disease. Further indication of the involvement of cholesterol has been shown in a cellular model of Parkinson's disease, where statins reduced α-synuclein aggregation and cholesterol supplementation increased α-synuclein aggregation (Bar-On et al., 2008). Similar to AD, it has also been claimed that cholesterol oxidation products are closely involved in α-synuclein fibrillization, of relevance to the development of Parkinson's and Lewy body disease (Bosco et al., 2006). Contrary to the above, Karube et al. (2008) have shown that the N-terminal region of α-synuclein is essential of fatty acid-induced oligomerization of this protein.

The involvement of cholesterol in the generation of other amyloid diseases is limited. Hou et al. (2008) presented evidence that cholesterol and anionic phospholipids are important for transthyretin fibrillogeneis and the resulting cytotoxicity of this protein, which is responsible for familial amyloidotic polyneuropathy. Calcitonin, a 32-aminoacid peptide involved in bone calcium metabolism, undergoes a structural transformation similar to other amyloidogenic proteins (Avidan-Shpalter and Gazit, 2006). The pore-like oligomers formed by calcitonin have an affinity for cholesterol containing rafts in membranes, "termed hydrophobicity-based toxicity" (Diociaiuti et al., 2006), which act as calcium channels. Similarly, in type II diabetes, amylin, the islet amyloid protein which leads to pancreatic fibrillar deposits, undergoes a cytotoxic membrane interaction (Jayasinghe and Langen, 2007), but the evidence presented indicated that phosphatidyl serine was the required lipid for the partly unfolded amylin hydrophobic β-sheet interaction within the membrane bilayer, rather than cholesterol. However, Cho et al. (2008) have produced evidence to suggest that cholesterol regulates amylin non-fibrillar aggregation and deposition within planar cholesterol-containing raft lipid membranes. These workers have recently extended their studies by using model membranes (Cho et al.,

2009), by showing amylin clustering and aggregation on cholesterol-containing membranes, whereas on cholesterol-depleted membranes amylin formed smaller oligomeric structures.

Acidic phospholipids were shown to be necessary for pore formation by human stefin B in model membranes (Rabzelj et al., 2008). Oxidized cholesterol aldehyde products, present in atherosclerotic lesions also have the ability to promote apolipoprotein C-II amyloid fibril formation (Stewart et al., 2007).

2.8 Conclusions

Whole animal and human metabolic studies, together with genetic, cellular and biochemical studies have provided a wealth of information supporting a link between cholesterol and Alzheimer's disease. Increasing evidence suggests that cholesterol interaction is involved in the oligomerization, membrane pore formation and fibrillogenesis of the Alzheimer's Aβ peptide and other amyloid peptides. Undoubtedly, further structural studies will have much to contribute, and are likely to correlate well with the other diverse approaches being used extensively for the study of Alzheimer's disease and other amyloidopathies.

References

Abildayeva, K., Jansen, P.J., Hirsch-Reinshagen, V., Bloks, V.W., Bakker, A.H. F., Ramaekers, F.C.S., de Vente, J., Groen, A.K., Wellington, C.L., Kuipers, F., Mulder, M., 2006, 24(S)-hydroxycholesterol participates in a liver X receptor-controlled pathway in astrocytes that regulates apolipoprotein E-mediated cholesterol efflux. *J. Biol. Chem.* **281**: 12799–12808.

Adalbert, R., Gilley, J., Coleman, M.P., 2007, Aβ, tau and ApoE4 in Alzheimer's disease: The axonal connection. *Trends Mol. Med.* **13**: 135–142.

Aizawa, Y., Fukatsu, R., Takamaru, Y., Tsuzuki, K., Chiba, H., Kobayashi, K., Fujii, N., Takahata, N., 1997, Amino-terminus truncated apolipoprotein E is the major species in amyloid deposits in Alzheimer's disease-affected brains: A possible role for apolipoprotein E in Alzheimer's disease. *Brain Res.* **768**: 208–214.

Avdulov, N.A., Chochina, S.V., Igbavboa, U., Warden, C.S., Vassiliev, A.V., Wood, W. G., 1997, Lipid binding to amyloid β-peptide aggregates: Preferential binding of cholesterol as compared with phosphatidylcholine and fatty acids. *J. Neurochem.* **69**: 1746–1752.

Avidan-Shpalter, C., Gazit, E., 2006, The early stages of amyloid formation: Biophysical and structural characterization of human calcitonin pre-fibrillar assemblies. *Amyloid* **13**: 216–225.

Bar-On, P., Crews, L., Koob, A.O., Mizuno, H., Adame, A., Spencer, B., Masliah, E., 2008, Statins reduce neuronal α-synuclein aggregation in *in vitro* models of Parkinson's disease. *J. Neurochem.* **105**: 1656–1667.

Behl, C., 2005, Oxidative stress in Alzheimer's disease: Implications for prevention and therapy. In: Harris, J.R. and Fahrenholz, F. (eds.), Alzheimer's Disease: Cellular and Molecular Aspects of Amyloid β, Springer, New York, pp. 65–78.

Bishop, G.M., Robinson, S.R., 2004, The amyloid paradox: Amyloid-β-metal complexes can be neurotoxic and neuroprotective. *Brain Pathol.* **14**: 448–452.

Björkhem, I., Cedazo-Minguez, A., Leoni, V., Meaney, S., 2009, Oxysterols and neurodegenerative diseases. *Mol. Asp. Med.* **30**: 171–179.

Borroni, B., Archetti, S., Agosti, C., Akkawi, N., Brambilla, C., Caimi, L., Caltragirone, C., DiLuca, M., Padovani, A., 2004, Intronic CYP46 polymorphism along with ApoE genotype in sporadic Alzheimer disease: From risk factors to disease modulators. *Neurobiol. Aging* **25**: 747–751.

Bosco, D.A., Fowler, D.M., Zhang, Q., Nieva, J., Powers, E.T., Wentworth, P., Lerner, R.A., Kelly, J. W., 2006, Elevated levels of oxidized cholesterol metabolites in Lewy body disease brains accelerate α-synuclein fibrillization. *Nat. Chem. Biol.* **2**: 249–253.

Brown, J., Theisler, C., Silberman, S., Magnuson, D., Gottardi-Littell, N., Lee, J.M., Yager, D., Crowley, J., Sambamurti, K., Rahman, M.M., Reiss, A.B., Eckman, C.R., Wolozin, B., 2004, Differential expression of cholesterol hydroxylases in Alzheimer's disease. *J. Biol. Chem.* **279**: 34674–34681.

Burns, M., Duff, K., 2002, Cholesterol in Alzheimer's disease and tauopathy. *Ann NY Acad. Sci.* **977**: 367–375.

Burns, M., Gaynor, K., Olm, V., Mercken, M., LaFrancois, J., Wang, L., Mathews, P.M., Noble, W., Matsuoka, Y., Duff, K., 2003a, Presenilin redistribution associated with aberrant cholesterol transport enhances β-amyloid production *in vivo*. *J. Neurosci.* **23**: 5645–5649.

Burns, M.P., Noble, W.J., Olm, V., Gaynor, K., Casey, E., LaFrancois, J., Wang, L., Duff, K., 2003b, Co-localization of cholesterol, apolipoprotein E and fibrillar Aβ in amyloid plaques. *Mol. Brain Res.* **110**: 119–125.

Burns, M.P., Igbavboa, U., Wang, L., Wood, W.G., Duff, K., 2006, Cholesterol distribution, not total levels, correlate with altered amyloid precursor protein processing in statin-treated mice. *Neuromol. Med.* **8**: 319–328.

Butterfield, D.A., Castegna, A., Lauderback, C.M., Drake, J., 2002, Evidence that amyloid beta-peptide-induced lipid peroxidation and its sequelae in Alzheimer's disease brain contribute to neuronal death. *Neurobiol. Aging* **23**: 655–664.

Canevari, L., Clark, J.B., 2007, Alzheimer's disease and cholesterol: The fat connection, *Neurochem. Res.* **32**: 739–750.

Carter, D.B., 2005, The interaction of amyloid-β with ApoE. In: Harris, J.R. and Fahrenholz, F. (eds.), Alzheimer's Disease: Cellular and Molecular Aspects of Amyloid β, Springer, New York, pp. 255–272.

Carter, C.J., 2007, Convergence of genes implicated in Alzheimer's disease on the cerebral cholesterol shuttle: APP, cholesterol, lipoproteins, and atherosclerosis. *Neurochem. Int.* **50**: 12–38.

Casserly, I., Topol, E., 2004, Convergence of atherosclerosis and Alzheimer's disease: inflammation, cholesterol, and misfolded proteins. *Lancet* **363**: 1139–1146.

Castaño, E.M., Prelli, F., Wisniewski, T., Golabek, A., Kumar, R.A., Soto, C., Frangione, B., 1995, Fibrillogenesis in Alzheimer's disease of amyloid β peptides and apolipoprotein E. *Biochem. J.* **306**: 599–604.

Cechetto, D.F., Hachinski, V., Whitehead, S.N., 2008, Vascular risk factors and Alzheimer's disease. *Expert Rev. Neurotherapeut.* **8**: 743–750.

Chen, X., Ghribi, O., Geiger, J.D., 2008, Rabbits fed cholesterol-enriched diets exhibit pathological features of inclusion body myositis, *Am. J. Physiol. Regulatory Integr. Comp. Physiol.* **294**: R829–R835.

Cheon, M., Chang, I., Mohanty, S., Luheshi, L.M., Dobson, C.M., Vendruscolo, M., Favrin, G., 2007, Structural reorganization and potential toxicity of oligomeric species formed during the assembly of amyloid fibrils. *PLoS Comp. Biol.* **3**: 1727–1738.

Cho, W.-J., Jena, B.P., Jeremie, A.M., 2008, Nano-scale imaging and dynamics of amylin-membrane interactions and its implications in type II diabetes mellitus. *Meth. Cell Biol.* **90**: 267–286.

Cho, W.-J., Trikha, S., Jeremic, A.M., 2009, Cholesterol regulates assembly of human islet amyloid polypeptide on model membranes. *J. Mol. Biol.* In press. Doi.10.0116/j.jmb.2009.08.055.

Chochina, S.V., Avdulov, N.A., Igbavboa, U., Cleary, J.P., O'Hare, E.O., Wood, W.G., 2001, Amyloid β-peptide$_{1-40}$ increases neuronal membrane fluidity: Role of cholesterol and brain region. *J. Lipid Res.* **42**: 1292–1297.

Christidis, D.S., Liberopoulos, E.N., Kakafika, A.I., Miltiadous, G.A., Cariolou, M., Ganotakis, E.S., Mikhailidis, D.P., Elisaf, M.S., 2006, The effect of apolipoprotein E polymorphism on the response to lipid-lowering treatment with atorvastatin or fenofibrate. *J. Cardiovasc. Pharmacol. Ther.* **11**(3): 211–221.

Crameri, A., Biondi, E., Kuehnle, K., Lutjohann, D., Thelen, K.M., Perga, S., Dotti, C.G., Nitsch, R.M., Dolores, M., Ledesma, D.M., Mohajeri, M.H., 2006, The role of seladin-1/DHCR24 in cholesterol biosynthesis, APP processing and Aβ generation in vivo. *EMBO J.* **25**: 432–443.

Crutcher, K.A., 2004, Apolipoprotein E is a prime suspect, not just and accomplice, in Alzheimer's disease. *J. Mol. Neurosci.* **23**: 181–188.

Cutler, R.G., Kelly, J., Storie, K., Pedersen, W.A., Tammara, A., Hatanpaa, K., Troncoso, J.C., Mattson, M.P., 2004, Involvement of oxidative stress-induced abnormalities in ceramide and cholesterol metabolism in brain aging and Alzheimer's disease. *Proc. Natl. Acad. Sci. USA* **101**: 2070–2075.

D'Errico, G., Vitiello, G., Ortona, O., Tedeschi, A., Ramunno, A., D'Ursi, A.M., 2008, Interaction between Alzheimer's Aβ(25-35) peptide and phospholipid bilayers: The role of cholesterol. *Biochim. Biophys. Acta* **1778**: 2710–2716.

Desai, P., DeKosky, S.T., Kamboh, M.I., 2002, Genetic variation in the cholesterol 24-hydroxylase (CYP46) gene and the risk of Alzheimer's disease. *Neurosci. Lett.* **328**: 9–12.

Diociaiuti, M., Polzi, L.Z., Valvo, L., Malchiodi-Albedi, F., Bombelli, C., Gaudiano, M.C., 2006, Calcitonin forms oligomeric pore-like structures in lipid membranes. *Biophys. J.* **91**: 2275–2281.

Distl, R., Meske, V., Ohm, T.G., 2001, Tangle-bearing neurons contain more free cholesterol than adjacent tangle-free neurons. *Acta Neuropathol.* **101**: 547–554.

Distl, R., Treiber-Held, S., Albert, F., Meske, V., Harzer, K., Ohm, T.G., 2003, Cholesterol storage and tau pathology in Neimann-Pick type C disease in the brain. *J. Pathol.* **200**: 104–111.

Eckert, G.P., Wood, W.G., Müller, W.E., 2005, Membrane disordering effects of β-amyloid peptides. In: Harris, J.R. and Fahrenholz, F. (eds.), Alzheimer's Disease: Cellular and Molecular Aspects of Amyloid β, Springer, New York, pp. 319–349.

Ehehalt, R., Keller, P., Haass, C., Thiele, C. Simons, K., 2003, Amyloidogenic processing of the Alzheimer beta-amyloid precursor protein depends on lipid rafts. *J. Cell Biol.* **160**: 113–123.

Fan, Q.-W., Wei, Y., Takao, S., Yanagisawa, K., Makoto, M., 2001, Cholesterol-dependent modulation of tau phosphorylation in cultured neurons. *J. Neurochem.* **76**: 391–400.

Fassbender, K., Simons, M., Bergmann, C., Stoick, M., Lüjohann, D., Keller, P., Runz, H., Kühl, S., Bertsch, T., von Bergmann, K., Hennerici, M., Beyreuther, K., Hartmann, T., 2001, Simvastatin strongly reduces levels of Alzheimer's disease β-amyloid peptides Aβ42 and Aβ40 *in vitro* and *in vivo*. *Proc. Natl. Acad. Sci. USA* **98**: 5856–5961.

Fernández, A., Llacuna, L., Fernández-Checa, J.C., Colell, A., 2009, *J. Neurosci.* **29**: 6394–6405.

Ferrera, P., Mercado-Gómez, O., Silva-Aguilar, M., Valverde, M., Arias, C., 2008, Cholesterol potentiates beta-amyloid-induced toxicity in human neuroblastoma cells: Involvement of oxidative stress. *Neurochem. Res.* **33**: 1509–1517.

Finch, C.E., 2005, Developmental origins of aging in brain and blood vessels: An overview. *Neurobiol. Aging* **26**: 281–291.

Fortin, D.L., Troyer, M.D., Nakamura, K., Kubo, S.-I., Anthony, M.D., Edwards, R.H., 2004, Lipid rafts mediate the synaptic localization of α-synuclein. *Neurobiol. Dis.* **24**: 6715–6723.

Fryer, J.D., Simmins, K., Paradanian, M., Bales, K.R., Paul, S.M., Sullivan, P.M., Holtzman, D.M., 2005, Human apolipoprotein E4 alters the amyloid-β 40:42 ratio and promotes the formation of cerebral amyloid angiopathy in an amyloid precursor protein transgenic model. *Neurobiol. Dis.* **25**: 2803–2810.

Galloway, S., Jian, L., Johnsen, R., Chew, S., Mamo, J.C.L., 2007, β-Amyloid or its precursor protein is found in epithelial cells of the small intestine and is stimulated by high-fat feeding. *J. Nutr. Biochem.* **18**: 279–284.

Gehman, J.D., O'Brien, C.C., Shabanpoor, F., Wade, J.D., Saparovic, F., 2008, Metal effects on the membrane interactions of amyloid-β-peptides. *Eur. Biophys. J.* **37:** 333–344.

Gellermann, G.P., Appel, T.R., Tannert, A., Radestock, A., Hortschansky, P., Schroeckh, V., Leisner, C., Lütkepohl, T., Shtrasburg, S., Röcken, C., Pras, M., Linke, R.P., Diekmann, S., Fändrich, M., 2005, Raft lipids as common components of human extracellular amyloid fibrils. *Proc. Natl. Acad. Sci. USA* **102:** 6297–6302.

Gellermann, G.P., Ullrich, K., Tannert, A., Unger, C., Habicht, G., Sauter, S.R.N., Hortschansky, P., Horn, U., Möllmann, U., Decker, M., Lehmann, J., Fändrich, M., 2006, Alzheimer-like plaque formation by human macrophages is reduced by fibrillation inhibitors and lovastatin. *J. Mol. Biol.* **360:** 251–257.

Ghribi, O., Larsen, B., Schrag, M., Herman, M.M., 2006, High cholesterol content in neurons increases BACE, β-amyloid, and phosphorylated tau levels in rabbit hippocampus. *Ext. Neurol.* **200:** 460–467.

Ghribi, O., Golovko, M.Y., Larsen, B., Schrag, M., Murphh, E.J., 2007, Deposition of iron and beta-amyloid plaques is associated with cortical cellular damage in rabbits fed with long-term cholesterol-enriched diets. *J. Neurochem.* **103:** 423–424.

Ghribi, O., Schommer, E., Prasanthi, J.R.P., 2009, 27-hydroxycholestrol as the missing link between circulating cholesterol and AD-like pathology. *Alzheimer's & Dementia: J. Alzheimer's Assoc* **5**: 180.

Gilbert, R.J.C., Rosjohn, J., Parker, M.W., Tweten, R.K., Morgan, P.J., Mitchell, T.J., Errington, N., Rowe, A.J., Andrew, P.W., Byron, O., 1998, Self-interaction of pneumolysin, the pore forming protein toxin of *Streptococcus pneumoniae*. *J. Mol Biol.* **284:** 1223–1237.

Gong, J.S., Kobayashi, M., Hayashi, H., Zou, K., Sawamura, N., Fujita, S.C., Yanagisawa, K., Michikawa, H., 2002, Apolipoprotein E (ApoE) isoform-dependent lipid release from astrocytes prepared from human ApoE3 and ApoE4 knock-in mice. *J. Biol. Chem.* **277:** 29919–29926.

Gong, J.S., Morita, S.Y., Handa, T., Fujita, S.C., Yanagisawa, K., Michikawa, M., 2007, Novel action of lipoprotein E (ApoE): ApoE isoform specifically inhibits lipid-particle-mediated cholesterol release from neurons. *Mol. Neurodegener.* **2:** 9. doi: 10.1 186/1750-1326-2-9.

Greeve, I., Hermans-Borgmeyer, I., Brellinger, C., Kasper, D., Gomez-Isla, T., Behl, C., Levkau, B., Nitsch, R.M., 2000, The human DIMINUTO/DWARF1 homolog seladin-1 confers resistance to Alzheimer's disease-associated neurodegeneration and oxidative stress. *J. Neurosci.* **20:** 7345–7352.

Gylys, K.H., Fein, J.A., Yang, F., Miller, C.A., Cole, G.M., 2007, Increased cholesterol in Aβ-positive nerve terminals from Alzheimer's disease cortex. *Neurobiol. Aging* **28:** 8–17.

Haberland, M.E., Reynolds, J.A., 1973, Self-association of cholesterol in aqueous solution. *Proc. Natl. Acad. Sci. USA* **70:** 2313–2316.

Harris, J.R., 1988, Electron microscopy of cholesterol. *Micron Microsc. Acta* **19:** 19–32.

Harris, J.R., 2002, *In vitro* fibrillogenesis of the amyloid β$_{1-42}$ peptide: cholesterol potentiation and aspirin inhibition. *Micron* **33:** 609–626.

Harris, J.R., 2008, Cholesterol binding to amyloid-β fibrils: A TEM study. *Micron* **39:** 1192–1196.

Hartmann, T., 2005, Cholesterol and Alzheimer's disease: Statins, cholesterol depletion in APP processing and Aβ generation. In: Harris, J.R. and Fahrenholz, F. (eds.), Alzheimer's Disease: Cellular and Molecular Aspects of Amyloid β, Springer, New York, pp. 365–380.

Hatters, D.M., Peters-Libeu, C.A., Weisgraber, K.W., 2006, Apolipoprotein E structure: Insights into function. *Trends Biochem. Sci.* **31:** 445.

Hirsch-Reinshagen, V., Wellington, C.L., 2007, Cholesterol metabolism, apolipoprotein E, adenosine triphosphate-binding cassette transporters, and Alzheimer's disease. *Curr. Opin. Lipidol.* **18:** 325–332.

Hofman, A., Ott, A., Breteler, M.M., Bots, M.L., Slooter, A.J., van Harskamp, F., van Duijn, C.N., van Broeckhoven, C., Grobbee, D.E., 1997, Atherosclerosis, apolipoprotein E, and prevalence of dementia and Alzheimer's disease in the Rotterdam Study. *Lancet* **349:** 151–154.

Hooijmans, C.R., Van der Zee, C.E.E.M., Dederen, P.J., Brouwer, K.M., Reijmer, Y.D., van Groen, T., Broersen, L.M., Lütjohann, D., Heerschap, A., Kiliaan, A.J., 2009, DHA and cholesterol containing diets influence Alzheimer-like pathology, cognition and cerebral vasculature in APP$_{SWE}$/PS1$_{dE9}$ mice. Neurobiol. Dis. **33**: 482–498.

Hou, X., Mechler, A., Martin, L.L., Aguilar, M.-I., Small, D.H., 2008, Cholesterol and anionic phospholipids increase the binding of amyloidogenic transthyretin to lipid membranes. *Biochim. Biophys. Acta* **1778**: 198–205.

Igbavboa, U., Pidcock, J.M., Johnson, L.N.A., Malo, T.M., Studniski, A.E., Yu, S., Sun, G.Y., Wood, W. G., 2003, Cholesterol distribution in the Golgi complex of DITNC1 astrocytes is differentially altered by fresh and aged amyloid β-peptide-(1-42). *J. Biol. Chem.* **278**: 17150–17167.

Igbavboa, U., Sun, G.Y., Weisman, G.A., He, Y., Wood, W.G., 2009, Amyloid β-protein stimulates trafficking of cholesterol and caveolin-1 from the plasma membrane to the Golgi complex in mouse primary astrocytes. *Neuroscience* **162**: 328–338.

Infante, J., Rodríguez-Rodríguez, E., Mateo, I., Llorca, J., Vázquez-Higuera, J.L., Berciano, J., Combarros, O., 2008, Gene-gene interaction between heme oxygenase-1 and liver X receptor-β and Alzheimer's disease risk. *Neurobiol. Aging*, In press, doi:10.1016/j.neurobiolaging.2008.05.025.

Ishiwata, H., Sato, S.B., Vertut-Doï, A., Hamashima, Y., Miyajima, K., 1997, Cholesterol derivative of poly(ethylene glycol) inhibits clathrin-independent, but not clathrin-dependent endocytosis. *Biochim. Biophys. Acta* **1359**: 123–135.

Jang, H., Zheng, J., Nussinov, R., 2007a, Models of β-amyloid ion channels in the membrane suggest that channel formation in the bilayer is a dynamic process. *Biophys. J.* **93**: 1938–1949.

Jang, H., Zheng, J., Lal, R., Nussinov, R., 2007b, New structures help the modelling of toxic amyloidβ ion channels. *Trends Biochem. Sci.* **33**: 91–100.

Jayasinghe, S.A., Langen, R., 2007, Membrane interaction of islet amyloid polypeptide. *Biochim. Biophys. Acta* **1768**: 2002–2009.

Kakio, A., Nishimoto, S., Yanaagisawa, K., Kozutsumi, Y., Matsuzaki, K., 2001, Cholesterol-dependent formation of GM1 ganglioside-bound amyloid beta-protein, an endogenous seed for Alzheimer amyloid. *J. Biol. Chem.* **276**: 24985–24990.

Kakio, A., Nishimoto, S., Yanagisawa, K., Kozutsumi, Y., Matsuzaki, K., 2002, Interactions of amyloid beta-protein with various gangliosides in raft-like membranes: Importance of GM1 ganglioside-bound form as an endogenous seed for Alzheimer amyloid. *Biochemistry* **41**: 7385–7390.

Kalman, J., Janka, Z., 2005, Cholesterol and Alzheimer's disease. *Orv. Hetil.* **146**: 1903–1911.

Kalvodova, L., Kahya, N., Schwille, P., Ehehalt, R., Verkade, P., Drechsel, D., Simons, K., 2005, Lipids as modulators of proteolytic activity of BACE: involvement of cholesterol, glycosphingolipids, and anionic phospholipids in vitro. *J. Biol. Chem.* **280**: 36815–36823.

Karube, H., Sakamoato, M., Arawaka, S., Hara, S., Sato, H., Ren, C.-H., Goto, S., Koyama, S., Wada, M., Kawanami, T., Kurita, K., Kato, T., 2008, N-terminal region of a-synuclein is essential for the fatty acid-induced oligomerization of the molecules. *FEBS Lett.* **582**: 3693–3700.

Kimura, T., Yanagisawa, K., 2007, Endosomal accumulation of GM1 ganglioside-bound amyloid beta-protein in neurons of aged monkey brains. *Neuroreport* **18**: 1669–1673.

Kirkitadze, M.D., Kowalska, A., 2005, Molecular mechanisms initiating amyloid β-fibril formation in Alzheimer's disease. *Acta Biochim. Pol.* **52**: 417–423.

Klinknera, A.M., Waites, C.R., Kerns, W.D., Bugelski, P.J., 1995, Evidence of foam cell and cholesterol crystal formation in macrophages incubated with oxidized LDL by fluorescence and electron microscopy. *J. Histochem. Cytochem.* **43**: 1071–1078.

Kölsch, H., Lütjohann, D., Ludwig, M., Schulte, A., Ptok, U., Jessen, F., von Bergmann, K., Rao, M.L., Maier, W., Heun, R., 2002, Polymorphism in the cholesterol 24S-hydroxylase gene is associated with Alzheimer's disease. *Mol. Psych.* **7**: 899–902.

Koppaka, V., Paul, C., Murray, I.V.J., Axelsen, P.H., 2003, Early synergy between Aβ42 and oxidatively damaged membranes in promoting amyloid fibril formation by Aβ40. *J. Biol. Chem.* **278:** 36277–36284.

Koudinov, A.R., Koudinova, N.V., 2001a, Brain cholesterol pathology is the cause of Alzheimer's disease. *Clin. Med. Health Res.* Clinmed/2001100005.

Koudinov, A.R., Koudinova, N.V., 2001b, Essential role for cholesterol in synaptic plasticity and neuronal degeneration. *FASEB J.* doi: 10.1096/fj.00-0815fje.

Krištofiková, Z., Kopecky, V., Hofbauerova, K., Hovorková, P., Řipová, D., 2008, Complex of amyloid β peptides with 24-hydroxycholesterol and its effect on hemicholinium-3 sensitive carriers. *Neurochem. Res.* **33:** 412–421.

Lashuel, H.A., Lansbury, P.T., 2006, Are amyloid diseases caused by protein aggregates that mimic bacterial pore-forming toxins. *Quart. Rev. Biophys.* **39:** 167–201.

Lau, T.-L., Gehman, J.D., Wade, J.D., Masters, C.L., Barnham, K.J., Separovic, F., 2007, Cholesterol and clioquinol modulation of Aβ(1-42) interaction with phospholipid bilayers and metals. *Biochim. Biophys. Acta* **1768:** 3135–3144.

Lebouvier, T., Perruchini, C., Panachal, M., Potier, M.C., Duyckaerts, C., 2009, Cholesterol in the senile plaque: Often mentioned, never seen. *Acta Neuropathol.* **117:** 31–34.

Ledesma, M.D., Dotti, C.G., 2006, Amyloid excess in Alzheimer's disease: What is cholesterol to be blamed for? *FEBS Lett.* **580:** 5525–5532.

Li, L., Cao, D., Garber, D.W., Kim, H., Fukuchi, K., 2003, Association of aortic atherosclerosis with cerebral β-amyloidosis and learning deficits in a mouse model of Alzheimer's disease. *Amer. J. Pathol.* **163:** 2155–2164.

Liao, G., Yao, Y., Liu, J., Yu, Z., Cheung, S., Xie, A., Liang, X., Bi, X., 2007, Cholesterol accumulation is associated with lysosomal dysfunction and autophagic stress in Npc1$^{-/-}$ mouse brain. *Amer. J. Pathol.* **171:** 962–975.

Lin, M.-S., Chen, L.-Y., Wang, S. S.-S., Chang, Y., Chen, W.-Y., 2008, Examining the levels of ganglioside and cholesterol in cell membrane on attenuating the cytotoxicity of beta-amyloid peptide. *Colloid. Surf. B: Biointerfaces* **65:** 172–177.

Liu, Y., Peterson, D.A., Schubert, D., 1998, Amyloid β peptide alters intracellular vesicle trafficking and cholesterol homeostasis, *Proc. Natl. Acad. Sci. USA* **95:** 13266–13271.

Lu, J., Wu, D., Zeng, Y., Sun, D., Hu, B., Shan, Q., Zhang, Z., Fan, S., 2009, Trace amounts of copper exacerbate beta amyloid-induced neourotoxicity in the cholesterol-fed mice through TNF-mediated inflammatory pathway. *Brain, Behav. Immunol.* **23:** 193–203.

Lukiw, W.J., Pappolla, M., Pelaez, R.P., Bazan, N.G., 2005, Alzheimer's disease – A dysfunction in cholesterol and lipid metabolism. *Cell. Mol. Neurobiol.* **25:** 475–483.

Malinchik, S.B., Inyouye, H., Szumowski, K.E., Kirschner, D.A., 1998, Structural analysis of the Alzheimer's β(1-40) amyloid: Protofilament assembly of tubular fibrils. *Biophys. J.* **74:** 537–545.

Martins, I.J., Hone, E., Foster, J.K., Sünram-Lea, S.I., Gnjec, A., Fuller, S.J., Nolan, D., Gandy, S.E., Martins, R.N., 2006, Apolipoprotein E, cholesterol metabolism, diabetes, and the convergence of risk factors for Alzheimer's disease and cardiovascular disease. *Mol. Psychiatry* **11:** 721–736.

Mason, R.P., Walter, M.F., Day, C.A., Jacob, R.F., 2006, Active metabolite of atorvastatin inhibits membrane cholesterol domain formation by an antioxidant mechanism. *J. Biol. Chem.* **281:** 9337–9345.

Mast, N., White, M.A., Bjorkhem, I., Johnson, E.F., Stout, C.D., 2008, Crystal structure of substrate-bound and substrate-free cytochrome P450 46A1, the principal cholesterol hydroxylase in the brain. *Proc. Natl. Acad. Sci. USA* **105:** 9546–9551.

McLaurin, J., Darabie, A.A., Morrison, M.R., 2002, Cholesterol, a modulator of membrane-associated amyloid-β fibrillogenesis. *Ann. NY Acad. Sci.* **977:** 376–383.

Micelli, S., Meleleo, D., Picciarelli, V., Gallucci, E., 2004, Effect of sterols on β-amyloid peptide (AβP 1-40) channel formation and their properties in planar lipid membranes. *Biophys. J.* **86:** 2231–2237.

Michikawa, M., 2004, Neurodegenerative disorders and cholesterol. *Curr. Alzh. Res.* **1:** 271–275.
Michikawa, M., 2006, Role of cholesterol in amyloid cascade: cholesterol-dependent modulation of tau phosphorylation and mitochondrial function. *Acta Neurol. Scand. Suppl.* **114:** 21–26.
Milton, N.G.N., Harris, J.R., 2009, Polymorphism of amyloid-β fibrils and its effects on human erythrocyte catalase binding. *Micron.* **40:** 800–810.
Mizuno, T., Nakata, M., Hironobu, N., Michikawa, M., Wang, R., Haass, C., Yanagisawa, K., 1999, Cholesterol-dependent generation of a seeding amyloid β-protein in cell culture. *J. Biol. Chem.* **274:** 15110–15114.
Mori, T., Paris, D., Town, T., Sparks, D.L., Delledonne, A., Crawford, F., Abdullah, L.I., Humphrey, J.A., Dickson, D.W., Mullan, M.J., 2001, Cholesterol accumulates in senile plaques of Alzheimer disease patients and in transgenic APP(SW) mice. *J. Neuropathol. Exp. Neurol.* **60:** 778–785.
Morishima-Kawashima, M., Han, X., Tanimura, Y., Hamanaka, H., Kobayashi, M., Sakurai, T., Yokayama, M., Wada, K., Nukina, N., Fujita, S.C., Ihara, Y., 2007, Effects of human apolipoprotein E isoforms on the amyloid beta-protein concentration and lipid composition of brain low-density membrane domains. *J. Neurochem.* **101:** 949–958.
Nelson, T.J., 2007, Cholesterol oxidation and β-amyloid. In: M.-K. Sun (Ed.), Research Progress in Alzheimer's Disease and Dementia, Vol. 2, Nova Science Publishers, New York, pp. 137–174.
Nelson, T.J., Alkon, D.L., 2005, Oxidation of cholesterol by amyloid precursor protein and beta-amyloid peptide. *J. Biol. Chem.* **280**: 7377–7387.
Nelson, T.J., Alkon, D.L., 2007, Protection against β-amyloid-induced apoptosis by peptides interacting with β-amyloid. *J. Biol. Chem.* **282:** 31238–31249.
Nixon, R.A., 2004, Niemann-Pick type C disease and Alzheimer's disease. *Am. J. Pathol.* **164:** 757–761.
Ohm, T.G., Treiber-Held, S., Distl, R., Glöckner, R., Schönheit, B., Tamanai, M., Meske, V., 2003, Cholesterol and tau protein – findings in Alzheimer's and Niemann Pick C's disease. *Pharmopsychiatry* **36:** 120–126.
Ong, W.Y., Goh, E.W., Lu, X.R., Farooqui, A.A., Patel, S.C., Halliwell, B., 2003, Increase in cholesterol and cholesterol oxidation products, and role of cholesterol oxidation products in kainite-induced neuronal injury. *Brain Pathol.* **13:** 250–262.
Olsen, O.F., Dragø, L., 2000, High density lipoprotein inhibits assembly of amyloid beta-pepptides into fibrils. *Biochem. Biophys. Res. Commun.* **270**: 62–66
Opazo, C., Huang, X., Cherny, R.A., Moir, R.D., Roher, A.E., White, A.R., Cappai, R., Masters, C.I., Tanzi, R.E., Inestrosa, N.C., Bush, A.I., 2002, Metalloenzyme-like activity of Alzheimer's disease β-amyloid. *J. Biol. Chem.* **277:** 40302–40308.
Orr, J.D., 2008, Statins in the spectrum of neurological diseases. *Curr. Atheroscler. Rep.* **10:** 11–18.
Panza, F., D'Introno, A., Colacicco, A.M., Capruso, C., Pichichero, G., Capruso, S.A., Capruso, A., Solfrizzi, V., 2005, Lipid metabolism in cognitive decline and dementia. *Brain Res. Rev.* **52:** 275–292.
Papassotiropoulos, A., Streffer, J., Tsolaki, M., Schmid, S., Thal, D., Nicosa, F., Iakovidou, V., Maddalena, A., Lütjohan, D., Ghebremedhin, E., Hegi, T., Pasch, T., Träxler, M., Brühl, A., Benussi, L., Boinetti, G., Braak, H., Mnitsch, R.M., Hock, C., 2003, Increased brain β-amyloid load, phosphorylated tau, and risk of Alzheimer's disease associated with an intronic *CYP46* polymorphism. *Arch. Neurol.* **60:** 29–35.
Papassotiropoulos, A., Wollmer, A., Tsolaki, M., Brunner, F., Molyva, D., Lütjohann, D., Nitsch, R.M., Hock, C., 2005, A cluster of cholesterol-related genes confers susceptibility for Alzheimer's disease. *J. Clin. Psych.* **66:** 940–947.
Pappolla, M.A., Smith, M.A., Bryant-Thomas, T., Bazan, N., Petanceska, S., Perry, G., Thal, L.J., Sano, M., Refolo, L.M., 2002, Cholesterol, oxidative stress and Alzheimer's disease: Expanding the horizons of pathogenesis. *Free Rad. Biol. Med.* **33:** 173–181.
Phillips, S.E., Woodruff, E.A., Liang, P., Patten, M., Broadie, K., 2008, Neuronal loss of *Drosophila* NCP1a causes cholesterol aggregation and age-progressive neurodegeneration. *J. Neurosci.* **28:** 6569–6582.

Prasanthi, J.R.P., Huls, A., Thomasson, S., Thompson, A., Schommer, E., Ghribi, O., 2009, Differential effects of 24-hydroxycholesterol and 27-hydroxycholesterolon β-amyloid precursor protein levels and processing in human neuroblastoma SH-SY5Y cells. *Mol. Neurodegen.* **4**:1, doi:10.1186/1750-1326-4-1

Prasanthi, J.R.P., Schommer, E., Thomasson, S., Thompson, A., Feist, G., Ghribi, O., 2008, Regulation of beta-amyloid levels in the brain of cholesterol-fed rabbit, a model system for sporadic Alzheimer's disease. *Mech. Ageing Dev.* **129:** 649-655.

Prasher, V.P., Airuehia, E., Patel, A., Haque, M.S., 2008, Total serum cholesterol levels and Alzheimer's dementia in patients with Down syndrome. *Int. J. Geriatr. Psychiatry.* **23**: 937–942.

Puglielli, L., Friedlich, A.L., Setchell, K.D.R., Nagano, S., Opazo, C., Cherny, R.A., Barnham, K.J., Wade, J.D., Melov, S., Kovacs, D.M., Bush, A.I., 2005, Alzheimer disease β-amyloid activity mimics cholesterol oxidase. *J. Clin. Invest.* **115**: 2556–2563.

Qui, L., Lewis, A., Como, J., Vaughn, M.W., Huang, J., Somerharju, P., Virtanen, J., Cheng, K.H., 2009, Cholesterol modulates the interaction of β-amyloid peptide with lipid bilayers. *Biophys. J.* **96**: 4299–4307.

Rabzelj, S., Viero, G., Gutierrez-Aguirre, I., Turk, V., Dalla Serra, M., Anderluh, G., Zerovnik, E., 2008, Interaction with model membranes and pore formation by human stefin B: Studying the native and prefibrillar states. *FEBS J.* **275:** 2455–2466.

Raffaï, R.L., Weisgraber, K.H., 2003, Cholesterol: From heart attacks to Alzheimer's disease. *J. Lipid Res.* **44**: 1423–1430.

Rahman, A., Akterin, S., Flores-Morales, A., Crisby, M., Kivipelto, M., Schultzberg, M., Cedazo-Minguez, A., 2005, High cholesterol diet induced tau hyperphosphorylation in apolipoprotein E deficient mice. *FEBS Lett.* **579**: 6411–6416.

Rapp, A., Gmeineer, B., Hüttinger, M., 2006, Implication of apoE isoforms in cholesterol metabolism by primary rat hippocampal neurons and astrocytes. *Biochimie* **88:** 473–483.

Rapp, J.H., Pan, X.M., Neumann, M., Hong, M., Hollenbeck, K., Liu, J., 2008, Microemboli composed of cholesterol crystals disrupt the blood-brain barrier and reduce cognition. *Stroke* **39**: 2354–2361.

Reid, P.C., Sakashita, N., Sguii, S., Ohno-Iwashita, Y., Shimada, Y., Hickey, W.F., Chang, T.-Y., 2004, A novel cholesterol stain reveals early neuronal cholesterol accumulation in the Niemann-Pick type C1 mouse brain. *J. Lipid. Res.* **45**: 582–591.

Reimann, E.M., Chen, K., Caselli, R.J., Alexander, G.E., Bandy, D., Adamson, J.L., Lee, W., Cannon, A., Stephan, E.A., Stephan, D.A., Papassotiropoulos, A., 2008, Cholesterol-related genetic scores are associated with hypometabolism in Alzheimer's-affected brain regions. *Neuroimage* **40:** 1214–1221.

Reiss, A.B., 2005, Cholesterol and apolipoprotein E in Alzheimer's disease. *Am. J. Alzheimer's Dis Other Demen*, **20:** 91–96.

Rodríguez-Rodríguez, E., Mateo, I., Infante, J., Llorca, J., García-Gorostiaga, I., Vázquez-Higuera, J.L., Sánchez-Juan, P., Berciano, J., Onofre, C., 2009 Interaction between HMGCR and ABCA1 cholesterol-related genes modulates Alzheimer's disease risk. *Brain Res.* **1280:** 1660171.

Saito, Y., Suzuki, K., Nanba, E., Tamamoto, T., Ohno, K., Murayama, S., 2002, Niemann-Pick type C disease: Accelerated neurofibrillary tangle formation and amyloid βdeposition associated with apolipoprotein E ϵ4 homozygosity. *Ann. Neurol.* **52:** 351–355.

Sathishkumar, K., Xi, X., Martin, R., Uppu, R.M., 2007, Cholesterol secoaldehyde, an oxonation product of cholesterol, induces amyloid aggregation and apoptosis in murine GT1-7 hypothalamic neurons. *J. Alzheimer's Dis.* **11:** 261–274.

Scheinost, J.C., Wang, H., Boldt, G.E., Offer, J., Wentworth, Jr., P., 2008, Cholesterol *seco*-sterol-induced aggregation of methylated amyloid-β peptides – Insights into aldehyde fibrillization of amyloid-β. *Angew. Chem.* **120**: 3983–3986.

Schneider, A., Schulz-Schaeffer, W., Hartmann, T., Schultz, J.B., Simons, M., 2006, Cholesterol depletion reduces aggregation of amyloid-beta peptide in hippocampal neurons. *Neurobiol. Dis.* **23:** 573–577.

Schöneich, C., 2002, Redox processes of methionine relevant to β-amyloid oxidation and Alzheimer's disease. *Arch. Biochem. Biophys.* **397**: 370–376.

Schulz, B., Liebisch, G., Grandl, M., Werner, T., Barlage, S., Schmitz, G., 2007, β-Amyloid (Aβ$_{40}$, Aβ$_{42}$) binding to modified LDL accelerates macrophage foam cell formation. *Biochim. Biophys. Acta* **1771**: 1335–1344.

Scott Kim, W., Chan, S.L., Hill, A.F., Guillemin, G.J., Garner, D., 2009, Impact of 27-hydroxycholesterol on amyloid-β production and ATP-binding cassette transporter expression in primary human neurons. *J. Alzheimer's Dis.* **16**: 121–131.

Shobab, L.A., Hsiung, G.-Y.R., Feldman, H.H., 2005, Cholesterol in Alzheimer's disease. *Lancet Neurol.* **4**: 841–852.

Simons, M., Keller, P., Dichgans, J., Schulz, J.B., 2001, Cholesterol and Alzheimer's disease. *Neurology* **57**: 1089–1093.

Simons, M., Keller, P., De Strooper, B., Beyruther, K., Dotti, C.G., Simons, K., 1998, Cholesterol depletion inhibits the generation of beta-amyloid in hippocampal neurons. *Proc. Natl. Acad. Sci.* USA **95**: 6460–6464.

Sinha, N., Tsai, C.-J., Nussinov, R., 2001, A proposed structural model for amyloid fibril elongation: Domain swapping form an interdigitating β-structure polymer. *Protein Eng.* **14**: 93–103.

Smith, M.E., 2001, Phagocytic properties of microglia in vitro: implications for a role in multiple sclerosis and EAE. *Microsc. Res. Tech.* **54**: 81–94.

Sparks, D.L., 2004, Cholesterol, copper, and accumulation of thioflavine S-reactive Alzheimer's-like amyloid β in rabbit brain. *J. Mol. Neurosci.* **24**: 97–104.

Sparks, D.L., 2007, Cholesterol metabolism and brain amyloidosis: Evidence for a role of copper in the clearance of Aβ through the liver. *Curr. Alz. Res.* **4**: 165–169.

Sparks, D.L., Martin, T.A., Gross, D.R., Hunsaker, J.C., 2000, Link between heart disease, cholesterol, and Alzheimer's disease: A review. *Microsc. Res. Techn.* **50**: 287–290.

Sparks, D.L., Schreurs, B.G., 2003, Trace amounts of copper in water induce β-amyloid plaques and learning deficits in a rabbit model of Alzheimer's disease. *Proc. Natl. Acad. Sci. USA* **100**: 11065–11069.

Srisailam, S., Wang, H.-M., Kumart, T.K.S., Rajalingam, D., Sivaraja, V., Sheu, H.-S., Chang, Y.-C., Yu, C., 2002, Amyloid-like fibril formation in an all β-barrel protein involves the formation of partially structured intermediate(s). *J. Biol. Chem.* **277**: 19027–19036.

Stadler, M., Phinney, A., Probst, A., Sommer, B., Staufenbiel, M., Jucker, M., 1999, Association of microglia with amyloid plaques in brains of APP23 transgenic mice. *Amer. J. Pathol.* **154**: 1673–1684.

Stanyer, L., Betteridge, D.J., Smith, C.C.T., 2002, An investigation into the mechanisms mediating plasma lipoprotein-potentiated β-amyloid fibrillogenesis. *FEBS Lett.* **518**: 72–78.

Stewart, C.R., Wilson, L.M., Zhang, Q., Pham, C.L.L., Waddington, L.J., Staples, M.K., Stapleton, D., Kelly, J.W., Howlett, G.J., 2007, Oxidized cholesterol metabolites found in human atherosclerotic lesions promote apolipoprotein C-II amyloid fibril formation. *Biochemistry* **46**: 5552–5561.

Streit, W.J., Sparks, D.L., 1997, Activation of microglia in the brains of humans with heart disease and hypercholesterolemic rabbits. *J. Mol. Med.* **75**: 1432–1440.

Stromer, T., Serpell, L.C., 2005, Structure and morphology of the Alzheimer's amyloid fibril. *Microsc. Res. Techn.* **67**: 210–217.

Subashinghe, S., Unabia, S., Barrow, C.J., Mok, S.S., Aguilar, M.-I., Small, D.H., 2003, Cholesterol is necessary both for the toxic effect of Aβ peptides on vascular smooth muscle cells and for Aβ binding to vascular smooth muscle cell membranes. *J. Neurochem.* **84**: 685–694.

Sullivan, P.M., Mace, B.E., Estrada, J.C., Schmechel, D.E., Alberts, M.J., 2008, Human apolipoprotein E4 targeted replacement mice show increased prevalence of intracerebral hemorrhage associated with vascular amyloid deposition. *J. Stroke Cerebrovasc. Dis.* **17:** 303–311.

Tamamizu-Kato, S., Cohen, J.K., Drake, C.B., Kosaraju, M.G., Drury, J., Narayanaswami, V., 2008, Interaction of amyloid β peptide compromises the lipid binding function of apolipoprotein E, *Biochemistry,* **47:** 5225–5234.

Tashima, Y., Oe, R., Lee, S., Sugihara, G., Chambers, E.J., Takahashi, M., Yamada, T., 2004, The effect of cholesterol and monosialoganglioside (GM1) on the release and aggregation of amyloid β-peptide from liposomes prepared from brain membrane-like lipids. *J. Biol. Chem.* **279:** 17587–17595.

Tedde, A., Rotondi, M., Cellini, E., Bagnoli, S., Muratore, L., Nacmias, B., Sorbi, S., 2006, Lack of association between the CYP46 gene polymorphism and Italian late-onset sporadic Alzheimer's disease. *Neurobiol. Aging* **27:** 773.e1-773.e3.

Thirumangalakudi, L., Prakasam, A., Zhang, R., Bimonte-Nelson, H., Sambamurti, K., Kindy, M.S., Bhat, N.R., 2008, High cholesterol-induced neuroinflammation and amyloid precursor protein processing correlate with loss of working memory in mice. *J. Neurochem.* **106:** 475–485.

Tong, X.-K., Nicolakakis, N., Fernandes, P., Ongali, B., Brouillette, J., Quirion, R., Hamel, E., 2009. Simvastatin improves cerebrovascular function and counters soluble amyloid-beta, inflammation and oxidative stress in aged APP mice. *Neurobiol. Dis.* **35:** 406–414.

Turkoglu, O.F., Erglu, H., Okutan, O., Gurcan, O., Bodur, E., Sargon, M.F., Öner, L., Beskonakh, E., 2008, Atorvastatin efficiency after traumatic brain injury in rats. *Surgical Neurol.* **72:** 146–152.

Vaya, J., Schipper, H.M., 2007, Oxysterols, cholesterol homeostasis, and Alzheimer's disease. *J, Neurochem.* **102:** 1727–1737.

Wakabayashi, M., Matsuzaki, K., 2007, Formation of amyloids by Aβ-(1-42) on NGF-differentiated PC12 cells: Roles of gangliosides and cholesterol. *J. Mol. Biol.* **371:** 924–933.

Wang, Y., Muneton, S., Sjövall, J., Jovanovic, J.N., Griffiths, W.J., 2008, The effect of 24S-hydroxycholesterol on cholesterol homeostasis in neurons: Quantitative changes to the cortical neuron proteome. *J. Proteome Res.* **7:** 1606–1614.

White, M.A., Mast, N., Bjorkhem, I., Johnson, E.F., Stout, C.D., Pikuleva, I.A., 2008, Use of complementary cation and anion heavy-atom salt derivatives to solve the structure of cytochrome P450 46A1. *Acta Cryst.* D **64:** 487–495.

Wolozin, B., 2004, Cholesterol and the biology of Alzheimer's disease, *Neuron* **41:** 7–10.

Wolozin, B., Kellman, W., Ruosseau, P., Celesia, G.G., Siegel, G., 2000, Decreased prevalence of Alzheimer disease associated with 3-hydroxy-3-methylglutaryl Coenzyme A reductase inhibitors. *Arch. Neurol.* **57:** 1439–1443.

Wood, W.G., Schroeder, F., Igbavboa, U., Avdulov, N.A., Chochina, S.V., 2002, Brain membrane cholesterol domains, aging and amyloid beta-peptides. *Neurobiol. Aging* **23:** 685–694.

Xiong, H., Callaghan, D., Jones, A., Walker, D.G., Lue, L.-F., Beach, T.G., Sue, L.I., Woulfe, J., Xu, H., Stanimirovic, D.B., Zhang, W., 2008, Cholesterol retention in Alzheimer's brain is responsible for high β- and γ-secretase activities and Aβ production. *Neurobiol. Dis.* **29:** 422–437.

Xue, Q.-S., Sparks, L.D., Streit, W.J., 2007, Microglial activation in the hippocampus of hypercholesterolemic rabbits occurs independent of increased amyloid production. *J. Neuroinflamm.* doi: 10.1186/1743-2094-4-20.

Yamazaki, T., Chang, T.Y., Haass, C., Ihara, Y., 2001, Accumulation and aggregation of amyloid beta-protein in late endosomes of Niemann-Pick type C cells. *J. Biol. Chem.* **276:** 4454–4460.

Yanagisawa, K., 2005, Cholesterol and amyloid β fibrillogenesis. In: Harris, J.R. and Fahrenholz, F. (eds.), Alzheimer's Disease: Cellular and Molecular Aspects of Amyloid β, Springer, New York, pp.179–202.

Yanagisawa, K., 2007, Role of gangliosides in Alzheimer's disease. *Biochim. Biophys. Acta* **1768:** 1943–1951.

Yao, Z.-X., Papadopoulos, V., 2002, Function of β-amyloid in cholesterol transport: A lead to neurotoxicity. *FASEB J.* **16:** 1677–1679.

Yao, Z.-X., Brown, R.C., Teper, G., Greeson, J., Papadopoulos, V., 2002, 22R-hydroxycholesterol protects neuronal cells from β-amyloid-induced cytotoxicity by binding to β-amyloid peptide. *J. Neurochem.* **83**: 1110–1119.

Yamamoto, N., Matsubara, T., Sato, T., Yanagisawa, K., 2008, Age-dependent high-density clustering of GM1 ganglioside at presynaptic neuritic terminals promotes amyloid beta protein fibrillogenesis. *Biochim. Biophys. Acta* **1778**: 2717–2726.

Yuyama, K., Yamamoto, N., Yanagisawa, K., 2008, Accelerated release of exosome-associated GM1 ganglioside (GM1) by endocytic pathway abnormality: Another putative pathway for GM1-induced amyloid fibril formation. *J. Neurochem.* **105**: 217–224.

Yip, C.M., McLaurin, J., 2001, Amyloid-β peptide assembly: A critical step in fibrillogenesis and membrane disruption. *Biophys. J.* **80**: 1359–1371.

Yip, C.M., Elton, E.A., Darabie, A.A., Morrison, M.R., McLaurin, J., 2001, Cholesterol, a modulator of membrane-associated Aβ-fibrillogenesis and neurotoxicity. *J. Mol. Biol.* **311**: 723–734.

Yoshiike, Y., Kayed, R., Milton, S.C., Takashima, A., Clabe, C.G., 2007, Pore-forming proteins share structural and functional homology with amyloid oligomers. *Neuromol. Med.* **9**: 270–275.

Yoshimoto, N., Tasaki, M., Shimanouchi, T., Umakoshi, H., Kuboi, R., 2005, Oxidation of cholesterol catalyzed by amyloid beta-peptide (Abeta)-Cu complex on lipid membrane. *J. Biosci. Bioeng.* **100**: 455–459.

Zamrini, E., McGwin, G., Roseman, J.M., 2004, Association between statin use and Alzheimer's disease. *Neuroepidemiology* **23**: 94–98.

Zatta, P., Zambenedetti, P., Stella, M.P., Licastro, F., 2002, Astrocytosis, microgliosis, metallothionein-I-II and amyloid expression in high cholesterol-fed rabbits, *J. Alzheimer's Dis.* **4**:1–9.

Zelcer, N., Khanlou, N., Clare, R., Jiang, Q., Reed-Geaghan, E.G., Landreth, G.E., Vinters, H.V., Tontonoz, P., 2007, Attenuation of neuroinflammation and Alzheimer's disease pathology by liver X receptors. *Proc. Natl. Acad. Sci. USA* **104**: 10601–10606.

Zhang, Q., Powers, E.T., Nieva, J., Huff, M.E., Dendle, M.A., Bieschke, J., Glabe, C.G., Eschenmoser, A., Wentworth, Jr., P., Lerner, R.A., Kelly, J.W., 2004, Metabolite-initiated protein misfolding may trigger Alzheimer's disease. *Proc. Natl. Acad. Sci. USA* **101**: 4752–4757.

Zinser, E.G., Hartmann, T., Grimm, M.O.W., 2007, Amyloid beta-protein and lipid metabolism. *Biochim. Biophys. Acta* **1768**: 1991–2001.

Chapter 3
Cholesterol-Binding Viral Proteins in Virus Entry and Morphogenesis

Cornelia Schroeder

Abstract Up to now less than a handful of viral cholesterol-binding proteins have been characterized, in HIV, influenza virus and Semliki Forest virus. These are proteins with roles in virus entry or morphogenesis. In the case of the HIV fusion protein gp41 cholesterol binding is attributed to a cholesterol recognition consensus (CRAC) motif in a flexible domain of the ectodomain preceding the trans-membrane segment. This specific CRAC sequence mediates gp41 binding to a cholesterol affinity column. Mutations in this motif arrest virus fusion at the hemifusion stage and modify the ability of the isolated CRAC peptide to induce segregation of cholesterol in artificial membranes.

Influenza A virus M2 protein co-purifies with cholesterol. Its proton translocation activity, responsible for virus uncoating, is not cholesterol-dependent, and the transmembrane channel appears too short for integral raft insertion. Cholesterol binding may be mediated by CRAC motifs in the flexible post-TM domain, which harbours three determinants of binding to membrane rafts. Mutation of the CRAC motif of the WSN strain attenuates virulence for mice. Its affinity to the raft–non-raft interface is predicted to target M2 protein to the periphery of lipid raft microdomains, the sites of virus assembly. Its influence on the morphology of budding virus implicates M2 as factor in virus fission at the raft boundary. Moreover, M2 is an essential factor in sorting the segmented genome into virus particles, indicating that M2 also has a role in priming the outgrowth of virus buds.

SFV E1 protein is the first viral type-II fusion protein demonstrated to directly bind cholesterol when the fusion peptide loop locks into the target membrane. Cholesterol binding is modulated by another, proximal loop, which is also important during virus budding and as a host range determinant, as shown by mutational studies.

C. Schroeder (✉)
Max Planck Institute for Molecular Cell Biology and Genetics, Pfotenhauerstr. 108, D-01307, Dresden, Germany
e-mail: cornelia.schroeder@mpi-cbg.de

Keywords HIV gp41 influenza M2 protein · Alphavirus E1 protein · Peripheral raft protein · Cholesterol binding site · Virus budding · Virus fusion · Virus fission · Filamentous virus particles

Abbreviation

aa	amino acid
CBPPA	cholesterol-protein binding blot assay
CCM	cholesterol consensus motif
CHS	cholesterol hemisuccinate
CRAC	cholesterol recognition amino acid consensus
DHSM	dihydrosphingomyelin
DRM	detergent-resistant membrane
DSC	differential scanning calorimetry
DV	dengue virus
FP	fusion peptide
FPLC	fast performance liquid chromatography
gp41	glycoprotein 41 (refers to molecular weight 41 kD)
GPCR	G protein-coupled receptor
HA	hemagglutinin
HIV	human immunodeficiency virus
K-D	Kyte-Doolittle (scale of hydrophobicity)
ld	liquid-disordered
LLP	lentivirus lytic peptide
lo	liquid-ordered
LUV	large unilamellar vesicles
Mab	monoclonal antibody
MAS	magic angle spinning
MBP	maltose binding protein
MLV	murine leukaemia virus
MPR	membrane proximal region
mβCD	methyl-β-cyclodextrin
NA	neuraminidase
NMR	nuclear magnetic resonance
pHtrans	pH of conformational transition
PIP3	phosphatidylinositol-3,4,5-triphosphate
PIP4,5P2	phosphatidylinositol-4,5-bisphosphate
PM	plasma membrane
POPC	palmitoyl oleyl phosphatidylcholine
pre-TM	pre-transmembrane
pre-TMp	pre-transmembrane peptide
RNP	ribonucleoprotein
S protein	spike protein
SARS	severe acute respiratory syndrome
SFV	Semliki Forest virus

SIN	Sindbis virus
SIV	Simian immunodeficiency virus
SM	sphingomyelin
SOPC	stearoyl oleyl phosphatidylcholine
β2AR	human β2-adrenergic receptor
TBE	tick-borne encephalitis virus
TM	transmembrane
TSPO	outer mitochondrial membrane translocator protein
TX-100	Triton X-100
Udorn	influenza A/Udorn/307/72
VSV	Vesicular stomatitis virus
WSN	influenza A/WSN/33
W-W	White-Wimley (scale of hydrophobicity)
XIP	exchanger inhibitory peptide

3.1 Introduction

Viruses cross membrane barriers in the process of infection and again during assembly and release through membranous compartments. The role of membrane lipid composition and protein-lipid binding for specific biological functions of viruses and organelles is an area of intense investigation. Proteins with functions in membrane domain organization, trafficking, fusion and fission often possess specific lipid binding sites. X-ray crystallography revealed the first cholesterol-binding site in the 3D-structure of a signalling protein (Cherezov et al., 2007, Hanson et al., 2008). The discovery of another cholesterol binding motif 10 years earlier in a mitochondrial protein (Li and Papadopoulos, 1998) has been influential for a number of studies of viral proteins carrying this motif.

Cholesterol is a class apart from the other membrane lipids. Unlike these it cannot form membranes on its own, and being the most compact of the lipids, it penetrates less deeply into the hydrophobic layer of membrane leaflets, while its miniature hydroxyl head-group barely projects into the interfacial zone. Cholesterol is also the most diffusible lipid, and the one most akin to small-molecule drugs in structure and function. Modulation of membrane cholesterol levels – physiological or induced by drug therapies or viral infection itself – can have significant consequences for virus replication. Cholesterol anchors on antiviral drugs and other inhibitors make these raftophilic and target and concentrate them into membrane rafts and to membrane trafficking pathways where they interfere most effectively with pathogenic processes (reviewed by Rajendran et al., 2010).

Membrane rafts are implicated in the entry and egress of many virus species (Nayak and Hui, 2004; Ono and Freed, 2005). Rafts are nanoscale dynamic, lateral membrane domains with a specific lipid and protein composition, enriched in cholesterol and sphingolipids, that form a liquid ordered-like phase separated from bulk membrane (Simons and Ikonen, 1997; Hancock, 2006). Signalling cycles at the plasma membrane involve reversible coalescence and disassembly of rafts driven by activation states of the raft proteins (Rajendran and Simons, 2005), whereas during virus morphogenesis rafts merge irreversibly into microscale platforms of assembly and budding (for reviews see Schmitt and Lamb, 2005; Waheed and Freed, 2009). The affinity of virus envelope proteins to rafts is not attributed to specific lipid binding sites but to a combination of acylation and long transmembrane (TM) segments (Scheiffele et al., 1997; Melkonian et al., 1999; Rousso et al., 2000). In addition, certain viral envelope proteins exhibit distinct affinity to cholesterol. It has proved challenging to correlate this property to specific cholesterol binding sites on the one hand and to biological function on the other. These issues are at the center of this review. Since viral cholesterol-binding proteins usually are components of the virion, the review begins with virus lipidomics.

3.1.1 Virus Lipidomics

The composition of influenza virus (IFV) and human immunodeficiency virus (HIV) envelopes was determined by liquid chromatography (Aloia et al., 1993; Zhang et al., 2000), while recent data for HIV and other viruses has been generated by mass spectroscopy (Brügger et al., 2006, 2007; Chan et al., 2008; Kalvodova et al., 2009; Lorizate et al., 2009).

The lipid compositions of the envelopes of 'non-raft' enveloped viruses (Semliki Forest virus, SFV, family Togaviridae, genus Alphavirus) and vesicular stomatitis virus – VSV, family Rhabdoviridae, genus Vesiculovirus) are remarkably similar and closely resemble that of the host cell plasma membrane (PM) from which they bud (Kalvodova et al., 2009). No significant differences between SFV and VSV were seen at the lipid class level, but saturated and mono-unsaturated glycerophospholipids were enriched in SFV as compared to VSV and differences in fatty acid chain length were seen. Compared to PM the viruses showed some selectivity for sphingomyelin (SM), especially, long chain and dihydrosphingomyelin – DHSM, and depletion of GM3 (Kalvodova et al., 2009).

IFV buds from apical PM with an envelope of cholesterol and sphingolipid-rich raft membrane (Scheiffele et al. 1999, Zhang et al., 2000). Raft association is intrinsically encoded in HA (Scheiffele et al., 1997), the most abundant envelope glycoprotein. The ability of viral envelope proteins to select a cognate lipid environment is uncovered in IFV mutants defective in raft association and budding. Thus, the envelope of a double mutant lacking the cytoplasmic tails of hemagglutinin (HA) and neuraminidase (NA) (HAt-/NAt-) incorporated three times more triglycerides and proportionally less raft lipids, cholesterol and SM (Zhang et al., 2000). Membrane rafts and specifically cholesterol are also involved in IFV entry. Sun and Whittaker (2003) demonstrated that depleting IFV envelope cholesterol by methyl-β-cyclodextrin (mβCD) extraction specifically blocked the fusion of infectious virus with the PM of pH 5-treated host cells, a process mimicking virus infection by fusion with the endosomal membrane. Similarly, the requirement of cholesterol in the HIV envelope for infection (Campbell et al., 2002; Guyader et al., 2002) correlates with the integrity of viral envelope rafts (Campbell et al., 2004).

Brügger et al. (2006) presented the first comprehensive HIV-1 lipidomics, extended by Chan et al. (2008) who added analysis of phosphoinositides and compared retroviruses, HIV-1 and 2 and murine leukaemia virus (MLV), as well as the PMs of three host cell lines. The lipid compositions of HIV-1 and 2 vary with that of the plasma membrane raft domain of their host cells (Chan et al., 2008). For example, the content of DHSM in HIV-1 from a macrophage cell line (MDM) is twice that of the T cell line H9. Similar differences in lipid composition were observed between HIV budding from MT4 versus 293T cells and correlated to membrane order of the virus envelope reported by the dye laurdan (Lorizate et al., 2009).

The HIV envelope protein gp160 seems to exert little influence on the envelope's lipid composition (Chan et al. 2008). Nevertheless, there are ways in which HIV proteins modify the lipid composition of the viral envelope. HIV accessory protein Nef enforces the raft character of the plasma membrane by significantly reducing polyunsaturated PC species and enriching SM (Brügger et al., 2007). Although Nef also boosts cholesterol synthesis (Zheng et al., 2003), cholesterol levels of the PM and the virus envelope are not altered in HIV infection (Brügger et al., 2007). Nef is a raft protein (Wang et al., 2000) but cholesterol binding by Nef itself (Zheng et al., 2003) has been contested (Brügger et al., 2007).

Gag is the main determinant of HIV raft association (Bhattacharya et al., 2006) and the driving force of particle formation and budding (Morita and Sundquist, 2004).

Phosphatidylinositol phosphates are the one raft lipid class preferentially incorporated into retroviral envelope over PM. Phosphatidylinositol-4,5-bisphosphate (PIP4,5P2) enrichment disappears upon deletion of the polybasic stretch at the head of the Gag protein MA domain (Chan et al. 2008), confirming its essential binding to PIP4,5P2 and via PIP4,5P2 to membrane (Ono et al., 2004; Murray et al., 2005; Chukkapalli et al., 2008). Enzymatic degradation of PIP4,5P2 also interferes with HIV budding (Chan et al., 2008). Similarly, PIP4,5P2 and PIP3 levels are increased during Respiratory syncytial virus (RSV) infection, and inhibiting their synthesis impaired formation of virus progeny (Yeo et al., 2009).

3.1.2 Cholesterol Binding Sites

In 1998 Papadopoulos and colleagues described a cholesterol recognition site VLNYYVWR in the outer mitochondrial membrane translocator protein TSPO, formerly known as peripheral-type benzodiazepine binding protein. Based on homology searches of other cholesterol binding proteins they proposed a cholesterol recognition amino acid consensus L/V-$(X)_{1-5}$-Y-$(X)_{1-5}$-R/K (CRAC; Li and Papadopoulos, 1998). TSPO is involved in cholesterol transport to cytochrome P450 which catalyzes the first steroidogenic reaction (Papadopoulos et al., 2007) (see also Chapter 15). The cholesterol-binding groove in a hydrophobic α-helix near the cytosolic C-terminus was confirmed by mutational studies and modelled (Li and Papadopoulos, 1998, Li et al., 2001, Jamin et al., 2005). Contributions from other TSPO α-helices to the binding site were predicted (Jamin et al., 2005; Murail et al.,

2008) and presented in a 3D homology model based on apolipophorin III, with the five TM α-helices surrounding one cholesterol molecule docked to the CRAC domain (Rone et al., 2009).

The crystal structure of a human β2-adrenergic receptor (β2AR), a G protein-coupled receptor (GPCR), revealed a cholesterol binding site formed by amino acids of α-helices IV and II (Cherezov et al., 2007; Hanson et al., 2008). These define a cholesterol consensus motif, CCM, conserved among human class A GPCRs (Hanson et al., 2008). For the purpose of comparison to CRAC, CCM is written as R/K-(X)$_{7-10}$-W/Y-(X)$_4$-I/V/L on one α-helix, and F/Y on the other. The motifs CCM and CRAC are obviously related by inversion. In three dimensions both motifs may determine similar binding grooves, however, β2AR in contrast to TSPO appears to bind two stacked molecules of cholesterol. Since CRAC and CCM are both quite degenerate, they occur frequently; not every occurrence will be a cholesterol recognition site. A common feature of CRAC and CCM is the α-helical secondary structure. It is reasonable to muster additional criteria for a cholesterol recognition site of this type, i.e. inclusion in or proximity to an α-helical amphiphilic or trans-membrane (TM) domain. Table 3.1 cites examples of CRAC motifs in viral proteins reviewed here, in comparison to CRAC motifs of cellular proteins. Also shown are CRAC motifs currently not implicated in cholesterol binding. For example, the influenza A M1 protein exhibits three such motifs, one of which is shown. It is part of the helix six domain which has affinity to membrane and to RNP (Ruigrok et al., 2000). Other short sequence motifs proposed in cholesterol binding (cf. Politowska et al., 2001; Yao and Papadopoulos, 2002) have not been analysed in virus proteins.

3.1.3 Methods Demonstrating Protein–Cholesterol Binding

Since cholesterol adheres to hydrophobic surfaces, evidence of binding specificity collected with independent methods is desirable. Table 3.2 lists approaches for probing the physical association of proteins with cholesterol, as reported for selected viral and cellular proteins. A comprehensive discussion of such methods is the subject of Chapter 1 of this book (Gimpl, 2010). The upper half of Table 3.2 lists ways of analysing cholesterol bound to proteins or peptides, purified or in membrane fractions, the lower part addresses binding-site mapping. The β2-adrenergic receptor, where X-ray crystallography revealed the cholesterol binding site belongs to the GPCR superfamily; previous studies on various GPCR have indicated a function of cholesterol in receptor activity (reviewed by Hanson et al., 2008; Paila et al., 2009). The natural variation of the cholesterol-binding site CCM in GPCRs will enable structure-function analysis. In Drosophila metabotropic glutamate receptor, ligand affinity increases with raft association; labelling with ^{3}H-photocholesterol (Thiele et al., 2000) demonstrated the sterol affinity of this particular GPCR (Eroglu et al., 2003). Semliki Forest Virus (SFV) E1 fusion protein was also labelled with photocholesterol (Umashankar et al., 2008).

Table 3.1 Occurrence and proposed function of CRAC-like motifs in viral and cellular proteins

Protein	Protein domain function	Motif	Localization protein domain/membrane face	References/comments
TSPO mouse	Cholesterol transport	$_{147}$VL$\underline{N}$$\underline{Y}$$\underline{Y}VW\underline{R}$$_{154}$	C-terminal TM α-helix/mitochondrial outer membrane	Li and Papadopoulos (1998); Li et al. (2001); Jamin et al. (2005)
TSPO human		$_{147}$T$\underline{L}$$\underline{N}$$\underline{Y}CVW\underline{R}$$_{154}$		Binds to CHS agarose (cp. Table 3.2) Vincent et al. (2002)
gp41, HIV-1	Membrane fusion	$_{679}$$\underline{L}W\underline{Y}I\underline{K}$$_{683}$	Pre-TM ecto-domain/outer leaflet of PM and viral envelope	No binding to CHS agarose (cp. Table 3.2) Vincent et al. (2002)
	Incorporation of Env into virus particles	$_{763}$$\underline{L}C\underline{L}FS\underline{Y}$$\underline{H}$$\underline{R}$$\underline{L}$$\underline{R}$$_{772}$ $_{833}$$\underline{V}QAA\underline{Y}$$\underline{R}A\underline{I}$$\underline{R}$$_{841}$	Endodomain: pre-LLP-2 and inside LLP-1/inner leaflet of PM and viral envelope	
M2 Influenza A virus	Assembly budding/pinching off	$_{44}$$\underline{D}$$\underline{R}$$\underline{L}$$\underline{F}$$\underline{F}KC\underline{I}$$\underline{Y}$$\underline{R}$$\underline{R}$$\underline{L}$$\underline{K}$$\underline{Y}G\underline{L}$$\underline{K}$$_{60}$[1] $_{44}$$\underline{D}$$\underline{R}$$\underline{L}$$\underline{F}$$\underline{F}KC\underline{I}$$\underline{Y}$$\underline{R}$$\underline{R}$$\underline{F}$$\underline{K}$$\underline{Y}G\underline{L}$$\underline{K}$$_{60}$[2] $_{44}$$\underline{D}$$\underline{R}$$\underline{L}$$\underline{F}$$\underline{F}KC\underline{I}$$\underline{Y}$$\underline{R}$$\underline{F}$$\underline{F}$$\underline{K}HG\underline{L}$$\underline{K}$$_{60}$[3] $_{44}$$\underline{D}$$\underline{R}$$\underline{L}$$\underline{F}$$\underline{F}KC\underline{I}$$\underline{Y}$$\underline{R}$$\underline{F}$$\underline{F}E\underline{H}G\underline{L}$$\underline{K}$$_{60}$[4] $_{55}$$\underline{L}$$\underline{K}$$\underline{Y}G\underline{L}$$\underline{K}$$_{60}$[1]	PM and viral envelope; endodomain/cytoplasmic	Schroeder et al. (2005)
M1 Influenza A virus	Assembly budding	$_{97}$$\underline{V}$$\underline{K}$$\underline{L}$$\underline{Y}$$\underline{R}$$\underline{K}$$\underline{L}$$\underline{K}$$_{104}$	Matrix protein 'helix six domain' interacts with membrane and RNP	(One of three incidental CRAC motifs)
Human cardiac Na+/Ca^{2+} exchanger[5]	Specific binding to PIP2 regulates activity	$_{253}$$\underline{D}$$\underline{R}$$\underline{R}$$\underline{L}$$\underline{F}$$\underline{Y}$$\underline{K}$$\underline{Y}$$\underline{V}$$\underline{Y}$$\underline{K}$$\underline{R}$$\underline{Y}RAG\underline{K}$$_{269}$ $_{261}$$\underline{V}$$\underline{Y}$$\underline{K}$$\underline{R}$$\underline{Y}RAG\underline{K}$$_{269}$	Endogenous XIP region	(Incidental CRAC motifs)
NAP-22 chicken, rodent CNS	Cholesterol-dependent sequestering of PtdIns(4,5)P2	$_{4}$$\underline{L}S\underline{K}$$\underline{K}$$\underline{K}$$\underline{K}G\underline{Y}$$\underline{N}$VNDE$\underline{K}$$_{17}$	N-terminal/PM rafts, nucleus	Terashita et al. (2002); Epand et al. (2004, 2005a)

Sequences of influenza A virus strains: [1] Germany/27 (H7N7); [2] WSN/33 (H3N2); [3] Singapore/1/57 (H2N2); [4] Udorn/307/72 (H3N2); cp. legend to Fig. 3.2; [5] accession number AAD26362

Table 3.2 Methods and parameters in the analysis of cholesterol binding

Method/parameter		Protein					
		HIV gp41	Influenza M2[5]	SFV E1	TSPO	GPCR	Aβ[a]
Binding properties	Cholesterol incorporation	–	+	–	+[7]	–	+[14]
	– stoichiometry cholesterol per subunit	–	0.5–1	–	–	2 [11,12]	–
	– binding constant	–	–	–	6.1 nM [8]	–	–
	– Filipin staining	–	+	–	–	–	–
	Affinity chromatography	+[1]	+	–	–	–	–
	Chemical shift of cholesterol carbon atoms (^{13}C MAS NMR)	+[2]	–	–	–	–	–
	Complexation with cholesterol crystallites	(+)[3]	–	–	–	–	+[15]
Binding site	CBPPA[b] mapping	–	–	–	–	–	+[14]
	Binding site transplantation	+[1]	–	–	+[9]	–	–
	Photo-affinity labelling with ^{3}H steroid	–	–	+[6]	+[9]	+[13]	–
	X-ray crystallography	–	–	–	–	+	–
	Mutant studies	(+)[4]	–	–	+[7,9,10]	–	–

[a] Aβ is the cleavage product of amyloid precursor protein
[b] CPBBA – Cholesterol–protein binding blot assay
[1] Affinity chromatography on cholesterol-hemisuccinate (CHS) agarose (Vincent et al., 2002)
[2] Epand et al. (2003)
[3] Experiments on short peptides, not full-length gp41
[4] Mutant studies on full-length protein biological function, not cholesterol binding (cf. Table 3.3)
[5] Schroeder and Lin (2005)
[6] Umashankar et al. (2008)
[7] Jamin et al. (2005)
[8] Lacapère et al. (2001)
[9] Li et al. (2001)
[10] Li and Papadopoulos (1998)
[11] Cherezov et al. (2007)
[12] Hanson et al. (2008)
[13] Eroglu et al. (2003)
[14] Yao and Papadopoulos (2002)
[15] Harris and Milton (2009)

The TSPO CRAC motif is currently the best-studied cholesterol-binding site, which perhaps explains why such motifs are being investigated in other proteins. Single mutants, where the signature residues Y and R of the CRAC motif were replaced by S and L abolished cholesterol uptake by bacteria expressing TSPO (Li and Papadopoulos, 1998). Transplantation of the CRAC motif into another protein, as done for TPSO and for HIV gp41, substantiated its assignment as a cholesterol-binding site. Photoaffinity labelling with ^{3}H-promegestone of the TPSO CRAC motif transplanted onto Tat, a cell-permeating HIV protein, was competed 1000-times more efficiently by cold cholesterol than by promegestone. The triple mutant V149G, Y152S, R156L of this construct could not be photoaffinity labelled (Li et al., 2001), and the TSPO single mutant Y152S no longer bound ^{3}H-cholesterol (Jamin et al., 2005). A cholesterol-protein binding blot assay enabled the delineation of a cholesterol-binding site in Aβ (Yao and Papadopoulos, 2002). Independent studies by electron microscopy also point to specific cholesterol binding of β-amyloid (Harris and Milton, 2009; see also Chapter 2). The analysis of the chemical shift of cholesterol carbon atoms in complexes with short peptides containing cholesterol-binding motifs was pioneered by Epand et al. (2003) and applied to gp41 CRAC. These authors also studied a number of sequence variations in short CRAC-containing peptides, discussed in Section 3.2.3.

3.2 Human Immunodeficiency Virus Fusion Protein gp41

In 2009 the notion that HIV is the paradigm of a virus that enters cells by fusion with the plasma membrane (reviewed by Gallo et al., 2003) was overturned. HIV enters the cell by receptor-mediated endocytosis, albeit at neutral pH (Miyauchi et al., 2009). HIV is transmitted either by free virus particles or via fusion of infected with non-infected cells, forming multi-nucleate syncytia. Both the Env protein clusters on the donor side, and primary and secondary receptors on the acceptor side, reside in raft membrane domains (reviewed by Waheed and Freed, 2009).

Interactions of HIV gp41 with cholesterol have been investigated more extensively than those of any other viral protein. HIV gp41 is derived by proteolytic cleavage from its precursor, the envelope glycoprotein 160. In complex with the other cleavage product gp120, gp41 forms the trimeric spikes of the virus particle. The gp120 subunit presents the receptor binding sites for the primary receptor CD4 and for secondary receptors and, with the gp41 subunit, functions as a class I fusion protein. Fusion is prepared by a sequence of events triggered by adsorption to the primary receptor (Fig. 3.1). Receptor binding sets off extensive restructuring of gp41 to expose and propel the N-terminal fusion peptide into the target membrane (reviewed by Gallo et al., 2003).

The ectodomain comprises defined sub-domains, which refold into different secondary structures during fusion. From the point of view of cholesterol binding, the pre-transmembrane (pre-TM) or membrane proximal region (MPR) (Fig. 3.1) of 20 amino acids (664–683) immediately preceding the TM segment has attracted special interest: DKWASLWNWFNITNWLWYIK. It forms an α-helix in lipid micelles, wherein four of the five tryptophan and the tyrosine residues align as a 'collar of aromatic residues' (Schibli et al., 2001). Analogous to tryptophan-rich antimicrobial peptides, the aromatic collar was predicted to engage with the aqueous interface of the membrane bilayer. The pre-TM terminates on LWIYK (679–683), the CRAC

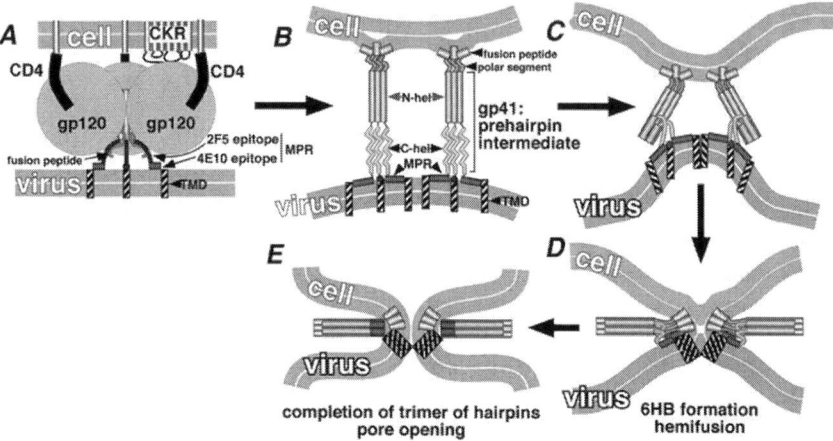

Fig. 3.1 Conformational transitions of HIV protein gp41 during the fusion cascade. **A** Release of the metastable state of the gp120-gp41 complex by binding to the primary and secondary receptors, CD4 and CKR. The membrane-proximal region MPR (pre-TM) is exposed adjacent to the virus envelope. The MPR-distal sequence occludes the fusion peptide. **B** gp120 trimers refold into extended α-helical structure and harpoon the fusion peptide into the target membrane. Coil-to-amphipathic helix transition of the MPR-distal sequence enables immersion of the pre-TM in the membrane interfacial zone. **C** Extended α-helices zip into a six-helix bundle (6HB) and clamp virus and cell membrane, causing (**D**) hemifusion and, by pulling the pre-TM into the trimer of hairpins, (**E**) fusion pore opening. Model of Bellamy-McIntyre et al. (2007), Figure 8 (modified), with permission from the American Society for Biochemistry and Molecular Biology

motif immediately proximal to the transmembrane domain. The role of the pre-TM has been studied at all levels of complexity, from mutational study of virus reproduction in the cell and effects on the various functions of gp41, to isolated proteins and peptides in artificial membrane systems.

3.2.1 Mutational Studies on the Pre-TM CRAC Motif in Virus-Cell Systems

Helseth et al. (1990) found that substituting the lysine of the LWYIK motif by isoleucine reduced syncytium formation by 95%. The mutation did not interfere with translation, processing and cell surface expression of Env, or with its binding to CD4, but this mutant Env expressed from a plasmid was completely unable to trans-complement the single-cycle replication of an Env-deleted virus. Salzwedel et al. (1999) explored the function of the pre-TM through substitution, deletion and insertion, and constructed a number of CRAC mutants – prior to the recognition of its cholesterol-binding significance. They found that the pre-TM is dispensable for maturation, trafficking, cell surface expression and CD4 binding, but is required for cell–cell fusion. The substitution WA within LWYIK was tolerated. In contrast, replacing K abrogated fusion. Deletion of LWYI inhibited incorporation of gp120 into virus particles, viral entry and syncytium formation (Salzwedel et al., 1999) while lipid mixing and small molecule transfer were only reduced by 50% (Muñoz-Barroso et al., 1999). Analogously, inserting nine amino acids between Y and K inhibited entry (Salzwedel et al., 1999) and fusion, but reduced lipid and small content mixing only by about 50%. Thus, a dysfunctional LWYIK motif appeared to allow fusion pore formation, but arrest fusion pore expansion.

Ten years later, a new mutational study has now focused on the CRAC motif (Chen et al., 2009). Three deletion mutants and three point mutations were studied: ΔLWYIK, ΔYI, ΔIK, KE, WA, YA. All mutant proteins underwent normal synthesis, oligomerization, cell surface expression, and incorporation with normally assembled Gag into mutant virus, i.e. budding was not impaired. However, multi-cycle replication of deletion mutants was significantly slowed, and virus infectivity and cell-cell fusion were strongly impaired (Table 3.3). The deletion mutants also interfered in *trans* with virus infectivity and to a lesser degree with fusion elicited by wild-type Env. The mutation KE suppressed infectivity in *trans* and eliminated fusion, and WA substitution was less disruptive than the other point mutations.

Overall, the recent study by Chen et al. (2009) shows that function of the LWYIK motif in virus infection is most sensitive to alteration of the CRAC consensus residues L, Y and K, confirming the earlier results of Salzwedel et al. (1999). ΔLWYIK gp41 remains susceptible to peptides blocking the formation of the six-helix bundle fusion intermediate (Fig. 3.1), which proves a degree of functional independence of these subdomains. Thus, the effects of the mutations may be attributed to local interactions of the LWYIK motif with lipid membranes. Of interest, none of the mutations of the CRAC motif influenced raft association of gp41. In a dye transfer assay between Env-expressing effector and CD4-expressing target cells pre-TM ΔLWYIK supported lipid mixing, tantamount to hemifusion, but inhibited small molecule content mixing (Chen et al. 2009). The Chen study

Table 3.3 Phenotypes of CRAC motif* mutations of HIV gp41

Mutant	Infectivity		Cell–cell fusion	Transdominant interference with		Lipid vs. content mixing
	Direct	Transcomplementation		Cell–cell fusion	Virus	
ΔLWYIK[1]	4	4	1	70	25	140 vs. 32
ΔYI[1]	8	8	2	66	30	–
ΔIK[1]	8	8	<1	60	35	–
LI[2]	–	–	44–70	–	–	–
WA[1]	37	43	90	100	90	–
YA[1]	46	48	35	–	–	–
KE[1]	40	17	< 10	–	–	–

% control; *consensus residues are underlined; – not done
after [1]Chen et al. (2009); [2]Epand et al. (2006)

confirms the conclusions of earlier studies (Muñoz-Barroso et al., 1999; Salzwedel et al., 1999), showing that LWYIK is required for the formation and dilation of fusion pores.

3.2.2 Studies on Full-Length gp160, gp41 and Polypeptide Constructs

One method for assaying the cholesterol affinity of a protein is by binding to cholesterol hemisuccinate (CHS) linked to agarose. Vincent et al. (2002) found that a soluble Env construct binds to CHS, but hardly at all to cholic acid agarose. As further controls, calmodulin agarose bound the construct via the Env calmodulin binding site and ConA-sepharose B via gp120-linked mannose. Maltose binding protein (MBP) fusions of gp41 sequences overlapping the pentapeptide LWYIK, even the minimal construct MBP-LWYIK, also attached to cholesterol hemisuccinate agarose, whereas others like the gp41 N-terminal fusion peptide, an immunodominant epitope or an endodomain fragment (aa752–856) did not. This sequence spans two CRAC motifs (Table 3.2), which are thus discounted. MBP fusions with complete or incomplete gp41, with or without TM, bound equally, as did a construct containing LWYIR. The study of Vincent et al. (2002) presented compelling evidence of cholesterol binding by the gp41 pre-TM CRAC motif, however, Chen et al. (2009) made the point that none of this work was replicated.

3.2.3 Peptide Studies and Modelling

The pre-TM has been analysed in detail by modelling and by experiments on isolated peptides and their membrane interaction. According to epitope mapping and hydropathy analyses pre-TM was subdivided into short defined sequence elements. N-terminally, pre-TM overlaps the epitope of the monoclonal antibody (Mab) 2F5. Residues 666–673 constitute interfacial subdomain I, 670–676 the epitope of Mab 4E10, and 677–683 interfacial subdomain II (reviewed by Lorizate et al., 2008). Different from the TM segment, which is hydrophobic according to the classical Kyte-Doolittle (K-D) scale, the pre-TM exhibits interfacial hydrophobicity as defined by White and Wimley (1999) (W-W). The K-D scale is based on phase partitioning of hydrophobic side chains, while the W-W scale reflects whole residue partitioning of oligopeptides into the bilayer interface of POPC. For the N-terminal fusion peptide W-W and K-D hydropathy overlap. In contrast, for pre-TM of HIV-1, 2 and SIV the distance between the W-W peak and the transmembrane K-D peak is 15–20 residues. Pre-TM interfacial hydrophobicity analysed in a narrow window of 5 aa exhibits two peaks, and mutations eliminating this bifurcation also interfere most with fusion (Sáez-Cirión et al., 2003), e.g. ΔLWYI, the CRAC deletion of Salzwedel et al. (1999).

Suárez et al. (2000a) tested the ability of the gp41 fusion (FP) and pre-TM peptides (pre-TMp) to permeabilize and fuse artificial membranes. Surprisingly, induction of membrane leakage by, and fusogenic activity of pre-TMp on large unilamellar vesicles (LUVs), is greater than the activity of the fusion peptide (FP). The validity of these peptide assays with respect to the earlier virus-cell studies (Salzwedel et al., 1999) was underscored by the inactivity of pre-TMp mutant W(1–3)A and its inability to cooperate with FP (Suárez et al., 2000a, b). Equimolar mixtures of wild-type pre-TMp and FP exhibited cooperativity and an increase in tryptophan fluorescence, indicative of physical interaction. Binary peptide mixtures exhibited higher reactivity to the Mab2F5 epitope of pre-TM (Fig. 3.1), and peptide hybrids exhibited even greater antibody affinity but less membrane destabilizing power. Their binary interaction was interpreted as a 'kinetic trap' stalling fusion, and it was inferred that in the metastable structure of gp120/gp41 the two membrane-active segments mutually mask their hydrophobic surfaces (Fig. 3.1A; Lorizate et al., 2006a, b).

In planar supported membrane bilayers with SM and cholesterol where liquid-ordered (lo) and –disordered (ld) lipid domains co-exist, pre-TMp clusters formed exclusively at the domain boundary (Sáez-Cirión et al., 2002). In a strictly cholesterol-dependent manner Mab4E10 bound and blocked liposome permeabilization by these pre-TMp clusters (Lorizate et al., 2006c). These findings are consistent with the pre-TM structure being embedded in the HIV envelope (Fig. 3.1B), as seen in the low resolution pre-fusion SIV spike 3D-structure (Zhu et al., 2006).

The Epand group investigated the potential of peptides derived from the pre-TM sequence to bind cholesterol and induce phase separation in membranes (see also Chapter 9). Differential scanning calorimetry (DSC) was used to monitor the enthalpy of acyl chain melting transition, which increases upon demixing of cholesterol (Epand et al., 2003). As a consequence of demixing, cholesterol crystallites form. Liposomes were prepared in the presence of peptide at high peptide-to-lipid ratio, 5 to 15 mol%. Introduction of LWYIK into multilamellar vesicles (MLV) of cholesterol admixed to SOPC or POPC increased the enthalpy of acyl chain melting transition and concomitantly induced cholesterol segregation into crystallites. Nuclear Overhauser effect spectroscopy indicated deeper penetration of the aromatic amino acids into the bilayer in the presence of cholesterol. This was corroborated by the increased quench of tryptophan fluorescence in the presence of cholesterol, of LWYIK (Epand et al., 2003) and LASWIK (Epand, 2004; Epand et al., 2005b), the analogous gp41 sequence of most HIV-2 strains. ^{13}C Magic angle spinning NMR suggested stronger interactions with the cholesterol A ring than with the interior of the leaflet. The complete pre-TMp was actually less prone to sequester cholesterol into domains than LWYIK itself, with LASWIK intermediate (Epand et al., 2005b). Cholesterol sequestration in SOPC/Cholesterol and enhancement of Trp fluorescence in the presence of cholesterol generally were strongest for wild-type CRAC and were diminished most by consensus-violating substitutions (Epand et al., 2006).

Altering the first CRAC residue L to V (within consensus) or A had a less profound effect than to I. IWYIK, unlike all other CRAC peptide variants, lowered the melting enthalpy of pure SOPC (Epand et al., 2006). The formation of cholesterol crystallites in SOPC/cholesterol was interpreted as displacement of cholesterol from positions adjacent to SOPC molecules rather than domain formation as with LWYIK (Greenwood et al., 2008). These three variants were also probed in the context of gp41 co-expressed with Tat from a plasmid. The mutation LI had similar effects as other single mutations of the CRAC motif (Table 3.3). A series of double mutants was made to alter the consensus or the intervening residues. Whereas GWGIK and LWGIG inhibited cell-cell fusion by > 60%, LGYGK inhibited only 25–30% (Vishwanathan et al., 2008a, b). Epand et al. (2006) also modelled a structure maximizing cholesterol-peptide interactions. All variants partitioned in the acyl chain-polar interface, but LWYIK was unique in that cholesterol OH was both H-bond acceptor to tyrosine OH, and H-bond donor to the lysine terminal CO. In the optimized model LWYIK enwraps the cholesterol A-ring and does not contact the hydrophobic bulk of the molecule.

3.3 Infuenza Virus M2 Protein

3.3.1 Influenza Virus Entry and Egress

Influenza virus has a segmented RNA genome packed as ribonucleoprotein (RNP) into a protein matrix, surrounded by an envelope carrying three transmembrane proteins, HA, NA and M2. The eight RNA segments are transcribed and replicated in the nucleus (reviewed by Whittaker et al., 2000). IFV invades the cell by adsorptive endocytosis (reviewed by Smith and Helenius, 2004), which delivers the virus to a perinuclear site (Lakadamyali et al., 2003). Here, the endosomal pH decreases to a threshold (pHtrans) triggering conformational transition of viral hemagglutinin and activation of the proton channel M2. The fusion peptide of HA is unburied and propelled into the endosomal membrane, launching fusion of the viral with the endosomal membrane (reviewed by Cross et al., 2001). Concomitant proton influx through the M2 ion channel dissociates the dense matrix ('uncoating') making RNP susceptible to primary transcription (cf. Whittaker et al., 2000). It is critical that all eight genome segments arrive in the nucleus. Likewise, at completion of the infectious cycle during virus assembly, the eight RNA segments must be sorted into the virus particle. The M2 protein plays a role both in packing and unpacking the genome.

3.3.2 M2 Protein Structural and Functional Domains

Influenza A M2 protein is a unique, multifunctional protein critical for initiation and completion of the infectious cycle. M2 is a tetrameric class III, single-pass TM protein (Holsinger and Lamb, 1991; Sugrue and Hay, 1991). Figure 3.2 depicts the structural and functional domains of influenza A M2. The 25 amino-acid

3   Cholesterol-Binding Viral Proteins

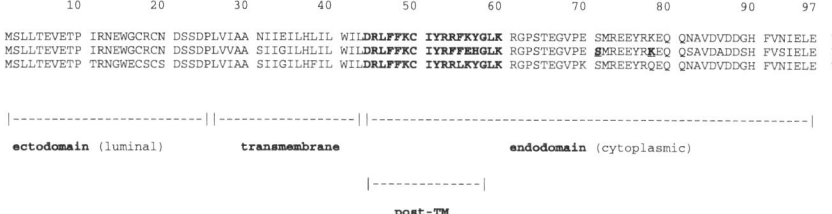

Fig. 3.2 Structural and functional domains of influenza A M2 protein. Influenza A strains: WS – WSN/33 (H1N1) – NCBI accession L25818.1; Ud – Udorn/307/72 – NCBI accession J02167.1 (H3N2); We – Weybridge/27 (H7N7) – EMBL accession AX006731.1. Bold print: post-TM. Bold printed, underlined residues interact with M1 protein

ectodomain is required for the incorporation of M2 into virus particles (Park et al., 1998) and forms the channel mouth. The α-helical 19 amino-acid TM segment (Lamb et al., 1985; Schnell and Chou, 2008; Stouffer et al., 2008) defines a minimal proton channel (Duff and Ashley, 1992). However, the following sequence up to at least residue 62 is also necessary for ion channel activity in vivo (Tobler et al., 1999). This amphipathic post-TM domain (D44–K60) comprises a sharp turn and a second alpha-helix according to solid-state NMR of the monomer (Tian et al., 2003). The 3D structure of the tetramer $M2_{18-60}$ elucidated by NMR in solution (Schnell and Chou, 2008) shows the TM and the post-TM α-helices connected by a loop of residues 47–50 (Fig. 3.3). The post-TM includes the palmitoylation site and CRAC motifs overlapping a sequence with predicted affinity to PIP4,5P2 (Table 3.1). Most of the M2 endodomain is involved in virus assembly, required

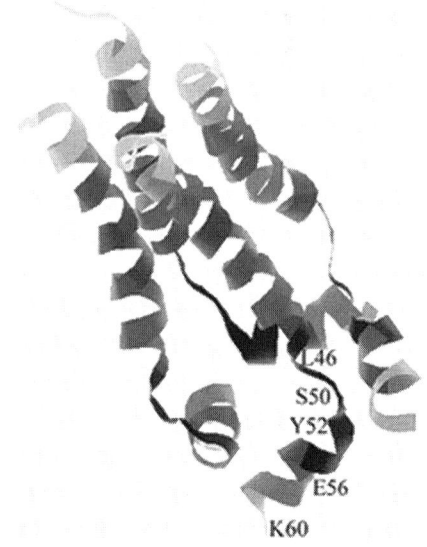

Fig. 3.3 3D-model of the tetrameric M2 transmembrane and post-TM structure (Schnell and Chou, 2008). The four helices forming the ion channel are in the upper left, the four post-TM helices in the lower right. Key residues are indicated: S50 marks the position of the palmitoylated C50 of the wild-type sequence. L46 is the first and Y52 the central residue of a common M2 CRAC motif. The position normally occupied by a basic residue is mutated in the Udorn strain (E56) (cp. Table 3.1, Fig. 3.2) (Redrawn after MMDB ID: 62125; PDB ID: 2RLF.)

both for the incorporation of M1 and genome packing (McCown and Pekosz, 2005, 2006; Iwatsuki-Horimoto et al., 2006; Chen et al., 2008).

3.3.3 Influenza Virus Membrane Rafts

HA and NA are integral raft proteins (Skibbens et al., 1989, Kurzchalia et al., 1992; Scheiffele et al., 1997; Zhang et al. 2000; Nayak and Barman, 2002) and the budding of influenza virus particles is raft-dependent (Scheiffele et al., 1999, Zhang et al., 2000). Raft trans-membrane proteins typically possess a long TM segment, since raft membranes are thicker than bulk membrane (Coxey et al., 1993; Ren et al., 1997). M2 differs from the large spike proteins HA and NA by its short 19 residue TM segment vs. 25–30 for the latter. Compared to HA and NA only a small amount of M2 is extracted into detergent-resistant membrane (DRM) by cold TX-100 (Zhang et al., 2000, Schroeder et al., 2005). Clusters of HA can be visualized in contours of the cell surface by immuno-gold labelling, whereas non-raft mutant HA is distributed more uniformly; these clusters were identified as raft micro-domains and platforms of virus budding (Takeda et al., 2003), which also contain NA and the raft marker GM1 (Leser and Lamb, 2005). Low density of M2 staining did not allow for assessment of its surface distribution in PM contours. In planar plasma membrane sheets HA appears in large 2-dimensional clusters (Hess et al., 2005) and these enable analysis of HA, M1 and M2 co-clustering (Chen et al., 2008). M2 protein co-clustered with M1, however, an alanine-scanning mutant $M2_{71}$ SMR → AAA (cp. Fig. 3.2) did not. Statistical analysis of co-clustering with HA revealed that this mutant still displayed significant long-range (> 200 nm) association with HA, albeit less than wild-type M2. Moreover, M1 in the background of this M2 mutation also remains associated with HA. This analysis suggested that different sequence elements of M2 are responsible for the association with M1 on the one side and HA on the other, and that association of M2 with HA is not mediated by M1 (Chen et al., 2008). Indeed, M1-binding sites in the M2 sequence are spatially separated from the post-TM (Fig. 3.2). We hypothesized that M2 attaches peripherally to the HA-studded membrane raft (Schroeder et al., 2005). Its short TM domain should lock M2 into non-raft membrane while post-TM lipid-binding determinants form a bridge into raft domains (see Section 3.6).

3.3.4 Cholesterol in the Apical Transport and Maturation of M2 Protein

In contrast to HA and NA, M2 is recycled between the PM and the TGN (Henkel and Weisz, 1998) where it equilibrates pH and protects acid-labile HA species from premature low-pH conformational transition (Sugrue et al., 1990; Grambas and Hay, 1992; Ciampor et al., 1992a, b; Ohuchi et al., 1994; Takeuchi and Lamb, 1994). HA and M2 are apparently co-transported to the PM, sharing the same transport pathways and vesicles. By peripherally inserting at the raft–non-raft interface, M2 may already associate to HA-bearing rafts during vesicular transport.

Cholesterol depletion slows down apical transport via sphingolipid-cholesterol rafts and causes mis-sorting of HA to the basolateral membrane (Keller and Simons, 1998). Cholesterol is also required for the maturation and stability of M2. This was first observed in a heterologous expression system. Extreme cholesterol depletion of insect cells altered the ultrastructure of the Golgi and interfered with cytotoxicity of expressed M2 (Cleverley et al., 1997). We studied cholesterol requirements of the ion channel function on M2 expressed in insect cells in the presence of cholesterol. The proton channel activity of liposome-reconstituted M2 was found to be independent of cholesterol (Lin and Schroeder, 2001). We also expressed M2 protein in E. coli which is intrinsically cholesterol-free. Irrespective of cholesterol content, different M2 preparations exhibited nearly the same activity and susceptibility to the antiviral drug rimantadine (Schroeder et al., 2005). While cholesterol is not directly required for ion channel activity it promotes tetramerization, a prerequisite of ion channel activity (Sakaguchi et al., 1997). Synthetic lipid bilayers of Golgi thickness (C16-C18 phospholipids) support tetramerization of the TM peptide $M2_{19-46}$ better than shorter phospholipids. Inclusion of cholesterol into the bilayers enhanced membrane thickness as well as M2 tetramerization. Judged by Scatchard analysis cholesterol did not directly bind to the transmembrane peptide (Cristian et al., 2003). Full-length M2 expressed in the absence of cholesterol in E. coli exhibited a higher dimer content and lower stability than M2 expressed in insect cells in the presence of serum (Schroeder et al., 2005). The lack of M2 activity in cholesterol-free insect cells (Cleverley et al., 1997) may therefore be attributed to a thinning of the Golgi membranes resulting in the failure of M2 to tetramerize. This is all the more likely as insect cells are cholesterol auxotroph and grow at 27°C, and their membranes are composed of shorter phospholipids than membranes of cells grown at 37°C (Rietveld et al., 1999). Introduction of cholesterol also causes the incorporation of longer chain phospholipids into insect cell membranes (Gimpl et al., 1995; Marheineke et al., 1998).

3.3.5 M2 Protein-Cholesterol Binding Experiments

M2 protein co-purifies with cholesterol which survives extensive detergent washes (Schroeder et al., 2005). Following expression and immunoaffinity purification from virus-infected chick embryo cells labelled with tritiated cholesterol the co-purified, extractable neutral lipid was subjected to thin-layer chromatography. About 69% of the extracted material coincided with the cholesterol spot. The cholesterol content of purified Weybridge M2 was 0.9 mol per M2 subunit. Sequence-identical M2 protein expressed and purified from insect cells by immunoaffinity FPLC contained 0.5 mol cholesterol per subunit. By prolonged treatment of solid-phase bound M2 with 40 mM 1-octyl-β-D-glucoside, most but not all cholesterol could be removed

(C.S., unpublished). This M2 preparation was captured by cholesteryl hemisuccinate agarose but not by unmodified agarose (Schroeder et al., 2005). Cholesterol co-purification with M2 isolated from homologous and heterologous expression systems and adsorption to cholesterol hemisuccinate indicated cholesterol binding by M2 (Table 3.2).

3.3.6 Membrane Raft Binding Determinants and CRAC Motifs in the Post-TM

The post-TM sequence of the M2 protein (D44–K60; see Figs. 3.2 and 3.3) exhibits interfacial hydrophobicity (White and Wimley, 1999) up to residue 57 (cf. Schroeder et al., 2005) and covers three overlapping determinants of lipid raft binding, palmitoylation at C50, one or two CRAC motifs, and an XIP-like motif (Table 3.1). The M2 protein of influenza A/Udorn 307/1972 (H3N2) (Udorn) lacks R54, K56 and L55 and therefore does not possess a bona-fide CRAC motif, but basic residues are present further downstream of Y52 (Table 3.1). The M2 Weybridge post-TM sequence is most closely homologous to the XIP region of Na/Ca exchangers (Table 3.1) that has specific affinity to PIP4,5P2 (He et al., 2000), a lipid species enriched in the cytoplasmic leaflet of raft membranes (Liu et al., 1998). NAP-22 is another example of a protein predicted to interact with cholesterol via an N-terminal motif similar to CRAC with longer spacers between the L, Y, and K residues (Terashita et al., 2002; Epand et al., 2004, 2005a; Table 3.1). Similar to M2 post-TM, this motif also exhibits a close overlap of elements determining raft binding, myristoylation, affinity to PIP4,5P2, and predicted cholesterol affinity. Epand et al. (2005a) determined that replacing the Y residue of this motif in a 19 residue NAP-22 peptide abolishes its ability to induce a cholesterol-depleted domain in LUVs.

Influenza B and C viruses also encode M2-like ion channel proteins, albeit less extensively studied, with analogous roles to influenza A M2, despite lack of sequence homology, involved in virus entry and capable of equilibrating pH gradients; BM2 also has a function in viral assembly and egress similar to AM2 (Hongo et al., 2004; Imai et al., 2004, 2008; Betakova and Hay, 2007; Pinto and Lamb, 2006). Both influenza B and C M2 sequences include a CRAC motif of unknown significance (not shown).

Stewart et al. (submitted) introduced mutations at CRAC motifs of M2 of the Udorn and WSN strains. The substitution R54F that restores a standard CRAC motif into Udorn M2 (see Table 3.1) neither influenced virus replication kinetics in vitro nor affected the formation of filamentous virus or the incorporation of matrix and envelope proteins into progeny virus. Likewise, alanine substitution of the key residues L46, Y52 and R54 eliminating the CRAC motif (WSN M2delCRAC) did not affect WSN replication in vitro, but caused attenuation of virulence. At a dose of 105 $TCID_{50}$ infection by WSN was lethal, whereas 80% of mice infected with WSN delCRAC survived; mutant R54F had an intermediate phenotype.

M2 palmitoylation at C50 is not conserved in all influenza virus strains, e.g. C50 is present in most H3N2 but only in a third of the H1N1 strains (Grantham et al., 2009). This is in contrast to HA where palmitoylation is conserved throughout the subtypes and improves virus budding (Jin et al., 1996), raft targeting (Melkonian et al., 1999) and assembly with the M1 protein in a strain-dependent manner (Chen et al., 2005). Judged by the behaviour of M2 C50S mutants, C50 palmitoylation was not required for the replication or filamentous budding of the Udorn and WSN strains in vitro, however, the WSN C50S virus mutant was attenuated in mice (Grantham et al., 2009). The robustness of IFV to mutations abolishing palmitoylation and the CRAC motif of the M2 post-TM implies functional redundancy and the need to disrupt more than one membrane-targeting determinant in the post-TM to see effects in vitro.

The properties of the short TM and amphiphilic post-TM sequence may encode dual affinity to raft and non-raft membrane, targeting M2 to the membrane domain border as a 'peripheral raft protein'. We proposed a simple model (Fig. 3.4) of the post-TM anchored in membrane rafts via palmitate. Palmitoylation, CRAC motifs and the PIP4,5P2-binding motif may support affinity to the periphery of rafts, while the short TM segment remains surrounded by non-raft membrane.

M2 post-TM features a conserved endocytic internalization motif at residues 52–55, YxxΦ, that often marks tight turns in the three-dimensional structures of internalized proteins (Collawn et al., 1990); also a kink at K60 was predicted by Saldanha et al. (2002). These elements may confer the structural flexibility required for a role of the post-TM region in membrane fission (see below). Figure 3.3 shows a cartoon of the 3D-model of $M2_{18-60}$ C50S (Schnell and Chou, 2008) indicating the position of the CRAC motif. Inclusion of C50 palmitoylation providing a second anchor perpendicular to the membrane plane should cause significant alterations to this model. Schnell and Chou (2008) address this issue: 'Modelling shows that extending the transmembrane helix to Phe 48 would place residue 50 facing the membrane, allowing for insertion of the palmitoyl acyl chain into the lipid bilayer.

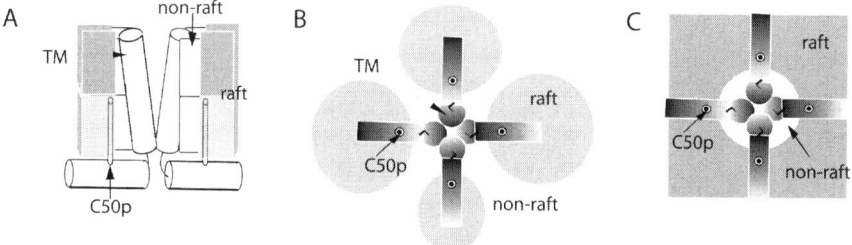

Fig. 3.4 Peripheral raft association of the M2 tetramer. (**a**) Cross-section of the membrane showing the TM and post-TM of two of the four subunits of the tetramer. TM is surrounded by non-raft membrane while post-TM connects to raft membrane via the palmitate bound to C50 (C50p) and other raft-targeting sequence elements. (**b**) Tetramer viewed from the endodomain. Subunits bridge separate rafts. (**c**) Merger of rafts, trapping the tetramer in small patch of non-raft membrane within raft domain. From Schroeder et al. (2005), with permission from Springer Publishers

This minor rearrangement would also move the amphipathic helices closer to the transmembrane domain.' Only structural studies on palmitoylated M2 will reveal whether and in which membrane environments such elongated TM segments may exist.

3.3.7 Morphogenesis and Budding

Influenza virus buds at the plasma membrane as spherical (80–120 nm in diameter) or filamentous particles up to > 10 μm long (for a recent review see Schmitt and Lamb, 2005). The latter may have a role in virus transmission in infected lung tissue by bridging infected and non-infected cells (Roberts and Compans, 1998), while spherical particles are expected to be more stable and suitable for aerosol transmission between hosts (Bourmakina and García-Sastre, 2003). Transmission electron microscopy of serially sectioned budding virus particles revealed seven RNP segments of different lengths surrounding the central eighth segment, projecting downwards from the top of the bud (Noda et al., 2006). Once the virus particle is pinched off the RNP segments become indistinguishable. An unknown packaging mechanism sorts the eight different RNP segments into each assembling virus (Fujii et al., 2003; Noda et al., 2006). The M2 protein is involved in this process (McCown and Pekosz, 2005, 2006).

Filamentous particles usually predominate irrespective of virus subtype, passage history or host species (cf. Elleman and Barclay, 2004). The filamentous phenotype is associated with gene segment 7 (Smirnov et al., 1991) encoding M2 and M1 protein and also requires a functional cortical actin microfilament array (Roberts and Compans, 1998; Simpson-Holley et al., 2002). Like spherical morphogenesis, the formation of virus filaments is raft-associated (Simpson-Holley et al., 2002). The outgrowth of virus filaments appears to bypass a decision point to complete spherical morphogenesis, as suggested by the following observations:

M2 is implicated in influenza virus morphogenesis, since a specific monoclonal antibody to the M2 ectodomain (Mab 14C2) suppressed the production of filamentous virus; strains unable to generate filamentous particles were not susceptible (Zebedee and Lamb, 1989; Elleman and Barclay, 2004). The antibody clusters the M2 protein on the PM and reduces its surface expression. Moreover, this antibody labelled M2 in spherical but not in filamentous virus particles (Hughey et al., 1995). Resistance to 14C2 mapped to the M2 endodomain, or to the M1 (matrix) protein of A/Udorn (Fig. 3.2; Zebedee and Lamb, 1989), resulting in distinct morphological phenotypes. M1 A41V generates exclusively spherical particles (Roberts et al., 1998; Elleman and Barclay, 2004). This substitution also occurs in high producer H1N1 laboratory strains PR8/34 and WSN/33 that have lost the morphotype switch, which confers no selective advantage for propagation in vitro. Mutations in Udorn M2 (S71Y or K78Q, Fig. 3.2) render filamentous particles less susceptible to antibody restriction but these have a much lower yield of infectious virus and may be defective in pinching-off (Hughey et al., 1995; Roberts et al., 1998). This data implicated interactions between the M2 endodomain region 71–78 with M1

31-41 (Zebedee and Lamb, 1989). Physical interaction of M1 with the M2 cytoplasmic tail has since been proven by immunoprecipitation and pull-down experiments (McCown and Pekosz, 2006; Chen et al., 2008). In the ensuing mutational studies of the M2 and M1 genes of filamentous Udorn and spherical WSN it turned out that mutations at these and other sites where the M1 s differ, or truncation of the M2 cytoplasmic tail, switch the morphotype (Bourmakina and García-Sastre, 2003; Elleman and Barclay, 2004; Burleigh et al., 2005; McCown and Pekosz, 2006; Iwatsuki-Horimoto et al., 2006; Chen et al., 2008).

Truncation of the M2 cytoplasmic tail (McCown and Pekosz, 2006) but also mutation at the extreme N-terminus of NA (Barman et al., 2004) led to 'daisy chains' of spherical, budding virus particles, defective in pinching-off (fission). A recent observation directly implicates membrane rafts in this process: Viperin is an interferon-induced protein that disperses lipid rafts (Wang et al., 2007) as evidenced by reduced copatching of HA with GM1, enhanced membrane fluidity, TX-100 extractability and lateral mobility of HA. Viperin binds and inhibits farnesyl diphosphate synthase upstream the cholesterol biosynthetic pathway. Remarkably, viperin expression also elicits 'daisy chain'-like IFV budding.

3.3.8 Incorporation of M2 into Virus Particles and the Process of Membrane Fission

Although M2 is expressed as abundantly as HA, the ratio of HA (trimer) to M2 (tetramer) in the virus envelope was estimated to be 500:15 (Zebedee and Lamb, 1988) and this sub-stoichiometric incorporation attributed to the exclusion of M2 from rafts (Zhang et al., 2000). The essential functions of M2 in virus uncoating (Kato and Eggers, 1969; Takeda et al., 2002) and genome packaging imply mechanisms specifically incorporating a few molecules of M2 into the envelope. As detailed in the previous sections: (1) The M2 endodomain at and beyond residue 71 physically interacts with matrix protein M1. (2) Although not an integral raft protein, M2 is associated specifically with raft-embedded HA, independent of M1 binding, apparently mediated by determinants for peripheral raft association in the pre-TM (Fig. 3.4). (3) M2 transits with HA through the trans-Golgi to the PM within the same transport vesicles. (4) M2 truncated at residue 70 still packages into virus particles (McCown and Pekosz, 2006), thus, the sequences responsible for M2 packaging include the post-TM. (5) The M2 post-TM is implicated as a factor in membrane fission (pinching-off) as are membrane rafts. We suggested that pinching-off occurs at the fault-line between raft and non-raft membrane at the budding pore (Schroeder et al., 2005; Schroeder and Lin, 2005), where M2 protein concentrates due to its affinity to the raft periphery. This gained further support from an experiment where transient cholesterol depletion, restricted to the process of pinching-off, actually increased the yield of spherical (WSN strain) particles (Barman and Nayak, 2007).

The proposed role of M2 in pinching-off is in agreement with electron micrographs showing immunogold-labelled M2 clustered at the neck of virus buds (Hughey et al., 1995; Lamb and Krug, 1996). The antibody 14C2 that stalls filamentous growth is not effective as Fab chains (Hughey et al., 1995), it must cross-link M2 to cluster it. We suggested that M2 cross-linking would accelerate pinching-off like a draw-string (Schroeder et al., 2005). The role of M2 in genome packaging (McCown and Pekosz, 2005, 2006) and the fact that the genome segments are coordinated to the top of the bud (Noda et al., 2006) suggests that M2 may also prime budding as an initial focus for RNP and M1 assembly. This would produce virus particles with distinct poles of M2.

In summary, the data indicates that the cholesterol affinity of the M2 protein is one of the functionally redundant elements targeting it to the periphery of membrane rafts. Association to the raft periphery appears to underlie its role during the budding and fission of virus particles.

3.4 Fusion Proteins of Alphavirus Species

The class II fusion proteins, E1 and E, of alpha- and flaviviruses power low pH-dependent fusion of the viral and the endosomal membrane during virus infection (reviewed by Heinz and Allison, 2000; Kielian, 1995, 2006). Alphavirus is a genus of togaviridae, flavivirus a genus of flaviviridae. In cholesterol-depleted insect cells alphavirus Semliki Forest virus (SFV) growth is 1000-fold restricted (Phalen and Kielian, 1991), whereas a less cholesterol-dependent point mutant, P226S, is only restricted 40-fold (Vashishtha et al., 1998). This mutation has arisen independently and repeatedly during appropriate selection conditions (Chatterjee et al., 2002). Likewise, the Sindbis alphavirus (SIN) is cholesterol-dependent for entry and egress (Lu et al., 1999). Fusion and infection of insect cells by SFV and Sindbis virus are stimulated by cholesterol. Despite similarity in fusion mechanism flaviviruses like yellow fever and several dengue virus (DV) strains do not require cholesterol.

Umashankar et al. (2008) demonstrated that an SFV E1 ectodomain protein (E1*) incubated with liposomes was labelled by photocholesterol (Table 3.2) at pH 5, the pH of fusion, whereas DV2 E* protein was not. A cholesterol-dependent cytolysin served as the positive control, which was also labelled by photocholesterol. In contrast, full length E1 membrane-inserted by its TM domain was not photoaffinity labelled, even at low pH. Labelling by photocholesterol indicates a specific interaction of the fusion peptide with cholesterol in the target membrane.

SFV E1* inserted into liposomes encompassing a liquid ordered SM and cholesterol-enriched phase could be extracted with mβCD, along with the cholesterol, while DV E* was not extracted (Umashankar et al., 2008), similar to E* of tick-borne encephalitis virus (TBE), another flavivirus (Stiasny et al., 2003). Both SFV and SIN, although requiring SM and cholesterol for infection, do not depend on membrane rafts (Waarts et al., 2002), and it is just the ectodomain during fusion but not the full length E1 protein that associates with rafts (Ahn et al., 2002).

The P226S mutation of SFV and SIN (Lu et al., 1999) is located to the ij loop which is apposed to the fusion peptide loop in the E protein 3D structure (Roussel et al., 2006) and apparently modulates cholesterol dependence of fusion. Recently the mutation A226V arose during a Chikungunya (alphavirus) epidemic. This mutation was associated with transmission by a new vector, *Aedes albopictus* and, at the same time, increased cholesterol dependence of the virus (Tsetsarkin et al., 2007).

SIN and SFV virus budding also exhibits a cholesterol requirement, attenuated by the same P226S mutation in E1 protein. In the absence of cholesterol, E1 protein is preferentially degraded rather than incorporated into progeny virus.

3.5 Other Cholesterol-Binding Virus Proteins

The first instance of a virus protein reported to bind cholesterol was Sendai virus class I fusion (F) protein (Asano and Asano, 1988). ^{3}H-cholesterol was added to a purified F protein preparation and a complex isolated by immuno-precipitation. The 3-OH group of cholesterol was not required for binding. Cholesterol labelled about 10% of the monomer. Cholesterol binding was blocked by a fusion inhibitory peptide. This work was apparently not followed up. Sendai virus F protein is extracted into DRM and the virus is proposed to bud from membrane rafts (Sanderson et al., 1995; Ali and Nayak, 2000).

The coronavirus spike (S) protein is a class I fusion protein, structurally and functionally similar to HIV Env. Analogous to HIV gp41, peptides representing the S protein pre-TM amphiphilic sequence are able to permeabilize and fuse membranes (Sainz et al., 2005). The pre-TM sequences of SARS and other coronaviruses, e.g. mouse hepatitis virus, harbor a CRAC motif (not shown). Cholesterol-binding studies have not been reported.

3.6 Conclusions

HIV gp41, influenza A M2 and SFV E1 protein are vastly different in structure and function. They have in common flexible domains which undergo conformational transition during the membrane restructuring processes and contain cholesterol binding sites. The type of studies performed and the amount of data available documenting cholesterol binding and its biological function vary greatly. For HIV gp41 cholesterol dependence of fusion is proven, a specific CRAC motif is shown to mediate cholesterol binding of gp41-derived peptides, and mutations to this motif arrest HIV infection at the hemifusion stage. The missing link in the chain of evidence is the demonstration at the virus-cell level that these gp41 mutations abrogate fusion due to impaired cholesterol binding. The influenza A virus M2 protein binds cholesterol which is, however, not required for its proton channel function. Mutation of a potential cholesterol binding CRAC motif attenuates virus virulence. This motif may be one of the lipid-binding determinants linking M2 peripherally to raft microdomains and predicted to play a role in virus budding and fission. Since this hypothesis was first published data has accrued consistent with peripheral raft targeting and supporting the role of the M2 cytoplasmic tail in morphogenesis. The specific role of cholesterol-binding in these processes requires further study. SFV is the first virus for which direct cholesterol binding by the fusion peptide was proven. A second-site locus modulating cholesterol affinity is related to host range and virulence in a number of alphavirus species.

Acknowledgements I am grateful to Prof. Andrew Pekosz (W. Harry Feinstone Department of Molecular Microbiology and Immunology, Johns Hopkins University, Baltimore, USA) for critically commenting the M2 (Section 3.3) and for sending his submitted manuscript and a paper in press.

References

Ahn, A., Gibbons, D.L., Kielian, M., 2002, The fusion peptide of Semliki Forest virus associates with sterol-rich membrane domain s. J. Virol. 76: 3267–3275.

Ali, A., Nayak, D.P., 2000, Assembly of Sendai virus: M protein interacts with F and HN proteins and with the cytoplasmic tail and transmembrane domain of F protein. Virology 276: 289–303.

Aloia, R.C., Tian, H., Jensen, F.C., 1993, Lipid composition and fluidity of the human immunodeficiency virus envelope and host cell plasma membranes. Proc. Natl. Acad. Sci. USA 90: 5181–5185.

Asano, K., Asano, A., 1988, Binding of cholesterol and inhibitory peptide derivatives with the fusogenic hydrophobic sequence of F-glycoprotein of HVJ (Sendai virus): possible implication in the fusion reaction. Biochemistry. 27: 1321–1329.

Barman, S., Adhikary, L., Chakrabarti, A.K., Bernas, C., Kawaoka, Y., Nayak, D.P., 2004, Role of transmembrane domain and cytoplasmic tail amino acid sequences of influenza a virus neuraminidase in raft association and virus budding. J. Virol. 78: 5258–5269.

Barman, S., Nayak, D.P.,2007, Lipid raft disruption by cholesterol depletion enhances influenza A virus budding from MDCK cells. J. Virol. 81: 12169–12178.

Bellamy-McIntyre, A.K., Lay, C.S., Bär, S., Maerz, A.L., Talbo, G.H., Drummer, H.E., Poumbourios, P., 2007, Functional links between the fusion peptide-proximal polar segment and membrane-proximal region of human immunodeficiency virus gp41 in distinct phases of membrane fusion. J. Biol. Chem. 282: 23104–23116.

Betakova, T., Hay, A.J., 2007, Evidence that the CM2 protein of influenza C virus can modify the pH of the exocytic pathway of transfected cells. J. Gen. Virol. 88: 2291–2296.

Bhattacharya, J., Repik, A., Clapham, P.R., 2006, Gag regulates association of human immunodeficiency virus type 1 envelope with detergent-resistant membranes. J Virol. 80: 5292–5300.

Bourmakina, S.V., García-Sastre, A., 2003, Reverse genetics studies on the filamentous morphology of influenza A virus. J. Gen. Virol. 84: 517–527.

Brügger, B., Glass, B., Haberkant, P., Leibrecht, I., Wieland, F.T., Kräusslich, H.G., 2006, The HIV lipidome: a raft with an unusual composition. Proc. Natl. Acad. Sci. USA 103: 2641–2646.

Brügger, B., Krautkrämer, E., Tibroni, N., Munte, C.E., Rauch, S., Leibrecht, I., Glass, B., Breuer, S., Geyer, M., Kräusslich, H.G., Kalbitzer, H.R., Wieland, F.T., Fackler, O.T., 2007, Human immunodeficiency virus type 1 Nef protein modulates the lipid composition of virions and host cell membrane microdomains. Retrovirology 4: 70.

Burleigh, L.M., Calder, L.J., Skehel, J.J., Steinhauer, D.A., 2005, Influenza a viruses with mutations in the M1 helix six domain display a wide variety of morphological phenotypes. J. Virol. 79: 1262–1270.

Campbell, S.M., Crowe, S.M., Mak, J., 2002, Virion-associated cholesterol is critical for the maintenance of HIV-1 structure and infectivity. AIDS 16: 2253–2261.

Campbell S., Gaus, K., Bittman, R., Jessup, W., Crowe, S., Mak, J., 2004, The raft-promoting property of virion-associated cholesterol, but not the presence of virion-associated Brij 98 rafts, is a determinant of human immunodeficiency virus type 1 infectivity. J. Virol. 78: 10556–10565.

Chan, R., Uchil, P.D., Jin, J., Shui, G., Ott, D.E., Mothes, W., Wenk, M.R., 2008, Retroviruses human immunodeficiency virus and murine leukemia virus are enriched in phosphoinositides. J. Virol. 82: 11228–11238.

Chatterjee, P.K., Eng, C.H., Kielian, M., 2002, Novel mutations that control the sphingolipid and cholesterol dependence of the Semliki Forest virus fusion protein. J. Virol. 76: 12712–12722.

Chen, B.J., Takeda, M., Lamb, R.A., 2005, Influenza virus hemagglutinin (H3 subtype) requires palmitoylation of its cytoplasmic tail for assembly: M1 proteins of two subtypes differ in their ability to support assembly. J. Virol. 79: 13673–13684

Chen, B.J., Leser, G.P., Jackson, D., Lamb, R.A., 2008, The influenza virus M2 protein cytoplasmic tail interacts with the M1 protein and influences virus assembly at the site of virus budding. J. Virol. 82: 10059–10070.

Chen, S.S., Yang, P., Ke, P.Y., Li, H.F., Chan, W.E., Chang, D.K., Chuang, C.K., Tsai, Y., Huang, S.C., 2009, Identification of the LWYIK motif located in the human immunodeficiency virus type 1 transmembrane gp41 protein as a distinct determinant for viral infection. J. Virol. 83: 870–883.

Cherezov, V., Rosenbaum, D.M., Hanson, M.A., Rasmussen, S.G., Thian, F.S., Kobilka, T.S., Choi, H.J., Kuhn, P., Weis, W.I., Kobilka, B.K., Stevens, R.C., 2007, High-resolution crystal structure of an engineered human beta2-adrenergic G protein-coupled receptor. Science 318: 1258–1265.

Chukkapalli, V., Hogue, I.B., Boyko, V., Hu, W.S., Ono, A., 2008, Interaction between the human immunodeficiency virus type 1 Gag matrix domain and phosphatidylinositol-(4,5)-bisphosphate is essential for efficient gag membrane binding. J. Virol. 82: 2405–2417.

Ciampor, F., Bayley, P.M., Nermut, M.V., Hirst, E.M.A., Sugrue, R.J., Hay, A.J., 1992a, Evidence that the amantadine-induced, M2-mediated conversion of influenza A virus haemagglutinin to the low pH conformation occurs in an acidic trans Golgi compartment. Virology 188: 14–24.

Ciampor, F., Thompson, C.A., Hay, A.J., 1992b, Regulation of pH by the M2 protein of influenza A viruses. Virus Res. 22: 247–258.

Cleverley, D.Z., Geller, H.M., Lenard, J., 1997, Characterization of cholesterol-free insect cells infectible by baculoviruses: effects of cholesterol on VSV fusion and infectivity and on cytotoxicity induced by influenza M2 protein. Exp. Cell Res. 233: 288–296.

Collawn, J.F., Stangel, M., Kuhn, L.A., Esekogwu, V., Jing, S., Trowbridge, I.S., Tainer, J.A., 1990, Transferrin receptor internalization sequence YXRF implicates a tight turn as the structural recognition motif for endocytosis. Cell 63: 1061–1072.

Coxey, R.A., Pentchev, P.G., Campbell, G., Blanchette-Mackie, E.J., 1993, Differential accumulation of cholesterol in Golgi compartments of normal and Niemann-Pick type C fibroblasts incubated with LDL: a cytochemical freeze-fracture study. J. Lipid Res. 34: 1165–1176.

Cristian, L., Lear, J.D., DeGrado, W.F., 2003, Use of thiol-disulfide equilibria to measure the energetics of assembly of transmembrane helices in phospholipid bilayers. Proc Natl. Acad. Sci. USA 100: 14772–14777.

Cross, K.J., Burleigh, L.M., Steinhauer, D.A., 2001, Mechanisms of cell entry by influenza virus. Expert Rev. Mol. Med. 3: 1–18.

Duff, K.C., Ashley, R.H., 1992, The transmembrane domain of influenza A M2 protein forms amantadine-sensitive proton channels in planar lipid bilayers. Virology 190: 485–489.

Elleman, C.J., Barclay, W.S., 2004, The M1 matrix protein controls the filamentous phenotype of influenza A virus. Virology 321: 144–153.

Epand, R.F., 2004, Do proteins facilitate the formation of cholesterol-rich domains? Biochim. Biophys. Acta 1666: 227–238.

Epand, R.F., Sayer, B.G., Epand, R.M., 2003, Peptide-induced formation of cholesterol-rich domains. Biochemistry 42: 14677–14689.

Epand, R.F., Sayer, B.G., Epand, R.M., 2005a, Induction of raft-like domains by a myristoylated NAP-22 peptide and its Tyr mutant. FEBS J. 272: 1792–1803.

Epand, R.F., Sayer, B.G., Epand, R.M., 2005b, The tryptophan-rich region of HIV gp41 and the promotion of cholesterol-rich domains. Biochemistry 44: 5525–5531.

Epand, R.F., Thomas, A., Brasseur, R., Vishwanathan, S.A., Hunter, E., Epand, R.M., 2006, Juxtamembrane protein segments that contribute to recruitment of cholesterol into domains. Biochemistry 45: 6105–6114.

Epand, R.M., Sayer, B.G., Vuong, P., Yip, C.M., Maekawa, S., Epand, R.F., 2004, Cholesterol-dependent partitioning of PtdIns(4,5)P2 into membrane domains by the N-terminal fragment of NAP-22 (neuronal axonal myristoylated membrane protein of 22 kDa). Biochem. J. 379: 527–532.

Eroglu, C., Brügger, B., Wieland, F., Sinning, I., 2003, Glutamate-binding affinity of Drosophila metabotropic glutamate receptor is modulated by association with lipid rafts. Proc. Natl. Acad. Sci. USA 100: 10219–10224.

Fujii, Y., Goto, H., Watanabe, T., Yoshida, T., Kawaoka, Y. (2003). Selective incorporation of influenza virus RNA segments into virions. Proc. Natl. Acad. Sci. USA 100: 2002–2007.

Gallo, S.A., Finnegan, C.M., Viard, M., Raviv, Y., Dimitrov, A., Rawat, S.S., Puri, A., Durell, S., Blumenthal, R., 2003, The HIV Env-mediated fusion reaction. Biochim. Biophys. Acta 1614: 36–50.

Gimpl, G., 2010, Cholesterol protein interaction: methods and cholesterol reporter molecules. In: Cholesterol Binding and Cholesterol Transport Proteins, Harris, J.R. (Ed.), Springer, pp. 1–45.

Gimpl, G., Klein, U., Reiländer, H., Fahrenholz, F., 1995, Expression of oxytocin receptor in baculovirus-infected insect cells: high affinity binding is induced by a cholesterol-cyclodextrin complex. Biochemistry 34: 13794–13801.

Grambas S., Hay A.J., 1992, Maturation of influenza A virus haemagglutinin – estimates of the pH encountered during transport and its regulation by the M2 protein. Virology 190: 11–18.

Grantham, M.L., Wu, W., Lalime, E.N., Lorenzo, M.E., Klein, S.L., Pekosz, A., 2009, Palmitoylation of the influenza A virus M2 protein is not required for virus replication in vitro but contributes to virus virulence. J. Virol. 83: 8655–8661.

Greenwood, A.I., Pan, J., Mills, T.T., Nagle, J.F., Epand, R.M., Tristram-Nagle, S., 2008, CRAC motif peptide of the HIV-1 gp41 protein thins SOPC membranes and interacts with cholesterol. Biochim. Biophys. Acta 1778: 1120–1130.

Guyader, M., Kiyokawa, E., Abrami, L., Turelli, P., Trono, D., 2002, Role for human immunodeficiency virus type 1 membrane cholesterol in viral internalization. J. Virol. 76: 10356–10364.

Hancock, J.F., 2006, Lipid rafts: contentious only from simplistic standpoints. Nat. Rev. Mol. Cell Biol. 7: 456–462.

Hanson, M.A, Cherezov, V., Griffith, M.T., Roth, C.B., Jaakola, V.P., Chien, E.Y., Velasquez, J., Kuhn, P., Stevens, R.C., 2008, A specific cholesterol binding site is established by the 2.8 Å structure of the human beta2-adrenergic receptor. Structure 16: 897–905.

Harris, J.R., Milton, N.G.N., 2010, Cholesterol in Alzheimer's disease and other amyloidogenic disorders. In: Cholesterol Binding and Cholesterol Transport Proteins, Harris, J.R. (Ed.), Springer pp. 47–76.

He, Z., Feng, S., Tong, Q., Hilgemann, D.W., Philipson, K.D., 2000, Interaction of PIP(2) with the XIP region of the cardiac Na/Ca exchanger. Am. J. Physiol. Cell Physiol. 278: C661–C666.

Heinz, F.X., Allison, S.L., 2000, Structures and mechanisms in flavivirus fusion. Adv. Virus Res. 55: 231–269.

Helseth, E., Olshevsky, U., Gabuzda, D., Ardman, B., Haseltine, W., Sodroski, J. 1990, Changes in the transmembrane region of the Human Immunodeficiency Virus Type 1 gp41 envelope glycoprotein affect membrane fusion. J. Virol. 64: 6314–6318.

Henkel, J.R., Weisz, O.A., 1998, Influenza M2 protein slows traffic along the secretory pathway; pH perturbation of acidified compartments affects early Golgi transport steps. J. Biol. Chem. 273: 6518–6524.

Hess, S.T., Kumar, M., Verma, A., Farrington, J., Kenworthy, A., Zimmerberg, J., 2005, Quantitative electron microscopy and fluorescence spectroscopy of the membrane distribution of influenza hemagglutinin. J. Cell Biol. 169: 965–976.

Holsinger, L.J., Lamb, R.A., 1991, Influenza virus M2 integral membrane protein is a homotetramer stabilized by formation of disulphide bonds. Virology 183: 32–43.

Hongo, S., Ishii, K., Mori, K., Takashita, E., Muraki, Y., Matsuzaki, Y., Sugawara, K., 2004, Detection of ion channel activity in Xenopus laevis oocytes expressing Influenza C virus CM2 protein. Arch. Virol. 149: 35–50.

Hughey, P.G., Roberts, P.C, Holsinger, L.J., Zebedee, S.L., Lamb, R.A., Compans, R.W., 1995, Effects of antibody to the influenza A virus M2 protein on M2 surface expression and virus assembly. Virology 212: 411–421.

Imai, M., Watanabe, S., Ninomiya, A., Obuchi, M., Odagiri, T., 2004, Influenza B virus BM2 protein is a crucial component for incorporation of viral ribonucleoprotein complex into virions during virus assembly. J. Virol. 78: 11007–11015.

Imai, M., Kawasaki, K., Odagiri, T., 2008, Cytoplasmic domain of influenza B virus BM2 protein plays critical roles in production of infectious virus. J. Virol. 82: 728–739.

Iwatsuki-Horimoto, K., Horimoto, T., Noda, T., Kiso, M., Maeda, J., Watanabe, S., Muramoto, Y., Fujii, K., Kawaoka, Y., 2006, The cytoplasmic tail of the influenza A virus M2 protein plays a role in viral assembly. J. Virol. 80: 5233–5240.

Jamin, N., Neumann, J.M., Ostuni, M.A., Vu, T.K., Yao, Z.X., Murail, S., Robert, J.C., Giatzakis, C., Papadopoulos, V., Lacapère, J.J., 2005, Characterization of the cholesterol recognition amino acid consensus sequence of the peripheral-type benzodiazepine receptor. Mol. Endocrinol. 19: 588–594.

Jin, H., Subbarao, K., Bagal, S., Leser, G.P., Murphy, B.R., Lamb, R.A., 1996, Palmitoylation of the influenza virus hemagglutinin (H3) is not essential for virus assembly or infectivity. J. Virol. 70: 1406–1414.

Kalvodova, L., Sampaio, J.L., Cordo, S., Ejsing, C.S., Shevchenko, A., Simons, K., 2009, The lipidomes of VSV, SFV and the host plasma membrane analyzed by quantitative shotgun mass spectrometry. J. Virol. doi:10.1128/JVI.00635-09.

Kato, N., Eggers, H.J., 1969, Inhibition of uncoating of fowl plague virus by 1-adamantamine hydrochloride. Virology 37: 632–641.

Keller, P., Simons, K., 1998, Cholesterol is required for surface transport of influenza virus hemagglutinin. J. Cell. Biol. 140: 1357–1367.

Kielian, M., 1995, Membrane fusion and the alphavirus life cycle. Adv. Virus Res. 45: 113–151.

Kielian, M., 2006, Class II virus membrane fusion proteins. Virology 344: 38–47.

Kurzchalia, T.V., Dupree, P., Parton, R.G., Kellner, R., Virta, H., Lehnert, M., Simons, K., 1992, VIP21 a 21-kD membrane protein is an integral component of trans-Golgi-network-derived transport vesicles. J. Cell. Biol. 118: 1003–1014.

Lacapère, J.J., Delavoie, F., Li, H., Péranzi, G., Maccario, J., Papadopoulos, V., Vidic, B., 2001, Structural and functional study of reconstituted peripheral benzodiazepine receptor. Biochem. Biophys. Res. Commun. 284: 536–541.

Lakadamyali, M., Rust, M.J., Babcock, H.P., Zhuang, X., 2003, Visualizing infection of individual influenza viruses. Proc. Natl. Acad. Sci. USA 100: 9280–9285.

Lamb, R.A., Krug, R.M., 1996, Orthomyxoviridae: the viruses and their replication. In: Fields Virology 3rd edn., Fields, B.N., Knipe, D.M., Howley, P.M., Chanock, R.M., Melnick, J.L., Monath, T.P., Roizman, B., Straus, S.E. (Eds.), Lippincott-Raven, Philadelphia, pp. 1353–1395.

Lamb, R.A., Zebedee, S.L., Richardson, C.D., 1985, Influenza virus M2 protein is an integral membrane protein expressed on the infected cell surface. Cell 40: 627–633.

Leser, G.P., Lamb, R.A., 2005, Influenza virus assembly and budding in raft-derived microdomains: a quantitative analysis of the surface distribution of HA, NA and M2 proteins. Virology 342: 215–227.

Li, H., Papadopoulos, V., 1998, Peripheral-type benzodiazepine receptor function in cholesterol transport. Identification of a putative cholesterol recognition/interaction amino acid sequence and consensus pattern. Endocrinology 139: 4991–4997.

Li, H., Yao, Z., Degenhardt, B., Teper, G., Papadopoulos, V., 2001, Cholesterol binding at the cholesterol recognition/interaction amino acid consensus (CRAC) of the peripheral-type benzodiazepine receptor and inhibition of steroidogenesis by an HIV TAT-CRAC peptide. Proc. Natl. Acad. Sci. USA 98: 1267–1272.

Lin, T., Schroeder, C., 2001, Definitive assignment of proton selectivity and attoampere unitary current to the M2 ion channel protein of influenza A virus. J. Virol. 75: 3647–3656.

Liu, Y., Casey, L., Pike, L.J., 1998, Compartmentalization of phosphatidylinositol 4,5-bisphosphate in low-density membrane domains in the absence of caveolin. Biochem. Biophys. Res. Commun. 245: 684–690.

Lorizate, M., Gómara, M.J., de la Torre, B.G., Andreu, D., Nieva, J.L., 2006a, Membrane-transferring sequences of the HIV-1 Gp41 ectodomain assemble into an immunogenic complex. J. Mol. Biol. 360: 45–55.

Lorizate, M., de la Arada, I., Huarte, N., Sánchez-Martínez, S., de la Torre, B.G., Andreu, D., Arrondo, J.L., Nieva, J.L., 2006b, Structural analysis and assembly of the HIV Gp41 amino-terminal fusion peptide and the pretransmembrane amphipathic-at-interface sequence. Biochemistry 45: 14337–14346.

Lorizate, M., Cruz, A., Huarte, N., Kunert, R., Pérez-Gil, J., Nieva, J.L., 2006c, Recognition and blocking of HIV-1 gp41 pre-transmembrane sequence by monoclonal 4E10 antibody in Raft-like membrane environment. J. Biol. Chem. 281: 39598–39606.

Lorizate, M., Huarte, N., Sáez-Cirión, A., Nieva, J.L., 2008, Interfacial pre-transmembrane domains in viral proteins promoting membrane fusion and fission. Biochim. Biophys. Acta 1778: 1624–1639.

Lorizate, M., Brügger, B., Akiyama, H., Glass, B., Mueller, B., Anderluh, G., Wieland, F.T., Kräusslich, H.G., 2009, Probing HIV-1 membrane liquid order by laurdan staining reveals producer cell dependent differences. J. Biol. Chem. doi/10.1074/jbc.M109.029256.

Lu, Y.E., Cassese, T., Kielian, M., 1999, The cholesterol requirement for sindbis virus entry and exit and characterization of a spike protein region involved in cholesterol dependence. J. Virol. 73: 4272–4278.

Marheineke, K., Grünewald, S., Christie, W., Reiländer, H., 1998, Lipid composition of Spodoptera frugiperda (Sf9) and Trichoplusia ni (T.n) insect cells used for baculovirus infection. FEBS Lett. 441: 49–52.

McCown, M.F., Pekosz, A., 2005, The influenza A virus M2 cytoplasmic tail is required for infectious virus production and efficient genome packaging. J. Virol. 79: 3595–3605.

McCown, M.F., Pekosz, A., 2006, Distinct domains of the influenza a virus M2 protein cytoplasmic tail mediate binding to the M1 protein and facilitate infectious virus production. J. Virol. 80: 8178–8189.

Melkonian, K.A., Ostermeyer, A.G., Chen, J.Z., Roth, M.G., Brown, D.A., 1999, Role of lipid modifications in targeting proteins to detergent-resistant membrane rafts. J. Biol. Chem. 274: 3910–3917.

Miyauchi, K., Kim, Y., Latinovic, O., Morozov, V., Melikyan, G.B., 2009, HIV enters cells via endocytosis and dynamin-dependent fusion with endosomes. Cell 137: 433–444.

Morita, E., Sundquist, W.I., 2004, Retrovirus budding. Annu. Rev. Cell Dev. Biol. 20: 395–425.

Muñoz-Barroso, I., Salzwedel, K., Hunter, E., Blumenthal, R., 1999, Role of the membrane-proximal domain in the initial stages of human immunodeficiency virus type 1 envelope glycoprotein-mediated membrane fusion. J. Virol. 73: 6089–6092.

Murail, S., Robert, J.C., Coïc, Y.M., Neumann, J.M., Ostuni, M.A., Yao, Z.X., Papadopoulos, V., Jamin. N., Lacapère, J.J., 2008, Secondary and tertiary structures of the transmembrane domains of the translocator protein TSPO determined by NMR. Stabilization of the TSPO tertiary fold upon ligand binding. B iochim. Biophys. Acta. 1778: 1375–1381.

Murray, P.S., Li, Z., Wang, J., Tang, C.L., Honig, B., Murray, D., 2005, Retroviral matrix domains share electrostatic homology: models for membrane binding function throughout the viral life cycle. Structure 13: 1521–1531.
Nayak, D.P., Barman, S., 2002, Role of lipid rafts in virus assembly and budding. Adv. Virus Res. 58: 1–28.
Nayak, D.P., Hui, E.K., 2004, The role of lipid microdomains in virus biology. Subcell. Biochem. 37: 443–491.
Noda, T., Sagara, H., Yen, A., Takada, A., Kida, H., Cheng, R.H., Kawaoka, Y., 2006, Architecture of ribonucleoprotein complexes in influenza A virus particles. Nature 439: 490–492.
Ohuchi, M., Cramer, A., Vey, M., Ohuchi, R., Garten, M., Klenk, H.D., 1994, Rescue of vector-expressed fowl plague virus hemagglutinin in biologically active form by acidotropic agents and coexpressed M2 protein. J. Virol. 68: 920–926.
Ono, A., Freed, E.O., 2005, Role of lipid rafts in virus replication. Adv. Virus Res. 64: 311–358.
Ono, A., Ablan, S.D., Lockett, S.J., Nagashima, K., Freed, E.O., 2004, Phosphatidylinositol (4,5) bisphosphate regulates HIV-1 Gag targeting to the plasma membrane. Proc. Natl. Acad. Sci. USA 101: 14889–14894
Paila, Y.D., Tiwari, S., Chattopadhyay, A., 2009, Are specific nonannular cholesterol binding sites present in G-protein coupled receptors? Biochim. Biophys. Acta 1788: 295–302.
Papadopoulos, V., Liu, J., Culty, M., 2007, Is there a mitochondrial signaling complex facilitating cholesterol import? Mol. Cell Endocrinol. 265–266: 59–64.
Park, E.K., Castrucci, M.R., Portner, A., Kawaoka,Y., 1998, The M2 Ectodomain Is important for its incorporation into Influenza A virions. J. Virol. 72: 2449–2455.
Phalen, T., Kielian, M., 1991, Cholesterol is required for infection by Semliki Forest virus. J. Cell Biol. 112: 615–623.
Pinto, L.H., Lamb, R.A., 2006, The M2 proton channels of influenza A and B viruses. J. Biol. Chem. 281: 8997–9000.
Politowska, E., Kaźmierkiewicz, R., Wiegand, V., Fahrenholz, F., Ciarkowski, J., 2001, Molecular modelling study of the role of cholesterol in the stimulation of the oxytocin receptor. Acta Biochim. Pol. 48: 83–93.
Rajendran, L., Simons, K., 2005, Lipid rafts and membrane dynamics. J. Cell Sci. 118: 1099–1102.
Rajendran, L., Knölker, H.J., Simons K., 2010, Subcellular targeting strategies for drug design and delivery. Nature Rev. Drug Disc. 9: 29–42.
Ren, J., Lew, S., Wang, Z., London, E., 1997, Transmembrane orientation of hydrophobic alpha-helices is regulated both by the relationship of helix length to bilayer thickness and by cholesterol concentration. Biochemistry 36: 10213–10220.
Rietveld, A., Neutz, S., Simons, K., Eaton, S., 1999, Association of sterol- and glycosylphosphatidylinositol-linked proteins with Drosophila raft lipid microdomains. J. Biol. Chem. 274: 12049–12054.
Roberts, P.C., Compans, R.W., 1998, Host cell dependence of viral morphology. Proc. Natl. Acad. Sci. USA 95: 5746–5751.
Roberts, P.C., Lamb, R.A., Compans, R.W., 1998, The M1 and M2 proteins of influenza virus are important determinants in filamentous particle formation. Virology 240: 127–137.
Rone, M.B., Fan, J., Papadopoulos, V., 2009, Cholesterol transport in steroid biosynthesis: role of protein-protein interactions and implications in disease states. Biochim. Biophys. Acta 1791: 646–658.
Roussel, A., Lescar, J., Vaney, M.C., Wengler, G., Wengler, G., Rey, F.A., 2006, Structure and interactions at the viral surface of the envelope protein E1 of Semliki Forest virus. Structure 14: 75–86.
Rousso, I., Mixon, M.B., Chen, B.K., Kim, P.S., 2000, Palmitoylation of the HIV-1 envelope glycoprotein is critical for viral infectivity. Proc. Natl. Acad. Sci. USA 97: 13523–13525.
Ruigrok, R.W., Barge, A., Durrer, P., Brunner, J., Ma, K., Whittaker, G.R., 2000, Membrane interaction of influenza virus M1 protein. Virology 267: 289–298.

Sáez-Cirión, A., Nir, S., Lorizate, M., Agirre, A., Cruz, A., Pérez-Gil, J., Nieva, J.L., 2002, Sphingomyelin and cholesterol promote HIV-1 gp41 pretransmembrane sequence surface aggregation and membrane restructuring. J. Biol. Chem. 277: 21776–21785.

Sáez-Cirión, A., Arrondo, J.L., Gómara, M.J., Lorizate, M., Iloro, I., Melikyan, G., Nieva, J.L., 2003, Structural and functional roles of HIV-1 gp41 pretransmembrane sequence segmentation. Biophys. J. 85: 3769–3780.

Sainz, B., Rausch, J.M., Gallaher, W.R, Garry, R.F., Wimley, W.C., 2005, The aromatic domain of the coronavirus class I viral fusion protein induces membrane permeabilization: putative role during viral entry. Biochemistry 44: 947–958.

Sakaguchi, T., Tu, Q., Pinto, L.H., Lamb, R.A., 1997, The active oligomeric state of the minimalistic influenza virus M2 ion channel is a tetramer. Proc. Natl. Acad. Sci. USA 94: 5000–5005.

Saldanha, J.W., Czabotar, P.E., Hay, A.J., Taylor, W.R., 2002, A model for the cytoplasmic domain of the influenza A virus M2 channel by analogy to the HIV-1 vpu protein. Protein Pept. Lett. 9: 495–502.

Salzwedel, K., West, J.T., Hunter, E., 1999, A conserved tryptophan-rich motif in the membrane-proximal region of the human immunodeficiency virus type 1 gp41 ectodomain is important for Env-mediated fusion and virus infectivity. J. Virol. 73: 2469–2480.

Sanderson, C.M., Avalos, R., Kundu, A., Nayak, DP, 1995, Interaction of Sendai viral F, HN and M proteins with host cytoskeletal and lipid components in Sendai virus-infected BHK cells. Virology 209: 701–707.

Scheiffele, P., Roth, M.G., Simons, K., 1997, Interaction of influenza virus haemagglutinin with sphingolipid-cholesterol membrane domains via its transmembrane domain. EMBO J. 16: 5501–5508.

Scheiffele, P., Rietveld, A., Wilk, T., Simons, K., 1999, Influenza viruses select ordered lipid domains during budding from the plasma membrane. J. B iol. Chem. 274: 2038–2044.

Schibli, D.J., Montelaro, R.C., Vogel, H.J., 2001, The membrane-proximal tryptophan-rich region of the HIV glycoprotein, gp41, forms a well-defined helix in dodecylphosphocholine micelles. Biochemistry 40: 9570–9578.

Schmitt, A.P., Lamb, R.A., 2005, Influenza virus assembly and budding at the viral budozone. Adv. Virus Res. 64: 383–416.

Schnell, J.R., Chou, J.J., 2008, Structure and mechanism of the M2 proton channel of influenza A virus. Nature 451: 591–595.

Schroeder, C., Heider, H., Möncke-Buchner, E., Lin, T., 2005, The influenza virus ion channel and maturation cofactor M2 is a cholesterol-binding protein. Eur. Biophys. J. 34: 52–66.

Schroeder, C., Lin, T., 2005, Influenza A virus M2 protein: proton selectivity of the ion channel, cytotoxicity, and a hypothesis on peripheral raft association and virus budding. In: Viral Membrane Proteins: Structure, Function, and Drug Design. Fischer, W. (Ed.), Kluwer Academic/Plenum Publishers, New York, pp. 113–130.

Simons, K., Ikonen, E., 1997, Functional rafts in cell membranes. Nature 387: 569–572.

Simpson-Holley, M., Ellis, D., Fisher, D., Elton, D., McCauley, J., Digard, P., 2002, A functional link between the actin cytoskeleton and lipid rafts during budding of filamentous influenza virions. Virology 301: 212–225.

Skibbens, J.E., Roth, M.G., Matlin, K.S., 1989, Differential extractability of influenza virus hemagglutinin during intracellular transport in polarized epithelial cells and nonpolar fibroblasts. J Cell. Biol. 108: 821–832.

Smirnov, Y.A., Kuznetsova, M.A., Kaverin, N.V., 1991, The genetic aspects of influenza virus filamentous particle formation. Arch. Virol. 118: 279–284.

Smith, A.E., Helenius, A., 2004, How viruses enter animal cells. Science 304: 237–242.

Stewart, S.M., Wu, W., Lalime E.N., Pekosz, A. The cholesterol recognition/interaction amino acid consensus motif of the influenza A virus M2 protein is not required for virus replication but contributes to virulence (submitted).

Stiasny, K., Koessl, C., Heinz, F.X., 2003, Involvement of lipids in different steps of the flavivirus fusion mechanism. J. Virol. 77: 7856–7862.

Stouffer, A.L., Acharya, R., Salom, D., Levine, A.S., Di Costanzo, L., Soto, C.S., Tereshko, V., Nanda, V., Stayrook, S., DeGrado, W.F., 2008, Structural basis for the function and inhibition of an influenza virus proton channel. Nature 451: 596–599.

Suárez, T., Gallaher, W.R., Agirre, A., Goñi, F.M., Nieva, J.L., 2000a, Membrane interface-interacting sequences within the ectodomain of the human immunodeficiency virus type 1 envelope glycoprotein: putative role during viral fusion. J. Virol. 74: 8038–8047.

Suárez, T., Nir, S., Goñi, F.M., Saéz-Cirión, A., Nieva, J.L., 2000b, The pre-transmembrane region of the human immunodeficiency virus type-1 glycoprotein: a novel fusogenic sequence. FEBS Lett. 477: 145–149.

Sugrue, R.J., Hay, A.J., 1991, Structural characteristics of the M2 protein of influenza A viruses: evidence that it forms a tetrameric channel. Virology 180: 617–624.

Sugrue, R.J., Bahadur, G., Zambon, M.C., Hall-Smith, M., Douglas, A.R., Hay, A.J., 1990, Specific structural alteration of the influenza haemagglutinin by amantadine. EMBO J. 9: 3469–3476.

Sun, X., Whittaker, G.R., 2003, Role for influenza virus envelope cholesterol in virus entry and infection. J Virol. 77: 12543–12551.

Takeda, M., Pekosz, A., Shuck, K., Pinto, L.H., Lamb, R.A., 2002, Influenza A virus M2 ion channel activity is essential for efficient replication in tissue culture. J. Virol. 76: 1391–1399.

Takeda, M., Leser, G.P., Russell, C.J., Lamb, R.A., 2003, Influenza virus hemagglutinin concentrates in lipid raft microdomains for efficient viral fusion. Proc Natl. Acad. Sci. USA 100: 14610–14617.

Takeuchi, K., Lamb, R.A., 1994, Influenza virus M2 protein ion channel activity stabilizes the native form of fowl plague virus hemagglutinin during intracellular transport. J. Virol. 68: 911–919.

Terashita, A., Funatsu, N., Umeda, M., Shimada, Y., Ohno-Iwashita, Y., Epand, R.M., Maekawa, S., 2002, Lipid binding activity of a neuron-specific protein NAP-22 studied in vivo and in vitro. J. Neurosci. Res. 70: 172–179.

Thiele, C., Hannah, M.J., Fahrenholz, F., Huttner, W., 2000, Cholesterol binds to synaptophysin and is required for biogenesis of synaptic vesicles. Nat. Cell Biol. 2: 42–49.

Tian, D., Gao, P.F., Pinto, L.H., Lamb, R.A., Cross, T.A., 2003, Initial structure and dynamic characterization of the M2 protein transmembrane and amphipathic helices in lipid bilayers. Protein Sci. 12: 2597–2605.

Tobler, K., Kelly, M.L., Pinto, L.H., Lamb, R.A. (1999). Effect of cytoplasmic tail truncations on the activity of the M2 ion channel of influenza A virus. J. Virol. 73: 9695–9701.

Tsetsarkin, K.A., Vanlandingham, D.L., McGee, C.E., Higgs, S., 2007, A single mutation in chikungunya virus affects vector specificity and epidemic potential. PLoS Pathog. 3: e201.

Umashankar, M., Sánchez-San Martín, C., Liao, M., Reilly, B., Guo, A., Taylor, G., Kielian, M., 2008, Differential cholesterol binding by class II fusion proteins determines membrane fusion properties. J. Virol. 82: 9245–9253.

Vashishtha, M., Phalen, T., Marquardt, M.T., Ryu, J.S., Ng, A.C., Kielian, M., 1998, A single point mutation controls the cholesterol dependence of Semliki Forest virus entry and exit. J. Cell Biol. 140: 91–99.

Vincent, N., Genin, C., Malvoisin, E., 2002, Identification of a conserved domain of the HIV-1 transmembrane protein gp41 which interacts with cholesteryl groups. Biochim. Biophys. Acta 1567: 157–164

Vishwanathan, S.A., Thomas, A., Brasseur, R., Epand, R.F., Hunter, E., Epand, R.M., 2008a, Hydrophobic substitutions in the first residue of the CRAC segment of the gp41 protein of HIV. Biochemistry 47: 124–130.

Vishwanathan, S.A., Thomas, A., Brasseur, R., Epand, R.F., Hunter, E., Epand, R.M., 2008b, Large changes in the CRAC segment of gp41 of HIV do not destroy fusion activity if the segment interacts with cholesterol. Biochemistry 47: 11869–11876.

Waarts, BL, Bittman, R, Wilschut, J., 2002, Sphingolipid and cholesterol dependence of alphavirus membrane fusion. Lack of correlation with lipid raft formation in target liposomes. J. Biol. Chem. 277: 38141–38147.

Waheed, A.A., Freed, E.O., 2009, Lipids and membrane microdomains in HIV-1 replication. Virus Res. doi:10.1016/j.virusres.2009.04.007

Wang, J.K, Kiyokawa, E., Verdin, E., Trono, D., 2000, The Nef protein of HIV-1 associates with rafts and primes T cells for activation. Proc. Natl. Acad. Sci. USA 97: 394–399.

Wang, X., Hinson, E.R., Cresswell, P., 2007, The interferon-inducible protein viperin inhibits influenza virus release by perturbing lipid rafts. Cell Host Microbe 2: 96–105.

White, S.H., Wimley, W.C., 1999, Membrane protein folding and stability: physical principles. Annu. Rev. Biophys. Biomol. Struct. 28: 319–365.

Whittaker, G.R., Kann, M., Helenius, A., 2000, Viral entry into the nucleus. Annu. Rev. Cell Dev. Biol. 16: 627–651.

Yao, Z.X., Papadopoulos, V., 2002, Function of beta-amyloid in cholesterol transport: a lead to neurotoxicity. FASEB J. 16: 1677–1679.

Yeo, D.S., Chan, R., Brown, G., Ying, L., Sutejo, R., Aitken, J., Tan, B.H., Wenk, M.R., Sugrue, R.J., 2009, Evidence that selective changes in the lipid composition of raft-membranes occur during respiratory syncytial virus infection. Virology 386: 168–182.

Zebedee, S.L., Lamb, R.A., 1989, Growth restriction of influenza A virus by M2 protein antibody is genetically linked to the M1 protein. Proc. Natl. Acad. Sci. USA 86: 1061–1065.

Zebedee, S.L., Lamb, R.A., 1988, Influenza A virus M2 protein: monoclonal antibody restriction of virus growth and detection of M2 in virions J. Virol. 62: 2762–2772.

Zhang, J., Pekosz, A., Lamb, R.A., 2000, Influenza virus assembly and lipid raft microdomains: a role for the cytoplasmic tails of the spike glycoproteins. J. Virol. 74: 4634–4644.

Zheng, Y.H., Plemenitas, A., Fielding, C.J., Peterlin, B.M., 2003, Nef increases the synthesis of and transports cholesterol to lipid rafts and HIV-1 progeny virions. Proc. Natl. Acad. S ci. USA 100: 8460–8465.

Zhu, P., Liu, J., Bess, J., Chertova, E., Lifson, J.D., Grisé, H., Ofek, G.A., Taylor, K.A., Roux, K.H., 2006, Distribution and three-dimensional structure of AIDS virus envelope spikes. Nature 441: 847–852.

Chapter 4
Sterol–Protein Interactions in Cholesterol and Bile Acid Synthesis

Emma De Fabiani, Nico Mitro, Federica Gilardi, and Maurizio Crestani

Abstract Cholesterol and other cholesterol related metabolites, oxysterols, and bile acids, establish specific interactions with enzymes and other proteins involved in cholesterol and bile acid homeostasis, triggering a variety of biological responses. The substrate-enzyme binding represents the best-characterized type of complementary interaction between proteins and small molecules. Key enzymes in the pathway that converts cholesterol to bile acids belong to the cytochrome P450 superfamily. In contrast to the majority of P450 enzymes, those acting on cholesterol and related metabolites exhibit higher stringency with respect to substrate molecules. This stringency, coupled with the specificity of the reactions, dictates the chemical features of intermediate metabolites (oxysterols) and end products (bile acids). Both oxysterols and bile acids have emerged in recent years as new signalling molecules due to their ability to interact and activate nuclear receptors, and consequently to regulate the transcription of genes involved in cholesterol and bile acid homeostasis and metabolism, but also in glucose and fatty acid metabolism. Interestingly, other proteins function as bile acid or sterol receptors. New findings indicate that bile acids also interact with a membrane G protein-coupled receptor, triggering a signalling cascade that ultimately promote energy expenditure. On the other end, cholesterol and side chain oxysterols establish specific interactions with different proteins residing in the endoplasmic reticulum that result in controlled protein degradation and/or trafficking to the Golgi and the nucleus. These regulatory pathways converge and contribute to adapt cholesterol uptake and synthesis to the cellular needs.

Keywords Cytochrome P450 · Insig proteins · Nuclear receptors · Sterol regulatory element binding protein · TGR5

E. De Fabiani (✉)
Department of Pharmacological Sciences, "Giovanni Galli" Laboratory of Biochemistry and Molecular Biology of Lipids, Via Balzaretti, 20133 Milan, Italy
e-mail: emma.defabiani@unimi.it

Abbreviations

SREBP	Sterol Regulatory Element Binding Protein
Scap	SREBP cleavage activating protein
HMG-CoA	3-hydroxy-3-methylglutaryl coenzyme A
LXR	liver X receptor
FXR	farnesoid X receptor;
CYP7A1	cholesterol 7α-hydroxylase
CYP27	sterol 27-hydroxylase
LBD	ligand binding domain
DBD	DNA binding domain
PXR	pregnane X receptor
VDR	vitamin D receptor
LDL	low density lipoprotein
ER	endoplasmic reticulum

4.1 Introduction

The aim of this chapter is to critically review recent publications dealing with the binding and/or interaction of cholesterol and other cholesterol derivatives, namely oxysterols, and bile acids, with enzymes and other proteins involved in cholesterol and bile acid homeostasis. The nature of sterol-protein interactions is crucial in numerous aspects: substrate specificity, as in the case of enzymes acting on cholesterol and/or its derivatives (i.e. cytochrome P450 s); intracellular trafficking, as in the case of Sterol regulatory element binding protein (SREBP) cleavage activating protein (SCAP); stability and susceptibility to degradation, as in the case of 3-hydroxy-3-methylglutaryl coenzyme A (HMG CoA) reductase and Insig proteins; ligand-specific conformational changes and consequently interactions with other protein complexes involved in the regulation of gene transcription, as in the case of nuclear receptors. The features of the interactions between sterols and nuclear receptors have been discovered and elucidated more recently but their molecular and metabolic consequences in the feedback regulation of cholesterol and bile acid synthesis, in the modulation of cholesterol transport, and in other key cellular functions, have immediately gained a primary level of attention.

Other sterol–protein interactions play a highly relevant role in cholesterol homeostasis, We refer in particular to proteins involved in cholesterol efflux from cells, for example the membrane transporters belonging to the ATP-binding cassette (ABC) family, and in its transport in the blood stream, specifically the apo-components of lipoprotein particles. Since these topics will be described in detail in other chapters, we will only cover a few concepts that are strictly linked to the subjects of the present chapter. The exit of cholesterol from non-hepatic cells, the initial event in the so-called reverse cholesterol transport, is carried out mainly by members of the ABC family (ABCA1 and ABCG1). In the extracellular compartment, Apolipoprotein (apo) E represents one of the most important cholesterol acceptors. The expression of these proteins is positively regulated at the level of gene transcription by the nuclear receptor activated by oxysterols (see below). A relevant fraction of blood cholesterol is transported by low density lipoprotein whose clearance is accomplished through a receptor-mediated pathway. The expression of this receptor is finely modulated through multiple mechanisms (see below) that adapt intracellular cholesterol levels to cellular needs and, at the same time, greatly influence the levels of circulating lipoprotein-associated cholesterol.

4.2 Brief Overview of Cholesterol and Bile Acid Biosynthesis

A complete overview of cholesterol and bile acid biosynthesis is far beyond the scope of this chapter and readers interested in a deeper description of this issue can find more complete information elsewhere (*see* (Goldstein and Brown, 1990) and related references for cholesterol synthesis and (Chiang, 2004, Russell, 2003), for bile acid synthesis).

Figure 4.1 summarizes the main steps of cholesterol and bile acid synthesis. Cholesterol synthesis, also referred to as mevalonate pathway, is a ubiquitous pathway and virtually every cell possesses the essential complement of enzymes and intracellular transport proteins to produce cholesterol and non sterol isoprenoids (Fig. 4.1A). Bile acid production is a typical liver function and the complete set of enzymes and proteins required for bile acid synthesis through the classic pathway is selectively expressed in hepatic cells (Fig. 4.1B). However, it should be pointed out that commitment of cholesterol to bile acid conversion can also commence in non-liver cells, through the so-called alternative pathway. In this pathway, the hydroxylation of cholesterol side-chain precedes the hydroxylation at position 7 of the B ring.

From a general point of view, the pathways leading to cholesterol and bile acid synthesis share some common features:

- Most of the key enzymes of both cholesterol and bile acid synthesis are transmembrane proteins residing in the endoplasmic reticulum (ER).
- The rate of both cholesterol and bile acid synthesis is regulated at an early stage, namely at the level of the conversion of HMG CoA to mevalonate and at the level of 7α-hydroxylation of cholesterol, respectively. Furthermore, the enzymes catalyzing the rate-limiting steps, HMG CoA reductase and cholesterol 7α-hydroxylase (CYP7A1), are mainly regulated at the transcriptional level.
- The end products of the pathways, and/or closely related metabolites, are the most potent effectors exerting a feedback control. In fact, HMG CoA reductase senses the negative effect of sterols and oxysterols, whereas CYP7A1 is down-regulated by bile acids. Both cholesterol and bile acids control their own synthesis through

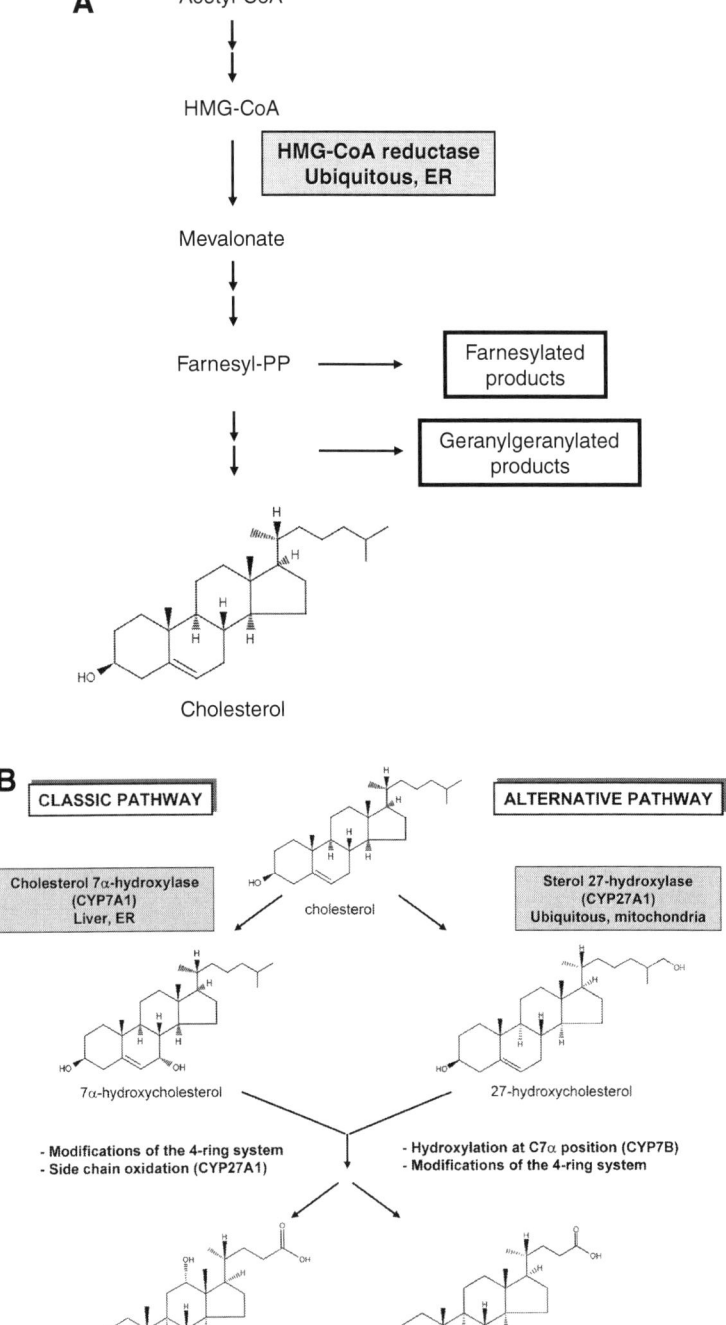

Fig. 4.1 (continued)

4.3 The Binding of (Chole)sterols to Cytochrome P450 Enzymes: Highly Stringent and Less Stringent Enzyme–Substrate Interactions

Several members of the cytochrome P450 (P450) superfamily utilize cholesterol or its derivatives as substrates for enzymatic reactions. These members include enzymes of the bile acid biosynthetic pathway and enzymes producing steroid hormones. In this chapter we will focus in particular on CYP7A1 and sterol 27-hydroxylase (CYP27A1).

It is well known that many P450 enzymes are involved in the biotransformation of xenobiotics and indeed, from a certain point of view, bile acid synthesis can be considered as a way to eliminate cholesterol from the body, following the classical steps used to remove highly lipophilic exogenous compounds. These steps include, firstly, introduction or generation of hydroxyl and carboxyl groups, and then conjugation with amino acids, taurine or glycine, to increase hydrophilicity and produce molecules suitable for excretion.

Hydroxylation at the 7α-position of cholesterol B ring is an essential feature of primary bile acids and is carried out by the liver-specific CYP7A1, in the classic pathway, and by the more widely expressed oxysterol 7α-hydroxylase, in the alternative pathway (Schwarz et al., 1998, Lathe, 2002). Since the early observations in the late 1950 s (Bergstrom, 1958), it was immediately clear that the 7α-hydroxylation of cholesterol represents a key step in the conversion of cholesterol to bile acids through the classic pathway, thus opening the way to a great deal of work aimed at discovering the biochemical features of the enzymatic reaction.

CYP7A1 is active almost exclusively on cholesterol and on its 5α-saturated analog cholestanol (Ogishima et al., 1987), thus showing a strict substrate specificity that is unusual for P450 enzymes. In addition, CYP7A1 displays high catalytic efficiency, a feature that well correlates with the ability of the liver to convert about

Fig. 4.1 Main steps of cholesterol and bile acid synthesis. **A** The synthesis of cholesterol is mainly regulated at the level of the conversion of HMG-CoA to mevalonate by the HMG-CoA reductase, an ER protein expressed in almost all tissues and cells. **B** The synthesis of bile acids starts in the liver (classic pathway) with the hydroxylation of cholesterol carried out by the P450 enzyme CYP7A1 and proceeds with modifications of the 4-ring system and the oxidation of the side chain by the mitochondrial P450 CYP27A1. The main primary bile acids found in humans are cholic acid and chenodeoxycholic acid. An alternative pathway has been described to take place in non-hepatic cells, according to which cholesterol is first hydroxylated at the side chain by CYP27A1. The hydroxyl group at the C7α position is introduced by the oxysterol 7α-hydroxylase (CYP7B1) and further reactions required for the conversion to bile acids are carried out in liver by the same enzymes of the classic pathway

600 mg of cholesterol a day, in comparison to the minute amounts (few milligrams) of other hydroxyl-derivatives of cholesterol produced by other sterol hydroxylases (Murtazina et al., 2002, Mast et al., 2004). CYP7A1 shows more stringent requirements for the substrate ring system than for the side chain, since it acts only on substrates carrying a free hydroxyl group at C3 and a *trans* or quasi-*trans* A/B ring configuration (Ogishima et al., 1987), but, on the other hand, has the ability to metabolize side-chain hydroxyl-cholesterols (Norlin et al., 2000a, b).

The molecular basis of the high substrate specificity of CYP7A1 were elegantly investigated in detail by the group of I. Pikuleva by using several complementary approaches: computer modeling, site-directed mutagenesis, and substrate-binding assays.

Mammalian P450 s are membrane-associated proteins that are anchored in the membrane by an N-terminal helix. Additional membrane binding sites were identified in some P450 s and crystallographic data suggest that these non-contiguous portions of the protein form a monofacial hydrophobic surface (Williams et al., 2000). In mammalian, as well as in bacterial P450 s, the F and G helices and the F/G-loop, which are flexible and undergo open/closed motions, are portions of the P450 polypeptide chain of particular importance for substrate access and catalysis. By using a theoretical approach, it was concluded that the F/G-loop is in contact with the membrane and it has been hypothesized that lipophilic substrates enter P450 protein from the membrane through an access path close to the F/G-loop and hydroxylated products leave the active site through another egress path, directly to the cytoplasm (Williams et al., 2000). To test whether this model also applies to CYP7A1, Nakayama and colleagues generated and analyzed a series of mutant CYP7A1 proteins. Mutations within the hydrophobic region of CYP7A1, comprising residues 214-227 and corresponding to the putative F/G-loop and the adjacent helical segments, yield to impaired interactions of recombinant enzyme with *Escherichia coli* membrane, reduced k_{cat}, increased K_m, but not K_d, for cholesterol (Nakayama et al., 2001). The results obtained with these mutations suggest that the strict substrate specificity of CYP7A1 is in part determined by the initial recognition that takes place on the surface of the molecule. Further studies by the same group provided evidence suggesting that Asn288 is a key residue for binding cholesterol, because its side chain interacts with the 3β-hydroxyl group either directly or via a water molecule (Mast et al., 2005). According to the view of these authors, Asn288 could form a stabilizing hydrogen bond network in the substrate-free protein, thus contributing to maintain a stable structure. However, when cholesterol is appropriately docked at the active site, the C3 β-hydroxyl group of the substrate would allow for an alternate network. The mutagenesis data reported by Mast et al. also suggest that other amino acids of CYP7A1 interact with cholesterol, thus strengthening the concept of a complementary fit between the cholesterol molecule and the enzyme active site, possibly due to reduced rotational freedom of the substrate inside the enzyme active site, resulting in a single binding orientation (Mast et al., 2005).

Hydroxylation of sterol side chain at position C27 is carried out by sterol 27-hydroxylase (CYP27A1), a mitochondrial P450 expressed in many tissues and cell types. This hydroxylase participates in bile acid synthesis, efficiently acting on bile acid intermediates in the classic pathway, or on cholesterol itself in the alternative pathway. In this respect, CYP27A1 would appear to exhibit less stringent requirements for substrate molecules in comparison to CYP7A1. However, it was observed that phospholipids stimulate enzyme activity to a greater extent when the substrate is cholesterol than when it acts on 5β-cholestane-3α,7α,12α-triol (Murtazina et al., 2004). To explain the molecular basis of this peculiar behaviour, I. Pikuleva and her group investigated how the two physiological substrates, cholesterol and 5β-cholestane-3α,7α,12α-triol, interact with CYP27A1, by combining computer modelling and site-directed mutagenesis (Mast et al., 2006). The computer models suggest that cholesterol and 5β-cholestane-3α,7α,12α-triol occupy different regions within the substrate-binding pocket and binds in different orientations. As a result, some of the active site residues interact with both substrates, although they are situated differently relative to each steroid, and some residues bind only one substrate. Mutation of the overlapping substrate-contact residues affected CYP27A1 binding and enzyme activity in a substrate-dependent manner and allowed identification of several important side chains. Threo110 is proposed to interact with the 12α-hydroxyl of 5β-cholestane-3α,7α,12α-triol, whereas Val367 seems to be crucial for correct positioning of the cholesterol C26 methyl group and for region-selective hydroxylation of this substrate. Dissecting the role of individual amino acids in the binding of physiological substrates to CYP27A1 active site may provide valuable insight for the understanding of phenotypic manifestations of cerebrotendinous xanthomatosis (CTX). CTX is a rare autosomal recessive disease caused by a deficiency of CYP27A1 (Cali et al., 1991) and characterized by a wide array of symptoms, tendon xanthomas, cataract, and complex neurologic impairment (Federico and Dotti, 2001). Since the cloning of the cDNA encoding the human gene, many mutations have been found and characterized. Pathologic allelic variants cause amino acid substitution, synthesis of truncated protein, and abnormal pre-mRNA splicing. Although a genotype-phenotype correlation has not been found so far, unravelling the structure of CYP27A1 might contribute to the understanding of its function in patho-physiology.

4.4 Bile Acid–Protein Interactions: The Molecular Mechanisms Underlying Bile Acid Synthesis Feedback Regulation and Beyond

Bile acids have been known in the past simply as cholesterol end-products, whose only reported function was the formation of mixed micelles because of their amphipathic nature. In fact, bile acids present specific conformational and physicochemical properties, displaying a concave hydrophilic face harbouring 2 (chenodeoxycholic acid) or 3 (cholic acid) hydroxyl groups, and a convex hydrophobic face, through which bile acids interact with other hydrophobic molecules.

An unexpected breakthrough in bile acid biology occurred with the discovery that physiological bile acids bind and activate a nuclear receptor, the farnesoid X receptor (FXR) (Makishima et al., 1999; Parks et al., 1999), thus emerging as signalling molecules that can modulate gene transcription. A few years later, the biological properties of bile acids expanded further with the demonstration that they can also interact with a G protein-coupled receptor (TGR5) (Kawamata et al., 2003).

Nuclear receptors are ligand-activated transcription factors (Chawla et al., 2001). FXR belongs to Class II, a group of nuclear receptors that are mainly localized in the nucleus, also when not activated by their ligand and depending on the receptor, they can bind DNA either as heterodimers, homodimers or, in some cases, as monomers. Almost all members of this nuclear receptor class present a conserved modular structure (Fig. 4.2). Essential elements of this structure are the A/B-domain, containing the ligand-independent transcription activation function (AF-1) at the N-terminus; the C-domain containing the DNA-binding domain (DBD), characterized by two typical zinc-fingers that are involved in the recognition of specific DNA consensus sequences; the D-domain, also known as hinge region, which confers flexibility to the receptor for dimerization and interaction to the DNA consensus sequences located on target genes; the D-domain can also interact with co-repressor proteins (Horlein et al., 1995; Chen and Evans, 1995); the E-domain comprising the ligand-binding domain (LBD), a hydrophobic pocket that can accommodate the receptor ligand, and the ligand-dependent transcription activation function (AF-2). The binding of the ligand promotes conformational changes that allow the physical interaction with coactivator proteins. A key role is played by helix 12 (H12) in the LDB of the nuclear receptor and by a short α-helical LxxLL sequence present on coactivator proteins, but it should be underlined that other helices, i.e. H3 and H4, of the nuclear receptor contribute to the contact area.

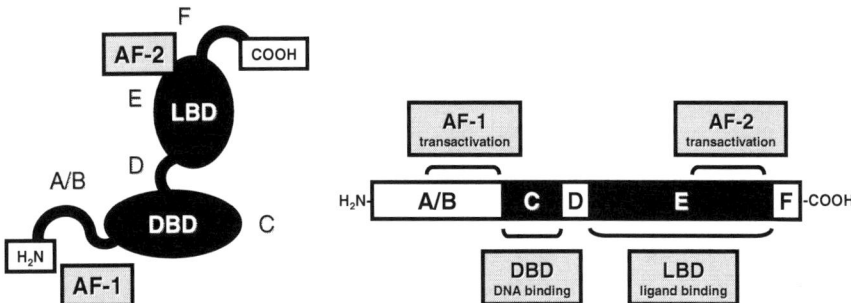

Fig. 4.2 Modular structure of nuclear receptors. Nuclear receptors present a conserved structure consisting of functional domains that are essential for their action. At the N-terminus, the A/B domain contains the ligand-independent transcription activation function (AF-1). The C domain contains the DNA binding domain (DBD) and is separated by a hinge region (D-domain) from the E-domain. This domain presents the ligand binding domain (LBD) and the ligand-dependent activation function 2 (AF-2). The F-domain is not conserved in all nuclear receptors

Hydrophobic bile acids, chenodeoxycholic acid, deoxycholic acid and lithocholic acid were the first reported ligands of FXR (Parks et al., 1999, Makishima et al., 1999), however, in the following years, the list of natural and synthetic compounds active as FXR ligands has increased tremendously. For a complete overview on FXR, its ligands and biological effects, interested readers are referred to a recent review by Lefebvre et al. (2009).

First of all, it should be pointed out that the ability of a given compound to bind and activate a nuclear receptor can be assessed by different means, using either cell-based and cell-free assays. Very briefly, in the cell-free fluorescence resonance energy transfer (FRET) assay, the binding of a molecule to the LBD is evaluated indirectly by measuring the recruitment of a cofactor peptide. Therefore the potency and efficacy of the ligand depend in part on the cofactor peptide used in the assay and consequently, when translating these observations in biological systems, the expression levels and availability of nuclear cofactors influence the final effect. On the other hand, the ability of a given compound to activate a nuclear receptor and gene transcription can be evaluated in transfection assays in cell cultures, usually employing standard reporting systems, i.e. promoters bearing responsive elements, often in multiple copies. In this case, the potency and efficacy may be influenced by the rate of uptake, modification and or sequestration of the ligand.

Although the approaches mentioned above are standardized procedures commonly used in the screening and characterization of nuclear receptor ligands, in this section we will focus and summarize the findings on the structural interactions between FXR and its natural ligands, bile acids, obtained by X-ray crystallography and/or by computer modelling. Although FXR binds to bile acids with a ~1000-fold weaker affinity than other steroid receptors bind to their cognate hormones, FXR displays a remarkable specificity for bile acids, the recognition being based on their unique non-planar shapes and amphipathic physicochemical properties.

Valuable hints on the interactions between the nuclear receptor and its cognate ligand/s have been obtained by solving the crystallographic structure of the LDB of FXR complexed with natural and synthetic agonists (Downes et al., 2003; Mi et al., 2003). These studies have revealed that similarly to other nuclear receptors, the FXR LDB is organized in a 12 α-helix bundle. Upon interaction with the ligand, it undergoes major structural transitions that enable the selective recruitment of coactivators by forming a charge clamp and a hydrophobic groove that function as an interaction interface with the LxxLL motifs present in coactivator proteins (Mi et al., 2003). Due to the fact that their skeleton is not planar, bile acids present a rounded shape that allows a close fit with respect to the pocket in FXR. Furthermore, it is noteworthy that bile acids occupy their pocket with their steroid nucleus positioned in the reverse orientation to that of all other steroids (Mi et al., 2003). In addition, the FXR ligand-binding pocket also utilizes the amphipathic properties of the bile acids to provide additional molecular recognition beyond their unique shape. In particular, the role and contribution to the binding of individual hydroxyl groups at position 3α, 7α, 12α, of the A/B ring conformation, and of the carboxyl terminus have been investigated in considerable detail. The side chain oxygen of Tyr366 establishes hydrogen bonds with the axial 7α-OH on ring B, and the lack of this hydroxyl in lithocholic acid lowers its affinity for the receptor. Similar results were obtained by modelling the molecule of chenodeoxycholic acid into the binding pocket of the human FXR (Downes et al., 2003). According to this model, potential bonds could occur between the hydroxyl groups of the bile acid with residues on helices 7 and 11, while hydrophobic interactions were predicted to secure helix 3 in an orientation allowing a compact conformation of helix 12 (activation function-2 domain), that enables stable interactions between the nuclear receptor and their coactivator partners. The influence of the hydrophobic interactions between the ligand and helix 3 on the activation state of the receptor is well demonstrated by the synthetic ligand fexaramine, that establishes a greater number of contacts and is a stronger activator in comparison to chenodeoxycholic acid (Downes et al., 2003). All naturally occurring bile acids contain a *cis*-oriented A/B ring juncture and a 3α-hydroxyl group in their A ring. The FXR structure shows the A ring and the 3-hydroxy group oriented toward helix 12 where they interact with a His residue on helix 10/11 (His444) and a Trp residue on helix 12 (Trp-466). Therefore the binding of ligand stabilizes the interaction between the indole ring of Trp466 (π electron system) and the Nϵ (cation) on the perpendicularly oriented His444 side chain. This molecular switch for nuclear receptor activation is peculiar of FXR and of the oxysterol receptor, as will be discussed below. By using a 3α-dehydroxylated bile acid, Mi et al. (2003) also demonstrated that the 3α-hydroxyl group is not indispensable to induce the optimal conformation change of FXR and that a correctly positioned ring A is the dominant factor mediating the agonist function of bile acids. On the other hand, bile acids lacking the hydroxyl group at 7 position have been predicted to interact only with helix 7, thus failing to bridge helix 3 to helix 7 as securely as chenodeoxycholic acid, a condition that in turn would affect the rigidity of helix 12. Moreover, as the ligand pocket does not provide any polar side chains to accommodate the 12α-OH group on ring C, the binding of cholic acid and deoxycholic acid would be energetically costly in comparison to chenodeoxycholic acid (Mi et al., 2003). The carboxylic extremity of bile acids is oriented towards the entry of the ligand binding pocket, thus explaining why conjugated derivatives still exhibit the ability to bind and activate FXR. It should be emphasized that the binding of different bile acids to FXR per se does not explain the complexity of the biological outcome, since the conformational changes caused by the ligand, which vary depending on the chemical structure of the ligand, induce the recruitment of cofactors and the assembly of complexes promoting gene transcription in a promoter- and tissue-specific fashion.

To complete the issue of the interactions between bile acids and nuclear receptors, it should be mentioned that the secondary bile acid lithocholic acid activates two other members of the nuclear receptor superfamily, the pregnane receptor, PXR (Staudinger et al., 2001, Xie et al., 2001) and the vitamin D receptor (Makishima et al., 2002). PXR is also activated by a variety of drugs and xenobiotics and indeed the crystal structure of the LBD has been investigated using ligands other than lithocholic acid. According to crystallographic studies, the PXR LBD presents a flexible and conformable ligand-binding pocket that adjusts its shape to accommodate ligands of distinct size and structure, establishing a combination of hydrophobic and polar interactions with PXR ligand-binding pocket residues (Watkins et al., 2003a, b; Watkins et al., 2001). These structural features account for the ligand binding promiscuity of PXR.

Surprisingly, lithocholic acid can also activate the vitamin D receptor (VDR), another member of the nuclear receptor superfamily, and studies of the structure-function relationships have allowed the identification of amino acids required for the specific interactions with 1α, 25-dihydroxyvitamin D_3 (high affinity ligand) and lithocholic acid (low affinity ligand), suggesting that VDR adopts distinct conformations in response to the binding with the two molecules (Adachi et al., 2004; Choi et al., 2003).

The discovery of bile acids as ligands for a novel G protein-coupled receptor (TGR5) was reported by two independent groups as a result of chemical library screenings (Maruyama et al., 2002; Kawamata et al., 2003). TGR5 is a cell-surface receptor associated with the intracellular accumulation of cyclic AMP, which is widely expressed in diverse cell types and whose downstream effects vary from attenuation of pro-inflammatory cytokine production by monocytes (Kawamata et al., 2003), to enhancement of energy expenditure in adipocytes and myocytes (Watanabe et al., 2006).

The initial observations indicated that the rank order of potency was correlated to hydrophobicity, lithocholic acid being the most hydrophobic and the most potent, and cholic acid the most hydrophilic and the weakest agonist, independent of the conjugation state (Kawamata et al., 2003; Maruyama et al., 2002). Notably, the 7β-hydroxyl epimer of chenodeoxycholic acid, ursodeoxycholic acid, and cholesterol were found to be only slightly active (Kawamata et al., 2003).

Taking advantage of their experience in structure-activity relationship studies on bile acid derivatives as FXR modulators, Pellicciari et al. (2007) investigated the impact of modifications in the side chain of chenodeoxycholic acid on FXR and TGR5 activation. Their results from docking experiments indicated that the binding pocket for bile acids is not entirely conserved between TGR5 and FXR since TGR5 displays an accessory binding pocket. Further studies showed that the tauro-conjugated derivatives as well as the sulfonic substituted of bile acids are more potent agonists of TGR5, probably due to the presence of the sulfonic moiety (Sato et al., 2008). In this regard it is noteworthy that binding pockets specifically recognizing sulfonic moieties in general do not contain positively charged residues (Macchiarulo and Pellicciari, 2007). On this basis the authors speculated that TGR5, converse to FXR, is endowed with a neutral charged binding site (Sato et al., 2008).

The proteins described in this section, FXR and TGR5, mediate most of the biological effects exerted by bile acids. The findings collected in recent years clearly indicate that bile acids are master regulators of bile acid metabolism and transport in liver and intestine (FXR activation); they contribute to the hepatic control of triglycerides and glucose homeostasis (FXR activation); they promote energy expenditure in skeletal muscle and adipose tissue (TGR5 activation); they may exert anti-inflammatory activity on macrophages (TGR5 activation).

Finally, it should not be forgotten that bile acids are likely to interact with other proteins/receptors, some of which are still unknown, through still undefined molecular interactions. These interactions are responsible for the activation of several protein kinases (*see for review* Hylemon et al., 2009) and also underlie processes such as the activation state of hepatic nuclear factor 4 (De Fabiani et al., 2001) and the nuclear-cytoplasmic shuttling of histone deacetylase 7 (Mitro et al., 2007a).

4.5 The Liver X Receptor: "Sterol" or "Non-sterol", This Is the Question

The liver X receptor (LXR) α and β are nuclear receptors belonging to class II, as well as FXR, and display most of the general features described in the previous section. LXRs have been defined as sterol sensors, essential for the maintenance of cholesterol homeostasis and protecting the cells from cholesterol overload. In fact, LXRs regulate the expression of proteins involved in cholesterol efflux and transport and, at least in rodents, metabolism to bile acids through CYP7A1 (Tontonoz and Mangelsdorf, 2003). The two isoforms are differently expressed in cells and tissues since the α isoform is more abundant in liver, kidney, intestine, fat tissue, macrophages, lung, spleen, while the β isoform is ubiquitously expressed. The two isoforms also share a high sequence identity (78%) and significantly, residue differences are located far away from the ligand-binding pocket (Alberti et al., 2000, Williams et al., 2003).

With the aim of identifying endogenous LXR ligands, Janowski et al. (1996) found that "oxysterol congeners", that is oxygenated forms of cholesterol, whose known function at that time was the repression of cholesterol synthesis (see below), were able to activate gene transcription through LXRα at concentrations not too far from physiology. Similar results were obtained soon after by other groups (Lehmann et al., 1997; Forman et al., 1997). Figure 4.3 shows a list of some oxysterols displaying the ability to activate LXRs and the corresponding enzymes responsible for their synthesis. In the following years a number of steroidal and non-steroidal compounds were synthesized and characterized, also because LXR activation appeared as an attractive pharmacological strategy to prevent atherosclerosis, due do their effects on ABC transporters and apoE (Millatt et al., 2003).

Initial studies on the structure-function relationship showed that the minimal pharmacophore for receptor activation is a sterol with a hydrogen bond acceptor at C24, such as that found in 24(*S*),25-epoxycholesterol (Spencer et al., 2001). According to the model, Trp443 in LXRα AF2 is essential for activation by oxysterols, since the side chain of 24(*S*),25-epoxycholesterol adopts a low-energy extended conformation that permits a hydrogen bond interaction between this residue and the epoxide oxygen. The model proposed by Spencer and colleagues also identified Arg305 as the amino acid that may interact with the C3 hydroxyl group of the sterol A-ring at the other end of the ligand-binding pocket (Spencer et al., 2001).

The X-ray crystal structures of LXRα/β LBD, published in 2003 by three independent groups, provided valuable insights on the structural determinants for ligand-dependent activation of these receptors. Svensson et al. reported the crystal

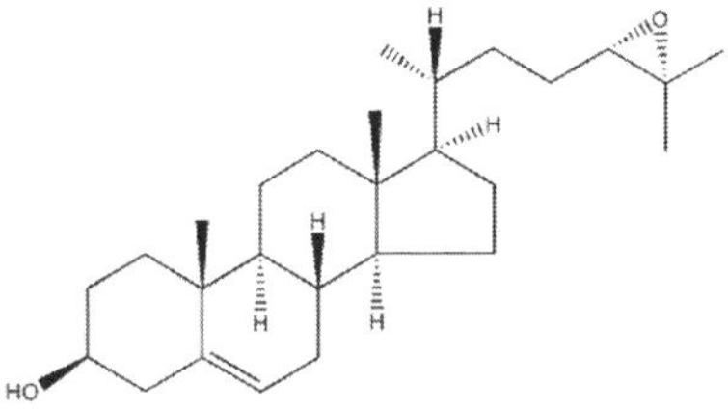

24(*S*),25-epoxycholesterol
2,3-oxidosqualene:lanosterol cyclase

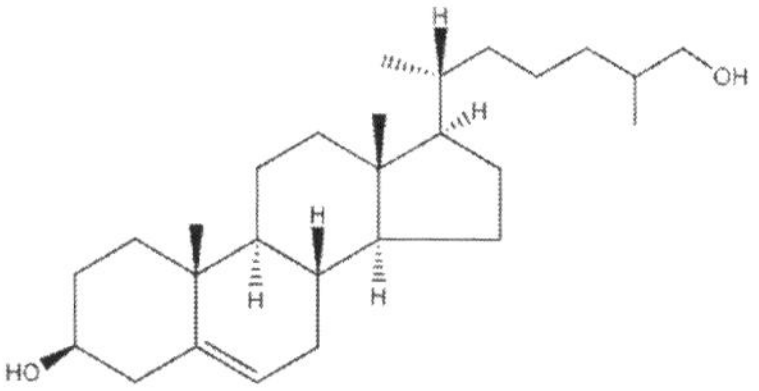

27-hydroxycholesterol
CYP27A1

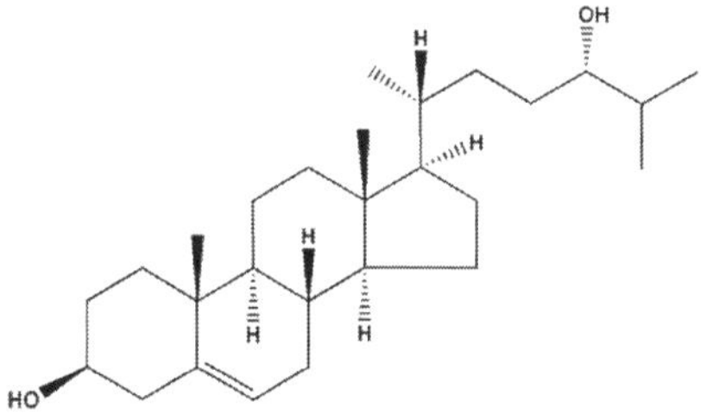

24-hydroxycholesterol
Cholesterol 24-hydroxylase (CYP46A1)

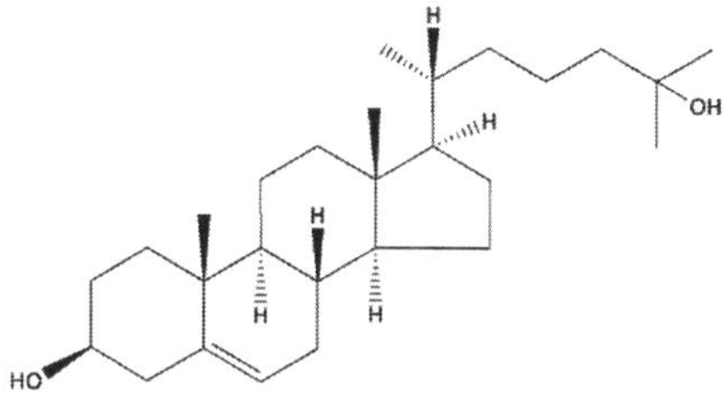

25-hydroxycholesterol
Cholesterol 25-hydroxylase

Fig. 4.3 Oxysterols activating LXRs. The chemical structures of side chain oxysterols exhibiting the ability to activate LXRs are shown. These molecules are enzymatic products and the enzymes responsible are indicated

structure of LXRα LBD complexed to a synthetic ligand and the docking of oxysterols on the obtained structure (Svensson et al., 2003), whereas Williams et al. and Farmegardh et al. reported the structure of LXRβ LBD complexed with natural and synthetic ligands (Farnegardh et al., 2003, Williams et al., 2003). The accessible volume of the LXRα ligand-binding pocket was estimated to be in the range 700-800 Å^3 (Svensson et al., 2003) while the cavity volume of the LXRβ ligand binding pocket was reported to vary depending on the size of the ligand (from a minimum of ~600 to a maximum of 1100 Å^3) (Farnegardh et al., 2003; Williams et al., 2003). In both cases, the volume is larger than those found in classic steroid hormone receptors. The analysis of the overall LBD structure of both receptors revealed a remarkable flexibility and capability to accommodate ligands of different structure, in accordance with the fact that LXRs can be activated by ligands with different chemical structures.

Svensson et al. (2003) showed that the ligand binding pocket of LXRα is predominantly hydrophobic, with only a few possible hydrogen bond interactions and the docking results suggested a common anchoring of the side chain hydroxyl/epoxy group to His421 and Trp443. The structural data were confirmed by functional analysis of mutated forms of LXRα LBD by transfection assays with reporter systems, since it was demonstrated that the transcriptional activation of oxysterols is strictly dependent on His421 and Trp443 (Svensson et al., 2003). It is noteworthy that both residues are present in corresponding positions in FXR and are most likely responsible for the cation-π interaction of this receptor with bile acids, as discussed above (Mi et al., 2003). As expected on the basis of the high degree of identity between the two isoforms, the crystal structure of LXRβ LBD complexed with the natural oxysterol 24(S),25-epoxycholesterol provided similar results. In particular Williams and colleagues showed that the sterol bound with the A ring oriented toward helix 1 and with the D-ring and epoxide tail oriented toward the C-terminal end of helix 10 (Williams et al., 2003). In particular, the epoxide oxygen atom, although adjacent to Trp457 (corresponding to Trp443 in the α isoform), actually makes its hydrogen bond with the imidazole ring of His435 (corresponding to His421 in the α isoform), in contrast with the model proposed by Spencer et al. (2001), discussed above. Histidine is unique among the naturally occurring amino acids in that it is able to function as either a hydrogen bond donor or acceptors by changing tautomers. Therefore His435 can donate a hydrogen bond to neutral oxysterols (i.e. 24(S),25-epoxycholesterol) whereas it may act as an acceptor with acidic ligands. The hydrogen bond present within the side chain of 24(S),25-epoxycholesterol affects the orientation adopted by the His435 imidazole, which is in turn crucial for the electrostatic interaction (cation-π) between the electropositive nuclei of His435 imidazole and the electronegative π-cloud of Trp457 side chain. This structural feature would explain the mechanism through which oxysterols hold the AF-2 helix of LXRβ in its active conformation. The importance of this mechanism in the recruitment of coactivator proteins and, ultimately, in the transactivation potential of the bound nuclear receptor, is underscored by the fact that the synthetic ligand T0901317, through it's acidic hydroxyl group, makes a shorter hydrogen bond with His435, thus leading to stronger electrostatic interactions with Trp457 (Williams et al., 2003) and exhibiting higher ability to recruit cofactors in functional assays, in comparison to natural oxysterols (Albers et al., 2006). Since the amino acids that line the ligand binding pocket are conserved in LXRα, the mechanism of ligand activation is almost certainly identical for the two isoforms, making the identification of α/β selective LXR ligands difficult.

Despite the large amount of evidence, there is no complete consensus that oxysterols may act in vivo as the true LXR endogenous ligands, owing to the fact that transgenic mouse models with markedly reduced or increased concentration of some specific oxysterols do not seem to present marked disturbances in cholesterol turnover and homeostasis (Bjorkhem, 2009); and, indeed, the attempt to demonstrate the relevance of endogenous oxysterols as LXR ligands by using a triple knock-out model deficient in sterol 27-hydroxylase, cholesterol 24-hydroxylase and cholesterol 25-hydroxylase, did not provide a clear-cut phenotype (Chen et al., 2007). On the other hand, overexpression of a sulfotransferase that specifically transfers a sulphate group to oxysterols that become inactive as LXR ligands, prevents the induction of LXR target genes in response to dietary cholesterol (Chen et al., 2007). Therefore, in our opinion it is likely that oxysterols might truly act as LXR ligands; however, the possibility that other endogenous metabolites may activate LXR should not be excluded.

In a recent paper Mitro and colleagues provided sound evidence that LXRs can be bound and transactivated by glucose, at concentrations that can be normally found in the liver after a meal (Mitro et al., 2007b). Data obtained in functional assays support the idea that glucose directly interacts with the receptors and promotes the recruitment of cofactors. The crystal structure of LXR LBD complexed with glucose is not yet available, however, the fact that addition of a synthetic LXR ligand to cultured cells potentiates glucose effects on gene expression, suggests that the receptor can be simultaneously activated by both ligands (Mitro et al., 2007b). This discovery is extremely important in order to fully appreciate the biological roles of LXRs. By sensing sterols, LXRs contribute to maintain cholesterol homeostasis, promoting its efflux from cells through the ATP-binding cassette transporters and apolipoprotein E (Tontonoz and Mangelsdorf, 2003). Furthermore, LXR activation also suppresses low-density lipoprotein (LDL) uptake through a newly identified mechanism by which LDL receptor (LDLR) is targeted to degradation by ubiquitination of its cytoplasmic domain carried out by an E3 ubiquitin ligase named Idol (Inducible Degrader of the LDLR), transcriptionally induced by LXR (Zelcer et al., 2009). On the other hand, by sensing glucose levels, LXRs participate in the regulation of both lipogenesis and carbohydrate metabolism.

4.6 Interactions Between Sterols and Sterol-Sensing Proteins Dictate Their Fate Toward Retention in the Endoplasmic Reticulum

Synthesis of cholesterol through the mevalonate pathway and its uptake via the low density lipoprotein (LDL) receptor pathway are finely tuned to meet the needs of cells. Cholesterol and oxysterols suppress cholesterol synthesis and uptake through multiple and complex mechanisms, whose existence has long been known (Brown and Goldstein, 1980, 2009).

In 1993, J. Goldstein and M. Brown reported the purification and characterization of a nuclear protein named sterol regulatory element binding protein (SREBP) due to its ability to bind a sequence of DNA mediating the effects of sterols on gene transcription (SRE), present in the LDL receptor promoter (Briggs et al., 1993; Wang et al., 1993). Since that discovery, Goldstein and Brown have worked extensively to dissect the multiple aspects of the SREBP pathway, transport from ER to the nucleus through the Golgi compartment, proteolytic activation, component of the sterol-sensing machinery. Briefly, SREBP proteins are oriented in ER membranes in a hairpin fashion with both the N-terminal domain and the C-terminal regulatory domain of the transcription factor facing the cytosol. Immediately after their synthesis on ER membranes, the SREBPs bind to SREBP cleavage activating protein (Scap) through an interaction between the C-terminal regulatory domain of the SREBP and the cytosolically-oriented C-terminal WD-repeat domain of Scap. Scap is embedded in ER membranes through its N-terminal domain. In sterol-depleted cells, the Scap/SREBP complex exits the ER in COPII-coated vesicles that bud

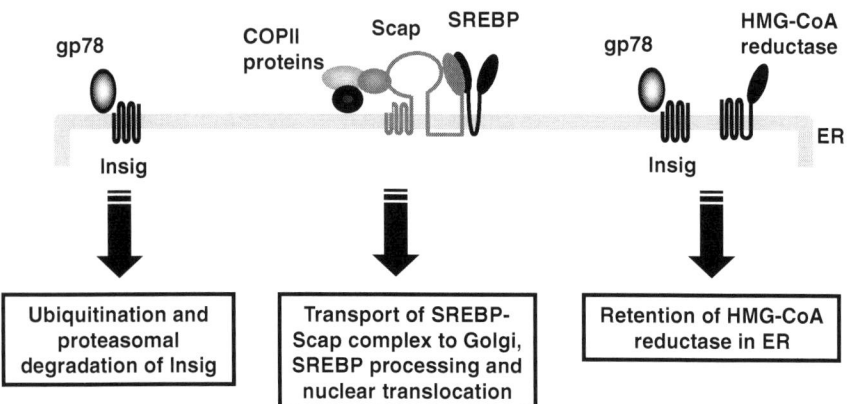

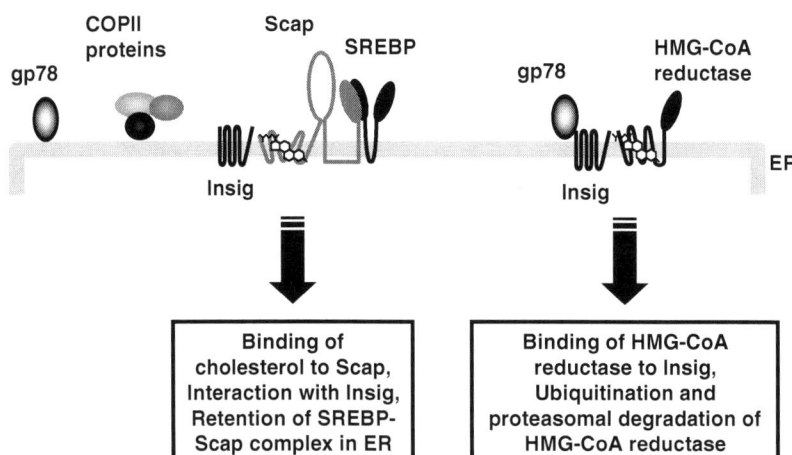

Fig. 4.4 (continued)

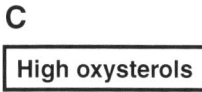

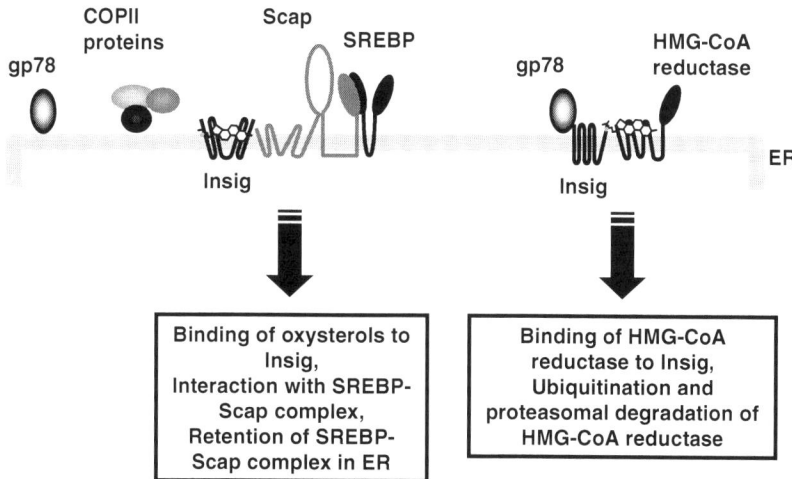

Fig. 4.4 Interactions between cholesterol and/or oxysterols ER proteins underlying the regulation of cholesterol homeostasis. **A** In the ER of cholesterol-deprived cells, the interaction between SREBP and Scap results in a conformational switch allowing the interaction with COPII proteins through the MELADL sequence on Scap. This association promotes the budding of vesicles and transport of the SREBP-Scap complex to Golgi where SREBP is processed to a mature form that can translocate to the nucleus and bind DNA at target promoters. Under these conditions, ER residing Insig proteins become accessible to the action of gp78 ubiquitin ligase, and undergo proteasomal degradation. HMG-CoA reductase is retained in the ER as well. In sterol-loaded cells, cholesterol interacts with a transmembrane region of Scap causing a conformational change that allows the association to Insig (**B**). This association is also promoted by interaction of Insig with oxysterols (**C**). Sterol-induced interaction of Scap with Insig blocks the transport of SREBP to Golgi and the downstream effects on gene transcription. The interaction between HMG-CoA reductase and cholesterol precursors (**B**) or oxysterols (**C**) promotes binding of Insig. The ubiquitin ligase gp78 associated to Insig acts upon HMG-CoA reductase directing it toward proteasomal degradation

from ER membranes (Goldstein et al., 2006) (Fig. 4.4A). Cholesterol and oxysterols inhibit SREBP processing by blocking the Scap-mediated transport of SREBP to the Golgi (Nohturfft et al., 2000). Hence Scap represents the central "sensor" through which the regulation of the proteolytic activation of SREBP occurs. The block of ER-Golgi transport of SREBP is, in turn, the consequence of sterol-induced interaction of Scap with other ER anchor proteins, Insig-1 and Insig-2 (Adams et al., 2003, 2004) (Figs. 4.4B,C). Notably, cholesterol and oxysterols induce the Scap-Insig interactions through different mechanisms that were investigated in great detail by the group of Goldstein and Brown. According to their results, Scap interacts with cholesterol through a "sterol-sensing" domain present in the transmembrane portion of the protein (Hua et al., 1996; Nohturfft et al., 1998); it specifically recognizes the 4-ring system and the 3β-hydroxyl group of cholesterol, independent of the presence of the side chain, but it does not bind a sterol containing a hydroxyl group in the side chain (Radhakrishnan et al., 2007). Based on these observations the authors suggested that Scap recognizes cholesterol in its usual orientation in the membrane, i.e. when its 3β-hydroxyl group is exposed at the surface and its side chain is buried in the hydrophobic bilayer. Hydroxylation of the side chain would most likely prevent such membrane insertion, thus making impossible the interaction with the sterol-sensing domain of Scap. It was then demonstrated that the sequence involved in the formation of the Scap-Insig complex is a tetrapeptide motif YIYF, located in the transmembrane segment of Scap (Yabe et al., 2002; Yang et al., 2002). On the contrary, Insig-2 binding has an absolute requirement for a sterol side chain with a hydroxyl group, that can be located in the side chain at position 22, 24, 25 or 27, thus suggesting that it recognizes a sterol that lies on the cytosolic surface of the membrane with its hydroxylated side chain exposed (Radhakrishnan et al., 2007). Either the hydroxyl group orients the oxysterol in relation to the membrane lipid bilayer so that Insig-2 can recognize it, or Insig-2 interacts indirectly with the hydroxyl group of the oxysterol through water-mediated interaction, as reported for the yeast oxysterol-binding protein Osh4 (Im et al., 2005).

Hence, the binding of cholesterol to Scap and of oxysterols to Insig have a common result, i.e. the triggering of Scap-Insig interaction that results in a conformation change of Scap. The following question can be posed: by which means does this conformational change block the exit of the SREBP-Scap complex from the ER?

First of all, the characterization of the type of vesicles containing SREBP-Scap, the dynamics of the process and the effect of oxysterols was investigated in great detail using isolated microsomal membranes, but also confirmed in living cells (Sun et al., 2005; Nohturfft et al., 2000; Espenshade et al., 2002). Taken together, the results obtained indicate that the SREBP-Scap complex is carried to the Golgi by means of COPII-coated vesicles (Espenshade et al., 2002), through the interaction of Scap with Sec24, one of the five proteins that, in complex with Sec23 (Sec23/24 heterodimer), form the COPII coat (Sun et al., 2005), and that sterols block the budding of the Scap-containing vesicles from the ER (Nohturfft et al., 2000). In particular, the binding site of Scap for Sec24 is presented as a hexapeptide sequence, Met-Glu-Leu-Ala-Asp-Leu (MELADL), located in the cytoplasmic loop between transmembrane α helices 6 and 7 (Sun et al., 2005). Cholesterol and oxysterols prevent the interaction between the MELADL sequence of Scap and Sec24 (Sun et al., 2005), thus ultimately blocking SREBP export from ER and cleavage. To further dissect the mechanism through which sterols affect this interaction, Sun and colleagues provided evidence suggesting that the binding of Insig to Scap induces a conformational change that moves the cytoplasmatic MELADL sequence of Scap closer to the membrane, sequestering it from Sec24 and therefore out of reach of the COPII proteins (Sun et al., 2007). In particular, mutational analysis revealed that the distance between the MELADL sequence of Scap and the membrane, rather than its absolute structure, is crucial for the binding of COPII proteins. Putting all these findings together, the model of sterol-mediated feedback of cholesterol synthesis and uptake via SREBP can be summarized as follows: the regulatory action of cholesterol and oxysterols initiate by binding to intracellular receptors, Scap for cholesterol, and Insig for oxysterols. Thereafter, their actions converge since both ligands cause Insig to bind to Scap, and this produces a single conformational change that switches the MELADL sorting signal in Scap to a new location with respect to the ER membrane, thereby precluding COPII protein binding.

4.7 Interactions Between Sterols and Sterol-Sensing Proteins Dictate their Fate Toward Degradation

The links between Insig proteins and the Scap-SREBP pathway are indeed more complex than those described in the preceding section. In fact, sterol deprivation, by causing the transfer of the Scap-SREBP complex to the Golgi, the proteolytic activation of SREBP and its nuclear translocation, ultimately results in activation of SREBP-regulated genes that include Insig-1 (Yang et al., 2002; Horton et al., 2002). On the contrary, Insig-2 is constitutively expressed. The levels of Insig-1 are also regulated at the level of protein stability since, in the absence of sterols, it dissociates from the Scap-SREBP complex, is then ubiquitinated on lysine residues, and rapidly degraded in proteasomes (Gong et al., 2006). Sterol-mediated reassociation of Insig-1 to the Scap-SREBP complex prevents its ubiquitination. Based on the discovery of gp78 as an Insig-1-associated protein (Song et al., 2005b), Lee et al. (2006) described in detail the mechanism of Insig-1 ubiquitination and the role of cholesterol in this process. In sterol-depleted cells, Insig-1 binds to a fraction of gp78, which transfers ubiquitin to Insig-1, targeting it for proteasomal degradation (Fig. 4.4A). In the presence of sterols, Scap binds to Insig-1 in a reaction that displaces gp78, thus preventing ubiquitination of Insig-1, which results in the stabilization of the protein (Fig. 4.4B).

Based on these findings, the sterol-modulated production and degradation of Insig-1 represents a "convergent feedback inhibition" mechanism that adapts the rate of cholesterol synthesis and uptake to the cellular needs. In summary, the production of new Insig-1 protein, induced by sterol-deprivation, must be followed by the availability of newly synthesized cholesterol in order to trigger the conformational change of Scap necessary to retain SREBP in the ER and inhibit its processing. This convergence ensures that SREBP processing will not be terminated before the cholesterol needs of the cell are met (Gong et al., 2006).

Protein stability is an important level of regulation of HMG-CoA reductase and it was shown that both sterol and non-sterol end-products of mevalonate metabolism contribute to accelerate degradation of the enzyme through a mechanism mediated by the ubiquitin-proteasome pathway (McGee et al., 1996; Ravid et al., 2000; Roitelman and Simoni, 1992). In all mammalian species investigated so far HMG-CoA reductase localizes to the ER membrane and consists of two domains: the cytoplasmic C-terminal domain, containing the catalytic domain, and the N-terminal domain that is integrated into membranes by virtue of eight membrane-spanning segments. Early studies had shown that expression of the truncated, cytosolic C-terminal domain of reductase produced a stable, catalytically active protein whose degradation was not influenced by sterols (Gil et al., 1985). Whereas a fusion protein containing the membrane domain of HMG-CoA reductase exhibited sterol-accelerated degradation, similar to the full length reductase (Skalnik et al., 1988). The HMG-CoA reductase protein, as well as Scap and other polytopic membrane proteins, contains a sequence of transmembrane α helices defined as the sterol-sensing domain (Kuwabara and Labouesse, 2002). It was then demonstrated that ubiquitination and degradation of HMG-CoA reductase was accelerated by the sterol-induced binding of its sterol-sensing domain to Insig-1 (Sever et al., 2003b) and mutational analysis revealed that the binding of HMG-CoA reductase to Insig-1 strictly depends on a YIYF sequence, located in the transmembrane segment of HMG-CoA reductase (Sever et al., 2003a), and also found in the transmembrane segment of Scap, as mentioned above. However, in contrast with Scap, that interacts selectively with cholesterol, the ubiquitination of the HMG-CoA reductase is stimulated by sterols others than cholesterol. In fact, HMG-CoA reductase ubiquitination in permeabilized cells is promoted by oxysterols, 25-hydroxycholesterol, 5-cholesten-3β,16β,27-triol, 24(S)-hydroxycholesterol, 27-hydroxycholesterol, 24(S),25-epoxycholesterol, and 19-hydroxycholesterol, at concentrations in the micromolar range (Song and DeBose-Boyd, 2004). Further studies demonstrated that also lanosterol and its metabolite 24,25-dihydrolanosterol, two intermediates in cholesterol biosynthesis, stimulate the ubiquitination and degradation of HMG-CoA reductase, although less potently than oxysterols (Song et al., 2005a). Experiments with isolated ER membranes incubated with sterols indicated that the mechanism of action is direct. The analysis of the structural requirements revealed that the 4,4-dimethyl moiety of lanosterol is most likely the major determinant, while the 3β-hydroxyl group and the C14 methyl group seems to play a secondary role in recognition (Song et al., 2005a). Therefore, the interaction of HMG-CoA reductase with sterols represents the indispensable trigger for the formation of the complex with Insig proteins and subsequent ubiquitination. As discussed above, ubiquitination of Insig proteins is made possible because they associate in ER membranes with the ubiquitin ligase gp78. Song et al. (2005b) investigated in great detail the interactions among all these ER proteins, and the effect of sterols on these interactions, by coupling affinity purification with tandem mass spectrometry. As mentioned above, they found that gp78 is an Insig-1-associated protein, that Insig-1 binds to the membrane domain of gp78 in the absence or presence of sterols, and moreover, that upon the addition of sterols, HMG-CoA reductase is recruited to the complex (Figs. 4.4B,C). Hence, the formation of this complex represents a key step that commits HMG-CoA reductase toward dislodgement from the ER and proteasomal degradation.

4.8 Summary

In the above sections we have described in detail the molecular interactions between individual proteins and cholesterol related molecules. However, these interactions should also be considered from a more general point of view, to appreciate the contributions of different pathways to a regulatory network, to highlight common mechanisms, and to identify effector molecules to which multiple pathways converge for the maintenance of homeostasis.

Cholesterol uptake: In most cells cholesterol uptake is mediated by the LDL receptor. In cholesterol-deprived cells the production of the receptor is transcriptionally induced by nuclear SREBP. Under these conditions, ER residing Insig proteins can no longer interact with Scap, become accessible to the action of the associated gp78 ubiquitin ligase, and undergo proteasomal degradation. In apparent contradiction, new Insig-1 protein molecules are produced under the transcriptional control of SREBP to attenuate the action of SREBP itself, once intracellular levels of cholesterol are restored. On the other hand, in sterol-loaded cells, cholesterol itself, by interacting directly with Scap, and side chain oxysterols, and by interacting directly with Insig, cause the retention of SREBP in the ER, thus reducing the transcription of LDL receptor gene. At the same time, side chain oxysterols, by activating LXRs, induce the expression of the ubiquitin ligase Idol and the proteasomal-mediated degradation of existing LDL receptor molecules.

Cholesterol synthesis: The rate limiting step in cholesterol synthesis is the reaction catalyzed by HMG-CoA reductase. It is regulated at the level of gene transcription by SREBP in a sterol-modulated fashion, similarly to the LDL receptor. In addition, existing HMG-CoA reductase is degraded via the proteasome with the involvement of Insig proteins and gp78, and the molecular trigger is represented by the direct interactions between its membrane domain and cholesterol, or closely related metabolites (side chain oxysterols and the precursor lanosterol).

Cholesterol transport: Side chain oxysterols, by activating LXRs, induce the expression of genes encoding cholesterol transporters and acceptors, thus favouring its efflux and protecting cells and tissues against abnormal cholesterol accumulation.

Bile acid synthesis: In mammals, one of the main pathways to eliminate cholesterol from the body is its conversion to bile acids. The key enzymes in this pathway, cholesterol 7α-hydroxylase and sterol 27-hydroxylase, are atypical P450 enzymes inasmuch as they exhibit stringent requirements for substrate molecules. This feature may be linked to the fact that cholesterol is the most abundant sterol in mammals. A major consequence of this substrate-specificity is the production of primary bile acids with a typical overall shape and the presence of hydroxyl groups at conserved key positions. Due to their features, primary bile acids, in particular chenodeoxycholic acid, can specifically interact and activate the nuclear receptor FXR, triggering a signalling cascade that has important effects on bile acid homeostasis: feedback downregulation of their synthesis, upregulation of bile acid transporters to promote their excretion.

Energy metabolism: Bile acid-dependent activation of FXR in the liver profoundly affects glucose metabolism. Furthermore, through the membrane receptor

TGR5, bile acids, in particular secondary bile acids, regulate energy metabolism in extra-hepatic tissues promoting energy expenditure. On the other hand, activation of LXR, results in increased hepatic synthesis of fatty acids, at least in rodents.

4.9 Conclusions

Cholesterol, its hydroxylated derivatives, oxysterols, and bile acids, are much more than intermediates or end products of metabolic pathways; they are indeed multi-tasking signalling molecules that, by interacting with different types of protein targets, exhibit regulatory functions on cholesterol metabolism and homeostasis, and beyond. Given the relevance of these regulatory mechanisms for understanding both the physiological chemistry and the bases of various diseases (hypercholesterolemia, cardiovascular disease, metabolic syndrome, diabetes), some issues should be carefully addressed.

Various distinct regulatory networks can be modulated by the same class of molecules, thus showing that chemical entities, similar for the core structure, but different for physico-chemical properties, also differ in their ability to interact with protein partners. For example, 25-hydrocholesterol is the most potent inhibitor of SREBP processing, but is one of the weakest oxysterol ligands of LXR. Therefore, we need to gain further insights into the structure-activity relationships of these molecules. Secondly, while there is no doubt about the role of primary bile acids in the physiological activation of FXR in liver and intestine, observations in several cell types suggest that different oxysterols may be produced in a cell and tissue-specific fashion, with different downstream effects on gene transcription. For example, 24-hydroxycholesterol is typically found in cells of the nervous system expressing high levels of cholesterol 24-hydroxylase, whereas in macrophages that express sterol 27-hydroxylase, 27-hydroxycholesterol is the physiological ligand of LXR.

Finally, by the use of global approaches, library screenings and "omics" technologies, it is likely that new partners and new functions for both sterols and bile acids will be discovered.

References

Adachi, R., Shulman, A. I., Yamamoto, K., Shimomura, I., Yamada, S., Mangelsdorf, D. J. and Makishima, M., 2004, Structural determinants for vitamin D receptor response to endocrine and xenobiotic signals. *Mol. Endocrinol.*, **18:** 43–52.

Adams, C. M., Goldstein, J. L. and Brown, M. S., 2003, Cholesterol-induced conformational change in SCAP enhanced by Insig proteins and mimicked by cationic amphiphiles. *Proc. Natl. Acad. Sci. USA*, **100:** 10647–10652.

Adams, C. M., Reitz, J., De Brabander, J. K., Feramisco, J. D., Li, L., Brown, M. S. and Goldstein, J. L., 2004, Cholesterol and 25-hydroxycholesterol inhibit activation of SREBPs by different mechanisms, both involving SCAP and Insigs. *J. Biol. Chem.*, **279:** 52772–52780.

Albers, M., Blume, B., Schlueter, T., Wright, M. B., Kober, I., Kremoser, C., Deuschle, U. and Koegl, M., 2006, A novel principle for partial agonism of liver X receptor ligands. Competitive recruitment of activators and repressors. *J. Biol. Chem.*, **281:** 4920–4930.

Alberti, S., Steffensen, K. R. and Gustafsson, J. A., 2000, Structural characterisation of the mouse nuclear oxysterol receptor genes LXRalpha and LXRbeta. *Gene,* **243:** 93–103.

Bergstrom, S., 1958, The formation and metabolism of bile acids under different conditions. In: Pincus, G. (Ed.) *Hormones and Atherosclerosis.* Brighton, Utah, Academic Press, New York.

Bjorkhem, I., 2009, Are side-chain oxidized oxysterols regulators also in vivo? *J. Lipid Res.,* **50 Suppl:** S213–218.

Briggs, M. R., Yokoyama, C., Wang, X., Brown, M. S. and Goldstein, J. L., 1993, Nuclear protein that binds sterol regulatory element of low density lipoprotein receptor promoter. I. Identification of the protein and delineation of its target nucleotide sequence. *J. Biol. Chem.,* **268:** 14490–14496.

Brown, M. S. and Goldstein, J. L., 1980, Multivalent feedback regulation of HMG CoA reductase, a control mechanism coordinating isoprenoid synthesis and cell growth. *J. Lipid Res.,* **21:** 505–517.

Brown, M. S. and Goldstein, J. L., 2009, Cholesterol feedback: from Schoenheimer's bottle to Scap's MELADL. *J. Lipid Res.,* **50 Suppl:** S15–S27.

Cali, J. J., Hsieh, C. L., Francke, U. and Russell, D. W., 1991 Mutations in the bile acid biosynthetic enzyme sterol 27-hydroxylase underlie cerebrotendinous xanthomatosis. *J. Biol. Chem.,* **266:** 7779–7783.

Chawla, A., Repa, J. J., Evans, R. M. and Mangelsdorf, D. J., 2001, Nuclear receptors and lipid physiology: opening the X-files. *Science,* **294:** 1866–1870.

Chen, J. D. and Evans, R. M., 1995, A transcriptional co-repressor that interacts with nuclear hormone receptors. *Nature,* **377:** 454–457.

Chen, W., Chen, G., Head, D. L., Mangelsdorf, D. J. and Russell, D. W., 2007, Enzymatic reduction of oxysterols impairs LXR signaling in cultured cells and the livers of mice. *Cell Metab.,* **5:** 73–79.

Chiang, J. Y., 2004, Regulation of bile acid synthesis: pathways, nuclear receptors, and mechanisms. *J. Hepatol.,* **40:** 539–551.

Choi, M., Yamamoto, K., Itoh, T., Makishima, M., Mangelsdorf, D. J., Moras, D., Deluca, H. F. and Yamada, S., 2003, Interaction between vitamin D receptor and vitamin D ligands: two-dimensional alanine scanning mutational analysis. *Chem. Biol.,* **10:** 261–270.

De Fabiani, E., Mitro, N., Anzulovich, A. C., Pinelli, A., Galli, G. and Crestani, M., 2001, The negative effects of bile acids and tumor necrosis factor-alpha on the transcription of cholesterol 7alpha-hydroxylase gene (CYP7A1) converge to hepatic nuclear factor-4. A novel mechanism of feedback regulation of bile acid synthesis mediated by nuclear receptors. *J. Biol. Chem.,* **276:** 30708–30716.

Downes, M., Verdecia, M. A., Roecker, A. J., Hughes, R., Hogenesch, J. B., Kast-Woelbern, H. R., Bowman, M. E., Ferrer, J. L., Anisfeld, A. M., Edwards, P. A., Rosenfeld, J. M., Alvarez, J. G., Noel, J. P., Nicolaou, K. C. and Evans, R. M., 2003, A chemical, genetic, and structural analysis of the nuclear bile acid receptor FXR. *Mol. Cell,* **11:** 1079–1092.

Espenshade, P. J., Li, W. P. and Yabe, D., 2002, Sterols block binding of COPII proteins to SCAP, thereby controlling SCAP sorting in ER. *Proc. Natl. Acad. Sci. USA,* **99:** 11694–11699.

Farnegardh, M., Bonn, T., Sun, S., Ljunggren, J., Ahola, H., Wilhelmsson, A., Gustafsson, J. A. and Carlquist, M., 2003, The three-dimensional structure of the liver X receptor beta reveals a flexible ligand-binding pocket that can accommodate fundamentally different ligands. *J. Biol. Chem.,* **278:** 38821–38828.

Federico, A. and Dotti, M. T., 2001, Cerebrotendinous xanthomatosis. *Neurology,* **57:** 1743.

Forman, B. M., Ruan, B., Chen, J., Schroepfer, G. J., Jr. and Evans, R. M., 1997, The orphan nuclear receptor LXRalpha is positively and negatively regulated by distinct products of mevalonate metabolism. *Proc. Natl. Acad. Sci. USA,* **94:** 10588–10593.

Gil, G., Faust, J. R., Chin, D. J., Goldstein, J. L. and Brown, M. S., 1985, Membrane-bound domain of HMG CoA reductase is required for sterol-enhanced degradation of the enzyme. *Cell,* **41:** 249–258.

Goldstein, J. L. and Brown, M. S., 1990, Regulation of the mevalonate pathway. *Nature,* **343:** 425–430.

Goldstein, J. L., Debose-Boyd, R. A. and Brown, M. S., 2006, Protein sensors for membrane sterols. *Cell,* **124:** 35–46.

Gong, Y., Lee, J. N., Lee, P. C., Goldstein, J. L., Brown, M. S. and Ye, J., 2006, Sterol-regulated ubiquitination and degradation of Insig-1 creates a convergent mechanism for feedback control of cholesterol synthesis and uptake. *Cell Metab.,* **3:** 15–24.

Horlein, A. J., Naar, A. M., Heinzel, T., Torchia, J., Gloss, B., Kurokawa, R., Ryan, A., Kamei, Y., Soderstrom, M., Glass, C. K. et al., 1995, Ligand-independent repression by the thyroid hormone receptor mediated by a nuclear receptor co-repressor. *Nature,* **377:** 397–404.

Horton, J. D., Goldstein, J. L. and Brown, M. S., 2002, SREBPs: activators of the complete program of cholesterol and fatty acid synthesis in the liver. *J. Clin. Inv.,* **109:** 1125–1131.

Hua, X., Nohturfft, A., Goldstein, J. L. and Brown, M. S., 1996, Sterol resistance in CHO cells traced to point mutation in SREBP cleavage-activating protein. *Cell,* **87:** 415–426.

Hylemon, P. B., Zhou, H., Pandak, W. M., Ren, S., Gil, G. and Dent, P., 2009, Bile acids as regulatory molecules. *J. Lipid Res.* **50:** 1509–1520.

Im, Y. J., Raychaudhuri, S., Prinz, W. A. and Hurley, J. H., 2005, Structural mechanism for sterol sensing and transport by OSBP-related proteins. *Nature,* **437:** 154–158.

Janowski, B. A., Willy, P. J., Devi, T. R., Falck, J. R. and Mangelsdorf, D. J., 1996, An oxysterol signalling pathway mediated by the nuclear receptor LXR alpha. *Nature,* **383:** 728–731.

Kawamata, Y., Fujii, R., Hosoya, M., Harada, M., Yoshida, H., Miwa, M., Fukusumi, S., Habata, Y., Itoh, T., Shintani, Y., Hinuma, S., Fujisawa, Y. and Fujino, M., 2003, A G protein-coupled receptor responsive to bile acids. *J. Biol. Chem.,* **278:** 9435–9440.

Kuwabara, P. E. and Labouesse, M., 2002, The sterol-sensing domain: multiple families, a unique role? *Trends Genet.,* **18:** 193–201.

Lathe, R., 2002 Steroid and sterol 7-hydroxylation: ancient pathways. *Steroids,* **67:** 967–977.

Lee, J. N., Song, B., Debose-Boyd, R. A. and Ye, J., 2006, Sterol-regulated degradation of Insig-1 mediated by the membrane-bound ubiquitin ligase gp78. *J. Biol. Chem.,* **281:** 39308–39315.

Lefebvre, P., Cariou, B., Lien, F., Kuipers, F. and Staels, B., 2009 Role of bile acids and bile acid receptors in metabolic regulation. *Physiol. Rev.,* **89:** 147–191.

Lehmann, J. M., Kliewer, S. A., Moore, L. B., Smith-Oliver, T. A., Oliver, B. B., Su, J. L., Sundseth, S. S., Winegar, D. A., Blanchard, D. E., Spencer, T. A. and Willson, T. M., 1997, Activation of the nuclear receptor LXR by oxysterols defines a new hormone response pathway. *J. Biol. Chem.,* **272:** 3137–3140.

Macchiarulo, A. and Pellicciari, R., 2007, Exploring the other side of biologically relevant chemical space: insights into carboxylic, sulfonic and phosphonic acid bioisosteric relationships. *J. Mol. Graph. Model,* **26:** 728–739.

Makishima, M., Lu, T. T., Xie, W., Whitfield, G. K., Domoto, H., Evans, R. M., Haussler, M. R. and Mangelsdorf, D. J., 2002, Vitamin D receptor as an intestinal bile acid sensor. *Science,* **296:** 1313–1316.

Makishima, M., Okamoto, A. Y., Repa, J. J., Tu, H., Learned, R. M., Luk, A., Hull, M. V., Lustig, K. D., Mangelsdorf, D. J. and Shan, B., 1999 Identification of a nuclear receptor for bile acids. *Science,* **284:** 1362–1365.

Maruyama, T., Miyamoto, Y., Nakamura, T., Tamai, Y., Okada, H., Sugiyama, E., Nakamura, T., Itadani, H. and Tanaka, K., 2002, Identification of membrane-type receptor for bile acids (M-BAR). *Biochem. Biophys. Res. Commun.,* **298:** 714–719.

Mast, N., Andersson, U., Nakayama, K., Bjorkhem, I. and Pikuleva, I. A., 2004, Expression of human cytochrome P450 46A1 in Escherichia coli: effects of N- and C-terminal modifications. *Arch. Biochem. Biophys.,* **428:** 99–108.

Mast, N., Graham, S. E., Andersson, U., Bjorkhem, I., Hill, C., Peterson, J. and Pikuleva, I. A., 2005, Cholesterol binding to cytochrome P450 7A1, a key enzyme in bile acid biosynthesis. *Biochemistry,* **44:** 3259–3271.

Mast, N., Murtazina, D., Liu, H., Graham, S. E., Bjorkhem, I., Halpert, J. R., Peterson, J. and Pikuleva, I. A., 2006, Distinct binding of cholesterol and 5beta-cholestane-3alpha,7alpha,12alpha-triol to cytochrome P450 27A1: evidence from modeling and site-directed mutagenesis studies. *Biochemistry,* **45:** 4396–4404.

Mcgee, T. P., Cheng, H. H., Kumagai, H., Omura, S. and Simoni, R. D., 1996, Degradation of 3-hydroxy-3-methylglutaryl-CoA reductase in endoplasmic reticulum membranes is accelerated as a result of increased susceptibility to proteolysis. *J. Biol. Chem.,* **271:** 25630–25638.

Mi, L. Z., Devarakonda, S., Harp, J. M., Han, Q., Pellicciari, R., Willson, T. M., Khorasanizadeh, S. and Rastinejad, F., 2003, Structural basis for bile acid binding and activation of the nuclear receptor FXR. *Mol. Cell,* **11:** 1093–1100.

Millatt, L. J., Bocher, V., Fruchart, J. C. and Staels, B., 2003 Liver X receptors and the control of cholesterol homeostasis: potential therapeutic targets for the treatment of atherosclerosis. *Biochim. Biophys. Acta,* **1631:** 107–118.

Mitro, N., Godio, C., De Fabiani, E., Scotti, E., Galmozzi, A., Gilardi, F., Caruso, D., Chacon, A. B. and Crestani, M., 2007a, Insights in the regulation of cholesterol 7alpha-hydroxylase gene reveal a target for modulating bile acid synthesis. *Hepatology,* **46:** 885–897.

Mitro, N., Mak, P. A., Vargas, L., Godio, C., Hampton, E., Molteni, V., Kreusch, A. and Saez, E., 2007b, The nuclear receptor LXR is a glucose sensor. *Nature,* **445:** 219–223.

Murtazina, D., Puchkaev, A. V., Schein, C. H., Oezguen, N., Braun, W., Nanavati, A. and Pikuleva, I. A., 2002, Membrane-protein interactions contribute to efficient 27-hydroxylation of cholesterol by mitochondrial cytochrome P450 27A1. *J. Biol. Chem.,* **277:** 37582–37589.

Murtazina, D. A., Andersson, U., Hahn, I. S., Bjorkhem, I., Ansari, G. A. and Pikuleva, I. A., 2004 Phospholipids modify substrate binding and enzyme activity of human cytochrome P450 27A1. *J. Lipid Res.,* **45:** 2345–2353.

Nakayama, K., Puchkaev, A. and Pikuleva, I. A., 2001, Membrane binding and substrate access merge in cytochrome P450 7A1, a key enzyme in degradation of cholesterol. *J. Biol. Chem.,* **276:** 31459–31465.

Nohturfft, A., Brown, M. S. and Goldstein, J. L., 1998, Sterols regulate processing of carbohydrate chains of wild-type SREBP cleavage-activating protein (SCAP), but not sterol-resistant mutants Y298C or D443N. *Proc. Natl. Acad. Sci. USA,* **95:** 12848–12853.

Nohturfft, A., Yabe, D., Goldstein, J. L., Brown, M. S. and Espenshade, P. J., 2000, Regulated step in cholesterol feedback localized to budding of SCAP from ER membranes. *Cell,* **102:** 315–323.

Norlin, M., Andersson, U., Bjorkhem, I. and Wikvall, K., 2000a, Oxysterol 7 alpha-hydroxylase activity by cholesterol 7 alpha-hydroxylase (CYP7A). *J. Biol. Chem.,* **275:** 34046–34053.

Norlin, M., Toll, A., Bjorkhem, I. and Wikvall, K., 2000b, 24-hydroxycholesterol is a substrate for hepatic cholesterol 7alpha-hydroxylase (CYP7A). *J. Lipid Res.,* **41:** 1629–1639.

Ogishima, T., Deguchi, S. and Okuda, K., 1987, Purification and characterization of cholesterol 7 alpha-hydroxylase from rat liver microsomes. *J. Biol. Chem.,* **262:** 7646–7650.

Parks, D. J., Blanchard, S. G., Bledsoe, R. K., Chandra, G., Consler, T. G., Kliewer, S. A., Stimmel, J. B., Willson, T. M., Zavacki, A. M., Moore, D. D. and Lehmann, J. M., 1999, Bile acids: natural ligands for an orphan nuclear receptor. *Science,* **284:** 1365–1368.

Pellicciari, R., Sato, H., Gioiello, A., Costantino, G., Macchiarulo, A., Sadeghpour, B. M., Giorgi, G., Schoonjans, K. and Auwerx, J., 2007, Nongenomic actions of bile acids. Synthesis and preliminary characterization of 23- and 6,23-alkyl-substituted bile acid derivatives as selective modulators for the G-protein coupled receptor TGR5. *J. Med. Chem.,* **50:** 4265–4268.

Radhakrishnan, A., Ikeda, Y., Kwon, H. J., Brown, M. S. and Goldstein, J. L., 2007, Sterol-regulated transport of SREBPs from endoplasmic reticulum to Golgi: oxysterols block transport by binding to Insig. *Proc. Natl. Acad. Sci. USA,* **104:** 6511–6518.

Ravid, T., Doolman, R., Avner, R., Harats, D. and Roitelman, J., 2000, The ubiquitin-proteasome pathway mediates the regulated degradation of mammalian 3-hydroxy-3-methylglutaryl-coenzyme A reductase. *J. Biol. Chem.,* **275:** 35840–35847.

Roitelman, J. and Simoni, R. D., 1992, Distinct sterol and nonsterol signals for the regulated degradation of 3-hydroxy-3-methylglutaryl-CoA reductase. *J. Biol. Chem.,* **267:** 25264–25273.

Russell, D. W., 2003, The enzymes, regulation, and genetics of bile acid synthesis. *Annu. Rev. Biochem.,* **72:** 137–174.

Sato, H., Macchiarulo, A., Thomas, C., Gioiello, A., Une, M., Hofmann, A. F., Saladin, R., Schoonjans, K., Pellicciari, R. and Auwerx, J., 2008, Novel potent and selective bile acid derivatives as TGR5 agonists: biological screening, structure-activity relationships, and molecular modeling studies. *J. Med. Chem.,* **51:** 1831–1841.

Schwarz, M., Lund, E. G. and Russell, D. W., 1998, Two 7 alpha-hydroxylase enzymes in bile acid biosynthesis. *Curr. Opin. Lipidol.,* **9:** 113–118.

Sever, N., Song, B. L., Yabe, D., Goldstein, J. L., Brown, M. S. and Debose-Boyd, R. A., 2003a, Insig-dependent ubiquitination and degradation of mammalian 3-hydroxy-3-methylglutaryl-CoA reductase stimulated by sterols and geranylgeraniol. *J. Biol. Chem.,* **278:** 52479–52490.

Sever, N., Yang, T., Brown, M. S., Goldstein, J. L. and Debose-Boyd, R. A., 2003b, Accelerated degradation of HMG CoA reductase mediated by binding of insig-1 to its sterol-sensing domain. *Mol. Cell,* **11:** 25–33.

Skalnik, D. G., Narita, H., Kent, C. and Simoni, R. D., 1988, The membrane domain of 3-hydroxy-3-methylglutaryl-coenzyme A reductase confers endoplasmic reticulum localization and sterol-regulated degradation onto beta-galactosidase. *J. Biol. Chem.,* **263:** 6836–6841.

Song, B. L. and Debose-Boyd, R. A., 2004, Ubiquitination of 3-hydroxy-3-methylglutaryl-CoA reductase in permeabilized cells mediated by cytosolic E1 and a putative membrane-bound ubiquitin ligase. *J. Biol. Chem.,* **279:** 28798–28806.

Song, B. L., Javitt, N. B. and Debose-Boyd, R. A., 2005a, Insig-mediated degradation of HMG CoA reductase stimulated by lanosterol, an intermediate in the synthesis of cholesterol. *Cell Metab.,* **1:** 179–189.

Song, B. L., Sever, N. and Debose-Boyd, R. A., 2005b, Gp78, a membrane-anchored ubiquitin ligase, associates with Insig-1 and couples sterol-regulated ubiquitination to degradation of HMG CoA reductase. *Mol. Cell,* **19:** 829–840.

Spencer, T. A., Li, D., Russel, J. S., Collins, J. L., Bledsoe, R. K., Consler, T. G., Moore, L. B., Galardi, C. M., Mckee, D. D., Moore, J. T., Watson, M. A., Parks, D. J., Lambert, M. H. and Willson, T. M., 2001, Pharmacophore analysis of the nuclear oxysterol receptor LXRalpha. *J. Med. Chem.,* **44:** 886–897.

Staudinger, J. L., Goodwin, B., Jones, S. A., Hawkins-Brown, D., Mackenzie, K. I., Latour, A., Liu, Y., Klaassen, C. D., Brown, K. K., Reinhard, J., Willson, T. M., Koller, B. H. and Kliewer, S. A., 2001, The nuclear receptor PXR is a lithocholic acid sensor that protects against liver toxicity. *Proc. Natl. Acad. Sci. USA,* **98:** 3369–3374.

Sun, L. P., Li, L., Goldstein, J. L. and Brown, M. S., 2005, Insig required for sterol-mediated inhibition of Scap/SREBP binding to COPII proteins in vitro. *J. Biol. Chem.,* **280:** 26483–26490.

Sun, L. P., Seemann, J., Goldstein, J. L. and Brown, M. S., 2007, Sterol-regulated transport of SREBPs from endoplasmic reticulum to Golgi: Insig renders sorting signal in Scap inaccessible to COPII proteins. *Proc. Natl. Acad. Sci. USA,* **104:** 6519–6526.

Svensson, S., Ostberg, T., Jacobsson, M., Norstrom, C., Stefansson, K., Hallen, D., Johansson, I. C., Zachrisson, K., Ogg, D. and Jendeberg, L., 2003, Crystal structure of the heterodimeric complex of LXRalpha and RXRbeta ligand-binding domains in a fully agonistic conformation. *EMBO J.,* **22:** 4625–4633.

Tontonoz, P. and Mangelsdorf, D. J., 2003, Liver X receptor signaling pathways in cardiovascular disease. *Mol. Endocrinol.,* **17:** 985–993.

Wang, X., Briggs, M. R., Hua, X., Yokoyama, C., Goldstein, J. L. and Brown, M. S., 1993, Nuclear protein that binds sterol regulatory element of low density lipoprotein receptor promoter. II. Purification and characterization. *J. Biol. Chem.,* **268:** 14497–14504.

Watanabe, M., Houten, S. M., Mataki, C., Christoffolete, M. A., Kim, B. W., Sato, H., Messaddeq, N., Harney, J. W., Ezaki, O., Kodama, T., Schoonjans, K., Bianco, A. C. and Auwerx, J., 2006, Bile acids induce energy expenditure by promoting intracellular thyroid hormone activation. *Nature,* **439:** 484–489.

Watkins, R. E., Davis-Searles, P. R., Lambert, M. H. and Redinbo, M. R., 2003a, Coactivator binding promotes the specific interaction between ligand and the pregnane X receptor. *J. Mol. Biol.,* **331:** 815–828.

Watkins, R. E., Maglich, J. M., Moore, L. B., Wisely, G. B., Noble, S. M., Davis-Searles, P. R., Lambert, M. H., Kliewer, S. A. and Redinbo, M. R., 2003b, 2.1 A crystal structure of human PXR in complex with the St. John's wort compound hyperforin. *Biochemistry*, **42:** 1430–1438.

Watkins, R. E., Wisely, G. B., Moore, L. B., Collins, J. L., Lambert, M. H., Williams, S. P., Willson, T. M., Kliewer, S. A. and Redinbo, M. R., 2001, The human nuclear xenobiotic receptor PXR: structural determinants of directed promiscuity. *Science*, **292:** 2329–2333.

Williams, P. A., Cosme, J., Sridhar, V., Johnson, E. F. and Mcree, D. E., 2000, Mammalian microsomal cytochrome P450 monooxygenase: structural adaptations for membrane binding and functional diversity. *Mol. Cell*, **5:** 121–131.

Williams, S., Bledsoe, R. K., Collins, J. L., Boggs, S., Lambert, M. H., Miller, A. B., Moore, J., Mckee, D. D., Moore, L., Nichols, J., Parks, D., Watson, M., Wisely, B. and Willson, T. M., 2003, X-ray crystal structure of the liver X receptor beta ligand binding domain: regulation by a histidine-tryptophan switch. *J. Biol. Chem.*, **278:** 27138–27143.

Xie, W., Radominska-Pandya, A., Shi, Y., Simon, C. M., Nelson, M. C., Ong, E. S., Waxman, D. J. and Evans, R. M., 2001, An essential role for nuclear receptors SXR/PXR in detoxification of cholestatic bile acids. *Proc. Natl. Acad. Sci. USA*, **98:** 3375–3380.

Yabe, D., Brown, M. S. and Goldstein, J. L., 2002, Insig-2, a second endoplasmic reticulum protein that binds SCAP and blocks export of sterol regulatory element-binding proteins. *Proc. Natl. Acad. Sci. USA*, **99:** 12753–12758.

Yang, T., Espenshade, P. J., Wright, M. E., Yabe, D., Gong, Y., Aebersold, R., Goldstein, J. L. and Brown, M. S., 2002, Crucial step in cholesterol homeostasis: sterols promote binding of SCAP to INSIG-1, a membrane protein that facilitates retention of SREBPs in ER. *Cell*, **110:** 489–500.

Zelcer, N., Hong, C., Boyadjian, R. and Tontonoz, P., 2009, LXR regulates cholesterol uptake through Idol-dependent ubiquitination of the LDL receptor. *Science*, **325:** 100–104.

Chapter 5
Cholesterol Oxidase: Structure and Function

Alice Vrielink

Abstract Cholesterol oxidase is a bacterial-specific flavoenzyme that catalyzes the oxidation and isomerisation of steroids containing a 3β hydroxyl group and a double bond at the Δ5–6 of the steroid ring system. The enzyme is a member of a large family of flavin-specific oxidoreductases and is found in two different forms: one where the flavin adenine dinucleotide (FAD) cofactor is covalently linked to the protein and one where the cofactor is non-covalently bound to the protein. These two enzyme forms have been extensively studied in order to gain insight into the mechanism of flavin-mediated oxidation and the relationship between protein structure and enzyme redox potential. More recently the enzyme has been found to play an important role in bacterial pathogenesis and hence further studies are focused on its potential use for future development of novel antibacterial therapeutic agents. In this review the biochemical, structural, kinetic and mechanistic features of the enzyme are discussed.

Keywords Cholesterol oxidase · Flavoenzyme · Enzyme mechanism · Redox catalysis · Oxygen channel · Protein structure

5.1 Introduction

Cholesterol oxidases are secreted bacterial enzymes that catalyze the first step in the degradation of cholesterol. They are flavoenzymes containing the redox cofactor, flavin adenine dinucleotide (FAD). The enzyme catalyzes three chemical conversions (*see* Fig. 5.1). In the first step, called the reductive half-reaction, the 3β-hydroxyl group of the steroid ring system is oxidized to the corresponding ketone. Key to this conversion is the FAD cofactor, which becomes reduced in the process. In the second step the enzyme catalyzes isomerization of the double bond in the oxidized steroid ring system from the Δ5–6 position to Δ4–5 position, to give

---

A. Vrielink (✉)
School of Biomedical Biomolecular and Chemical Sciences, University of Western Australia, 35 Stirling Highway, Crawley, WA 6009, Australia
e-mail: alice.vrielink@uwa.edu.au

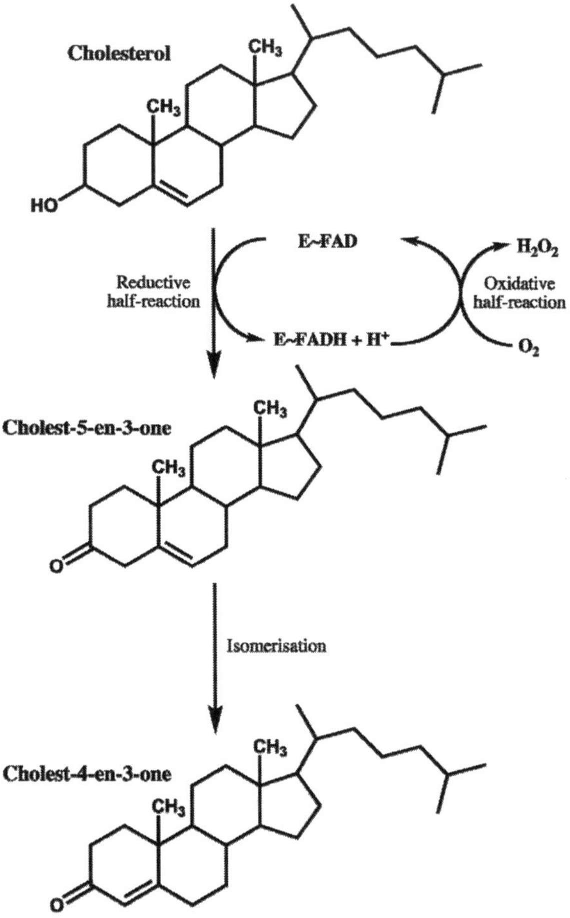

Fig. 5.1 Reaction catalyzed by the enzyme cholesterol oxidase showing the substrate, cholesterol, the product, cholest-4-en-3one and the intermediate formed after oxidation and prior to isomerisation, cholest-5-en-3-one. The cofactor, FAD (flavin adenine dinucleotide) remains bound to the enzyme and is reoxidized in the oxidative half reaction by dioxygen to form hydrogen peroxide

the final steroid product, cholest-4-en-3-one. In the final step of the enzyme, called the oxidative half-reaction, the reduced cofactor reacts with dioxygen and is thus reoxidized while O_2 is reduced to H_2O_2.

5.2 Forms of Cholesterol Oxidase

The enzyme is found only in microorganisms and studies have shown that for many bacteria the enzyme expression can be induced by the presence of cholesterol in the growth medium. Reports in the literature indicate that some bacteria, such as *Mycobacterium*, *Rhodococcus* and *Nocardia* produce an intracellular form of the enzyme that is membrane associated (Buckland et al., 1976; Zajaczkowska et al., 1988; Zajaczkowska and Sedlaczek, 1988; Wilmanska et al., 1995) while the enzyme from *Arthrobacter, Schizopyllum, Streptoverticillium Brevibacterium* and *Streptomyces* is found in the extracellular fraction (Fukuyama and Miyake, 1979; Inouye et al., 1982; Ishizaki et al., 1989; Kamei et al., 1978; Lartillot and Kedziora; 1990, Uwajima et al., 1973; Wilmanska and Sedlaczek, 1988). In addition to the differences in location of the enzyme, two molecular forms are also known and have been extensively characterized.

The majority of known examples of the enzyme contain the FAD cofactor tightly but non-covalently bound to the protein (CO-1). The second form has the cofactor covalently attached to the protein chain via a bond linking the 8-methyl group of the isoalloxazine portion of the FAD to a histidine side chain of the polypeptide chain (CO-2). The amino acid sequences of these two forms of the enzyme differ substantially and indeed their structure, biochemical and kinetic properties are also highly divergent (*see below* for further discussion). There is one reported case of the enzyme containing covalently bound flavin mononucleotide (FMN) as the cofactor, however evidence does not appear to be conclusive (Iwaki et al., 2005).

Cholesterol oxidases are part of a unique class of enzymes that are soluble proteins although they interact with highly insoluble substrates. The natural substrate, cholesterol, is an important membrane component exhibiting low solubility in the aqueous medium of the cell, hence the enzyme must interact with the lipid bilayer in order to for the substrate to partition out of the membrane and undergo oxidation at the enzyme active site. In this regard cholesterol oxidase is an interfacial enzyme as it binds transiently to the membrane surface during catalysis and can only access the substrate from the membrane phase. Other examples of interfacial enzymes include phospholipases (Berg et al., 2001; Gelb et al., 1995; Jain et al., 1995).

Early studies on the substrate specificity by the enzyme from *Nocardia erythropolis* were carried out by Smith and Brooks (Brooks and Smith, 1975, 1980; Smith and Brooks, 1974, 1975, 1977). Since then, further studies on the enzyme from different bacterial sources have added to our understanding of the steroid specificity. In general, the enzyme exhibits a broad specificity for steroid substrates with the main feature being the presence of a hydroxyl group in a β configuration at the C3 center of the steroid A ring and a *trans* A–B ring junction (*see* Fig. 5.1); hydroxyl groups bonded to the A ring in α-linkage are not oxidized by the enzyme. However, the presence of a steroid skeleton is not essential for oxidation; low molecular weight alcohols, including methanol and propan-2-ol, can also be oxidized by the enzyme (albeit that higher concentrations are required, indicating poorer binding affinity for smaller substrates) (Pollegioni et al., 1999). This highlights the notion that cholesterol oxidases may have been adapted from alcohol oxidases to accommodate the large and bulky steroid ring system, while still retaining oxidation activity.

In addition, features needed for isomerization have evolved within the same protein such that the enzyme is able to carry out this added catalytic step. Interestingly, bacterial-specific ketosteroid isomerases are known that carry out the isomerisation step independently of steroid oxidation (Linden and Benisek, 1986; Talalay and Wang, 1955). The reasons for having bifunctional (oxidation and isomerisation reactions are carried out by the same enzyme) as well as monofunctional (when only isomerisation chemistry is carried out) enzymes in bacteria are not well understood.

In eukaryotes, steroid oxidation and isomerisation are important steps in the synthesis of a wide variety of steroid hormones and are carried out by NAD^+ dependent 3β-hydroxysteroid dehydrogenases as membrane-bound proteins located in the endoplasmic reticulum and mitochondria (Cherradi et al., 1993, 1994, 1997; Luu The et al., 1989; Sauer et al., 1994; Thomas et al., 1998). Hence flavin mediated cholesterol oxidation is a process unique to microorganisms.

5.3 Applications of Cholesterol Oxidase

Cholesterol oxidase has a number of important commercial applications. Initial studies on the enzyme focused on its use for the determination of cholesterol in serum, HDL or LDL (*see* review, Smith and Brooks, 1976). Serum cholesterol levels are determined using a three-enzyme assay, including cholesterol esterase, cholesterol oxidase and peroxidase (Richmond, 1973; Allain et al., 1974; Richmond, 1976). More recently electrochemical biosensors with immobilized cholesterol oxidase have been employed to determine cholesterol levels in serum or food (Arya et al., 2007; Basu et al., 2007; Vidal et al., 2004). Determining the serum cholesterol levels is critical for the assessment of a variety of diseases including hypercholesterolemia, coronary heart disease and lipid disorders for estimating the risk of thrombosis and myocardial infarction.

The enzyme has also been used extensively as a probe for studying biological membranes (Bittman et al., 1994; Slotte, 1992a, b, 1995; Ohvo-Rekila et al., 1998; Barenholz et al., 1978; Lange, 1992; Lange et al., 1984, 2007; el Yandouzi and Le Grimellec, 1993) and the role cholesterol–lipid interactions play in the formation of membrane microdomains such as caveolae and lipid rafts (Lange and Steck, 2008; Le Lay et al., 2009; Gimpl and Gehrig-Burger, 2007). These studies have helped in the study of eukaryotic membranes and the function of lipid rafts, which are implicated in many cellular processes such as signal transduction, protein and lipid sorting events and viral budding.

In addition the enzyme has been found to exhibit insecticidal properties against cotton boll weevil larvae (Purcell et al., 1993) as well as tobacco budworms, corn earworms and pink bollworm (Greenplate et al., 1995). Due to these observations, genetic approaches are being used to produce transgenic plants designed to express the enzyme as an effective in situ insecticide (Corbin et al., 1994, 1998). The lethal effect exhibited by the enzyme on larvae has been attributed to the oxidation of cholesterol in the epithelial membrane of the organism's midgut, thus resulting in the physical and functional destruction of the membrane (Purcell et al., 1993). Corbin and colleagues have shown that the enzyme is expressed in transgenic tobacco plants and that insecticidal activity against boll weevil larvae was observed (Corbin et al., 2001). Furthermore they showed that when the enzyme was produced in the cytosol, low levels of saturated sterols were found and the plant exhibited developmental aberrations, whereas if the enzyme was targeted to chloroplasts a larger accumulation of saturated sterols occurred and the plant appeared more normal in terms of development and phenotype.

More recently cholesterol oxidase has been found to play a role in bacterial pathogenesis due to its membrane disrupting activity (Navas et al., 2001). A number of microorganisms, including fast-growing mycobacteria, are able to metabolize cholesterol and use it as a carbon and energy source (Martin, 1977; Sedlaczek, 1988; Pandey and Sassetti, 2008). A gene cluster containing enzymes involved in cholesterol catabolism has been identified in *Mycobacterium sp.* and found to be upregulated during bacterial survival in macrophages (Van der Geize et al., 2007). Indeed these results correlated well with earlier studies that showed cholesterol to be essential for the uptake of mycobacteria by macrophages (Gatfield and Pieters, 2000). Further studies by Brzostek and co-workers on *Mycobacterium tuberculosis* indicate that cholesterol oxidase is essential for the survival of the bacteria during infection and hence plays a role in pathogenesis of *M. tuberculosis* (Brzostek et al., 2007). Similarly, the bacterium, *Rhodococcus equi* was originally identified as a horse pathogen, however, more recently it has been found to be pathogenic in humans, affecting especially immune compromised individuals such as those infected with HIV (Weinstock and Brown, 2002). Like mycobacteria, these bacteria also infect host macrophages through cholesterol oxidase-mediated membrane lysis (Fuhrmann et al., 2002). The uptake of the bacterium in macrophages was correlated to increased survival of the bacteria and oxidation of cholesterol in macrophages. Furthermore, cholesterol oxidation was increased in the presence of the sphingomyelinase-producing bacterium, *Corynebacterium pseudotuberculosis* (Linder and Bernheimer, 1982). The sphingomyelinase activity results in sublytic damage to the membrane, thus allowing cholesterol oxidase to more easily reach its target substrate. The 3-hydroxyl group of cholesterol is thought to mediate sterol-phospholipid interaction and, once oxidized, disrupts the membrane structure (Linder, 1984). Thus the observed virulence activity occurs in a cooperative manner resulting in membrane damage, similarly to what has been reported with hemolysin and phospholipases in *Listeria* (Vazquez-Boland et al., 2001). Importantly, since there is no equivalent enzyme in eukaryotes, the enzyme constitutes a potential target for development of new antibacterial therapeutic agents to treat immune compromised patients.

The enzyme from *Streptomyces natalensis* has been shown to play a key role in the biosynthesis of the polyene macrolide pimaricin, an important antifungal antibiotic used in the food industry to prevent mold contamination (Aparicio et al., 2003, 2004). A cluster of genes coding encoding enzymes involved in pimaricin biosynthesis includes 13 polyketide synthases, incorporated into 5 multifunctional enzymes as well as 12 other proteins that control the modification of the polyketide skeleton, in addition to proteins involved in export and regulation of gene expression (Anton et al., 2007; Mendes et al., 2001, 2005). One of the genes (*pimE*) within this cluster encodes a cholesterol oxidase. Gene inactivation studies showed that the enzyme adversely affected pimaricin biosynthesis, suggesting that it plays a role as a signaling protein in the production of the antifungal agent (Mendes et al., 2007).

The antifungal activity of pimaricin is due to its interaction with membrane sterols thus resulting in membrane disruption and cell damage.

5.4 Redox Properties of Cholesterol Oxidase

Flavoproteins exhibit a unique absorption spectrum due to the presence of the FAD cofactor. This absorption spectrum provides information on the electronic state of the isoalloxazine moiety, the chromophore of the cofactor, and hence, on the redox state and the ionization state of the flavin. The precise details of the spectra are sensitive to the microenvironment of the cofactor and thus give information on the active site region of the enzyme (Ghisla and Massey, 1986). The absorption spectra of both CO-1 and CO-2 are shown in Fig. 5.2 in the oxidized and fully reduced enzyme states. In addition the spectrum of free FAD is shown. Evident from these spectra are the precise location of the two maxima at the near UV (360 nm) and visible (450 nm) wavelength regions for the oxidized forms of the cofactor and the absence of these peaks in the reduced states. The precise locations of the maxima can vary by up to ~20 nm depending on the microenvironment around the isoalloxazine moiety. As the maxima shift towards longer wavelengths (red shift) compared to the free FAD, the flavin moiety within the protein is in a more apolar environment; in

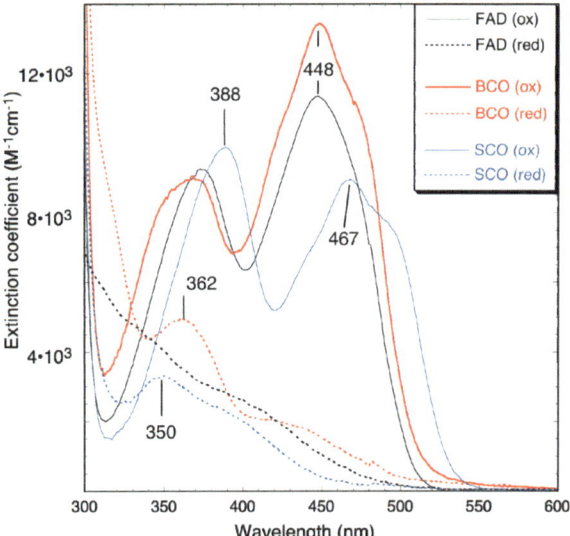

Fig. 5.2 The absorption spectra of cholesterol oxidase containing FAD non-covalently bound to the enzyme, *Streptomyces hygroscopicus*, (SCO) is shown in blue lines and the enzyme containing FAD covalently bound to the enzyme from *Brevibacterium sterolicum* (BCO) is shown in red lines. The spectrum for free FAD is shown in black lines. In each case the spectra include the oxidized (ox) and reduced (ref) forms of the cofactor. Figure reproduced with permission from Vrielink and Ghisla, 2009

contrast when the maxima shift towards shorter wavelengths (blue shift) than free FAD, the flavin is in a more polar environment. The reduced flavin spectra also reveal the electronic state of the reduced cofactor.

Both forms of the enzyme catalyze the dehydrogenation of the steroid substrate via a single 2-electron step resulting in a fully reduced FAD cofactor ($FADH_2$ or $FADH^- + H^+$). The redox properties of the two enzyme forms differ however in terms of their kinetic and thermodynamic behavior. Wild-type CO-2 exhibits a redox potential, $E_m = -101$ mV however a H121A mutant of the enzyme where the covalent linkage between the flavin and the polypeptide chain has been removed exhibits a redox potential $E_m = -204$ mV (Lim et al., 2006). This mutation shows that the isoalloxazine moiety of the flavin cofactor adopts a more planar conformation than in the wild-type enzyme structure suggesting that the flavin ring conformation plays a role in modulating the reduction potential of the cofactor. The redox potentials for two different non-covalent forms of the enzyme have been studied and show some degree of difference with reported values of $E_m = -217$ mV for the enzyme from *Streptomyces hygroscopicus* (Gadda et al., 1997) and $E_m = -131$ mV for *Streptomyces* SA-COO (Chen et al., 2008). In the case of the *Streptomyces* SA-COO cholesterol oxidase, both sequence and structural information can provide insight into the factors that mediate redox potential activity. For the *Streptomyces hygroscopicus* enzyme there is currently no known structure or sequence information, thus the factors that lead to the redox potential are not as clearly understood.

Extensive structural and kinetic work has been carried out for both the covalent form of cholesterol oxidase (from *Brevibacterium sterolicum*) and the non-covalent enzyme (from *Streptomyces* SA-COO and *Rodococcus equi*). These studies have provided extensive insights into the mechanism of catalysis by each form of the enzyme. The structural and mechanistic features will be described further below. The sequences of the covalent and non-covalent forms of the enzyme differ significantly (15% identity between the two sequences) suggesting that the topologies of the enzyme forms will differ as well. Indeed, this was shown to be the case from their high resolution crystal structures (Coulombe et al., 2001; Vrielink et al., 1991; Yue et al., 1999).

5.5 Structure Characterization

5.5.1 Non-covalent Enzyme Structure

The structures of two non-covalent forms of the enzyme from *Streptomyces* SA-COO and *Rhodococcus equi* have been determined by crystallographic methods (Vrielink et al., 1991; Yue et al., 1999). The enzyme comprises two domains: an FAD-binding domain and a substrate-binding domain (*see* Fig. 5.3a). The FAD-binding domain is responsible for anchoring the FAD cofactor in the enzyme; it is composed of a 5-stranded beta pleated sheet, sandwiched between two sets of alpha helices. This domain contains a consensus sequence of glycine residues (GXGXXG)

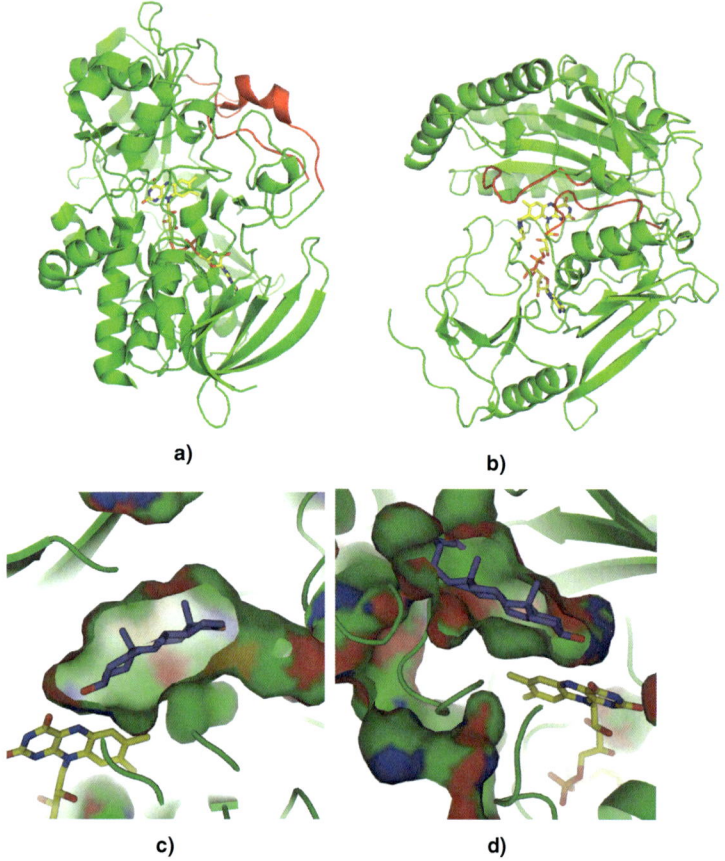

Fig. 5.3 Three dimensional structures of (**a**) cholesterol oxidase from *Streptomyces sp* SA-COO and (**b**) *Brevibacterium sterolicum*. The secondary structure elements are shown and the loop regions important for substrate access to the active site and proposed to be involved in interaction with the membrane where the substrate is located are coloured red. The FAD cofactor is shown in yellow bonds. Close up solvent accessible surface representations of the substrate binding site for (**c**) the enzyme from *Streptomyces sp* SA-COO in complex with the bound substrate, dehydroisoandosterone, (PDB id 1COY) (Li et al., 1993) and (**d**) the enzyme from *Brevibacterium sterolicum* with a cholesterol molecule bound in the active site cavity. The surfaces are computed using a sphere with radius 1.4 Å equivalent to the radius of a water molecule. The electrostatics of the protein have been mapped onto the surface. Red indicates negatively charged regions and blue indicates positively charged regions of the protein. The bound steroid molecules are shown with blue bonds

located at a loop region between the N terminus of the first alpha helix and a beta strand in the structure. This motif of conserved glycine residues has been seen in many nucleotide binding proteins (Eventoff and Rossmann, 1975; Ohlsson et al., 1974) and accommodates a close approach of the negatively charged diphosphate moiety of the cofactor to the protein chain in such as way as to stabilize the negative charge by the positively charged electrostatic field at the amino terminus of the helix. Further interactions are seen between the 2' and 3' hydroxyl groups of the cofactor ribose moiety and a conserved glutamate/aspartate residue located at the C terminal region of the first beta strand and approximately 20 residues away from the glycine-rich loop.

The second domain is called the substrate-binding domain. It contains a 6-stranded beta pleated sheet, forming a "roof" over the substrate-binding cavity with alpha helices on one side of the sheet. The isoalloxazine moiety of the FAD cofactor is positioned on one side of the substrate-binding cavity. A series of loops close over a buried cavity, which is large enough to accommodate the 4 steroid ring structure but not the hydrocarbon moiety at C17 of the steroid D ring. This was confirmed by a structure of an enzyme/steroid complex using dehydroisoandosterone as the substrate (Li et al., 1993) (Fig. 5.3c). The loops that close the substrate-binding cavity from the bulk solvent are proposed to act as the entrance route for cholesterol and play a role in interaction with the steroid C17 hydrocarbon "tail" as well as with the membrane in order to extract the steroid from the bilayer into the buried enzyme active site cavity. Differences in the loops are evident between the two non-covalent structures (*Rhodococcus equi* and *Streptomyces* SA-COO) that are correlated with substrate binding affinity. In the *Streptomyces* SA-COO structure one of the loops contains an amphipathic helical turn; in the *Rhodococcus* structure the loop is an extended coiled structure and exhibits higher temperature factors. The decreased rigidity in the loop for the *Rhodococcal* structure is correlated with higher observed binding constants (K_m) for cholesterol and dehydroisoandosterone compared to that observed for the *Streptomyces* enzyme (Sampson et al., 1998). It is thought that the rigidity of the loops pre-orients the amino acid residues needed for binding the steroid C17 hydrocarbon tail and thus increases the efficiency of the enzyme for catalysis.

5.5.2 Covalent Enzyme Structure

The covalent enzyme structure also comprises two domains, however the topology lacks the characteristic nucleotide binding fold seen in most FAD binding proteins (*see* Fig. 5.3b). The cofactor is covalently linked to a histidine residue (His121) located in a loop region between the third and fourth beta-strand of a four stranded beta pleated sheet (Coulombe et al., 2001). In addition, the phosphate oxygen atoms of the cofactor make hydrogen bond interactions with main chain nitrogen atoms of this loop region, however there are no hydrogen bond interactions between the protein and the 2' and 3' ribose hydroxyl groups of the cofactor. Despite the covalent linkage the cofactor remains bound to the protein even when the histidine residue needed for covalent linkage is mutated to alanine, hence the covalent bond is not an absolute requirement for FAD binding. Structural and redox studies on this mutant have however shown that it is involved in modulating the redox potential of the cofactor; mutation to alanine decreases the redox potential by ∼100 mV relative to wild-type enzyme (Motteran et al., 2001). The structure of the mutant enzyme confirms that the cofactor is no longer covalently linked to the protein (Lim et al., 2006). Furthermore this mutant structure revealed differences in the conformation of the isoalloxazine ring system relative to the wild-type enzyme that adversely affect redox chemistry. Studies on the apoenzyme form of this mutant indicate reduced protein stability relative to the holoenzyme form, suggesting a link between enzyme stability and cofactor linkage (Caldinelli et al., 2005, 2008).

As observed for the non-covalent enzyme, the substrate binding domain in the covalent enzyme is also composed of a large beta pleated sheet which forms a "roof" over the steroid binding cavity buried within the structure of the protein. Two extended loops exhibiting higher thermal motion relative to the remainder of the enzyme are proposed to form the entrance to the steroid binding cavity.

The active site cavity is large enough to contain the entire cholesterol substrate including the C17 hydrocarbon moiety, as shown in a model of the enzyme/cholesterol complex (Fig. 5.3d). The isoalloxazine moiety of the cofactor is located at the base of the cavity. Its position provides a basis for modelling of the steroid substrate in the cavity and hence insight into the residues that may be implicated in catalysis (*see* further discussion below).

5.6 Catalytic Mechanism

In order for the cholesterol oxidase to efficiently carry out both oxidation and isomerization catalysis a number of important features are required. The substrate must be oriented correctly relative to the flavin ring system to allow for: 1) hydride transfer from the steroid C3 to N5 of the cofactor and 2) rearrangement of the double bond to give the final conjugated ketone product. Furthermore, to facilitate both steps of steroid oxidation and isomerization a base is required. In the steroid dehydrogenation step, the base is needed to deprotonate the steroid C3–OH proton; both the high pK_a of the unactivated steroid hydroxyl group (>15) and the concerted mechanism of O–H and C–H bond breakage strongly indicate the need for a base in the steroid dehydrogenation step. Furthermore, as the isomerization step requires both a C–H bond breaking and a C–H bond making step, an acid/base residue is likely to be needed.

5.6.1 Non-covalent Enzyme Mechanism

The structure of an enzyme/steroid substrate (Fig. 5.3c) complex for the non-covalent enzyme has given important initial insights into binding mode of the substrate and the catalytic residues that are likely to be involved in the mechanisms of oxidation and isomerisation (Li et al., 1993). Hypotheses regarding the roles of residues in catalysis, based on this complex structure have been further tested by mutagenesis, kinetic analyses and further structural work. Originally the base for oxidation chemistry in the non-covalent enzyme was thought to be His447 (Kass and Sampson, 1998a; Li et al., 1993; Yue et al., 1999; Yamashita et al., 1998) however high resolution structural studies have shown that, in the active pH range of the enzyme, this histidine is protonated at NE2 precluding its ability to act as a base. Reinterpretation of the mutagenesis and kinetic results directed at studies of this residue suggested it acts as a hydrogen bond donor to the steroid hydroxyl oxygen atom thereby allowing correct orbital alignment for deprotonation of the steroid C3–OH proton and hydride transfer of the C3–H as to the cofactor (Lario et al., 2003; Lyubimov et al., 2006).

Further residues that may be important in oxidation chemistry in the non-covalent enzyme include Asn485 and Tyr446. Asn485 acts to stabilize the reduced cofactor through a movement of the side chain closer to the reduced cofactor, allowing hydrogen bond interactions between the amide side chain NH_2 group and the π system of the cofactor pyrimidine ring. Mutagenesis of Asn485 to leucine or aspartate decreases the reduction potential relative to the wild-type enzyme thereby adversely affecting the ability of the cofactor to be reduced (Lyubimov et al., 2009; Yin et al., 2001).

Recent structural work on a complex of the H447Q/E361Q double mutant, exhibiting much slower oxidation chemistry than the wild-type enzyme, bound to a substrate analogue revealed differences in the conformation of the isoalloxazine ring moiety due to steric pressure exerted on the ring by alternate conformations of the side chain of Tyr446 (Lyubimov et al., 2007) (*see* Fig. 5.4). The altered isoalloxazine ring conformation, induced by substrate binding at the active site and the movement of Tyr446 is thought to modulate the reduction potential of the cofactor and thus affect oxidation chemistry. Further studies on the role of Tyr446 in oxidation chemistry and in substrate binding need to be carried out.

Identification of the base for deprotonation of the substrate C3–OH proton is still under some debate. Glu361 may act as the base, (*see* proposed mechanism on Fig. 5.5) however this residue is also required as the base for the isomerisation step (*see* below). If Glu361 acts as the base for both oxidation and isomerisation the proton extracted from the steroid during oxidation would need to be transferred away from the glutamate side chain before the isomerisation step could occur. The proton extracted is likely to be transferred to oxygen during the oxidative half reaction when the cofactor is reoxidized. However the exact mechanism and kinetics of this step is not yet clearly understood.

The isomerisation step, where the intermediate cholest-5-en-3-one is converted to the final product, cholest-4-en-3-one, occurs through the transfer of a hydrogen atom from the C4 to the C6 position on the steroid by a single base (Kass and Sampson, 1995). The reaction is facilitated by activation of the C–H group at C4 by the neighboring C3 carbonyl group formed during the previous oxidation step. As mentioned above, Glu361 is the base required to carry out this chemistry. It is positioned in the structure on the "roof" of the active site cavity lying above the C4–C6 locus of the steroid ring system. Furthermore, the side chain exhibits higher temperature factors in the crystal structure (Vrielink et al., 1991; Yue et al., 1999) suggesting it is highly mobile as needed to transfer the proton from C4 to C6 of the steroid substrate (*see* proposed mechanism in Fig. 5.6). Studies by Sampson

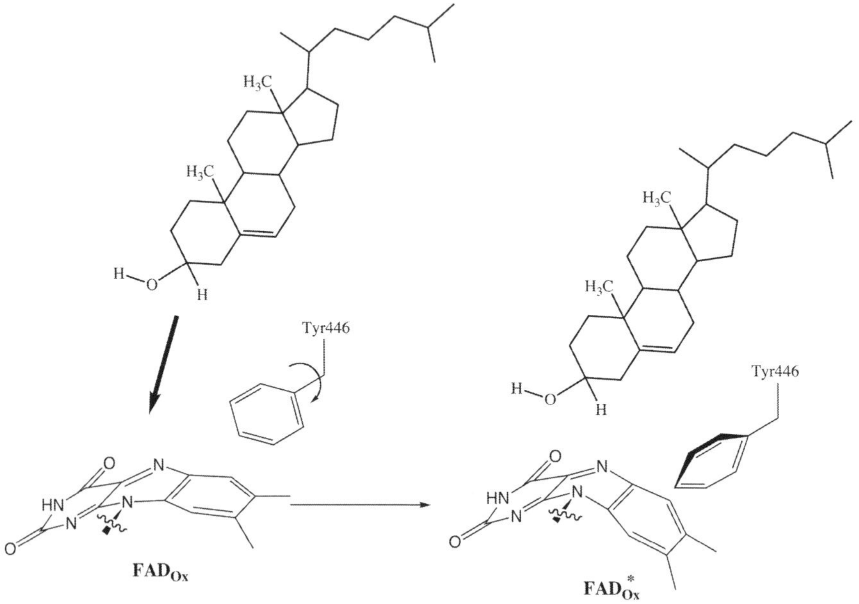

Fig. 5.4 Depiction of the proposed induced fit mechanism upon substrate binding observed for the non-covalent enzyme from *Streptomyces sp* SA-COO. The isoalloxazine moiety of the FAD cofactor is shown. In the absence of the substrate the isoalloxazine ring system is non planar (FAD_{Ox}). Upon binding of the substrate, the side chain of Tyr446 reorients placing steric pressure on the dimethylbenzene portion of the isoalloxazine ring system and resulting in further bending. The increased distortion of the isoalloxazine moiety modulates the redox potential of the cofactor thus facilitating substrate oxidation (FAD_{Ox}^*)

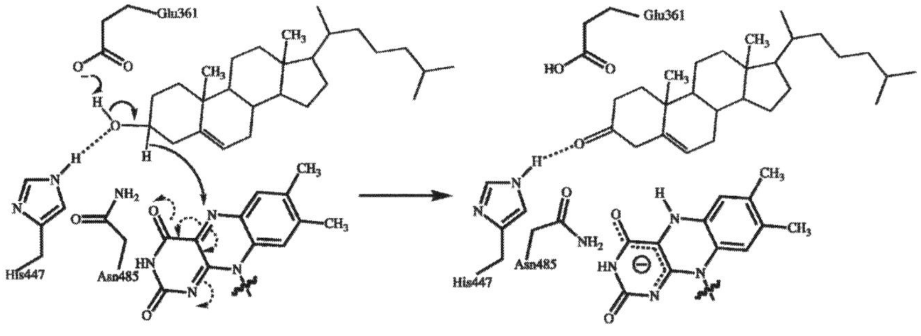

Fig. 5.5 Proposed mechanism of oxidation for the non-covalent enzyme from *Streptomyces sp* SA-COO. Glu361 acts as the base for abstraction of the 3β-hydroxyl proton and His447 acts to orient the hydroxyl proton through a hydrogen bond. Asn485 adopts two conformations, one with the side chain amide nitrogen pointing away from the pyrimidine ring of the isoalloxazine moiety when the cofactor is oxidized and the second conformation pointing towards the pyrimidine ring system when the cofactor is reduced. This second conformation stabilizes the reduced cofactor

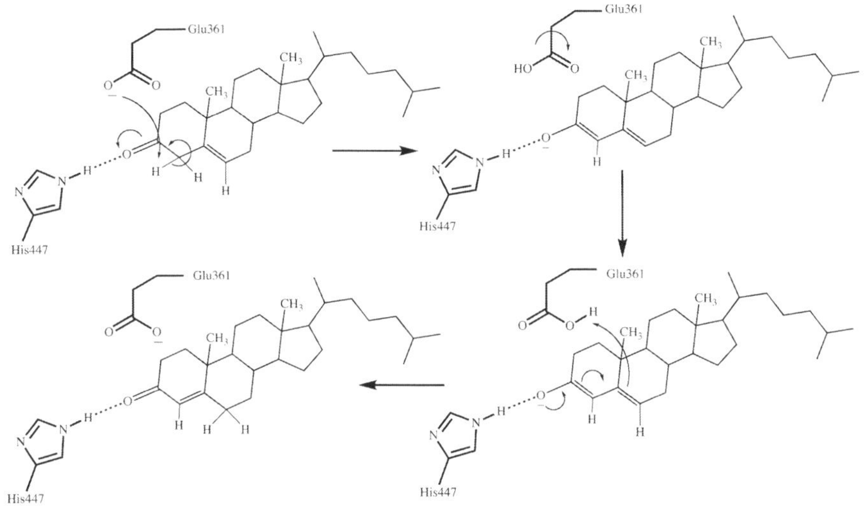

Fig. 5.6 Proposed mechanism for isomerisation of the steroid intermediate, cholest-5-ene-3-one after oxidation. Glu361 acts as the base extracting the proton from C4 of the steroid A ring to generate a negatively charged transition state stabilized by interaction with His447. The side chain of Glu361 reorients to reprotonate the steroid at C6 to give the final cholest-4-en-3-one product

and co-workers have shown that the isomerisation reaction proceeds stereospecifically on the β-face of the steroid ring system by Glu361 (Kass and Sampson, 1995, 1998b).

5.6.2 Covalent Enzyme Mechanism

Although numerous studies have been carried out to characterize the catalytic mechanism of the non-covalent enzyme, there has been less work on the covalent form. The steroid binding pocket for this form is much more hydrophilic in nature than the non-covalent enzyme. Two charged side chain (Glu475 and Arg477) adopt multiple conformations in the crystal structure correlated with one another, suggesting they move in a concerted fashion. The covalent enzyme has not been crystallized in the presence of a steroid ligand, thus the mode of substrate binding is only speculative and based on an *in silico* enzyme/substrate model (Coulombe et al., 2001) (Fig. 5.3d). However the positions of side chains relative to the flavin moiety provide some basis for hypotheses as to the roles of specific residues in oxidation and isomerisation chemistry.

Based on the enzyme/substrate model, it was initially proposed that Glu475 acts as the base for extraction of the steroid C3–OH proton in the reductive half reaction (Coulombe et al., 2001). In addition, the positively charged Arg477 was thought to stabilize the reduced cofactor in a similar manner to Asn485 in the non-covalent enzyme. Alternatively this residue may play a similar role to that of His447 in the non-covalent enzyme by positioning the substrate hydroxyl hydrogen appropriately for correct orbital alignment to allow deprotonation of the C3–OH and hydride transfer of the C3–H to the cofactor. Furthermore, Glu475 was also thought to be the base for the isomerisation step, hence the same residue may act as the base for both oxidation and isomerisation, as has recently also been proposed for the non-covalent enzyme. If this residue is needed as the base for both steps the proton extracted from the steroid C3–OH group must be transferred to another site after oxidation and prior to isomerisation. Based on an inspection of the residues that lie near to Glu475, Glu311 may play the role of a proton shuttle residue, accepting the proton from Glu475. Interestingly Glu311 is also located near to the proposed oxygen entry channel and hence may transfer the proton onto an incoming O_2 molecule (see below).

Mutational and kinetic analyses have been carried out for Glu311, Glu475 and Arg477 (Piubelli et al., 2008). These studies fail to show a marked decrease in either substrate oxidation or intermediate isomerisation for mutations of Glu475 which remove its basic properties, hence the authors concluded that this residue is not likely to be essential for catalysis. However their studies did show a pronounced decrease in both substrate oxidation and isomerisation for mutations of Glu311, supporting the hypothesis that it plays a role in catalysis. Perhaps the base needed for oxidation and isomerisation is Glu311. As in the non-covalent enzyme, the finite identification of the base for proton abstraction remains elusive and will require further study.

5.7 Oxygen Channel

Cholesterol oxidases require the binding and oxidation of molecular oxygen to complete the catalytic cycle of the enzyme (*see* Fig. 5.1). Molecular oxygen accepts the electrons from the reduced FAD cofactor in order to reoxidize it. This oxidative half reaction results in the formation of H_2O_2. The kinetics of O_2 binding to the enzyme has been the subject of some studies for the covalent enzyme from *Brevibacterium sterolicum* and for the non-covalent enzyme from *Streptomyces hygroscopicus* (Pollegioni et al., 1999). These studies reveal that reaction with dioxygen occurs very differently for the two enzyme forms. In the covalent enzyme, reaction with O_2 proceeds with two phases: an initial rapid phase that shows a saturation dependence on $[O_2]$ followed by a slow phase, independent of $[O_2]$. It was proposed that this behavior results from the interconversion of two enzyme states, each with different reactivities toward O_2. The proposed scheme for oxygen reactivity is shown in Fig. 5.7. The reduced enzyme converts between two different forms: E~FAD_{red} and E~FAD^*_{red}. One form (denoted E~FAD_{red} in Fig. 5.7) is more reactive to oxygen and thus is the preferred pathway for flavin reoxidation (formation of E~FAD_{ox} and H_2O_2).

In the case of the non-covalent enzyme from *Streptomyces hygroscopicus*, reoxidation of reduced FAD by O_2 in the absence of bound ligand occurs in a monophasic

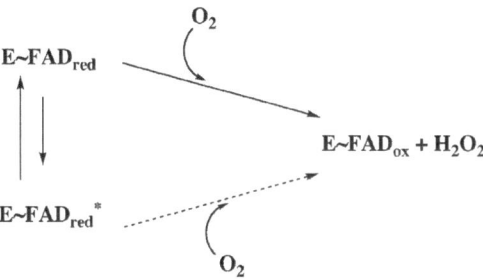

Fig. 5.7 Proposed mechanism for reaction of reduced FAD in cholesterol oxidase from *Brevibacterium sterolicum* with dioxygen. E~FAD_{red} and E~FAD_{red}* depict different conformational forms of the reduced enzyme. E~FAD_{ox} denotes the oxidized form of the cofactor. Dioxygen reacts more readily with E~FAD_{red} than with E~FAD_{red}*

manner, similarly to what was been observed for the majority of flavoprotein oxidases (Ghisla and Massey, 1989). However the enzyme reduction kinetics is different in the presence of the bound steroid product, cholest-4-en-3-one, where the reaction exhibits a dependence on [O_2] (Pollegioni et al., 1999) suggesting some similarities with the covalent enzyme. Flavin-dependent hydroxylases are known to form an intermediate (such as a 4a-flavin hydroperoxide) during the oxidative half reaction (Ghisla and Massey, 1989).

The structures of both the covalent and non-covalent enzyme forms give intriguing keys into the oxidative half reaction and the mechanism of dioxygen binding. In the non-covalent enzyme initial structural studies at 1.8 Å resolution did not give a clear indication of a unique oxygen binding pathway into the enzyme active site. It was supposed that molecular oxygen entered the active site through the surface loops which also allowed entry of the steroid substrate. However, when structural studies were carried out to atomic resolution (0.95 Å), a narrow channel was visible, large enough to accommodate a dioxygen molecule (Lario et al., 2003) (Figs. 5.8a,b). This channel is located at the interface between the FAD and substrate-binding domains, extending from the surface of the protein, opposite to the steroid entry loops, towards the isoalloxazine moiety in the buried binding cavity. The channel is visible because of the ability to resolve multiple side chain conformations for discrete residues in the active site and along the channel surface at higher resolution. Two mutually exclusive side chain populations are evident; when the side chains adopt conformer A, access from the exterior to the active site through the channel is blocked. In contrast, when the side chains adopt conformer B, there is sufficient space in the channel to allow a dioxygen molecule to move from the exterior of the protein to the active site (Fig. 5.8b). Based on these observations it was proposed that the channel functioned as the entry route for dioxygen in the oxidative half reaction. Gating of the channel is accomplished by Asn485, the same residue shown to be important for stabilization of the reduced cofactor (Lyubimov et al., 2009; Yin et al., 2001). Based on these observations it was suggested that cofactor reduction was correlated to oxygen entry through the movement of Asn485 and that once this

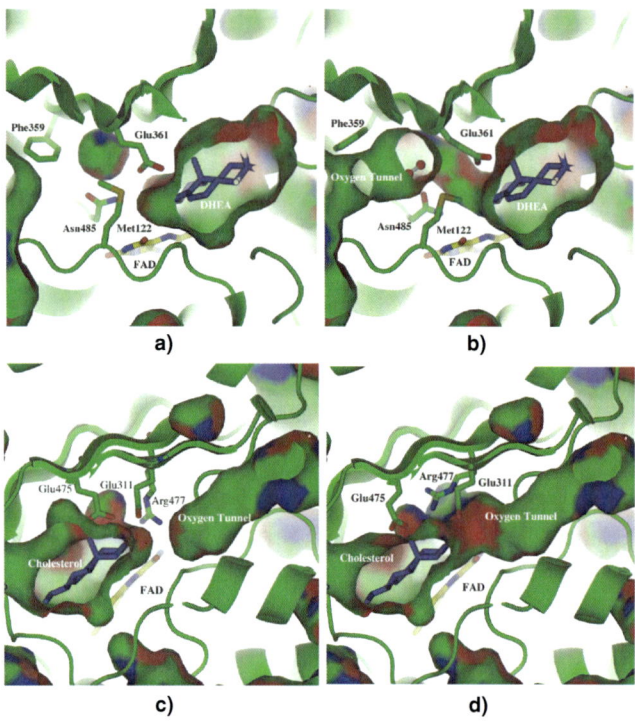

Fig. 5.8 Close up of the active site regions of both forms of cholesterol oxidase depicting the gating residues and the proposed oxygen channel formed. The non-covalent enzyme is depicted in (**a**) channel closed conformation and (**b**) channel open conformation. In (**b**) a dioxygen molecule is seen bound inside the channel. Residues involved in channel gating are labeled. The covalent enzyme is depicted in (**a**) channel closed conformation and (**b**) channel open conformation. Models of the bound substrate are shown with blue bonds. The surface was computed as described in Fig. 5.3

side chain moved to stabilize the reduced cofactor, other side chains were also able to move allowing the oxygen entry channel to open. Mutation and kinetic analysis of the gate residue (Asn485Asp), as well as residues that line the channel (Phe359Trp and Gly347Asn) show that these residues, decrease the overall catalytic efficiency of oxidation by the enzyme and suggests that oxygen binds to the enzyme to form a complex before chemically reacting with the flavin. Furthermore, a rate-limiting conformational change occurs for the binding of oxygen to the enzyme (Chen et al., 2008). Indeed the tunnel-mutant forms of the enzyme exhibit kinetic cooperativity with respect to dioxygen. This is not observed in the wild-type enzyme suggesting that a rate-limiting conformational change occurs for binding of oxygen and the release of hydrogen peroxide in the non-covalent enzyme. These studies support the role of the channel in oxygen binding and access to the enzyme.

In the covalent enzyme the presence of a narrow channel between the FAD binding domain and the substrate binding domain is evident in the structure at 1.7 Å resolution (Coulombe et al., 2001). Gating is also evident in the structure through the alternate positions of Glu475 and Arg477 (Figs. 5.8c,d). Studies by Pollegioni and co-workers have shown that mutations of Glu311 affect reaction of the reduced enzyme with dioxygen by altering the Glu475/Arg477 gate (Piubelli et al., 2008).

In general the observed channels in the structures of both forms of cholesterol oxidases and the effect mutations of residues have, either in gating the channel, or by affecting the channel gating residues, indicate that dioxygen must enter the active site through specific engineered routes and that control of oxygen entry is mediated by the protein. These studies add to the increasing evidence that enzymes specific for reactivity with oxygen have engineered routes to allow these small gaseous substrate to access the reactive centres and they do not simply diffuse into the proteins (Saam et al., 2007; Furse et al., 2006; Hofacker and Schulten, 1998; Soulimane et al., 2000; Brunori, 2000; Deng et al., 2007; Moustafa et al., 2006).

The extensive biochemical and kinetic studies on cholesterol oxidase have provided important insights into flavin-mediated redox chemistry. These studies have also revealed features necessary for modulating the flavin reduction potential and hence the reactivity of the enzyme. Importantly, studies of this enzyme have also revealed gated access routes for molecular oxygen to enter protein active sites resulting in new views regarding the mechanisms by which the protein controls oxygen access and reactivity. Further studies should perhaps focus more on understanding the detailed features of protein recognition of the substrate in the context of the membrane environment where cholesterol is found and the role that this enzyme plays in bacterial pathogenesis. Indeed, such knowledge may provide guide future work on the enzyme as a target for novel therapeutic agents as inhibitors to treat virulent bacterial infections such as tuberculosis.

References

Allain, C.C., Poon, L.S., Chan, C.S., Richmond, W., and Fu, P.C., 1974, Enzymatic determination of total serum cholesterol. *Clin. Chem.* **20:** 470–475.

Anton, N., Santos-Aberturas, J., Mendes, M.V., Guerra, S.M., Martin, J.F., and Aparicio, J.F., 2007, PimM, a PAS domain positive regulator of pimaricin biosynthesis in Streptomyces natalensis. *Microbiology* **153:** 3174–3183.

Aparicio, J.F., Caffrey, P., Gil, J.A., and Zotchev, S.B., 2003, Polyene antibiotic biosynthesis gene clusters. *Appl. Microbiol. Biotechnol.* **61:** 179–188.

Aparicio, J.F., Mendes, M.V., Anton, N., Recio, E., and Martin, J.F., 2004, Polyene macrolide antibiotic biosynthesis. *Curr. Med. Chem.* **11:** 1645–1656.

Arya, S.K., Datta, M., Singh, S.P., and Malhotra, B.D., 2007, Biosensor for total cholesterol estimation using N-(2-aminoethyl)-3-aminopropyltrimethoxysilane self-assembled monolayer. *Anal. Bioanal. Chem.* **389:** 2235–2242.

Barenholz, Y., Patzer, E.J., Moore, N.F., and Wagner, R.R., 1978, Cholesterol oxidase as a probe for studying membrane composition and organization. *Adv. Exp. Med. Biol.* **101:** 45–56.

Basu, A.K., Chattopadhyay, P., Roychoudhuri, U., and Chakraborty, R., 2007, Development of cholesterol biosensor based on immobilized cholesterol esterase and cholesterol oxidase on oxygen electrode for the determination of total cholesterol in food samples. *Bioelectrochemistry* **70:** 375–379.

Berg, O.G., Gelb, M.H., Tsai, M.D., and Jain, M.K., 2001, Interfacial enzymology: the secreted phospholipase A(2)-paradigm. *Chem. Rev.* **101**: 2613–2654.

Bittman, R., Kasireddy, C.R., Mattjus, P., and Slotte, J.P., 1994, Interaction of cholesterol with sphingomyelin in monolayers and vesicles. *Biochemistry* **33**: 11776–11781.

Brooks, C.J., and Smith, A.G., 1975, Cholesterol oxidase. Further studies of substrate specificity in relation to the analytical characterisation of steroids. *J. Chromatogr.* **112**: 499–511.

Brooks, C.J., and Smith, A.G., 1980, More on substrate specificity of cholesterol oxidase. *Clin. Chem.* **26**: 1918.

Brunori, M., 2000, Structural dynamics of myoglobin. *Biophys. Chem.* **86**: 221–230.

Brzostek, A., Dziadek, B., Rumijowska-Galewicz, A., Pawelczyk, J., and Dziadek, J., 2007, Cholesterol oxidase is required for virulence of *Mycobacterium tuberculosis*. *FEMS Microbiol. Lett.* **275**: 1106–1112.

Buckland, B.C., Lilly, M.D., and Dunnill, P., 1976, The kinetics of cholesterol oxidase synthesis by *Nocardia rhodocrous*. *Biotechnol. Bioeng.* **18**: 601–621.

Caldinelli, L., Iametti, S., Barbiroli, A., Bonomi, F., Fessas, D., Molla, G., Pilone, M.S., and Pollegioni, L., 2005, Dissecting the structural determinants of the stability of cholesterol oxidase containing covalently bound flavin. *J. Biol. Chem.* **280**: 22572–22581.

Caldinelli, L., Iametti, S., Barbiroli, A., Fessas, D., Bonomi, F., Piubelli, L., Molla, G., and Pollegioni, L., 2008, Relevance of the flavin binding to the stability and folding of engineered cholesterol oxidase containing noncovalently bound FAD. *Protein Sci.* **17**: 409–419.

Chen, L., Lyubimov, A., Brammer, L., Vrielink, A., and Sampson, N.S., 2008, The binding and release of oxygen and hydrogen peroxide are directed by a hydrophobic tunnel in cholesterol oxidase. *Biochemistry* **47**: 5368–5377.

Cherradi, N., Defaye, G., and Chambaz, E.M., 1993, Dual subcellular localization of the 3 beta-hydroxysteroid dehydrogenase isomerase: characterization of the mitochondrial enzyme in the bovine adrenal cortex. *J. Steroid Biochem. Mol. Biol.* **46**: 773–779.

Cherradi, N., Defaye, G., and Chambaz, E.M., 1994, Characterization of the 3 beta-hydroxysteroid dehydrogenase activity associated with bovine adrenocortical mitochondria. *Endocrinology* **134**: 1358–1364.

Cherradi, N., Rossier, M.F., Vallotton, M.B., Timberg, R., Friedberg, I., Orly, J., Wang, X.J., Stocco, D.M., and Capponi, A.M., 1997, Submitochondrial distribution of three key steroidogenic proteins (steroidogenic acute regulatory protein and cytochrome p450scc and 3beta-hydroxysteroid dehydrogenase isomerase enzymes) upon stimulation by intracellular calcium in adrenal glomerulosa cells. *J. Biol. Chem.* **272**: 7899–7907.

Corbin, D.R., Grebenok, R.J., Ohnmeiss, T.E., Greenplate, J.T., and Purcell, J.P., 2001, Expression and chloroplast targeting of cholesterol oxidase in transgenic tobacco plants. *Plant Physiol.* **126**: 1116–1128.

Corbin, D.R., Greenplate, J.T., and Purcell, J.P., 1998, The identification and development of proteins for control of insects in genetically modified crops. *Hort. Sci.* **33**: 614–617.

Corbin, D.R., Greenplate, J.T., Wong, E.Y., and Purcell, J.P., 1994, Cloning of an insecticidal cholesterol oxidase gene and Its expression in bacteria and in plant protoplasts. *Appl. Environ. Microbiol.* **60**: 4239–4244.

Coulombe, R., Yue, K.Q., Ghisla, S., and Vrielink, A., 2001, Oxygen access to the active site of cholesterol oxidase through a narrow channel is gated by an Arg-Glu pair. *J. Biol. Chem.* **276**: 30435–30441.

Deng, P., Nienhaus, K., Palladino, P., Olson, J.S., Blouin, G., Moens, L., Dewilde, S., Geuens, E., and Nienhaus, G.U., 2007, Transient ligand docking sites in *Cerebratulus lacteus* mini-hemoglobin. *Gene* **398**: 208–223.

El Yandouzi, E.H., and Le Grimellec, C., 1993, Effect of cholesterol oxidase treatment on physical state of renal brush border membranes: evidence for a cholesterol pool interacting weakly with membrane lipids. *Biochemistry* **32**: 2047–2052.

Eventoff, W., and Rossmann, M.G., 1975, The evolution of dehydrogenases and kinases. *CRC Crit. Rev. Biochem.* **3**: 111–140.

Fuhrmann, H., Dobeleit, G., Bellair, S., and Guck, T., 2002, Cholesterol oxidase and resistance of Rhodococcus equi to peroxidative stress in vitro in the presence of cholesterol. *J. Vet. Med. B Infect. Dis. Vet. Public Health* **49:** 310–321.

Fukuyama, M., and Miyake, Y., 1979, Purification and some properties of cholesterol oxidase from *Schizophyllum commune* with covalently bound flavin. *J. Biochem. (Tokyo)* **85:** 1183–1193.

Furse, K.E., Pratt, D.A., Schneider, C., Brash, A.R., Porter, N.A., and Lybrand, T.P., 2006, Molecular dynamics simulations of arachidonic acid-derived pentadienyl radical intermediate complexes with COX-1 and COX-2: Insights into oxygenation regio- and stereoselectivity. *Biochemistry* **45:** 3206–3218.

Gadda, G., Wels, G., Pollegioni, L., Zucchelli, S., Ambrosius, D., Pilone, M.S., and Ghisla, S., 1997, Characterization of cholesterol oxidase from *Streptomyces hydroscopicus* and *Brevibacterium sterolicum*. *Eur. J. Biochem.* **250:** 369–376.

Gatfield, J., and Pieters, J., 2000, Essential role for cholesterol in entry of mycobacteria into macrophages. *Science* **288:** 1647–1650.

Gelb, M.H., Jain, M.K., Hanel, A.M., and Berg, O.G., 1995, Interfacial enzymology of glycerolipid hydrolases: lessons from secreted phospholipases A2. *Ann. Rev. Biochem.* **64:** 653–688.

Ghisla, S., and Massey, V., 1986, New flavins for old: artificial flavins as active site probes of flavoproteins. *Biochem. J.* **239:** 1–12.

Ghisla, S., and Massey, V., 1989, Mechanism of flavoprotein-catalysed reactions. *Eur. J. Biochem.* **181:** 1–17.

Gimpl, G., and Gehrig-Burger, K., 2007, Cholesterol reporter molecules. *Biosci. Rep.* **27:** 335–358.

Greenplate, J.T., Duck, N.B., Pershing, J.C., and Purcell, J.P., 1995, Cholesterol oxidase – an oostatic and larvicidal agent active against the cotton boll weevil, *Anthonomus grandis*. *Entomol. Exp. Appl.* **74:** 253–258.

Hofacker, I., and Schulten, K., 1998, Oxygen and proton pathways in cytochrome c oxidase. *Proteins* **30:** 100–107.

Inouye, Y., Taguchi, K., Fuki, A., Ishimaru, K., Snakamura, S., and Nomi, R., 1982, Purification and characterisation of extracellular 3beta-hydroxysteroid oxidase produced by *Streptoverticillium cholesterolicum*. *Chem. Pharm. Bull. (Tokyo)* **30:** 951–958.

Ishizaki, R., Hirayama, N., Shinkawa, H., Nimi, O., and Murooka, Y., 1989, Nucleotide sequence of the gene for cholesterol oxidase from a *Streptomyces sp*. *J. Bacteriol.* **171:** 596–601.

Iwaki, M., Yakovlev, G., Hirst, J., Osyczka, A., Dutton, P.L., Marshall, D., and Rich, P.R., 2005, Direct observation of redox-linked histidine protonation changes in the iron-sulfur protein of the cytochrome bc(1) complex by ATR-FTIR spectroscopy. *Biochemistry* **44:** 4230–4237.

Jain, M.K., Gelb, M.H., Rogers, J., and Berg, O.G., 1995, Kinetic basis for interfacial catalysis by phospholipase A2. *Meth. Enzymol.* **249:** 567–614.

Kamei, T., Takiguchi, Y., Suzuki, H., Matsuzaki, M., and Nakamura, S., 1978, Purification of 3beta-hydroxysteroid oxidase of *Streptomyces violascens* origin by affinity chromatography on cholesterol. *Chem. Pharm. Bull. (Tokyo)* **26:** 2799–2804.

Kass, I.J., and Sampson, N.S., 1995, The isomerization catalyzed by *Brevibacterium sterolicum* cholesterol oxidase proceeds stereospecifically with one base. *Biochem. Biophys. Res. Commun.* **206:** 688–693.

Kass, I.J., and Sampson, N.S., 1998a, Evaluation of the role of His447 in the reaction catalyzed by cholesterol oxidase. *Biochemistry* **37:** 17990–18000.

Kass, I.J., and Sampson, N.S., 1998b, The importance of Glu^{361} position in the reaction catalyzed by cholesterol oxidase. *Bioorg. Med. Chem. Lett.* **8:** 2663–2668.

Lange, Y., 1992, Tracking cell cholesterol with cholesterol oxidase. *J. Lipid Res.* **33:** 315–321.

Lange, Y., Matthies, H., and Steck, T.L., 1984, Cholesterol oxidase susceptibility of the red cell membrane. *Biochim. Biophys. Acta* **769:** 551–562.

Lange, Y., and Steck, T.L., 2008, Cholesterol homeostasis and the escape tendency (activity) of plasma membrane cholesterol. *Prog. Lipid Res.* **47:** 319–332.

Lange, Y., Ye, J., and Steck, T.L., 2007, Scrambling of phospholipids activates red cell membrane cholesterol. *Biochemistry* **46**: 2233–2238.

Lario, P.I., Sampson, N., and Vrielink, A., 2003, Sub-atomic resolution crystal structure of cholesterol oxidase: what atomic resolution crystallography reveals about enzyme mechanism and the role of the FAD cofactor in redox activity. *J. Mol. Biol.* **326**: 1635–1650.

Lartillot, S., and Kedziora, P., 1990, Production, purification and some properties of cholesterol oxidase from a *Streptomyces sp. Prep. Biochem.* **20**: 51–62.

Le Lay, S., Li, Q., Proschogo, N., Rodriguez, M., Gunaratnam, K., Cartland, S., Rentero, C., Jessup, W., Mitchell, T., and Gaus, K., 2009, Caveolin-1-dependent and -independent membrane domains. *J. Lipid Res.* **50**: 1609–1620.

Li, J., Vrielink, A., Brick, P., and Blow, D.M., 1993, Crystal structure of cholesterol oxidase complexed with a steroid substrate: Implications for flavin adenine dinucleotide dependent alcohol oxidases. *Biochemistry* **32**: 11507–11515.

Lim, L., Molla, G., Guinn, N., Ghisla, S., Pollegioni, L., and Vrielink, A., 2006, Structural and kinetic analysis of the H121A mutant of cholesterol oxidase. *Biochem. J.* **400**: 13–22.

Linden, K.G., and Benisek, W.F., 1986, The amino acid sequence of a delta 5-3-oxosteroid isomerase from *Pseudomonas putida* biotype B. *J. Biol. Chem.* **261**: 6454–6460.

Linder, R., 1984, Alteration of mammalian membranes by the cooperative and antagonistic actions of bacterial proteins. *Biochim. Biophys. Acta* **779**: 432–435.

Linder, R., and Bernheimer, A.W., 1982, Enzymatic oxidation of membrane cholesterol in relation to lysis of sheep erythrocytes by corynebacterial enzymes. *Arch. Biochem. Biophys.* **213**: 395–404.

Luu The, V., Lachance, Y., Labrie, C., Leblanc, G., Thomas, J.L., Strickler, R.C., and Labrie, F., 1989, Full length cDNA structure and deduced amino acid sequence of human 3 beta-hydroxy-5-ene steroid dehydrogenase. *Mol. Endocrinol.* **3**: 1310–1312.

Lyubimov, A.Y., Chen, L., Sampson, N.S., and Vrielink, A., 2009, A hydrogen-bonding network is important for oxidation and isomerization in the reaction catalyzed by cholesterol oxidase. *Acta Crystallogr. D* **65**: 1221–1231.

Lyubimov, A.Y., Heard, K., Tang, H., Sampson, N.S., and Vrielink, A., 2007, Distortion of flavin geometry linked to ligand binding in cholesterol oxidase. *Prot. Sci.* **16**: 2647–2656.

Lyubimov, A.Y., Lario, P.I., Moustafa, I., and Vrielink, A., 2006, Atomic resolution crystallography reveals how changes in pH shape the protein microenvironment. *Nat. Chem. Biol.* **2**: 259–264.

Martin, C.K., 1977, Microbial cleavage of sterol side chains. *Adv. Appl. Microbiol.* **22**: 29–58.

Mendes, M.V., Anton, N., Martin, J.F., and Aparicio, J.F., 2005, Characterization of the polyene macrolide P450 epoxidase from Streptomyces natalensis that converts de-epoxypimaricin into pimaricin. *Biochem. J.* **386**: 57–62.

Mendes, M.V., Recio, E., Anton, N., Guerra, S.M., Santos-Aberturas, J., Martin, J.F., and Aparicio, J.F., 2007, Cholesterol oxidases act as signaling proteins for the biosynthesis of the polyene macrolide pimaricin. *Chem. Biol.* **14**: 279–290.

Mendes, M.V., Recio, E., Fouces, R., Luiten, R., Martin, J.F., and Aparicio, J.F., 2001, Engineered biosynthesis of novel polyenes: a pimaricin derivative produced by targeted gene disruption in *Streptomyces natalensis*. *Chem. Biol.* **8**: 635–644.

Motteran, L., Pilone, M.S., Molla, G., Ghisla, S., and Pollegioni, L., 2001, Cholesterol oxidase from *Brevibacterium sterolicum* – The relationship between covalent flavinylation and redox properties. *J. Biol. Chem.* **276**: 18024–18030.

Moustafa, I., Foster, S., Lyubimov, A.Y., and Vrielink, A., 2006, Crystal structure of LAAO from *Calloselasma rhodostoma* with an L-phenylalanine substrate: insights into structure and mechanism. *J. Mol. Biol.* **364**: 991–1002.

Navas, J., Gonzalez-Zorn, B., Ladron, N., Garrido, P., and Vazquez-Boland, J.A., 2001, Identification and mutagenesis by allelic exchange of choE, encoding a cholesterol oxidase from the intracellular pathogen *Rhodococcus equi*. *J. Bacteriol.* **183**: 4796–4805.

Ohlsson, I., Nordstrom, B., and Branden, C.I., 1974, Structural and functional similarities within the coenzyme binding domains of dehydrogenases. *J. Mol. Biol.* **89**: 339–354.

Ohvo-Rekila, H., Mattjus, P., and Slotte, J.P., 1998, The influence of hydrophobic mismatch on androsterol/phosphatidylcholine interactions in model membranes. *Biochim. Biophys. Acta* **1372:** 331–338.

Pandey, A.K., and Sassetti, C.M., 2008, Mycobacterial persistence requires the utilization of host cholesterol. *Proc. Natl. Acad. Sci. USA* **105:** 4376–4380.

Piubelli, L., Pedotti, M., Molla, G., Feindler-Boeckh, S., Ghisla, S., Pilone, M.S., and Pollegioni, L., 2008, On the oxygen reactivity of flavoprotein oxidases: an oxygen access tunnel and gate in *Brevibacterium sterolicum* cholesterol oxidase. *J. Biol. Chem.* **283:** 24738–24747.

Pollegioni, L., Wels, G., Pilone, M.S., and Ghisla, S., 1999, Kinetic mechanisms of cholesterol oxidase from *Streptomyces hygroscopicus* and *Brevibacterium sterolicum*. *Eur. J. Biochem.* **263:** 1–13.

Purcell, J.P., Greenplate, J.T., Jennings, M.G., Ryerse, J.S., Pershing, J.C., Sims, S.R., Prinsen, M.J., Corbin, D.R., Tran, M., Sammons, R.D., and Stonard, R.J., 1993, Cholesterol oxidase – a potent insecticidal protein active against boll weevil larvae. *Biochem. Biophys. Res. Commun.* **196:** 1406–1413.

Richmond, W., 1973, Preparation and properties of a cholesterol oxidase from *Nocardia sp.* and its application to the enzymatic assay of total cholesterol in serum. *Clin. Chem.* **19:** 1350–1356.

Richmond, W., 1976, Use of cholesterol oxidase for assay of total and free cholesterol in serum by continuous-flow analysis. *Clin. Chem.* **22:** 1579–1588.

Saam, J., Ivanov, I., Walther, M., Holzhutter, H., and Kuhn, H., 2007, Molecular dioxygen enters the active site of 12/15-lipoxygenase via dynamic oxygen access channels. *Proc. Natl. Acad. Sci. USA* **104:** 13319–13324.

Sampson, N.S., Kass, I.J., and Ghoshroy, K.B., 1998, Assessment of the role of an Ω loop of cholesterol oxidase: A truncated loop mutant has altered substrate specificity. *Biochemistry* **37:** 5770–5778.

Sauer, L.A., Chapman, J.C., and Dauchy, R.T., 1994, Topology of 3 beta-hydroxy-5-ene-steroid dehydrogenase/delta 5-delta 4-isomerase in adrenal cortex mitochondria and microsomes. *Endocrinology* **134:** 751–759.

Sedlaczek, L., 1988, Biotransformations of steroids. *Crit. Rev. Biotechnol.* **7:** 187–236.

Slotte, J.P., 1992a, Enzyme-catalyzed oxidation of cholesterol in mixed phospholipid monolayers reveals the stoichiometry at which free cholesterol clusters disappear. *Biochemistry* **31:** 5472–5477.

Slotte, J.P., 1992b, Substrate specificity of cholesterol oxidase from *Streptomyces cinnamomeus* – a monolayer study. *J. Steroid Biochem. Mol. Biol.* **42:** 521–526.

Slotte, J.P., 1995, Direct observation of the action of cholesterol oxidase in monolayers. *Biochim. Biophys. Acta* **1259:** 180–186.

Smith, A.G., and Brooks, C.J., 1974, Application of cholesterol oxidase in the analysis of steroids. *J. Chromatogr.* **101:** 373–378.

Smith, A.G., and Brooks, C.J., 1975, Studies of the substrate specificity of cholesterol oxidase from *Nocardia erythropolis* in the oxidation of 3-hydroxy steroids. *Biochem. Soc. Trans.* **3:** 675–677.

Smith, A.G., and Brooks, C.J., 1976, Cholesterol oxidases: properties and applications. *J. Steroid Biochem.* **7:** 705–713.

Smith, A.G., and Brooks, C.J., 1977, The substrate specificity and stereochemistry, reversibility and inhibition of the 3-oxo steroid delta 4-delta 5-isomerase component of cholesterol oxidase. *Biochem. J.* **167:** 121–129.

Soulimane, T., Buse, G., Bourenkov, G.P., Bartunik, H.D., Huber, R., and Than, M.E., 2000, Structure and mechanism of the aberrant ba(3)-cytochrome c oxidase from *Thermus thermophilus*. *EMBO J.* **19:** 1766–1776.

Talalay, P., and Wang, V.S., 1955, Enzymic isomerization of delta5-3-ketosteroids. *Biochim. Biophys. Acta* **18:** 300–301.

Thomas, J.L., Evans, B.W., Blanco, G., Mercer, R.W., Mason, J.I., Adler, S., Nash, W.E., Isenberg, K.E., and Strickler, R.C., 1998, Site-directed mutagenesis identifies amino acid residues associated with the dehydrogenase and isomerase activities of human type I (placental) 3beta-hydroxysteroid dehydrogenase/isomerase. *J. Steroid Biochem. Mol. Biol.* **66:** 327–334.

Uwajima, T., Yagi, H., Nakamurs, S., and Terada, O., 1973, Isolation and crystallization of extracellular 3β-hydroxysteroid oxidase of *Brevibacterium sterolicum* nov. sp. *Agr. Biol. Chem.* **37:** 2345–2350.

Van Der Geize, R., Yam, K., Heuser, T., Wilbrink, M.H., Hara, H., Anderton, M.C., Sim, E., Dijkhuizen, L., Davies, J.E., Mohn, W.W., and Eltis, L.D., 2007, A gene cluster encoding cholesterol catabolism in a soil actinomycete provides insight into *Mycobacterium tuberculosis* survival in macrophages. *Proc. Natl. Acad. Sci. USA* **104:** 1947–1952.

Vazquez-Boland, J.A., Kuhn, M., Berche, P., Chakraborty, T., Dominguez-Bernal, G., Goebel, W., Gonzalez-Zorn, B., Wehland, J., and Kreft, J., 2001, Listeria pathogenesis and molecular virulence determinants. *Clin. Microbiol. Rev.* **14:** 584–640.

Vidal, J.C., Espuelas, J., and Castillo, J.R., 2004, Amperometric cholesterol biosensor based on in situ reconstituted cholesterol oxidase on an immobilized monolayer of flavin adenine dinucleotide cofactor. *Anal. Biochem.* **333:** 88–98.

Vrielink, A., and Ghisla, S., 2009, Cholesterol oxidase: Biochemistry and structural features. *FEBS J.* **276:** 6826–6843.

Vrielink, A., Lloyd, L.F., and Blow, D.M., 1991, Crystal structure of cholesterol oxidase from *Brevibacterium sterolicum* refined at 1.8 Å resolution. *J. Mol. Biol.* **219:** 533–554.

Weinstock, D.M., and Brown, A.E., 2002, *Rhodococcus equi*: an emerging pathogen. *Clin. Infect. Dis.* **34:** 1379–1385.

Wilmanska, D., Dziadek, J., Sajduda, A., Milczarek, K., Jaworski, A., and Murooka, Y., 1995, Identification of cholesterol oxidase from fast-growing Mycobacterial strains and *Rhododoccus sp. J. Ferment. Bioeng.* **29:** 119–124.

Wilmanska, D., and Sedlaczek, L., 1988, The kinetics of biosynthesis and some properties of an extracellular cholesterol oxidase produced by *Arthrobacter sp.* IM 79. *Acta Microbiol. Polon.* **37:** 45–51.

Yamashita, M., Toyama, M., Ono, H., Fujii, I., Hirayama, N., and Murooka, Y., 1998, Separation of the two reactions, oxidation and isomerization, catalyzed by *Streptomyces* cholesterol oxidase. *Protein Eng.* **11:** 1075–1081.

Yin, Y., Sampson, N.S., Vrielink, A., and Lario, P.I., 2001, The presence of a hydrogen bond between asparagine 485 and the pi system of FAD modulates the redox potential in the reaction catalyzed by cholesterol oxidase. *Biochemistry* **40:** 13779–13787.

Yue, Q.K., Kass, I.J., Sampson, N.S., and Vrielink, A., 1999, Crystal structure determination of cholesterol oxidase from *Streptomyces* and structural characterization of key active site mutants. *Biochemistry* **38:** 4277–4286.

Zajaczkowska, E., Bartoszko-Tyczkowska, A., and Sedlaczek, L., 1988, Microbiological degradation of sterols. II. Isolation of *Rhodococcus sp.* IM 58 mutants with a block of the cholesterol degradation pathway. *Acta Microbiol. Polon.* **37:** 39–44.

Zajaczkowska, E., and Sedlaczek, L., 1988, Microbiolgical degradation of sterols. I. Selective induction of enzyme of the cholesterol side chain cleavage in *Rhodococcus sp.* IM 58. *Acta Microbiol. Polon.* **37**.

Chapter 6
Oxysterol-Binding Proteins

Neale D. Ridgway

Abstract In eukaryotic cells, membranes of the late secretory pathway contain a disproportionally large amount of cholesterol in relation to the endoplasmic reticulum, nuclear envelope and mitochondria. At one extreme, enrichment of the plasma membrane with cholesterol and sphingolipids is crucial for formation of liquid ordered domains (rafts) involved in cell communication and transport. On the other hand, regulatory machinery in the endoplasmic reticulum is maintained in a relatively cholesterol-poor environment, to ensure appropriate rapid responses to fluctuations in cellular sterol levels. Thus, cholesterol homeostasis is absolutely dependent on its distribution along an intracellular gradient. It is apparent that this gradient is maintained by a combination of sterol-lipid interactions, vesicular transport and sterol-binding/transport proteins. Evidence for rapid, energy-independent transport between organelles has implicated transport proteins, in particular the eukaryotic oxysterol binding protein (OSBP) family. Since the founding member of this family was identified more than 25 years ago, accumulated evidence implicates the 12-member family of OSBP and OSBP-related proteins (ORPs) in sterol signalling and/or sterol transport functions. The OSBP/ORP gene family is characterized by a conserved β-barrel sterol-binding fold but is differentiated from other sterol-binding proteins by the presence of additional domains that target multiple organelle membranes. Here we will discuss the functional and structural characteristics of the mammalian OSBP/ORP family that support a 'dual-targeting' model for sterol transport between membranes.

Keywords Cholesterol · Endoplasmic reticulum · Golgi apparatus · Oxysterols · Sterol transport

Abbreviations

ABC	ATP-binding cassette
ACAT	acyl-CoA:cholesterol acyltransferase

N.D. Ridgway (✉)
Departments of Pediatrics and Biochemistry & Molecular Biology, Atlantic Research Centre, Dalhousie University, 5849 University Av. Halifax, Nova Scotia B3H 4H7, Canada
e-mail: nridgway@dal.ca

APP	amyloid precursor protein
CERT	ceramide transfer protein
ER	endoplasmic reticulum
FFAT	two phenylalanines in an acidic tract
LDL	low density lipoprotein
LXR	liver X receptor
NPC	Niemann-Pick C
OHD	OSBP-homology domain
OSBP	oxysterol binding protein
ORP	OSBP-related protein
NVJ	nuclear-vacuolar junction
PDK	phosphoinositide-dependent kinase
PIP	phosphatidylinositol phosphate
PI4P	phosphatidylinositol 4-phosphate
PI4,5P_2	phosphatidylinositol 4,5-*bis*phosphate
PH	pleckstrin homology
PM	plasma membrane
PP2A	protein phosphatase 2A
RNAi	RNA interference
SREBP	sterol-regulatory element binding protein
SM	sphingomyelin
START	steroidogenic acute regulatory transport
VAP	vesicle-associated membrane protein-associated protein

6.1 Introduction

Cellular and extracellular cholesterol homeostasis is primarily controlled by the sterol-regulatory element binding protein (SREBP)/SREBP-cleavage activating protein/Insig complex (Goldstein et al., 2006) and nuclear liver X receptors (LXR) (Schmitz and Langmann, 2005), transcriptional regulatory circuits that physically interact with cholesterol or its oxygenated derivatives (oxysterols) in the endoplasmic reticulum (ER) and nucleus to impart negative and positive feed-back regulation of cholesterol synthesis, catabolism, uptake and efflux. The ER and nuclear envelope contain <1% of total cholesterol (Lange and Steck, 1997). However, culminating with the plasma membrane (PM), which contains 25-90% of total cellular sterols, organelles along the secretory pathway are progressively enriched in cholesterol. Hence transport between cholesterol-rich compartments and the ER are imperative in this regulatory scheme. It is increasingly apparent that de novo synthesized and lipoprotein-derived cholesterol is redistributed along the cellular cholesterol gradient by non-vesicular mechanisms involving transport proteins, and by association with sphingolipids in laterally segregated, liquid-ordered membrane domains termed 'rafts' (Ikonen, 2008).

6.2 Pathways of Intracellular Cholesterol Transport

Low density lipoprotein (LDL)-derived and de novo synthesized cholesterol traverses a complex intracellular pathway in membrane transport vesicles or complexed with carrier proteins (Ikonen, 2008; Liscum and Munn, 1999). Cholesterol synthesized in the ER is rapidly exported to the PM and late Golgi compartments by a non-vesicular, energy-dependent, cytoskeleton-independent pathway with kinetics that are consistent with involvement of transport proteins (Heino et al., 2000; Lange et al., 1991; Urbani and Simoni, 1990). Although controversial, recent evidence using fluorescence analogues indicates that 60-70% of cholesterol that arrives at the PM is associated with the cytosolic leaflet (Mondal et al., 2009). PM cholesterol is rapidly redistributed to the endosome recycling compartment with a $T_{1/2}$ of 2.5 min, indicative of a protein carrier(s) (Hao et al., 2002; Maxfield and Wustner, 2002). On the other hand, cholesterol is transported from the endosome recycling compartment to the PM by an energy-dependent mechanism with kinetics that suggest vesicular and tubulo-vesicular carriers (Hao et al., 2002). Cholesterol transported from the PM to the ER inhibits processing of SREBP and is esterified by acyl-CoA:cholesterol acyltransferase (ACAT). Transport is sensitive to hydrophobic amines and disruption of intermediate filaments but insensitive to energy poisons, indicating soluble carriers are involved that transit the lysosomes/endosomes (Liscum and Munn, 1999; Underwood et al., 1996). The sphingolipid content of the PM also dictates the rate of cholesterol desorption and delivery to PM. Removal of sphingomyelin (SM) with sphingomyelinase results in rapid influx of cholesterol to the ER where it is esterified and inhibits SREBP processing (Porn and Slotte, 1990; Scheek et al., 1997; Slotte et al., 1990). Sphingomyelinase-mediated cholesterol efflux to the ER is energy independent and unaffected by inhibitors of vesicular trafficking (Skiba et al., 1996).

The trafficking itinerary of LDL-derived cholesterol is relatively well understood due to studies on the Niemann-Pick C (NPC) cholesterol/sphingolipid storage disease (*see* Chapter 11). The cholesterol esters in endocytosed LDL are hydrolyzed by acid lipase in late endosomes (Chang et al., 2006). Cholesterol then appears to have two fates; transport to the PM and subsequent redistribution to the ER and other membranes (Cruz et al., 2000; Wojtanik and Liscum, 2003) or transport to the Golgi and ER (Underwood et al., 1998; Urano et al., 2008). NPC disease is caused by mutations in two genes products (NPC1 and NPC2) that promote egress of cholesterol and sphingolipids from late endosomes (Vance, 2006). NPC1 is a polytopic protein in the endosomes that binds oxysterols and cholesterol (Infante et al., 2008a; Ohgami et al., 2004), while NPC2 is a 16 kDa soluble glycoprotein in the lumen of the endosomes/lysosomes that binds and transfers cholesterol in vitro (Cheruku et al., 2006; Millat et al., 2001). Because of phenotypic similarities resulting from NPC1 and NPC2 deficiencies, it is possible that the two proteins function in a common pathway to promote cholesterol efflux from endosomes (Infante et al., 2008a,b).

6.3 Cholesterol Transfer by Soluble Binding Proteins

Although many cholesterol trafficking pathways involve transfer proteins, technical obstacles, such as high rates of spontaneous cholesterol transfer, lack of suitable probes and functional redundancy, have hampered their identification. Despite these problems, cholesterol transfer pathways have been identified that involve: 1) caveolin-stabilized lipid particles (Smart et al., 1996; Uittenbogaard et al., 1998), 2) ATP-binding cassette (ABC) transporters (Oram and Vaughan, 2006) and 3) soluble sterol-binding proteins. Although cholesterol has an appreciable rate of spontaneous exchange between membranes, transfer is significantly enhanced by soluble binding proteins that shield cholesterol from the aqueous environment by accommodation in hydrophobic pockets (Frolov et al., 1996; McLean and Phillips, 1981). Two large gene families that encode proteins with unique, high-affinity lipid binding folds have been implicated in sterol transfer; steroidogenic acute regulatory transport (START) proteins and OSBP/ORPs. The 15-member mammalian START family has helix-grip folds that, in the case of STARD1/StAR, STARD3/MLN64, STARD4 and STARD5, bind and/or transfer cholesterol (Kishida et al., 2004; Miller, 2007; Soccio et al., 2005). For a detailed overview of the START domain family, the reader is referred to a recent review (Alpy and Tomasetto, 2005) and Chapter 15 herein.

6.4 Oxysterol-Binding Protein (OSBP) and OSBP-Related Proteins (ORPs)

Oxysterol-binding protein (OSBP), the founder of a 12-member mammalian gene family, was identified in the 1980s as a high affinity receptor for a variety of side-chain and ring oxides of cholesterol (termed oxysterols) (Kandutsch and Shown, 1981). Although originally touted as a mediator of oxysterol suppression of cholesterol synthesis and uptake, recent studies indicate that the OSBP family has sterol transfer and/or sterol-sensing activities that serve to integrate cholesterol and lipid homeostasis with other cellular activities. The nomenclature for the mammalian family follows two related designations (Fig. 6.1); OSBP1, OSBP2, OSBPL1, OSBPL2, OSBPL3 and OSBPL5-OSBPL11 (Jaworski et al., 2001) or OSBP and ORP1-ORP11. The majority of studies use the latter designations and thus it will be used in this review and clarified where necessary. Since yeast OSBP homologues (OSH) were recently reviewed (Prinz, 2007; Schulz and Prinz, 2007), the emphasis here will be on structural and functional characterization of mammalian OSBP/ORPs, and evidence supporting the signalling and/or sterol transfer activity for this family.

6.5 Phylogenetic Distribution of OSBP/ORPs

OSBP/ORPs are restricted to eukaryotic lineages, with numerous homologues identified in animals, plants, fungi and protists (Beh et al., 2001; Lehto and Olkkonen, 2003; Skirpan et al., 2006; Zeng and Zhu, 2006). For example, the *S. cerevisiae*

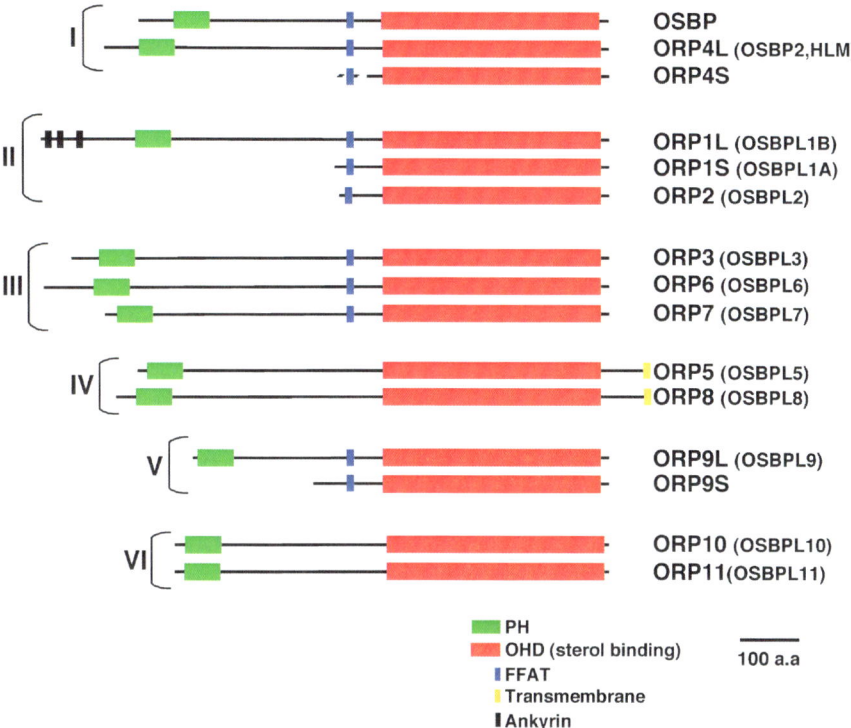

Fig. 6.1 Structural organization of the human OSBP/ORP family. Human OSBP/ORP family members are arranged into families I–VI (Lehto and Olkkonen, 2003). Alternate nomenclature is indicated in brackets next to each OSBP/ORP. Included in the figure are only those truncated or 'short' variants whose protein expression was verified in cell lines or tissue. Individual domain are colour-coded (see key) as follows; *black*, ankyrin; *green*, pleckstrin; *blue*, FFAT; *red*, OHD (sterol-binding); *yellow*, *trans*-membrane. The hatched line at the N-terminus of ORP4S indicates two alternate translation start sites.

genome encodes seven Osh proteins while mammalian genomes encode 12 members that can be subdivided into six subfamilies based on sequence similarity (Jaworski et al., 2001; Lehto and Olkkonen, 2003; Schulz and Prinz, 2007) (Fig. 6.1). OSBP/ORP expression does not strictly correlate with sterol biosynthetic activity since *D. malanogaster* and *C. elegans*, which do not synthesize but require sterols for survival, express four OSBP homologues (Jaworski et al., 2001) (Flybase and Wormbase websites). In an extreme case, parasitic *apicomplexan* protists have minimal synthetic capacity for lipids and sterols but express two ORPs that bind various negatively charged glycerolipids and possibly sterols (Zeng and Zhu, 2006).

Phylogenetic relationships between 120 taxa revealed support for clustering on the basis of type rather than taxonomic distribution. The clustering of OSBP/ORPs from numerous taxa into related groups suggests a set of early evolutionary ancestors (Jaworski et al., 2001; Zeng and Zhu, 2006). However, groupings of two to three mammalian and fungal OSBP/ORPs within each cluster also indicate more recent duplication events. Presently, there is insufficient data on individual OSBP/ORPs to determine whether groupings are indicative of a common function(s), but it is noteworthy that the presence or absence of domains in several human and fungal isoforms predicts their grouping into subfamilies.

6.6 Structural Organization of the OSBP/ORP Family

6.6.1 Ligand Binding Domain

All OSBP/ORPs share a highly conserved C-terminal 300–350 amino acid OSBP-homology domain (OHD) that binds cholesterol and/or oxysterols (Lehto et al., 2001; Ridgway et al., 1992) (Fig. 6.1). ORPs exist as either full-length, multi-domain proteins or truncated versions containing only the OHD. Recently, a 1.5 Å resolution structure of the OHD-only yeast homologue Osh4p was solved, to reveal a 270 amino acid, 19-strand β-barrel capped by a flexible α-helical lid (Im et al., 2005). Interestingly, the interior of the β-barrel is lined with hydrophobic residues but also contains as many as 15 water molecules, some of which form hydrogen bonds between the 3-hydroxyl of cholesterol, side-chain hydroxyls of oxysterols and amino acids lining the binding fold. The sterol 3-hydroxyl group is positioned at the bottom of the tunnel via hydrogen bonding to two water molecules and polar amino acids. The α-helical lid assumes a closed conformation by van der Waals interactions between the sterol ligand and aromatic residues in the lid and leucine residues in the lid hinge. The absence of direct ligand-protein interactions, structural similarity to other known binding proteins and similar affinity for cholesterol and oxysterols led the authors to conclude that Osh4p is a sterol transporter and not a signalling protein. This is also supported by a proposed sterol-binding mechanism, which involves a series of four basic residues around the entrance to the binding tunnel that in the apo conformation facilitate binding to anionic lipids and extraction of sterols (Im et al., 2005). This binding and release mechanism for sterol transfer between membranes is supported by molecular dynamics simulations (Canagarajah et al., 2008), and analysis of sterol binding and transfer by Osh4p mutants in vitro (Raychaudhuri et al., 2006).

The high degree of sequence conservation in the OHD suggests that other family members bind sterols in a similar manner to Osh4p. OSBP, ORP9, ORP4, ORP1 and ORP8 bind oxysterols (K_D 20–50 nM) and/or cholesterols (K_D 200–400 nM) (Ngo and Ridgway, 2009; Wang et al., 2002, 2008; Yan et al., 2007a, 2008) with similar affinity to Osh4p. The affinity and specificity of other ORPs for sterols is uncertain; all were derivatized to varying extents with photo-activated cholesterol and/or 25-hydroxycholesterol, but not with photo-activated phosphatidylcholine, in vitro and/or when expressed in COS cells (Suchanek et al., 2007). Osh4p, as well as the OHD-only proteins ORP2, ORP1S, ORP9S and ORP10S, bound phosphatidylinositol phosphates (PIP) and other anionic phospholipids through surface electrostatic interactions, possibly as part of a sterol extraction mechanism. However, Osh4p also catalyzed the transfer of phosphatidylinositol 4,5-*bis*phosphate (PI4,5P$_2$) and phosphatidylserine between vesicles in vitro, suggesting that lipid acyl-chains could be accommodated in the sterol binding fold (Raychaudhuri et al., 2006). The absence of direct contacts between sterol ligands and amino acid residues lining the Osh4p binding fold supports the concept of relaxed binding specificity, however non-sterol ligands have yet to be identified.

6.6.2 Organelle-Specific Targeting Domains

In addition to the highly conserved ligand-binding domain, OSBP/ORPs contain other protein modules-pleckstrin homology (PH), two phenylalanines in an acid tract (FFAT), Golgi dynamics and ankyrin domains-that bind lipids and proteins (Fig. 6.1). The unique combination and specificities of these domains facilitates the differential interaction of ORPs with organelle membranes as part of a sterol-transfer or signalling function. PH domains are found in a majority of OSBP/ORP and facilitate interaction with PIPs that are enriched in specific organelle membranes (Lemmon and Ferguson, 2001). For example, OSBP and ORP9L bind phosphatidylinositol 4-phosphate (PI4P) with relatively high specificity resulting in Golgi localization (Levine and Munro, 1998; Ngo and Ridgway, 2009), while N-terminal PH domains in ORP3, ORP6 and ORP7 facilitate localization to the PM (Lehto et al., 2004). A comprehensive analysis in yeast demonstrated that Osh1p and Osh2p were two of only six PH domain-containing proteins that were relatively non-specific for PIPs when assayed in vitro, but displayed high affinity for subcellular membranes, especially the Golgi apparatus (Yu et al., 2004). The lack of specificity for individual PIPs indicates that other factors also participate in interaction of OSBP/ORP PH domains with membranes, such as the small GTP-binding protein Arf1 (Godi et al., 2004; Levine and Munro, 2002; Roy and Levine, 2004). PH domains are present in ORPs across all taxa indicating this is a conserved mechanism for membrane binding.

Eight mammalian ORPs have FFAT domains (EFFDAxE) that interact with vesicle-associated membrane protein-associated protein (VAP) (Loewen and Levine, 2005; Wyles et al., 2002; Wyles and Ridgway, 2004). VAP is a type II integral membrane protein consisting of a N-terminal major sperm protein domain, an internal coiled-coiled domain and a C-terminal transmembrane domain (Skehel et al., 1995). Mammalian VAP-A and B genes encode isoforms that share 60% identity, localize to the ER and appear to be functionally redundant with respect to interaction with ORPs. Structural similarity to and interaction with v- and t-SNAREs indicates that VAP is involved in vesicular trafficking events (Soussan et al., 1999; Weir et al., 1998, 2001). However, results from these overexpression and in vitro studies could be due to non-specific sequestration of SNARES by VAP and inhibition of subsequent vesicular trafficking events.

VAP has a pleiotropic role in lipid transport and regulation by anchoring numerous proteins, including OSBP/ORPs, to the ER. FFAT motifs interact with a VAP consensus motif imbedded in the major sperm protein domain in a 2:2 stoichiometry

(Kaiser et al., 2005). Since FFAT domains are found in other lipid metabolic and regulatory proteins (Loewen et al., 2003), such as Nir2 and ceramide transfer protein (CERT), VAP is a docking site for other lipid regulatory and transfer proteins at the ER. A mutation in VAP (P56S) was shown to be the causative factor in a rare form of amyotrophic lateral sclerosis, by inappropriately inducing an unfolded protein response in the ER (Park et al., 2002). Whether mistargeting of OSBP/ORPs to the ER plays a role in the etiology of this form of amyotrophic lateral sclerosis is unknown.

Only ORP1L contains ankyrin motifs, a ubiquitous protein module involved in protein–protein interactions (Li et al., 2006). Ankyrin motifs in ORP1L mediate interaction with rab7 on the cytoplasmic aspect of endosomes. Interestingly, yeast Osh1p and Osh2p also have N-terminal ankyrin motifs, which in the case of Osh1p is required for localization to the nuclear-vacuolar junction (NVJ) (Levine and Munro, 2001), the site of piecemeal microautophagy of the nucleus. Recruitment of Osh1p to the NVJ occurs in response to nutrient deprivation and could be a mechanism to sequester Osh1p away from the Golgi apparatus, where it is involved in sterol-dependent transport of permeases to the PM (Kvam and Goldfarb, 2006, 2007).

A myriad of ORP mRNAs are produced as a result of alternate splicing and promoters, a feature that potentially impacts on the distribution of organelle-specific targeting domains and functions of individual ORPs. While data on these truncated variants is limited, truncated or 'short' ORP1, ORP4 and ORP9 (denoted by S) that encompass the C-terminal sterol binding domain and are missing the PH domains (Fig. 6.1), have activities that are distinct from the full-length or 'long' versions (denoted by L) (Johansson et al., 2003; Wang et al., 2002; Wyles and Ridgway, 2004).

6.7 Role of Mammalian OSBP/ORPs in Sterol Transport and Signalling

Mammalian OSBP/ORPs have been implicated in cell signalling (Wang et al., 2005, 2008), cytoskeletal organization (Johansson et al., 2007; Wang et al., 2002; Wyles et al., 2007), lipid homeostasis (Laitinen et al., 2002; Yan et al., 2007a,b), LXR/ABCA1 regulation (Bowden and Ridgway, 2008; Johansson et al., 2003; Yan et al., 2008), cell adhesion (Lehto et al., 2008) and SM metabolism (Bowden and Ridgway, 2008; Perry and Ridgway, 2006; Ridgway, 1995). A major challenge has been to explain how these diverse outcomes could be related to a common activity. For instance, how is OSBP regulation of CERT and ABCA1 (Bowden and Ridgway, 2008; Perry and Ridgway, 2006), related to the observation that OSBP is a cholesterol-dependent scaffold for phosphatases that regulate extracellular signal-regulated kinase (ERK) activity (Wang et al., 2005)? The apparent complexity has lead to the mutually inclusive views that OSBP/ORPs are sterol-signalling and/or transport proteins. Starting with OSBP, we will review the state

6.7.1 OSBP

OSBP was originally identified in the mid-1980s by Andrew Kandutsch and co-workers as a soluble, high affinity receptor for oxysterols such as 25-hydroxycholesterol. (Kandutsch and Shown, 1981; Kandutsch et al., 1984). Purification, cDNA cloning and expression studies revealed a unique 809 amino acid peptide that bound a variety of oxysterols but not DNA (Dawson et al., 1989a,b; Ridgway et al., 1992). It was only after searchable cDNA and genomic DNA databases were established that it became apparent that OSBP was the founding member of a large family of related sterol-binding proteins.

The PH domain of OSBP has affinity for a variety of phosphorylated PIs, but has preference for PI4P and PI4,5P$_2$ (Levine and Munro, 1998, 2002; Ngo and Ridgway, 2009). In vivo, OSBP interacts with the *trans* Golgi/TGN in a PI4P-specific manner, but also requires additional factors such as Arf1 (Levine and Munro, 2002). The FFAT domain of OSBP interacts with VAP in the ER (Loewen et al., 2003; Wyles et al., 2002). Thus the function(s) of OSBP are intimately related to 'dual targeting' to the ER and Golgi apparatus, which is modulated by sterol-binding (Ridgway et al., 1992; Wyles and Ridgway, 2004), phosphorylation status (Mohammadi et al., 2001; Wyles et al., 2007) and cellular SM content (Mohammadi et al., 2001).

In vitro, OSBP binds 25-hydroxycholesterol and cholesterol with high affinity (K_D s of 10 and 173 nM, respectively) (Dawson et al., 1989a; Wang et al., 2008), and catalyzes the PI4P-dependent transfer of cholesterol between liposomes (Ngo and Ridgway, 2009). Initial deletion mutations showed the OSBP sterol-binding domain encompassed amino acids 455–805 (Ridgway et al., 1992), the location of the predicted β-barrel from the structure of Osh4p (Im et al., 2005). However, a recent deletion analysis identified a minimal region between amino acids 408 and 459 that was sufficient for cholesterol and 25-hydroxycholesterol binding (Wang et al., 2008). This short region contains a putative cholesterol recognition/interaction amino acid consensus (CRAC) motif (Epand, 2006) that encompasses just the α-helical lid and β-sheets 1–3 of the predicted binding fold (Im et al., 2005). Moreover, CRAC domains are generally involved in sterol binding by integral proteins, and it is difficult to rationalize how this motif is involved in sterol-binding by a soluble protein such as OSBP.

Overexpression studies showed that OSBP increased cholesterol synthesis, reduced cholesterol esterification (Lagace et al., 1997) and enhanced oxysterol-dependent activation of SM synthesis (Lagace et al., 1999; Ridgway, 1995). The effects on cholesterol regulation could not be reproduced by RNA interference (RNAi) experiments and were likely related to sterol sequestration (Nishimura et al., 2005; Perry and Ridgway, 2006). Because eight OSBP family members have FFAT motifs and interact with the ER, it was not unexpected that ACAT activity and SREBP processing were unaffected by the loss of OSBP. However, the other site(s) for dual localization of OSBP/ORPs are unique and functions in these compartments are sensitive to silencing of individual genes. This implies a situation where numerous OSBP/ORPs collect sterols at the ER and deliver these ligands to unique compartments throughout the cell.

The activation of SM synthesis by 25-hydroxycholesterol was blocked by RNAi depletion of OSBP, and required sterol-dependent translocation of the ceramide transfer protein (CERT) to the Golgi apparatus. Like OSBP, CERT has PH and FFAT domains that mediate vectoral delivery of ceramide, bound by a C-terminal START domain, from the ER to SM synthase 1 in the *trans* Golgi/TGN (Hanada et al., 2003). Mechanistically this involves the VAP, PH and sterol-binding activities of OSBP, but not a physical interaction with CERT. Recent in vitro studies showing that OSBP extracts and transfers cholesterols imply that the effect on CERT is mediated indirectly by altering the sterol environment in the Golgi compartment (Ngo and Ridgway, 2009). Indeed, a TGN/endosomal PI4-kinase IIα is activated by OSBP, possibly by altering its sterol environment, leading to increased PI4P synthesis and CERT recruitment to the Golgi apparatus (Banerji and Ridgway, unpublished results). Thus OSBP is a sterol-transfer protein that regulates the cholesterol and/or oxysterol environment of the late Golgi compartment, an activity that is coupled to CERT-dependent ceramide delivery and SM synthesis in the same compartment (Fig. 6.2). This provides a plausible mechanism for coordinated raft assembly, which is known to occur in the *trans* Golgi/TGN (Klemm et al., 2009).

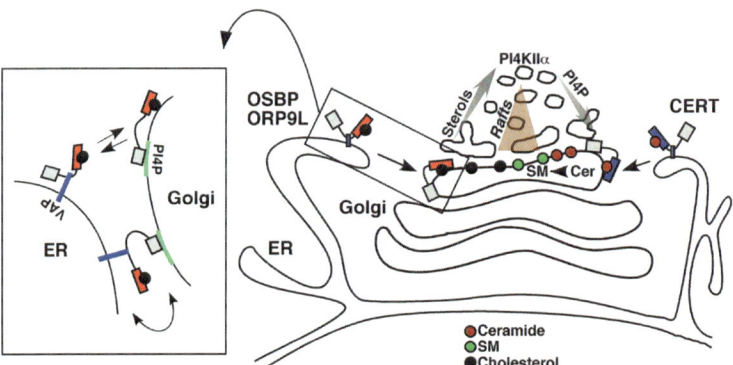

Fig. 6.2 Functional consequences of OSBP- and ORP9L-dependent sterol transfer between the ER and Golgi apparatus. Directional transfer of cholesterol and/or oxysterol by OSBP and ORP9L would maintain sterol homeostasis in the ER to *trans*-Golgi/TGN as well as downstream organelles. In the case of OSBP, sterol transfer would optimize the membrane environment of PI4KIIα resulting in increased PI4P synthesis, recruitment of CERT and increased ceramide delivery for SM synthesis. The coordinated transfer and synthesis of sterols and SM in elements of the late Golgi apparatus and endosomes would then maintain lipid raft assembly. In the case of ORP9L, cholesterol delivery to the Golgi apparatus affects secretion and sterol levels in post-Golgi compartments. The boxed insert shows that the mechanism of OSBP/ORP9L sterol transfer could involve: (1) sequential binding to donor and acceptor membranes or (2) tethering at membranes contact sites through simultaneous PH and FFAT domain interactions (Levine, 2004).

In addition to sterol-transfer activity, OSBP also imparts sterol-dependent regulation of cellular signalling pathways. In HEK293 cells, OSBP is a scaffold for two phosphatases that regulate the ERK pathway (Wang et al., 2005). Binding of cholesterol to OSBP recruits protein phosphatase 2A (PP2A) and the phosphotyrosine phosphatase HePTP resulting in ERK dephosphorylation and reduced kinase activity. Depletion of cholesterol or exogenous oxysterols dissembled the complex resulting in ERK hyper-phosphorylation. OSBP overexpression also stabilized the complex and promoted ERK dephosphorylation. Viral overexpression of OSBP in liver and hepatoma cells caused a similar decrease in phospho-ERK levels that was correlated with increased nuclear expression of SREPB-1c (Yan et al., 2007b). Deletion mapping and pull-down experiments showed that PP2A and HePTP bind at non-overlapping sites in the C-terminal 400 amino acids of OSBP (Wang et al., 2008).

In an apparently unrelated role, stimulation of pro-atherogenic profillin-1 expression in endothelial cells by 7-ketocholesterol is preceded by JAK2 phosphorylation of Y394 in OSBP, and recruitment and phosphorylation of STAT3 (Romeo and Kazlauskas, 2008). This provides a mechanism for oxysterol-mediated dysfunction of the cytoskeleton in endothelial cells but seems at odds with the relatively low affinity of OSBP for side-chain hydroxylated sterols (Dawson et al., 1989b; Taylor et al., 1984).

OSBP is also phosphorylated on serines 381, 384 and 387 adjacent to the FFAT motif in response to changes in SM and cholesterol metabolism (Mohammadi et al., 2001). Based on consensus sequences, S381 could be a protein kinase A site while S384 and S387 are casein kinase I sites. Depletion of cellular SM and cholesterol (>6 h) results in OSBP dephosphorylation and enhanced Golgi localization (Mohammadi et al., 2001; Ridgway et al., 1998b; Storey et al., 1998). Disruption of the Golgi with brefeldin A also promoted OSBP dephosphorylation (Ridgway et al., 1998a). This suggests that Golgi-localized OSBP is dephosphorylated and activated while the phospho-form is cytoplasmic or ER-localized. This model fits with recent studies of CERT, which is phosphorylated at a similar serine-rich site by protein kinase D and casein kinase I, rendering the enzyme cytoplasmic and inactive (Fugmann et al., 2007; Kumagai et al., 2007; Tomishige et al., 2009).

It is not surprising given the potential for functional redundancy within a multi-member gene family that manipulation of OSBP expression has only minor effects on SREBP regulation in the ER. In fact, OSBP appears to predominately affect cholesterol-regulated functions at the Golgi apparatus and in post-Golgi compartments. The activity and expression of the cholesterol and lipid efflux pump ABCA1 was increased by RNAi silencing of OSBP (Bowden and Ridgway, 2008). Interestingly, this did not involve transcriptional activation by LXR but rather stabilization of ABCA1 protein by a mechanism that involves the OSBP sterol-binding domain. This suggests that OSBP works in opposition to LXR by exposing ABCA1 to a membrane environment that increases protease degradation, possibly by internalization to endosomal compartments where calpains reside (Martinez et al., 2003; Wang et al., 2003). OSBP expression is also negatively correlated with processing of amyloid precursor protein (APP) to the amyloidogenic Aβ peptide (Zerbinatti et al.,

2008). Since APP processing by secretases is cholesterol-regulated (Grimm et al., 2005; Wolozin, 2001), OSBP could affect the sterol composition of membranes where Aβ processing occurs. Collectively, these data point to a model wherein OSBP controls cholesterol or oxysterol distribution in post-Golgi compartments, thus regulating the activity of cholesterol sensitive functions, including secretases, PI 4-kinases and ABCA1.

6.7.2 ORP1

Two ORP1 variants are expressed from separate promoters; a full-length protein containing ankyrin, PH, FFAT and sterol binding domains (ORP1L), and a truncated variant containing the FFAT and sterol-binding domains (ORP1S) (Johansson et al., 2003) (Fig. 6.1). ORP1L binds 25- and 22(R)-hydroxycholesterol, and modulates LXR transcriptional activity when overexpressed in macrophages (Yan et al., 2007a) and COS cells (Johansson et al., 2003). The PH domain of ORP1L has low specificity and affinity for PIPs and does not localize to organelle membranes when expressed in isolation. However, like full-length ORP1L, GFP-fusions with the ankyrin motifs are localized to the limiting membranes of endosomes. ORP1S is primarily found in the cytoplasm, binds anionic phospholipids and regulates vesicle transport from the Golgi apparatus (Xu et al., 2001).

ORP1L interaction with rab7 and the rab7-interacting lysosomal protein (RILP) has been implicated in positioning and maturation of endosomes. The dynein-dynactin motor is recruited to late endosomal membranes by interaction of RILP with p150^{Glued} and ORP1L with βIII spectrin (Johansson et al., 2005, 2007). This complex is then competent for minus-end transport of late-endosomes along microtubules. The cholesterol content in the outer membranes of late endosomes regulates the recruitment of the dynein-dynactin/p150^{glued} complex by a sensing function of the ORP1L sterol-binding domain (Rocha et al., 2009). When endosomal cholesterol is elevated, ORP1L binds cholesterol and promotes minus end transport and clustering of endosomes around the microtubule organizing center. Depletion of endosomal cholesterol changes the confirmation of ORP1L to enhance interaction with the ER partner VAP and subsequent displacement of the dynein/dynactin/p150^{glued}, a situation that favours dispersion of the late-endosomes.

Regulation of endosomal positioning and maturation could potentially affect cholesterol egress to other organelles and the sterol regulatory machinery in the ER or at other sites. While this has not been directly addressed, OPR1L enhanced (Johansson et al., 2005, 2007) or repressed (Yan et al., 2007a) the activity LXR and expression of some target genes, indicating a cell-specific effect at the levels of a sterol-transfer or signalling function.

6.7.3 ORP2

ORP2 is the only mammalian isoform expressed exclusively as a truncated 'short' version missing the PH but harbouring a FFAT motif. ORP2 bound

22(R)-hydroxycholesterol with a KD of 17 nM, but affinity for 25-and 7-keto-cholesterol are substantially less (Hynynen et al., 2009). ORP2 does not have a PH domain, but reportedly binds anionic phospholipids in a region of the sterol-binding domain analogous to the yeast homologue Osh4p (Li et al., 2002). In the case of Osh4p, anionic lipid binding stimulates cholesterol-transport activity and itself could also be a direct substrate for transfer (Raychaudhuri et al., 2006).

PIP binding activity could also be responsible for targeting of ORP2 to the Golgi apparatus where it disrupted ER-Golgi trafficking when overexpressed in CHO cells (Laitinen et al., 2002; Xu et al., 2001). However, overexpression studies in A431 cells showed that ORP2 was localized to the surface of lipid droplets and dissociated to the cytoplasm upon addition of exogenous 22(R)-hydroxycholesterol (Hynynen et al., 2009). Localization of endogenous ORP2 has yet to be verified, but its presence on lipid droplets is consistent with an inhibitory role in triglyceride, phospholipid and cholesterol ester metabolism (Hynynen et al., 2009; Kakela et al., 2005; Laitinen et al., 2002).

6.7.4 ORP3, ORP6 and ORP7

These three ORPs share extensive organizational and sequence similarity and are grouped in subfamily III (Lehto and Olkkonen, 2003) (Fig. 6.1). ORP3 has a complex pattern of alternate splicing that produces species with intact sterol-binding and PH domains, as well as variants with C-and N-terminal truncations that remove most of the sterol-binding and PH domains, respectively (Collier et al., 2003; Lehto et al., 2004). The kidney, lymph nodes and thymus have the highest level of ORP3 expression (Lehto et al., 2004). The sterol-binding activity of ORP3 has not been fully characterized but it was derivatized with photoactive 25-hydroxycholesterol and cholesterol (Suchanek et al., 2007). The ORP3-PH domain interacts with PI 3-phosphate, PI 3,4-*bis*phosphate and $PI4,5P_2$, and is localized to the PM when expressed as a tandem fusion with GFP (Lehto et al., 2005). This, combined with efficient targeting to the ER by its FFAT motif, suggests that ORP3 facilitates sterol transfer or communication between the PM and ER. The only functional analysis of ORP3 demonstrated effects of overexpression on cell spreading, adhesion and cytoskeleton organization that were consistent with negative regulation of ras activity (Lehto et al., 2008). ORP3 is also phosphorylated in response to loss of cell adhesion indicating regulation by inside out signalling via cell adhesion receptors. It remains to be determined how sterol binding by ORP3 is involved in ras regulation at the PM.

ORP6 is expressed highly in the brain and skeletal muscle, while ORP7 expression is limited to the stomach and intestinal tract (Lehto et al., 2004). ORP6 and ORP7 bound photo-activated 25-hydroxycholesterol but not photo-cholesterol (Suchanek et al., 2007). C- and N-terminal splice variants were identified that remove the PH and sterol binding domains, however the functional significance of these truncations is unknown. Interestingly, ORP6 is the only family member to be up-regulated in cholesterol-loaded macrophages (Lehto et al., 2001).

6.7.5 ORP4

ORP4 has a restricted expression profile with the highest mRNA levels in the brain, retina, heart and kidney (Moreira et al., 2001; Wang et al., 2002). Expression of OSBP and ORP4L in the retina suggests that these two receptors could protect against oxysterols in that tissue (Moreira et al., 2001). ORP4 (termed *HLM*) was also detected in the peripheral blood of patients with different solid tumours and in various cancer cell lines, and could be a prognostic marker for poor clinical outcome (Fournier et al., 1999).

ORP4 shares the greatest degree of sequence similarity and sterol-binding properties with OSBP, yet appears to be functionally unrelated. The ORP4 gene is expressed from alternate promoters to produce full-length ORP4L and an ORP4S variant missing the PH domain (Wang et al., 2002). Both forms bound 25-hydroxycholesterol (K_D 48 nM) and cholesterol (K_D 267 nM) in vitro (Wyles et al., 2007), but were insensitive to sterol-induced translocation to organelle membranes when expressed in CHO cells. Instead, ORP4S constitutively interacted with and reorganized or 'bundled' the vimentin intermediate filament network, such that it collapsed around the nucleus of CHO cells (Wang et al., 2002). The vimentin binding site is in the sterol-binding region of ORP4, but the two binding sites appeared to be non-interacting in vitro (Wyles et al., 2007). A leucine repeat adjacent to the FFAT motif in ORP4 was required for normal vimentin organization; overexpression of leucine repeat mutants of ORP4L or ORP4S caused reorganization and bundling of the vimentin filament network. Overexpression of ORP4L and ORP4S inhibits cholesterol esterification (Wang et al., 2002), consistent with the role of vimentin in mobilization of LDL-derived cholesterol from endosomes to the ER (Evans, 1994; Styers et al., 2004). Interestingly, ORP4L heterodimerized with OSBP, and mutations in the corresponding leucine repeat of OSBP also disrupted the vimentin network in an ORP4L-dependent manner (Wyles et al., 2007). Thus ORP4 could regulate cholesterol homeostasis by either directly modifying vimentin/endosome interaction or by utilizing vimentin as a scaffold for directed sterol transport.

6.7.6 ORP5 and ORP8

These two family members have C-terminal transmembrane domains and N-terminal PH domains, but lack FFAT motifs (Fig. 6.1). The transmembrane domain anchors ORP8 to the ER, presumably with the N-terminal region extending into the cytoplasm (Yan et al., 2008), thus dispensing with the need for the FFAT motif. The C-terminal region of ORP8 (a.a. 242–828) excluding the transmembrane domain, bound 25-hydroxycholesterol but not 24(S)- or 7-keto-cholesterol (Yan et al., 2008).

Gene silencing by RNAi revealed that ORP8 negatively regulates ABCA1 expression in macrophages. Depletion of ORP8 in THP-1 macrophages increased ABCA1 expression via up-regulation of LXR promoter activity due to: (1) sequestration of LXR ligands at the ER or (2) an effect on ER sterol metabolism that impacts indirectly on LXR activity in the nucleus (Yan et al., 2007b). This is similar to the observed negative regulation of ABCA1 expression by OSBP (Bowden and Ridgway, 2008), except that effect was at the post-transcriptional level. The domain structure of ORP5 is similar to ORP8 but its characteristics and functions have not been explored.

6.7.7 ORP9

ORP9 is expressed as a full-length (ORP9L) and N-terminal truncated variant expressed from an alternate promoter (ORP9S) (Wyles and Ridgway, 2004). ORP9L mRNA and protein are expressed highly in liver, kidney, heart and skeletal muscle (Wyles and Ridgway, 2004). Expression of the ORP9S mRNA was restricted to liver, lung, heart and prostate, and the protein is absent from most commonly used immortalized cell lines (unpublished results).

The ORP9L PH domain is specific for PI4P in the *trans* and *medial* Golgi (Ngo and Ridgway, 2009), while the FFAT domain binds VAP in the ER (Wyles et al., 2002; Wyles and Ridgway, 2004). Interestingly, ORP9L does not bind 25-hydroxycholesterol or cholesterol when the sterols are presented in aqueous dispersions but efficiently extracts and transfers cholesterol between phospholipid liposomes (Ngo and Ridgway, 2009). This is in contrast to OSBP, which binds sterols under both conditions. Cholesterol extraction and transfer by ORP9L was stimulated by PI4P and required intact sterol-binding and PH domains. Interestingly, transfer was only stimulated when PI4P was in donor or donor and acceptor liposomes, suggesting that the PH domain ligand must be in the correct context with the cholesterol ligand (Ngo and Ridgway, 2009).

In accordance with dual localization to the Golgi/ER and cholesterol transfer activity, a role for ORP9 in cholesterol homeostasis in the early secretory pathway was proposed (Ngo and Ridgway, 2009). Knockdown of ORP9L with siRNAs caused fragmentation of the Golgi and a modest reduction in ER–Golgi protein trafficking. In ORP9L knockdown CHO cells, there was a significant increase in cholesterol accumulation in the endosomal compartment (measured by filipin fluorescence) and 10–20% increase in total cellular cholesterol. This data suggests that ORP9L regulates cholesterol in the ER and Golgi apparatus, disruption of which causes abnormal cholesterol accumulation in the endosomes and defective vesicular transport. OSBP and ORP9L have non-overlapping functions in the ER/Golgi since ORP9L did not effect SM synthesis (Ngo and Ridgway, 2009) and OSBP did not affect ER-Golgi transport or Golgi organization (Perry and Ridgway, 2006). A recent study identified an up-regulated microRNA that modifies cholesterol uptake and represses ORP9 expression in oxidized-LDL-stimulated macrophages, further evidence for a role in cellular cholesterol transport and regulation (Chen et al., 2009).

When inducibly overexpressed in CHO cells, ORP9S caused reversible cessation of cell growth and ER-Golgi protein transport, and promoted Golgi fragmentation (Ngo and Ridgway, 2009). These effects were dependent on sterol binding and FFAT domains, indicating that ORP9S is a dominant inhibitor of ORP9L (and possibly other ORPs) at the ER. A negative regulatory role in ER–Golgi secretion is supported by the observation that ORP9S complemented the inhibitory effects of Osh4p on Golgi secretion in yeast (Fairn and McMaster, 2005).

ORP9L is phosphorylated at serine 473, a phosphoinositide-dependent kinase (PDK) 2 site that is required for the activation of a variety of AGC kinases (Lessmann et al., 2007). Phosphorylation of ORP9S at the corresponding site was increased in antigen-stimulated bone marrow-derived mast cells, and was sensitive to PKC inhibitors. Although not linked directly to function, depletion of OPR9L in CHO cells resulted in reduced phosphorylation of the AGC kinase Akt at a PDK2 site that is required for activation of the PI 3-kinase pathway. In light of recent evidence implicating ORP9 in cholesterol trafficking and ER-Golgi secretion, we suspect that Akt phosphorylation is linked to altered cellular sterol distribution.

6.7.8 ORP10 and ORP11

These final two members are grouped in the poorly characterized subfamily VI (Fig. 6.1). ORP10 is expressed in a number of tissues, including liver, kidney and lung, while ORP11 is highly expressed in lung and spleen (Lehto et al., 2001). Both proteins were labelled with photo-25-hydroxycholesterol, but not photo-cholesterol (Suchanek et al., 2007), and a truncated version of ORP10 missing the PH domain bound PI-3P that was immobilized on nitrocellulose (Fairn and McMaster, 2005). A potential role for ORP11 in lipid metabolism was highlighted by elevated mRNA expression and significant linkage between single nucleotide polymorphisms and LDL-cholesterol, hyperglycemia/diabetes and other abnormalities associated with Metabolic Syndrome (Bouchard et al., 2007, 2009).

6.8 Summary and Conclusions

Recent evidence from our lab and others suggest that many functions of fungal and mammalian OSBP/ORPs can be explained by a sterol transfer activity that alters the membrane sterol environment and thus activity of associated proteins and pathways. In this context, one can envision how this could be viewed as a sterol-signalling function, particularly when the sequence of events between sterol-binding and the observed phenotypic event are not identified. Dual targeting to organellar membranes, coupled with high affinity sterol binding, would allow OSBP/ORPs to selectively transfer cholesterol or oxysterols between donor and acceptor membranes in the following manner (Fig. 6.2). In the case of OSBP and ORP9L, binding to ER membranes by VAP recognition would initiate sterol extraction. The sterol-loaded form would then interact with an acceptor Golgi membrane containing PI4P and release the bound sterol. Transfer could take place by: (1) a diffusional mechanism, wherein the liganded ORP completely disengages from the donor membrane prior to interaction with the target membrane, or (2) bridging of closely apposed donor and acceptor membranes at contact sties by simultaneous binding of VAP and PIPs (Levine, 2004) (Fig. 6.2). Cholesterol delivery from the ER to the late Golgi/TGN by OSBP and ORP9L could be important to maintain the sterol gradient within the secretory pathway and coordinate this with SM to synthesis facilitate raft assembly.

However, before concluding that all ORPs are sterol transfer proteins it is necessary to resolve a number of outstanding issues. Important among these will be to assess the sterol-binding and transfer specificity of individual ORPs in vitro and in vivo. The measurement of in vivo transfer will be challenging owing to potential redundancy within the family and technical problems associated with measuring sterol transfer between organelles. The concept of redundancy was demonstrated in yeast, where lethality caused by deletion of all 7 OSH genes was rescued by expression of any one OSH gene (Beh and Rine, 2004). Yeast missing all 7 OSH genes displayed increased ergosterol levels (Beh et al., 2001), and a >7-fold reduction in cholesterol and ergosterol transfer from the PM to ER (Raychaudhuri et al., 2006). Defective sterol transfer was partially rescued by expression of Osh3p, Osh5p or Osh4p (Raychaudhuri et al., 2006), but results suggested that additional Osh proteins are also involved.

The ligand specificity of individual ORPs has also not been fully explored nor has the relationship between cholesterol versus oxysterol binding. Some ORPs bind oxysterols with an affinity that is approximately 5-fold greater than that for cholesterol (Im et al., 2005; Wang et al., 2008; Wyles et al., 2007). However, the concentration of cholesterol in a cell far exceeds that of oxysterols (approximately 10^3-fold), suggesting that the primary ligand is cholesterol. With this in mind, oxysterols could: (1) preferentially bind to ORPs in the cytoplasmic compartment or (2) alter the conformation and transfer activity due to differential interaction in the ORP sterol-binding pocket.

Unlike other sterol-binding and transfer proteins, the OSBP family of proteins possess high affinity sterol-binding coupled with auxiliary domains that target these proteins to organelle membranes. This combination designates them as important factors in the regulation of cholesterol distribution between membranes, which impacts on cholesterol synthesis, esterification and efflux. However, diverse expression, ligand specificity and localization ensure that the complex function(s) of this gene family will not be solved without considerable future effort.

References

Alpy, F., and Tomasetto, C., 2005, Give lipids a START: the StAR-related lipid transfer (START) domain in mammals. *J. Cell Sci.* **118**:2791–801.

Beh, C.T., Cool, L., Phillips, J., and Rine, J., 2001, Overlapping functions of the yeast oxysterol-binding protein homologues. *Genetics.* **157**:1117–40.

Beh, C.T., and Rine, J., 2004, A role for yeast oxysterol-binding protein homologs in endocytosis and in the maintenance of intracellular sterol-lipid distribution. *J. Cell Sci.* **117**:2983–96.

Bouchard, L., Faucher, G., Tchernof, A., Deshaies, Y., Marceau, S., Lescelleur, O., Biron, S., Bouchard, C., Perusse, L., and Vohl, M.C., 2009, Association of OSBPL11 gene polymorphisms with cardiovascular disease risk factors in obesity. *Obesity* **17**:1466–72.

Bouchard, L., Tchernof, A., Deshaies, Y., Marceau, S., Lescelleur, O., Biron, S., and Vohl, M.C., 2007, ZFP36: a promising candidate gene for obesity-related metabolic complications identified by converging genomics. *Obes. Surg.* **17**:372–82.

Bowden, K., and Ridgway, N.D., 2008, OSBP negatively regulates ABCA1 protein stability. *J. Biol. Chem.* **283**:18210–7.

Canagarajah, B.J., Hummer, G., Prinz, W.A., and Hurley, J.H., 2008, Dynamics of cholesterol exchange in the oxysterol binding protein family. *J. Mol. Biol.* **378**:737–48.

Chang, T.Y., Chang, C.C., Ohgami, N., and Yamauchi, Y., 2006, Cholesterol sensing, trafficking, and esterification. *Annu. Rev. Cell. Dev. Biol.* **22**:129–57.

Chen, T., Huang, Z., Wang, L., Wang, Y., Wu, F., Meng, S., and Wang, C., 2009, MicroRNA-125a-5p partly regulates the inflammatory response, lipid uptake, and ORP9 expression in oxLDL-stimulated monocyte/macrophages. *Cardiovasc. Res.* **83**:131–139.

Cheruku, S.R., Xu, Z., Dutia, R., Lobel, P., and Storch, J., 2006, Mechanism of cholesterol transfer from the Niemann-Pick type C2 protein to model membranes supports a role in lysosomal cholesterol transport. *J. Biol. Chem.* **281**:31594–604.

Collier, F.M., Gregorio-King, C.C., Apostolopoulos, J., Walder, K., and Kirkland, M.A., 2003, ORP3 splice variants and their expression in human tissues and hematopoietic cells. *DNA Cell Biol.* **22**:1–9.

Cruz, J.C., Sugii, S., Yu, C., and Chang, T.Y., 2000, Role of Niemann-Pick type C1 protein in intracellular trafficking of low density lipoprotein-derived cholesterol. *J. Biol. Chem.* **275**:4013–21.

Dawson, P.A., Ridgway, N.D., Slaughter, C.A., Brown, M.S., and Goldstein, J.L., 1989a, cDNA cloning and expression of oxysterol-binding protein, an oligomer with a potential leucine zipper. *J Biol Chem.* **264**:16798–803.

Dawson, P.A., Van der Westhuyzen, D.R., Goldstein, J.L., and Brown, M.S., 1989b, Purification of oxysterol binding protein from hamster liver cytosol. *J. Biol. Chem.* **264**:9046–52.

Epand, R.M., 2006, Cholesterol and the interaction of proteins with membrane domains. *Prog. Lipid Res.* **45**:279–94.

Evans, R.M., 1994, Intermediate filaments and lipoprotein cholesterol. *Trends Cell Biol.* **4**: 149–51.

Fairn, G.D., and McMaster, C.R., 2005, The roles of the human lipid-binding proteins ORP9S and ORP10S in vesicular transport. *Biochem. Cell Biol.* **83**:631–6.

Fournier, M.V., Guimaraes da Costa, F., Paschoal, M.E., Ronco, L.V., Carvalho, M.G., and Pardee, A.B., 1999, Identification of a gene encoding a human oxysterol-binding protein- homologue: a potential general molecular marker for blood dissemination of solid tumors. *Cancer Res.* **59**:3748–53.

Frolov, A., Woodford, J.K., Murphy, E.J., Billheimer, J.T., and Schroeder, F., 1996, Spontaneous and protein-mediated sterol transfer between intracellular membranes. *J. Biol. Chem.* **271**:16075–83.

Fugmann, T., Hausser, A., Schoffler, P., Schmid, S., Pfizenmaier, K., and Olayioye, M.A., 2007, Regulation of secretory transport by protein kinase D-mediated phosphorylation of the ceramide transfer protein. *J. Cell Biol.* **178**:15–22.

Godi, A., Di Campli, A., Konstantakopoulos, A., Di Tullio, G., Alessi, D.R., Kular, G.S., Daniele, T., Marra, P., Lucocq, J.M., and De Matteis, M.A., 2004, FAPPs control Golgi-to-cell-surface membrane traffic by binding to ARF and PtdIns(4)P. *Nat. Cell Biol.* **6**:393–404.

Goldstein, J.L., DeBose-Boyd, R.A., and Brown, M.S., 2006, Protein sensors for membrane sterols. *Cell* **124**:35–46.

Grimm, M.O., Grimm, H.S., Patzold, A.J., Zinser, E.G., Halonen, R., Duering, M., Tschape, J.A., De Strooper, B., Muller, U., Shen, J., and Hartmann, T., 2005, Regulation of cholesterol and sphingomyelin metabolism by amyloid-β and presenilin. *Nat. Cell Biol.* **7**:1118–23.

Hanada, K., Kumagai, K., Yasuda, S., Miura, Y., Kawano, M., Fukasawa, M., and Nishijima, M., 2003, Molecular machinery for non-vesicular trafficking of ceramide. *Nature* **426**: 803–9.

Hao, M., Lin, S.X., Karylowski, O.J., Wustner, D., McGraw, T.E., and Maxfield, F.R., 2002, Vesicular and non-vesicular sterol transport in living cells. The endocytic recycling compartment is a major sterol storage organelle. *J. Biol. Chem.* **277**:609–17.

Heino, S., Lusa, S., Somerharju, P., Ehnholm, C., Olkkonen, V.M., and Ikonen, E., 2000, Dissecting the role of the golgi complex and lipid rafts in biosynthetic transport of cholesterol to the cell surface. *Proc. Natl. Acad. Sci. U S A* **97**:8375–80.

Hynynen, R., Suchanek, M., Spandl, J., Back, N., Thiele, C., and Olkkonen, V.M., 2009, OSBP-related protein 2 (ORP2) is a sterol receptor on lipid droplets that regulates the metabolism of neutral lipids. *J. Lipid. Res.* **50**:1305–1315.

Ikonen, E., 2008, Cellular cholesterol trafficking and compartmentalization. *Nat. Rev. Mol. Cell Biol.* **9**:125–38.

Im, Y.J., Raychaudhuri, S., Prinz, W.A., and Hurley, J.H., 2005, Structural mechanism for sterol sensing and transport by OSBP-related proteins. *Nature* **437**:154–8.

Infante, R.E., Abi-Mosleh, L., Radhakrishnan, A., Dale, J.D., Brown, M.S., and Goldstein, J.L., 2008a, Purified NPC1 protein. I. Binding of cholesterol and oxysterols to a 1278-amino acid membrane protein. *J. Biol. Chem.* **283**:1052–63.

Infante, R.E., Wang, M.L., Radhakrishnan, A., Kwon, H.J., Brown, M.S., and Goldstein, J.L., 2008b, NPC2 facilitates bidirectional transfer of cholesterol between NPC1 and lipid bilayers, a step in cholesterol egress from lysosomes. *Proc. Natl. Acad. Sci. U S A* **105**: 15287–92.

Jaworski, C.J., Moreira, E., Li, A., Lee, R., and Rodriguez, I.R., 2001, A family of 12 human genes containing oxysterol-binding domains. *Genomics* **78**:185–96.

Johansson, M., Bocher, V., Lehto, M., Chinetti, G., Kuismanen, E., Ehnholm, C., Staels, B., and Olkkonen, V.M., 2003, The two variants of oxysterol binding protein-related protein-1 display different tissue expression patterns, have different intracellular localization, and are functionally distinct. *Mol. Biol. Cell* **14**:903–15.

Johansson, M., Lehto, M., Tanhuanpaa, K., Cover, T.L., and Olkkonen, V.M., 2005, The oxysterol-binding protein homologue ORP1L interacts with Rab7 and alters functional properties of late endocytic compartments. *Mol. Biol. Cell* **16**:5480–92.

Johansson, M., Rocha, N., Zwart, W., Jordens, I., Janssen, L., Kuijl, C., Olkkonen, V.M., and Neefjes, J., 2007, Activation of endosomal dynein motors by stepwise assembly of Rab7-RILP-p150Glued, ORP1L, and the receptor betaIII spectrin. *J. Cell Biol.* **176**:459–71.

Kaiser, S.E., Brickner, J.H., Reilein, A.R., Fenn, T.D., Walter, P., and Brunger, A.T., 2005, Structural basis of FFAT motif-mediated ER targeting. *Structure* **13**:1035–45.

Kakela, R., Tanhuanpaa, K., Laitinen, S., Somerharju, P., and Olkkonen, V.M., 2005, Overexpression of OSBP-related protein 2 (ORP2) in CHO cells induces alterations of phospholipid species composition. *Biochem. Cell Biol.* **83**:677–83.

Kandutsch, A.A., and Shown, E.P., 1981, Assay of oxysterol-binding protein in a mouse fibroblast, cell-free system. Dissociation constant and other properties of the system. *J. Biol. Chem.* **256**:13068–73.

Kandutsch, A.A., Taylor, F.R., and Shown, E.P., 1984, Different forms of the oxysterol-binding protein. Binding kinetics and stability. *J. Biol. Chem.* **259**:12388–97.

Kishida, T., Kostetskii, I., Zhang, Z., Martinez, F., Liu, P., Walkley, S.U., Dwyer, N.K., Blanchette-Mackie, E.J., Radice, G.L., and Strauss, J.F., 3rd., 2004, Targeted mutation of the MLN64 START domain causes only modest alterations in cellular sterol metabolism. *J. Biol. Chem.* **279**:19276–85.

Klemm, R.W., Ejsing, C.S., Surma, M.A., Kaiser, H.J., Gerl, M.J., Sampaio, J.L., de Robillard, Q., Ferguson, C., Proszynski, T.J., Shevchenko, A., and Simons, K., 2009, Segregation of sphingolipids and sterols during formation of secretory vesicles at the trans-Golgi network. *J. Cell Biol.* **185**:601–12.

Kumagai, K., Kawano, M., Shinkai-Ouchi, F., Nishijima, M., and Hanada, K., 2007, Interorganelle trafficking of ceramide is regulated by phosphorylation-dependent cooperativity between the PH and START domains of CERT. *J. Biol. Chem.* **282**:17758–66.

Kvam, E., and Goldfarb, D.S., 2006, Nucleus-vacuole junctions in yeast: anatomy of a membrane contact site. *Biochem. Soc. Trans.* **34**:340–2.

Kvam, E., and Goldfarb, D.S., 2007, Nucleus-vacuole junctions and piecemeal microautophagy of the nucleus in S. cerevisiae. *Autophagy* **3**:85–92.

Lagace, T.A., Byers, D.M., Cook, H.W., and Ridgway, N.D., 1997, Altered regulation of cholesterol and cholesteryl ester synthesis in Chinese-hamster ovary cells overexpressing the oxysterol-binding protein is dependent on the pleckstrin homology domain. *Biochem. J.* **326**:205–13.

Lagace, T.A., Byers, D.M., Cook, H.W., and Ridgway, N.D., 1999, Chinese hamster ovary cells overexpressing the oxysterol binding protein (OSBP) display enhanced synthesis of sphingomyelin in response to 25-hydroxycholesterol. *J. Lipid Res.* **40**:109–16.

Laitinen, S., Lehto, M., Lehtonen, S., Hyvarinen, K., Heino, S., Lehtonen, E., Ehnholm, C., Ikonen, E., and Olkkonen, V.M., 2002, ORP2, a homolog of oxysterol binding protein, regulates cellular cholesterol metabolism. *J. Lipid Res.* **43**:245–55.

Lange, Y., Echevarria, F., and Steck, T.L., 1991, Movement of zymosterol, a precursor of cholesterol, among three membranes in human fibroblasts. *J. Biol. Chem.* **266**:21439–43.

Lange, Y., and Steck, T.L., 1997, Quantitation of the pool of cholesterol associated with acyl-CoA:cholesterol acyltransferase in human fibroblasts. *J. Biol. Chem.* **272**:13103–8.

Lehto, M., Hynynen, R., Karjalainen, K., Kuismanen, E., Hyvarinen, K., and Olkkonen, V.M., 2005, Targeting of OSBP-related protein 3 (ORP3) to endoplasmic reticulum and plasma membrane is controlled by multiple determinants. *Exp. Cell Res.* **310**:445–62.

Lehto, M., Laitinen, S., Chinetti, G., Johansson, M., Ehnholm, C., Staels, B., Ikonen, E., and Olkkonen, V.M., 2001, The OSBP-related protein family in humans. *J. Lipid Res.* **42**:1203–13.

Lehto, M., Mayranpaa, M.I., Pellinen, T., Ihalmo, P., Lehtonen, S., Kovanen, P.T., Groop, P.H., Ivaska, J., and Olkkonen, V.M., 2008, The R-Ras interaction partner ORP3 regulates cell adhesion. *J. Cell Sci.* **121**:695–705.

Lehto, M., and Olkkonen, V.M., 2003, The OSBP-related proteins: a novel protein family involved in vesicle transport, cellular lipid metabolism, and cell signalling. *Biochim. Biophys. Acta* **1631**:1–11.

Lehto, M., Tienari, J., Lehtonen, S., Lehtonen, E., and Olkkonen, V.M., 2004, Subfamily III of mammalian oxysterol-binding protein (OSBP) homologues: the expression and intracellular localization of ORP3, ORP6, and ORP7. *Cell Tissue Res.* **315**:39–57.

Lemmon, M.A., and Ferguson, K.M., 2001, Molecular determinants in pleckstrin homology domains that allow specific recognition of phosphoinositides. *Biochem. Soc. Trans.* **29**:377–84.

Lessmann, E., Ngo, M., Leitges, M., Minguet, S., Ridgway, N.D., and Huber, M., 2007, Oxysterol-binding protein-related protein (ORP) 9 is a PDK-2 substrate and regulates Akt phosphorylation. *Cell Signal.* **19**:384–92.

Levine, T., 2004, Short-range intracellular trafficking of small molecules across endoplasmic reticulum junctions. *Trends Cell Biol.* **14**:483–90.

Levine, T.P., and Munro, S., 1998, The pleckstrin homology domain of oxysterol-binding protein recognises a determinant specific to Golgi membranes. *Curr. Biol.* **8**:729–39.

Levine, T.P., and Munro, S., 2001, Dual targeting of Osh1p, a yeast homologue of oxysterol-binding protein, to both the Golgi and the nucleus-vacuole junction. *Mol. Biol. Cell.* **12**:1633–44.

Levine, T.P., and Munro, S., 2002, Targeting of Golgi-specific pleckstrin homology domains involves both PtdIns 4-kinase-dependent and -independent components. *Curr. Biol.* **12**:695–704.

Li, J., Mahajan, A., and Tsai, M.D., 2006, Ankyrin repeat: a unique motif mediating protein-protein interactions. *Biochemistry* **45**:15168–78.

Li, X., Rivas, M.P., Fang, M., Marchena, J., Mehrotra, B., Chaudhary, A., Feng, L., Prestwich, G.D., and Bankaitis, V.A., 2002, Analysis of oxysterol binding protein homologue Kes1p function in regulation of Sec14p-dependent protein transport from the yeast Golgi complex. *J. Cell Biol.* **157**:63–77.

Liscum, L., and Munn, N.J., 1999, Intracellular cholesterol transport. *Biochim. Biophys. Acta* **1438**:19–37.

Loewen, C.J., and Levine, T.P., 2005, A highly conserved binding site in vesicle-associated membrane protein-associated protein (VAP) for the FFAT motif of lipid-binding proteins. *J. Biol. Chem.* **280**:14097–104.

Loewen, C.J., Roy, A., and Levine, T.P., 2003, A conserved ER targeting motif in three families of lipid binding proteins and in Opi1p binds VAP. *EMBO J.* **22**:2025–35.

Martinez, L.O., Agerholm-Larsen, B., Wang, N., Chen, W., and Tall, A.R., 2003, Phosphorylation of a pest sequence in ABCA1 promotes calpain degradation and is reversed by ApoA-I. *J. Biol. Chem.* **278**:37368–74.

Maxfield, F.R., and Wustner, D., 2002, Intracellular cholesterol transport. *J Clin. Invest.* **110**:891–8.

McLean, L.R., and Phillips, M.C., 1981, Mechanism of cholesterol and phosphatidylcholine exchange or transfer between unilamellar vesicles. *Biochemistry* **20**:2893–900.

Millat, G., Chikh, K., Naureckiene, S., Sleat, D.E., Fensom, A.H., Higaki, K., Elleder, M., Lobel, P., and Vanier, M.T., 2001, Niemann-Pick disease type C: spectrum of HE1 mutations and genotype/phenotype correlations in the NPC2 group. *Am. J. Hum. Genet.* **69**:1013–21.

Miller, W.L., 2007, Steroidogenic acute regulatory protein (StAR), a novel mitochondrial cholesterol transporter. *Biochim. Biophys. Acta* **1771**:663–76.

Mohammadi, A., Perry, R.J., Storey, M.K., Cook, H.W., Byers, D.M., and Ridgway, N.D., 2001, Golgi localization and phosphorylation of oxysterol binding protein in Niemann-Pick C and U18666A-treated cells. *J. Lipid Res.* **42**:1062–71.

Mondal, M., Mesmin, B., Mukherjee, S., and Maxfield, F.R., 2009, Sterols are mainly in the cytoplasmic leaflet of the plasma membrane and the endocytic recycling compartment in CHO cells. *Mol. Biol. Cell* **20**:581–8.

Moreira, E.F., Jaworski, C., Li, A., and Rodriguez, I.R., 2001, Molecular and biochemical characterization of a novel oxysterol-binding protein (OSBP2) highly expressed in retina. *J. Biol. Chem.* **276**:18570–8.

Ngo, M., and Ridgway, N.D., 2009, Oxysterol binding protein-related Protein 9 (ORP9) is a cholesterol transfer protein that regulates Golgi structure and function. *Mol. Biol. Cell* **20**:1388–99.

Nishimura, T., Inoue, T., Shibata, N., Sekine, A., Takabe, W., Noguchi, N., and Arai, H., 2005, Inhibition of cholesterol biosynthesis by 25-hydroxycholesterol is independent of OSBP. *Genes Cells* **10**:793–801.

Ohgami, N., Ko, D.C., Thomas, M., Scott, M.P., Chang, C.C., and Chang, T.Y., 2004, Binding between the Niemann-Pick C1 protein and a photoactivatable cholesterol analog requires a functional sterol-sensing domain. *Proc. Natl. Acad. Sci. U S A* **101**:12473–8.

Oram, J.F., and Vaughan, A.M., 2006, ATP-Binding cassette cholesterol transporters and cardiovascular disease. *Circ. Res.* **99**:1031–43.

Park, Y.U., Hwang, O., and Kim, J., 2002, Two-hybrid cloning and characterization of OSH3, a yeast oxysterol-binding protein homolog. *Biochem. Biophys. Res. Commun.* **293**:733–40.

Perry, R.J., and Ridgway, N.D., 2006, Oxysterol-binding protein and vesicle-associated membrane protein-associated protein are required for sterol-dependent activation of the ceramide transport protein. *Mol. Biol. Cell* **17**:2604–16.

Porn, M.I., and Slotte, J.P., 1990, Reversible effects of sphingomyelin degradation on cholesterol distribution and metabolism in fibroblasts and transformed neuroblastoma cells. *Biochem. J.* **271**:121–6.

Prinz, W.A., 2007, Non-vesicular sterol transport in cells. *Prog. Lipid Res.* **46**:297–314.

Raychaudhuri, S., Im, Y.J., Hurley, J.H., and Prinz, W.A., 2006, Nonvesicular sterol movement from plasma membrane to ER requires oxysterol-binding protein-related proteins and phosphoinositides. *J. Cell Biol.* **173**:107–19.

Ridgway, N.D., 1995, 25-Hydroxycholesterol stimulates sphingomyelin synthesis in Chinese hamster ovary cells. *J. Lipid Res.* **36**:1345–58.

Ridgway, N.D., Badiani, K., Byers, D.M., and Cook, H.W., 1998a, Inhibition of phosphorylation of the oxysterol binding protein by brefeldin A. *Biochim. Biophys. Acta* **1390**:37–51.

Ridgway, N.D., Dawson, P.A., Ho, Y.K., Brown, M.S., and Goldstein, J.L., 1992, Translocation of oxysterol binding protein to Golgi apparatus triggered by ligand binding. *J. Cell Biol.* **116**: 307–19.

Ridgway, N.D., Lagace, T.A., Cook, H.W., and Byers, D.M., 1998b, Differential effects of sphingomyelin hydrolysis and cholesterol transport on oxysterol-binding protein phosphorylation and Golgi localization. *J. Biol. Chem.* **273**:31621–8.

Rocha, N., Kuijl, C., van der Kant, R., Janssen, L., Houben, D., Janssen, H., Zwart, W., and Neefjes, J., 2009. Cholesterol sensor ORP1L contacts the ER protein VAP to control Rab7-RILP-p150^{glued} and late endosome positioning. *J. Cell Biol.* **185**:1209–25.

Romeo, G.R., and Kazlauskas, A., 2008, Oxysterol and diabetes activate STAT3 and control endothelial expression of profilin-1 via OSBP1. *J. Biol. Chem.* **283**:9595–605.

Roy, A., and Levine, T.P., 2004, Multiple pools of PtdIns 4-phosphate detected using the pleckstrin homology domain of Osh2p. *J. Biol. Chem.* **279**:44683–9.

Scheek, S., Brown, M.S., and Goldstein, J.L., 1997, Sphingomyelin depletion in cultured cells blocks proteolysis of sterol regulatory element binding proteins at site 1. *Proc. Natl. Acad. Sci. U S A* **94**:11179–83.

Schmitz, G., and Langmann, T., 2005, Transcriptional regulatory networks in lipid metabolism control ABCA1 expression. *Biochim. Biophys. Acta* **1735**:1–19.

Schulz, T.A., and Prinz, W.A., 2007, Sterol transport in yeast and the oxysterol binding protein homologue (OSH) family. *Biochim. Biophys. Acta* **1771**:769–80.

Skehel, P.A., Martin, K.C., Kandel, E.R., and Bartsch, D., 1995, A VAMP-binding protein from Aplysia required for neurotransmitter release. *Science* **269**:1580–3.

Skiba, P.J., Zha, X., Maxfield, F.R., Schissel, S.L., and Tabas, I., 1996, The distal pathway of lipoprotein-induced cholesterol esterification, but not sphingomyelinase-induced cholesterol esterification, is energy-dependent. *J. Biol. Chem.* **271**:13392–400.

Skirpan, A.L., Dowd, P.E., Sijacic, P., Jaworski, C.J., Gilroy, S., and Kao, T.H., 2006, Identification and characterization of PiORP1, a Petunia oxysterol-binding-protein related protein involved in receptor-kinase mediated signalling in pollen, and analysis of the ORP gene family in Arabidopsis. *Plant Mol. Biol.* **61**:553–65.

Slotte, J.P., Harmala, A.S., Jansson, C., and Porn, M.I., 1990, Rapid turn-over of plasma membrane sphingomyelin and cholesterol in baby hamster kidney cells after exposure to sphingomyelinase. *Biochim. Biophys. Acta* **1030**:251–7.

Smart, E.J., Ying, Y., Donzell, W.C., and Anderson, R.G., 1996, A role for caveolin in transport of cholesterol from endoplasmic reticulum to plasma membrane. *J. Biol. Chem.* **271**: 29427–35.

Soccio, R.E., Adams, R.M., Maxwell, K.N., and Breslow, J.L., 2005, Differential gene regulation of StarD4 and StarD5 cholesterol transfer proteins. Activation of StarD4 by sterol regulatory element-binding protein-2 and StarD5 by endoplasmic reticulum stress. *J. Biol. Chem.* **280**:19410–8.

Soussan, L., Burakov, D., Daniels, M.P., Toister-Achituv, M., Porat, A., Yarden, Y., and Elazar, Z., 1999, ERG30, a VAP-33-related protein, functions in protein transport mediated by COPI vesicles. *J. Cell Biol.* **146**:301–11.

Storey, M.K., Byers, D.M., Cook, H.W., and Ridgway, N.D., 1998, Cholesterol regulates oxysterol binding protein (OSBP) phosphorylation and Golgi localization in Chinese hamster ovary cells: correlation with stimulation of sphingomyelin synthesis by 25-hydroxycholesterol. *Biochem. J.* **336**:247–56.

Styers, M.L., Salazar, G., Love, R., Peden, A.A., Kowalczyk, A.P., and Faundez, V., 2004, The endo-lysosomal sorting machinery interacts with the intermediate filament cytoskeleton. *Mol. Biol. Cell* **15**:5369–82.

Suchanek, M., Hynynen, R., Wohlfahrt, G., Lehto, M., Johansson, M., Saarinen, H., Radzikowska, A., Thiele, C., and Olkkonen, V.M., 2007, The mammalian oxysterol-binding protein-related proteins (ORPs) bind 25-hydroxycholesterol in an evolutionarily conserved pocket. *Biochem. J.* **405**:473–80.

Taylor, F.R., Saucier, S.E., Shown, E.P., Parish, E.J., and Kandutsch, A.A., 1984, Correlation between oxysterol binding to a cytosolic binding protein and potency in the repression of hydroxymethylglutaryl coenzyme A reductase. *J. Biol. Chem.* **259**:12382–7.

Tomishige, N., Kumagai, K., Kusuda, J., Nishijima, M., and Hanada, K., 2009, Casein kinase Iγ2 down-regulates trafficking of ceramide in the synthesis of sphingomyelin. *Mol. Biol. Cell* **20**:348–57.

Uittenbogaard, A., Ying, Y., and Smart, E.J., 1998, Characterization of a cytosolic heat-shock protein-caveolin chaperone complex. Involvement in cholesterol trafficking. *J. Biol Chem.* **273**:6525–32.

Underwood, K.W., Andemariam, B., McWilliams, G.L., and Liscum, L., 1996, Quantitative analysis of hydrophobic amine inhibition of intracellular cholesterol transport. *J. Lipid Res.* **37**:1556–68.

Underwood, K.W., Jacobs, N.L., Howley, A., and Liscum, L., 1998, Evidence for a cholesterol transport pathway from lysosomes to endoplasmic reticulum that is independent of the plasma membrane. *J. Biol. Chem.* **273**:4266–74.

Urano, Y., Watanabe, H., Murphy, S.R., Shibuya, Y., Geng, Y., Peden, A.A., Chang, C.C., and Chang, T.Y., 2008, Transport of LDL-derived cholesterol from the NPC1 compartment to the ER involves the trans-Golgi network and the SNARE protein complex. *Proc. Natl. Acad. Sci. U S A* **105**:16513–8.

Urbani, L., and Simoni, R.D., 1990, Cholesterol and vesicular stomatitis virus G protein take separate routes from the endoplasmic reticulum to the plasma membrane. *J. Biol. Chem.* **265**:1919–23.

Vance, J.E., 2006, Lipid imbalance in the neurological disorder, Niemann-Pick C disease. *FEBS Lett.* **580**:5518–24.

Wang, C., JeBailey, L., and Ridgway, N.D., 2002, Oxysterol-binding-protein (OSBP)-related protein 4 binds 25-hydroxycholesterol and interacts with vimentin intermediate filaments. *Biochem. J.* **361**:461–72.

Wang, N., Chen, W., Linsel-Nitschke, P., Martinez, L.O., Agerholm-Larsen, B., Silver, D.L., and Tall, A.R., 2003, A PEST sequence in ABCA1 regulates degradation by calpain protease and stabilization of ABCA1 by apoA-I. *J. Clin. Invest.* **111**:99–107.

Wang, P.Y., Weng, J., and Anderson, R.G., 2005, OSBP is a cholesterol-regulated scaffolding protein in control of ERK 1/2 activation. *Science* **307**:1472–6.

Wang, P.Y., Weng, J., Lee, S., and Anderson, R.G., 2008, The N terminus controls sterol binding while the C terminus regulates the scaffolding function of OSBP. *J. Biol. Chem.* **283**:8034–45.

Weir, M.L., Klip, A., and Trimble, W.S., 1998, Identification of a human homologue of the vesicle-associated membrane protein (VAMP)-associated protein of 33 kDa (VAP-33): a broadly expressed protein that binds to VAMP. *Biochem. J.* **333**:247–51.

Weir, M.L., Xie, H., Klip, A., and Trimble, W.S., 2001, VAP-A binds promiscuously to both v- and tSNAREs. *Biochem. Biophys. Res. Commun.* **286**:616–21.

Wojtanik, K.M., and Liscum, L., 2003, The transport of low density lipoprotein-derived cholesterol to the plasma membrane is defective in NPC1 cells. *J. Biol. Chem.* **278**:14850–6.

Wolozin, B., 2001, A fluid connection: cholesterol and Aβ. *Proc. Natl. Acad. Sci. U S A* **98**:5371–3.

Wyles, J.P., McMaster, C.R., and Ridgway, N.D., 2002, Vesicle-associated membrane protein-A (VAP-A) interacts with the oxysterol-binding protein to modify export from the endoplasmic reticulum. *J. Biol. Chem.* **277**:29908–18.

Wyles, J.P., Perry, R.J., and Ridgway, N.D., 2007, Characterization of the sterol-binding domain of oxysterol-binding protein (OSBP)-related protein 4 reveals a novel role in vimentin organization. *Exp. Cell Res.* **313**:1426–37.

Wyles, J.P., and Ridgway, N.D., 2004, VAMP-associated protein-A regulates partitioning of oxysterol-binding protein-related protein-9 between the endoplasmic reticulum and Golgi apparatus. *Exp. Cell Res.* **297**:533–47.

Xu, Y., Liu, Y., Ridgway, N.D., and McMaster, C.R., 2001, Novel members of the human oxysterol-binding protein family bind phospholipids and regulate vesicle transport. *J. Biol Chem.* **276**:18407–14.

Yan, D., Jauhiainen, M., Hildebrand, R.B., Willems van Dijk, K., Van Berkel, T.J., Ehnholm, C., Van Eck, M., and Olkkonen, V.M., 2007a, Expression of human OSBP-related protein 1L in macrophages enhances atherosclerotic lesion development in LDL receptor-deficient mice. *Arterioscler. Thromb. Vasc. Biol.* **27**:1618–24.

Yan, D., Lehto, M., Rasilainen, L., Metso, J., Ehnholm, C., Yla-Herttuala, S., Jauhiainen, M., and Olkkonen, V.M., 2007b, Oxysterol binding protein induces upregulation of SREBP-1c and enhances hepatic lipogenesis. *Arterioscler. Thromb. Vasc. Biol.* **27**:1108–14.

Yan, D., Mayranpaa, M.I., Wong, J., Perttila, J., Lehto, M., Jauhiainen, M., Kovanen, P.T., Ehnholm, C., Brown, A.J., and Olkkonen, V.M., 2008, OSBP-related protein 8 (ORP8) suppresses ABCA1 expression and cholesterol efflux from macrophages. *J. Biol. Chem.* **283**:332–340.

Yu, J.W., Mendrola, J.M., Audhya, A., Singh, S., Keleti, D., DeWald, D.B., Murray, D., Emr, S.D., and Lemmon, M.A., 2004, Genome-wide analysis of membrane targeting by S. cerevisiae pleckstrin homology domains. *Mol. Cell* **13**:677–88.

Zeng, B., and Zhu, G., 2006, Two distinct oxysterol binding protein-related proteins in the parasitic protist Cryptosporidium parvum (Apicomplexa). *Biochem. Biophys. Res. Commun.* **346**: 591–99.

Zerbinatti, C.V., Cordy, J.M., Chen, C.D., Guillily, M., Suon, S., Ray, W.J., Seabrook, G.R., Abraham, C.R., and Wolozin, B., 2008, Oxysterol-binding protein-1 (OSBP1) modulates processing and trafficking of the amyloid precursor protein. *Mol. Neurodegener.* **3**:5.

Chapter 7
High Density Lipoprotein Structure–Function and Role in Reverse Cholesterol Transport

Sissel Lund-Katz and Michael C. Phillips

Abstract High density lipoprotein (HDL) possesses important anti-atherogenic properties and this review addresses the molecular mechanisms underlying these functions. The structures and cholesterol transport abilities of HDL particles are determined by the properties of their exchangeable apolipoprotein (apo) components. ApoA-I and apoE, which are the best characterized in structural terms, contain a series of amphipathic α-helical repeats. The helices located in the amino-terminal two-thirds of the molecule adopt a helix bundle structure while the carboxy-terminal segment forms a separately folded, relatively disorganized, domain. The latter domain initiates lipid binding and this interaction induces changes in conformation; the α-helix content increases and the amino-terminal helix bundle can open subsequently. These conformational changes alter the abilities of apoA-I and apoE to function as ligands for their receptors. The apoA-I and apoE molecules possess detergent-like properties and they can solubilize vesicular phospholipid to create discoidal HDL particles with hydrodynamic diameters of ~10 nm. In the case of apoA-I, such a particle is stabilized by two protein molecules arranged in an anti-parallel, double-belt, conformation around the edge of the disc. The abilities of apoA-I and apoE to solubilize phospholipid and stabilize HDL particles enable these proteins to be partners with ABCA1 in mediating efflux of cellular phospholipid and cholesterol, and the biogenesis of HDL particles. ApoA-I-containing nascent HDL particles play a critical role in cholesterol transport in the circulation whereas apoE-containing HDL particles mediate cholesterol transport in the brain. The mechanisms by which HDL particles are remodeled by lipases and lipid transfer proteins, and interact with SR-BI to deliver cholesterol to cells, are reviewed.

Keywords HDL · Cholesterol · Lipoprotein · apoA-I · apoE

M.C. Phillips (✉)
The Children's Hospital of Philadelphia, Philadelphia, PA 19104-4318, USA
e-mail: phillipsmi@email.chop.edu

Abbreviations

ANS	8-anilino-1-napthalenesulfonic acid
AP	acute phase
apo	apolipoprotein
CE	cholesterol ester
CETP	cholesteryl ester transfer protein
DMPC	dimyristoyl PC
EL	endothelial lipase
FC	free (unesterified) cholesterol
HDL	high density lipoprotein
HL	hepatic lipase
LCAT	lecithin-cholesterol acyltransferase
LDL	low density lipoprotein
LUV	large unilamellar vesicle
MLV	multilamellar vesicle
PC	phosphatidylcholine
PL	phospholipid
PLTP	phospholipid transfer protein
RCT	reverse cholesterol transport
SAA	serum amyloid A
SUV	small unilamellar vesicle
TAG	triacylglycerol
VLDL	very low density lipoprotein

7.1 Introduction

Human serum lipoproteins are soluble complexes of proteins (apolipoproteins) and lipids that represent the major cholesterol transport vehicles in both the intravascular and extravascular compartments. Lipoprotein particles are synthesized by the liver and intestine and mediate lipid transport from the intestine to the liver, and between the liver and cells in the periphery of the body. Mature lipoproteins are microemulsion or emulsion particles (diameter range = 7–600 nm) containing a core of neutral lipids (triacylglycerol (TAG), cholesteryl ester (CE) and cholesterol) stabilized by a surface monomolecular film of phospholipids (PL), cholesterol and apolipoproteins (apo). Lipoprotein particles are traditionally fractionated on the basis of their densities (Jonas and Phillips, 2008).

Apolipoproteins are part of a multi-gene family (Li et al., 1988). ApoB100 is the principal protein constituent of low density lipoprotein (LDL) particles and, by acting as a ligand for the low density lipoprotein receptor, it targets TAG and cholesterol for delivery to cells. Because of this activity, apoB100-containing lipoprotein particles are atherogenic (*see* Chapter 8 for details). The focus in this chapter is on the anti-atherogenic lipoprotein, high density lipoprotein (HDL), which mediates the efflux of cellular cholesterol. HDL participates in the reverse cholesterol transport process whereby excess cholesterol in cells in the periphery is transported to the liver and ultimately excreted from the body in the feces (Cuchel and Rader, 2006). Apolipoprotein-mediated interactions of HDL particles with cell surface receptors and lipid transporters are critical for this process. HDL contains exchangeable apolipoproteins of the A, C and E families that evolved from a common ancestral gene and are structurally similar. These protein molecules contain 22-amino acid tandem repeats that are often separated by a proline residue (Li et al., 1988). The repeating 22-amino acid segments form amphipathic α-helices (Segrest et al., 1992) thereby enabling these apolipoprotein molecules to bind well to PL-water surfaces and stabilize lipoprotein particles. Lipoprotein particles in the circulation are highly dynamic and, besides acting as ligands for cell surface receptors, apolipoproteins also mediate lipoprotein remodeling. In the case of HDL, the apolipoproteins participate in particle remodeling by interacting with lipid transfer proteins and enzymes that modify lipids (Lund-Katz et al., 2003). These reactions are critical for HDL metabolism and effective reverse cholesterol transport (RCT).

In this consideration of apolipoproteins as cholesterol transport proteins, the focus is on human HDL and two of its constituent proteins, apoA-I and apoE, that are the best characterized in structural terms. The structural basis for the multiple functions of these protein molecules and how they influence the contributions of HDL to RCT are addressed in the following sections.

7.2 Structures of ApoA-I and ApoE in the Lipid-Free State

7.2.1 Primary and Secondary Structures

The genes of apoA-I and apoE contain four exons and three introns, with similar locations of intron–exon boundaries and similar intron and exon lengths for the first three exons. The differences in the total length of the mRNA are due to variations in the length of the fourth exon. Exons three and four in the apoA-I and apoE genes encode the entire mature protein sequence. In both cases, exon 4 codes for a primary structure of 11- and 22-amino acids tandem repeats that span residues 44–243 in apoA-I and 62–299 in apoE (Fig. 7.1A) (Li et al., 1988). Each of these repeats has the periodicity of an amphipathic α-helix and these helices are often separated by a proline residue (Li et al., 1988; Segrest et al., 1992). The amphipathic α-helices have been classified into several distinct classes according to the distribution of charged residues around the axis of the helices (Segrest et al., 1994). The class A amphipathic helix is a major lipid-binding motif in exchangeable apolipoproteins and is characterized by the location of basic residues near the hydrophilic/hydrophobic interface and acidic residues at the center of the polar face (Fig. 7.1B). Class G* and Y amphipathic α-helices are also present in exchangeable apolipoproteins (Segrest et al., 1992). The class G* helix is similar to the amphipathic α-helices present in water-soluble globular proteins and possesses a random radial arrangement of positive and negative amino acids in the polar face. Sometimes the amphipathic α-helix

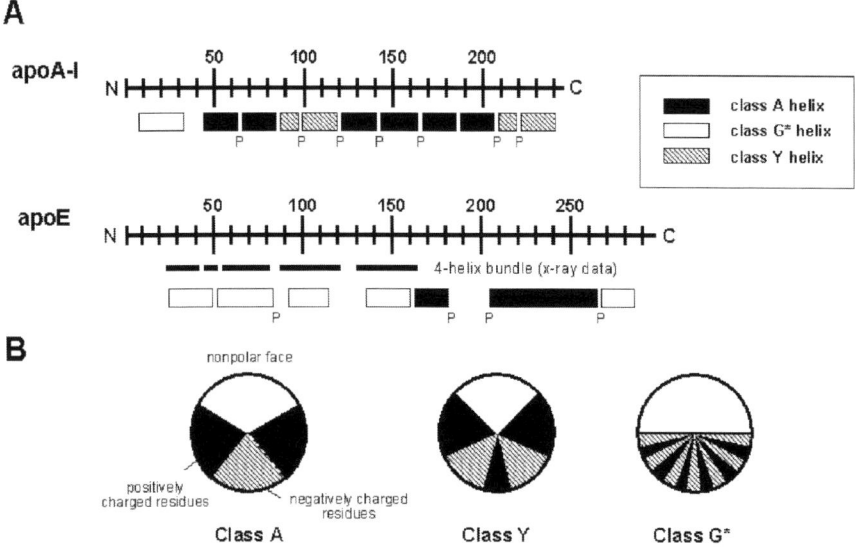

Fig. 7.1 **A** Distribution of amphipathic α-helices in the human exchangeable apolipoproteins, apoA-I and apoE. The letter P below the rectangles indicates positions of all proline residues. **B** Amphipathic helix classes found in the exchangeable apolipoproteins. Classification is based on the distribution of charged residues (see Section 2.1). These figures were adapted from Segrest et al. (1992)

is characterized by the presence of a Y-shaped cluster of basic amino acids in the polar face (Fig. 7.1B) giving a class Y helix (Segrest et al., 1994).

Human apoA-I is a 243 amino acid protein (molecular mass = 28.1 kDa) in which the region coded by exon 4 is predicted to contain eight 22-mer and two 11-mer amphipathic α-helices with most of the helices being punctuated by prolines (Fig. 7.1A) (Brouillette et al., 2001; Segrest et al., 1992). The predicted α-helices shown in Fig. 7.1A for human apoA-I include approximately 80% of the amino acids and represent the maximal helix content; for comparison, the lipid-free protein is about 50% α-helical in dilute solution, as revealed by circular dichroism measurements (Saito et al., 2003b). Comparison of sequences between mammals indicates that the N-terminal region of apoA-I is highly conserved while the central and C-terminal regions show conservative substitutions between species (Brouillette et al., 2001; Frank and Marcel, 2000). Studies of synthetic peptides corresponding to each of the 22-residue amphipathic α-helices of human apoA-I have shown that the first (residues 44–65) and last (residues 220–241) repeat helices have the greatest lipid affinity (Palgunachari et al., 1996). Hydropathy analysis of the amino acid sequence indicates that the C-terminal region of human apoA-I is very hydrophobic (Saito et al., 2004b), consistent with this region having significant lipid binding ability. Studies of both natural and engineered mutations in the human apoA-I molecule have revealed that the C-terminal region is indeed important for lipid binding (Brouillette et al., 2001; Saito et al., 2004b) and that the central region corresponding to residues 121–186 is important for activation of the enzyme lecithin: cholesterol acyltransferase (LCAT) (Frank and Marcel, 2000; Sorci-Thomas and Thomas, 2002).

Human apoE is a 299 amino acid protein with a molecular mass of 34.2 kDa (Weisgraber, 1994), that is predicted to contain amphipathic α-helices along its length. In contrast to apoA-I, a high proportion of these amphipathic α-helices are class G* (Fig. 7.1A). The distribution of 22-residue repeats and the proline punctuation are less regular than occurs in the apoA-I molecule (Fig. 7.1A). The predicted α-helices include about 70% of the amino acids which is somewhat higher than the value of ~60% measured by circular dichroism for lipid-free apoE in dilute solution (Morrow et al., 2000). There is a high degree of sequence conservation across species with the exceptions of the N- and C-termini: homology begins in the vicinity of residue 26 in the human sequence and continues to approximately residue 288 (Weisgraber, 1994). Hydropathy analysis of the human apoE sequence shows that the C-terminus as well a central region (residues 192–215) are relatively hydrophobic (Saito et al., 2004b); the C-terminal region plays a critical role in lipid binding (Saito et al., 2004b; Weisgraber, 1994). Human apoE exists as three major isoforms, apoE2, apoE3 and apoE4, each differing by cysteine and arginine at positions 112 and 158. ApoE3, the most common form, contains cysteine and arginine at these positions, respectively, whereas apoE2 contains cysteine and apoE4 contains arginine at both sites (Weisgraber, 1994). These apoE isoforms are associated with different levels of disease risk, most notably for atherosclerosis (Davignon et al., 1988; Getz and Reardon, 2009) and Alzheimer's disease (Mahley et al., 2009; Mahley and Huang, 1999) (*see* also Chapter 2). Comparisons of apoE2, apoE3 and apoE4 together with studies of other natural apoE mutations have led to the identification of a cluster of basic amino acids in the regions spanning residues 136–150 as the recognition site responsible for the binding of apoE to the low density lipoprotein (LDL) receptor (Hatters et al., 2006; Weisgraber, 1994).

7.2.2 Tertiary Structure

A variety of studies using protein engineering techniques have provided important insights into the lipid-free structures of apoA-I and apoE (for reviews, *see* Brouillette et al., 2001; Davidson and Thompson, 2007; Saito et al., 2004b; Weisgraber, 1994). Proteolysis analysis (Roberts et al., 1997) and deletion mutagenesis studies (Davidson et al., 1996; Saito et al., 2003b) have suggested that the lipid-free apoA-I molecule is organized into two structural domains; the N-terminal and central parts form a helix bundle whereas the C-terminal α-helices form a separate, less organized structure. The helix bundle organization in the N-terminal domain is also supported by fluorescence studies of single tryptophan mutants of human (Brouillette et al., 2005; Davidson et al., 1999) and chicken apoA-I (Kiss et al., 1999). The guanidine-induced denaturation curve of apoA-I is monophasic (Reijngoud and Phillips, 1982) whereas that for apoE is biphasic (Morrow et al., 2000) indicating that apoE also adopts a two-domain tertiary structure, and that the

N- and C-terminal domains unfold independently. The helix bundle motif of the N-terminal domain is similar in apoA-I and apoE except for it being less organized and less stable in apoA-I. A thermal unfolding study has demonstrated that the apoA-I molecule exhibits a loosely folded, molten globular-like structure (i.e. the α-helices do not occupy fixed positions with respect to one another and are not organized into a unique tertiary structure) (Gursky and Atkinson, 1996).

ApoA-I is the only intact human apolipoprotein in the lipid-free state for which a high resolution structure is available to date. The crystal structure (Ajees et al., 2006) reveals an N-terminal anti-parallel four-helix bundle domain and a separate two-helix C-terminal domain (Fig. 7.2A). Apparently, the conditions used for crystallization induced helix formation because some 80% of the residues in this structure are in α-helices, whereas the helix content is closer to 50% for monomeric apoA-I in dilute solution. Nevertheless, cross-linking/mass spectrometry experiments (Silva et al., 2005) indicate that the apoA-I molecule in solution is still folded into two domains, with residues 1–189 forming the helix bundle and residues 190–243 folding separately. In the case of apoE, the structure of the 22-kDa N-terminal fragment has been solved by crystallographic (Wilson et al., 1991) and

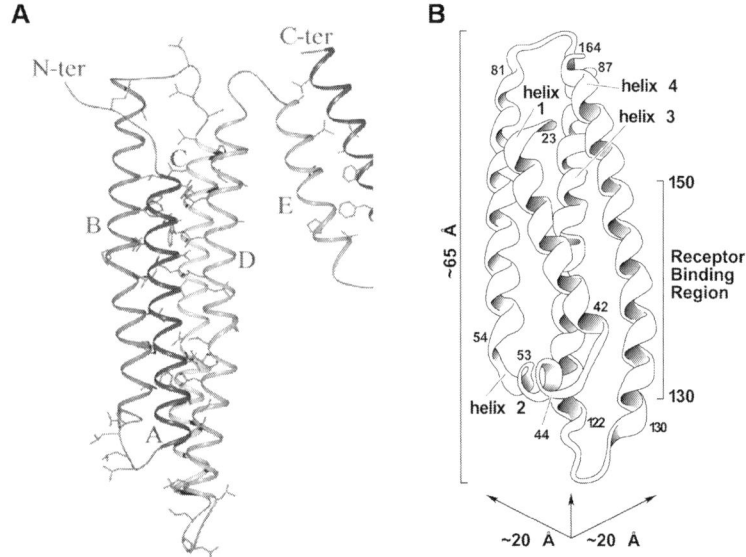

Fig. 7.2 Crystal structures of human apolipoproteins in the lipid-free state. **A** The six α-helices in human apoA-I are shown (*from* Ajees et al., 2006, with permission). The N-terminal anti-parallel four-helix bundle contains helices A (residues 10–39), B (50–84), C (97–137), and D (147–187). The C-terminal domain is formed by the two α-helices E (residues 196–213) and F (219–242). Hydrophobic residues located in the interior of the helix bundles are shown as sticks. **B** Ribbon model of the structure of the 22-kDa N-terminal domain fragment of human apoE3 (from Weisgraber (1994), with permission). Four of the five helices are arranged in an anti-parallel four-helix bundle. The residues spanned by each helix, together with the region in helix 4 recognized by the LDL receptor, are indicated

NMR (Sivashanmugam and Wang, 2009) methods (Fig. 7.2B). As occurs with apoA-I, this domain forms a globular bundle of four elongated α-helices in which the helices are arranged in an anti-parallel manner with their hydrophobic faces oriented towards the interior of the bundle. This structure shares the same basic architecture as the helix bundle of insect apolipophorin III (Narayanaswami and Ryan, 2000). The N-terminal fragments of all three apoE isoforms adopt such a four-helix bundle motif, but subtle differences in the side-chain conformations and salt-bridge arrangements of the isoforms affect their functions and characteristics (Dong et al., 1994, 1996; Wilson et al., 1994). Studies of guanidine (Morrow et al., 2000) and thermal (Acharya et al., 2002) denaturation revealed that the N-terminal fragments of the apoE isoforms differ in stability (apoE4 < apoE3 < apoE2), indicating that replacing cysteine residues by arginine results in a progressive decrease in hydrophobicity and stability of the helix bundle such that in apoE4 this domain exhibits molten globule characteristics (Morrow et al., 2002).

The structural organization of the C-terminal domain in both apoA-I and apoE is not well characterized. Fluorescence measurements with 8-anilino-1-naphthalenesulfonic acid (ANS) have suggested that the C-terminal domain of apoE forms a solvent-exposed, less organized structure (Saito et al., 2003a). A model has been proposed in which the helices in the C-terminal domain form an intermolecular coiled-coil helix structure (Choy et al., 2003). The polymorphism at position 112 in the N-terminal domain of apoE affects the properties of the C-terminal domain because of interactions between the domains (Hatters et al., 2006). In apoE4, the N- and C-terminal domains interact differently than they do in the other isoforms: Arg-112 causes a rearrangement of the Arg-61 side chain in the N-terminal domain of apoE4, allowing it to interact with Glu-255 in the C-terminal domain (Dong et al., 1994; Dong and Weisgraber, 1996). This domain interaction in human apoE4 leads to a less organized structure in the C-terminal domain, leading to preferential association with VLDL rather than HDL, which is contrary to the behavior of apoE3 (Weisgraber, 1990). The domain interaction in apoE4 has been suggested to contribute to the accelerated catabolism of this isoform and, consequently, the increased cholesterol and LDL levels in the plasma of individuals with this genotype (Davignon et al., 1988; Mahley et al., 1999). Fluorescence energy transfer measurements indicate that the N- and C-terminal domains of lipid-free apoE3 are close to one another and interact, probably through weak hydrophobic interaction (Narayanaswami et al., 2001); the two domains are closer together in apoE4 than in apoE3 (Hatters et al., 2005). In the case of apoA-I, an electron paramagnetic resonance spectroscopic study indicated that the helices in the C-terminal domain form a compact anti-parallel alignment with residues 188–205 existing as a flexible loop (Oda et al., 2003), whereas fluorescence resonance energy transfer measurements suggested an extended conformation (Behling Agree et al., 2002). Regardless of its exact conformation, the C-terminal domain appears to be relatively disorganized because it contains an exposed hydrophobic surface that is accessible to ANS binding (Saito et al., 2003b). Analogous to the situation with apoE, the N- and C-terminal domains in the apoA-I molecule interact with each other to maintain the overall structure of the protein (Fang et al., 2003; Tricerri et al., 2000). Human and mouse apoA-I both adopt a two-domain tertiary structure implying that apoA-I from all higher mammals is organized similarly (Tanaka et al., 2008). Human apoA-I functions with a relatively stable N-terminal helix bundle and a hydrophobic C-terminal domain, whereas mouse apoA-I functions with an unstable helix bundle domain and a polar C-terminal domain.

7.2.3 Quaternary Structure

It is well established that lipid-free apoA-I and apoE tend to self-associate in aqueous solution because of hydrophobic interaction between amphipathic α-helices. Lipid-free apoA-I self-associates reversibly as a function of protein concentration and forms oligomers at concentrations > 0.1 mg/ml (Donovan et al., 1987; Vitello and Scanu, 1976). Sedimentation equilibrium ultracentrifugation and gel filtration chromatography data suggest a model of apoA-I self-association that involves equilibration between monomer–dimer–tetramer–octamer states. Studies on C-terminal truncation mutants demonstrate that the oligomerization is less pronounced (Davidson et al., 1996), indicating that self-association of apoA-I is mediated by hydrophobic interactions between α-helices in the C-terminal domain (Section 2.2). Compared to human apoA-I, lipid-free mouse apoA-I undergoes only minimal concentration-dependent self-association (Gong et al., 1994), because its C-terminal domain is relatively polar (Tanaka et al., 2008).

ApoE is known to exist as a tetramer in the lipid-free state (Yokoyama et al., 1985) and this tetramerization is thought to be mediated by the C-terminal domain, because the N-terminal 22-kDa fragment is monomeric whereas the 10-kDa C-terminal fragment is tetrameric in aqueous solution (Aggerbeck et al., 1988). Sedimentation velocity experiments provide direct evidence for heterogeneous solution structures of apoE3 and apoE4 (Perugini et al., 2000). In a lipid-free environment, both proteins exist as a slow equilibrium mixture of monomers, tetramers, octamers and a small proportion of higher oligomers. Experiments with mixtures of apoE and phospholipid micelles indicate that apoE oligomers undergo a phospholipid-induced dissociation, suggesting that the monomeric form predominates on lipoprotein particle surfaces (Perugini et al., 2000). Sedimentation velocity studies using recombinant apoE isoforms fused with an amino-terminal extension of 43 amino acids confirmed that monomers, dimers and tetramers are the major species of lipid-free apoE2, apoE3 and apoE4 (Barbier et al., 2006). Analysis of apoE C-terminal truncation variants indicates that the extreme C-terminal residues 267–299, which are known to be very hydrophobic by hydropathy analysis (Saito et al., 2004b), are responsible for the self-association of apoE (Westerlund and Weisgraber, 1993). A segment containing residues 218–266 in the apoE C-terminal domain possesses a high propensity to form a coiled-coil helix structure and an apoE construct containing C-terminal residues 201–299 gives circular dichroism spectra characteristic of coiled-coil helices (Choy et al., 2003). This apoE C-terminal domain construct predominantly forms dimeric and tetrameric species in aqueous solution. A monomeric, biologically active apoE C-terminal domain mutant has been prepared using protein engineering techniques (Fan et al., 2004). In this model, five bulky hydrophobic residues (F257, W264, V269, L279, and V287) are replaced with either smaller hydrophobic or polar/charged residues. Cross-linking experiments indicate that this mutant is 100% monomeric even at concentrations as high as 5 mg/ml. A monomeric, biologically active, full-length apoE has also been generated in similar fashion (Zhang et al., 2007). This mutant is nearly 95–100% monomeric even at 20 mg/ml, consistent with interactions between hydrophobic residues in the C-terminal domain playing a key role in the self-association process. To understand the molecular basis for the different degrees of self-association of the apoE isoforms, the effects of progressive truncation of the C-terminal domain in human apoE3 and apoE4 on their lipid-free structures have been compared (Sakamoto et al., 2008). In contrast to previously reported findings, gel filtration chromatography experiments demonstrate that the monomer–tetramer distribution is different for the two isoforms with apoE4 being more monomeric than apoE3; removal of the C-terminal helices in both isoforms favors the monomeric state.

Overall, it can be inferred that self-association of apoA-I and apoE promotes stabilization of the potential α-helical segments of the C-terminal domain (Sections 7.2.1 and 7.2.2) that are less organized in the monomeric form.

7.3 Interaction of ApoA-I and ApoE with Lipids

7.3.1 Molecular Mechanism of Lipid-Binding

It is well established that the C-terminal domain in the apoA-I and apoE molecules is critical for lipid-binding (for reviews, see Weisgraber, 1994; Saito et al., 2004b; Hatters et al., 2006). In the case of apoA-I, an early model proposed a multi-step mechanism in which the initial binding to lipid occurs through the C-terminal region followed by a conformational switch of residues 1–43 which unmasks a latent lipid-binding domain comprising residues 44–65 (Rogers et al., 1998). By using a series of deletion mutants that progressively lacked different regions along the molecule, we showed that the binding of apoA-I to lipids is modulated by reorganization of the N-terminal helix bundle structure (Saito et al., 2003b). Recent fluorescence experiments with pyrene-labelled apoA-I indicate the helix bundle can open upon interaction with the surface of a lipid particle (Kono et al., 2008). This concept is consistent with earlier experiments with apoE demonstrating that upon binding to lipids, the four-helix bundle in the N-terminal domain undergoes a conformational reorganization to expose the hydrophobic faces of its amphipathic helices for interaction with lipid molecules (Weisgraber, 1994). Molecular area measurements at an air-water interface indicated that the N-terminal domain occupies a larger surface area than can be accounted for by its globular four-helix bundle conformation, suggesting adoption of an open conformation by the helix bundle (Weisgraber et al., 1992; Weisgraber, 1994). Subsequent studies using infrared spectroscopy (Raussens et al., 1998), fluorescence resonance energy transfer (Fisher and Ryan, 1999), and inter-helical disulfide mutants of the apoE N-terminal domain (Lu et al., 2000) confirmed that the four-helix bundle undergoes conformational opening when apoE is complexed with PL. This conformational rearrangement is associated with changes in functionality (Saito et al., 2004b). The helix bundle can adopt either open (recognized by the LDL receptor) or closed (not recognized by the LDL receptor) conformations, depending upon the concentration of apoE bound to the PL particle surface (Saito et al., 2001). The rate of interfacial rearrangement is affected by the stability of the helix bundle domain. Thus, apoE4 rearranges more rapidly than apoE3 upon binding to a lipoprotein surface (Nguyen et al., 2009) because, as mentioned in Section 7.2.2, the apoE4 N-terminal domain is relatively unstable. The different domain-domain interaction in apoE4 compared to apoE3 (Section 7.2.2) causes the C-terminal domain in apoE4 to be relatively disorganized, leading to a higher lipid affinity for this apoE isoform (Saito et al., 2003a).

The similar two-domain tertiary structures of apoA-I and apoE (Section 7.2.2) led to a common two-step mechanism describing the binding of both proteins to spherical lipid particles (Fig. 7.3) (Saito et al., 2003b, 2004b). In this model, the apolipoprotein initially binds to a lipid surface through amphipathic α-helices in the C-terminal domain; this process is accompanied by an increase in α-helicity in this domain (Oda et al., 2003; Saito et al., 2003b). Subsequently, the helix bundle in the N-terminal domain undergoes a conformational opening, converting hydrophobic helix-helix interactions to helix-lipid interactions. The conformational transition from random coil to α-helix upon lipid-binding provides the energetic source to drive the lipid interaction of apolipoproteins (Massey et al., 1979). Calorimetric measurements of binding of apoA-I (Arnulphi et al., 2004; Saito et al., 2003b),

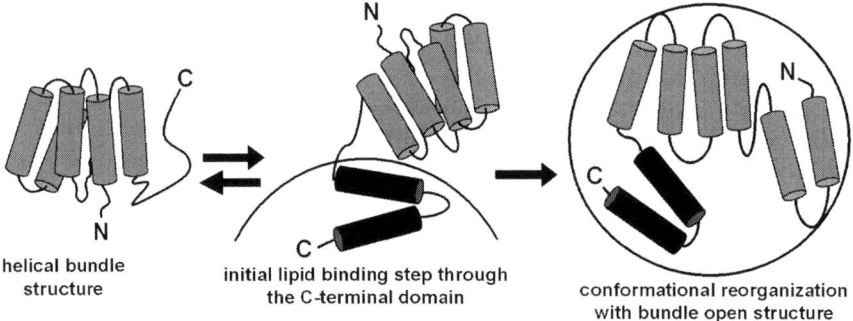

Fig. 7.3 Model of the two-step mechanism of apoA-I binding to a spherical lipid particle. In the lipid-free state, apoA-I is organized into two structural domains in which the N-terminal domain forms a helix bundle whereas the C-terminal domain forms a separate, less organized structure. Initial lipid binding occurs through amphipathic α-helices in the C-terminal domain accompanied by an increase in α-helicity probably in the region including residues 187–220. Subsequently, the helix bundle in the N-terminal domain undergoes a conformational opening, converting hydrophobic helix-helix interactions to the helix–lipid interaction; this second step is only slowly reversible. Reproduced with permission from Saito et al. (2003b)

apoE (Saito et al., 2001) and apoA-I model peptides (Gazzara et al., 1997) to lipid particles, indicate that lipid-binding of these proteins is accompanied by a large release of exothermic heat, consistent with the lipid-binding of apolipoproteins being enthalpically driven. The contribution of α-helix formation to the enthalpy of binding of apoA-I to egg PC small unilamellar vesicles is -1.1 kcal/α-helical residue (Saito et al., 2004a), is in agreement with a prior estimate (-1.3 kcal/α-helical residue) for plasma apolipoproteins using different membrane systems (Massey et al., 1979). α-Helix formation in apoA-I also contributes to an increase in the favourable free energy of binding to lipid (-42 cal/α-helical residue), leading to an approximately two-order of magnitude increase in the affinity of binding (Saito et al., 2004b). This indicates that the transition of random coil to α-helix plays a critical role in driving apoA-I to interact with a lipid surface. This phenomenon probably explains why many exchangeable apolipoproteins in the lipid-free state contain random coil structure especially in the lipid-binding domain, such as the C-terminal domain of apoA-I and apoE.

The binding of apoA-I and apoE to PL is affected by the composition of the lipid particle. For example, addition of either 20–40 mol% cholesterol or 33 mol% egg yolk phosphatidylethanolamine to 100 nm egg yolk phosphatidylcholine (PC) large unilamellar vesicles (LUV) increases the amount of apoA-I binding by increasing the PL polar headgroup space in the relatively flat surface of the LUV (Saito et al., 1997). In contrast, addition of cholesterol to either 100 nm TAG-PC emulsion particles (Saito et al., 1997) or to 20 nm PC small unilamellar vesicles (SUV) (Arnulphi et al., 2004) decreases apoA-I binding; this effect apparently occurs because cholesterol condenses the PL packing and decreases the space available in the highly curved SUV surface.

7.3.2 Apolipoprotein Conformation in Discoidal and Spherical HDL Particles

It is difficult to obtain high resolution structures for HDL particles, because they are not suitable for study by X-ray crystallography and NMR. However, progress has been made by applying alternative methods to study homogeneous, reconstituted, HDL particles containing apoA-I as the sole protein. The study of such particles is relevant because apoA-I is the major protein component of both nascent discoidal and circulating spherical HDL particles (Lund-Katz et al., 2003). The reconstituted HDL discs comprise a segment of PL bilayer surrounded at the edge by apoA-I molecules arranged in a belt-like fashion. The generally accepted double-belt model for the structure of a discoidal HDL particle containing two apoA-I molecules (Segrest et al., 1999) is depicted in Fig. 7.4A. The details of the helix organization in this structure have been resolved by chemical cross-linking and mass spectrometry methods (Davidson and Silva, 2005; Thomas et al., 2006). The two apoA-I molecules are aligned in an anti-parallel fashion so that the amphipathic α-helix spanning residues 121–142 in one apoA-I molecule is opposite the same

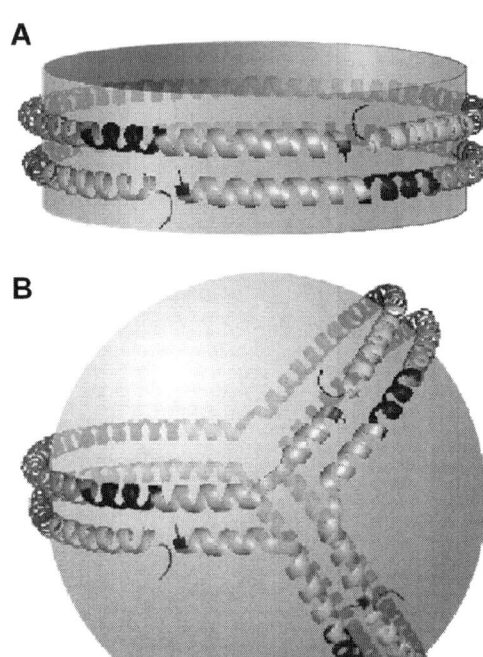

Fig. 7.4 ApoA-I conformation on discoidal and spherical HDL particles. The apoA-I molecules are organized as a double-belt in discoidal particles and as a trefoil in spherical particles. All helix–helix interactions between the two molecules of apoA-I in the disc double-belt arrangement are also present between the three apoA-I molecules in the trefoil organization on the surface of a spherical HDL particle. Reproduced with permission from Silva et al. (2008)

helix in the other apoA-I molecule. Salt-bridges between the two apoA-I molecules help to stabilize this structure. While there is general agreement about the basic structure shown in Fig. 7.4A, there are uncertainties about the organization of the N- and C-terminal ends of the two apoA-I molecules (Davidson and Thompson, 2007; Thomas et al., 2008). The size of the discoidal HDL particle is determined primarily by the number of apoA-I molecules per particle. However, several discrete-sized particles can be formed in complexes containing a constant number of apoA-I molecules (Li et al., 2004). The conformationally flexible apoA-I molecule (Section 7.3.1) adjusts to different particle sizes by certain segments in the protein desorbing from the particle surface and looping into the aqueous phase. Neither the location within the apoA-I molecule nor the precise nature of the protruding segments are well established (Davidson and Thompson, 2007; Thomas et al., 2008).

Given the similarities in apoA-I and apoE structure (Section 7.2), it is unsurprising that apoE also complexes with PL to make HDL particles. As occurs in discoidal HDL particles containing apoA-I, the α-helices in apoE molecules also align perpendicular to the acyl chains of the PL molecules (Narayanaswami et al., 2004; Schneeweis et al., 2005). However, the apoE-PL interaction is more complex than that seen with apoA-I in that some apoE helices seem to be situated among the PL polar headgroups on the faces of the disc. X-ray diffraction data for dipalmitoyl

PC complexes containing two apoE molecules (Peters-Libeu et al., 2006, 2007) indicate that, unlike apoA-I, apoE does not simply form discoidal particles. Rather, the apoE particles are quasi-spheroidal and the apoE molecules are folded into a helical hairpin and interact primarily with the PL polar headgroups (Hatters et al., 2009; Peters-Libeu et al., 2006, 2007).

In vivo, discoidal nascent HDL particles created by the activity of ABCA1 (Section 7.5.2) are remodelled by various factors (Section 7.3.3) to yield the mature spherical HDL particles found in the circulation (Jonas and Phillips, 2008; Lund-Katz et al., 2003). These spherical HDL particles contain a neutral lipid core composed of CE and TAG surrounded by a monolayer of PL and cholesterol molecules and, unlike a discoidal particle, do not have a particle edge to constrain the apolipoproteins. However, the protein-protein interactions that occur in the double-belt model of a discoidal particle are maintained in a spherical, apoA-I containing, HDL particle (Fig. 7.4B) (Silva et al., 2008). All three apoA-I molecules in the trefoil model of a spherical HDL particle are in identical conformations and the inter-molecular salt bridges are the same as those that exist in the double-belt model of a discoidal particle. The apoA-I molecules in the trefoil arrangement are thought to provide the structural scaffold that stabilizes the spherical HDL particle. However, there is another pool of HDL-associated apoA-I molecules that readily exchanges on and off the particle surface. These apoA-I molecules are likely to be bound with only their C-terminal domain in contact with the particle surface and their N-terminal helix bundle domain in the closed conformation protruding into the aqueous phase (Kono et al., 2008) (cf. Fig. 7.3). ApoE molecules associated with spherical HDL particles also presumably adopt either the helix bundle-closed or -open conformation (cf. Section 7.3.2).

7.3.3 Remodeling of HDL Particles

Remodeling by plasma factors is a critical part of HDL metabolism and underlies the dynamic nature of HDL particles. Thus, the various subpopulations of HDL particles that exist in human plasma (Lund-Katz et al., 2003; Zannis et al., 2006) are continually interconverting due to the lipolytic and lipid transfer activities of the plasma factors listed in Table 7.1. As shown in Fig. 7.7 in Section 7.5.1, the changes in HDL particle shape and size caused by these plasma factors are also central to the

Table 7.1 Plasma factors involved in remodeling of HDL

HDL conversion	Plasma factors
disc → sphere	LCAT
large sphere → small sphere + free apo A-I	CETP and HL
large sphere → small sphere	EL
sphere → larger and smaller spheres + free apo A-I	PLTP

participation of HDL in the RCT pathway. The key function of lecithin-cholesterol acyltransferase (LCAT) is to form CE while the other enzymes, hepatic lipase (HL) and endothelial lipase (EL), are involved in releasing fatty acids from PL and TAG. The lipid transfer proteins, cholesterol ester transfer protein (CETP) and phospholipid transfer protein (PLTP), are involved in moving CE, TAG and PL molecules among HDL and other lipoprotein particles (Masson et al., 2009; Rye et al., 2009). As shown in Table 7.1, the interconversion of HDL particles is frequently accompanied by the release of apoA-I molecules into the aqueous phase; rearrangements of HDL particles that are accompanied by a decrease in net surface area lead to desorption of apoA-I molecules and vice versa. Such cycling of apoA-I molecules between HDL particles and the aqueous phase is critical for HDL metabolism (Pownall and Ehnholm, 2006; Rye and Barter, 2004).

LCAT is secreted by the liver in humans and is the major enzyme responsible for the esterification of free unesterified cholesterol (FC) present in circulating lipoproteins (Santamarina-Fojo et al., 2000; Zannis et al., 2006). The protein comprises a single polypeptide chain of 416 amino acids and it is glycosylated at four sites giving a molecular mass of ~60 kDa. This lipase contains an α/β-hydrolase fold and an Asp-His-Ser catalytic triad, with Ser 181 being in the active site (Jonas, 2000). The enzyme catalyses a transesterification reaction in which an unsaturated fatty acid is transferred from the sn-2 position of PC to the hydroxyl group of cholesterol via Ser 181, generating CE and lyso-PC. ApoA-I is the major activator of LCAT in plasma so PC and FC molecules in HDL particles are the preferred substrates. Amphipathic α-helices located between residues 143–187 in the apoA-I molecule apparently mediate the binding of LCAT to the HDL particle, with three arginine residues (R149, R153, R160) playing a critical role (Roosbeek et al., 2001). LCAT binds with high affinity to apoA-I-containing discoidal HDL particles (cf. Section 7.3.2) and converts the FC in them to CE. The CE molecules produced are poorly soluble in a PL bilayer and form a neutral lipid core; the resultant spherical, microemulsion-like, particle is stabilized by a mixed PL/apoA-I monolayer (cf. Section 7.3.2). As expected, in LCAT-knockout mice the levels of spherical CE-enriched HDL particles are reduced and the levels of FC-enriched discoidal HDL particles are increased (Santamarina-Fojo et al., 2000).

CETP and PLTP are members of the lipopolysaccharide-binding/lipid transfer protein family and each contains 476 residues in a single polypeptide chain; both proteins are glycosylated and the apparent molecular masses are 74 and 84 kDa, respectively (Masson et al., 2009; Rye et al., 2009). The primary structure of PLTP is ~25% identical to that of CETP and both proteins can be predicted to have a similar tertiary structure. The crystal structure of CETP indicates that the molecule contains a 6 nm-long tunnel that can be filled with two CE molecules and plugged at each end by a PC molecule (Qiu et al., 2007). It is suggested that when CETP binds to a lipoprotein particle, PL molecules bound at the end of the tunnel merge into the PL monolayer at the particle surface and allow neutral lipid molecules to enter and exit the tunnel. Since CETP can bind CE and TAG molecules, it is possible that the protein transfers lipids between lipoprotein particles by a shuttle mechanism. However, CETP can bridge two HDL particles to form a ternary complex and induce particle fusion, thereby changing HDL particle size (Rye et al., 1997) (cf. Table 7.1). CETP-mediated exchange of CE and TAG between HDL and VLDL enriches the HDL particles with TAG, which destabilizes them and promotes dissociation of apoA-I molecules (Rye et al., 2009; Sparks et al., 1995). TAG-enriched HDL particles are also good substrates for HL, the activity of which further reduces the particle size and enhances apoA-I dissociation (Table 7.1). PLTP transfers PL molecules between VLDL and HDL, as well as between different HDL particles (Huuskonen et al., 2001; Van Tol, 2002). As indicated in Table 7.1, PLTP remodels HDL into large and small particles and promotes the dissociation of apoA-I. This remodelling process involves HDL particle fusion and the remodelling is faster with TAG-enriched HDL (Settasatian et al., 2001). The activities of both CETP and PLTP impact significantly on plasma HDL levels. Thus, PLTP-knockout mice exhibit ~50% reductions in plasma HDL levels, whereas inhibition of CETP in both animals and humans increases HDL levels (Masson et al., 2009).

HL and EL are members of the lipase gene family that also includes pancreatic lipase and lipoprotein lipase; members of this family have similar tertiary structures (Wong and Schotz, 2002). Both HL and EL hydrolyze HDL lipids, although the substrate specificities are different. HDL TAG and PL are substrates for HL whereas EL is primarily a phospholipase (Jaye and Krawiec, 2004; Thuren, 2000). Presumably, this variation in substrate specificity is the reason that HL action on TAG-enriched HDL leads to dissociation of apoA-I (Clay et al., 1991), whereas EL action does not (Jahangiri et al., 2005); lipolysis reduces HDL particle size in both cases (Table 7.1). Both lipases affect plasma HDL levels although to different extents; EL-knockout mice have ~50% elevated HDL cholesterol levels (Badellino and Rader, 2004), whereas HL-knockout mice exhibit much smaller increases in HDL cholesterol levels (Homanics et al., 1995).

Plasma HDL consists of particles that contain only apoA-I and particles that contain both apoA-I and apoA-II (Lund-Katz et al., 2003). The latter particles are formed by LCAT-induced fusion of nascent, discoidal HDL particles that contain either apoA-I or apoA-II (Clay et al., 2000). Compared to apoA-I, apoA-II is more lipophilic and dissociates less readily from HDL particles (Pownall and Ehnholm, 2006). The presence of apoA-II stabilizes the HDL particle and reduces its ability to be remodeled. Thus, apoA-II inhibits the ability of CETP to shrink spherical HDL particles and prevents apoA-I dissociation; the latter effect occurs because of stabilizing protein-protein interactions between apoA-I and apoA-II in the HDL particle surface (Rye et al., 2008).

7.4 Lipid Solubilizing Properties of ApoA-1 and ApoE

7.4.1 Historical Perspective

The fact that the apolipoproteins of HDL can stabilize ~10 nm lipid microemulsion particles with the structure depicted in Fig. 7.4B for a spherical HDL particle, indicates that an exchangeable apolipoprotein such as apoA-I possesses detergent-like properties and can solubilize lipids. This was first demonstrated directly in the 1960 s by Scanu (1967), who reported that the apoproteins from human HDL can form stable, water-soluble, lipid–protein complexes with HDL PL when the mixture is sonicated. Subsequently, electron microscopic examination (Forte et al., 1971) indicated that such complexes are disc-shaped and circular dichroism measurements (Lux et al., 1972) showed that the interaction with lipid increases the α-helix content of the apolipoprotein. Scanu and colleagues also used the sonication procedure to show that (a) apo HDL can complex with a range of synthetic phospholipids when the lipids are in a liquid-crystalline state (Kruski and Scanu, 1974) and (b) spherical HDL particles can be reassembled in vitro when the HDL neutral lipids (CE and TAG) are included (Hirz and Scanu, 1970).

HDL particles reconstituted with either natural or synthetic lipids have been the subject of extensive investigations since the early 1970s. The first detailed characterization in 1974 was of the disc-shaped complexes formed by dimyristoyl PC (DMPC) and porcine apo HDL (primarily apoA-I). Co-sonication of this mixture under conditions where all the apoprotein is complexed yields lipoprotein discs consisting of two apoprotein molecules and about 200 DMPC molecules; the major diameter is 8–9 nm and the thickness is 4.4 nm, in good agreement with the thickness of a DMPC bilayer (Hauser et al., 1974). The acyl chain melting transition of the DMPC in such a reconstituted HDL particle is broadened, indicating that the apoprotein molecules modify the packing of the lipid molecules (Barratt et al., 1974). The fundamental organization of the discoidal particle as a small segment of DMPC bilayer stabilized by apoprotein molecules was confirmed by small-angle X-ray scattering (Atkinson et al., 1976), and it was found that the apoprotein amphipathic α-helices reduce the cooperativity of DMPC acyl chain motions, but do not bind the lipid molecules tightly (Andrews et al., 1976). Importantly, it was shown in these studies with porcine apo HDL that sonication is not required and that the highly surface-active protein molecules can spontaneously solubilize multilamellar vesicles (MLV) of DMPC to form discoidal HDL particles of the type described above (also see Fig. 7.5) (Hauser et al., 1974). In 1977 it was demonstrated that pure

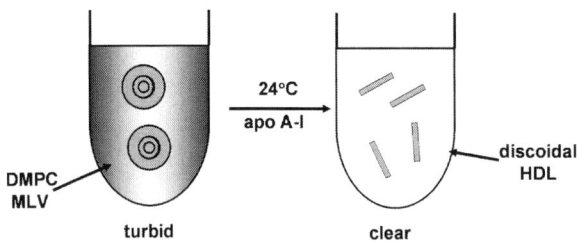

Fig. 7.5 Schematic representation of the spontaneous solubilization of dimyristoyl phosphatidylcholine (DMPC) multilamellar vesicles (MLV) by apoA-I. When apoA-I is incubated with the turbid MLV suspension at 24°C (the melting temperature of the DMPC acyl chains), the solution becomes optically clear because the DMPC bilayers are solubilized in a few minutes to create discoidal HDL particles that are too small to scatter visible light

apoA-I, either human or bovine, can form discoidal complexes with DMPC (Jonas et al., 1977) and that the particle size increases with increasing DMPC/apoprotein ratio (Tall et al., 1977); the latter finding is consistent with the predominant location of the apoprotein being an annulus around the perimeter of the disc (Tall et al., 1977) (cf. Fig. 7.4A).

7.4.2 Mechanism of Solubilization Reaction

The first detailed study of the kinetics of solubilization of DMPC by human apoA-I was conducted by Pownall and colleagues in 1978 (Pownall et al., 1978). They used turbidimetric measurements (Fig. 7.5) to show that the rate of conversion of DMPC MLV into discoidal apoA-I/DMPC complexes is highest at the gel to liquid-crystal phase transition temperature (24°C) of DMPC. The enhanced rate at the phase transition is due to the presence of lattice defects that create sites to which apoA-I molecules can bind (with an associated increase in α-helix content), thereby destabilizing the DMPC bilayer and promoting rearrangement into discoidal, HDL-like, particles (Pownall et al., 1979, 1981). These fundamental concepts concerning the mechanism of PL solubilization by apolipoproteins are now well accepted. The laboratories of Pownall (Pownall et al., 1987) and Jonas (1992) have contributed much of our detailed understanding of the solubilization process. Generally, the reaction is kinetically controlled. Factors such as the physical state of the PL bilayer, the stability of the bilayer, the curvature of the vesicles and their cholesterol content (see below) are important. Regarding the apolipoprotein component, its molecular weight, hydrophobicity and state of self-association affect the rate of solubilization.

Analysis of the clearance kinetics in longer-term incubations of apoA-I with DMPC MLV at 24°C shows that the reaction is second order and consists of two simultaneous kinetic phases (Segall et al., 2002). As summarized in Fig. 7.6, the two kinetic components are proposed to arise from two distinct types of binding site (with or without lattice defects) for apoA-I on MLV surfaces. If the initial apoA-I/MLV contact occurs at a packing defect in the DMPC bilayer surface, then the reaction proceeds directly to stage 2. Initial contact at a non-defect site requires diffusion of the apoA-I molecule over the surface to a defect site, giving rise to a slow kinetic phase in obtaining stage 2. After adsorption of apoprotein molecules to DMPC bilayer defects, the rate of arrival at stage 3 depends upon the rate at which these molecules can rearrange and insert their α-helices into the defects (Fig. 7.6). As proposed earlier (Pownall et al., 1987), once a critical concentration of α-helices absorbed in the defects is attained, the lipid bilayer becomes unstable and rearranges to form a bilayer disc (stage 3 in Fig. 7.6). Apoprotein molecules at stage 2 of the reaction are bound to the DMPC vesicle surface reversibly whereas the vesicle to disc conversion in stage 3 is irreversible. The reaction of apoA-I with either SUV or MLV of DMPC to form discoidal complexes proceeds similarly. Detailed examination of this system (Jonas et al., 1980; Jonas and Drengler, 1980) has shown that apoA-I binding to saturate the vesicle surface (stage 1) is rapid and that the bilayer disruption is relatively slow. The nature of any intermediates in the vesicle

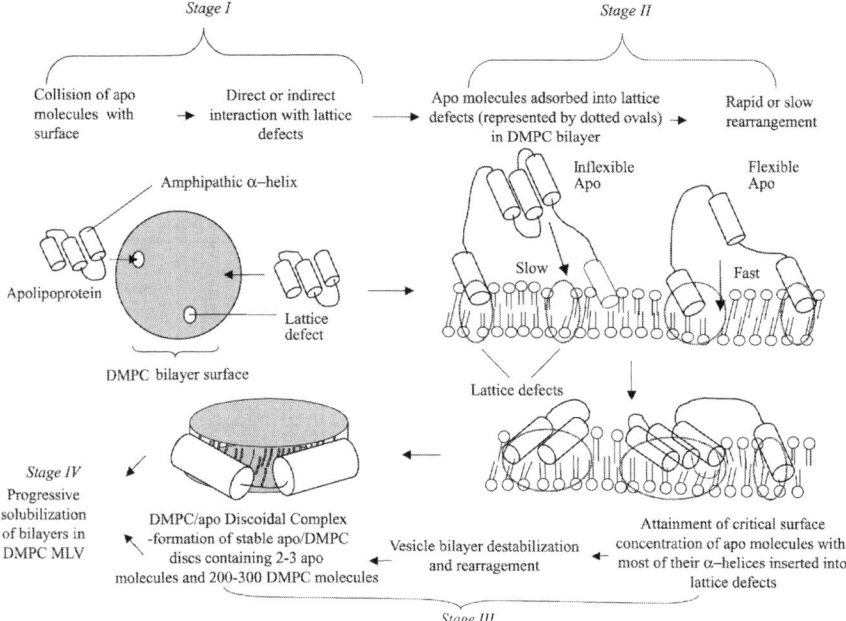

Fig. 7.6 Molecular model explaining the two simultaneous kinetic phases of DMPC solubilization by apolipoproteins. The solubilization of DMPC MLV by apolipoprotein molecules involves four states (as indicated). Completion of the first and second of these stages can each occur by two simultaneous alternative pathways, one more rapid than the other, whereas stages three and four comprise a common pathway. Flexible apolipoprotein molecules react more rapidly than inflexible ones. See Segall et al. (2002) for more details. Reproduced with permission from Segall et al. (2002)

to disc conversion step (stage 3 in Fig. 7.6) is not well understood. In the case of a DMPC SUV (~20 nm in diameter), when the surface is saturated with apoA-I the DMPC/apoA-I stoichiometry is ~1000/1 mol/mol, as compared to a ratio of ~ 100/1 in the discs that are formed (Jonas et al., 1980). Thus, it follows that an entire DMPC SUV cannot convert directly in one step into discoidal product without either incorporation of additional apoA-I molecules or the loss of excess DMPC to other particles. Intermediate complexes have not been detected in the DMPC SUV/apoA-I reaction (Jonas et al., 1980) but the formation of small discs proceeds through the formation of a large disc intermediate when DMPC LUV (~100 nm diameter) are solubilized by apoA-I (Zhu et al., 2007).

At temperatures other than the DMPC gel to liquid/crystal transition temperature of 24°C, apoA-I reacts slowly with DMPC vesicles because the bilayer contains relatively few packing defects (Pownall et al., 1978, 1987). However, the addition of cholesterol to the DMPC bilayer accelerates the rate of solubilization. The reaction is fastest at 12 mol% cholesterol and it becomes less temperature sensitive (Pownall et al., 1979). The combined effects of cholesterol and temperature can alter the rate of reaction by more than three orders of magnitude because of large variations in the number of lattice defects in the DMPC-cholesterol mixed bilayer. All the cholesterol is solubilized by apoA-I when the DMPC vesicle contains less than about 20 mol% cholesterol but higher levels of cholesterol inhibit the formation of discoidal complexes. The sizes of the HDL particles formed are dependent upon the initial level of cholesterol in the DMPC MLV (Massey and Pownall, 2008); increasing the bilayer cholesterol concentration results in the formation of larger sized discoidal particles that contain more apoA-I molecules.

In contrast to the DMPC system, MLV prepared from a PL such as an egg PC obtained from natural sources, are not solubilized in a matter of minutes at room temperature by apoA-I. The reason for this low reactivity is that the natural PC molecules contain unsaturated acyl chains so that their gel to liquid-crystal phase transition temperatures are well below room temperature and the bilayers contain few lattice defects into which apoA-I molecules can insert (cf. Fig. 7.6). However, a complex lipid mixture representing the lipids of a mammalian plasma membrane forms MLV that are solubilized by apoA-I at 37°C (Vedhachalam et al., 2007a). The rate of ~10% solubilization per hour is much slower than that typically observed with DMPC MLV, presumably because the numbers of lattice defects are fewer in the membrane lipid bilayer. Such a membrane bilayer solubilization process is integral to the mechanism of nascent HDL particle formation by ABCA1 (cf. Section 7.5.2).

7.4.3 Influence of Apolipoprotein Structure

All exchangeable apolipoproteins exhibit some ability to solubilize PL vesicles and create discoidal HDL particles which is consistent with the amphipathic α-helix, rather than a specific amino acid sequence, being the structural motif responsible for this activity (cf. Section 7.3). In agreement with this concept, fragments of the human apoA-I molecule corresponding to residues 1–86 and 149–243 contain amphipathic α-helices and can form similar discoidal DMPC complexes to the intact apoA-I molecule (Vanloo et al., 1991). Since the hydrophobic C-terminal domain (residues 190–243) of human apoA-I molecule is critical for lipid binding (Fig. 7.3), deletion of this segment reduces the ability of the protein to solubilize DMPC MLV (Vedhachalam et al., 2007a). For the same reasons, either deletion or disruption of the C-terminal α-helix (residues 221–243) has the same effect. Deletion of residues 1–43 or 44–65 in the N-terminal helix bundle domain has a less marked effect on the ability of the protein to solubilize DMPC (Vedhachalam et al., 2007a).

Human apoE also solubilizes DMPC vesicles to create discoidal particles (Innerarity et al., 1979). However, apoE reacts more slowly than apoA-I and this is attributed to the less flexible structure of the apoE molecule which reduces the rate of stage 2 in the solubilization reaction (see Fig. 7.6) (Segall et al., 2002). As observed with human apoA-I, removal of the lipid-binding C-terminal domain (residues 192–299) greatly reduces the rate at which apoE solubilizes DMPC MLV. The isolated and flexible C-terminal domain of apoE solubilizes DMPC at a similar rate to apoA-I. The reaction rates of the helix bundle domains of the three commons apoE isoforms vary inversely with the stabilities of these fragments. Overall, it seems that flexibility in an apolipoprotein molecule increases the time-average exposure of hydrophobic surface area, thereby increasing the rate of PL solubilization (*see* Figs. 7.5 and 7.6; Segall et al. (2002)).

7.5 HDL and Reverse Cholesterol Transport (RCT)

7.5.1 Overview of RCT Pathway – HDL Species and Receptors Involved

Figure 7.7 summarizes the RCT pathway in which HDL mediates the movement of cholesterol from peripheral cells to the liver for excretion from the body (Fielding and Fielding, 1995; Oram and Heinecke, 2005). It is clear from the pathway shown in Fig. 7.7 that apoA-I is involved in all stages of RCT, including the formation of

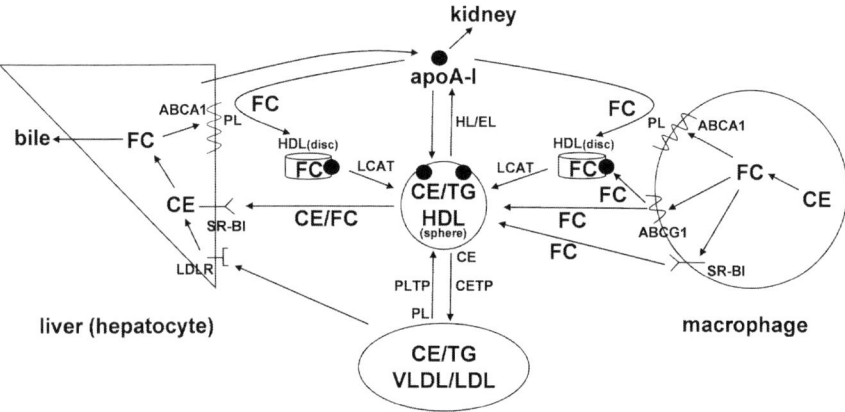

Fig. 7.7 Schematic overview of the major pathways involved in HDL-mediated macrophage cholesterol efflux and reverse cholesterol transport to the liver. ApoA-I is produced by the liver and acquires free cholesterol (FC) and phospholipid (PL) from liver and peripheral cells (including macrophages) via the ABCA1 transporter to form nascent (discoidal) HDL particles. Non-lipidated apoA-I is cleared by the kidney. FC efflux from macrophages to HDL particles is also promoted by the ABCG1 transporter and SR-BI. As discussed in Section 7.3.3, the FC in discoidal HDL particles is converted to CE by LCAT activity leading to the formation of mature, spherical HDL particles. PLTP mediates transfer of PL from VLDL into HDL thereby providing PL for the LCAT reaction. Mature HDL particles can be remodeled to smaller particles with the release of apoA-I by the actions of hepatic lipase (HL) and endothelial lipase (EL) which hydrolyze HDL TAG and PL, respectively. In humans, but not rodents, HDL-CE can be transferred to the VLDL/LDL pool by CETP and taken up by endocytosis into hepatocytes via interaction with the LDL receptor (LDLR). HDL-CE and FC are also transferred directly to hepatocytes via SR-BI-mediated selective uptake. Cholesterol taken up by the liver can be recycled back into the ABCA1 pathway, secreted into bile as either FC or bile acids, or assembled into lipoprotein particles that are secreted back into the circulation (not shown)

nascent HDL particles, HDL remodeling by LCAT and delivery of HDL cholesterol to the liver via scavenger receptor class B, type 1 (SR-BI). The finding that the severe HDL deficiency associated with the genetic disorder, Tangier disease, is caused by mutations in the ATP-binding cassette transporter A1 (ABCA1) demonstrated that this transporter plays a critical role in HDL production (Oram, 2000). The bulk of the HDL particles in the circulation are produced by ABCA1 expressed in the liver and intestine, with the former being the major contributor (Lee and Parks, 2005). Removal of cholesterol from macrophages in the walls of blood vessels via RCT is critical for preventing the development of atherosclerotic plaque (Cuchel and Rader, 2006).

Studies of macrophage RCT in mice have demonstrated that both apoA-I and ABCA1 play essential roles in promoting RCT; it follows that they are anti-atherogenic proteins (Wang et al., 2007a; Zhang et al., 2003). These two proteins interact to mediate the first step in RCT, the efflux of cellular cholesterol (Fig. 7.7). Efflux of cholesterol from macrophages involves both active and passive processes (Yancey et al., 2003). In addition to the active ABCA1-mediated efflux of cellular PL and FC to lipid-free/poor apoA-I, a related transporter, ABCG1, can promote FC efflux to HDL particles. Passive FC efflux from macrophages also occurs by the so-called aqueous diffusion pathway (Adorni et al., 2007). This simple diffusion process involves desorption of FC molecules from the PL bilayer of the plasma membrane, followed by their diffusion in the aqueous phase and collision-mediated absorption into HDL particles (Phillips et al., 1987). Efflux of cellular FC from macrophages to HDL can also be facilitated by SR-BI (Adorni et al., 2007).

As can be seen from Fig. 7.7, besides being located in macrophages, SR-BI is located on the surface of hepatocytes. SR-BI is abundantly expressed in the liver where it mediates the selective uptake of FC and CE from HDL (Trigatti et al., 2003; van Eck et al., 2005; Zannis et al., 2006). This FC and CE is then released into bile as either FC or bile acid and then, in the last step of RCT, excreted from the body in feces (Fig. 7.7). The hepatic expression of SR-BI positively promotes the flux of cholesterol from macrophages through the RCT pathway (Zhang et al., 2005). In contrast, the SR-BI expressed in macrophages, which can promote efflux of cellular cholesterol (*see* Section 7.5.3), does not promote macrophage RCT in vivo (Wang et al., 2007b). However, at later stages of atherosclerotic lesion development in mice, SR-BI in macrophages plays an anti-atherogenic role (van Eck et al., 2005). Consistent with this atheroprotective effect, inactivation of the SR-BI gene in apoE-null mice accelerates coronary atherosclerosis (Trigatti et al., 1999). The loss of SR-BI activity is associated with an increase in total plasma cholesterol and the incidence of abnormally large, apoE-enriched HDL particles (Rigotti et al., 1997). The fact that the incidence of atherosclerosis in these animals is associated with higher HDL cholesterol levels indicates that the quality and not the quantity of HDL particles is critical in preventing coronary artery disease. The HDL particles that are present must be able to maintain the appropriate flux of cholesterol through the RCT pathway (Fig. 7.7). Enhancement of RCT from plaque to liver can enhance rapid regression of atherosclerosis (Williams et al., 2007).

7.5.2 ABCA1

ABCA1 is a member of the ATP binding cassette (ABC) family of membrane transporters and its ability to translocate PL across the plasma membrane of cells leads to the formation of nascent HDL particles when lipid-free/poor apoA-I is present in the extracellular medium. The preferred substrate for translocation by ABCA1 has not been established unambiguously, but it is clear that phosphatidylserine can be pumped across the membrane (Alder-Baerens et al., 2005; Chambenoit et al., 2001). ABCA1 is a 2261 amino acid integral membrane protein consisting of two halves of similar structure (Oram and Heinecke, 2005). Each half contains a six-helix transmembrane domain together with a cytosolic nucleotide-binding domain that mediates the ATPase activity. The topology in a cell plasma membrane is predicted to involve a cytosolic N-terminus and some extracellular loops that are heavily glycosylated (Dean et al., 2001). The two transmembrane helical domains form a chamber in which PL molecules are translocated. The molecular mechanism of this pumping action in ABCA1 has not been elucidated, but is likely to be similar to that of a related microbial lipid transporter whose crystal structure is known (Dawson and Locher, 2006).

The expression of ABCA1 is increased by loading cells with cholesterol because the consequent increase in oxysterol level activates the nuclear liver X receptor (LXR) (Oram and Heinecke, 2005). The transcription of ABCA1 is also induced by ligands for the retinoid X receptor and, in the case of murine macrophages, also by cyclic AMP. ABCA1 is degraded rapidly after transcription (half-life of 1–2 h) and its cellular level is sensitive to the presence of an apolipoprotein such as apoA-I, because apoA-I binds to ABCA1 and stabilizes it by modulating its phosphorylation, thereby protecting the transporter from calpain-mediated proteolysis (Wang and Tall, 2003; Yokoyama, 2006). ABCA1 recycles rapidly between the plasma membrane and late endosomal/lysosomal compartments (Neufeld et al., 2001) and is degraded at an intracellular site during this trafficking. When apoA-I is bound to ABCA1 (Oram et al., 2000; Vedhachalam et al., 2007b; Wang et al., 2000), the endocytosis is unaffected whereas the intracellular degradation is reduced which leads to higher levels of the transporter recycling back to the plasma membrane (Lu et al., 2008). This effect leads to enhanced HDL biogenesis because the ABCA1-mediated assembly of nascent HDL particles occurs at the cell surface (Denis et al., 2008; Faulkner et al., 2008).

The nascent HDL products of the apoA-I/ABCA1 reaction are primarily discoidal particles (cf. Section 7.3.2) containing 2, 3 or 4 apoA-I molecules (Duong et al., 2006; Krimbou et al., 2006; Liu et al., 2003). These particles are not only heterogeneous with respect to diameter, but also with respect to lipid composition in that they have different FC contents and PL compositions (Duong et al., 2006). This creation of variable nascent HDL species underlies the heterogeneity in the population of HDL particles present in the plasma compartment (Lund-Katz et al., 2003). The ABCA1/apoA-I reaction also leads to the production of some monomeric apoA-I molecules that are associated with 3–4 PL molecules; this lipid-poor apoA-I (preβ1-HDL) (Chau et al., 2006) is a product of the reaction but also a substrate in that it can react further and be converted into larger discoidal particles (Duong et al., 2008). Different apolipoproteins such as apoE (Krimbou et al., 2004; Vedhachalam et al., 2007c), and peptides containing amphipathic α-helices (Remaley et al., 2003), can react with ABCA1 to create nascent HDL particles and the particle sizes are dependent upon the protein structure. A striking example of such a protein structural requirement is the observation that removal of the C-terminal α-helix of human apoA-I drastically reduces the level of PL and FC efflux (Vedhachalam et al., 2004), prevents formation of normal size nascent HDL particles and causes formation of very large HDL particles (Liu et al., 2003). As summarized in Sections 7.3 and 7.4.1, the C-terminal α-helix of apoA-I plays a critical role in lipid-binding and lipid-solubilization. It can be inferred that, since this helix is also essential for ABCA1-mediated HDL particle biogenesis, plasma membrane microsolubilization occurs during HDL biogenesis (Gillotte et al., 1999).

As summarized above, much has been learned about the ways in which ABCA1 contributes to the biogenesis of HDL particles. There has been great interest in elucidating the mechanism by which the transporter and apoA-I react to create nascent HDL particles and various models have been proposed (*for reviews see* Oram and Heinecke, 2005; Yokoyama, 2006; Zannis et al., 2006). Figure 7.8 summarizes a reaction scheme we have proposed recently, that integrates key findings from the literature and our laboratory (Vedhachalam et al., 2007a). A central feature of this mechanism is that membrane PL translocation via ABCA1 induces bending of the membrane bilayer to create high curvature sites to which apoA-I can bind and solubilize membrane PL and FC to create nascent HDL particles. Step 1 involves binding of apoA-I (in a lipid-free/poor state but not in a fully lipidated state) to the ABCA1 molecule probably via interaction of an amphipathic α-helix with a site on an extracellular loop of ABCA1 (Fitzgerald et al., 2002). As mentioned above, this association with apoA-I stabilizes the transporter in the plasma membrane, which leads to enhanced PL translocation and asymmetric PL packing across the bilayer. The resultant membrane strain is relieved by formation of highly curved exovesiculated domains in the plasma membrane to which apoA-I molecules can bind with high affinity (Step 2). The high curvature disorders the molecular packing in the bilayer and creates spaces between the PL polar groups into which apoA-I amphipathic α-helices can penetrate. The binding of apoA-I to the exovesiculated plasma membrane domains creates conditions for the formation of nascent HDL particles, which follows in Step 3.

Step 3 requires the spontaneous solubilization (cf. Section 7.4) of the membrane lipid bilayer to create nascent HDL. We have defined this process, which leads to simultaneous release of cellular PL and FC, as membrane microsolubilization (Gillotte-Taylor et al., 2002; Gillotte et al., 1998, 1999). As discussed in Section 7.4.1, apoA-I can spontaneously solubilize bilayers comprised of the PL typically found in the plasma membrane of mammalian cells to create discoidal HDL particles (Vedhachalam et al., 2007a). The apolipoprotein content of these particles is determined by apoA-I/PL/FC interactions in which ABCA1 is not directly involved and by the structural properties of the apoA-I molecule. Step 3 in the reaction scheme summarized in Fig. 7.8 is the slowest and, therefore, rate-limiting. This fact

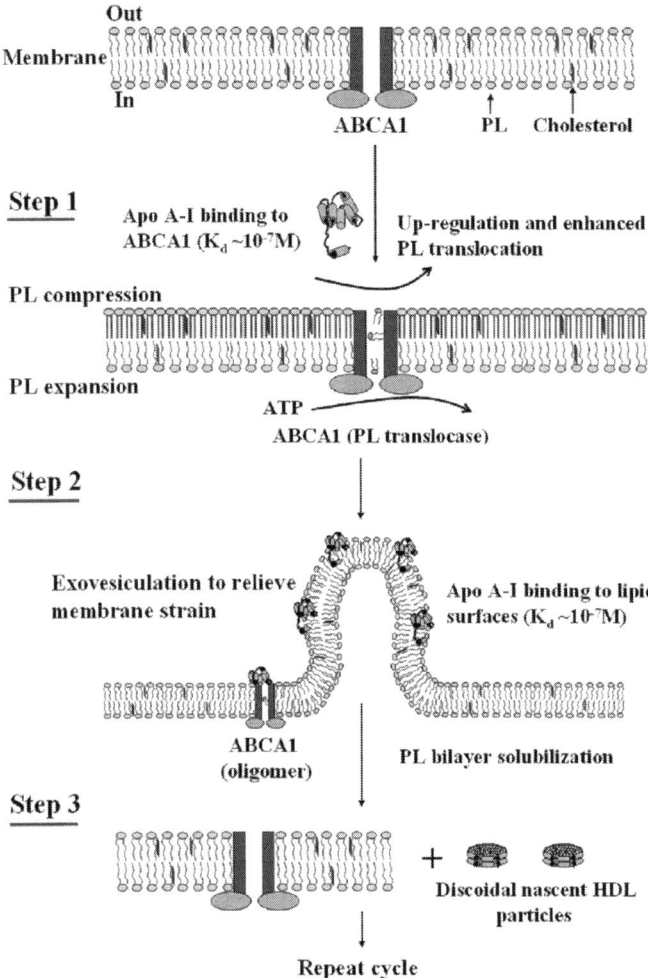

Fig. 7.8 Mechanism of interaction of apoA-I with ABCA1 and efflux of cellular phospholipids and cholesterol. The reaction in which apoA-I binds to ABCA1 and membrane lipids to create discoidal nascent HDL particles contains three steps. Step 1 involves the high affinity binding of a small amount of apoA-I to ABCA1 located in the plasma membrane PL bilayer; this regulatory pool of apoA-I up-regulates ABCA1 activity, thereby enhancing the active translocation of membrane PL from the cytoplasmic to exofacial leaflet. This translocase activity leads to lateral compression of the PL molecules in the exofacial leaflet and expansion of those in the cytoplasmic leaflet. Step 2 involves the bending of the membrane to relieve the strain induced by the unequal molecular packing density across the membrane and the formation of an exovesiculated domain to which apoA-I can bind with high affinity. This interaction with the highly curved membrane surface involves apoA-I/membrane lipid interactions and creates a relatively large pool of bound apoA-I. Step 3 involves the spontaneous solubilization by the bound apoA-I of membrane PL and cholesterol in the exovesiculated domains to create discoidal HDL particles containing two, three or four apoA-I molecules/particle. In the diagram, the two transmembrane six-helix domains of ABCA1 are represented as rectangles, whereas the two ATPase domains are shown as ovals. The space between the two rectangles represents the chamber in which translocation of PL molecules occurs. Reproduced with permission from Vedhachalam et al. (2007a)

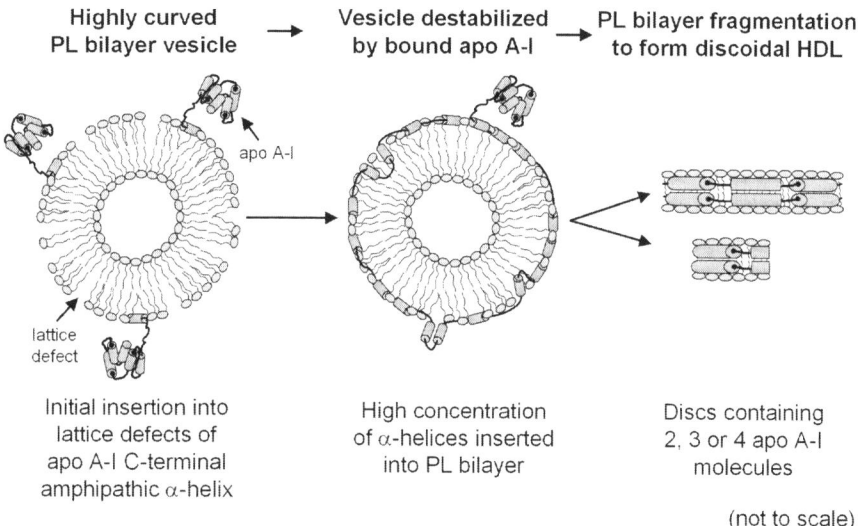

Fig. 7.9 Suggested molecular mechanism for the solubilization of PL bilayers by apoA-I to create discoidal HDL particles. This process is envisaged to underlie the solubilization of DMPC MLV depicted in Figs. 7.5 and 7.6, and Step 3 in the formation of nascent HDL particles in the apoA-I/ABCA1 reaction depicted in Fig. 7.8

is exemplified by the observation that apoA-I structural alterations that modulate the rate of model membrane bilayer solubilization have a similar effect on the rate of PL and FC efflux from cells via ABCA1 (Vedhachalam et al., 2007a). Figure 7.9 summarizes a molecular mechanism for Step 3 in which vesiculated membrane bilayers are fragmented into small segments to form discoidal HDL particles. Overall, the mechanism depicted in Fig. 7.8 is consistent with the known properties of ABCA1 and apoA-I.

As shown in the summary of the RCT pathway (Fig. 7.7), the removal of excess cholesterol from macrophages can occur via both ABCA1 and ABCG1 pathways. The nascent HDL particles created by the ABCA1 reaction contain some FC (Duong et al., 2006, 2008), but they are able to acquire additional cholesterol by participation in the ABCG1 pathway. Thus, ABCA1 and ABCG1 can act sequentially to mediate cellular cholesterol export to apoA-I (Gelissen et al., 2008; Vaughan and Oram, 2006). This synergy between lipid transporters is important for regulating cholesterol efflux and maintaining the appropriate cholesterol levels in macrophages (Jessup et al., 2006; Marcel et al., 2008).

7.5.3 SR-BI

SR-BI was the first recognized HDL receptor and it mediates the flux of FC and CE between bound HDL particles and the cell plasma membrane (Acton et al., 1996). Of particular note, SR-BI mediates the selective uptake of CE from HDL, a process whereby HDL lipids are taken up preferentially by cells through a non-endocytic mechanism without either degradation of apolipoproteins or whole particle uptake (Connelly and Williams, 2004; Pittman et al., 1987; Zannis et al., 2006). SR-BI is an 82-kDa membrane glycoprotein containing a large extracellular domain (408 residues) and two transmembrane domains with short cytoplasmic N- and C-terminal domains. The extracellular domain plays a critical role in mediating the selective uptake process (Connelly et al., 1999, 2001; Gu et al., 1998). SR-BI contains a PDZK1 binding motif in its C-terminal domain and interaction with this scaffold protein controls the abundance and localization in the plasma membrane of hepatic SR-BI (Fenske et al., 2008). A minor, alternatively spliced, form called SR-BII has a different C-terminal domain that does not bind to PDZK1, so that this isoform is mostly located in the cell interior (Eckhardt et al., 2004). SR-BI self-associates into dimers and tetramers but this process is not dependent upon the C-terminal domain (Sahoo et al., 2007). SR-BI-mediated lipid uptake does not require endocytosis, indicating that the uptake occurs at the plasma membrane (Harder et al., 2006; Nieland et al., 2005). Furthermore, interactions with other proteins or specific cellular structures are not required for this activity as it has been shown that the purified protein reconstituted into a model membrane is functional (Liu and Krieger, 2002).

As befits a scavenger receptor, SR-BI binds to a range of ligands besides HDL; these include VLDL, LDL, modified LDL and PL vesicles (Connelly and Williams, 2004; Trigatti et al., 2003; Zannis et al., 2006). The amphipathic α-helix is the recognition motif for SR-BI and HDL binds to the receptor via the multiple amphipathic α-helical repeats in the apoA-I molecule (Williams et al., 2000). A specific amino acid sequence in apoA-I is not required for the interaction and there is more than one apolipoprotein binding site on the SR-BI molecule (Thuahnai et al., 2003). The binding of HDL to SR-BI is influenced by the conformation of apoA-I so that larger (~10 nm diameter) HDL particles bind better than smaller (~8 nm) particles (de Beer et al., 2001a; Thuahnai et al., 2004). The presence of apoA-II in HDL particles attenuates the binding to SR-BI (de Beer et al., 2001b). The dependence on HDL concentration of binding to SR-BI and CE selective uptake is similar, indicating that the two processes are linked (Rodrigueza et al., 1999). It is apparent that the mechanism of HDL CE selective uptake by SR-BI involves a two-step process in which the initial binding of an HDL particle to the receptor is followed by the transfer of CE molecules from the bound HDL particle into the cell plasma membrane (Gu et al., 1998; Rodrigueza et al., 1999). The rate of CE selective uptake from the donor HDL particle is proportional to the amount of CE initially present in the particle, indicating a mechanism in which CE moves down its concentration gradient from HDL particles docked on SR-BI into the cell plasma membrane. The activation energy for this process is ~9 kcal/mol, which is consistent with the uptake occurring by a non-aqueous pathway. As depicted in the model shown in Fig. 7.10, HDL binding to SR-BI allows access CE molecules to a channel formed by the extracellular domain of the receptor, from which water is excluded and along which HDL CE molecules move down their concentration gradient into the cell plasma membrane (Rodrigueza et al., 1999). An alternative model proposed a hemi-fusion event between the PL monolayer of an HDL particle and the external leaflet of the plasma

7 HDL and Cholesterol Transport

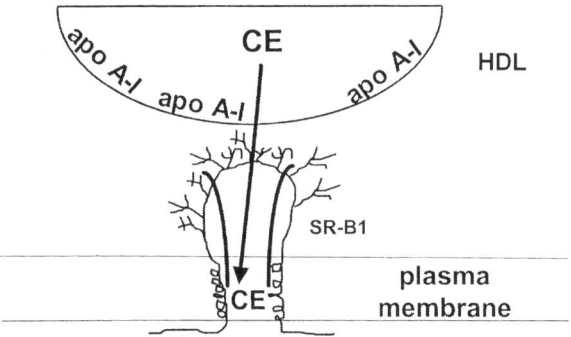

Fig. 7.10 Model of SR-BI-mediated selective uptake of cholesteryl ester (CE) from HDL. This model proposes that SR-BI contains a non-aqueous channel, which excludes water, and serves as a conduit for hydrophobic CE molecules diffusing from bound HDL down their concentration gradient to the cell plasma membrane. The scheme depicts a channel formed by a single SR-BI molecule, but it is possible that self-association of SR-BI is required to create the channel. Reproduced with permission from Rodrigueza et al. (1999)

membrane, thereby allowing lipid transfer to occur (Gu et al., 1998). However, the fact that the relative rates of selective uptake of CE, FC, PC and sphingomyelin from HDL are different (Rodrigueza et al., 1999; Thuahnai et al., 2001) argues against this model. The efficient transfer of lipid molecules from HDL to the cell plasma membrane requires an optimal alignment of the apoA-I-containing HDL/SR-BI complex (Liu et al., 2002; Thuahnai et al., 2004). The presence of an apolipoprotein such as apoA-I in the donor particle is required for the selective uptake of CE to occur because in the absence of apolipoprotein-mediated binding to SR-BI, the lipid components of the donor fuse with the cell membrane giving stoichiometric lipid uptake (Thuahnai et al., 2001). This phenomenon suggests that there is a fusogenic motif in the extracellular domain of SR-BI.

Because the SR-BI-mediated transfer of lipid molecules between bound HDL and the cell plasma membrane is a passive process, the rate and direction of net transfer are sensitive to the concentration gradient existing between the HDL particle and the plasma membrane. Thus, in the case of FC, SR-BI facilitates the bi-directional flux between HDL and the plasma membrane (Yancey et al., 2003). The concentration gradient for FC is sensitive to the PL content of HDL so that the net transfer of cholesterol out of the cell is promoted by PL-enrichment of HDL (Pownall, 2006; Yancey et al., 2000). Different kinds of cells exhibit large differences in the rate of FC efflux to PL-containing acceptors (Rothblat et al., 1986) and these variations are due to differences in the expression levels of SR-BI (Ji et al., 1997; Jian et al., 1998). The fact that SR-BI-mediated FC efflux to HDL is dependent upon HDL binding to the receptor (Gu et al., 2000) and that the appropriate apoA-I/SR-BI complex must be formed (Liu et al., 2002), point to a similar mechanism operating for SR-BI-mediated HDL CE selective uptake and cellular FC efflux. Depending upon the concentration of HDL present in the extracellular medium, SR-BI-facilitated FC efflux can occur by pathways dependent on and independent of

HDL binding to the receptor (Thuahnai et al., 2004). FC efflux is binding-dependent at low concentrations of HDL, where binding to SR-BI is not saturated; under this condition (Fig. 7.10), FC molecules diffuse from the plasma membrane to the bound HDL particle via the hydrophobic channel created by SR-BI. At saturation concentration of HDL, FC efflux is independent of HDL binding to SR-BI. Under this condition, efflux occurs by the aqueous diffusion mechanism whereby FC particles desorb from the plasma membrane and diffuse through the aqueous phase and collide with HDL acceptor particles. The receptor enhances the rate of FC efflux via aqueous diffusion because it perturbs the packing of FC molecules in the plasma membrane; this reorganization and activation of plasma membrane FC molecules is evident from the increased pool of FC accessible to cholesterol oxidase in SR-BI-containing membranes (Kellner-Weibel et al., 2000; Llera-Moya et al., 1999). Similar to CE selective uptake, SR-BI-mediated FC efflux is dependent upon HDL particle size so that, at the same particle concentration, large HDL promotes more FC efflux than small HDL.

7.5.4 ApoE-HDL

In the plasma compartment, apoE influences the lipid concentration because it affects the levels of all lipoproteins by modulating their receptor-mediated clearance into cells. In addition, apoE affects the production of hepatic VLDL and the lipolytic processing of TAG-rich lipoproteins. ApoE, one of the least abundant plasma apolipoproteins, is present on all lipoprotein particles, with the exception of LDL (Mahley and Rall, 1995). The RCT pathway (Section 7.5.1) directs excess cholesterol in peripheral tissues through HDL acceptors to the liver for elimination (Fig. 7.7). Although the major apolipoprotein in HDL_2 and HDL_3 is apoA-I, HDL_2 also contains significant amounts of apoE. ApoE uniquely facilitates RCT by allowing CE-rich core expansion in HDL (Mahley et al., 2006) after LCAT converts FC to CE (Section 7.3.3). ApoA-I-containing HDL can accommodate only a limited amount of CE in its core, resulting in a limited size expansion, whereas the size and CE content of HDL can be increased significantly when apoE is present. Since apoE is a ligand for the LDL receptor, these apoE-enriched HDLs can deliver cholesterol acquired from the periphery to the liver via hepatic LDL receptors (Mahley, 1988). After secretion from the liver as a component of VLDL, apoE redistributes to enrich chylomicrons, remnant lipoprotein particles and HDL (Fazio et al., 2000). ApoE associated with lipoprotein particles serves additional functions such as binding to heparan sulfate proteoglycans to facilitate internalization of the lipoprotein by cells (Mahley et al., 2006). The rapid removal of chylomicron remnant particles requires the presence of a cellular pool of apoE. ApoE internalized by hepatocytes is partially protected from lysosomal degradation and recycles through the Golgi apparatus, suggesting that recycled apoE may have a physiological role in lipoprotein assembly, remnant removal and cholesterol efflux (Fazio et al., 2000). After receptor-mediated endocytosis, the intracellular fate of the TAG-rich lipoproteins is very complex and differs from the degradation pathway of LDL (Heeren et al.,

2006). The majority of TAG-rich lipoprotein-derived apoE remains in peripheral recycling endosomes. This pool of apoE is then mobilized by HDL or HDL-derived apoA-I to be recycled back to the plasma membrane; this apoE is secreted into the extracellular medium where it can participate in the formation of apoE-HDL particles. The recycling of apoE in murine macrophages occurs via ABCA1 and, in the presence of apoA-I, recycled apoE exits the cells on HDL-like particles (Hasty et al., 2005). On the other hand, apoA-I stimulates secretion of apoE independently of cholesterol efflux in a human macrophage cell line; this effect reflects a novel, ABCA1-independent, positive feedback pathway for stimulation of apoE secretion (Kockx et al., 2004). Intracellular trafficking of recycling apoE in Chinese hamster ovary cells appears to be linked to cellular cholesterol removal via the endosomal recycling compartment and PL-containing receptors in an alternative pathway to that of the ABCA1/apoA-I route (Braun et al., 2006). The relationships between apoE and ABCA1 for regulating cellular sterol efflux have been examined in macrophages expressing both apoE and ABCA1. ABCA1 expression is required for apoE-mediated cholesterol efflux when endogenously synthesized apoE accumulates extracellularly. However, an ABCA1-independent pathway for lipid efflux that requires the intracellular synthesis and/or transport of apoE also exists (Huang et al., 2006). Overall, it is clear that the secretion of apoE from macrophages is a regulated process and that both ABCA1-dependent and -independent pathways exist (Kockx et al., 2008).

ApoE is the major apolipoprotein in cerebrospinal fluid and it plays a pivotal role in maintaining cholesterol homeostasis in the brain. The most cholesterol-rich organ in the body is the brain, which contains some 25% of total body cholesterol; this cholesterol is a major component of myelin and neuronal and glial cell membranes (Dietschy and Turley, 2001). Nearly all brain cholesterol is synthesized in situ and it is transported on apoE-HDL particles that circulate in the cerebrospinal fluid (Pitas et al., 1987; LaDu et al., 2000). ABCA1 and ABCG1 are involved in regulation of lipid and lipoprotein metabolism in the brain. ABCA1 mediates the initial lipidation of nascent apoE particles (cf. Section 7.5.2), which are secreted from both astrocytes and microglia (Hirch-Reinshagen and Wellington, 2007). These discoidal particles are then thought to acquire further lipids from either neurons or glia through an active transport process mediated by ABCG1 (Kim et al., 2006). This step, together with the subsequent maturation of the circulating lipoprotein, results in the spherical apoE-containing particles observed in cerebrospinal fluid. ApoE has been implicated in the clearance and deposition of amyloid-β in the brain. When complexed with amyloid-β, apoE facilitates the cellular uptake of amyloid-β via apoE receptors (Tokuda et al., 2000). ApoE is required for the extracellular deposition of amyloid-β as amyloid and it delays the clearance of amyloid-β across the blood-brain barrier (Hirch-Reinshagen and Wellington, 2007). It is probable that one of the major functions of apoE in the central nervous system is to mediate neuronal repair, remodelling and protection with apoE4 being less effective than the apoE2 and apoE3 isoforms (Mahley and Huang, 1999). ApoE has been linked to the pathogenesis of Alzheimer's disease with apoE polymorphism having significant effects (Roses, 1996). The apoE4 allele is a major susceptibility gene that is associated with

40–65% of cases of sporadic and familial Alzheimer's disease; individuals carrying this isoform have an increased occurrence and reduced age of onset of the disease (Corder et al., 1993). The apoE2 allele may be even more protective than apoE3 against the development of Alzheimer's disease (Corder et al., 1994). The mechanisms by which apoE is involved in the pathogenesis of Alzheimer's disease remain incompletely understood. Cholesterol homeostasis and trafficking is currently recognized to participate in aspects of amyloid-β metabolism, but how this may be related to apoE as a risk factor for Alzheimer's disease requires further clarification.

7.6 HDL and Inflammation

Inflammation, such as is caused by direct tissue injury induces the acute phase (AP) response in which the liver increases the synthesis of a number of particular proteins including serum amyloid A (SAA) (O'Brien and Chait, 2006). Macrophages are mobilized to the injury site where cell death has occurred to remove the cell debris, such as membranes rich in PL and cholesterol. This movement of macrophages to the injury site is assisted by HDL, which becomes proinflammatory in the AP response and can induce monocyte chemotaxis (Ansell et al., 2007). The HDL is also required to mediate RCT and remove the excess cholesterol from the macrophages. The conversion of HDL from its usual anti-inflammatory condition to a proinflammatory state during the AP response is caused by the activity of the AP proteins, SAA and Group IIa secretory phospholipase A_2 (Ansell et al., 2007; Kontush and Chapman, 2006; van der Westhuijzen et al., 2007). The enhanced expression of these proteins leads to extensive HDL remodelling and lower circulating levels of HDL cholesterol and apoA-I (Jahangiri et al., 2009; van der Westhuijzen et al., 2007). These changes in HDL are presumably beneficial during the 24 hours or so of the AP response. However, the presence of proinflammatory HDL during the chronic inflammation associated with atherosclerosis (Kontush and Chapman, 2006; O'Brien and Chait 2006) is harmful (Ansell et al., 2007). Although SAA is a major player in the HDL remodeling that occurs during the AP response, current understanding of the role of SAA in the inflammatory response and of how this protein affects HDL metabolism and cholesterol transport is limited.

7.6.1 Serum Amyloid A

SAA consists of four isoforms and, of these, isoforms SAA1 and SAA2 are major AP response proteins whose plasma concentrations increase by some three orders of magnitude after injury. Most of the plasma SAA is bound to HDL so that SAA is the major protein in AP HDL (Tam et al., 2002; van der Westhuijzen et al., 2007). Human SAA1 and SAA2 each contain 122 amino acids and structure prediction analysis suggests that about 80% of these residues are located in an N-terminal helix bundle, with the remainder of the C-terminus being disordered (Stevens, 2004); this structure is analogous to that of apoA-I (*see* Section 7.2.2). Furthermore, SAA contains amphipathic α-helices and can bind to PL to create SAA-PL complexes in an analogous fashion to apoA-I (Bausserman et al., 1983; Segrest et al., 1976). However, unlike apoA-I, SAA can directly bind cholesterol in solution with high affinity to form an equimolar complex (Liang and Sipe, 1995); the N-terminal region is involved in the binding site (Liang et al., 1996). The lipid-binding capabilities of the SAA molecule allow it to replace apoA-I on HDL particles and thereby modify HDL-mediated cholesterol transport (van der Westhuijzen et al., 2007).

Currently, it is thought that the SAA in AP HDL particles is involved in enhancing removal of cholesterol from macrophages at sites of injury thereby playing a protective role (Tam et al., 2002; van der Westhuijzen et al., 2007). Consistent with this concept, like apoA-I, SAA can participate in the various pathways for FC efflux from macrophages (*see* Fig. 7.7). Thus, SAA in a lipidated state can mediate cellular FC efflux via SR-BI (Cai et al., 2005; Marsche et al., 2007) (cf. Section 7.5.3) while lipid-free SAA promotes ABCA1-dependent FC and PL efflux and nascent HDL particle assembly (Abe-Dohmae et al., 2006; Stonik et al., 2004) (cf. Section 7.5.2). AP HDL promotes more FC efflux from cholesterol-loaded macrophages than normal HDL and this is thought to occur because SAA2 promotes the availability for export of FC in cells (Tam et al., 2002; van der Westhuijzen et al., 2007). Thus, peptides from SAA2 have been reported to operate inside macrophages to reduce acyl CoA-acyltransferase (ACAT) activity and increase neutral CE hydrolase activity (Tam et al., 2002). These various abilities of SAA to promote cellular cholesterol efflux lead to the retention of cholesterol efflux capacity in AP plasma despite marked decreases in HDL cholesterol and apoA-I levels (Jahangiri et al., 2009).

7.7 Summary and Conclusions

Significant progress is being made in understanding the structures of human exchangeable apolipoproteins, especially apoA-I and apoE. It is established that the amphipathic α-helical repeats in these proteins are the key structural elements responsible for the functions of these molecules. The application of protein engineering techniques together with a range of physical-biochemical methods has shown that apoA-I and apoE adopt two-domain tertiary structures. The α-helices located in the N-terminal two-thirds of the molecule adopt a helix-bundle conformation while the C-terminal region forms a separately folded, relatively disordered, domain. ApoA-I and apoE bind to lipid surfaces with high affinity, and this binding is initiated by the C-terminal domain, which becomes more α-helical upon interaction with the lipid. Subsequently, the N-terminal helix bundle domain can open, allowing helix–helix contacts to be replaced by helix-lipid interactions. ApoA-I and apoE possess detergent-like properties in that their lipid-binding capabilities permit them to solubilize vesicular phospholipid and form discoidal HDL particles. The apoA-I molecules in such particles adopt a 'double-belt' conformation with defined protein-protein contacts that are also maintained in spherical HDL particles. These structural models are proving valuable for understanding HDL function, but atomic level resolution structures are required to elucidate the detailed molecular mechanisms underlying the participation of HDL in the RCT pathway. The heterogeneity of HDL with regard to particle size, particle shape, protein composition and lipid composition complicates attempts to derive high resolution structural information. The lipid-binding and lipid-solubilizing properties of apoA-I and apoE underlie their abilities to interact with ABCA1, efflux cellular lipids and create nascent HDL particles. Detailed understanding of why a heterogeneous population of nascent HDL particles is formed will require more knowledge of the structure and lipid translocase activity of ABCA1, as well as of the membrane microenvironment in which it resides. Good progress is being made in understanding the ways in which apoA-I-containing HDL particles participate in the reverse transport of cholesterol from peripheral cells to the liver. Less is known about how apoE-HDL mediates cholesterol transport in the brain. At this stage, relatively little in structural and mechanistic terms is known about the effects of natural mutations and oxidative modifications on the functionalities of apoA-I, apoE and the HDL particles they form.

Acknowledgements We are indebted to all our colleagues for their valuable contributions to the studies from our laboratory described here. Our research reported here was supported by NIH Grants HL22633 and HL 56083.

References

Abe-Dohmae, S., Kato, K. H., Kumon, Y., Hu, W., Ishigami, H., Iwamoto, N., Okazaki, M., Wu, C. A., Ueda, K. and Yokoyama, S., 2006, Serum amyloid A generates high density lipoprotein with cellular lipid in an ABCA1- or ABCA7-dependent manner. *J. Lipid Res.* **47:** 1542–1550.

Acharya, P., Segall, M. L., Zaiou, M., Morrow, J. A., Weisgraber, K., Phillips, M. C., Lund-Katz, S. and Snow, J. W., 2002, Comparison of the stabilities and unfolding pathways of human apolipoprotein E isoforms by differential scanning calorimetry and circular dichroism. *Biochim. Biophys. Acta* **1584:** 9–19.

Acton, S., Rigotti, A., Landschulz, K. T., Xu, S., Hobbs, H. H. and Krieger, M., 1996, Identification of scavenger receptor SR-BI as a high density lipoprotein receptor. *Science* **271:** 518–520.

Adorni, M. P., Zimetti, F., Billheimer, J. T., Wang, N., Rader, D. J., Phillips, M. C. and Rothblat, G. H., 2007, The roles of different pathways in the release of cholesterol from macrophages, *J. Lipid Res.* **48:** 2453–2462.

Aggerbeck, L. P., Wetterau, J. R., Weisgraber, K. H., Wu, C. S. C. and Lindgren, F. T., 1988, Human apolipoprotein E3 in aqueous solution. II. Properties of the amino- and carboxyl-terminal domains. *J. Biol. Chem.* **263:** 6249–6258.

Ajees, A. A., Anantharamaiah, G. M., Mishra, V. K., Hussain, M. M. and Murthy, S., 2006, Crystal structure of human apolipoprotein A-I: Insights into its protective effect against cardiovascular diseases. *Proc. Natl. Acad. Sci. USA* **103:** 2126–2131.

Alder-Baerens, N., Muller, P., Pohl, A., Korte, T., Hamon, Y., Chimini, G., Pomorski, T., and Herrmann, A., 2005, Headgroup-specific exposure of phospholipids in ABCA1-expressing cells. *J. Biol. Chem.* **280:** 26321–26329.

Andrews, A. L., Atkinson, D., Barratt, M. D., Finer, E. G., Hauser, H., Henry, R., Leslie, R. B., Owens, N. L., Phillips, M. C., and Robertson, R. N., 1976, Interaction of apoprotein from porcine high-density lipoprotein with dimyristoyl lecithin. 2. Nature of lipid-protein interaction. *Eur. J. Biochem.* **64:** 549–563.

Ansell, B. J., Fonarow, G. C., and Fogelman, A. M., 2007, The paradox of dysfunctional high-density lipoprotein. *Curr. Opin. Lipidol.* **18:** 427–434.

Arnulphi, C., Jin, L., Tricerri, M. A., and Jonas, A., 2004, Enthalpy-driven apolipoprotein A-I and lipid bilayer interaction indicating protein penetration upon lipid binding. *Biochemistry* **43**: 12258–12264.

Atkinson, D., Smith, H. M., Dickson, J., and Austin, J. P., 1976, Interaction of apoprotein from porcine high-density lipoprotein with dimyristoyl lecithin. 1. The structure of the complexes. *Eur. J. Biochem.* **64**: 541–547.

Badellino, K. O. and Rader, D. J., 2004, The role of endothelial lipase in high-density lipoprotein metabolism. *Curr. Opin. Lipidol.* **19**: 392–395.

Barbier, A., Clement-Collin, V., Dergunov, A. D., Visvikis, A., Siest, G., and Aggerbeck, L. P., 2006, The structure of human apolipoprotein E2, E3 and E4 in solution 1. Tertiary and quaternary structure. *Biophys. Chem.* **20**: 158–169.

Barratt, M. D., Badley, R. A., and Leslie, R. B., 1974, The interaction of apoprotein from porcine high-density lipoprotein with dimyristoyl phosphatidylcholine. *Eur. J. Biochem.* **48**: 595–601.

Bausserman, L. L., Herbert, P. N., Forte, T., Klausner, R. D., McAdam, K. P. W. J., Osborne, J. C., Jr., and Rosseneu, M., 1983. Interaction of the serum amyloid A proteins with phospholipid. *J. Biol. Chem.* **258**: 10681–10688.

Behling Agree, A. K., Tricerri, M. A., Arnvig-McGuire, K., Tian, S. M., and Jonas, A., 2002, Folding and stability of the C-terminal half of apolipoprotein A-I examined with a Cys-specific fluorescence probe. *Biochim. Biophys. Acta* **1594**: 286–296.

Braun, N. A., Mohler, P. J., Weisgraber, K. H., Hasty, A. H., Linton, M. F., Yancey, P. G., Su, Y. R., Fazio, S., and Swift, L. L., 2006, Intracellular trafficking of recycling apolipoprotein E in Chinese hamster ovary cells. *J. Lipid Res.* **47**: 1176–1186.

Brouillette, C. G., Anantharamaiah, G. M., Engler, J. A., and Borhani, D. W., 2001, Structural models of human apolipoprotein A-I: a critical analysis and review. *Biochim. Biophys. Acta* **1531**: 4–46.

Brouillette, C. G., Dong, W. J., Yang, Z. W., Ray, M. J., Protasevich, I. I., Cheung, H. C., and Engler, J. A., 2005, Forster resonance energy transfer measurements are consistent with a helical bundle model for lipid-free apolipoprotein A-I. *Biochemistry* **44**: 16413–16425.

Cai, L., de Beer, M. C., de Beer, F. C., and van der Westhuijzen, D. R., 2005, Serum amyloid A is a ligand for scavenger receptor class B type I and inhibits high density lipoprotein binding and selective lipid uptake. *J. Biol. Chem.* **280**: 2954–2961.

Chambenoit, O., Hamon, Y., Margue, D., Rigneult, H., Rosseneu, M., and Chimini, G., 2001, Specific docking of apolipoprotein A-I at the cell surface requires a functional ABCA1 transporter. *J. Biol. Chem.* **276**: 9955–9960.

Chau, P., Nakamura, Y., Fielding, C. J., and Fielding, P. E., 2006, Mechanism of prebeta-HDL formation and activation. *Biochemistry* **45**: 3981–3987.

Choy, N., Raussens, V., and Narayanaswami, V., 2003, Inter-molecular coiled-coil formation in human apolipoprotein E C-terminal domain. *J. Mol. Biol.* **334**: 527–539.

Clay, M. A., Newnham, H. H., and Barter, P. J., 1991, Hepatic lipase promotes a loss of apolipoprotein A-I from triglyceride-enriched human high density lipoproteins during incubation in vitro. *Arterioscler. Thromb.* **11**: 415–422.

Clay, M. A., Pyle, D. H., Rye, K. A., and Barter, P. J., 2000, Formation of spherical, reconstituted high density lipoproteins containing both apolipoproteins A-I and A-II is mediated by lecithin: cholesterol acyltransferase. *J. Biol. Chem.* **275**: 9019–9025.

Connelly, M. A., Klein, S. M., Azhar, S., Abumrad, N. A., and Williams, D. L., 1999, Comparison of class B scavenger receptors, CD36 and scavenger receptor BI (SR-BI), shows that both receptors mediate high density lipoprotein-cholesteryl ester selective uptake but SR-BI exhibits a unique enhancement of cholesteryl ester uptake. *J. Biol. Chem.* **274**: 41–47.

Connelly, M. A., Llera-Moya, M., Monzo, P., Yancey, P., Drazul, D., Stoudt, G., Fournier, N., Klein, S. M., Rothblat, G., and Williams, D. L., 2001, Analysis of chimeric receptors shows that multiple distinct functional activities of scavenger receptor, class B, type I (SR-BI), are localized to the extracellular receptor domain. *Biochemistry* **40**: 5249–5259.

Connelly, M. A. and Williams, D. L., 2004, Scavenger receptor BI: A scavenger receptor with a mission to transport high density lipoprotein lipids. *Curr. Opin. Lipidol.* **15**: 287–295.

Corder, E. H., Saunders, A. M., Risch, N. J., Strittmatter, W. J., Schmechel, D. E., Gaskell, P. C., Jr., Rimmler, J. B., Locke, P. A., Conneally, P. M., Schmader, K. E., Small, G. W., Roses, A. D., Haines, J. L., and Pericak-Vance, M. A., 1994, Protective effect of apolipoprotein E type 2 allele for late onset Alzheimer's disease. *Nat. Genet.* **7**: 180–184.

Corder, E. H., Saunders, A. M., Strittmatter, W. J., Schmechel, D. E., Gaskell, P. C., Small, G. W., Roses, A. D., Haines, G. L., and Pericak-Vance, M. A., 1993, Gene dose of apolipoprotein E type 4 allele and the risk of Alzheimer's disease in late onset families. *Science* **261**: 921–923.

Cuchel, M. and Rader, D. J., 2006, Macrophage reverse cholesterol transport. Key to the regression of atherosclerosis? *Circulation* **113**: 2548–2555.

Davidson, W. S., Arnvig-McGuire, K., Kennedy, A., Kosman, J., Hazlett, T., and Jonas, A., 1999, Structural organization of the N-terminal domains of apolipoprotein A-I: Studies of tryptophan mutants. *Biochemistry* **38**: 14387–14395.

Davidson, W. S., Hazlett, T., Mantulin, W. M., and Jonas, A. 1996, The role of apolipoprotein AI domains in lipid binding. *Proc. Natl. Acad. Sci. USA* **93**: 13605–13610.

Davidson, W. S. and Silva, R. A. G. 2005, Apolipoprotein structural organization in high density lipoproteins: Belts, bundles, hinges and hairpins. *Curr. Opin. Lipidol.* **16**: 295–300.

Davidson, W. S. and Thompson, T. B. 2007, The structure of apolipoprotein A-I in high density lipoproteins. *J. Biol. Chem.* **282**: 22249–22253.

Davignon, J., Gregg, R. E., and Sing, C. F. 1988, Apolipoprotein E polymorphism and atherosclerosis. *Arteriosclerosis* **8**: 1–21.

Dawson, R. J. P. and Locher, K. P. 2006, Structure of a bacterial multidrug ABC transporter, *Nature* **443**: 180–185.

de Beer, M. C., Durbin, D. M., Cai, L., Jonas, A., and de Beer, F. C. 2001a, Apolipoprotein A-I conformation markedly influences HDL interaction with scavenger receptor BI. *J. Lipid Res.* **42**: 309–313.

de Beer, M. C., Durbin, D. M., Cai, L., Mirocha, N., Jonas, A., Webb, N. R., de Beer, F. C., and van der Westhuyzen, D. R. 2001b, Apolipoprotein A-II modulates the binding and selective lipid uptake of reconstituted high density lipoprotein by scavenger receptor BI. *J. Biol. Chem.* **276**: 15832–15839.

Dean, M., Hamon, Y., and Chimini, G. 2001, The human ATP-binding cassette (ABC) transporter superfamily. *J. Lipid Res.* **42**: 1007–1017.

Denis, M., Landry, Y. D., and Zha, X. 2008, ATP-binding cassette A1-mediated lipidation of apolipoprotein A-I occurs at the plasma membrane and not in the endocytic compartments. *J. Biol. Chem.* **283**: 16178–16186.

Dietschy, J. M. and Turley, S. D. 2001, Cholesterol metabolism in the brain. *Curr. Opin. Lipidol.* **12**: 105–112.

Dong, L. M., Parkin, S., Trakhanov, S. D., Rupp, B., Simmons, T., Arnold, K. S., Newhouse, Y. M., Innerarity, T. L., and Weisgraber, K. H. 1996, Novel mechanism for defective receptor binding of apolipoprotein E2 in type III hyperlipoproteinemia. *Nat. Struct. Biol.* **3**: 718–722.

Dong, L. M. and Weisgraber, K. H. 1996, Human apolipoprotein E4 domain interaction. Arginine 61 and glutamic acid 255 interact to direct the preference for very low density lipoproteins. *J. Biol. Chem.* **271**: 19053–19057.

Dong, L. M., Wilson, C., Wardell, M. R., Simmons, T., Mahley, R. W., Weisgraber, K. H., and Agard, D. A. 1994, Human apolipoprotein E. Role of arginine 61 in mediating the lipoprotein preferences of the E3 and E4 isoforms. *J. Biol. Chem.* **269**: 22358–22365.

Donovan, J. M., Benedek, G. B., and Carey, M. C. 1987, Self-association of human apolipoproteins A-I and A-II and interactions of apolipoprotein A-I with bile salts: Quasi-elastic light scattering studies. *Biochemistry* **26**: 8116–8125.

Duong, P. T., Collins, H. L., Nickel, M., Lund-Katz, S., Rothblat, G. H., and Phillips, M. C. 2006, Characterization of nascent HDL particles and microparticles formed by ABCA1-mediated efflux of cellular lipids to apoA-I. *J. Lipid Res.* **47**: 832–843.

Duong, P. T., Weibel, G. L., Lund-Katz, S., Rothblat, G. H., and Phillips, M. C. 2008, Characterization and properties of pre beta-HDL particles formed by ABCA1-mediated cellular lipid efflux to apoA-I. *J. Lipid Res.* **49:** 1006–1014.

Eckhardt, E. R. M., Cai, L., Sun, B., Webb, N. R., and van der Westhuyzen, D. R. 2004, High density lipoprotein uptake by scavenger receptor SR-BII. *J. Biol. Chem.* **279:** 14372–14381.

Fan, D., Li, Q., Korando, L., Jerome, W. G., and Wang, J. 2004, A monomeric human apolipoprotein E carboxyl-terminal domain. *Biochemistry* **43:** 5055–5064.

Fang, Y., Gursky, O., and Atkinson, D. 2003, Structural studies of N- and C-terminally truncated human apolipoprotein A-I. *Biochemistry* **42:** 6881–6890.

Faulkner, L. E., Panagotopulos, S. E., Johnson, J. D., Woollett, L. A., Hui, D. Y., Witting, S. R., Maiorano, J. N., and Davidson, W. S. 2008, An analysis of the role of a retroendocytosis pathway in ABCA1-mediated cholesterol efflux from macrophages. *J. Lipid Res.* **49:** 1322–1332.

Fazio, S., Linton, M. F., and Swift, L. L. 2000, The cell biology and physiologic relevance of apoE recycling. *Trends Cardiovasc. Med.* **10:** 23–30.

Fenske, S. A., Yesilaltay, A., Pal, R., Daniels, K., Rigotti, A., Krieger, M., and Kocher, O. 2008, Overexpression of the PDZ1 domain of PDZK1 blocks the activity of hepatic scavenger receptor, class B, type I (SR-BI) by altering its abundance and cellular localization. *J. Biol. Chem.* **283:** 22097–22104.

Fielding, C. J. and Fielding, P. E. 1995, Molecular physiology of reverse cholesterol transport. *J. Lipid Res.* **36:** 211–228.

Fisher, C. A. and Ryan, R. O. 1999, Lipid binding-induced conformational changes in the N-terminal domain of human apolipoprotein E. *J. Lipid Res.* **40:** 93–99.

Fitzgerald, M. L., Morris, A. L., Rhee, J. S., Andersson, L. P., Mendez, A. J., and Freeman, M. W. 2002, Naturally occurring mutations in the largest extracellular loops of ABCA1 can disrupt its direct interaction with apolipoprotein A-I. *J. Biol. Chem.* **277:** 33178–33187.

Forte, T. M., Nichols, A. V., Gong, E. L., Lux, S., and Levy, R. I. 1971, Electron microscopic study on reassembly of plasma high density apoprotein with various lipids. *Biochim. Biophys. Acta* **248:** 381–386.

Frank, P. G. and Marcel, Y. L. 2000, Apolipoprotein A-I: Structure-function relationships. *J. Lipid Res.* **41:** 853–872.

Gazzara, J. A., Phillips, M. C., Lund-Katz, S., Palgunachari, M. N., Segrest, J. P., Anantharamaiah, G. M., and Snow, J. W. 1997, Interaction of class A amphipathic helical peptides with phospholipid unilamellar vesicles. *J. Lipid Res.* **38:** 2134–2146.

Gelissen, I. C., Harris, M., Rye, K. A., Quinn, C., Brown, A. J., Kockx, M., Cartland, S., Packianathan, M., Kritharides, L., and Jessup, W. 2008, ABCA1 and ABCG1 synergize to mediate cholesterol export to apoA-I. *Arterioscler. Thromb. Vasc. Biol.* **26:** 534–540.

Getz, G. S. and Reardon, C. A. 2009, Apoprotein E as a lipid transport and signaling protein in the blood, liver, and artery wall. *J. Lipid Res.* **50:** S156–S161.

Gillotte, K. L., Davidson, W. S., Lund-Katz, S., Rothblat, G. H., and Phillips, M. C. 1998, Removal of cellular cholesterol by pre-beta-HDL involves plasma membrane microsolubilization. *J. Lipid Res.* **39:** 1918–1928.

Gillotte, K. L., Zaiou, M., Lund-Katz, S., Anantharamaiah, G. M., Holvoet, P., Dhoest, A., Palgunachari, M. N., Segrest, J. P., Weisgraber, K. H., Rothblat, G. H., and Phillips, M. C. 1999, Apolipoprotein-mediated plasma membrane microsolubilization. *J. Biol. Chem.* **274:** 2021–2028.

Gillotte-Taylor, K. L., Nickel, M., Johnson, W. J., Francone, O., Holvoet, P., Lund-Katz, S., Rothblat, G. H., and Phillips, M. C. 2002, Effects of enrichment of fibroblasts with unesterified cholesterol on the efflux of cellular lipids to apolipoprotein A-I. *J. Biol. Chem.* **277:** 11811–11820.

Gong, E., Tan, C. S., Shoukry, M. I., Rubin, E. M., and Nichols, A. V. 1994, Structural and functional properties of human and mouse apolipoprotein A-I. *Biochim. Biophys. Acta* **1213:** 335–342.

Gu, X., Kozarsky, K., and Krieger, M. 2000, Scavenger receptor class B, type I-mediated [3-H]cholesterol efflux to high and low density lipoproteins is dependent on lipoprotein binding to the receptor. *J. Biol. Chem.* **275:** 29993–30001.

Gu, X., Trigatti, B., Xu, S., Acton, S., Babitt, J., and Krieger, M. 1998, The efficient cellular uptake of high density lipoprotein lipid via scavenger receptor class B type I requires not only receptor-mediated surface binding but also receptor-specific lipid transfer mediated by its extracellular domain. *J. Biol. Chem.* **273:** 26338–26348.

Gursky, O. and Atkinson, D. 1996, Thermal unfolding of human high-density apolipoprotein A-I: implications for a lipid-free molten globular state. *Proc. Natl. Acad. Sci. USA* **93:** 2991–2995.

Harder, C. J., Vassiliou, G., McBride, H. M., and McPherson, R. 2006, Hepatic SR-BI-mediated cholesteryl ester selective uptake occurs with unaltered efficiency in the absence of cellular energy. *J. Lipid Res.* **47:** 492–503.

Hasty, A. H., Plummer, M. R., Weisgraber, K. H., Linton, M. F., Fazio, S., and Swift, L. L. 2005, The recycling of apolipoprotein E in macrophages: influence of HDL and apolipoprotein A-I. *J. Lipid Res.* **46:** 1433–1439.

Hatters, D. M., Budamagunta, M. S., Voss, J. C., and Weisgraber, K. H. 2005, Modulation of apolipoprotein E structure by domain interaction. Differences in lipid-bound and lipid-free forms. *J. Biol. Chem.* **280:** 34288–34295.

Hatters, D. M., Peters-Libeu, C. A., and Weisgraber, K. H. 2006, Apolipoprotein E structure: Insights into function. *Trends Biochem. Sci.* **31:** 445–454.

Hatters, D. M., Voss, J. C., Budamagunta, M. S., Newhouse, Y. N., and Weisgraber, K. H. 2009, Insight on the molecular envelope of lipid-bound apolipoprotein E from electron paramagnetic resonance spectroscopy. *J. Mol. Biol.* **386:** 261–271.

Hauser, H., Henry, R., Leslie, R. B., and Stubbs, J. M. 1974, The interaction of apoprotein from porcine high-density lipoprotein with dimyristoyl phosphatidylcholine. *Eur. J. Biochem.* **48:** 583–594.

Heeren, J., Beisiegel, U., and Grewal, T. 2006, Apolipoprotein E recycling. Implications for dyslipidemia and atherosclerosis. *Arterioscler. Thromb. Vasc. Biol.* **26:** 442–448.

Hirch-Reinshagen, V. and Wellington, C. L. 2007, Cholesterol metabolism, apolipoprotein E, adenosine triphosphate-binding cassette transporters, and Alzheimer's disease. *Curr. Opin. Lipidol.* **18:** 325–332.

Hirz, R. and Scanu, A. M. 1970, Reassembly in vitro of a serum high-density lipoprotein. *Biochim. Biophys. Acta* **207:** 364–367.

Homanics, G. E., de Silva, H. V., Osada, J., Zhang, S. H., Wong, H., Borensztajn, J., and Maeda, N. 1995, Mild dyslipidemia in mice following targeted inactivation of the hepatic lipase gene. *J. Biol. Chem.* **270:** 2974–2980.

Huang, Z. H., Fitzgerald, M. L., and Mazzone, T. 2006, Distinct cellular loci for the ABCA1-dependent and ABCA1-independent lipid efflux mediated by endogenous apolipoprotein E expression. *Arterioscler. Thromb. Vasc. Biol.* **26:** 157–162.

Huuskonen, J., Olkkonen, V. M., Jauhiainen, M., and Ehnholm, C. 2001, The impact of phospholipid transfer protein (PLTP) on HDL metabolism. *Atherosclerosis* **155:** 269–281.

Innerarity, T. L., Pitas, R. E., and Mahley, R. W. 1979, Binding of arginine-rich (E) apoprotein after recombination with phospholipid vesicles to the low density lipoprotein receptors of fibroblasts. *J. Biol. Chem.* **254:** 4186–4190.

Jahangiri, A., de Beer, M. C., Noffsinger, V., Tannock, L. R., Ramaiah, C., Webb, N. R., van der Westhuijzen, D. R., and de Beer, F. C. 2009, HDL remodeling during the acute phase response. *Arterioscler. Thromb. Vasc. Biol.* **29:** 261–267.

Jahangiri, A., Rader, D. J., Marchadier, D., Curtiss, L. K., Bonnet, D. J., and Rye, K. A. 2005, Evidence that endothelial lipase remodels high density lipoproteins without mediating the dissociation of apolipoprotein A-I. *J. Lipid Res.* **46:** 896–903.

Jaye, M. and Krawiec, J. 2004, Endothelial lipase and HDL metabolism. *Curr. Opin. Lipidol.* **15:** 183–189.

Jessup, W., Gelissen, I. C., Gaus, K., and Kritharides, L. 2006, Roles of ATP binding cassette transporters A1 and G1, scavenger receptor BI and membrane lipid domains in cholesterol export from macrophages. *Curr. Opin. Lipidol.* **17:** 247–257.

Ji, Y., Jian, B., Wang, N., Sun, Y., Llera Moya, M., Phillips, M. C., Rothblat, G. H., Swaney, J. B., and Tall, A. R. 1997, Scavenger receptor BI promotes high density lipoprotein-mediated cellular cholesterol efflux. *J. Biol. Chem.* **272:** 20982–20985.

Jian, B., Llera-Moya, M., Ji, Y., Wang, N., Phillips, M. C., Swaney, J. B., Tall, A. R., and Rothblat, G. H. 1998, Scavenger receptor class B type I as a mediator of cellular cholesterol efflux to lipoproteins and phospholipid acceptors. *J. Biol. Chem.* **273:** 5599–5606.

Jonas, A. 1992, Lipid-binding properties of apolipoproteins, in *Structure and Function of Apolipoproteins*. M. Rosseneu, ed., CRC Press, Boca Raton, pp. 217–250.

Jonas, A. 2000, Lecithin cholesterol acyltransferase. *Biochim. Biophys. Acta* **1529:** 245–256.

Jonas, A. and Drengler, S. M. 1980, Kinetics and mechanism of apolipoprotein A-I interaction with L-alpha-dimyristoylphosphatidylcholine vesicles. *J. Biol. Chem.* **255:** 2190–2194.

Jonas, A., Drengler, S. M., and Patterson, B. W. 1980, Two types of complexes formed by the interaction of apolipoprotein A-I with vesicles of L-alpha-dimyristoylphosphatidylcholine. *J. Biol. Chem.* **255:** 2183–2189.

Jonas, A., Krajnovich, D. J., and Patterson, B. W. 1977, Physical properties of isolated complexes of human and bovine A-I apolipoproteins with L-alpha-dimyristoyl phosphatidylcholine. *J. Biol. Chem.* **252:** 2200–2205.

Jonas, A. and Phillips, M. C. 2008, Lipoprotein Structure, in *Biochemistry of Lipids, Lipoproteins and Membranes*. 5th edn., D. E. Vance and J. E. Vance, eds., Elsevier, pp. 485–506.

Kellner-Weibel, G., Llera-Moya, M., Connelly, M. A., Stoudt, G., Christian, A. E., Haynes, M. P., Williams, D. L., and Rothblat, G. H. 2000, Expression of scavenger receptor BI in COS-7 cells alters cholesterol content and distribution. *Biochemistry* **39:** 221–229.

Kim, W. S., Rahmanto, A. S., Kamili, A., Rye, K. A., Guillemin, G. J., Gelissen, I. C., Jessup, W., Hill, A. F., and Garner, B. 2006, Role of ABCG1 and ABCA1 in regulation of neuronal cholesterol efflux to apolipoprotein-E discs and suppression of amyloid-β peptide generation. *J. Biol. Chem.* **282:** 2851–2861.

Kiss, R. S., Kay, C. M., and Ryan, R. O. 1999, Amphipathic alpha-helix bundle organization of lipid-free chicken apolipoprotein A-I. *Biochemistry* **38:** 4327–4334.

Kockx, M., Jessup, W., and Kritharides, L. 2008, Regulation of endogenous apolipoprotein E secretion by macrophages. *Arterioscler. Thromb. Vasc. Biol.* **28:** 1060–1067.

Kockx, M., Rye, K. A., Gaus, K., Quinn, C. M., Wright, J., Sloane, T., Sviridov, D., Fu, Y., Sullivan, D., Burnett, J. R., Rust, S., Assmann, G., Anantharamaiah, G. M., Palgunachari, M. N., Lund-Katz, S., Phillips, M. C., Dean, R. T., Jessup, W., and Kritharides, L. 2004, Apolipoprotein A-I-simulated apolipoprotein E secretion from human macrophages is independent of cholesterol efflux. *J. Biol. Chem.* **279:** 25966–25977.

Kono, M., Okumura, Y., Tanaka, M., Nguyen, D., Dhanasekaran, P., Lund-Katz, S., Phillips, M. C., and Saito, H. 2008, Conformational flexibility of the N-terminal domain of apolipoprotein A-I bound to spherical lipid particles. *Biochemistry* **47:** 11340–11347.

Kontush, A. and Chapman, M. J. 2006, Functionally defective high-density lipoprotein: A new therapeutic target at the crossroads of dyslipidemia, inflammation, and atherosclerosis. *Pharmacol. Rev.* **58:** 342–374.

Krimbou, L., Denis, M., Haidar, B., Carrier, M., Marcil, M., and Genest, J., Jr. 2004, Molecular interactions between apoE and ABCA1: impact on apoE lipidation. *J. Lipid Res.* **45:** 839–848.

Krimbou, L., Marcil, M., and Genest, J. 2006, New insights into the biogenesis of human high-density lipoprotein. *Curr. Opin. Lipidol.* **17:** 257–267.

Kruski, A. W. and Scanu, A. M. 1974, Interaction of human serum high density lipoprotein apoprotein with phospholipids. *Chem. Phys. Lipids* **13:** 27–48.

LaDu, M. J., Reardon, C., Van Eldick, L., Fagan, L. M., Bu, G., Holtzman, D., and Getz, G. S. 2000, Lipoproteins in the central nervous system. *Ann. NY Acad. Sci.* **903:** 167–175.

Lee, J. Y. and Parks, J. S. 2005, ATP-binding cassette transporter AI and its role in HDL formation. *Curr. Opin. Lipidol.* **16:** 19–25.

Li, L., Chen, J., Mishra, V. K., Kurtz, J. A., Cao, D., Klon, A. E., Harvey, S. C., Anantharamaiah, G. M., and Segrest, J. P. 2004, Double belt structure of discoidal high density lipoproteins: Molecular basis for size heterogeneity. *J. Mol. Biol.* **343:** 1293–1311.

Li, W., Tanimura, M., Luo, C., Datta, S., and Chan, L. 1988, The apolipoprotein multigene family: biosynthesis, structure, structure-function relationships, and evolution. *J. Lipid Res.* **29:** 245–271.

Liang, J. S., Schreiber, B. M., Salmona, M., Phillip, G., Gonnerman, W. A., de Beer, F. C., and Sipe, J. D. 1996, Amino terminal region of acute phase, but not constitutive, serum amyloid A (apoSAA) specifically binds and transports cholesterol into aortic smooth muscle and HepG2 cells. *J. Lipid Res.* **37:** 2109–2116.

Liang, J. S. and Sipe, J. D. 1995, Recombinant human serum amyloid A (apoSSAp) binds cholesterol and modulates cholesterol flux *J. Lipid Res.* **36:** 37–46.

Liu, B. and Krieger, M. 2002, Highly purified scavenger receptor class B, type I reconstituted into phosphatidylcholine/cholesterol liposomes mediates high affinity high density lipoprotein binding and selective lipid uptake. *J. Biol. Chem.* **277:** 34125–34135.

Liu, L., Bortnick, A. E., Nickel, M., Dhanasekaran, P., Subbaiah, P. V., Lund-Katz, S., Rothblat, G. H., and Phillips, M. C. 2003, Effects of apolipoprotein A-I on ATP-binding cassette transporter A1-mediated efflux of macrophage phospholipid and cholesterol. *J. Biol. Chem.* **278:** 42976–42984.

Liu, T., Krieger, M., Kan, H. Y., and Zannis, V. I. 2002, The effects of mutations in helices 4 and 6 of apo A-I on SR-BI-mediated cholesterol efflux suggest that formation of a productive complex between reconstituted HDL and SR-BI is required for efficient lipid transport. *J. Biol. Chem.* **277:** 21576–21584.

Llera-Moya, M., Rothblat, G. H., Connelly, M. A., Kellner-Weibel, G., Sakr, S. W., Phillips, M. C., and Williams, D. L. 1999, Scavenger receptor BI (SR-BI) mediates free cholesterol flux independently of HDL tethering to the cell surface. *J. Lipid Res.* **40:** 575–580.

Lu, B., Morrow, J. A., and Weisgraber, K. H. 2000, Conformational reorganization of the four-helix bundle of human apolipoprotein E in binding to phospholipid. *J. Biol. Chem.* **275:** 20775–20781.

Lu, R., Arakawa, R., Ito-Osumi, C., Iwamoto, N., and Yokoyama, S. 2008, ApoA-I facilitates ABCA1 recycle/accumulation to cell surface by inhibiting its intracellular degradation and increases HDL generation. *Arterioscler. Thromb. Vasc. Biol.* **28:** 1820–1824.

Lund-Katz, S., Liu, L., Thuahnai, S. T., and Phillips, M. C. 2003, High density lipoprotein structure. *Frontiers in Bioscience* **8:** d1044–d1054.

Lux, S. E., Hirz, R., Shrager, R. I., and Gotto, A. M. 1972, The influence of lipid on the conformation of human plasma high density apolipoproteins. *J. Biol. Chem.* **247:** 2598–2606.

Mahley, R. W. 1988, Apolipoprotein E: Cholesterol transport protein with expanding role in cell biology. *Science* **240:** 622–630.

Mahley, R. W. and Huang, Y. 1999, Apolipoprotein E: From atherosclerosis to Alzheimer's disease and beyond. *Curr. Opin. Lipidol.* **10:** 207–217.

Mahley, R. W., Huang, Y., and Rall, S. C. 1999, Pathogenesis of type III hyperlipoproteinemia (dysbetalipoproteinemia): Questions, quandaries, and paradoxes. *J. Lipid Res.* **40:** 1933–1949.

Mahley, R. W., Huang, Y., and Weisgraber, K. H. 2006, Putting cholesterol in its place: apoE and reverse cholesterol transport. *J. Clin. Invest.* **116:** 1226–1229.

Mahley, R. W. and Rall, J. 1995, Type III hyperlipoproteinemia (dysbetalipoproteinemia): The role of apolipoprotein E in normal and abnormal lipoprotein metabolism. In *The Metabolic and Molecular Bases of Inherited Disease*, 7th edn, C. R. Scriver et al., eds., McGraw-Hill, Inc., New York, pp. 1953–1980.

Mahley, R. W., Weisgraber, K. H., and Huang, Y. 2009, Apolipoprotein E: Structure determines function – from atherosclerosis to Alzheimer's disease to AIDS. *J. Lipid Res.* **50:** S183–S188.

Marcel, Y. L., Ouimet, M., and Wang, M. D. 2008, Regulation of cholesterol efflux from macrophages. *Curr. Opin. Lipidol.* **19**: 455–461.

Marsche, G., Frank, S., Raynes, J. G., Kozarsky, K. F., Sattler, W., and Malle, E. 2007, The lipidation status of acute-phase protein serum amyloid A determines cholesterol mobilization via scavenger receptor class B, type I. *Biochem. J.* **402**: 117–124.

Massey, J. B., Gotto, A. M. J., and Pownall, H. J. 1979, Contribution of α helix formation in human plasma apolipoproteins to their enthalpy of association with phospholipids. *J. Biol. Chem.* **254**: 9359–9361.

Massey, J. B. and Pownall, H. J. 2008, Cholesterol is a determinant of the structures of discoidal high density lipoproteins formed by the solubilization of phospholipid membranes by apolipoprotein A-I. *Biochim. Biophys. Acta* **1781**: 245–253.

Masson, D., Jiang, X. C., Lagrost, L., and Tall, A. R. 2009, The role of plasma lipid transfer proteins in lipoprotein metabolism and atherogenesis. *J. Lipid Res.* **50**: S201–S206.

Morrow, J. A., Hatters, D. M., Lu, B., Hocht, P., Oberg, K. A., Rupp, B., and Weisgraber, K. H. 2002, Apolipoprotein E4 forms a molten globule. *J. Biol. Chem.* **277**: 50380–50385.

Morrow, J. A., Segall, M. L., Lund-Katz, S., Phillips, M. C., Knapp, M., Rupp, B., and Weisgraber, K. H. 2000, Differences in stability among the human apolipoprotein E isoforms determined by the amino-terminal domain. *Biochemistry* **39**: 11657–11666.

Narayanaswami, V., Maiorano, J. N., Dhanasekaran, P., Ryan, R. O., Phillips, M. C., Lund-Katz, S., and Davidson, W. S. 2004, Helix orientation of the functional domains in apolipoprotein E in discoidal high density lipoprotein particles. *J. Biol. Chem.* **279**: 14273–14279.

Narayanaswami, V. and Ryan, R. O. 2000, Molecular basis of exchangeable apolipoprotein function. *Biochim. Biophys. Acta* **1483**: 15–36.

Narayanaswami, V., Szeto, S. S., and Ryan, R. O. 2001, Lipid association-induced N- and C-terminal domain reorganization in human apolipoprotein E3. *J. Biol. Chem.* **276**: 37853–37860.

Neufeld, E. B., Remaley, A. T., Demosky, S. J., Stonik, J. A., Cooney, A., Comly, M., Dwyer, N. K., Zhang, M., Blanchette-Mackie, E. J., Santamarina-Fojo, S., and Brewer, H. B. J. 2001, Cellular localization and trafficking of the human ABCA1 transporter. *J. Biol. Chem.* **276**: 27584–27590.

Nguyen, D., Dhanasekaran, P., Phillips, M. C., and Lund-Katz, S. 2009, Molecular mechanism of apolipoprotein E binding to lipoprotein particles. *Biochemistry* **48**: 3025–3032.

Nieland, T. J., Ehrlich, M., Krieger, M., and Kirchhausen, T. 2005, Endocytosis is not required for the selective lipid uptake mediated by murine SR-BI. *Biochim. Biophys. Acta* **1734**: 44–51.

O'Brien, K. D. and Chait, A. 2006, Serum amyloid A: The other inflammatory protein. *Curr. Athero. Rep.* **8**: 62–68.

Oda, M. N., Forte, T. M., Ryan, R. O., and Voss, J. C. 2003, The C-terminal domain of apolipoprotein A-I contains a lipid-sensitive conformational trigger. *Nat. Struct. Biol.* **10**: 455–460.

Oram, J. F. 2000, Tangier Disease and ABCA1. *Biochim. Biophys. Acta* **1529**: 321–330.

Oram, J. F. and Heinecke, J. W. 2005, ATP-binding cassette transporter A1: A cell cholesterol exporter that protects against cardiovascular disease. *Physiol. Rev.* **85**: 1343–1372.

Oram, J. F., Lawn, R. M., Garvin, M. R., and Wade, D. P. 2000, ABCA1 is the cAMP-inducible apolipoprotein receptor that mediates cholesterol secretion from macrophages. *J. Biol. Chem.* **275**: 34508–34511.

Palgunachari, M. N., Mishra, V. K., Lund-Katz, S., Phillips, M. C., Adeyeye, S. O., Alluri, S., Anantharamaiah, G. M., and Segrest, J. P. 1996, Only the two end helixes of eight tandem amphipathic helical domains of human apo A-I have significant lipid affinity. *Arterioscler. Thromb. Vasc. Biol.* **16**: 328–338.

Perugini, M. A., Schuck, P., and Howlett, G. J. 2000, Self-association of human apolipoprotein E3 and E4 in the presence and absence of phospholipid. *J. Biol. Chem.* **275**: 36758–36765.

Peters-Libeu, C. A., Newhouse, Y., Hall, S. C., Witkowska, H. E., and Weisgraber, K. H. 2007, Apolipoprotein E dipalmitoylphosphatidylcholine particles are ellipsoidal in solution. *J. Lipid Res.* **48**: 1035–1044.

Peters-Libeu, C. A., Newhouse, Y., Hatters, D. M., and Weisgraber, K. H. 2006, Model of biologically active apolipoprotein E bound to dipalmitoylphosphatidylcholine. *J. Biol. Chem.* **281:** 1073–1079.

Phillips, M. C., Johnson, W. J., and Rothblat, G. H. 1987, Mechanisms and consequences of cellular cholesterol exchange and transfer. *Biochim. Biophys. Acta* **906:** 223–276.

Pitas, R. E., Boyles, J. K., Lee, S. H., Hui, D., and Weisgraber, K. H. 1987, Lipoproteins and their receptors in the central nervous system. *J. Biol. Chem.* **262:** 14352–14360.

Pittman, R. C., Knecht, T. P., Rosenbaum, M. S., and Taylor, C. A., Jr. 1987, A nonendocytotic mechanism for the selective uptake of high density lipoprotein-associated cholesteryl esters. *J. Biol. Chem.* **262:** 2443–2450.

Pownall, H., Pao, Q., Hickson, D., Sparrow, J. T., Kusserow, S. K., and Massey, J. B. 1981, Kinetics and mechanism of association of human plasma apolipoproteins with dimyristoylphosphatidylcholine: Effect of protein structure and lipid clusters on reaction rates. *Biochemistry* **20:** 6630–6635.

Pownall, H. J. 2006, Detergent-mediated phospholipidation of plasma lipoproteins increases HDL cholesterophilicity and cholesterol efflux via SR-BI. *Biochemistry* **45:** 11514–11522.

Pownall, H. J. and Ehnholm, C. 2006, The unique role apolipoprotein A-I in HDL remodeling and metabolism. *Curr. Opin. Lipidol.* **17:** 209–213.

Pownall, H. J., Massey, J. B., Kusserow, S. K., and Gotto, A. M. 1978, Kinetics of lipid-protein interactions: interaction of apolipoprotein A-I from human plasma high density lipoproteins with phosphatidylcholine. *Biochemistry* **17:** 1183–1188.

Pownall, H. J., Massey, J. B., Kusserow, S. K., and Gotto, A. M., Jr. 1979, Kinetics of lipid-protein interactions: Effect of cholesterol on the association of human plasma high-density apolipoprotein A-I with L-alpha-dimyristoylphosphatidylcholine. *Biochemistry* **18:** 574–579.

Pownall, H. J., Massey, J. B., Sparrow, J. T., and Gotto, A. M. 1987, Lipid-protein interactions and lipoprotein reassembly, in *Plasma Lipoproteins*. A. M. Gotto, ed., Elsevier Science Publishers B.V., Amsterdam, pp. 95–127.

Qiu, X., Mistry, A., Ammirati, M. J., Chrunyk, B. A., Clark, R. W., Cong, Y., Culp, J. S., Danley, D. E., Freeman, T. B., Goeghegan, K. F., Griffor, M. C., Hawrylik, S. J., Hayward, C. M., Hensley, P., Hoth, L. R., Karma, G. A., Lira, M. E., Lloyd, D. B., McGrath, K. M., Stutzman-Engwall, K. J., Subashi, A. K., Subashi, T. A., Thompson, J. F., Wang, I. K., Zhao, H., and Seddon, A. P. 2007, Crystal structure of cholesteryl ester transfer protein reveals a long tunnel and four bound lipid molecules. *Nat. Struct. Mol. Biol.* **14:** 106–113.

Raussens, V., Fisher, C. A., Goormaghtigh, E., Ryan, R. O., and Ruysschaert, J. M. 1998, The low density lipoprotein receptor active conformation of apolipoprotein E. *J. Biol. Chem.* **273:** 25825–25830.

Reijngoud, D. J. and Phillips, M. C. 1982, Mechanism of dissociation of human apolipoprotein A-I from complexes with dimyristoylphosphatidylcholine as studied by guanidine hydrochloride denaturation. *Biochemistry* **21:** 2969–2976.

Remaley, A. T., Thomas, F., Stonik, J. A., Demosky, S. J., Bark, S. E., Neufeld, E. B., Bocharov, A. V., Vishnyakova, T. G., Patterson, A. P., Eggerman, T. L., Santamarina-Fojo, S., and Brewer, B. H. 2003, Synthetic amphipathic helical peptides promote lipid efflux from cells by an ABCA1-dependent and an ABCA1-independent pathway. *J. Lipid Res.* **44:** 828–836.

Rigotti, A., Trigatti, B., Penman, M., Rayburn, H., Herz, J., and Krieger, M. 1997, A targeted mutation in the murine gene encoding the high density lipoprotein (HDL) receptor scavenger receptor class B type I reveals its key role in HDL metabolism. *Proc. Natl. Acad. Sci. USA* **94:** 12610–12615.

Roberts, L. M., Ray, M. J., Shih, T. W., Hayden, E., Reader, M. M., and Brouillette, C. G. 1997, Structural analysis of apolipoprotein A-I: limited proteolysis of methionine-reduced and -oxidized lipid-free and lipid-bound human apoA-I. *Biochemistry* **36:** 7615–7624.

Rodrigueza, W. V., Thuahnai, S. T., Temel, R. E., Lund-Katz, S., Phillips, M. C., and Williams, D. L. 1999, Mechanism of scavenger receptor class B type I-mediated selective uptake

of cholesteryl esters from high density lipoprotein to adrenal cells. *J. Biol. Chem.* **274:** 20344–20350.
Rogers, D. P., Roberts, L., Lebowitz, J., Datta, G., Anantharamaiah, G. M., Engler, J. A., and Brouillette, C. G. 1998, The lipid-free structure of apolipoprotein A-I: Effects of amino-terminal deletions. *Biochemistry* **37:** 11714–11725.
Roosbeek, S., Vanloo, B., Duverger, N., Caster, H., Breyne, J., De Beun, I., Patel, H., Vandekerckhove, J., Shoulders, C., Rosseneu, M., and Peelman, F. 2001, Three arginine residues in apolipoprotein A-I are critical for activation of lecithin: cholesterol acyltransferase. *J. Lipid Res.* **42:** 31–40.
Roses, A. D. 1996, Apolipoprotein E alleles as risk factors in Alzheimer's disease. *Ann. Rev. Med.* **47:** 387–400.
Rothblat, G. H., Bamberger, M., and Phillips, M. C. 1986, Reverse cholesterol transport. *Meth. Enzymol.* **129:** 628–644.
Rye, K. A. and Barter, P. J. 2004, Formation and metabolism of prebeta-migrating, lipid-poor apolipoprotein A-I. *Arterioscler. Thromb. Vasc. Biol.* **24:** 421–428.
Rye, K. A., Bursill, C. A., Lambert, G., Tabet, F., and Barter, P. J. 2009, The metabolism and anti-atherogenic properties of HDL. *J. Lipid Res.* **50:** S195–S200.
Rye, K. A., Hime, N. J., and Barter, P. J. 1997, Evidence that cholesteryl ester transfer protein-mediated reductions in reconstituted high density lipoprotein size involve particle fusion. *J. Biol. Chem.* **272:** 3953–3960.
Rye, K. A., Wee, K., Curtiss, L. K., Bonnet, D. J., and Barter, P. J. 2008, Apolipoprotein A-II inhibits high density lipoprotein remodeling and lipid-poor apolipoprotein A-I formation. *J. Biol. Chem.* **278:** 22530–22536.
Sahoo, D., Darlington, Y. F., Pop, D., Williams, D. L., and Connelly, M. A. 2007, Scavenger receptor class B type I (SR-BI) assembles into detergent-sensitive dimers and tetramers. *Biochim. Biophys. Acta* **1771:** 807–817.
Saito, H., Dhanasekaran, P., Baldwin, F., Weisgraber, K., Lund-Katz, S., and Phillips, M. C. 2001, Lipid binding-induced conformational change in human apolipoprotein E. *J. Biol. Chem.* **276:** 40949–40954.
Saito, H., Dhanasekaran, P., Baldwin, F., Weisgraber, K. H., Phillips, M. C., and Lund-Katz, S. 2003a, Effects of polymorphism on the lipid interaction of human apolipoprotein E. *J. Biol. Chem.* **278:** 40723–40729.
Saito, H., Dhanasekaran, P., Nguyen, D., Deridder, E., Holvoet, P., Lund-Katz, S., and Phillips, M. C. 2004a, Alpha-helix formation is required for high affinity binding of human apolipoprotein A-I to lipids. *J. Biol. Chem.* **279:** 20974–20981.
Saito, H., Dhanasekaran, P., Nguyen, D., Holvoet, P., Lund-Katz, S., and Phillips, M. C. 2003b, Domain structure and lipid interaction in human apolipoproteins A-I and E: A general model. *J. Biol. Chem.* **278:** 23227–23232.
Saito, H., Lund-Katz, S., and Phillips, M. C. 2004b, Contributions of domain structure and lipid interaction to the functionality of exchangeable human apolipoproteins. *Progress in Lipid Research* **43:** 350–380.
Saito, H., Miyako, Y., Handa, T., and Miyajima, K. 1997, Effect of cholesterol on apolipoprotein A-I binding to lipid bilayers and emulsions. *J. Lipid Res.* **38:** 287–294.
Sakamoto, T., Tanaka, M., Vedhachalam, C., Nickel, M., Nguyen, D., Dhanasekaran, P., Phillips, M. C., Lund-Katz, S., and Saito, H. 2008, Contributions of the carboxyl-terminal helical segment to the self-association and lipoprotein preferences of human apolipoprotein E3 and E4 isoforms. *Biochemistry* **47:** 2968–2977.
Santamarina-Fojo, S., Lambert, G., Hoeg, J. M., and Brewer, H. B. J. 2000, Lecithin-cholesterol acyltransferase: role in lipoprotein metabolism, reverse cholesterol transport and atherosclerosis. *Curr. Opin. Lipidol.* **11:** 267–275.
Scanu, A. 1967, Binding of human serum high density lipoprotein apoprotein with aqueous dispersions of phospholipids. *J. Biol. Chem.* **242:** 711–719.

Schneeweis, L. A., Koppaka, V., Lund-Katz, S., Phillips, M. C., and Axelsen, P. H. 2005, Structural analysis of lipoprotein E particles. *Biochemistry* **44**: 12525–12534.

Segall, M. L., Dhanasekaran, P., Baldwin, F., Anantharamaiah, G. M., Weisgraber, K., Phillips, M. C., and Lund-Katz, S. 2002, Influence of apoE domain structure and polymorphism on the kinetics of phospholipid vesicle solubilization. *J. Lipid Res.* **43**: 1688–1700.

Segrest, J. P., Garber, D. W., Brouillette, C. G., Harvey, S. C., and Anantharamaiah, G. M. 1994, The amphipathic alpha-helix: a multifunctional structural motif in plasma apolipoproteins. *Adv. Protein Chem.* **45**: 303–369.

Segrest, J. P., Jones, M. K., De Loof, H., Brouillette, C. G., Venkatachalapathi, Y. V., and Anantharamaiah, G. M. 1992, The amphipathic helix in the exchangeable apolipoproteins: a review of secondary structure and function. *J. Lipid Res.* **33**: 141–166.

Segrest, J. P., Jones, M. K., Klon, A. E., Sheldahl, C. J., Hellinger, M., DeLoof, H., and Harvey, S. C. 1999, A detailed molecular belt model for apolipoprotein A-I in discoidal high density lipoprotein. *J. Biol. Chem.* **274**: 31755–31758.

Segrest, J. P., Pownall, H. J., Jackson, R. L., Glenner, G. G., and Pollock, P. S. 1976, Amyloid A: Amphipathic helixes and lipid binding. *Biochemistry* **15**: 3187–3191.

Settasatian, N., Duong, M., Curtiss, L. K., Ehnholm, C., Jauhiainen, M., Huuskonen, J., and Rye, K. A. 2001, The mechanism of the remodeling of high density lipoproteins by phospholipid transfer protein. *J. Biol. Chem.* **276**: 26898–26905.

Silva, R. A. G., Hilliard, G. M., Fang, J., Macha, S., and Davidson, W. S. 2005, A three-dimensional molecular model of lipid-free apolipoprotein A-I determined by cross-linking/mass spectrometry and sequence tracking. *Biochemistry* **44**: 2759–2769.

Silva, R. A. G., Huang, R., Morris, J., Fang, J., Gracheva, E. O., Ren, G., Kontush, A., Jerome, W. G., Rye, K. A., and Davidson, W. S. 2008, Structure of apolipoprotein A-I in spherical high density lipoproteins of different sizes. *Proc. Natl. Acad. Sci. USA* **105**: 12176–12181.

Sivashanmugam, A. and Wang, J. 2009, A unified scheme for initiation and conformational adaptation of human apolipoprotein E N-terminal domain upon lipoprotein-binding and for receptor-binding activity. *J. Biol. Chem.* **284**: 14657–14666.

Sorci-Thomas, M. and Thomas, M. J. 2002, The effects of altered apolipoprotein A-I structure on plasma HDL concentration. *Trends Cardiovasc. Med.* **12**: 121–128.

Sparks, D. L., Davidson, W. S., Lund-Katz, S., and Phillips, M. C. 1995, Effects of the neutral lipid content of high density lipoprotein on apolipoprotein A-I structure and particle stability. *J. Biol. Chem.* **270**: 26910–26917.

Stevens, F. J. 2004, Hypothetical structure of human serum amyloid A protein. *Amyloid. J. Protein Folding. Disord.* **11**: 71–80.

Stonik, J. A., Remaley, A. T., Demosky, S. J., Neufeld, E. B., Bocharov, A., and Brewer, H. B. 2004, Serum amyloid A promotes ABCA1-dependent and ABCA1-independent lipid efflux from cells. *Biochem. Biophys. Res. Commun.* **321**: 936–941.

Tall, A. R., Small, D. M., Deckelbaum, R. J., and Shipley, G. G. 1977, Structure and thermodynamic properties of high density lipoprotein recombinants. *J. Biol. Chem.* **252**: 4701–4711.

Tam, S. P., Flexman, A., Hulme, J., and Kisilevsky, R. 2002, Promoting export of macrophage cholesterol: the physiological role of a major acute-phase protein, serum amyloid A 2.1. *J. Lipid Res.* **43**: 1410–1420.

Tanaka, M., Koyama, M., Dhanasekaran, P., Nguyen, D., Nickel, M., Lund-Katz, S., Saito, H., and Phillips, M. C. 2008, Influence of tertiary structure domain properties on the functionality of apolipoprotein A-I. *Biochemistry* **47**: 2172–2180.

Thomas, M. J., Bhat, S., and Sorci-Thomas, M. G. 2006, The use of chemical cross-linking and mass spectrometry to elucidate the tertiary conformation of lipid-bound apolipoprotein A-I. *Curr. Opin. Lipidol.* **17**: 214–220.

Thomas, M. J., Bhat, S., and Sorci-Thomas, M. G. 2008, Three-dimensional models of HDL apoA-I: Implications for its assembly and function. *J. Lipid Res.* **49**: 1875–1883.

Thuahnai, S. T., Lund-Katz, S., Anantharamaiah, G. M., Williams, D. L., and Phillips, M. C. 2003, A quantitative analysis of apolipoprotein binding to scavenger receptor class B, type I (SR-BI): multiple binding sites for lipid-free and lipid-associated apolipoproteins. *J. Lipid Res.* **44**: 1132–1142.

Thuahnai, S. T., Lund-Katz, S., Dhanasekaran, P., Llera-Moya, M., Connelly, M. A., Williams, D. L., Rothblat, G. H., and Phillips, M. C. 2004, SR-BI-mediated cholesteryl ester selective uptake and efflux of unesterified cholesterol: Influence of HDL size and structure. *J. Biol. Chem.* **279**: 12448–12455.

Thuahnai, S. T., Lund-Katz, S., Williams, D. L., and Phillips, M. C. 2001, Scavenger receptor class B, type I-mediated uptake of various lipids into cells. *J. Biol. Chem.* **276**: 43801–43808.

Thuren, T. 2000, Hepatic lipase and HDL metabolism. *Curr. Opin. Lipidol.* **11**: 277–283.

Tokuda, T., Calero, M., Matsubara, E., Vidal, R., Kumar, A., Permanne, B., Ziokovic, B., Smith, J. D., LaDu, M. J., Rostagno, A., Frangione, B., and Ghiso, J. 2000, Lipidation of apolipoprotein E influences its isoform-specific interaction with Alzheimer's amyloid β peptides. *Biochem. J.* **348**: 359–365.

Tricerri, M. A., Behling Agree, A. K., Sanchez, S. A., and Jonas, A. 2000, Characterization of apolipoprotein A-I structure using a cysteine-specific fluorescence probe. *Biochemistry* **39**: 14682–14691.

Trigatti, B., Rayburn, H., Vinals, M., Braun, A., Miettinen, H., Penman, M., Hertz, M., Schrenzel, M., Amigo, L., Rigotti, A., and Krieger, M. 1999, Influence of the high density lipoprotein receptor SR-BI on reproductive and cardiovascular pathophysiology. *Proc. Natl. Acad. Sci. USA* **96**: 9322–9327.

Trigatti, B. L., Krieger, M., and Rigotti, A. 2003, Influence of the HDL receptor SR-BI on lipoprotein metabolism and atherosclerosis. *Arterioscler. Thromb. Vasc. Biol.* **23**: 1732–1738.

van der Westhuijzen, D. R., de Beer, F. C., and Webb, N. R. 2007, HDL cholesterol transport during inflammation. *Curr. Opin. Lipidol.* **18**: 147–151.

van Eck, M., Pennings, M., Hoekstra, M., Out, R., and Van Berkel, T. J. C. 2005, Scavenger receptor BI and ATP-binding cassette transporter A1 in reverse cholesterol transport and atherosclerosis. *Curr. Opin. Lipidol.* **16**: 307–315.

Van Tol, A. 2002, Phospholipid transfer protein. *Curr. Opin. Lipidol.* **13**: 135–139.

Vanloo, B., Morrison, J., Fidge, N., Lorent, G., Lins, L., Brasseur, R., Ruysschaert, J. M., Baert, J., and Rosseneu, M. 1991, Characterization of the discoidal complexes formed between apoA-I-CNBr fragments and phosphatidylcholine. *J. Lipid Res.* **32**: 1253–1264.

Vaughan, A. M. and Oram, J. F. 2006, ABCA1 and ABCG1 or ABCG4 act sequentially to remove cellular cholesterol and generate cholesterol-rich HDL. *J. Lipid Res.* **47**: 2433–2443.

Vedhachalam, C., Duong, P. T., Nickel, M., Nguyen, D., Dhanasekaran, P., Saito, H., Rothblat, G. H., Lund-Katz, S., and Phillips, M. C. 2007a, Mechanism of ATP-binding cassette transporter A1-mediated cellular lipid efflux to apolipoprotein A-I and formation of high density lipoprotein particles. *J. Biol. Chem.* **282**: 25123–25130.

Vedhachalam, C., Ghering, A. B., Davidson, W. S., Lund-Katz, S., Rothblat, G. H., and Phillips, M. C. 2007b, ABCA1-induced cell surface binding sites for apoA-I. *Arterioscler. Thromb. Vasc. Biol.* **27**: 1603–1609.

Vedhachalam, C., Liu, L., Nickel, M., Dhanasekaran, P., Anantharamaiah, G. M., Lund-Katz, S., Rothblat, G., and Phillips, M. C. 2004, Influence of apo A-I structure on the ABCA1-mediated efflux of cellular lipids. *J. Biol. Chem.* **279**: 49931–49939.

Vedhachalam, C., Narayanaswami, V., Neto, N., Forte, T. M., Phillips, M. C., Lund-Katz, S., and Bielicki, J. K. 2007c, The C-terminal lipid-binding domain of apolipoprotein E is a highly efficient mediator of ABCA1-dependent cholesterol efflux that promotes the assembly of high-density lipoproteins. *Biochemistry* **46**: 2583–2593.

Vitello, L. B. and Scanu, A. M. 1976, Studies on human serum high density lipoproteins. Self-association of apolipoprotein A-I in aqueous solutions. *J. Biol. Chem.* **251**: 1131–1136.

Wang, N., Silver, D. L., Costet, P., and Tall, A. R. 2000, Specific binding of ApoA-I, enhanced cholesterol efflux, and altered plasma membrane morphology in cells expressing ABC1. *J. Biol. Chem.* **275**: 33053–33058.

Wang, N. and Tall, A. R. 2003, Regulation and mechanisms of ATP-binding cassette transporter A1-mediated cellular cholesterol efflux. *Arterioscler. Thromb. Vasc. Biol.* **23**: 1178–1184.

Wang, X., Collins, H. L., Ranalletta, M., Fuki, I. V., Billheimer, J. T., Rothblat, G. H., Tall, A. R., and Rader, D. J. 2007a, Macrophage ABCA1 and ABCG1, but not SR-BI, promote macrophage reverse cholesterol transport in vivo. *J. Clin. Invest.* **117**: 2216–2224.

Wang, X., Collins, H. L., Ranalletta, M., Fuki, I. V., Billheimer, J. T., Rothblat, G. H., Tall, A. R., and Rader, D. J. 2007b, Macrophage ABCA1 and ABCG1, but not SR-BI, promote macrophage reverse cholesterol transport in vivo. *J. Clin. Invest.* **117**: 2216–2224.

Weisgraber, K. H. 1990, Apolipoprotein E distribution among human plasma lipoproteins: role of the cysteine-arginine interchange at residue 112. *J. Lipid Res.* **31**: 1503–1511.

Weisgraber, K. H. 1994, Apolipoprotein E: Structure-function relationships. *Adv. Protein Chem.* **45**: 249–302.

Weisgraber, K. H., Lund-Katz, S., and Phillips, M. C. 1992, Apolipoprotein E: Structure-function correlations, in *High Density Lipoproteins and Atherosclerosis III*. N. E. Miller and A. R. Tall, eds., Elsevier Science Publications., pp. 175–181.

Westerlund, J. A. and Weisgraber, K. H. 1993, Discrete carboxyl-terminal segments of apolipoprotein E mediate lipoprotein association and protein oligomerization. *J. Biol. Chem.* **268**: 15745–15750.

Williams, D. L., Llera-Moya, M., Thuahnai, S. T., Lund-Katz, S., Connelly, M. A., Azhar, S., Anantharamaiah, G. M., and Phillips, M. C. 2000, Binding and cross-linking studies show that scavenger receptor bi interacts with multiple sites in apolipoprotein A-I and identify the class A amphipathic alpha-helix as a recognition motif. *J. Biol. Chem.* **275**: 18897–18904.

Williams, K. J., Feig, J. E., and Fisher, E. A. 2007, Cellular and molecular mechanisms for rapid regression of atherosclerosis: from bench top to potentially achievable clinical goal. *Curr. Opin. Lipidol.* **18**: 443–450.

Wilson, C., Mau, T., Weisgraber, K., Wardell, M. R., Mahley, R. W., and Agard, D. A. 1994, Salt bridge relay triggers defective LDL receptor binding by a mutant apolipoprotein. *Structure* **2**: 713–718.

Wilson, C., Wardell, M. R., Weisgraber, K. H., Mahley, R. W., and Agard, D. A. 1991, Three-dimensional structure of the LDL receptor-binding domain of human apolipoprotein E. *Science* **252**: 1817–1822.

Wong, H. and Schotz, M. C. 2002, The lipase gene family. *J. Lipid Res.* **43**: 993–999.

Yancey, P. G., Bortnick, A. E., Kellner-Weibel, G., Llera-Moya, M., Phillips, M. C., and Rothblat, G. H. 2003, Importance of different pathways of cellular cholesterol efflux. *Arterioscler. Thromb. Vasc. Biol.* **23**: 712–719.

Yancey, P. G., Llera-Moya, M., Swarnaker, S., Monzo, P., Klein, S. M., Connelly, M. A., Johnson, W. J., Williams, D. L., and Rothblat, G. H. 2000, High density lipoprotein phospholipid composition is a major determinant of the bi-directional flux and net movement of cellular free cholesterol mediated by scavenger receptor BI. *J. Biol. Chem.* **275**: 36596–36604.

Yokoyama, S. 2006, Assembly of high-density lipoprotein. *Arterioscler. Thromb. Vasc. Biol.* **26**: 20–27.

Yokoyama, S., Kawai, Y., Tajima, S., and Yamamoto, A. 1985, Behavior of human apolipoprotein E in aqueous solutions and at interfaces. *J. Biol. Chem.* **260**: 16375–16382.

Zannis, V. I., Chroni, A., and Krieger, M. 2006, Role of apoA-I, ABCA1, LCAT, and SR-BI in the biogenesis of HDL. *J. Mol. Med.* **84**: 276–294.

Zhang, Y., da Silva, J. R., Reilly, M., Billheimer, J. T., Rothblat, G. H., and Rader, D. J. 2005, Hepatic expression of scavenger receptor class B type I (SR-BI) is a positive regulator of macrophage reverse cholesterol transport in vivo. *J. Clin. Invest.* **115**: 2870–2874.

Zhang, Y., Vasudevan, S., Sojitrawala, R., Zhao, W., Cui, C., Xu, C., Fan, D., Newhouse, Y., Balestra, R., Jerome, W. G., Weisgraber, K., Li, Q., and Wang, J. 2007, A monomeric, biologically active, full-length human apolipoprotein E. *Biochemistry* **46:** 10722–10732.

Zhang, Y., Zanotti, I., Reilly, M., Glick, J. M., Rothblat, G. H., and Rader, D. J. 2003, Overexpression of apoA-I promotes reverse cholesterol transport of cholesterol from macrophages to feces. *Circulation* **108:** 661–663.

Zhu, K., Brubaker, G., and Smith, J. D. 2007, Large disk intermediate precedes formation of apolipoprotein A-I-dimyristoylphosphatidylcholine small disks. *Biochemistry* **46:** 6299–6307.

Chapter 8
Lipoprotein Modification and Macrophage Uptake: Role of Pathologic Cholesterol Transport in Atherogenesis

Yury I. Miller, Soo-Ho Choi, Longhou Fang, and Sotirios Tsimikas

Abstract Low-density lipoprotein (LDL) is a major extracellular carrier of cholesterol and, as such, plays important physiologic roles in cellular function and regulation of metabolic pathways. However, under pathologic conditions of hyperlipidemia, oxidative stress and/or genetic disorders, specific components of LDL become oxidized or otherwise modified, and the transport of cholesterol by modified LDL is diverted from its physiologic targets toward excessive cholesterol accumulation in macrophages and the formation of macrophage "foam" cells in the vascular wall. This pathologic deposition of modified lipoproteins and the attendant pro-inflammatory reactions in the artery wall lead to the development of atherosclerotic lesions. Continued accumulation of immunogenic modified lipoproteins and a pro-inflammatory milieu result in the progression of atherosclerotic lesions, which may obstruct the arterial lumen and/or eventually rupture and thrombose, causing myocardial infarction or stroke. In this review, we survey mechanisms of LDL modification and macrophage lipoprotein uptake, including results of recent in vivo experiments, and discuss unresolved problems and controversial issues in this growing field. Future directions in studying foam cell formation may include introducing novel animal models, such as hypercholesterolemic zebrafish, enabling dynamic in vivo observation of macrophage lipid uptake.

Keywords Low density lipoprotein · Macrophage · Cholesterol transport · Foam cell · Atheroscleosis

Abbreviations

apoA	apolipoprotein A
CE	cholesteryl ester
CAD	coronary artery disease
FFA	free fatty acids
Hp	haptoglobin

Y.I. Miller (✉)
Department of Medicine, University of California, San Diego, La Jolla, CA, 92037-0682, USA
e-mail: yumiller@ucsd.edu

Hb	hemoglobin
HDL	high density lipoprotein
FH	familial hypercholesterolemia
LDL	low density lipoprotein
LDL-C	LDL cholesterol
LDLR	LDL receptor
12/15LO	12/15-lipoxygenase
LRP-1	LDL receptor related protein-1
LOX-1	lectin-like oxidized LDL receptor-1
LXR	liver X receptors
M-CSF	macrophage colony stimulating factor
MPO	myeloperoxidase
Nox	NADPH oxidase
OxLDL	oxidized LDL
(PRRs)	pattern recognition receptors
PL	phospholipid
PUFA	polyunsaturated fatty acids
SR-A	scavenger receptor type A
RAGE	receptor for advanced glycation end products
SR-BI	scavenger receptor type BI
SREC-I	scavenger receptor expressed by endothelial cells
SR-PSOX/CXCL16	scavenger receptor for phosphatidylserine and oxidized lipoprotein/chemokine (C-X-C motif) ligand 16
Syk	spleen tyrosine kinase
VLDL	very low density lipoprotein

8.1 Introduction

8.1.1 Physiologic Role of LDL in Cholesterol Transport

Cholesterol metabolism is highly regulated and includes dietary delivery, endogenous synthesis as well as numerous pathways for cholesterol utilization. Dietary lipids are absorbed in the intestine where chylomicrons are synthesized to deliver nutrient lipids to tissues and the liver. The liver stores cholesterol and fatty acids and synthesizes VLDL, which is metabolized to LDL intravascularly. LDL then delivers cholesterol and other lipids to all tissues in the body. One LDL particle carries approximately 600 molecules of free cholesterol, 1600 molecules of cholesteryl esters, 700 molecules of phospholipids, and 185 molecules of triglycerides. LDL is recognized by the LDL receptor (LDLR), which is ubiquitously expressed and mediates LDL internalization. HDL and apoAI then participate in "reverse cholesterol transport" from cells, including macrophages in the vessel wall, to the liver. Abnormal regulation or genetic disorders of VLDL, LDL and HDL metabolism are etiologic factors in many diseases of lipid metabolism and constitute the leading cause of atherosclerosis.

8.1.2 Atherosclerosis and the LDL Paradox. Modified LDL

It is now widely accepted that elevated plasma levels of LDL cholesterol (LDL-C) are the major pathogenic factor in the development of atherosclerosis. Primary evidence for this is the development of atherosclerosis and myocardial infarction in children with homozygous familial hypercholesterolemia (FH) (Goldstein et al., 1983) and the reduction in the incidence of all cause mortality and recurrent cardiovascular events with statin therapy, which reduces endogenous synthesis of cholesterol, upregulates LDL receptors and lowers plasma levels of LDL-C (Steinberg et al., 2008). Statin therapy results in 20–30% reductions in LDL-C levels and 20–30% reductions in the incidence of coronary artery disease (CAD). Even more strikingly, recent studies demonstrated that a 28% reduction in the LDL-C levels sustained through life, as in patients with loss-of-function mutations in PCSK-9, was associated with an 88% decrease in the risk of CAD (Horton et al., 2009). Plasma and intracellular PCSK-9 promote degradation of the LDLR, and the PCSK-9 deficiency leads to a higher abundance of the LDLR on the cell surface, more efficient LDL removal and reduced LDL-C levels. In contrast, loss-of-function mutations in the LDLR gene, as in FH patients, result in elevated levels of LDL-C. Homozygous FH patients have LDL-C levels of 600–800 mg/dl and often develop myocardial infarction in the first two decades of life (Goldstein et al., 1983).

Although patients with homozygous FH have very few or no LDL receptors, yet they accumulate cholesterol in subcutaneous tissues and tendon xanthomas and in arterial lesions. Therefore, cholesterol accumulation must be occurring by a pathway other than the LDL receptor. Furthermore, incubation of monocyte/macrophages with native LDL in vitro does not lead to significant accumulation of cellular cholesterol (Goldstein et al., 1979).

This apparent "LDL paradox" was explained by demonstrating the presence of oxidized or modified LDL, which is taken up by macrophages via several mechanisms and receptors other than the LDLR and which accumulate in the arterial walls (Fig. 8.1). Goldstein and Brown (Goldstein et al., 1979) initially described the "acetyl LDL receptor", which binds LDL modified in vitro with acetic anhydride. Unlike the LDL receptor, which is downregulated as the cell cholesterol content increases, the acetyl LDL receptor was not downregulated, but continued to be fully active even as the cell cholesterol content increased markedly. Subsequently, the acetyl LDL receptor was cloned and sequenced, and re-designated the scavenger receptor, type A, or SR-A (Kodama et al., 1990). Further studies identified a number of other scavenger receptors as well as non-scavenger receptor-mediated mechanisms of uptake of lipoproteins by macrophages. It was also apparent from the in vitro experiments that LDL needed to be oxidized or otherwise modified and/or aggregated, in order to be internalized by macrophages. This led to formulating the oxidation hypothesis of atherosclerosis, summarized by Steinberg, Witztum and colleagues in 1989 in their classic article "Beyond cholesterol: Modifications of low density lipoprotein that increase its atherogenicity" (Steinberg et al., 1989), suggesting that oxidative modifications of LDL promote LDL atherogenicity. In this article, we will describe mechanisms of LDL oxidation and non-oxidative modification,

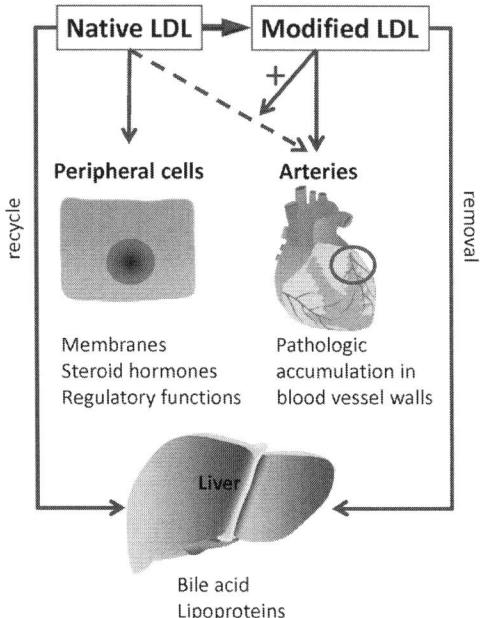

Fig. 8.1 Physiologic and pathologic metabolism of LDL. Native LDL delivers cholesterol and other lipid nutrients to peripheral cells and is recycled in the liver. These processes are mediated by the LDL receptor (LDLR) ubiquitously expressed on many cell types. Hypercholesterolemia leads to LDL accumulation in blood vessel walls. Unlike native LDL, modified LDL is poorly recognized by the LDLR, but is internalized mostly by macrophages via scavenger receptors-mediated uptake and other mechanisms discussed in this review. Modified LDL also facilitates the macrophage uptake of native LDL, like in mmLDL-induced macropinocytosis

including in vitro models as well as in vivo evidence of LDL modification, and discuss the mechanisms of macrophage lipoprotein uptake. Important mechanisms of reverse cholesterol transport from lipid-laden macrophages to the liver will be discussed elsewhere in this journal issue.

8.2 Mechanisms of LDL Oxidation and Enzymatic Degradation

There are now many lines of evidence that oxidation of lipoproteins does occur in vivo and that this process is quantitatively important. This evidence is summarized in Box 8.1. The LDL particle is uniquely sensitive to oxidative damage due to its complex lipid-protein composition. Each LDL particle contains approximately 700 molecules of phospholipids, 185 of triglycerides and 1600 of cholesteryl esters. The polyunsaturated acyl chains of each of these lipid classes are vulnerable to oxidation, as are the sterol of cholesteryl esters and 600 molecules of free cholesterol. LDL contains one molecule of apolipoprotein B-100, made of 4536 amino acid residues, with many exposed tyrosines and lysines, which can be directly oxidized or modified by lipid oxidation products.

> **Box 8.1 Evidence that LDL undergoes oxidation in vivo**
>
> - LDL gently extracted from the atherosclerotic tissue of rabbits and humans has all of the physical, biological, and immunologic properties observed with LDL oxidized *in vitro*
> - A small fraction of circulating LDL particles display a number of chemical indices consistent with early stages of LDL oxidation
> - Subtle modifications of LDL render autologous LDL immunogenic; "oxidation-specific" epitopes are present in atherosclerotic lesions, and "oxidation-specific" antibodies avidly bind to atherosclerotic lesions
> - Autoantibodies to a variety of epitopes of OxLDL can be found in plasma of experimental animals with atherosclerosis
> - The presence of OxLDL in the vessel wall can be imaged in vivo using radiolabeled oxidation-specific antibodies

8.2.1 LDL Oxidation by Copper

Overnight exposure of LDL to copper sulfate leads to its profound oxidative degradation (Steinbrecher et al., 1984). This type of Cu^{2+}-catalyzed oxidative attack on the polyunsaturated fatty acids (PUFA) in the *sn*-2 position of phospholipids may lead to degradation of 40% of the phosphatidylcholine and 50–75% of the PUFA (Esterbauer et al., 1987; Reaven et al., 1993). The apoB also undergoes drastic alterations, to a degree where it is no longer recognized by the LDLR. The recognition of oxidized LDL (OxLDL) by scavenger receptors depends in part upon the generation of neoepitopes created by the masking of epsilon amino groups of lysine residues by aldehyde fragments generated from the PUFA. Non-enzymatic oxidation catalyzed by Cu^{2+} is believed to depend upon the presence of lipid hydroperoxides in the starting material (Esterbauer et al., 1992). These hydroperoxides are degraded to peroxy radicals and alkoxy radicals by Cu^{2+} and in turn, these radicals initiate a chain reaction that generates many more hydroperoxides. The fatty acid side chains of cholesterol esters are susceptible to oxidative damage and the polycyclic sterol ring structure of the cholesterol molecule is also subject to oxidative attack. Incubation of LDL with Cu^{2+} for even a few hours, or with 15-lipoxygenase, is sufficient to oxidize it to the point that it develops important new biological properties (Berliner et al., 1990; Sigari et al., 1997; Miller et al., 2003b). This form of LDL, designated mmLDL, or "minimally oxidized LDL", is still recognized by the LDLR and it is not, at this stage of oxidation, a ligand for the scavenger receptors (Berliner et al., 1990; Navab et al., 1996). In vitro experiments have indicated a large number of biological properties that could in principle make mmLDL proatherogenic (Berliner et al., 2001; Miller et al., 2003a, 2005; Choi et al., 2009; Bae et al., 2009). Although copper-oxidized LDL is a convenient in vitro model of OxLDL, copper ions are unlikely to significantly contribute to LDL oxidation in vivo.

8.2.2 LDL Oxidation by Heme

Divalent iron cations (Fe^{2+}) can also induce LDL oxidation but to a lesser degree than the Cu^{2+}. However, heme, an iron complex with protoporphyrin IX, is a strong LDL-oxidizing agent, particularly when activated with low concentrations of peroxides (Miller et al., 1995). Heme is the oxygen-binding prosthetic group of hemoglobin. LDL oxidation by hemoglobin (Hb) results in apoB-apoB crosslinking and in Hb-apoB crosslinking in blood, as well as in robust lipid peroxidation (Miller et al., 1996; Ziouzenkova et al., 1999). Small amounts of Hb are constantly leaking from damaged erythrocytes, particularly in the vascular regions with turbulent flow, such as vessel bifurcations and aortic curvatures. These processes are exacerbated in hemodialysis patients with high rates of hemolysis, and Hb-induced LDL oxidation has been suggested to significantly contribute to the increased levels of OxLDL found in the plasma of the patients on hemodialysis (Sevanian and Asatryan, 2002; Ziouzenkova et al., 2002).

In plasma, the oxidative damage by free Hb is prevented by Hb binding to haptoglobin (Miller et al., 1997). In plasma, haptoglobin (Hp) exists as a homodimer, and each of its monomers could be one of two common allelic variants, Hp1 or Hp2. The Hp2 variant is less effective than Hp1 in preventing Hb-induced oxidation, and the anti-oxidative efficiency of Hp2 is further reduced when it is complexed with glycated Hb1Ac, present in diabetic patients (Asleh et al., 2003). Remarkably, subjects with the Hp2-2 phenotype, found in up to 37% of Caucasians, have a higher risk of cardiovascular events than the Hp1-1 or Hp2-1 populations (Asleh et al., 2005). The odds ratio of having CVD in diabetic patients with the Hp2-2 phenotype is 5 times greater than in the patients with the Hp1-1 phenotype. An intermediate risk of CVD is associated with the Hp2-1 phenotype (Levy et al., 2002; Roguin et al., 2003; Burbea et al., 2004).

Furthermore, transgenic Hp2-2 mice on an apoE$^{-/-}$ background and Hp2-2 diabetic patients have more iron deposits and lipid peroxidation products, as well as the macrophage accumulation in atherosclerotic lesions compared to respective Hp1-1 genotypes (Levy et al., 2007; Moreno et al., 2008). Remarkably, antioxidant vitamin E supplements reduced the incidence of myocardial infarction, stroke and cardiovascular death in diabetic Hp2-2 patients by more than 50%, which led to early termination of the clinical trial due to obvious beneficial effect of the treatment (Milman et al., 2008).

Even intact erythrocytes can be a source of catalytically active heme. Hemoglobin catabolism yields low levels of free hemin (Fe^{3+}), which accumulates in the erythrocyte membrane. Under normal circumstances, hemopexin and albumin clear hemin from the erythrocyte membrane. However, an in vitro study of the kinetics of hemin clearance demonstrates that under conditions of hyperlipidemia and inflammation, LDL can transiently bind hemin in whole blood, resulting in LDL oxidation (Miller and Shaklai, 1999).

8.2.3 Enzymatic and Cell-Mediated Oxidation of LDL

Incubation of LDL with several cell types in vitro accelerates its oxidative modification. Included among these are endothelial cells, smooth muscle cells and monocyte/macrophages, i.e. all of the cell types that are found in an atherosclerotic lesion. LDL is oxidized not only within the artery wall but also at peripheral sites of inflammation (Liao et al., 1994). There are many postulated mechanisms by which LDL could become oxidized within the artery wall or in plasma. A number of different enzyme systems, such as lipoxygenases, myeloperoxidase, NADPH oxidases, and nitric oxide synthases, have been shown to have the potential of contributing to the oxidation of LDL. Endothelial cells (EC), vascular smooth muscle cells (VSMC), macrophages and neutrophils express one or several of these enzymes. While macrophages may not be required to initiate LDL oxidation, they are likely to amplify oxidative reactions in macrophage-rich areas of atherosclerotic lesions.

8.2.3.1 12/15-Lipoxygenase

Among several mechanisms suggested to explain how LDL is oxidized in vivo, 12/15-lipoxygenase (12/15LO) has been proposed to play a major role (Cyrus et al., 1999, 2001; Glass and Witztum, 2001; George et al., 2001; Reilly et al., 2004; Huo et al., 2004). The family of 12/15LO enzymes is conserved between various animal and plant species and includes human 15LO, mouse 12/15LO (both expressed in macrophages), soybean lipoxygenase (SLO) and others (Yamamoto, 1992; Liavonchanka and Feussner, 2006). The classic in vitro reaction of 12/15LO is the oxygenation of arachidonic acid at carbons 12 and/or 15 (hence the name of the LO enzyme). 12/15LO is capable of oxygenating not only free fatty acids (FFA) but also polyunsaturated acyl chains in phospholipids (PL) and cholesteryl esters (CE) (Yamamoto, 1992). This is in contrast to 5LO, which oxygenates only FFA. An incubation of LDL with isolated 12/15LO or with the cells expressing 12/15LO produces hydroperoxides of three classes, FFA-OOH, PL-OOH and, most profusely, CE-OOH (Yamamoto, 1992; Ezaki et al., 1995; Belkner et al.,

1998; Harkewicz et al., 2008). Yoshimoto and co-workers suggested a highly plausible mechanism of how intracellular 12/15LO mediates oxidation of extracellular LDL and particularly of it's CE residing in the hydrophobic core of the lipoprotein (Zhu et al., 2003; Takahashi et al., 2005). According to their hypothesis, LDL binds to macrophage LDL receptor related protein-1 (LRP-1), which in turn induces 12/15LO translocation from the cytosol to the cell membrane, to the site of LDL–LRP-1 binding. This would be compatible with plasma membrane translocation of 12/15LO in macrophages during phagocytosis to sites where apoptotic cells were bound (Miller et al., 2001). At the site of the LDL–LRP-1 complex, LRP-1 initiates an exchange of CE between LDL and the cell, leading to 12/15LO-mediated oxygenation of the CE in an LRP-1-dependent manner. Further, LRP-1 contributes to the efflux of oxidized CE back to the LDL particle. This mechanism agrees well with the known preferential oxygenation of CE by 12/15LO expressing cells (Ezaki et al., 1995). Accumulation of CE hydroperoxides has been documented in human atherosclerotic lesions and in the lesions of apoE$^{-/-}$ mice fed a high-fat diet (Letters et al., 1999; Upston et al., 2002; Leitinger, 2003).

The importance of 12/15LO in the development of diet-induced atherosclerosis has been established in several murine models, including 12/15LO knockout and transgenic mice (Cyrus et al., 1999, 2001; George et al., 2001; Reilly et al., 2004; Huo et al., 2004; Poeckel et al., 2009). The 12/15LO$^{-/-}$, apoE$^{-/-}$ double knockout mice fed a high-fat diet have less atherosclerosis, significantly lower titers of autoantibodies against OxLDL in plasma and lower isoprostane levels in urine as compared to apoE$^{-/-}$ mice, indicating that 12/15LO is important in LDL oxidation in vivo (Cyrus et al., 2001). Overexpression of 15LO in the rabbit or in mouse macrophages paradoxically reduced atherosclerosis, which may be due to increased synthesis of 15LO-dependent anti-inflammatory eicosanoids (Shen et al., 1996; Merched et al., 2008).

8.2.3.2 Myeloperoxidase

Myeloperoxidase (MPO) is a heme enzyme secreted by neutrophils and monocyte/macrophages that generates a number of oxidants, including hypochlorous acid and peroxynitrite, which can initiate lipid and protein oxidation and produce chlorinated and nitrated LDL. Reactive nitrogen species convert LDL into a high affinity ligand for CD36, which mediates their uptake by macrophages (Podrez et al., 2000). More recently, it has been demonstrated that MPO-catalyzed carbamylation of LDL converts it into a SR-A ligand (Wang et al., 2007). MPO has been identified in human atherosclerotic lesions and is of particular interest because modifications found in human atherosclerosis bear similarities to hypochlorous acid-mediated derivation of lipoprotein components in vitro (Daugherty et al., 1994). MPO has been recently shown to specifically bind to HDL within human atherosclerotic lesions, with selective targeting of apoAI for site-specific chlorination and nitration by MPO-generated reactive oxidants in vivo. One apparent consequence of MPO-catalyzed apoAI oxidation includes the functional impairment of the ability of HDL to promote cellular cholesterol efflux, thereby generating dysfunctional HDL (Nicholls et al., 2005; Shao et al., 2006). However, in bone marrow transplantation experiments in which $LDLR^{-/-}$ mice received MPO-deficient bone marrow progenitor cells, larger lesions were observed than in $LDLR^{-/-}$ mice transplanted with wild type progenitor cells. Similar results were seen when MPO-deficient mice were crossed into $LDLR^{-/-}$ mice. This could be explained by the absence of MPO from murine lesions and the types of MPO-dependent oxidation products found in human lesions were not present in murine lesions (Brennan et al., 2001), suggesting that murine MPO could not be directly related to lesion formation in mice. Overexpression of human MPO in macrophages transplanted into $LDLR^{-/-}$ mice resulted in increased atherosclerosis burden (McMillen et al., 2005).

8.2.3.3 Endothelial and Inducible Nitric Oxide Synthases and NADPH Oxidases

Isolated nitric oxide (product of eNOS or iNOS), superoxide and hydrogen peroxide (products of NADPH oxidases [Nox]) are relatively weak oxidizers. However, nitric oxide and superoxide form peroxynitrite, a strong oxidizing agent, and hydrogen peroxide is needed for Fenton-type and peroxidase reactions, catalyzed by iron, peroxidase enzymes, like MPO, hemoglobin or hemin. In addition to their role in lipoprotein oxidation, many studies suggest that NOS and Nox enzymes play important signaling roles in many cell types. For example, we have recently demonstrated that mmLDL activates macrophage Nox2, which in turn mediates secretion of pro-inflammatory cytokines and VSMC migration (Bae et al., 2009).

Although one would predict that Nox2 would be proinflammatory, and thus proatherogenic, conclusive experimental data to support such a role in vivo is not currently available. The role of Nox components p47phox or gp91phox have been evaluated in various murine models by examining whole body knockouts in the background of apoE deficiency (Kirk et al., 2000; Hsich et al., 2000; Barry-Lane et al., 2001). In one study, the knockout of p47phox decreased lesion formation in the whole aorta, but not at the aortic root (Barry-Lane et al., 2001). In a second study, disruption of p47phox led to no changes in lesion formation at the aortic ring (Hsich et al., 2000). Yet, in a third study, gp91phox knockout led to lowered plasma cholesterol levels, but no measured decreases in atherosclerosis (Kirk et al., 2000), which in the setting of lowered plasma cholesterol, might even be interpreted as enhanced lesion formation. The reasons for these differences are not clear, but could be related to a differential impact of Nox on different stages of lesion development, or to its role in different cell types (Cathcart, 2004). Tissue specific

8.2.4 Non-oxidative Enzymatic Modifications of LDL

Subendothelial retention of LDL is a very early step in the development of atherosclerotic lesions (Tabas et al., 2007). LDL entrapped by extracellular matrix, particularly by proteoglycans, is vulnerable to hydrolysis by lipases and proteases. It has been demonstrated that lipoprotein lipase (LpL), secretory sphingomyelinase (S-SMase), secretory phospholipase A_2 (sPLA$_2$), cholesteryl ester hydrolase (CEH), matrix metalloproteinases (MMP), and plasmin are capable of LDL degradation and produce atherogenic LDL forms that are rapidly internalized by macrophages and/or activate inflammatory responses and the complement system (Bhakdi et al., 1995; Torzewski et al., 2004; Tabas et al., 2007; Fenske et al., 2008; Boyanovsky and Webb, 2009).

Recent experimental animal studies have convincingly demonstrated the role of acid SMase and sPLA$_2$ in the development of atherosclerosis. Deletion of the acid SMase gene *Asm*, which gives rise to both lysosomal and soluble SMase, in either apoE$^{-/-}$ or LDLR$^{-/-}$ mice resulted in 40–50% decrease in early foam cell aortic root lesion area and dramatically reduced subendothelial LDL retention (Devlin et al., 2008). Two isoforms of sPLA$_2$, group IIa and group V, have been shown to accelerate atherosclerosis in mouse models (Ivandic et al., 1999; Bostrom et al., 2007). Like group V sPLA$_2$, the group X enzyme exhibits the highest phospholipolytic activity toward LDL in vitro (Webb and Moore, 2007), but its role in the development of murine atherosclerosis has not yet been demonstrated.

8.3 Macrophage Uptake of Modified LDL

Under atherogenic conditions, normal cholesterol transport by LDL to cells and tissues is partially diverted toward excessive accumulation of cholesterol in macrophages within the blood vessel walls. Unlike the LDLR, which surface expression is tightly regulated by intracellular cholesterol levels, expression of scavenger receptors responsible for the uptake of modified LDL lacks this negative feedback mechanism, therefore mediating massive cholesterol accumulation in the macrophage. Some forms of modified LDL stimulate phagocytosis and macropinocytosis, which also contribute to lipoprotein accumulation in macrophages. In addition, modified LDL is immunogenic and/or recognized by the complement system, and is internalized by macrophages as immune complexes via FcRγ or via complement receptors (Box 8.2).

> **Box 8.2 Major pathways mediating uptake of modified LDL by macrophages**
>
> **Scavenger Receptor family**
>
> • CD36 • LOX-1 • SR-PSOX/CXCL16
> • SR-A • RAGE • SREC-I
>
> **Macropinocytosis**
>
> • TLR4-dependent
> • Constitutive
>   (in M-CSF differentiated macrophages)
>
> **Phagocytosis**
>
> Uptake of LDL aggregates
>
> **LDL Receptor family**
>
> • LDLR • LRP-1 • VLDLR
>
> **FcR and Complement Receptors**
>
> Modified LDL complexes with antibodies, CRP, and complement

8.3.1 Pattern-Recognition Receptors

Pattern recognition receptors (PRRs) are capable of recognizing pathogen-associated molecular patterns (PAMPs) rather than being strictly specific to individual ligands. PRRs have been proposed to play physiologic roles in the recognition and clearance of microbial pathogens and apoptotic cells. It appears that modified LDL has chemical moieties exposed on its surface that precisely match or closely resemble PAMPs and, accordingly, are recognized by PRRs.

8.3.1.1 CD36, SR-A and Other Scavenger Receptors

Macrophages express on their surface a large repertoire of PRRs at high densities, including so-called scavenger receptors that mediate binding and uptake of OxLDL via actin-independent mechanisms. CD36 and SR-A have the highest affinity for OxLDL and acetylated LDL and are responsible for up to 90% of their uptake by macrophages in vitro (Kunjathoor et al., 2002). The CD36-mediated OxLDL uptake depends on the activation of Src and JNK kinases (Rahaman et al., 2006). A comparison of monocyte/macrophages from patients with a total deficiency of CD36 with normal monocyte/macrophages suggests that about 50% of the in vitro uptake of OxLDL is attributable to this receptor, under the conditions studied (Yamashita et al., 2007). However, no complete human data correlating the CD36 deficiency with atherosclerosis are yet available, and mouse model studies produced mixed results. Using hypercholesterolemic CD36$^{-/-}$ and SR-A$^{-/-}$ mouse models, several groups suggested that SR-A and CD36 play quantitatively important roles in mediating uptake of OxLDL and promoting the development of atherosclerosis in apoE$^{-/-}$ mice (Suzuki et al., 1997; Sakaguchi et al., 1998; Febbraio et al., 2000; Kuchibhotla et al., 2008). In contrast, a different group demonstrated that SR-A$^{-/-}$, apoE$^{-/-}$ and CD36$^{-/-}$, apoE$^{-/-}$ double knockout mice, although having significant reductions in peritoneal macrophage lipid accumulation in vivo, had increased atherosclerosis or no change in the lesion size (Moore et al., 2005). The follow up study by this group confirmed that even the combined CD36/SR-A deficiency in apoE$^{-/-}$ mice had no effect on foam cells in atherosclerotic lesions or the lesion size, but revealed reduced complexity and the size of necrotic areas in the lesions, suggesting the role for CD36 and SR-A in cell death (Manning-Tobin et al., 2009). The discrepancy between the results received in different laboratories underscores the complexity of the mechanisms of macrophage lipid accumulation and atherosclerosis (Witztum, 2005; Webb and Moore, 2007).

Recently, new functions for CD36 related to the development of atherosclerosis have been suggested. It has been reported that CD36 is involved in oxidized lipid-induced platelet activation and that it mediates a pro-thrombotic phenotype under dislipidemic conditions (Podrez et al., 2007). In addition, CD36 has been shown to mediate cytoskeletal and Nox activity in macrophages in response to oxidized phospholipids and beta-amyloid (Moore et al., 2002; Park et al., 2009).

Other scavenger receptors reported to mediate OxLDL internalization include lectin-like oxidized LDL receptor-1 (LOX-1), scavenger receptor for phosphatidylserine and oxidized lipoprotein (SR-PSOX/CXCL16), scavenger receptor expressed by endothelial cells (SREC-I), scavenger receptor type BI (SR-BI), CD68, and receptor for advanced glycation end products (RAGE) (Witztum, 2005; Webb and Moore, 2007). While inhibition of macrophage scavenger receptor activity could potentially provide the basis of an anti-atherogenic therapy, it may be the case that several classes of proteins will have to be targeted simultaneously. Since these receptors are also involved in clearing microorganisms, enhanced susceptibility to specific infectious pathogens may also occur. It is also possible that inhibition of scavenger receptor function could have deleterious effects if they play an important role in the clearance of apoptotic cells or other critical functions. For example, although SR-BI binds OxLDL, its major function is to mediate reverse cholesterol transport by HDL, and SR-BI deletion in apoE$^{-/-}$ mice causes severe atherosclerosis with evidence of plaque rupture and acute myocardial infarction, complications that are rare in other murine models of atherosclerosis (Braun et al., 2002; Huby et al., 2006).

8.3.1.2 TLR4 and Macropinocytosis

Toll-like receptors (TLRs) are an important class of PRRs, sensing the presence of many microbial PAMPs and activating diverse programs of pro-inflammatory gene expression and cytoskeletal responses. The work from our laboratory has demonstrated that minimally oxidized LDL (mmLDL) binds CD14, a co-receptor of TLR4, and activates macrophages via TLR4/MD-2 (Miller et al., 2003b). The Tabas laboratory has reported that acetylated LDL also signals via TLR4 (Seimon et al., 2006). mmLDL and its active components, polyoxygenated cholesteryl ester hydroperoxides, induce extensive membrane ruffling and cytoskeletal rearrangements in macrophages in a TLR4-dependent manner (Harkewicz et al., 2008; Choi et al., 2009). The membrane ruffles eventually close to form large pinosomes, the process known as fluid phase uptake, or macropinocytosis (Fig. 8.2). We have demonstrated that the signalling of mmLDL-induced macropinocytosis includes the recruitment of spleen tyrosine kinase (Syk) to a TLR4 signaling complex, TLR4 and Syk phosphorylation, activation of a Vav1-Ras-Raf-MEK-ERK1/2 signaling cascade, phosphorylation of paxillin, and activation of Rac, Cdc42 and Rho. These mmLDL-induced and TLR4- and Syk-dependent signalling events and cytoskeletal rearrangements lead to enhanced uptake of small molecules, dextran and, most importantly, of both native and oxidized LDL, resulting in intracellular lipid accumulation. Remarkably, TLR4-dependent lipoprotein uptake occurs not only in differentiated macrophages but also in peripheral blood monocytes. An intravenous injection of fluorescently labelled mmLDL in wild type mice resulted in its rapid accumulation in circulating monocytes, which was significantly attenuated in TLR4-deficient mice (Choi et al., 2009). As TLR4 is highly expressed on the surface of circulating monocytes in patients with cardiovascular disease, and cholesteryl ester hydroperoxides are present and stable in plasma, lipid uptake by monocytes in circulation may contribute to the pathological roles of monocytes in atherogenesis.

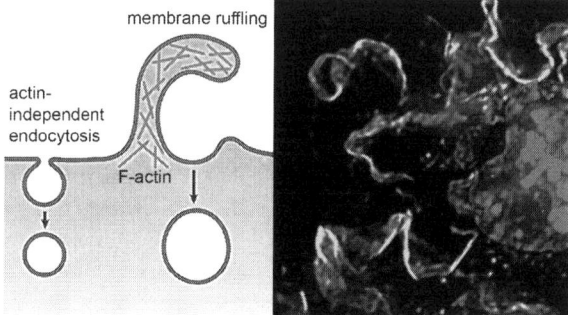

Fig. 8.2 Actin-dependent and –independent mechanisms of modified LDL uptake. Actin-dependent endocytosis is characterized by robust actin polymerization and formation of membrane ruffles, which eventually close and form large pinosomes, the process known as fluid phase uptake, or macropinocytosis. On the right, mmLDL-induced ruffling in macrophages; staining with FITC-phalloidin (F-actin) and Hoechst (nucleus)

8.3.1.3 Immune and Complement Complexes

Soluble PRRs, such as IgM and C-reactive protein (CRP), which bind oxidized and enzymatically modified LDL in plasma and in atherosclerotic lesions, can in turn bind complement and undergo enhanced binding via complement receptors (Bhakdi et al., 1995; Hartvigsen et al., 2009). In addition, oxidized and other modified LDL are immunogenic, and OxLDL/IgG complexes are found in plasma. These immune complexes can be internalized via Fc receptors (Virella and Lopes-Virella, 2008).

8.3.2 Constitutive Macropinocytosis

Monocyte differentiation into macrophages in atherosclerotic lesions requires macrophage colony stimulating factor (M-CSF). The osteopetrotic mouse, which carries a naturally occurring mutation in the gene encoding M-CSF and exhibits a near complete absence of macrophages, is extremely resistant to the development of atherosclerosis when bred to apoE-deficient mice, despite an increase in circulating cholesterol levels (Qiao et al., 1997). Because M-CSF-dependent differentiation of monocytes in vitro leads to the development of a macrophage phenotype characterized by constitutive macropinocytosis, Kruth, Jones and colleagues have suggested that foam cell formation in atherosclerotic lesions can occur due to increased uptake of native LDL mediated by scavenger receptor-independent macropinocytosis (Jones et al., 2000; Kruth et al., 2005; Zhao et al., 2006). Macropinocytosis in M-CSF-differentiated macrophages is inhibited by agonists of liver X receptors (LXR), suggesting an additional mechanism by which LXR agonists may inhibit macrophage cholesterol accumulation and atherosclerosis (Buono et al., 2007). Importantly, a macropinocytosis-type uptake of fluorescent nanoparticles, which are the size of LDL, by macrophages, has been convincingly demonstrated to occur in murine atherosclerotic lesions (Buono et al., 2009).

8.3.3 Phagocytosis and Patocytosis of Aggregated LDL

Disruption of LDL integrity often leads to its aggregation. For experimental purposes, LDL aggregation can be achieved simply by a vigorous shaking (vortexing) of a test tube for 1 minute. More relevant to atherogenesis, aggregation is often a result of enzymatic hydrolysis of LDL by sPLA$_2$, S-SMase, and proteases or by their combination, or extensive oxidation of LDL, as well as of the LDL retention by extracellular matrix in the subendothelial space (Oorni et al., 2000; Sakr et al., 2001). Aggregated LDL is taken up by macrophages via a phagocytic mechanism, which, similarly to macropinocytosis, requires intact actin cytoskeleton (Khoo et al., 1988). An alternative mechanism is patocytosis, which involves degradation of aggregated LDL by macrophage-released plasmin and subsequent uptake of cleaved fragments (Kruth, 2002). It has been reported that group V sPLA$_2$-modified LDL binds syndecan-4, which in turn mediates PI3K-dependent uptake of modified LDL

(Boyanovsky et al., 2008). Because sPLA$_2$-modified LDL readily aggregates and the PI3K activity is needed for both macropinocytosis and phagocytosis, it is difficult to discern which of these two actin-dependent processes are more relevant to the sPLA$_2$-modified LDL uptake.

8.3.4 LDLR Family Receptor-Mediated Uptake

Aggregated LDL as well as VLDL and triglyceride-rich remnant lipoprotein bind to the LDLR, LRP-1, and VLDLR on the cell surface, and mediate selective uptake of components of these lipoprotein particles. Thus, similarly to the mechanism of LDL oxidation by 12/15LO (*see* Section 8.2.3.1), LRP-1 mediates selective uptake of cholesteryl esters from aggregated LDL, without any involvement of cytoskeletal proteins (Llorente-Cortes et al., 2006). These cholesteryl esters are stored in lipid droplets, cannot be removed by HDL-mediated reverse cholesterol transport, and induce expression of adipocytes differentiation-related protein (Llorente-Cortes et al., 2006, 2007).

8.4 Future Directions

Oxidative and/or enzymatic modifications of LDL, uptake of modified LDL by macrophages and the formation of cholesterol-laden macrophage foam cells are rate-limiting processes of atherogenesis. Arguably, targeting foam cells in fatty streaks early would effectively treat atherosclerosis and prevent its subsequent complications (Steinberg et al., 2008). Although several molecular mechanisms leading to foam cell formation, as outlined in this review, have been proposed, controversy still surrounds their importance. Even less is known about the temporal and quantitative contributions of such mechanisms in fatty streak formation. The problem with interpreting the results of in vitro experiments studying modified LDL uptake by macrophages is that the relative abundance of a specific modified LDL species in vivo is unknown and may be different at different stages during the progression of atherosclerosis. Thus, given a variety of the uptake mechanisms, it is difficult to quantify their significance in atherogenesis. New approaches need to be developed to address these problems.

8.4.1 Zebrafish Model for Studying Early Events in Atherogenesis

One of the approaches that could be helpful in understanding the quantitative significance of different mechanisms of macrophage lipoprotein uptake would be to determine individual rates of in vivo lipid uptake via specific mechanisms and then build a cumulative model of foam cell formation. This challenging task could be accomplished using a novel hypercholesterolemic zebrafish model, as we suggested in a recent paper (Stoletov et al., 2009). In this work, we transplanted genetically modified mouse macrophages into zebrafish larvae in which a high-cholesterol diet has induced the formation of fatty streaks. Because zebrafish larvae are transparent and we used fluorescently labelled cells and dietary lipids, we were able to monitor macrophage lipid uptake in vivo, directly in the environment of a fatty streak. Using this model, we found that the rate of in vivo lipid uptake by TLR4-deficient macrophages was significantly lower compared to the uptake by wild type macrophages, supporting the results of our in vitro experiments (Choi et al., 2009; Stoletov et al., 2009). Similar measurements with macrophages lacking other receptors and signalling molecules discussed in this review will produce a set of rate constants that could be used in putting together a computational model of foam cell formation. Because the majority of the cellular aspects of macrophage biology are conserved from worms to mammals, such a quantitative model will be useful in understanding the processes of atherogenesis in humans and enable informed design of therapies targeting foam cells. An advantage of the optically transparent zebrafish model is that, unlike in cell culture experiments in which lipoproteins are modified in vitro, in these settings, macrophages are exposed to a multitude of lipoprotein modifications occurring in vivo.

8.5 Summary

Pathologic cholesterol transport by modified LDL results in cholesterol accumulation in the arterial walls and in a variety of pro-inflammatory pathways that lead to development of atherosclerotic lesions and, in humans, clinical events manifested by death, myocardial infarction, stroke and peripheral arterial disease. These multifaceted processes involve a variety of mechanisms of LDL modification and complex pathways of lipoprotein macrophage uptake. Understanding the most significant elements of these processes and designing efficient therapies targeting excessive cholesterol accumulation in macrophages will have important implications for the treatment of atherosclerosis.

Acknowledgments: Work described in this review contributed by the authors was supported by NIH grants HL081862 and GM069338 (Y.I.M.), and a grant from the Leducq Fondation (Y.I.M. and S.T.).

References

Asleh, R., Marsh, S., Shilkrut, M., Binah, O., Guetta, J., Lejbkowicz, F., Enav, B., Shehadeh, N., Kanter, Y., Lache, O., Cohen, O., Levy, N.S., and Levy, A.P., 2003, Genetically determined heterogeneity in hemoglobin scavenging and susceptibility to diabetic cardiovascular disease. *Circ. Res.* **92:** 1193–1200.

Asleh, R., Guetta, J., Kalet-Litman, S., Miller-Lotan, R., and Levy, A.P., 2005, Haptoglobin genotype- and diabetes-dependent differences in iron-mediated oxidative stress in vitro and in vivo. *Circ. Res.* **96:** 435–441.

Bae, Y.S., Lee, J.H., Choi, S.H., Kim, S., Almazan, F., Witztum, J.L., and Miller, Y.I., 2009, Macrophages generate reactive oxygen species in response to minimally oxidized low-density lipoprotein: Toll-like receptor 4- and spleen tyrosine kinase-dependent activation of NADPH oxidase 2. *Circ. Res.* **104:** 210–218.

Barry-Lane, P.A., Patterson, C., van der Merwe, M., Hu, Z., Holland, S.M., Yeh, E.T.H., and Runge, M.S., 2001, p47phox is required for atherosclerotic lesion progression in ApoE-/- mice. *J. Clin. Invest.* **108:** 1513–1522.

Belkner, J., Stender, H., and Kuhn, H., 1998, The rabbit 15-lipoxygenase preferentially oxygenates LDL cholesterol esters, and this reaction does not require vitamin E. *J. Biol. Chem.* **273:** 23225–23232.

Berliner, J.A., Subbanagounder, G., Leitinger, N., Watson, A.D., and Vora, D., 2001, Evidence for a role of phospholipid oxidation products in atherogenesis. *Trends Cardiovasc. Med.* **11:** 142–147.

Berliner, J.A., Territo, M.C., Sevanian, A., Ramin, S., Kim, J.A., Bamshad, B., Esterson, M., and Fogelman, A.M., 1990, Minimally modified low density lipoprotein stimulates monocyte endothelial interactions. *J. Clin. Invest.* **85:** 1260–1266.

Bhakdi, S., Dorweiler, B., Kirchmann, R., Torzewski, J., Weise, E., Tranum-Jensen, J., Walev, I., and Wieland, E., 1995, On the pathogenesis of atherosclerosis: enzymatic transformation of human low density lipoprotein to an atherogenic moiety. *J. Exp. Med.* **182:** 1959–1971.

Bostrom, M.A., Boyanovsky, B.B., Jordan, C.T., Wadsworth, M.P., Taatjes, D.J., de Beer, F.C., and Webb, N.R., 2007, Group v secretory phospholipase A2 promotes atherosclerosis: evidence from genetically altered mice. *Arterioscler. Thromb. Vasc. Biol.* **27:** 600–606.

Boyanovsky, B.B., Shridas, P., Simons, M., van der Westhuyzen, D.R., and Webb, N.R., 2008, Syndecan-4 mediates macrophage uptake of group V secretory phospholipase A2-modified low density lipoprotein. *J. Lipid Res.* **50:** 641–650.

Boyanovsky, B.B. and Webb, N.R., 2009, Biology of secretory phospholipase A2. *Cardiovasc. Drugs Ther.* **23:** 61–72.

Braun, A., Trigatti, B.L., Post, M.J., Sato, K., Simons, M., Edelberg, J.M., Rosenberg, R.D., Schrenzel, M., and Krieger, M., 2002, Loss of SR-BI expression leads to the early onset of occlusive atherosclerotic coronary artery disease, spontaneous myocardial infarctions, severe cardiac dysfunction, and premature death in apolipoprotein E-deficient mice. *Circ. Res.* **90:** 270–276.

Brennan, M.L., Anderson, M.M., Shih, D.M., Qu, X.D., Wang, X., Mehta, A.C., Lim, L.L., Shi, W., Hazen, S.L., Jacob, J.S., Crowley, J.R., Heinecke, J.W., and Lusis, A.J., 2001, Increased atherosclerosis in myeloperoxidase-deficient mice. *J. Clin. Invest.* **107:** 419–430.

Buono, C., Anzinger, J.J., Amar, M., and Kruth, H.S., 2009, Fluorescent pegylated nanoparticles demonstrate fluid-phase pinocytosis by macrophages in mouse atherosclerotic lesions. *J. Clin. Invest.* **119:** 1373–1381.

Buono, C., Li, Y., Waldo, S.W., and Kruth, H.S., 2007, Liver X receptors inhibit human monocyte-derived macrophage foam cell formation by inhibiting fluid-phase pinocytosis of LDL. *J. Lipid Res.* **48:** 2411–2418.

Burbea, Z., Nakhoul, F., Zoabi, R., Hochberg, I., Levy, N.S., Benchetrit, S., Weissgarten, J., Tovbin, D., Knecht, A., Iaina, A., Herman, M., Kristal, B., and Levy, A.P., 2004, Haptoglobin phenotype as a predictive factor of mortality in diabetic haemodialysis patients. *Ann. Clin. Biochem.* **41:** 469–473.

Cathcart, M.K., 2004, Regulation of Superoxide Anion Production by NADPH Oxidase in Monocytes/Macrophages: Contributions to Atherosclerosis. *Arterioscler. Thromb. Vasc. Biol.* **24:** 23–28.

Choi, S.-H., Harkewicz, R., Lee, J.H., Boullier, A., Almazan, F., Li, A.C., Witztum, J.L., Bae, Y.S., and Miller, Y.I., 2009, Lipoprotein accumulation in macrophages via toll-like receptor-4-dependent fluid phase uptake. *Circ. Res.* **104:** 1355–1363.

Cyrus, T., Pratico, D., Zhao, L., Witztum, J.L., Rader, D.J., Rokach, J., FitzGerald, G.A., and Funk, C.D., 2001, Absence of 12/15-lipoxygenase expression decreases lipid peroxidation and atherogenesis in apolipoprotein e-deficient mice. *Circulation* **103**: 2277–2282.

Cyrus, T., Witztum, J.L., Rader, D.J., Tangirala, R., Fazio, S., Linton, M.F., and Funk, C.D., 1999, Disruption of the 12/15-lipoxygenase gene diminishes atherosclerosis in apo E-deficient mice. *J. Clin. Invest.* **103**: 1597–1604.

Daugherty, A., Dunn, J.L., Rateri, D.L., and Heinecke, J.W., 1994, Myeloperoxidase, a catalyst for lipoprotein oxidation, is expressed in human atherosclerotic lesions. *J. Clin. Invest.* **94**: 437–444.

Detmers, P.A., Hernandez, M., Mudgett, J., Hassing, H., Burton, C., Mundt, S., Chun, S., Fletcher, D., Card, D.J., Lisnock, J., Weikel, R., Bergstrom, J.D., Shevell, D.E., Hermanowski-Vosatka, A., Sparrow, C.P., Chao, Y.S., Rader, D.J., Wright, S.D., and Pure, E., 2000, Deficiency in inducible nitric oxide synthase results in reduced atherosclerosis in apolipoprotein E-deficient mice. *J. Immunol.* **165**: 3430–3435.

Devlin, C.M., Leventhal, A.R., Kuriakose, G., Schuchman, E.H., Williams, K.J., and Tabas, I., 2008, Acid sphingomyelinase promotes lipoprotein retention within early atheromata and accelerates lesion progression. *Arterioscler. Thromb. Vasc. Biol.* **28**: 1723–1730.

Esterbauer, H., Gebicki, J., Puhl, H., and Jurgens, G., 1992, The role of lipid peroxidation and antioxidants in oxidative modification of LDL. *Free Radic. Biol. Med.* **13**: 341–390.

Esterbauer, H., Jurgens, G., Quehenberger, O., and Koller, E., 1987, Autoxidation of human low density lipoprotein: loss of polyunsaturated fatty acids and vitamin E and generation of aldehydes. *J. Lipid Res.* **28**: 495–509.

Ezaki, M., Witztum, J.L., and Steinberg, D., 1995, Lipoperoxides in LDL incubated with fibroblasts that overexpress 15-lipoxygenase. *J. Lipid Res.* **36**: 1996–2004.

Febbraio, M., Podrez, E.A., Smith, J.D., Hajjar, D.P., Hazen, S.L., Hoff, H.F., Sharma, K., and Silverstein, R.L., 2000, Targeted disruption of the class B scavenger receptor CD36 protects against atherosclerotic lesion development in mice. *J. Clin. Invest.* **105**: 1049–1056.

Fenske, D., Dersch, K., Lux, C., Zipse, L., Suriyaphol, P., Dragneva, Y., Han, S.R., Bhakdi, S., and Husmann, M., 2008, Enzymatically hydrolyzed low-density lipoprotein modulates inflammatory responses in endothelial cells. *Thromb. Haemost.* **100**: 1146–1154.

George, J., Afek, A., Shaish, A., Levkovitz, H., Bloom, N., Cyrus, T., Zhao, L., Funk, C.D., Sigal, E., and Harats, D., 2001, 12/15-Lipoxygenase gene disruption attenuates atherogenesis in LDL receptor-deficient mice. *Circulation* **104**: 1646–1650.

Glass, C.K. and Witztum, J.L., 2001, Atherosclerosis. The road ahead. *Cell* **104**: 503–516.

Goldstein, J.L., Ho, Y.K., Basu, S.K., and Brown, M.S., 1979, Binding site on macrophages that mediates uptake and degradation of acetylated low density lipoprotein, producing massive cholesterol deposition. *Proc. Natl. Acad. Sci. USA* **76**: 333–337.

Goldstein, J.L., Kita, T., and Brown, M.S., 1983, Defective lipoprotein receptors and atherosclerosis. Lessons from an animal counterpart of familial hypercholesterolemia. *N. Engl. J. Med.* **309**: 288–296.

Harkewicz, R., Hartvigsen, K., Almazan, F., Dennis, E.A., Witztum, J.L., and Miller, Y.I., 2008, Cholesteryl ester hydroperoxides are biologically active components of minimally oxidized LDL. *J. Biol. Chem.* **283**: 10241–10251.

Hartvigsen, K., Chou, M.Y., Hansen, L.F., Shaw, P.X., Tsimikas, S., Binder, C.J., and Witztum, J.L., 2009, The role of innate immunity in atherogenesis. *J. Lipid Res.* **50**: S388–S393.

Horton, J.D., Cohen, J.C., and Hobbs, H.H., 2009, PCSK9: A convertase that coordinates LDL catabolism. *J. Lipid Res.* **50**: S172–S177.

Hsich, E., Segal, B.H., Pagano, P.J., Rey, F.E., Paigen, B., Deleonardis, J., Hoyt, R.F., Holland, S.M., and Finkel, T., 2000, Vascular effects following homozygous disruption of p47phox: An essential component of NADPH oxidase. *Circulation* **101**: 1234–1236.

Huby, T., Doucet, C., Dachet, C., Ouzilleau, B., Ueda, Y., Afzal, V., Rubin, E., Chapman, M.J., and Lesnik, P., 2006, Knockdown expression and hepatic deficiency reveal an atheroprotective role for SR-BI in liver and peripheral tissues. *J. Clin. Invest.* **116**: 2767–2776.

Huo, Y., Zhao, L., Hyman, M.C., Shashkin, P., Harry, B.L., Burcin, T., Forlow, S.B., Stark, M.A., Smith, D.F., Clarke, S., Srinivasan, S., Hedrick, C.C., Pratico, D., Witztum, J.L., Nadler, J.L., Funk, C.D., and Ley, K., 2004, Critical role of macrophage 12/15-lipoxygenase for atherosclerosis in apolipoprotein E-deficient mice. *Circulation* **110**: 2024–2031.

Ihrig, M., Dangler, C.A., and Fox, J.G., 2001, Mice lacking inducible nitric oxide synthase develop spontaneous hypercholesterolaemia and aortic atheromas. *Atherosclerosis* **156**: 103–107.

Ivandic, B., Castellani, L.W., Wang, X.P., Qiao, J.H., Mehrabian, M., Navab, M., Fogelman, A.M., Grass, D.S., Swanson, M.E., De Beer, M.C., de Beer, F., and Lusis, A.J., 1999, Role of group II secretory phospholipase A2 in atherosclerosis : 1. Increased atherogenesis and altered lipoproteins in transgenic mice expressing group IIa phospholipase A2. *Arterioscler. Thromb. Vasc. Biol.* **19**: 1284–1290.

Jones, N.L., Reagan, J.W., and Willingham, M.C., 2000, The pathogenesis of foam cell formation: Modified LDL stimulates uptake of co-incubated LDL via macropinocytosis. *Arterioscler. Thromb. Vasc. Biol.* **20**: 773–781.

Khoo, J.C., Miller, E., McLoughlin, P., and Steinberg, D., 1988, Enhanced macrophage uptake of low density lipoprotein after self-aggregation. *Arteriosclerosis* **8**: 348–358.

Kirk, E.A., Dinauer, M.C., Rosen, H., Chait, A., Heinecke, J.W., and LeBoeuf, R.C., 2000, Impaired superoxide production due to a deficiency in phagocyte NADPH oxidase fails to inhibit atherosclerosis in mice. *Arterioscler. Thromb. Vasc. Biol.* **20**: 1529–1535.

Kodama, T., Freeman, M., Rohrer, L., Zabrecky, J., Matsudaira, P., and Krieger, M., 1990, Type I macrophage scavenger receptor contains alpha-helical and collagen-like coiled coils. *Nature* **343**: 531–535.

Kruth, H.S., 2002, Sequestration of aggregated low-density lipoproteins by macrophages. *Curr. Opin. Lipidol.* **13**: 483–488.

Kruth, H.S., Jones, N.L., Huang, W., Zhao, B., Ishii, I., Chang, J., Combs, C.A., Malide, D., and Zhang, W.Y., 2005, Macropinocytosis is the endocytic pathway that mediates macrophage foam cell formation with native low density lipoprotein. *J. Biol. Chem.* **280**: 2352–2360.

Kuchibhotla, S., Vanegas, D., Kennedy, D.J., Guy, E., Nimako, G., Morton, R.E., and Febbraio, M., 2008, Absence of CD36 protects against atherosclerosis in ApoE knock-out mice with no additional protection provided by absence of scavenger receptor A I/II. *Cardiovasc. Res.* **78**: 185–196.

Kunjathoor, V.V., Febbraio, M., Podrez, E.A., Moore, K.J., Andersson, L., Koehn, S., Rhee, J.S., Silverstein, R., Hoff, H.F., and Freeman, M.W., 2002, Scavenger receptors class A-I/II and CD36 are the principal receptors responsible for the uptake of modified low density lipoprotein leading to lipid loading in macrophages. *J. Biol. Chem.* **277**: 49982–49988.

Leitinger, N., 2003, Cholesteryl ester oxidation products in atherosclerosis. *Mol. Aspects Med.* **24**: 239–250.

Letters, J.M., Witting, P.K., Christison, J.K., Eriksson, A.W., Pettersson, K., and Stocker, R., 1999, Time-dependent changes to lipids and antioxidants in plasma and aortas of apolipoprotein E knockout mice. *J. Lipid Res.* **40**: 1104–1112.

Levy, A.P., Hochberg, I., Jablonski, K., Resnick, H.E., Lee, E.T., Best, L., and Howard, B.V., 2002, Haptoglobin phenotype is an independent risk factor for cardiovascular disease in individuals with diabetes: The strong heart study. *J. Am. Coll. Cardiol.* **40**: 1984–1990.

Levy, A.P., Levy, J.E., Kalet-Litman, S., Miller-Lotan, R., Levy, N.S., Asaf, R., Guetta, J., Yang, C., Purushothaman, K.R., Fuster, V., and Moreno, P.R., 2007, Haptoglobin genotype is a determinant of iron, lipid peroxidation, and macrophage accumulation in the atherosclerotic plaque. *Arterioscler. Thromb. Vasc. Biol.* **27**: 134–140.

Liao, F., Andalibi, A., Qiao, J.H., Allayee, H., Fogelman, A.M., and Lusis, A.J., 1994, Genetic evidence for a common pathway mediating oxidative stress, inflammatory gene induction, and aortic fatty streak formation in mice. *J. Clin. Invest.* **94**: 877–884.

Liavonchanka, A. and Feussner, I., 2006, Lipoxygenases: Occurrence, functions and catalysis. *J. Plant Physiol.* **163**: 348–357.

Llorente-Cortes, V., Otero-Vinas, M., Camino-Lopez, S., Costales, P., and Badimon, L., 2006, Cholesteryl esters of aggregated LDL are internalized by selective uptake in human vascular smooth muscle cells. *Arterioscler. Thromb. Vasc. Biol.* **26:** 117–123.

Llorente-Cortes, V., Royo, T., Juan-Babot, O., and Badimon, L., 2007, Adipocyte differentiation-related protein is induced by LRP1-mediated aggregated LDL internalization in human vascular smooth muscle cells and macrophages. *J. Lipid Res.* **48:** 2133–2140.

Manning-Tobin, J.J., Moore, K.J., Seimon, T.A., Bell, S.A., Sharuk, M., varez-Leite, J.I., de Winther, M.P.J., Tabas, I., and Freeman, M.W., 2009, Loss of SR-A and CD36 activity reduces atherosclerotic lesion complexity without abrogating foam cell formation in hyperlipidemic mice. *Arterioscler. Thromb. Vasc. Biol.* **29:** 19–26.

McMillen, T.S., Heinecke, J.W., and LeBoeuf, R.C., 2005, Expression of human myeloperoxidase by macrophages promotes atherosclerosis in mice. *Circulation* **111:** 2798–2804.

Merched, A.J., Ko, K., Gotlinger, K.H., Serhan, C.N., and Chan, L., 2008, Atherosclerosis: Evidence for impairment of resolution of vascular inflammation governed by specific lipid mediators. *FASEB J.* **22:** 3595–3606.

Miller, Y.I., Altamentova, S.M., and Shaklai, N., 1997, Oxidation of low-density lipoprotein by hemoglobin stems from a heme-initiated globin radical: antioxidant role of haptoglobin 248. *Biochemistry* **36:** 12189–12198.

Miller, Y.I., Chang, M.K., Binder, C.J., Shaw, P.X., and Witztum, J.L., 2003a, Oxidized low density lipoprotein and innate immune receptors. *Curr. Opin. Lipidol.* **14:** 437–445.

Miller, Y.I., Chang, M.K., Funk, C.D., Feramisco, J.R., and Witztum, J.L., 2001, 12/15-Lipoxygenase translocation enhances site-specific actin polymerization in macrophages phagocytosing apoptotic cells. *J. Biol. Chem.* **276:** 19431–19439.

Miller, Y.I., Felikman, Y., and Shaklai, N., 1995, The involvement of low-density lipoprotein in hemin transport potentiates peroxidative damage. *Biochim. Biophys. Acta* **1272:** 119–127.

Miller, Y.I., Felikman, Y., and Shaklai, N., 1996, Hemoglobin induced apolipoprotein B crosslinking in low-density lipoprotein peroxidation. *Arch. Biochem. Biophys.* **326:** 252–260.

Miller, Y.I. and Shaklai, N., 1999, Kinetics of hemin distribution in plasma reveals its role in lipoprotein oxidation. *Biochim. Biophys. Acta* **1454:** 153–164.

Miller, Y.I., Viriyakosol, S., Binder, C.J., Feramisco, J.R., Kirkland, T.N., and Witztum, J.L., 2003b, Minimally modified LDL binds to CD14, induces macrophage spreading via TLR4/MD-2, and inhibits phagocytosis of apoptotic cells. *J. Biol. Chem.* **278:** 1561–1568.

Miller, Y.I., Viriyakosol, S., Worrall, D.S., Boullier, A., Butler, S., and Witztum, J.L., 2005, Toll-like receptor 4-dependent and -independent cytokine secretion induced by minimally oxidized low-density lipoprotein in macrophages. *Arterioscler. Thromb. Vasc. Biol.* **25:** 1213–1219.

Milman, U., Blum, S., Shapira, C., Aronson, D., Miller-Lotan, R., Anbinder, Y., Alshiek, J., Bennett, L., Kostenko, M., Landau, M., Keidar, S., Levy, Y., Khemlin, A., Radan, A., and Levy, A.P., 2008, Vitamin E supplementation reduces cardiovascular events in a subgroup of middle-aged individuals with both type 2 diabetes mellitus and the haptoglobin 2-2 genotype: a prospective double-blinded clinical trial. *Arterioscler. Thromb. Vasc. Biol.* **28:** 341–347.

Moore, K.J., El Khoury, J., Medeiros, L.A., Terada, K., Geula, C., Luster, A.D., and Freeman, M.W., 2002, A CD36-initiated signaling cascade mediates inflammatory effects of beta-amyloid. *J. Biol. Chem.* **277:** 47373–47379.

Moore, K.J., Kunjathoor, V.V., Koehn, S.L., Manning, J.J., Tseng, A.A., Silver, J.M., McKee, M., and Freeman, M.W., 2005, Loss of receptor-mediated lipid uptake via scavenger receptor A or CD36 pathways does not ameliorate atherosclerosis in hyperlipidemic mice. *J. Clin. Invest.* **115:** 2192–2201.

Moreno, P.R., Purushothaman, K.R., Purushothaman, M., Muntner, P., Levy, N.S., Fuster, V., Fallon, J.T., Lento, P.A., Winterstern, A., and Levy, A.P., 2008, Haptoglobin genotype is a major determinant of the amount of iron in the human atherosclerotic plaque. *J. Am. Coll. Cardiol.* **52:** 1049–1051.

Navab, M., Berliner, J.A., Watson, A.D., Hama, S., Territo, M.C., Lusis, A.J., Shih, D.M., Van Lenten, B.J., Frank, J.S., Demer, L.L., Edwards, P.A., and Fogelman, A.M., 1996, The Yin and Yang of oxidation in the development of the fatty streak. *Arterioscler. Thromb. Vasc. Biol.* **16:** 831–842.

Nicholls, S.J., Zheng, L., and Hazen, S.L., 2005, Formation of dysfunctional high-density lipoprotein by myeloperoxidase. *Trends Cardiovasc. Med.* **15:** 212–219.

Oorni, K., Pentikainen, M.O., Ala-Korpela, M., and Kovanen, P.T., 2000, Aggregation, fusion, and vesicle formation of modified low density lipoprotein particles: Molecular mechanisms and effects on matrix interactions. *J. Lipid Res.* **41:** 1703–1714.

Park, Y.M., Febbraio, M., and Silverstein, R.L., 2009, CD36 modulates migration of mouse and human macrophages in response to oxidized LDL and may contribute to macrophage trapping in the arterial intima. *J. Clin Invest.* **119:** 136–145.

Podrez, E.A., Byzova, T.V., Febbraio, M., Salomon, R.G., Ma, Y., Valiyaveettil, M., Poliakov, E., Sun, M., Finton, P.J., Curtis, B.R., Chen, J., Zhang, R., Silverstein, R.L., and Hazen, S.L., 2007, Platelet CD36 links hyperlipidemia, oxidant stress and a prothrombotic phenotype. *Nat. Med.* **13:** 1086–1095.

Podrez, E.A., Febbraio, M., Sheibani, N., Schmitt, D., Silverstein, R.L., Hajjar, D.P., Cohen, P.A., Frazier, W.A., Hoff, H.F., and Hazen, S.L., 2000, Macrophage scavenger receptor CD36 is the major receptor for LDL modified by monocyte-generated reactive nitrogen species. *J. Clin. Invest.* **105:** 1095–1108.

Poeckel, D., Zemski Berry, K.A., Murphy, R.C., and Funk, C.D., 2009, Dual 12/15- and 5-lipoxygenase deficiency in macrophages alters arachidonic acid metabolism and attenuates peritonitis and atherosclerosis in APOE knockout mice. *J. Biol. Chem.* **284:** 21077–21089.

Qiao, J.H., Tripathi, J., Mishra, N.K., Cai, Y., Tripathi, S., Wang, X.P., Imes, S., Fishbein, M.C., Clinton, S.K., Libby, P., Lusis, A.J., and Rajavashisth, T.B., 1997, Role of macrophage colony-stimulating factor in atherosclerosis: studies of osteopetrotic mice. *Am. J. Pathol.* **150:** 1687–1699.

Rahaman, S.O., Lennon, D.J., Febbraio, M., Podrez, E.A., Hazen, S.L., and Silverstein, R.L., 2006, A CD36-dependent signaling cascade is necessary for macrophage foam cell formation. *Cell Metab.* **4:** 211–221.

Reaven, P., Parthasarathy, S., Grasse, B.J., Miller, E., Steinberg, D., and Witztum, J.L., 1993, Effects of oleate-rich and linoleate-rich diets on the susceptibility of low density lipoprotein to oxidative modification in mildly hypercholesterolemic subjects. *J. Clin. Invest.* **91:** 668–676.

Reilly, K.B., Srinivasan, S., Hatley, M.E., Patricia, M.K., Lannigan, J., Bolick, D.T., Vandenhoff, G., Pei, H., Natarajan, R., Nadler, J.L., and Hedrick, C.C., 2004, 12/15-Lipoxygenase activity mediates inflammatory monocyte/endothelial interactions and atherosclerosis in vivo. *J. Biol. Chem.* **279:** 9440–9450.

Roguin, A., Koch, W., Kastrati, A., Aronson, D., Schomig, A., and Levy, A.P., 2003, Haptoglobin genotype is predictive of major adverse cardiac events in the 1-year period after percutaneous transluminal coronary angioplasty in individuals with diabetes. *Diabetes Care* **26:** 2628–2631.

Sakaguchi, H., Takeya, M., Suzuki, H., Hakamata, H., Kodama, T., Horiuchi, S., Gordon, S., van der Laan, L.J., Kraal, G., Ishibashi, S., Kitamura, N., and Takahashi, K., 1998, Role of macrophage scavenger receptors in diet-induced atherosclerosis in mice. *Lab. Invest.* **78:** 423–434.

Sakr, S.W., Eddy, R.J., Barth, H., Wang, F., Greenberg, S., Maxfield, F.R., and Tabas, I., 2001, The uptake and degradation of matrix-bound lipoproteins by macrophages require an intact actin cytoskeleton, Rho family GTPases, and myosin ATPase activity. *J. Biol. Chem.* **276:** 37649–37658.

Seimon, T.A., Obstfeld, A., Moore, K.J., Golenbock, D.T., and Tabas, I., 2006, Combinatorial pattern recognition receptor signaling alters the balance of life and death in macrophages. *Proc. Natl. Acad. Sci. USA* **103:** 19794–19799.

Sevanian, A. and Asatryan, L., 2002, LDL modification during hemodialysis. Markers for oxidative stress. *Contrib. Nephrol.* 386–395.

Shao, B., Oda, M.N., Bergt, C., Fu, X., Green, P.S., Brot, N., Oram, J.F., and Heinecke, J.W., 2006, Myeloperoxidase impairs ABCA1-dependent cholesterol efflux through methionine oxidation and site-specific tyrosine chlorination of apolipoprotein A-I. *J. Biol. Chem.* **281:** 9001–9004.

Shen, J., Herderick, E., Cornhill, J.F., Zsigmond, E., Kim, H.S., Kuhn, H., Guevara, N.V., and Chan, L., 1996, Macrophage-mediated 15-lipoxygenase expression protects against atherosclerosis development. *J. Clin. Invest.* **98:** 2201–2208.

Shi, W., Wang, X., Shih, D.M., Laubach, V.E., Navab, M., and Lusis, A.J., 2002, Paradoxical reduction of fatty streak formation in mice lacking endothelial nitric oxide synthase. *Circulation* **105:** 2078–2082.

Sigari, F., Lee, C., Witztum, J.L., and Reaven, P.D., 1997, Fibroblasts that overexpress 15-lipoxygenase generate bioactive and minimally modified LDL. *Arterioscler. Thromb. Vasc. Biol.* **17:** 3639–3645.

Steinberg, D., Glass, C.K., and Witztum, J.L., 2008, Evidence mandating earlier and more aggressive treatment of hypercholesterolemia. *Circulation* **118:** 672–677.

Steinberg, D., Parthasarathy, S., Carew, T.E., Khoo, J.C., and Witztum, J.L., 1989, Beyond cholesterol: Modifications of low-density lipoprotein that Increase its atherogenicity. *New Engl. J. Med.* **320:** 915–924.

Steinbrecher, U.P., Parthasarathy, S., Leake, D.S., Witztum, J.L., and Steinberg, D., 1984, Modification of low density lipoprotein by endothelial cells involves lipid peroxidation and degradation of low density lipoprotein phospholipids. *Proc. Natl. Acad. Sci. USA* **81:** 3883–3887.

Stoletov, K., Fang, L., Choi, S.H., Hartvigsen, K., Hansen, L.F., Hall, C., Pattison, J., Juliano, J., Miller, E.R., Almazan, F., Crosier, P., Witztum, J.L., Klemke, R.L., and Miller, Y.I., 2009, Vascular lipid accumulation, lipoprotein oxidation, and macrophage lipid uptake in hypercholesterolemic zebrafish. *Circ. Res.* **104:** 952–960.

Suzuki, H., Kurihara, Y., Takeya, M., Kamada, N., Kataoka, M., Jishage, K., Ueda, O., Sakaguchi, H., Higashi, T., Suzuki, T., Takashima, Y., Kawabe, Y., Cynshi, O., Wada, Y., Honda, M., Kurihara, H., Aburatani, H., Doi, T., Matsumoto, A., Azuma, S., Noda, T., Toyoda, Y., Itakura, H., Yazaki, Y., and Kodama, T., 1997, A role for macrophage scavenger receptors in atherosclerosis and susceptibility to infection. *Nature* **386:** 292–296.

Tabas, I., Williams, K.J., and Boren, J., 2007, Subendothelial lipoprotein retention as the initiating process in atherosclerosis: update and therapeutic implications. *Circulation* **116:** 1832–1844.

Takahashi, Y., Zhu, H., Xu, W., Murakami, T., Iwasaki, T., Hattori, H., and Yoshimoto, T., 2005, Selective uptake and efflux of cholesteryl linoleate in LDL by macrophages expressing 12/15-lipoxygenase. *Biochem. Biophys. Res. Commun.* **338:** 128–135.

Torzewski, M., Suriyaphol, P., Paprotka, K., Spath, L., Ochsenhirt, V., Schmitt, A., Han, S.R., Husmann, M., Gerl, V.B., Bhakdi, S., and Lackner, K.J., 2004, Enzymatic modification of low-density lipoprotein in the arterial wall: A new role for plasmin and matrix metalloproteinases in atherogenesis. *Arterioscler. Thromb. Vasc. Biol.* **24:** 2130–2136.

Upston, J.M., Niu, X., Brown, A.J., Mashima, R., Wang, H., Senthilmohan, R., Kettle, A.J., Dean, R.T., and Stocker, R., 2002, Disease stage-dependent accumulation of lipid and protein oxidation products in human atherosclerosis. *Am. J. Pathol.* **160:** 701–710.

Virella, G. and Lopes-Virella, M.F., 2008, Atherogenesis and the humoral immune response to modified lipoproteins. *Atherosclerosis* **200:** 239–246.

Wang, Z., Nicholls, S.J., Rodriguez, E.R., Kummu, O., Horkko, S., Barnard, J., Reynolds, W.F., Topol, E.J., DiDonato, J.A., and Hazen, S.L., 2007, Protein carbamylation links inflammation, smoking, uremia and atherogenesis. *Nat. Med.* **13:** 1176–1184.

Webb, N.R. and Moore, K.J., 2007, Macrophage-derived foam cells in atherosclerosis: Lessons from murine models and implications for therapy. *Curr. Drug Targets* **8:** 1249–1263.

Witztum, J.L., 2005, You are right too! *J Clin Invest* **115:** 2072–2075.

Yamamoto, S., 1992, Mammalian lipoxygenases: Molecular structures and functions. *Biochim. Biophys. Acta* **1128:** 117–131.

Yamashita, S., Hirano, K., Kuwasako, T., Janabi, M., Toyama, Y., Ishigami, M., and Sakai, N., 2007, Physiological and pathological roles of a multi-ligand receptor CD36 in atherogenesis; insights from CD36-deficient patients. *Mol. Cell. Biochem.* **299:** 19–22.

Zhao, B., Li, Y., Buono, C., Waldo, S.W., Jones, N.L., Mori, M., and Kruth, H.S., 2006, Constitutive receptor-independent low density lipoprotein uptake and cholesterol accumulation by macrophages differentiated from human monocytes with macrophage-colony-stimulating factor (M-CSF). *J. Biol. Chem.* **281:** 15757–15762.

Zhu, H., Takahashi, Y., Xu, W., Kawajiri, H., Murakami, T., Yamamoto, M., Iseki, S., Iwasaki, T., Hattori, H., and Yoshimoto, T., 2003, Low density lipoprotein receptor-related protein-mediated membrane translocation of 12/15-lipoxygenase is required for oxidation of low density lipoprotein by macrophages. *J. Biol. Chem.* **278:** 13350–13355.

Ziouzenkova, O., Asatryan, L., Akmal, M., Tetta, C., Wratten, M.L., Loseto-Wich, G., Jurgens, G., Heinecke, J., and Sevanian, A., 1999, Oxidative cross-linking of ApoB100 and hemoglobin results in low density lipoprotein modification in blood. Relevance to atherogenesis caused by hemodialysis. *J. Biol. Chem.* **274:** 18916–18924.

Ziouzenkova, O., Asatryan, L., Tetta, C., Wratten, M.L., Hwang, J., and Sevanian, A., 2002, Oxidative stress during ex vivo hemodialysis of blood is decreased by a novel hemolipodialysis procedure utilizing antioxidants. *Free Radic. Biol. Med.* **33:** 248–258.

Chapter 9
Cholesterol Interaction with Proteins That Partition into Membrane Domains: An Overview

Richard M. Epand, Annick Thomas, Robert Brasseur, and Raquel F. Epand

Abstract Biological membranes are complex structures composed largely of proteins and lipids. These components have very different structural and physical properties and consequently they do not form a single homogeneous mixture. Rather components of the mixture are more enriched in some regions than in others. This can be demonstrated with simple lipid mixtures that spontaneously segregate components so as to form different lipid phases that are immiscible with one another. The segregation of molecular components of biological membranes also involves proteins. One driving force that would promote the segregation of membrane components is the preferential interaction between a protein and certain lipid components. Among the varied lipid components of mammalian membranes, the structure and physical properties of cholesterol is quite different from that of other major membrane lipids. It would therefore be expected that in many cases proteins would have very different energies of interaction with cholesterol vs. those of other membrane lipids. This would be sufficient to cause segregation of components in membranes. The factors that facilitate the interaction of proteins with cholesterol are varied and are not yet completely understood. However, there are certain groups that are present in some proteins that facilitate interaction of the protein with cholesterol. These groups include saturated acyl chains of lipidated proteins, as well as certain amino acid sequences. Although there is some understanding as to why these particular groups favour interaction with cholesterol, our knowledge of these molecular features is not sufficiently developed to allow for the design of agents that will modify such binding.

Keywords Phase · Raft · Domain · CRAC segment · HIV · gp41 · Sterol sensing domain · SCAP · Cholesterol

R.M. Epand (✉)
Department of Biochemistry and Biomedical Sciences, McMaster University, Hamilton, Ontario, L8N 3Z5, Canada
e-mail: epand@mcmaster.ca

9.1 Lipid Mixtures That Spontaneously Segregate into Cholesterol-Rich Domains

9.1.1 Lipid Mixtures Exhibiting Liquid–Liquid Phase Immiscibility

In the last several years, the phase properties of a mixture of equimolar concentrations of three lipids, dioleoyl phosphatidylcholine (DOPC), cholesterol and a high melting lipid of bovine brain sphingomyelin or dipalmitoyl phosphatidylcholine (DPPC) have been studied (Veatch et al., 2004; Veatch and Keller, 2005a,b). When sphingomyelin is used as the high melting lipid, this lipid mixture corresponds to the major lipid components of the outer leaflet of mammalian cell plasma membranes. It is also similar to the mixture of lipids found in the low density detergent insoluble fraction of these membranes (Brown and London, 2000). It has been pointed out that the lipid in this detergent insoluble fraction, also called detergent resistant membranes (DRM), is not necessarily equivalent to a pre-existing domain in the membrane (Lichtenberg et al., 2005).

9.1.2 The Liquid Ordered Phase

Lipid mixtures of the type described above that exhibit liquid-liquid immiscibility have two physically different kinds of liquid phases – a common kind of liquid disordered phase as well as a liquid ordered phase that may be particular for phases enriched with a sterol (Ipsen et al., 1987; Zuckermann et al., 2004). The difference between the two kinds of liquid phases is that the acyl chains of the phospholipids in the liquid ordered state are largely extended, resembling the acyl chain conformation observed in the solid or gel phase. However, both the liquid ordered and liquid disordered phases are liquid phases and exhibit rapid lateral mobility, a characteristic of a liquid phase. The putative "raft" domains in biological membranes (Simons and Ikonen, 1997) are thought to be in the liquid ordered state (Lingwood et al., 2008; Risselada and Marrink, 2008).

9.1.3 Comparison Between the Domains Formed in Simple Lipid Mixtures and Those of Biological Membranes

One obvious difference between the domains in model membranes and those in biological membranes is the presence of proteins in the later system. In addition, most model system studies are done with membrane bilayers that do not exhibit transbilayer asymmetry, while surface membranes of mammalian cells exhibit large differences between the extracellular and cytoplasmic leaflets. However, there is also a more subtle difference that is yet to be fully explained and that is the size of the membrane domains that are formed. It has recently been suggested that cytoskeletal obstacles may cause domains to break up into small size units (Yethiraj and Weisshaar, 2007). An alternative explanation of the formation of transient small domains is that they represent critical fluctuations (Honerkamp-Smith et al., 2009). In the case of the model membranes showing liquid-liquid immiscibility, the domains are of the order of microns in size and can be easily visualized by light microscopy. However, unless a cell is stimulated or molecular components of rafts are crosslinked, mammalian cells do not exhibit raft-like domains that are detectible by light microscopy. This finding has led to a prolonged controversy as to whether "raft" domains exist in biological membranes. There is little controversy regarding the question of whether there are domains in biological membranes. There is much evidence to indicate a large transmembrane asymmetry, with both the lipid and protein compositions being different for each monolayer. In addition, many studies have shown the presence of lateral inhomogeneity in the membrane, indicating the presence of domains within each monolayer of the membrane. However, a more difficult question is whether any of these domains should be classified as membrane "rafts". Part of the difficulty comes from an evolving definition of what constitutes a "raft" domain. A recent version of the definition of rafts is that they are small, highly dynamic microdomains of eukaryotic plasma membranes that function to compartmentalize several membrane-associated cellular processes (Pike, 2006). In practice, a membrane raft has often been defined empirically by the criterion used for its identification. A simple criterion that had been used defined a raft as the low density membrane fraction that was insoluble in 1% Triton X-100 at 4°C (Brown and Rose, 1992). The highly empirical nature of this criterion is illustrated by the fact that variation of the concentration of the detergent, the nature of the detergent or the temperature of solubilization, caused a change in the fraction of the membrane that was detergent insoluble. In addition, a more fundamental criticism was raised that the detergent caused a rearrangement of molecular components of the membranes such that the detergent insoluble fraction was formed as a consequence of the addition of the Triton and was not pre-existing in the membrane (Heerklotz, 2002).

Although the nature and even the very existence of "raft" domains in membranes is not completely resolved, there is a microdomain with many of the characteristics of "rafts" that is well described and understood in greater detail. These are the caveolae. Caveolae are domains that, like rafts, are enriched in sphingomyelin and cholesterol. They differ from rafts by concentrating the protein caveolin into this domain. Caveolin plays an important function in maintaining the integrity of this domain (Parton and Simons, 2007). This protein also is likely responsible for the characteristic morphology of this domain 50–80 nm diameter, flask-shaped invaginations in the plasma membrane. These domains are not present in all cells but are specific for cells that have a high capacity for endocytotic or transcytotic transport, such as endothelial, epithelial or phagocytic cells. Caveolae, like rafts (Simons and Toomre, 2000), are believed to function in signal transduction but are also involved in a separate pathway for endocytosis, including the uptake of certain pathogens (Parton and Simons, 2007; Harris et al., 2002).

9.1.4 Detection of Rafts in Biological Membranes

The detailed characterization of "floating rafts", i.e. cholesterol and sphingomyelin-rich domains that are not part of caveolae, but rather are free floating in the membrane, is a theme that is currently an active area of investigation. We describe the current state of progress.

Although individual raft domains are too small to visualize by light microscopy, there are conditions in which domains coalesce to form larger, micron-size structures. One of the common ways of visualizing raft domains by fluorescence microscopy is with the use of a fluorescently-labelled B-subunit of cholera toxin that has specificity for binding to the ganglioside GM1, a marker lipid for raft domains. Since cholera toxin is pentavalent, it causes the formation of patches of GM1 resulting in the aggregation of rafts to form structures that can be visualized by light microscopy (Brown and London, 2000; Harder et al., 1998; Kusumi and Suzuki, 2005). An alternative approach to labelling individual molecules in membranes in order to avoid the probe inducing the formation of domains or promoting the merger of existing domains is to use a monovalent antibody directed against a membrane component. Such monovalent antibodies can be produced in *Llama glama*, an animal that makes antibodies devoid of light chains and therefore do not aggregate antigens (Conrath et al., 2003). Such antibodies have been used for FLIM studies to evaluate the enlargement of membrane domains following the activation of the EGF receptor (Hofman et al., 2008). FLIM has also provided evidence for the presence of submicron lipid domains in biological membranes (Stockl et al., 2008).

Since the cytoplasmic leaflet of the plasma membrane is often the monolayer that is more directly involved in signal transduction pathways (Simons and Ikonen, 1997), it is important to show that raft domains are present in both leaflets of the plasma membrane. Evidence for the presence of cholesterol-rich domains on the cytoplasmic leaflet comes from studies using labelled perfringolysin O (*see also* Chapter 22), a cytolysin that binds specifically to cholesterol. The toxin could be expressed within the cell as a fusion protein with the Green Fluorescent Protein. Its punctate binding to the plasma membrane was evidence for the presence of cholesterol-rich domains on the plasma membrane (Hayashi et al., 2006).

9.1.5 Properties of Rafts in Biological Membranes

Phospholipids in the liquid ordered state are considered to have lateral mobility similar to that found in the liquid phase but have extended and more ordered acyl chains, resembling the gel phase. Despite the rapid lateral mobility expected in rafts, about a two-fold difference can be detected between the diffusion constant of lipids in the liquid ordered and that in the liquid disordered phase in model membranes (Dietrich et al., 2001). In addition to this slower rate of diffusion, it can be observed by single particle tracking that both raft associated lipids as well as proteins in the membranes of murine fibroblasts do not diffuse freely through the membrane but are trapped in regions termed transient confinement zones (TCZ) (Dietrich et al., 2002). The size of these TCZs were found to be ~700 nm with an average residence time of about 13 seconds (Schutz et al., 2000), although there was some heterogeneity between raft domains (Drbal et al., 2007). It should be realized that the formation of a TCZ can have several causes including being located in rafts, interacting with the actin cytoskeleton (Langhorst et al., 2007) or with other proteins (Douglass and Vale, 2005). Gold-labelled GPI-anchored proteins have been found to be transiently confined to a smaller area of ~100 nm diameter (Dietrich et al., 2002). A laser trap was employed to confine the motion of a bead bound to a raft protein to an area of about 100 nm in diameter. Local diffusion of this protein was then followed by high resolution single particle tracking. A domain size of ~25 nm was measured, from which the protein did not leave over the time course of the experiment of 10 minutes (Pralle et al., 2000). Recently, stimulated emission depletion (STED) far-field fluorescence nanoscopy was employed to image very small and transient raft domains (Eggeling et al., 2009). It was found that sphingolipids and GPI-anchored proteins are transiently trapped for ~10–20 ms in cholesterol-dependent domains that are <20 nm in diameter.

In addition to the imaging methods described above, it is possible to use fluorescent properties to obtain information about domains at a shorter length scale. These methods include fluorescence quenching, Forster resonance energy transfer (FRET), and fluorescence lifetime imaging microscopy (FLIM). In model systems the detailed analysis of FRET and FLIM results indicate the presence of small domains in certain lipid systems, with diameters approaching that which are expected for lipid rafts in biological membranes (Loura et al., 2009; de Almeida et al., 2009). Using FRET data, Meyer et al. (2006) concluded that G-protein coupled receptors were concentrated 80-fold into domains whose size was below 10 nm diameter. In another study FRET was measured between CFP and YFP proteins. It was shown that the extent of FRET was not sensitive to the aggregation state of the individual proteins and that fixing the cells significantly increases intramolecular FRET efficiency (Anikovsky et al., 2008). FRET has also been used to demonstrate the cholesterol-dependence of the interactions of proteins in the case of the interaction of the tetanus neurotoxin with Thy-1 (Herreros et al., 2001) and between BACE, the β site of amyloid precursor protein-cleaving enzyme, and the low density lipoprotein receptor-related protein (von Arnim et al., 2005).

9.1.6 Transbilayer Coupling and Rafts

As mentioned above, biological membranes exhibit marked transbilayer asymmetry. Some of the important functions ascribed to rafts involve proteins that reside on the cytoplasmic leaflet of the membrane and are involved in signal transduction events within the cell. However, the prominent lipid components of rafts are cholesterol and sphingomyelin. In the plasma membrane, sphingomyelin is primarily on the extracellular leaflet of the membrane. How then are the sphingomyelin-rich raft domains of the extracellular leaflet related spatially and functionally with the signal transduction proteins on the inner leaflet, whose function is also thought to be dependent on being in a raft domain?

Workers in the laboratory of Lukas Tamm have greatly advanced our understanding of the relationship of membrane domains and transbilayer asymmetry (Kiessling et al., 2009). They have overcome several technical difficulties in order to study this issue with the use of asymmetric supported bilayers. To measure lipid asymmetry, Tamm and colleagues adopted the method of fluorescence interference contrast microscopy to obtain rates of lipid flip-flop (Crane et al., 2005). They showed that flip-flop rates could be slowed and asymmetry maintained with the use of a polymer cushion between the asymmetric bilayer and the solid support (Kiessling and Tamm, 2003; Wagner and Tamm, 2000). Using known lipid compositions for each monolayer in the supported bilayer system that corresponded to the lipid compositions of the cytoplasmic and extracellular leaflets of the plasma membrane of eukaryotic cells, these workers showed a colocalization of liquid-ordered domains on each of the monolayers (Wan et al., 2008). This work demonstrated, in a simple controlled model system, that there could be transbilayer coupling in a raft-like domain. Such a property explains how signal transduction across the cell membrane could occur in the raft domains of biological membranes.

9.2 Perturbation of Phase Behaviour by Proteins

The phase behaviour of lipid mixtures have been extensively studied, and mixtures have been identified that exhibit liquid-liquid phase immiscibility. However, to relate this information to the properties of biological membranes one must also take into account the presence of proteins in the membrane. Proteins would in general be expected to interact with different affinities to different lipids. This is particularly true for the relative affinities of the protein to cholesterol compared with phospholipids. Cholesterol has a very different chemical structure to phospholipids, as well as different physical properties, being generally more hydrophobic and also having a more rigid, less flexible structure. Proteins will promote the formation of cholesterol-rich domains if they preferentially interact with cholesterol. However, the formation of cholesterol-rich domains will also be facilitated in cases in which proteins have a lower tendency to bind cholesterol compared with phospholipids (Epand et al., 2004). One can think of this as being analogous to the role of polyunsaturated acyl chains in promoting the formation of raft domains by excluding cholesterol (Soni et al., 2008). With regard to proteins and cholesterol in membranes, in general one would anticipate that proteins would be excluded from cholesterol-rich domains because such domains are more tightly packed and rigid and therefore it is more difficult for other materials to mix with this region of the membrane. Thus, in general one would anticipate that membrane proteins would facilitate the formation of cholesterol-rich domains by being excluded from interactions with cholesterol.

9.3 Proteins Favouring Colocalization with Cholesterol

However, there are also proteins that have favourable interactions with cholesterol and are found in cholesterol-rich domains to stabilize them. This may have an important functional role to group interacting proteins together in a more concentrated form within a domain. The motifs that promote interaction of proteins with cholesterol are not completely understood. There certainly are proteins that preferentially partition into raft domains. For example, the transmembrane segment of the influenza virus hemagglutinin has been found to sequester to raft domains (Scolari et al., 2009). However, the features of transmembrane segments that favour partitioning into raft domains are not known. There are also roles for segments near the membrane interface that recognize cholesterol. Some of the features of these proteins favouring interaction with cholesterol will be discussed below.

9.3.1 Lipidated Proteins

Just as polyunsaturated acyl chains disfavour interaction with cholesterol, so do certain lipid moieties that are covalently linked to proteins. In general the partitioning of lipidated proteins between cholesterol-rich and cholesterol-poor domains can be rationalized in terms of the compatibility of the lipid chain to pack together with cholesterol in the membrane. Thus, polyunsaturated acyl chains are highly flexible and disordered and hence mix poorly with the more rigid domains containing cholesterol with its fused ring system. Acylation is a post-translational modification of proteins that adds rigid saturated acyl chains, either as myristic acid on the N-terminal amino group of Gly or as palmitic acid on cysteine residues (Brown and London, 2000; Resh, 2004; Chakrabandhu et al., 2007; Resh, 2006). These saturated acyl chains behave in a manner opposite to the polyunsaturated acyl chains and favour interaction with the rigid cholesterol ring. Another example is the covalent attachment of cholesterol moieties that because of their smooth, regular surface would also partition with cholesterol-rich domains (Karpen et al., 2001). The opposite situation occurs with prenylated proteins that are excluded from raft domains because of the rough contour of their lipid chain with protruding methyl groups (Melkonian et al., 1999). GPI-linkage is another motif of lipidated proteins that partition into raft domains (Morandat et al., 2002; Sharom and Lehto, 2002; Milhiet et al., 2002; Wang et al., 2002; Sharma et al., 2004). The structure of the GPI linkage is that it has a phosphatidylinositol (PtnIns) moiety that is further glycosylated on the inositol sugar of the PtnIns headgroup. The most abundant form of PtnIns in mammalian membranes is 18:0/20:4. The arachidonoyl chain at the *sn-2* position of glycerol would inhibit partitioning into a raft domain. However, it has recently been found that there is remodelling of the acyl chains of PtnIns when it is converted to the final GPI anchor that is then linked with proteins (Houjou et al., 2007; Maeda et al., 2007; Kinoshita et al., 2008). The arachidonoyl (20:4) chain of PtnIns is replaced by a stearoyl (18:0) group in the final form of the GPI anchor that would be expected to partition well into raft domains. Some GPI groups have an ether linkage in the sn-1 position as a 1-alkyl, 2-acyl-PI derivative (McConville and Ferguson, 1993; Redman et al., 1994). Thus the acyl chain composition of the GPI anchor in nucleated cells is uncommon, by having a saturated acyl chain at the *sn-2* position. This acyl chain replacement takes place in the Golgi apparatus (Fujita and Jigami, 2008). The importance of acyl chain remodelling of GPI-proteins is demonstrated by the fact that knockout mice defective in this function died quickly after birth and had developmental defects (Ueda et al., 2007).

An exception to this general motif of GPI groups with two saturated acyl chains has been found with the GPI-proteins of human erythrocytes. The GPI group in these proteins maintains an unsaturated acyl chain in the *sn-2* position but in addition has a third acyl chain as a palmitoyl group linked to the 2-position of the inositol ring. Thus, these proteins have two saturated acyl chains and one unsaturated one (Roberts et al., 1988; Rudd et al., 1997).

9.3.2 CRAC Motif

Apart from the roles of lipidation, there has been limited progress in identifying the properties that determine the partitioning of proteins among different lipids. The first difficulty in formulating general rules about which protein sequences would favour interaction with cholesterol in biological membranes, is to divide membrane proteins into those that associate with "rafts" and those that do not. A simple way of doing this is by use of the DRMs, but this criterion for raft association has its own caveats. Nevertheless, there are proteins for which there is good evidence that they interact with cholesterol and that cholesterol may even be required for the functioning of these proteins. Among these proteins there has been identified a common motif that has been termed the cholesterol recognition/interaction amino acid consensus (CRAC) motif (Li and Papadopoulos, 1998). A CRAC motif is defined as one with the pattern $-L/V-X_{1-5}-Y- X_{1-5}-R/K-$. There is only one position, the Y residue, which is a unique residue. The first residue can be either L or V and the last residue, either of the two basic amino acids, R or K. One of the difficulties about this pattern is that there are two variable segments, X_{1-5}, each of which can have any length between 1 and 5 residues, as well as any kind of sequence. The looseness of this definition has been pointed out by Palmer (Palmer, 2004). He noted that the genome of *Streptococcus agalactiae* encodes 2094 known and hypothetical proteins. Since bacteria do not synthesize or contain cholesterol, almost all of these proteins will probably have no functional relationship with cholesterol whatsoever. The CRAC sequence occurs 5,737 times among these proteins, corresponding to 2.7 occurrences per protein, or to 1 occurrence in every 112 amino acids. The situation is similar with *Staphylococcus aureus* and of *Escherichia coli*. This makes clear the fact that the CRAC algorithm has limited predictive value in determining which proteins sequester into raft domains. Nevertheless, among proteins that are known to interact with cholesterol, many of them are found to contain CRAC segments. In such cases these segments likely play a role in the interaction with cholesterol. In addition, if one also requires that the CRAC segment be juxtaposed at the membrane interface, generally by being adjacent to a transmembrane segment, then the correlation between the presence of this segment and cholesterol interaction is found to be more reliable.

Two lines of evidence have been used to test the relationship of CRAC segments with function. One is to mutate the CRAC segment in the protein and determine if this alters function and/or interaction with cholesterol-rich domains. An alternative strategy is to utilize a peptide corresponding to the CRAC domain and test if the peptide has the ability to preferentially interact with cholesterol. The use of such peptides has advantage that their interactions with membranes of defined lipid composition can be accurately determined. Of course, the two approaches are not mutually exclusive and mutations in the CRAC domain in an intact protein can also be studied by using peptides with modified sequences.

Several proteins have been identified as interacting with cholesterol and having a CRAC segment adjacent to a transmembrane helix (Epand, 2006). We will mention two recently identified examples. One of them is the major protein of peripheral myelin, the P0 protein. This protein undergoes a cholesterol-dependent conformation change (Luo et al., 2007). This protein has a single transmembrane domain. It has a CRAC motif on both sides of the transmembrane segment. The sequence VTLYVFEK fulfils the requirements of a CRAC domain and is located on the amino terminal side of the transmembrane segment, which would correspond to the extracellular side of the membrane. On the carboxy-terminal side of the transmembrane segment is the sequence LFYLIR that is also a CRAC motif. Cholesterol is required for the formation and compaction of myelin. It was also demonstrated by labelling with a photo-activatable cholesterol that there is direct interaction between cholesterol and the P0 protein (Saher et al., 2009). Cholesterol also plays a role in the exiting of the P0 protein from the endoplasmic reticulum (ER), although cholesterol is not required for the exit of another major protein of peripheral myelin, the myelin-associated glycoprotein (MAG). We also discuss below the role of cholesterol in preventing escape of another protein, SCAP, from the ER. Mutating the cytoplasmic CRAC domain of the P0 protein from LFYLIR to LFSLIL resulted in the exit from the ER being independent of the presence of cholesterol (Saher et al., 2009).

Another recent example of the role of a CRAC segment in the interaction with cholesterol is the cytolethal distending toxin C of *Aggregatibacter actinomycetemcomitans* (CdtC). This toxin is not an integral membrane protein and does not have any transmembrane segments. Nevertheless, it has been shown that this toxin sequesters to membrane microdomains (Boesze-Battaglia et al., 2006; Shenker et al., 2005). In addition, the localization of CdtC to these domains is prevented by cholesterol depletion, as is the cytotoxicity of this protein (Boesze-Battaglia et al., 2009). These workers also identified a CRAC segment on the toxin with the sequence LIDYKGK. The toxin binds preferentially to liposomes containing cholesterol and mutation of the required Y residue to P resulted in reduced binding to cholesterol-containing membranes as well as to the surface of target cells, resulting in lower toxicity (Boesze-Battaglia et al., 2009). This CRAC domain is part of an amphoteric protein and thus the CRAC domain is not adjacent to a transmembrane segment, but is rather exposed on the surface of the protein, making it accessible for binding to cholesterol in target membranes.

There are several cases of fusion proteins of enveloped viruses that have CRAC domains adjacent to a transmembrane segment. One example that has been more extensively studied is the gp41 fusion protein of HIV-1. This protein has a CRAC domain on the extracellular side of the membrane that is adjacent to the transmembrane segment of this protein. This region of the gp41 has been suggested to be important for membrane fusion (Munoz-Barroso et al., 1999; Salzwedel et al., 1999; Lorizate et al., 2008; Saez-Cirion et al., 2003, 2002; Suarez et al., 2000; Apellaniz et al., 2009; Vishwanathan and Hunter, 2008) and it is also a target for developing vaccines (Huarte et al., 2008a,b; Lorizate et al., 2006; Sun et al., 2008).

Cholesterol was found to be required for HIV infection (*see also* Chapter 2) (Liao et al., 2001; Sarin et al., 1985; Schaffner et al., 1986) as well as for fusion promoted by synthetic peptides that include the membrane proximal region (Liao et al., 2001; Shnaper et al., 2004). In addition, depletion of cholesterol from target membranes results in the loss of HIV infectivity (Liao et al., 2003; Viard et al., 2002; Graham et al., 2003). An indication that cholesterol is important for the membrane interactions of HIV is the finding that the cholesterol/phospholipid ratios in the viral membrane of HIV are generally higher than that of the host membranes (Aloia et al., 1993), although the exact increase varies from strain to strain and also depends on other factors including the growth conditions and the type of target cell used. It has been suggested that the HIV envelope has a lipid composition that would have raft-like properties (Brugger et al., 2006). HIV viral membrane proteins have also been shown to sequester into cholesterol-rich raft domains in the membrane during viral assembly (Leung et al., 2008). Cell membrane raft domains are important for viral entry and assembly (Luo et al., 2008). Cholesterol may be particularly important for viral fusion and internalization since these processes are inhibited by lowering viral membrane cholesterol even though membrane binding is not inhibited (Guyader et al., 2002). Another indication of the importance of cholesterol for viral fusion is that removing the receptor for HIV, CD4, from raft domains by mutating the receptor results in the inhibition of HIV-1 infection (Del Real et al., 2002). The chemokine co-receptor of HIV-1 also has to be colocalized into raft domains (Popik et al., 2002).

The CRAC sequence that is adjacent to the transmembrane domain in the gp41 protein is LWYIK. This small segment has been demonstrated to interact with cholesterol by measuring the binding of a fusion protein of a segment of gp41 that is N-terminal to the transmembrane segment and the maltose binding protein binding with cholesterol-hemisuccinate agarose (Vincent et al., 2002). This work further demonstrated that the short peptide, LWYIK, was inhibitory to this interaction. In addition, we have shown that N-acetyl-LWYIK-amide can induce the segregation of cholesterol into cholesterol-rich domains in mixtures with 1-stearoyl-2-oleoyl phosphatidylcholine (SOPC) and that the aromatic groups of this peptide will penetrate more deeply into bilayers containing cholesterol than to membranes of pure phosphatidylcholine (Epand et al., 2003). A recent report has shown that a mutation of gp41 that deletes the LWYIK segment still appears to translocate to cholesterol domains (Chen et al., 2009). However, the criterion for raft association used in this work is that the protein be located in the detergent resistant membranes (DRM). This is not a very good criterion and has been put into question in recent years. Even using this criterion, Chen's results would be hard to explain based on sequestration of the protein to a particular domain. Why would deleting LWYIK cause a very large increase of this protein in DRM? All the mutants show increased DRM partitioning. It suggests that deleting residues in this region facilitates protein aggregation in the presence of Triton so that the mutated protein is not solubilized. In addition, however, this study demonstrates the lower fusogenic ability of the mutant proteins and suggests that the LWYIK segment promotes the enlargement of fusion pores and is involved in postfusion events (Chen et al., 2009).

There is thus good evidence that the segment LWYIK interacts with cholesterol and has a functional role in the fusion of HIV-1. The segment LWYIK fulfils the requirements of a CRAC motif. We have used the segment LWYIK to test how well the CRAC algorithm predicts the functioning of variations of this sequence. There is only one residue that is uniquely required in the CRAC motif, the Y residue. However, there are variants of HIV-1 known in which this Y (residue 681) of gp41 is substituted with S, making it a non-CRAC sequence. It is possible that this substitution makes the virus bind less well to cholesterol, but it still retains some infectivity. It should also be pointed out that both Y and S have OH groups in the side-chain that may have similar H-bonding functions. In addition to these mutant strains of HIV-1, there is also HIV-2 and several strains of SIV that have the segment LASWIK in place of the CRAC domain of HIV-1. This sequence has some cholesterol-sequestering activity, but less than that of LWYIK (Epand et al., 2005). This suggests that HIV-2 may have less dependence on the presence of raft domains in target membranes compared with HIV-1. This is supported by the observation that glycosphingolipids, components of raft domains, are required for the entry of HIV-1 (Hug et al., 2000; Rawat et al., 2004; Viard et al., 2004), while HIV-2 does not require the presence of glycosphingolipids for fusion (Hug et al., 2000). Since HIV-2 fusion is independent of the raft lipids, glycosphingolipids, it is likely also independent of the presence of rafts. Thus raft domains may be less important for the infectivity of HIV-2 compared with HIV-1.

We further evaluated how changes in the sequence LWYIK affected the interaction of peptides with membranes with and without cholesterol. In many cases we also compared these peptides with the consequences of the corresponding changes in sequence of the intact gp41 protein of HIV on the efficiency of membrane fusion. The three positions are required by the CRAC motif are L/V, Y and K/R. For LWYIK this corresponds to the first, third and fifth residue. We compared a group of peptides in which we did not replace the W or I residues. We made a series of N-Acetyl-peptide-amides in which the first, third and fifth residues of LWYIK are substituted. Among the peptides used the native sequence of this CRAC domain, N-acetyl-LWYIK-amide, is the most potent in segregating cholesterol followed by N-acetyl-LWFIK-amide, and then N-acetyl-LWYIR-amide and N-acetyl-LWYIH-amide that have comparable potencies (Epand et al., 2006). N-acetyl-LWWIK-amide and N-acetyl-LWLIK-amide have still weaker potency and

N-acetyl-IWYIK-amide is the poorest in forming regions in the membrane devoid of cholesterol. Differences in membrane partitioning among the peptides did not account for these results. The behavior of these peptides is not completely predicted by the CRAC algorithm. The peptide N-acetyl-LWFIK-amide is not formally a CRAC peptide and it is not as selective as N-acetyl-LWYIK-amide in sequestering cholesterol at low concentration, however it is more potent than another peptide that does correspond to a CRAC motif, N-acetyl-LWYIR-amide. Interestingly, the weakest peptide in sequestering cholesterol is N-acetyl-IWYIK-amide, which is not a CRAC motif. The two peptides that are the most different from the average are the LWYIK that sequesters cholesterol the best and IWYIK that is weakest in sequestering cholesterol. These results support the concept of CRAC sequences being required to sequester cholesterol. However, the other peptides do not quantitatively fit in with the CRAC algorithm, suggesting that these rules are suggestive, but have limited predictive power.

Even though LWYIK and IWYIK have such large differences sequestering cholesterol, the two peptides differ only in the movement of a CH_3 group from one carbon atom to another. We extended this comparison to include the other amino acids with aliphatic side chains, i.e. A and V. Substitution of either of these two amino acids in the first position of LWYIK produced a peptide (all as N-acetyl-peptide-amides, but for simplicity only the sequence of the peptide portion is indicated) with intermediate cholesterol-sequestering ability between that of LWYIK and IWYIK (Vishwanathan et al., 2008a). This also shows the limitations of the CRAC algorithm in being a quantitative predictive method, since VWYIK is also a CRAC sequence, like LWYIK, but is much less active. The results also illustrate that the behaviour of these peptides is not simply the result of differences in hydrophobicity since VWYIK and AWYIK are at opposite ends of the hydrophobicity scale for this group of peptides, yet they have about the same behaviour. It appears that the native sequence, LWYIK, is the most effective in sequestering cholesterol.

One of the factors that may contribute to decreased cholesterol sequestering ability is conformational flexibility since residues with a branched chain at the β-carbon, i.e. I or V in the first position of the CRAC segment result in poor cholesterol sequestering activity. We therefore systematically substituted either G, P or A in positions 2 or 4 of LWYIK to replace W and I (Vishwanathan et al., 2008b). These are the variable positions of the CRAC sequence; hence all of these peptides were CRAC motifs. The residues were chosen so as to maximize conformational flexibility with G, an amino acid with only a H-atom on the α-carbon; or to minimize conformational flexibility with P, an amino acid with limited conformational flexibility because of a ring structure implicating the backbone. The peptide with the greatest conformational flexibility had the largest effect in sequestering cholesterol.

A search was made for the most favorable conformation of LWYIK and analogs using PepLook (Thomas et al., 2006). The energy of peptide conformations was calculated by an all atom description of structures with the addition of van der Waals, electrostatic, internal and external hydrophobicity energy terms. When structures were calculated in water, the contribution of solvent was accounted for by an external hydrophobicity energy term where the solvent-accessible surface of atoms was calculated (Thomas et al., 2004; Brasseur, 1995; Lins et al., 2003). From the 500,000 calculated models, generally 99 or in some cases 999 of the most stable conformations were saved. Calculations were run either in conditions of implicit water, lipids or at a membrane interface. We found that the mean force potential (MFP) of the LWYIK 3D models at the interface were the largest, supporting the conclusion that this peptide has the best intrinsic possibilities of stability especially in a membrane interface. All 3D models of the peptides LWYIK, IWYIK, AWYIK and VWYIK prefer being at the membrane interface rather than in water or in the lipid core. The most stable at the interface is IWYIK, the difference with LWYIK being small. Interestingly, IWYIK has the largest structure diversity at the interface, more than LWYIK, AWYIK or VWYIK. IWYIK shows 5 different structures at the interface whereas there are only 2 for LWYIK and 3 for AWYIK and VWYIK. Next we analyzed the cholesterol-peptide interaction: all peptides generated electrostatic interactions with the OH group of cholesterol, at least as a H-donor or as a H-acceptor, but none was simultaneously a H-acceptor and donor as found for LWYIK. This double interaction satisfies all polar possibilities of cholesterol. Cholesterol in a membrane normally interacts as an H-donor with PC. The presence of LWYIK in a membrane will impair that possibility and thus should facilitate the segregation of cholesterol away from PC. A possibility that would explain the calorimetric data (Vishwanathan et al., 2008a) is that the structural diversity of IWYIK allows it to bind either PC or cholesterol polar headgroups, whereas the optimal binding cavity of LWYIK for the polar head of cholesterol with a well positioned H-donor and H-acceptor will prevent the cholesterol from any other specific interaction with PC or other lipids (Fig. 9.1). Hence, LWYIK selectivity would be linked to a preferential interaction with cholesterol resulting in a displacement of the cholesterol binding capacities. This analysis extends to the double mutants at positions 2 and 4

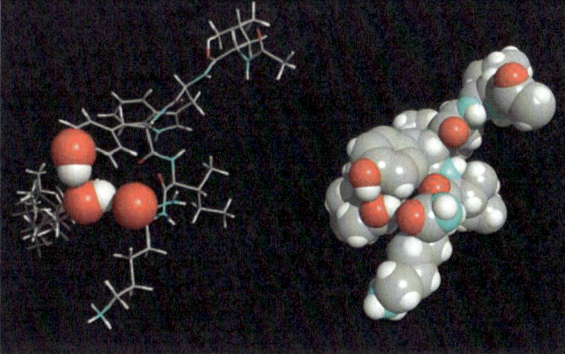

Fig. 9.1 One of the low energy complexes of LWYIK with cholesterol. Left is stick model with the dual H-bonding donor and acceptor interactions of the peptide with the OH group of cholesterol. Right is space-filling model illustrating the potential favorable packing of an aromatic side chain of the peptide with the cholesterol ring system

of LWYIK. We calculated that as expected LGYGK has the largest plasticity with a RMS deviation of the 999 PepLook models of up to 4 Å, the most constrained peptide being LPYPK. When we sorted the PepLook models of lower energy at the membrane interface and ranked them by delta energy we noticed that LWYIK has wider energy pits than LPYPK. Therefore, LPYPK is both structurally and energetically restricted, LGYGK has wider structural and energy possibilities, and LWYIK with intermediate structural possibilities has the best capacity of energy stabilization. This again demonstrates that LWYIK is the peptide that shows the greatest stability at a membrane interface.

Computer modeling studies show that the major factors contributing to the differences among these peptides include their structural flexibility, their position at the interface and their mechanism of interactions with cholesterol. However, in some conditions, more frequent with other peptides, aromatic groups stack with the A ring of cholesterol. The Y residue of the CRAC motif of the benzodiazepine receptor is also important for interaction with cholesterol (Jamin et al., 2005). A model was presented for the interaction of the CRAC motif of the peripheral-type benzodiazepine receptor and cholesterol. The modelling of LWYIK indicates that aromatic groups are likely important for stacking with cholesterol, but the conformation of the peptide and its presentation to the membrane are also important in determining its ability to preferentially bind and to increase the possibility of forming cholesterol-rich regions in the membrane. The central Y residue in LWYIK may have a particular role in stabilizing interactions with cholesterol by forming H-bonds. However, as expected because of the short length and flexibility of these peptides, this feature is necessary for cholesterol selectivity but is not sufficient. Changing K for R decreases this capacity because the polar head of R tends to force the peptide outward towards the water and decreases the structural flexibility at the interface.

In collaboration with Eric Hunter and Sundaram A. Vishwanathan at the Emory Vaccine Research Center, Atlanta, GA, we have also made mutations in the gp41 fusion protein of HIV-1 that corresponded to some of the peptides that we had studied with regard to preferential interaction with cholesterol. The juxtamembrane CRAC domain in the gp41 protein of HIV-1, LWYIK, corresponds to residues 679 to 683 of the intact protein. The fusion efficiency of L679I, L679A and L679V were all between 69 and 79% that of the wild-type protein, as assessed by the luciferase assay. The L679I mutant showed the least fusion, while the L679A and L679V showed similar patterns (A being slightly higher than V). The β-galactosidase assay gave similar results. The average number of nuclei per syncytium is lowest for I, and highest for A (Vishwanathan et al., 2008a). The results agree with another report qualitatively showing reduced fusogenicity of the L679A mutant (Zwick et al., 2005). There is only a small reduction in cell surface expression of any of the mutants, relative to the wild-type protein taken as 100%. Expression levels for the mutant (as % of wild type) were 79 ± 13 for L679A; 77 ± 12 for L679V; 73 ± 11 for L679I. The difference in cell surface expression among the mutants is small and within experimental error. However, the mutants have a slightly lower level of expression than the wild-type protein. This could make a small contribution to, but not fully account for, the weaker fusogenic activity of the mutants compared with the wild-type gp41. Previous studies from the Hunter laboratory demonstrated that in order to obtain a 50% decrease in fusion from the Env of SIV a 10-fold difference in expression level was required (Lin et al., 2003). In the present study a decrease of approximately 20% in the expression level leads to almost a 50% decrease in the number of nuclei per syncytia demonstrating that the reduction in fusion cannot be accounted for by the small differences in expression level.

There is a very good phenomenological correlation between the behavior of these protein segments with model membranes with regard to interaction with cholesterol and the fusogenic potency of mutant forms of the gp41 protein of HIV expressed in cells. One would not expect an exact agreement, as other factors in addition to cholesterol interactions undoubtedly affect fusogenic activity. This correlation holds reasonably well within this set. We can also compare the mutations and peptides expected to have a high degree of conformational flexibility. The fusion efficiency of W680G, I682G was somewhat less than the wild type, but was much greater than for L679G, Y681G or Y681G, K683G. All three of these mutations corresponded to replacing two residues of the LWYIK segment with G. The W680G, I 682G is the only one of the three peptides that has a CRAC domain and it also is potent in sequestering cholesterol. However, the interaction with cholesterol is not the only factor determining the fusogenic activity of the mutant forms of gp41. Thus, the L679I mutant has about the same fusogenic activity as does the W680G, I682G mutant. Nevertheless, the peptide LGYGK is much more effective in sequestering cholesterol (Vishwanathan et al., 2008b) than is the peptide IWYIK (Vishwanathan et al., 2008a). This supports the CRAC hypothesis for cholesterol interactions since LGYGK is a CRAC sequence but IWYIK is not. Although LGYGK has a stronger interaction with cholesterol than IWYIK, L679I and the W680G, I682G gp41 mutants have comparable fusogenic activity. Changing L to I involves only moving a methyl group from one carbon atom to another, while the W680G, I682G mutant replaces two hydrophobic, bulky residues with two that are small, conformationally flexible and not hydrophobic in a region of gp41 that is invariant. It is thus remarkable that this drastic change in the LWYIK segment of gp41 results in retaining 70% of the fusion activity, the same amount that is seen with a simple shift of a methyl group. There are undoubtedly other factors, in addition to the interaction with cholesterol that will determine the biological properties of these mutants. Nevertheless, within a similar series of mutations the correlation between the peptide affecting cholesterol distribution and the fusogenic activity of the mutant gp41 is good. In addition, the fact that we can make a drastic change in the LWYIK segment of gp41, and still retain considerable fusogenic activity as well as cholesterol sequestering ability (with the W680G, I682G mutant), is good evidence that interactions of this region of the protein with cholesterol play an important, but not unique role, in viral fusion.

It is also interesting to consider how cholesterol could affect the interaction of the intact HIV with membranes since there has been some imaging of the native structure of Env on the surface of the virion. One study proposed a splayed-legs model for Env wherein the Env trimer positions itself in the viral membrane like a tripod (Zhu et al., 2006). This model suggests extensive interaction of gp41 molecules with the viral membrane, and possible interaction of the Env with cholesterol-rich membrane domain on the target cell membrane. This could affect Env stability and subsequent internalization. However, there is an alternate model (Zanetti et al., 2006) in which a compact stalk-like Env structure interacts with membranes. In this model there is less interaction of gp41 monomers with the membrane and more with each other. For this model it is difficult to visualize the possible affects of modifications of the CRAC segment on viral entry. Thus, our results indicate that the CRAC domain is important for gp41-induced cell-cell fusion. It is not certain what its importance is for viral-cell fusion. Based on the model presented by (Zhu et al., 2006), it would be likely that this lipid interaction also has a role in virus–cell fusion.

9.3.3 Sterol-Sensing Domains

A type of cholesterol recognition domain that is very different to the short interfacial segments of the CRAC motif is the sterol-sensing domain. This domain contains five transmembrane helices and is found in several proteins that are involved in cholesterol homeostasis, including the enzymes 3-hydroxy-3-methylglutaryl coenzyme A reductase (HMG-CoA reductase) and 7-dehydrocholesterol reductase that are required for the biosynthesis of cholesterol (Kuwabara and Labouesse, 2002). The enzyme HMG-CoA reductase catalyzes the first step in cholesterol biosynthesis and it is the target for statins, a group of cholesterol-lowering drugs. The rate of cholesterol biosynthesis is also regulated at the level of transcription of cholesterol-synthesizing enzymes. This regulation is indirectly determined by the level of cholesterol in the ER (Goldstein et al., 2006). The cholesterol sensor in the ER is the SCAP protein (SREBP-cleavage activating protein) that contains a sterol-sensing domain (Levine, 2004). SCAP binds to cholesterol with a high degree of cooperativity resulting in a sensitive control of the cholesterol concentration in the ER (Radhakrishnan et al., 2008). Cholesterol in the ER is also responsive to changes in the level of cholesterol in other cell membranes, in particular the plasma membrane that has a much higher concentrations of cholesterol. It has been suggested that the concentration of cholesterol normally in the plasma membrane corresponds to a region in the phase diagram that is very sensitive to cholesterol concentration, which may result in the plasma membrane concentration of cholesterol regulating the rate of transport of this sterol to the ER (Radhakrishnan and McConnell, 2000). SCAP functions by carrying the sterol response element binding protein (SREBP) to the Golgi where SREBP is cleaved. A portion of SREBP then translocates to the nucleus where it acts as a transcriptional regulator of cholesterol-synthesizing enzymes. Cholesterol levels in the ER regulate the exit of this protein to the Golgi together with the transcription factors. When the cholesterol concentration is high, SCAP binds to cholesterol and adopts a conformation that allows it to bind to INSIG-1, an integral membrane protein of the ER, retaining the complex in the ER. This interaction is also sensitive to increased levels of expression of INSIG that can lower the concentration of cholesterol required to retain SCAP in the ER

(Radhakrishnan et al., 2008). In this way the cholesterol level in the ER, by binding to SCAP, can turn off its own synthesis.

There are several residues in SCAP that are important for its biological function. In particular, there is a segment YIYF corresponding to residues 298–301 of SCAP that is required for binding of SCAP to INSIG (Yang et al., 2002; Yabe et al., 2002). When SCAP is mutated to replace the Y-298 with C, the protein is not prevented by high cholesterol concentrations from escorting SREBP to function as a transcription factor, resulting in greater cholesterol synthesis (Brown et al., 2002). The importance of this segment is also indicated by the finding that the two Y residues in YIYF are conserved from human, hamster, *C. elegans* and *D. melanogaster*. The segment YIYF also occurs in HMG-CoA reductase, another protein with a sterol-sensing domain. In both cases, the YIYF is present at the C-terminal end of a transmembrane helix. In the case of SCAP the topology of the protein puts YIYF at the cytoplasmic side of the ER, while with HMG-CoA reductase it is on the luminal side. Interestingly, although both SCAP and HMG-CoA reductase bind to INSIG, their fate is quite different. HMG-CoA reductase is ubiquitinated and rapidly degraded, while SCAP is retained in the ER. SCAP exhibits a high degree of cooperativity in binding to cholesterol (Radhakrishnan et al., 2008), perhaps as a consequence of this protein having a tendency to oligomerize into a tetramer (Radhakrishnan et al., 2004), resulting in cholesterol biosynthesis being turned on and off over a narrow range of cholesterol concentrations.

In order for SCAP to dissociate from INSIG and be escorted out of the ER, the cholesterol concentration must be low, which results in a conformational change in SCAP, as detected by the appearance of new proteolytic cleavage sites (Adams et al., 2004; Brown et al., 2002). In sterol-depleted cells a small GTP binding protein, Sar1, binds to the ER and attracts Sec23 and Sec24. The Sec24 then binds to SCAP by interacting with a site on a cytoplasmic loop between helices 6 and 7 (Sun et al., 2005). Interestingly, the segment YIYF, that we suggested above interacts with cholesterol, is on an adjacent cytoplasmic loop between helices 2 and 3. Low cholesterol concentrations would result in the dissociation of cholesterol from binding to the YIYF segment and thereby freeing access to the adjacent cytoplasmic loop to bind the segment of SCAP with the sequence MELADL (Sun et al., 2005). The binding of Sec24 to this site mediates the binding of COPII coat proteins resulting in SCAP and SREBP exiting the ER in the form of COPII-coated vesicles that bud from the ER membrane (Fig. 9.2).

In support of the role of the YIYF segment in SCAP for binding to cholesterol, we have studied the ability of the peptide N-acetyl-YIYF-amide to form cholesterol-rich domains in bilayers of SOPC and cholesterol using DSC. We find that the peptide has no effect on the phase transition of pure SOPC, but it markedly raises the enthalpy of the gel to liquid crystalline phase transition of this phospholipid in mixtures with 30 or 40 mol% cholesterol. We interpret these findings to indicate that the peptide is very potent in recruiting cholesterol into a domain, leaving the remainder of the membrane partially depleted of cholesterol. In addition, this peptide promotes the crystallization of anhydrous cholesterol even in mixtures of SOPC with only 30 mol% cholesterol and 5 mol% N-acetyl-YIYF-amide (unpublished

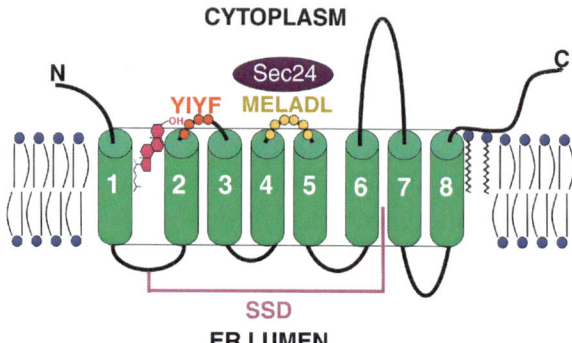

Fig. 9.2 A cartoon model of the polytopic protein SCAP with its 8 transmembrane helices. The protein is embedded the ER membrane and the cytoplasmic and ER Lumen sides are indicated. The 6 transmembrane helices corresponding to the sterol-sensing domain (SSD) are indicated. A cholesterol molecule is shown between helices 1 and 2 to illustrate its potential interaction with the YIYF sequence in the extramembranous loop between helices 2 and 3. Binding of cholesterol to SCAP causes the protein to be retained in the ER by binding to INSIG. Export of SCAP requires the binding of other proteins, beginning with the binding of Sec24 to the segment MELADL in the extramembranous loop between helices 4 and 5. We suggest that the juxtaposition of the loop between helices 2 and 3 with the loop between helices 4 and 5 prevents Sec24 and cholesterol binding simultaneously to SCAP

observations). Thus, the segment YIYF segment can itself contribute to cholesterol sequestration. This sequence occurs at the end of a transmembrane helix, a property shared with many CRAC domains that have been identified as functioning in cholesterol binding.

9.4 Summary and Future Perspectives

The interaction of proteins with cholesterol results in biological consequences of great significance. These include the determination of the organization of the membrane and the rearrangement of molecular components into domains, regulation of the rate of cholesterol biosynthesis, the transfer of cholesterol within the cell and the functioning of a class of bacterial toxin that is dependent on interactions with cholesterol for its action. In the present review we have focused on two examples of very different kinds of cholesterol binding motifs. One of these is the CRAC motif of short protein segments that interact with cholesterol at the membrane interface. The second is the sterol-sensing domain that is a large protein segment that includes five transmembrane helices.

The CRAC domain is a small segment of a protein that interacts preferentially with cholesterol over phospholipids. Although the CRAC algorithm predicts a very large number of potential cholesterol-binding sequences and may therefore be of limited predictive value, there are examples of many proteins that have CRAC domains that are important for interaction with cholesterol and for the control of protein function. We use as a specific example the juxtamembrane CRAC domain of the gp41 fusion protein of HIV-1. Analysis of this segment by studying native and modified peptides interacting with lipid membranes, the mutation of the intact viral gp41 protein and its effect on membrane fusion as well as in silico modelling studies have contributed to understanding some of the factors by which this CRAC domain interacts with cholesterol.

We discuss the sterol-sensing domain in the context of the SCAP protein. Although this domain is large, it appears that certain small critical regions play major roles in cholesterol recognition. Important functional regions of this protein are the adjacent extramembranous loops on the cytoplasmic side of the ER membrane. One of these loops contains the sequence YIYF that is required for interaction with cholesterol while the adjacent loop contains the sequence MELADL that is the site of binding of Sec24. We suggest that there is interaction between these two sites resulting in competitive binding. When cholesterol is bound on the extramembranous loop between helix 2 and helix 3 it prevents Sec24 from binding at the adjacent loop. In this state SCAP binds to INSIG and is retained in the ER. When cholesterol is low and is not bound to SCAP, then Sec24 binds to the extramembranous loop between helix 4 and helix 5 and SCAP resulting in SCAP leaving the ER together with SREBP. The sterol-sensing domain is large and has other functions in addition to binding to cholesterol.

References

Adams, C. M., Reitz, J., De Brabander, J. K., Feramisco, J. D., Li, L., Brown, M. S., and Goldstein, J. L., 2004, Cholesterol and 25-hydroxycholesterol inhibit activation of SREBPs by different mechanisms, both involving SCAP and Insigs. *J. Biol. Chem.* **279:** 52772–52780.

Aloia, R. C., Tian, H., and Jensen, F. C., 1993, Lipid composition and fluidity of the human immunodeficiency virus envelope and host cell plasma membranes. *Proc. Natl. Acad. Sci. USA* **90:** 5181–5185.

Anikovsky, M., Dale, L., Ferguson, S., and Petersen, N., 2008, Resonance energy transfer in cells: a new look at fixation effect and receptor aggregation on cell membrane. *Biophys. J.* **95:** 1349–1359.

Apellaniz, B., Nir, S., and Nieva, J. L., 2009, Distinct mechanisms of lipid bilayer perturbation induced by peptides derived from the membrane-proximal external region of HIV-1 gp41. *Biochemistry* **48:** 5320–5331.

Boesze-Battaglia, K., Besack, D., McKay, T., Zekavat, A., Otis, L., Jordan-Sciutto, K., and Shenker, B. J., 2006, Cholesterol-rich membrane microdomains mediate cell cycle arrest induced by Actinobacillus actinomycetemcomitans cytolethal-distending toxin. *Cell Microbiol.* **8:** 823–836.

Boesze-Battaglia, K., Brown, A., Walker, L., Besack, D., Zekavat, A., Wrenn, S., Krummenacher, C., and Shenker, B. J., 2009, Cytolethal distending toxin-induced cell cycle arrest of lymphocytes is dependent upon recognition and binding to cholesterol. *J. Biol. Chem.* **284:** 10650–10658.

Brasseur, R., 1995, Simulating the folding of small proteins by use of the local minimum energy and the free solvation energy yields native-like structures. *J. Mol. Graph.* **13:** 312–322.

Brown, A. J., Sun, L., Feramisco, J. D., Brown, M. S., and Goldstein, J. L., 2002, Cholesterol addition to ER membranes alters conformation of SCAP, the SREBP escort protein that regulates cholesterol metabolism. *Mol. Cell* **10:** 237–245.

Brown, D. A. and London, E., 2000, Structure and function of sphingolipid- and cholesterol-rich membrane rafts. *J. Biol. Chem.* **275**: 17221–17224.

Brown, D. A. and Rose, J. K., 1992, Sorting of GPI-anchored proteins to glycolipid-enriched membrane subdomains during transport to the apical cell surface. *Cell* **68**: 533–544.

Brugger, B., Glass, B., Haberkant, P., Leibrecht, I., Wieland, F. T., and Krausslich, H. G., 2006, The HIV lipidome: a raft with an unusual composition. *Proc. Natl. Acad. Sci. USA* **103**: 2641–2646.

Chakrabandhu, K., Herincs, Z., Huault, S., Dost, B., Peng, L., Conchonaud, F., Marguet, D., He, H. T., and Hueber, A. O., 2007, Palmitoylation is required for efficient Fas cell death signaling. *EMBO J.* **26**: 209–220.

Chen, S. S. L., Yang, P., Ke, P. Y., Li, H. F., Chan, W. E., Chang, D. K., Chuang, C. K., Tsai, Y., and Huang, S. C., 2009, Identification of the LWYIK motif located in the human immunodeficiency virus type 1 transmembrane gp41 protein as a distinct determinant for viral infection. *J. Virol.* **83**: 870–883.

Conrath, K. E., Wernery, U., Muyldermans, S., and Nguyen, V. K., 2003, Emergence and evolution of functional heavy-chain antibodies in Camelidae. *Dev. Comp. Immunol.* **27**: 87–103.

Crane, J. M., Kiessling, V., and Tamm, L. K., 2005, Measuring lipid asymmetry in planar supported bilayers by fluorescence interference contrast microscopy. *Langmuir* **21**: 1377–1388.

de Almeida, R. F., Loura, L. M., and Prieto, M., 2009, Membrane lipid domains and rafts: current applications of fluorescence lifetime spectroscopy and imaging. *Chem. Phys. Lipids* **157**: 61–77.

Del Real, G., Jimenez-Baranda, S., Lacalle, R. A., Mira, E., Lucas, P., Gomez-Mouton, C., Carrera, A. C., Martinez, A., and Manes, S., 2002, Blocking of HIV-1 infection by targeting CD4 to nonraft membrane domains. *J. Exp. Med.* **196**: 293–301.

Dietrich, C., Bagatolli, L. A., Volovyk, Z. N., Thompson, N. L., Levi, M., Jacobson, K., and Gratton, E., 2001, Lipid rafts reconstituted in model membranes. *Biophys. J.* **80**: 1417–1428.

Dietrich, C., Yang, B., Fujiwara, T., Kusumi, A., and Jacobson, K., 2002, Relationship of lipid rafts to transient confinement zones detected by single particle tracking. *Biophys. J.* **82**: 274–284.

Douglass, A. D. and Vale, R. D., 2005, Single-molecule microscopy reveals plasma membrane microdomains created by protein-protein networks that exclude or trap signaling molecules in T cells. *Cell* **121**: 937–950.

Drbal, K., Moertelmaier, M., Holzhauser, C., Muhammad, A., Fuertbauer, E., Howorka, S., Hinterberger, M., Stockinger, H., and Schutz, G. J., 2007, Single-molecule microscopy reveals heterogeneous dynamics of lipid raft components upon TCR engagement. *Int. J. Immunol.* **19**: 675–684.

Eggeling, C., Ringemann, C., Medda, R., Schwarzmann, G., Sandhoff, K., Polyakova, S., Belov, V. N., Hein, B., von, M. C., Schonle, A., and Hell, S. W., 2009, Direct observation of the nanoscale dynamics of membrane lipids in a living cell. *Nature* **457**: 1159–1162.

Epand, R. F., Sayer, B. G., and Epand, R. M., 2005, The tryptophan-rich region of HIV gp41 and the promotion of cholesterol-rich domains. *Biochemistry* **44**: 5525–5531.

Epand, R. F., Thomas, A., Brasseur, R., Vishwanathan, S. A., Hunter, E., and Epand, R. M., 2006, Juxtamembrane protein segments that contribute to recruitment of cholesterol into domains. *Biochemistry* **45**: 6105–6114.

Epand, R. M., 2006, Cholesterol and the interaction of proteins with membrane domains. *Prog. Lipid Res.* **45**: 279–294.

Epand, R. M., Epand, R. F., Sayer, B. G., Melacini, G., Palgulachari, M. N., Segrest, J. P., and Anantharamaiah, G. M., 2004, An apolipoprotein AI mimetic peptide: membrane interactions and the role of cholesterol. *Biochemistry* **43**: 5073–5083.

Epand, R. M., Sayer, B. G., and Epand, R. F., 2003, Peptide-induced formation of cholesterol-rich domains. *Biochemistry* **42**: 14677–14689.

Fujita, M. and Jigami, Y., 2008, Lipid remodeling of GPI-anchored proteins and its function. *Biochim. Biophys. Acta* **1780**: 410–420.

Goldstein, J. L., Bose-Boyd, R. A., and Brown, M. S., 2006, Protein sensors for membrane sterols. *Cell* **124**: 35–46.

Graham, D. R., Chertova, E., Hilburn, J. M., Arthur, L. O., and Hildreth, J. E., 2003, Cholesterol depletion of human immunodeficiency virus type 1 and simian immunodeficiency virus with beta-cyclodextrin inactivates and permeabilizes the virions: evidence for virion-associated lipid rafts. *J. Virol.* **77:** 8237–8248.

Guyader, M., Kiyokawa, E., Abrami, L., Turelli, P., and Trono, D., 2002, Role for human immunodeficiency virus type 1 membrane cholesterol in viral internalization. *J. Virol.* **76:** 10356–10364.

Harder, T., Scheiffele, P., Verkade, P., and Simons, K., 1998, Lipid domain structure of the plasma membrane revealed by patching of membrane components. *J. Cell Biol.* **141:** 929–942.

Harris, J., Werling, D., Hope, J. C., Taylor, G., and Howard, C. J., 2002, Caveolae and caveolin in immune cells: distribution and functions. *Trends Immunol.* **23:** 158–164.

Hayashi, M., Shimada, Y., Inomata, M., and Ohno-Iwashita, Y., 2006, Detection of cholesterol-rich microdomains in the inner leaflet of the plasma membrane. *Biochem. Biophys. Res. Commun.* **351:** 713–718.

Heerklotz, H., 2002, Triton promotes domain formation in lipid raft mixtures. *Biophys. J.* **83:** 2693–2701.

Herreros, J., Ng, T., and Schiavo, G., 2001, Lipid rafts act as specialized domains for tetanus toxin binding and internalization into neurons. *Mol. Biol. Cell* **12:** 2947–2960.

Hofman, E. G., Ruonala, M. O., Bader, A. N., van den, H. D., Voortman, J., Roovers, R. C., Verkleij, A. J., Gerritsen, H. C., and van Bergen En Henegouwen PM, 2008, EGF induces coalescence of different lipid rafts. *J. Cell Sci.* **121:** 2519–2528.

Honerkamp-Smith, A. R., Veatch, S. L., and Keller, S. L., 2009, An introduction to critical points for biophysicists; observations of compositional heterogeneity in lipid membranes. *Biochim. Biophys. Acta* **1788:** 53–63.

Houjou, T., Hayakawa, J., Watanabe, R., Tashima, Y., Maeda, Y., Kinoshita, T., and Taguchi, R., 2007, Changes in molecular species profiles of glycosylphosphatidylinositol anchor precursors in early stages of biosynthesis. *J. Lipid Res.* **48:** 1599–1606.

Huarte, N., Lorizate, M., Maeso, R., Kunert, R., Arranz, R., Valpuesta, J. M., and Nieva, J. L., 2008a, The broadly neutralizing anti-HIV-1 4E10 monoclonal antibody is better adapted to membrane-bound epitope recognition and blocking than 2F5. *J. Virol.* **82:** 8986–8996.

Huarte, N., Lorizate, M., Kunert, R., and Nieva, J. L., 2008b, Lipid modulation of membrane-bound epitope recognition and blocking by HIV-1 neutralizing antibodies. *FEBS Lett.* **582:** 3798–3804.

Hug, P., Lin, H. M., Korte, T., Xiao, X., Dimitrov, D. S., Wang, J. M., Puri, A., and Blumenthal, R., 2000, Glycosphingolipids promote entry of a broad range of human immunodeficiency virus type 1 isolates into cell lines expressing CD4, CXCR4, and/or CCR5. *J. Virol.* **74:** 6377–6385.

Ipsen, J. H., Karlstrom, G., Mouritsen, O. G., Wennerstrom, H., and Zuckermann, M. J., 1987, Phase equilibria in the phosphatidylcholine-cholesterol system. *Biochim. Biophys. Acta* **905:** 162–172.

Jamin, N., Neumann, J. M., Ostuni, M. A., Vu, T. K., Yao, Z. X., Murail, S., Robert, J. C., Giatzakis, C., Papadopoulos, V., and Lacapere, J. J., 2005, Characterization of the cholesterol recognition amino acid consensus sequence of the peripheral-type benzodiazepine receptor. *Mol. Endocrinol.* **19:** 588–594.

Karpen, H. E., Bukowski, J. T., Hughes, T., Gratton, J. P., Sessa, W. C., and Gailani, M. R., 2001, The sonic hedgehog receptor patched associates with caveolin-1 in cholesterol-rich microdomains of the plasma membrane. *J. Biol. Chem.* **276:** 19503–19511.

Kiessling, V. and Tamm, L. K., 2003, Measuring distances in supported bilayers by fluorescence interference-contrast microscopy: polymer supports and SNARE proteins. *Biophys. J.* **84:** 408–418.

Kiessling, V., Wan, C., and Tamm, L. K., 2009, Domain coupling in asymmetric lipid bilayers. *Biochim. Biophys. Acta* **1788:** 64–71.

Kinoshita, T., Fujita, M., and Maeda, Y., 2008, Biosynthesis, remodelling and functions of mammalian GPI-anchored proteins: recent progress. *J. Biochem. (Tokyo)* **144:** 287–294.

Kusumi, A. and Suzuki, K., 2005, Toward understanding the dynamics of membrane-raft-based molecular interactions. *Biochim. Biophys. Acta* **1746**: 234–251.

Kuwabara, P. E. and Labouesse, M., 2002, The sterol-sensing domain: multiple families, a unique role? *Trends Genet.* **18**: 193–201.

Langhorst, M. F., Solis, G. P., Hannbeck, S., Plattner, H., and Stuermer, C. A., 2007, Linking membrane microdomains to the cytoskeleton: regulation of the lateral mobility of reggie-1/flotillin-2 by interaction with actin. *FEBS Lett.* **581**: 4697–4703.

Leung, K., Kim, J. O., Ganesh, L., Kabat, J., Schwartz, O., and Nabel, G. J., 2008, HIV-1 assembly: viral glycoproteins segregate quantally to lipid rafts that associate individually with HIV-1 capsids and virions. *Cell Host. Microbe.* **3**: 285–292.

Levine, T., 2004, SSD: sterol-sensing direct. *Dev. Cell.* **7**: 152–153.

Li, H. and Papadopoulos, V., 1998, Peripheral-type benzodiazepine receptor function in cholesterol transport. Identification of a putative cholesterol recognition/interaction amino acid sequence and consensus pattern. *Endocrinology* **139**: 4991–4997.

Liao, Z., Cimakasky, L. M., Hampton, R., Nguyen, D. H., and Hildreth, J. E., 2001, Lipid rafts and HIV pathogenesis: host membrane cholesterol is required for infection by HIV type 1. *AIDS Res. Hum. Retroviruses* **17**: 1009–1019.

Liao, Z., Graham, D. R., and Hildreth, J. E., 2003, Lipid rafts and HIV pathogenesis: virion-associated cholesterol is required for fusion and infection of susceptible cells. *AIDS. Res. Hum. Retroviruses* **19**: 675–687.

Lichtenberg, D., Goñi, F. M., and Heerklotz, H., 2005, Detergent-resistant membranes should not be identified with membrane rafts. *Trends Biochem. Sci.* **30**: 430–436.

Lin, X., Derdeyn, C. A., Blumenthal, R., West, J., and Hunter, E., 2003, Progressive truncations C terminal to the membrane-spanning domain of simian immunodeficiency virus Env reduce fusogenicity and increase concentration dependence of Env for fusion. *J. Virol.* **77**: 7067–7077.

Lingwood, D., Ries, J., Schwille, P., and Simons, K., 2008, Plasma membranes are poised for activation of raft phase coalescence at physiological temperature. *Proc. Natl. Acad. Sci USA* **105**: 10005–10010.

Lins, L., Thomas, A., and Brasseur, R., 2003, Analysis of accessible surface of residues in proteins. *Protein Sci.* **12**: 1406–1417.

Lorizate, M., Cruz, A., Huarte, N., Kunert, R., Perez-Gil, J., and Nieva, J. L., 2006, Recognition and blocking of HIV-1 gp41 pre-transmembrane sequence by monoclonal 4E10 antibody in a raft-like membrane environment. *J. Biol. Chem.* **281**: 39598–39606.

Lorizate, M., Huarte, N., Saez-Cirion, A., and Nieva, J. L., 2008, Interfacial pre-transmembrane domains in viral proteins promoting membrane fusion and fission. *Biochim. Biophys. Acta* **1778**: 1624–1639.

Loura, L. M., de Almeida, R. F., Silva, L. C., and Prieto, M., 2009, FRET analysis of domain formation and properties in complex membrane systems. *Biochim. Biophys. Acta* **1788**: 209–224.

Luo, C., Wang, K., Liu, d. Q., Li, Y., and Zhao, Q. S., 2008, The functional roles of lipid rafts in T cell activation, immune diseases and HIV infection and prevention. *Cell. Mol. Immunol.* **5**: 1–7.

Luo, X., Sharma, D., Inouye, H., Lee, D., Avila, R. L., Salmona, M., and Kirschner, D. A., 2007, Cytoplasmic domain of human myelin protein zero likely folded as beta-structure in compact myelin. *Biophys. J.* **92**: 1585–1597.

Maeda, Y., Tashima, Y., Houjou, T., Fujita, M., Yoko-o T, Jigami, Y., Taguchi, R., and Kinoshita, T., 2007, Fatty acid remodeling of GPI-anchored proteins is required for their raft association. *Mol. Biol. Cell* **18**: 1497–1506.

McConville, M. J. and Ferguson, M. A., 1993, The structure, biosynthesis and function of glycosylated phosphatidylinositols in the parasitic protozoa and higher eukaryotes. *Biochem. J.* **294 (Pt 2)**: 305–324.

Melkonian, K. A., Ostermeyer, A. G., Chen, J. Z., Roth, M. G., and Brown, D. A., 1999, Role of lipid modifications in targeting proteins to detergent-resistant membrane rafts. Many raft proteins are acylated, while few are prenylated. *J. Biol. Chem.* **274**: 3910–3917.

Meyer, B. H., Segura, J. M., Martinez, K. L., Hovius, R., George, N., Johnsson, K., and Vogel, H., 2006, FRET imaging reveals that functional neurokinin-1 receptors are monomeric and reside in membrane microdomains of live cells. *Proc. Natl. Acad. Sci USA* **103**: 2138–2143.

Milhiet, P. E., Giocondi, M. C., and Le Grimellec, C., 2002, Cholesterol is not crucial for the existence of microdomains in kidney brush-border membrane models. *J. Biol. Chem.* **277**: 875–878.

Morandat, S., Bortolato, M., and Roux, B., 2002, Cholesterol-dependent insertion of glycosylphosphatidylinositol-anchored enzyme. *Biochim. Biophys. Acta* **1564**: 473–478.

Munoz-Barroso, I., Salzwedel, K., Hunter, E., and Blumenthal, R., 1999, Role of the membrane-proximal domain in the initial stages of human immunodeficiency virus type 1 envelope glycoprotein-mediated membrane fusion. *J. Virol.* **73**: 6089–6092.

Palmer, M., 2004, Cholesterol and the activity of bacterial toxins. *FEMS Microbiol. Lett.* **238**: 281–289.

Parton, R. G. and Simons, K., 2007, The multiple faces of caveolae. *Nat. Rev. Mol. Cell Biol.* **8**: 185–194.

Pike, L. J., 2006, Rafts defined: a report on the Keystone Symposium on Lipid Rafts and Cell Function. *J Lipid Res.* **47**: 1597–1598.

Popik, W., Alce, T. M., and Au, W. C., 2002, Human immunodeficiency virus type 1 uses lipid raft-colocalized CD4 and chemokine receptors for productive entry into CD4(+) T cells. *J. Virol.* **76**: 4709–4722.

Pralle, A., Keller, P., Florin, E. L., Simons, K., and Horber, J. K., 2000, Sphingolipid-cholesterol rafts diffuse as small entities in the plasma membrane of mammalian cells. *J. Cell Biol.* **148**: 997–1008.

Radhakrishnan, A., Goldstein, J. L., McDonald, J. G., and Brown, M. S., 2008, Switch-like control of SREBP-2 transport triggered by small changes in ER cholesterol: a delicate balance. *Cell Metab.* **8**: 512–521.

Radhakrishnan, A. and McConnell, H. M., 2000, Chemical activity of cholesterol in membranes. *Biochemistry* **39**: 8119–8124.

Radhakrishnan, A., Sun, L. P., Kwon, H. J., Brown, M. S., and Goldstein, J. L., 2004, Direct binding of cholesterol to the purified membrane region of SCAP: mechanism for a sterol-sensing domain. *Mol. Cell.* **15**: 259–268.

Rawat, S. S., Eaton, J., Gallo, S. A., Martin, T. D., Ablan, S., Ratnayake, S., Viard, M., KewalRamani, V. N., Wang, J. M., Blumenthal, R., and Puri, A., 2004, Functional expression of CD4, CXCR4, and CCR5 in glycosphingolipid-deficient mouse melanoma GM95 cells and susceptibility to HIV-1 envelope glycoprotein-triggered membrane fusion. *Virology* **318**: 55–65.

Redman, C. A., Thomas-Oates, J. E., Ogata, S., Ikehara, Y., and Ferguson, M. A., 1994, Structure of the glycosylphosphatidylinositol membrane anchor of human placental alkaline phosphatase. *Biochem. J.* **302 (Pt 3)**: 861–865.

Resh, M. D., 2004, Membrane targeting of lipid modified signal transduction proteins. *Subcell. Biochem.* **37**: 217–232.

Resh, M. D., 2006, Palmitoylation of ligands, receptors, and intracellular signaling molecules. *Sci. STKE* **2006**: re14.

Risselada, H. J. and Marrink, S. J., 2008, The molecular face of lipid rafts in model membranes. *Proc. Natl. Acad. Sci USA* **105**: 17367–17372.

Roberts, W. L., Myher, J. J., Kuksis, A., Low, M. G., and Rosenberry, T. L., 1988, Lipid analysis of the glycoinositol phospholipid membrane anchor of human erythrocyte acetylcholinesterase. Palmitoylation of inositol results in resistance to phosphatidylinositol-specific phospholipase C. *J. Biol. Chem.* **263**: 18766–18775.

Rudd, P. M., Morgan, B. P., Wormald, M. R., Harvey, D. J., van den Berg, C. W., Davis, S. J., Ferguson, M. A., and Dwek, R. A., 1997, The glycosylation of the complement regulatory protein, human erythrocyte CD59. *J. Biol. Chem.* **272:** 7229–7244.

Saez-Cirion, A., Arrondo, J. L., Gomara, M. J., Lorizate, M., Iloro, I., Melikyan, G., and Nieva, J. L., 2003, Structural and functional roles of HIV-1 gp41 pretransmembrane sequence segmentation. *Biophys. J.* **85:** 3769–3780.

Saez-Cirion, A., Nir, S., Lorizate, M., Agirre, A., Cruz, A., Perez-Gil, J., and Nieva, J. L., 2002, Sphingomyelin and cholesterol promote HIV-1 gp41 pretransmembrane sequence surface aggregation and membrane restructuring. *J. Biol. Chem.* **277:** 21776–21785.

Saher, G., Quintes, S., Mobius, W., Wehr, M. C., Kramer-Albers, E. M., Brugger, B., and Nave, K. A., 2009, Cholesterol regulates the endoplasmic reticulum exit of the major membrane protein P0 required for peripheral myelin compaction. *J. Neurosci.* **29:** 6094–6104.

Salzwedel, K., West, J. T., and Hunter, E., 1999, A conserved tryptophan-rich motif in the membrane-proximal region of the human immunodeficiency virus type 1 gp41 ectodomain is important for Env-mediated fusion and virus infectivity. *J. Virol.* **73:** 2469–2480.

Sarin, P. S., Gallo, R. C., Scheer, D. I., Crews, F., and Lippa, A. S., 1985, Effects of a novel compound (AL 721) on HTLV-III infectivity in vitro. *N. Engl. J. Med.* **313:** 1289–1290.

Schaffner, C. P., Plescia, O. J., Pontani, D., Sun, D., Thornton, A., Pandey, R. C., and Sarin, P. S., 1986, Anti-viral activity of amphotericin B methyl ester: inhibition of HTLV-III replication in cell culture. *Biochem. Pharmacol.* **35:** 4110–4113.

Schutz, G. J., Kada, G., Pastushenko, V. P., and Schindler, H., 2000, Properties of lipid microdomains in a muscle cell membrane visualized by single molecule microscopy. *EMBO J.* **19:** 892–901.

Scolari, S., Engel, S., Krebs, N., Plazzo, A. P., De Almeida, R. F. M., Prieto, M., Veit, M., and Herrmann, A., 2009, Lateral distribution of the transmembrane domain of influenza virus hemagglutinin revealed by time-resolved fluorescence imaging. *J. Biol. Chem.* **284:** 15708–15716.

Sharma, P., Varma, R., Sarasij, R. C., Ira, Gousset, K., Krishnamoorthy, G., Rao, M., and Mayor, S., 2004, Nanoscale organization of multiple GPI-anchored proteins in living cell membranes. *Cell* **116:** 577–589.

Sharom, F. J. and Lehto, M. T., 2002, Glycosylphosphatidylinositol-anchored proteins: structure, function, and cleavage by phosphatidylinositol-specific phospholipase C. *Biochem. Cell Biol.* **80:** 535–549.

Shenker, B. J., Besack, D., McKay, T., Pankoski, L., Zekavat, A., and Demuth, D. R., 2005, Induction of cell cycle arrest in lymphocytes by Actinobacillus actinomycetemcomitans cytolethal distending toxin requires three subunits for maximum activity. *J. Immunol.* **174:** 2228–2234.

Shnaper, S., Sackett, K., Gallo, S. A., Blumenthal, R., and Shai, Y., 2004, The C- and the N-terminal regions of glycoprotein 41 ectodomain fuse membranes enriched and not enriched with cholesterol, respectively. *J. Biol. Chem.* **279:** 18526–18534.

Simons, K. and Ikonen, E., 1997, Functional rafts in cell membranes. *Nature* **387:** 569–572.

Simons, K. and Toomre, D., 2000, Lipid rafts and signal transduction. *Nat. Rev. Mol. Cell Biol.* **1:** 31–39.

Soni, S. P., LoCascio, D. S., Liu, Y., Williams, J. A., Bittman, R., Stillwell, W., and Wassall, S. R., 2008, Docosahexaenoic acid enhances segregation of lipids between raft and nonraft domains: 2H-NMR study. *Biophys. J.* **95:** 203–214.

Stockl, M., Plazzo, A. P., Korte, T., and Herrmann, A., 2008, Detection of lipid domains in model and cell membranes by fluorescence lifetime imaging microscopy of fluorescent lipid analogues. *J. Biol. Chem.* **283:** 30828–30837.

Suarez, T., Nir, S., Goni, F. M., Saez-Cirion, A., and Nieva, J. L., 2000, The pre-transmembrane region of the human immunodeficiency virus type-1 glycoprotein: a novel fusogenic sequence. *FEBS Lett.* **477:** 145–149.

Sun, L. P., Li, L., Goldstein, J. L., and Brown, M. S., 2005, Insig required for sterol-mediated inhibition of Scap/SREBP binding to COPII proteins *in vitro*. *J. Biol. Chem.* **280**: 26483–26490.

Sun, Z. Y., Oh, K. J., Kim, M., Yu, J., Brusic, V., Song, L., Qiao, Z., Wang, J. H., Wagner, G., and Reinherz, E. L., 2008, HIV-1 broadly neutralizing antibody extracts its epitope from a kinked gp41 ectodomain region on the viral membrane. *Immunity* **28**: 52–63.

Thomas, A., Deshayes, S., Decaffmeyer, M., Van Eyck, M. H., Charloteaux, B., and Brasseur, R., 2006, Prediction of peptide structure: how far are we? *Proteins* **65**: 889–897.

Thomas, A., Milon, A., and Brasseur, R., 2004, Partial atomic charges of amino acids in proteins. *Proteins* **56**: 102–109.

Ueda, Y., Yamaguchi, R., Ikawa, M., Okabe, M., Morii, E., Maeda, Y., and Kinoshita, T., 2007, PGAP1 knock-out mice show otocephaly and male infertility. *J. Biol. Chem.* **282**: 30373–30380.

Veatch, S. L. and Keller, S. L., 2005a, Miscibility phase diagrams of giant vesicles containing sphingomyelin. *Phys. Rev. Lett.* **94**: 148101.

Veatch, S. L. and Keller, S. L., 2005b, Seeing spots: Complex phase behavior in simple membranes. *Biochim. Biophys. Acta Mol. Cell Res.* **1746**: 172–185.

Veatch, S. L., Polozov, I. V., Gawrisch, K., and Keller, S. L., 2004, Liquid domains in vesicles investigated by NMR and fluorescence microscopy. *Biophys. J.* **86**: 2910–2922.

Viard, M., Parolini, I., Rawat, S. S., Fecchi, K., Sargiacomo, M., Puri, A., and Blumenthal, R., 2004, The role of glycosphingolipids in HIV signaling, entry and pathogenesis. *Glycoconj. J.* **20**: 213–222.

Viard, M., Parolini, I., Sargiacomo, M., Fecchi, K., Ramoni, C., Ablan, S., Ruscetti, F. W., Wang, J. M., and Blumenthal, R., 2002, Role of cholesterol in human immunodeficiency virus type 1 envelope protein-mediated fusion with host cells. *J. Virol.* **76**: 11584–11595.

Vincent, N., Genin, C., and Malvoisin, E., 2002, Identification of a conserved domain of the HIV-1 transmembrane protein gp41 which interacts with cholesteryl groups. *Biochim. Biophys. Acta* **1567**: 157–164.

Vishwanathan, S. A. and Hunter, E., 2008, Importance of the membrane-perturbing properties of the membrane-proximal external region of human immunodeficiency virus type 1 gp41 to viral fusion. *J. Virol.* **82**: 5118–5126.

Vishwanathan, S. A., Thomas, A., Brasseur, R., Epand, R. F., Hunter, E., and Epand, R. M., 2008a, Hydrophobic substitutions in the first residue of the CRAC segment of the gp41 protein of HIV. *Biochemistry* **47**: 124–130.

Vishwanathan, S. A., Thomas, A., Brasseur, R., Epand, R. F., Hunter, E., and Epand, R. M., 2008b, Large changes in the CRAC segment of gp41 of HIV do not destroy fusion activity if the segment interacts with cholesterol. *Biochemistry*. **47**: 11869–11876.

von Arnim, C. A., Kinoshita, A., Peltan, I. D., Tangredi, M. M., Herl, L., Lee, B. M., Spoelgen, R., Hshieh, T. T., Ranganathan, S., Battey, F. D., Liu, C. X., Bacskai, B. J., Sever, S., Irizarry, M. C., Strickland, D. K., and Hyman, B. T., 2005, The low density lipoprotein receptor-related protein (LRP) is a novel beta-secretase (BACE1) substrate. *J. Biol. Chem.* **280**: 17777–17785.

Wagner, M. L. and Tamm, L. K., 2000, Tethered polymer-supported planar lipid bilayers for reconstitution of integral membrane proteins: silane-polyethyleneglycol-lipid as a cushion and covalent linker. *Biophys. J.* **79**: 1400–1414.

Wan, C., Kiessling, V., and Tamm, L. K., 2008, Coupling of cholesterol-rich lipid phases in asymmetric bilayers. *Biochemistry* **47**: 2190–2198.

Wang, J., Gunning, W., Kelley, K. M., and Ratnam, M., 2002, Evidence for segregation of heterologous GPI-anchored proteins into separate lipid rafts within the plasma membrane. *J. Membr. Biol.* **189**: 35–43.

Yabe, D., Xia, Z. P., Adams, C. M., and Rawson, R. B., 2002, Three mutations in sterol-sensing domain of SCAP block interaction with insig and render SREBP cleavage insensitive to sterols. *Proc. Natl. Acad. Sci. USA* **99**: 16672–16677.

Yang, T., Espenshade, P. J., Wright, M. E., Yabe, D., Gong, Y., Aebersold, R., Goldstein, J. L., and Brown, M. S., 2002, Crucial step in cholesterol homeostasis: sterols promote binding of

SCAP to INSIG-1, a membrane protein that facilitates retention of SREBPs in ER. *Cell* **110:** 489–500.

Yethiraj, A. and Weisshaar, J. C., 2007, Why are lipid rafts not observed in vivo? *Biophys. J.* **93:** 3113–3119.

Zanetti, G., Briggs, J. A., Grunewald, K., Sattentau, Q. J., and Fuller, S. D., 2006, Cryo-electron tomographic structure of an immunodeficiency virus envelope complex in situ. *PLoS Pathog.* **2:** e83.

Zhu, P., Liu, J., Bess, J., Jr., Chertova, E., Lifson, J. D., Grise, H., Ofek, G. A., Taylor, K. A., and Roux, K. H., 2006, Distribution and three-dimensional structure of AIDS virus envelope spikes. *Nature* **441:** 847–852.

Zuckermann, M. J., Ipsen, J. H., Miao, L., Mouritsen, O. G., Nielsen, M., Polson, J., Thewalt, J., Vattulainen, I., and Zhu, H., 2004, Modeling lipid-sterol bilayers: applications to structural evolution, lateral diffusion, and rafts. *Meth. Enzymol* **383:** 198–229.

Zwick, M. B., Jensen, R., Church, S., Wang, M., Stiegler, G., Kunert, R., Katinger, H., and Burton, D. R., 2005, Anti-human immunodeficiency virus type 1 (HIV-1) antibodies 2F5 and 4E10 require surprisingly few crucial residues in the membrane-proximal external region of glycoprotein gp41 to neutralize HIV-1. *J. Virol.* **79:** 1252–1261.

Chapter 10
Caveolin, Sterol Carrier Protein-2, Membrane Cholesterol-Rich Microdomains and Intracellular Cholesterol Trafficking

Friedhelm Schroeder, Huan Huang, Avery L. McIntosh, Barbara P. Atshaves, Gregory G. Martin, and Ann B. Kier

Abstract While the existence of membrane lateral microdomains has been known for over 30 years, interest in these structures accelerated in the past decade due to the discovery that cholesterol-rich microdomains serve important biological functions. It is increasingly appreciated that cholesterol-rich microdomains in the plasma membranes of eukaryotic cells represent an organizing nexus for multiple cellular proteins involved in transmembrane nutrient uptake (cholesterol, fatty acid, glucose, etc.), cell-signaling, immune recognition, pathogen entry, and many other roles. Despite these advances, however, relatively little is known regarding the organization of cholesterol itself in these plasma membrane microdomains. Although a variety of non-sterol markers indicate the presence of microdomains in the plasma membranes of living cells, none of these studies have demonstrated that cholesterol is enriched in these microdomains in living cells. Further, the role of cholesterol-rich membrane microdomains as targets for intracellular cholesterol trafficking proteins such as sterol carrier protein-2 (SCP-2) that facilitate cholesterol uptake and transcellular transport for targeting storage (cholesterol esters) or efflux is only beginning to be understood. Herein, we summarize the background as well as recent progress in this field that has advanced our understanding of these issues.

Keywords Cholesterol · Membrane · Domains · Caveolae · Sterol carrier protein-2

Abbreviations

ABCA1, ABCG1, ABCG5, and ABCG8   refer to ATP-binding cassette proteins A1, G1, G5, and G8 respectively
BC-θ   biotinylated Cθ–toxin

F. Schroeder (✉)
Department of Physiology and Pharmacology, Texas A&M University, TVMC College Station, TX 77843-4466, USA
e-mail: fschroeder@cvm.tamu.edu

J.R. Harris (ed.), *Cholesterol Binding and Cholesterol Transport Proteins*, Subcellular Biochemistry 51, DOI 10.1007/978-90-481-8622-8_10,
© Springer Science+Business Media B.V. 2010

CT-B	cholera toxin subunit B
DChol	6-dansyl-cholestanol
DHE	dehydroergosterol
DiD	1,1′-dioctadecyl-3,3,3′,3′-tetramethylindodicarbocyanine, 4-chlorobenzenesulfonate salt(DiIC$_{18}$(5))
DRM	detergent resistant membranes
FRET	fluorescence resonance energy transfer
HDL	high density lipoprotein
L-FABP	liver fatty acid binding protein
LSCM	laser scanning confocal microscopy
MPLSM	multiphoton laser scanning microscopy
MβCD	Methyl-β-cyclodextrin
NBD-cholesterol	22-(N-(7-nitrobenz-2-oxa-1,3-diazol-4-yl)amino)-23,24-bisnor-5-cholen-3β -ol
Pgp	P-glycoprotein
PM	plasma membrane
N-Rh-DOPE	1,2-Dioleoyl-sn-Glycero-3-Phosphoethanolamine-N-(Lissamine Rhodamine B Sulfonyl) (Ammonium Salt)
SCP-2	sterol carrier protein-2
SRB1	scavenger receptor B1

10.1 Introduction

Cholesterol is the driving force for lipid-rich microdomain formation in plasma membranes (Bretscher and Munro, 1993), providing a platform for several proteins known to regulate uptake and transport of cholesterol (Fig. 10.1) (Everson and Smart, 2005; Parathath et al., 2004; Connelly and Williams, 2004; Yancey et al., 2003), fatty acids (Ortegren et al., 2007), and glucose (Saltiel and Pessin, 2003; Vainio et al., 2005; Ikonen and Vainio, 2005). The microdomain concept, despite continually evolving details, provides a framework for understanding location and function of receptors, transport/translocase proteins, and downstream signaling molecules. Despite the importance of cholesterol for the very existence of microdomains, little is known regarding the properties of cholesterol therein—either in peripheral cells serving as cholesterol donors in reverse cholesterol transport or in hepatocytes, which are key for the net removal of cholesterol from the body as well as in transport and metabolism of both fatty acids and glucose. Likewise, the role of intracellular cholesterol-binding proteins in mediating cholesterol transfer from plasma membrane cholesterol-rich microdomains is only beginning to be resolved.

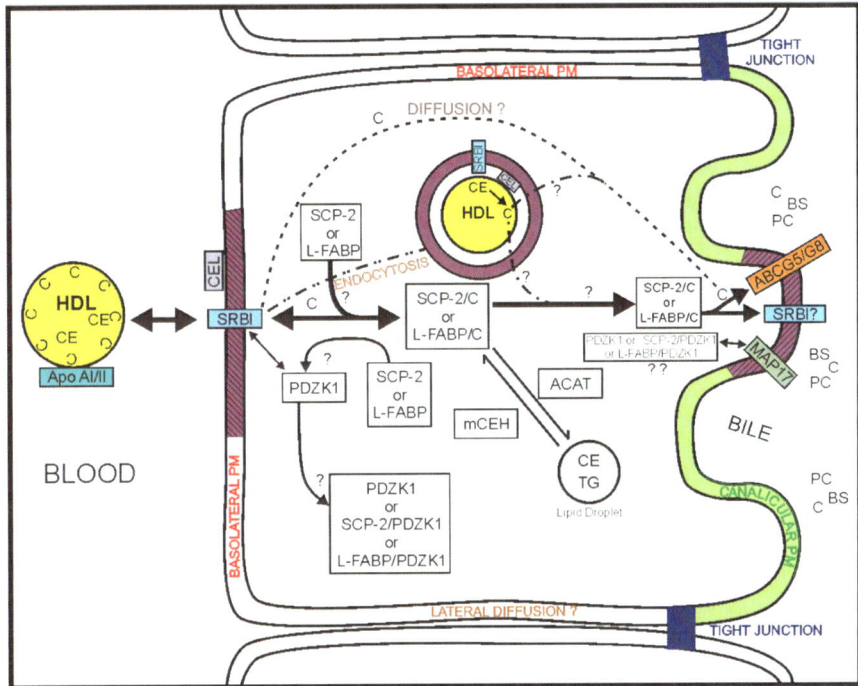

Fig. 10.1 Proposed mechanisms for transhepatocyte transport of HDL-derived cholesterol for efflux into bile. Spontaneous diffusion through aqueous cytosol (dotted line), vesicular transport (endocytosis), lateral diffusion through plasma membranes, and protein-mediated transport (SCP-2, L-FABP). Basolateral membranes, white; canalicular membranes, green; cholesterol-rich microdomains, purple; endocytosed cholesterol-rich microdomain surrounding HDL, purple sphere; HDL, yellow; lipid droplet, gray. CEL, carboxyl ester lipase; ACAT, acyl CoA cholesterol acyltransferase; mCEH, neutral cholesterol ester hydrolase; PM, plasma membrane; C, cholesterol; CE, cholesterol ester; TG, triacylglycerol; PC, phosphatidylcholine; BS, bile salts

10.2 What Are "Cholesterol-Rich and -Poor Microdomains"?

As evidenced by over 5,000 publications in little over a decade, lipid-rich microdomains have become a major focus in biomembrane research (Schroeder et al., 2005; Pike, 2006). Lipid-rich microdomains are defined as small, highly dynamic, cholesterol- and sphingolipid-rich domains that compartmentalize cellular processes and can be stabilized to form larger platforms through protein-protein and protein-lipid interactions (Pike, 2006). These domains, also rich in saturated or monounsaturated fatty acylated phospholipids, are physically distinct, liquid-ordered structures intermediate between fluid liquid-crystalline and rigid gel phases (Schroeder et al., 2005; Gallegos et al., 2006; Shahedi et al., 2006; Atshaves et al., 2007c). Lipid-rich microdomains range in size from 1-10 nm (single annular shells of cholesterol surrounding lipids or proteins) to >1 μm (including optically visible microvilli, filopodia, pseudopodia) (Schroeder et al., 2005). Although caveolin-1 was initially thought to be required for directing formation of remarkably stable cholesterol-rich microdomains enriched with a variety of proteins involved in cholesterol uptake/efflux, signaling, and other processes, similarly-sized microdomains also exist in cells lacking caveolin-1—suggesting that "caveolae" are a subset of cholesterol-rich microdomains within the plasma membrane (Smart and van der Westhuyzen, 1998; Schroeder et al., 2005; Pike, 2006). Although biochemical cell fractionation techniques have been used to isolate cholesterol-rich microdomains, the isolation methods are critical: (i) Quantity and purity vary dramatically depending on isolation method; (ii) Detergent (detergent resistant membranes, DRM) or high pH carbonate buffer-based preparations contain up to 30-75% contaminating non-microdomain and intracellular proteins. DRMs also exhibit artifactual sterol crystals, abnormal sterol flux, are unresponsive to intracellular cholesterol transfer proteins, and should not be identified with cholesterol-rich microdomains; (iii) Detergent-free, high pH carbonate-free methods for isolating cholesterol-rich microdomains from purified plasma membranes (e.g. affinity chromatography) are the least perturbing and simultaneously resolve highly purified cholesterol-rich and -poor microdomains (Gallegos et al., 2006; Schroeder et al., 2005; Storey et al., 2007a; Foster et al., 2003; Babiychuk and Draeger, 2006; Feul-Lagerstedt et al., 2007; Lichtenberg et al., 2005). Little is known about cholesterol-poor microdomains, since only recently were techniques developed that isolate these microdomains in sufficient purity and quantity (Atshaves et al., 2003, 2007a; Gallegos et al., 2004; Storey et al., 2007a). With regards to the existence of cholesterol-rich and -poor microdomains in plasma membranes of living cells, evidence of such was previously only inferred through indirect markers. However, recent real-time imaging studies of fluorescent sterols provide evidence for the existence of cholesterol-rich microdomains in plasma membranes of living cells.

10.3 How Is Cholesterol Organized Within Plasma Membranes?

Within model membranes, cholesterol can form small non-crystalline clusters (10–25 nm) (Stillwell and Wassall, 2003; Tulenko et al., 1998; Troup et al., 2003; McIntosh et al., 2003) whose hexagonal phase may account for the rapid spontaneous transbilayer migration of cholesterol ($t_{1/2}$=sec to min) (Hayakawa et al., 1998; Lange et al., 2007; Muller and Herrmann, 2002; Schroeder and Nemecz, 1990; Schroeder et al., 1991b; Haynes et al., 2000). Whether plasma membrane cholesterol-rich microdomains are specifically enriched in these cholesterol clusters is unknown.

Although plasma membrane transbilayer cholesterol gradients exist in non-hepatocytes (Schroeder et al., 1991b; Wood et al., 1990; Kier et al., 1986;

Schroeder and Sweet, 1988; Mondal et al., 2009), transbilayer distribution of cholesterol within hepatocyte membranes is an enigma. The cytofacial leaflet of non-hepatocyte plasma membranes is enriched in cholesterol (Sweet and Schroeder, 1988b; Schroeder, 1988; Pitto et al., 2000; Mondal et al., 2009). The cholesterol enrichment in the cytofacial versus exofacial leaflet results in a transbilayer fluidity gradient with more rigid cytofacial leaflet and more fluid exofacial leaflet (Schroeder and Sweet, 1988; Sweet and Schroeder, 1988a). However, enrichment of plasma membrane phospholipids with n-3 polyunsaturated fatty acids (PUFAs) significantly alters transbilayer cholesterol distribution and the transbilayer difference in structure (Sweet and Schroeder, 1988b; Schroeder, 1988; Pitto et al., 2000; Mondal et al., 2009). The functional importance of a transbilayer fluidity gradient is shown by Na^+,K^+-ATPase, whose activity is "optimal" at a specific lipid fluidity in the cytofacial (but not exofacial) leaflet, where the ATP hydrolyzing site is localized (Sweet and Schroeder, 1988a). ABC transporters involved in cholesterol efflux (e.g. ABCA1, ABCG1, ABCG5, and ABCG8) also have cytofacial ATP hydrolysis sites, and activities of several are sensitive to cholesterol content.

While plasma membrane lateral cholesterol-rich microdomains exhibit a unique liquid-ordered phase (Schroeder et al., 2005; Gallegos et al., 2006; Pike, 2006), transbilayer structure of lateral cholesterol-rich and -poor microdomains is only beginning to be resolved (Atshaves et al., 2007c; Gallegos et al., 2004; Gallegos et al., 2006). The cytofacial leaflet phospholipid fatty acyl chains of cholesterol-rich microdomains from non-hepatocytes and hepatocytes are more rigid than cytofacial acyl chains (Schroeder et al., 1989b; Jefferson et al., 1990; Wood and Schroeder, 1992; Igbavboa et al., 1996). Unexpectedly, despite the fact that the bulk lipid (i.e. acyl chain) fluidity of cholesterol-rich microdomains is lower than that of cholesterol-poor microdomains, cholesterol itself is less rigidly structured than the bulk lipids therein (Atshaves et al., 2007c; Gallegos et al., 2004, 2006). Cholesterol headgroup aqueous exposure in cholesterol-rich and -poor microdomains, especially biological membranes, is not known (Harroun et al., 2008). Model membranes exhibit two pools of cholesterol (Nemecz and Schroeder, 1988; Schroeder and Nemecz, 1989; Schroeder et al., 1988a, b, 1991a): (i) a small, more aqueous-exposed pool of exchangeable sterol; (ii) a large, less aqueous-exposed pool of very slowly transferable sterol—consistent with the tighter packing, higher order, and greater thickness of this region (Shaw et al., 2006).

10.4 What Are the Dynamics of Cholesterol Efflux in Cholesterol-Rich Versus -Poor Microdomains?

Cholesterol efflux from non-hepatocyte plasma membranes resolves into fast ($t_{1/2}$=min–hr) and slow ($t_{1/2}$=days) components (Schroeder et al., 1991a, 1996) which are both lateral microdomains since cholesterol transbilayer migration is rapid (sec–min) (Hayakawa et al., 1998; Lange et al., 2007; Muller and Herrmann, 2002; Schroeder and Nemecz, 1990; Schroeder et al., 1991b; Haynes et al., 2000).

Cholesterol translocation in model membranes relates directly to fluidity, suggesting that rapidly transferable cholesterol would be associated with more fluid cholesterol-poor microdomains (Schroeder et al., 1991a; Jessup et al., 2006). However, cholesterol efflux is actually faster from cholesterol-rich microdomains (more rigid) in non-hepatocytes (Atshaves et al., 2003; Gallegos et al., 2001a, 2004, 2006, 2008; Storey et al., 2007a), indicating that cholesterol in biological microdomains may not be organized in the same way as in model membrane microdomains.

10.5 Does Membrane Lipid Composition Affect Cholesterol Dynamics in Cholesterol-Rich and -Poor Microdomains?

Although regulation of cholesterol-rich microdomains has only been examined in detail in model membrane and non-hepatocyte systems, much less is known about microdomain regulation in hepatocytes (Schroeder et al., 2005; Atshaves et al., 2007c).

Cholesterol content: Despite many studies in model membranes, little is known regarding the role of cholesterol in regulating the structure of cholesterol-rich and -poor microdomains in biological membranes. In non-hepatocytes, methylcyclodextrin and statins preferentially deplete the rapid (<1hr) cholesterol efflux pool associated with cholesterol-rich microdomains (Jessup et al., 2006). While caveolar structure/function in non-hepatocytes as well as SRB1 and P-gp localization are exquisitely sensitive to cholesterol perturbation, ABCA1 activity is not (Martin and Parton, 2005; Parton and Simons, 2007; Parton et al., 2006; Harder et al., 2007; Orlowski et al., 2006).

Phospholipid content: Cholesterol uptake/efflux is sensitive to cholesterol-rich microdomain phospholipid species. Sphingomyelin (SM) depletion inhibits cholesterol flux mediated by caveolin-1 and SRB1 by displacing cholesterol and caveolin-1 from cholesterol-rich microdomains (Jessup et al., 2006; Yu et al., 2005; Gulbins and Li, 2006).

Dietary n-3 versus n-6 polyunsaturated fatty acids (PUFAs): Anthropological, epidemiological, and molecular studies indicate that humans evolved on a diet with a ratio of n-6/n-3 PUFA near 1:1 (Simopoulos, 2006; Morris, 2007a, b). However, since the common use of hot milling (higher PUFA degradation, especially n-3 PUFA) for preparing flour began in the early 1900 s, the total PUFA content in the modern Western diet has decreased, and the n-6/n-3 ratio has risen to 15:1 (Simopoulos, 2006; Morris, 2007a, b). Low dietary n-3 PUFAs as in Western diets particularly correlates with the pathogenesis of age related diseases including diabetes, accelerated cardiovascular disease, stroke, atherosclerosis, and chronic inflammatory diseases (Simopoulos, 2006; Morris, 2007a; Morris, 2007b). Since n-6 PUFAs [linoleic acid (18:2n-6), arachidonic acid (20:4n-6)] may be metabolized to both pro- and anti-inflammatory mediators, a recent AHA Science Advisory examining extensive literature concluded that high dietary n-6 PUFAs [linoleic acid (18:2n-6), arachidonic acid (20:4n-6)] may also exert cardiovascular/antiinflammatory benefits and may improve insulin resistance, effects likely to be dependent on the relative proportion of the various mediators produced (Harris et al., 2009). However, model membrane studies suggest cholesterol-rich microdomains as a likely target for effects of n-3 PUFAs, more so than n-6 PUFAs (Stillwell and Wassall, 2003). Cholesterol-rich microdomains of non-hepatocytes have much less n-3 PUFA-containing phospholipids than cholesterol-poor microdomains (Atshaves et al., 2003, 2007a; Pike et al., 2002), consistent with model membrane studies where cholesterol preferentially interacts with phospholipids containing saturated fatty acyl chains, but less so with n-3 PUFA acyl chains (Stillwell and Wassall, 2003). Supplementation of non-hepatocytes with n-3, but not n-6, PUFAs results in preferential accumulation of n-3 PUFA in phospholipids in the plasma membrane cytofacial leaflet (phosphatidylethanolamine, phosphatidylserine), which in turn is thought to enhance cholesterol translocation from cytofacial to exofacial leaflet, a location optimal for cholesterol transfer to/from HDL (Stillwell and Wassall, 2003; Sweet and Schroeder, 1988b). Incorporation of n-3, but not n-6, PUFAs into detergent resistant membrane (DRM) phospholipids of non-hepatocytes displaces SM and/or cholesterol from DRMs, is consistent with increased cholesterol flux (Stulnig et al., 2001; Ma et al., 2004; Li et al., 2007). In di-n-3 PUFA phospholipid model membranes, the OH group of cholesterol is parallel to the bilayer plane (opposite to normal perpendicular orientation), suggesting that n-3 PUFA phospholipids in cholesterol-rich microdomains may facilitate transbilayer cholesterol movement (Harroun et al., 2008).

10.6 How May Plasma Membrane Proteins Regulate Cholesterol Dynamics?

Cholesterol distribution and dynamics in cholesterol-rich versus -poor microdomains is also likely to be dependent upon interaction with proteins within these microdomains. Key among these are reverse cholesterol transport (RCT) proteins. RCT proteins such as caveolin-1, ABCA1, ABCG1, SRB1, and P-gp are selectively distributed into cholesterol-rich microdomains of non-hepatocytes (Everson and Smart, 2005; Jessup et al., 2006; Karten et al., 2006; Bourret et al., 2006; Kamau et al., 2005) and their activities are sensitive to cholesterol concentration and distribution. These RCT proteins also induce formation of more aqueous-exposed, cholesterol-oxidase-accessible cholesterol (Smart and van der Westhuyzen, 1998; Parathath et al., 2004; Vaughan and Oram, 2005; de la Llera-Moya et al., 1999; Baldan et al., 2006). SRB1 increases the membrane cholesterol pool available to cyclodextrin acceptors (Kellner-Weibel et al., 2000), and caveolin-1 and SRB1, but not ABCA1, directly bind cholesterol and/or are cross-linked by photoactivatable cholesterol (Thiele et al., 2000; Assanasen et al., 2005; Wanaski et al., 2003). In contrast, since hepatocytes are essentially deficient in caveolin-1 (Vainio et al., 2002; Atshaves et al., 2007c), microdomains and dynamics of cholesterol and RCT proteins could be significantly different to that in other cell types. Hepatocyte basolateral plasma membranes contain SRB1 for cholesterol uptake/efflux while canalicular membranes contain ABCA5, ABCA8, Pgp, and possibly SRB1 for biliary excretion (Fig. 10.1). In non-hepatocytes, caveolin-1 binds cholesterol and caveolin-1 overexpression increases cholesterol content of cholesterol-rich microdomains (Schroeder et al., 2005; Everson and Smart, 2005). SRB1 also binds cholesterol, increases aqueous accessibility of plasma membrane cholesterol, facilitates ATP-independent, bidirectional, free cholesterol flux across membranes, and increases non-lysosomal, selective endocytic uptake of HDL cholesterol ester (Everson and Smart, 2005; Assanasen et al., 2005; Kellner-Weibel et al., 2000). SRB1 overexpression in non-hepatocytes elicits accumulation of PC species with longer mono- or PUFA acyl chains, consistent with decreased PC/cholesterol interactions and increased cholesterol efflux to HDL (Parathath et al., 2004). Since hepatocytes contain little if any caveolin-1, SRB1 location likely is more complex, depending on HDL-cholesterol load, expression of co-regulator proteins PDZK1 (CLAMP) and MAP17 (Atshaves et al., 2007c; Silver, 2004) (Fig. 10.1). While some ABC transporters (ABCA1) do not interact with cholesterol, others such as ABCG1 increase cholesterol content in cholesterol-rich microdomains of non-hepatocytes (Karten and et al., 2006; Baldan et al., 2006). Both affinity chromatography (not requiring detergents or high pH carbonate buffers) and detergent-based methods suggest the existence of microdomains in hepatocyte basolateral plasma membranes (Balbis et al., 2004; Mazzone et al., 2006), while detergent-based methods suggest the existence of cholesterol-rich microdomains containing ABCG5 (important for biliary cholesterol efflux), ABCC2, ABCB4, and the bile salt export pump in hepatocyte canalicular plasma membranes (Ismair et al., 2009).

10.7 How May Intracellular Cholesterol-Binding Proteins Regulate Cholesterol Dynamics?

In non-hepatocytes, cholesterol movement from plasma membrane caveolae microdomains through cytosol to intracellular sites is thought to be mediated by slower vesicular transport and/or more rapidly via soluble cholesterol binding proteins (Schroeder et al., 2005; Everson and Smart, 2005). By contrast, in hepatocytes transhepatocyte cholesterol movement is a very rapid ($t_{1/2}$ of 1–3 min), protein-mediated process (Fig. 10.1) (Schroeder et al., 2005). This rapid trans-hepatocyte transport of cholesterol is much too fast to be accounted for by spontaneous diffusion, vesicular transfer, or lateral diffusion in membranes (Robins and Fasulo, 1999; Ji et al., 1999; Wustner et al., 2002). Instead, several families of intracellular proteins are thought to facilitate cholesterol intracellular movement in hepatocytes and other cells/tissues where they are highly expressed.

10.7.1 Sterol Carrier Protein-2 (SCP-2)

The SCP-2/SCP-x gene structure exhibits alternate transcription sites that encode for two markedly different proteins (Gallegos et al., 2001b): (i) The shorter SCP-2/SCP-x gene product, 15 kDa pro-SCP-2 (SCP-2 with a 20aa N-terminal presequence), undergoes complete post-translational cleavage to 13 kDa SCP-2. While the presence of the 20aa presequence facilitates SCP-2 targeting to peroxisomes, post-translational cleavage outside the peroxisome results in a more extraperoxisomal SCP-2 distribution (Gallegos et al., 2001b). As a result, nearly half of total SCP-2 is extraperoxisomal (Schroeder et al., 2000; Gallegos et al., 2001b). SCP-2 is a ubiquitous protein found in nearly all mammalian cells/tissues examined (Gallegos et al., 2001b). Tissues in which SCP-2 is most highly expressed are those most active in cholesterol uptake/transport/metabolism such as liver, steroidogenic tissues (ovary, testis, adrenal), and intestine. SCP-2 has high affinity for cholesterol (K_d 4–11 nM) (Schroeder et al., 1998, 2000; Gallegos et al., 2001b; Martin et al., 2008b). SCP-2 also binds to cholesterol-rich microdomains in anionic phospholipid-rich/cholesterol-rich model membranes and to caveolin-1 in plasma membrane caveolae (Huang et al., 1999a, b; Parr et al., 2007; Zhou et al., 2004). These features make SCP-2 an excellent candidate for facilitating cholesterol transfer between membranes, a possibility born out by SCP-2's ability to alter cholesterol-rich microdomain structure and enhance cholesterol dynamics (Atshaves et al., 2003, 2007c; Gallegos et al., 2001a). Overexpression of SCP-2 in cultured cells enhances sterol uptake/efflux (Moncecchi et al., 1996; Murphy and Schroeder, 1997; Atshaves et al., 2000). SCP-2 enhances cholesterol transfer from plasma membranes to intracellular sites in intact non-hepatocytes and enhances cytosolic diffusion of other bound ligands (fatty acids) in hepatocytes (Murphy and Schroeder, 1997; Murphy, 1998; McArthur et al., 1999). (ii) The other SCP-2/SCP-x gene product is the 58 kDa SCP-x (a branched-chain ketoacyl thiolase important in bile acid synthesis). SCP-x is an exclusively peroxisomal protein that undergoes partial post-translational cleavage to 13 kDa SCP-2 and a 46 kDa thiolase therein (Gallegos et al., 2001b). SCP-x catalyzes cholesterol side-chain oxidation required for bile acid synthesis (Gallegos et al., 2001b).

10.7.2 Liver Fatty Acid-Binding Protein (L-FABP)

L-FABP is a soluble, 14 kDa protein highly expressed in liver, intestine, and kidney (McArthur et al., 1999). Levels in liver cytosol are very high (200-400 µM) and can be upregulated as much as 4-5 fold in response to diet or reduction in SCP-2 (McArthur et al., 1999; Fuchs et al., 2001; Seedorf et al., 1998; Atshaves et al., 2007c). L-FABP also binds cholesterol and other sterols (NBD-cholesterol, DHE, photoactivatable FCBP), but with lower affinity than SCP-2 (Fischer et al., 1985; Sams et al., 1991; Schroeder et al., 1998, 2000; Stolowich et al., 1999; Martin et al., 2008b; Avdulov et al., 1999; Schroeder et al., 1998; Martin et al., 2009a). However,

L-FABP is 40-fold more prevalent than SCP-2 in hepatocytes (McArthur et al., 1999; Gallegos et al., 2001b; Stolowich et al., 2002; Schroeder et al., 1996) and results obtained with purified plasma membranes (Nemecz and Schroeder, 1991; Schroeder et al., 1996; Woodford et al., 1993, 1995a) and transfected cells (Jefferson et al., 1991; Incerpi et al., 1991, 1992) suggest this protein may play a role not only in intracellular trafficking of fatty acids (McArthur et al., 1999; Weisiger, 2005; Schroeder et al., 2008), but also of cholesterol.

10.7.3 Steroidogenic Acute Regulatory Related (START) Proteins

Many of the 15-member START protein family participate in intracellular lipid trafficking and signaling. START proteins are primarily membrane-bound (Kanno et al., 2007; Murcia et al., 2006; Soccio and Breslow, 2003; Miller, 2007; Holthuis and Levine, 2005). While StAR (released from ER, primarily in steroidogenic tissues), MLN64 (endosomal membrane), and StarD5 (cytosolic) bind cholesterol, all are reported mainly to facilitate cholesterol transfer to mitochondria for oxidation (Tsujishita and Hurley, 2000; Murcia et al., 2006; Petrescu et al., 2001; Rodriguez-Agundo et al., 2005). StARD4 (not yet shown to bind cholesterol; predicted to be cytosolic) also facilitates cholesterol delivery to mitochondria (Soccio and Breslow, 2003; Miller, 2007). PCTP binds and transfers phosphatidylcholine (not cholesterol) in cytosol and indirectly regulates phosphatidylcholine efflux into bile (Kanno et al., 2007). Roles for START proteins in direct cytoplasmic transfer of cholesterol for biliary efflux have not yet been established.

10.7.4 Oxysterol Related Proteins (ORP)

The 12-member ORP family participates in intracellular lipid trafficking (*see also* Chapter 6). Cytosolic ORPs bind oxysterols, then target to intracellular membranes, and influence transcriptional and post-transcriptional regulation of cholesterol metabolism via LXR and SREBP (Holthuis and Levine, 2005; Laitinen et al., 2002; Perry and Ridgway, 2006; Yan et al., 2007; Anniss et al., 2002). OSBP, ORP1L, ORP3, ORP5, and ORP8 also bind cholesterol (Laitinen et al., 2002; Wang et al., 2002, 2005; Lagace et al., 1997; Holthuis and Levine, 2005; Suchanek et al., 2007). OSBP, ORP2, or ORP4 over-expression enhances cholesterol efflux to apoA1 and decreases cholesterol esterification (Laitinen et al., 2002; Wang et al., 2002; Lagace et al., 1997; Holthuis and Levine, 2005; Suchanek et al., 2007). Whether ORPs promote rapid transhepatic cholesterol transfer is unknown.

10.7.5 Niemann Pick C (NPC) Proteins

Niemann Pick C (NPC1, NPC2, NPCL1) proteins facilitate intracellular lipid trafficking (e.g. LDL-cholesterol), and NPC disease elicits hepatocyte cholesterol accumulation and death (Ohgami et al., 2004; Chang et al., 2005; Beltroy et al., 2005; Gallegos et al., 2001b) (*see also* Chapter 11). Ablation of NPC1 (late endosomal membrane protein; binds oxysterol>cholesterol) (Subramanian and Balch, 2008) decreases SCP-2/SCP-x and increases lysosomal free cholesterol (Roff et al., 1992). NPC2 (soluble in lysosomes, late endosomes) binds cholesterol but not oxysterol (Subramanian and Balch, 2008). NPC1L (plasma membrane protein primarily in intestinal microvillus and in bile canaliculus of polarized hepatocytes) facilitates intestinal cholesterol uptake, transports cholesterol from bile back into hepatocytes, and is inhibited by ezetimibe (Zetia) (Petersen et al., 2008; Hui et al., 2008). Their membrane (lysosomal, plasma membrane) and intralysosomal localizations suggest that the NPC proteins are not directly involved in mediating rapid cholesterol transfer across hepatocytes or other cells.

10.7.6 Other Intracellular Proteins

Acyl CoA cholesterol acyltransferase (ACAT) and neutral cholesterol ester hydrolase (CEH) are enzymes that contribute to regulating the amount of free cholesterol available for biliary efflux (Cohen, 1999; Zanlungo et al., 2004). ACAT2 is the major form of ACAT in murine liver, while low levels of ACAT1 are also present (Chang et al., 1997, 2001). However, there is no evidence that these proteins are involved in directly binding and transporting cholesterol rapidly through the cytoplasm of hepatocytes or other cells.

In summary, cholesterol represents a driving force in formation of cholesterol-rich, liquid-ordered lipid microdomains that provide a platform to localize receptors, transport/ translocase proteins, and downstream signaling molecules that regulate cellular uptake and efflux of cholesterol as well as uptake of fatty acids and glucose in non-hepatocytes. Transhepatocyte cholesterol movement is very rapid, and not readily explained by spontaneous diffusion, vesicular transfer, or lateral diffusion in membranes. While many intracellular protein families (NPC, START, ORP, ACAT) facilitate intracellular cholesterol trafficking/metabolism, most transfer cholesterol to mitochondria, are membrane-bound, do not bind cholesterol, facilitate vesicular transfer, or regulate free cholesterol level. Based upon these data and their ability to enhance cytosolic transport of other lipids, SCP-2 and L-FABP are hypothesized to be likely candidate proteins for mediating the rapid transhepatocyte transfer of HDL free cholesterol from cholesterol-rich microdomains in the basolateral membrane to cholesterol-rich microdomains in the canalicular membrane for biliary efflux.

10.8 Sterol Carrier Protein-2 Facilitates Intermembrane Cholesterol Transfer In Vitro

The ability of SCP-2 to enhance transfer of cholesterol between model membranes or between plasma membranes in vitro has been previously established and reviewed in detail (Schroeder et al., 1989a, 1998, 1991a, 1996, 2001b, 2005; Gallegos et al., 2001b, 2008; McIntosh et al., 2008a;). Key conclusions of these studies were that: (i) SCP-2 enhances intermembrane cholesterol transfer by increasing the size of the exchangeable sterol fraction in membranes and by increasing the size of the soluble pool of cholesterol; (ii) SCP-2 directly interacts with the membrane anionic phospholipids (rich in the cytofacial facing side of membranes) and with caveolin-1 in plasma membrane caveolae (Huang et al., 1999a, b, 2002). Direct interaction of SCP-2 with membranes is required to mediate intermembrane cholesterol transfer (Woodford et al., 1995b); (iii) SCP-2 preferentially enhanced trafficking of cholesterol from lysosomes to plasma membranes and from plasma membranes to intracellular membranes in vitro (Schoer et al., 2000; Schroeder et al., 2001a; Frolov et al., 1996a, b).

10.9 Do Cholesterol-Rich and -Poor Microdomains Exist in the Plasma Membranes of Living Cells: Real-Time Multiphoton Imaging of a Naturally Fluorescent Sterol (Dehydroergosterol, DHE)?

While increasing evidence suggests the existence of cholesterol-rich and -poor microdomains in model membranes and biological membranes, especially plasma membranes, most of this evidence was indirect: (i) chemical analysis of cholesterol-rich and -poor microdomains isolated by biochemical fractionation without the use of detergents or high pH carbonate; (ii) imaging of non-sterol markers (Smart et al., 1995; Schroeder et al., 1982, 2005; Atshaves et al., 2003, 2007a; Storey et al., 2007a, b; Gallegos et al., 2006). Thus, until recently a key remaining question is whether cholesterol-rich and -poor microdomains exist in the plasma membranes of living cells. While real-time imaging cholesterol itself in live cells can be accomplished by some physical methods (e.g. NMR of cholesterol labelled with deuterium or carbon 13), the resolution of such approaches is insufficient to resolve cholesterol-rich and -poor microdomains. Consequently, most investigators in the field have turned to more sensitive fluorescence techniques utilizing fluorescent sterol probes. Proper application of fluorescent sterols requires: (i) selective non-perturbing "tags" for cholesterol; (ii) non-invasive, non-perturbing real-time visualization; and (iii) a mathematical basis for assigning a totally random versus clustered pattern. These issues were addressed by developing use of fluorescent sterols (Schroeder et al., 1998; McIntosh et al., 2003, 2007, 2008a, b, c; Hao et al., 2001; Mukherjee et al., 1998; Wustner et al., 2002, 2004, 2005). The first fluorescent sterol used for real-time imaging of cholesterol distribution in the plasma membranes of living cells was the naturally-occurring fluorescent sterol, dehydroergosterol (DHE). Although the use of DHE for in vitro studies examining cholesterol-protein interaction as well as distribution and dynamics in natural and model membranes is well established (Schroeder and Nemecz, 1990; Schroeder et al., 1991a, 1996, 1998), UV excitation of DHE results in significant photobleaching that precludes confocal imaging (Schroeder et al., 1998). Therefore, multiphoton laser scanning microscopy (MPLSM) was developed to examine sterol distribution without significant photobleaching (McIntosh et al., 2003, 2007, 2008a, b; Zhang et al., 2005; Schroeder et al., 2005). DHE and cholesterol share many structural properties, and DHE replaces up to 90% of cholesterol in cultured cells without altering composition of other lipids or activity of cholesterol-sensitive enzymes and receptors (McIntosh et al., 2008a, b, c). DHE codistributes with cholesterol in all cellular membranes and in plasma membrane cholesterol-rich and -poor microdomains of non-hepatocytes (L-cells, MDCK cells) (Hale and Schroeder, 1982; Schoer et al., 2000; Atshaves et al., 2003; Gallegos et al., 2006; Storey et al., 2007a). DHE is esterified with a similar fatty acid pattern (Schroeder et al., 1996; Frolov et al., 2000; Holtta-Vuori et al., 2008), and effluxes from L-cells to HDL with similar kinetics to cholesterol (Atshaves et al., 2000). Quantitative algorithms to analyze MPLSM images of DHE pattern, proportion, and size distribution (McIntosh et al., 2003, 2007, 2008a, b; Zhang et al., 2005; Gallegos et al., 2008) showed that DHE in plasma membranes of living non-hepatocytes (McIntosh et al., 2003, 2007, 2008a, b; Zhang et al., 2005; Schroeder et al., 2005) and hepatocytes (not shown) is clustered, rather than spatially random (i.e. homogenous Poisson) or regular. These sterol-rich domains comprised nearly 30% of plasma membrane imaging sections (Zhang et al., 2005), consistent with the proportion of cholesterol-rich microdomains resolved from purified plasma membranes by ConA-affinity chromatography—and were 200–565 nm diameter, which is in the lower quartile of sizes reported with other cholesterol-rich microdomain/caveolae markers (Anderson and Jacobson, 2002; Schroeder et al., 2005). Since caveolae are 50-100 nm in size and are known to cluster in many cell types, including fibroblasts, the DHE-rich regions may represent clustered caveolae (non-hepatocytes) or cholesterol-rich microdomains (hepatocytes) (Roper et al., 2000; Schroeder et al., 2005; Zhang et al., 2005; McIntosh et al., 2007, 2008a, b; Gallegos et al., 2008). Thus, MPLSM imaging of DHE provided the first evidence supporting the existence of plasma membrane microdomains in living cells.

10.10 Can the Existence of Cholesterol-Rich and -Poor Microdomains in Plasma Membranes of Living Cells Be Confirmed by Other Real-Time Approaches Using Synthetic Fluorescent Sterols Suitable for Confocal Imaging?

Several synthetic fluorescence (Dansyl-cholestanol; BODIPY-cholesterol, BCθ) probe approaches have been examined for selective probing of cholesterol-rich and poor microdomains (Huang et al., 2009; McIntosh et al., 2008b; Ohno-Iwashita et al., 2004; Shimada et al., 2002; Sugii et al., 2003). These probes are excited at longer wavelengths suitable for laser scanning confocal microscopy (LSCM) and well suited for fluorescence resonance energy transfer (FRET) with many commercially available fluorophores. Dansyl-cholestanol (DChol), synthesized by our lab, was taken up and esterified similarly as cholesterol (non-hepatocytes) (Huang et al., 2009): (i) L-cells incubated with DChol/ or cholesterol/methylcyclodextrin complex for 30 min, washed and incubated with serum-containing medium for 24 h, esterified DChol 19.4±0.7% and 14.5±1.0% respectively; (ii) L-cells incubated overnight with DChol or ^{3}H-cholesterol added to serum-containing medium esterified DChol and ^{3}H-cholesterol similarly, 3.4±0.8% and 3.9±0.2%, respectively; and (iii) DChol and cholesterol ester fatty acid distributions were the same (GC/MS, not shown). When L-cells were incubated for 30 min or less, <0.2% of DChol or cholesterol was esterified, and both DChol and cholesterol were distributed nearly 4-fold more into cholesterol-rich versus -poor microdomains resolved by affinity chromatography of purified plasma membranes (Huang et al., 2009). To further establish that the DChol-rich and -poor clusters obtained by imaging correlated with the biochemically-isolated DChol-rich and -poor clusters, a series of colocalization and fluorescence resonance energy transfer (FRET) studies were performed with additional markers of lipid-rich, cholesterol-rich microdomains (Huang et al., 2009).

10.10.1 Colocalization of DChol with the Lipid-Rich Microdomain Marker GM_1

GM_1, a known marker of cholesterol-rich domains (Atshaves et al., 2003; Schroeder et al., 2005; Gallegos et al., 2006), was labelled by binding to Alexa Fluor 594 CT-B,

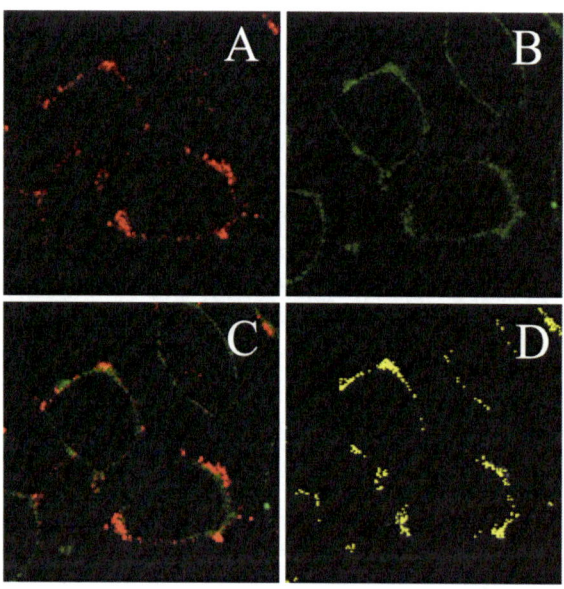

Fig. 10.2 Co-localization of DChol and GM_1 in L-cells. GM1, a cholesterol-rich domain maker, was labeled by incubating the cells with Alexa Fluor 594 CT-B. The cells were then incubated with DChol-MβCD at 10 μg/ml PBS for 10 min. Emission of Alexa Fluor 594 CT-B (*red, Panel* A) and DChol (*green, Panel* B) were simultaneously imaged through separate photomultipliers. *Panel* C shows the superposition of A and B. *Panel* D shows only the colocalized pixels in yellow

cholera toxin subunit B as described earlier (Huang et al., 2009). L-ells were incubated with Alexa Fluor 594 CT-B, then with DChol-MβCD, followed by real time imaging by LSCM. The emission of DChol and Alexa Fluor 594 CT-B were imaged simultaneously though two separate photomultipliers. Alexa Fluor 594 CT-B was distributed most intensely at the plasma membrane, but not uniformly (Fig. 10.2A). After 10 min incubation of L-cells with DChol-MβCD, DChol was taken up and distributed mostly at the plasma membrane, in a non-uniform pattern (Fig. 10.2B). Superposition of the simultaneously acquired images of Alexa Fluor 594 CT-B (red) and DChol (green), yielded a image with many pixels that were intermediate in color, suggesting colocalization (Fig. 10.2C). When only the colocalized yellow pixels were shown (Fig. 10.2D), it was apparent that DChol significantly colocalized with Alexa Fluor 594 CT-B bound to GM_1 at the plasma membrane and this distribution was not uniform, but clustered. Confocal imaging colocalization and statistical analysis of DChol codistribution with Alexa Fluor 594 CT-B in plasma membranes of living L-cells showed that DChol strongly codistributed with cholesterol-rich microdomain probe Alexa Fluor 594 CT-B (Huang et al., 2009).

10.10.2 DChol Colocalization and Fluorescence Resonance Energy Transfer (FRET) with DiD, a Liquid Ordered Phase Lipid-Rich Microdomain Marker

DiD is a fluorescent probe preferentially localized to plasma membrane liquid ordered phase (Spink et al., 1990; Schram and Thompson, 1997), a characteristic of lipid-rich (Atshaves et al., 2007c; Gallegos et al., 2004, 2006; Storey et al., 2007a), cholesterol-rich microdomains (also known as lipid rafts), rather than to plasma membrane liquid disordered phase, a characteristic of lipid-poor, cholesterol-poor microdomains (also known as non-rafts) (Atshaves et al., 2007c; Gallegos et al., 2004, 2006; Storey et al., 2007a). Imaging and FRET experiments were carried out to determine if DChol colocalized with DiD, and whether they localized in close proximity (Huang et al., 2009). L-cells were labelled with DiD and DChol, and the two probes were imaged simultaneously. DiD was distributed in the plasma membrane in a non-uniform, patchy pattern and also in the intracellular sites (Fig. 10.3A). DChol (Fig. 10.3B) mainly distributed in the plasma membrane, also in a non-uniform, patchy pattern. Colocalization (Fig. 10.3C) and displaying of only colocalized pixels (Fig. 10.3D) demonstrated that significant amount of DChol colocalized with DiD in the liquid ordered cholesterol-rich domains. Confocal imaging colocalization and statistical analysis of DChol codistribution with DiD in plasma membranes of living L-cells showed that DChol strongly codistributed with cholesterol-rich, liquid-ordered microdomain probe DiD (Huang et al., 2009).

The above co-localization results obtained by imaging had a resolution near 2000 Å and many cholesterol-rich domains are thought to be much smaller (Pike, 2004; Schroeder et al., 2005). Fluorescence resonance energy transfer (FRET) experiments were performed to improve the resolution (Huang et al., 2009),

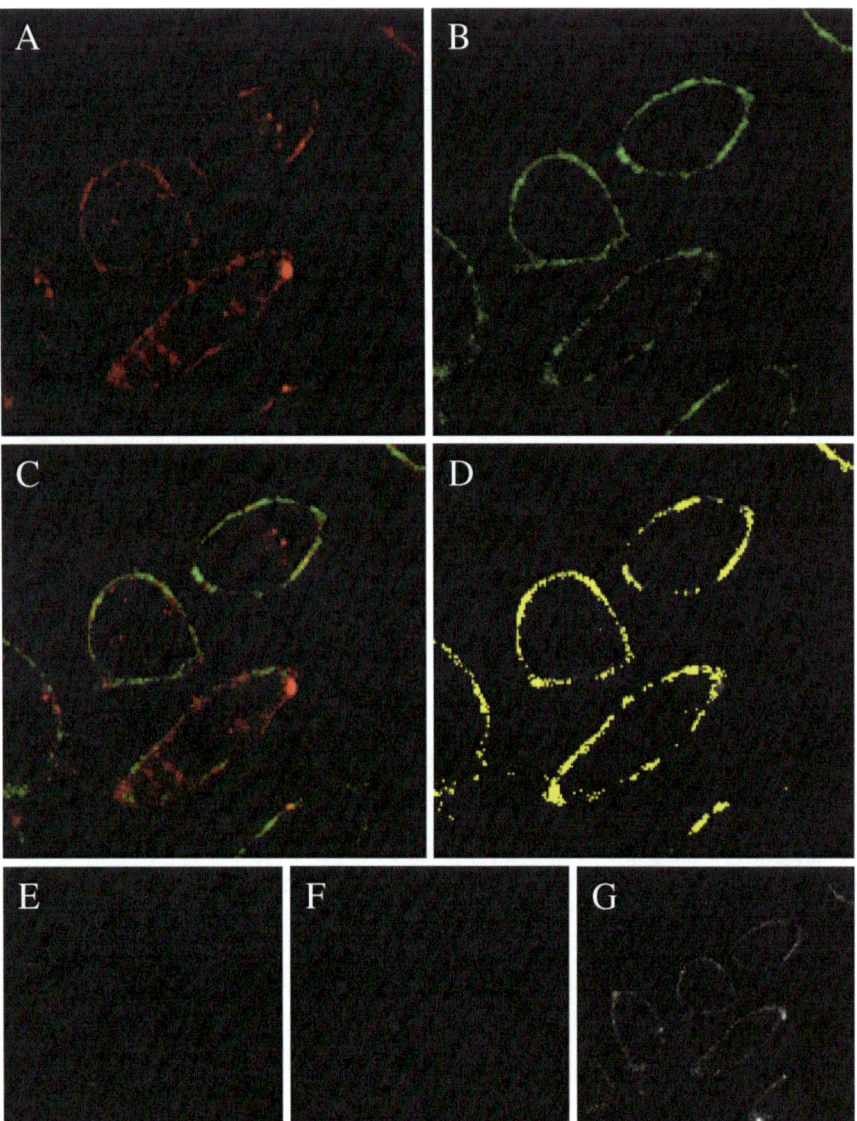

Fig. 10.3 Images of colocalization and FRET between DChol and DiD, a gel phase marker, in living cells. L-cells were incubated with DiD (25 μM) in culture medium for 10 min at 37°C, and then incubated with DChol-MβCD in PBS (DChol concentration 10 μg/ml) at room temperature for 15 min. Colocalization of DChol and DiD are shown as following, *Panel* A, DiD fluorescence; *Panel* B, DChol fluorescence; *Panel* C, DiD colocalization with DChol; *Panel* D, yellow colocalized pixels taken from *Panel* C. *Panels* (E–G) show the FRET between DChol and DiD. The images were taken by excitation of donor DChol with 408 nm laser and monitoring emission of acceptor DiD with 680/32 filter. *Panel* E, image of cells labeled with donor DChol only, no significant fluorescence signal was detected. *Panel* F, image of cells labeled with acceptor DiD only, no significant fluorescence signal was observed. *Panel* G, cells were labeled with both DChol and DiD; enhanced DiD emission was observed, indicating DiD and DChol were in close proximity to permit energy transfer from DChol to DiD

because FRET occurs only when donor and acceptor are located in close proximity (~10–100 Å apart) (Lakowicz, 2006). The significant overlap between DChol emission spectrum and DiD excitation spectrum (not shown) made them a potential donor/acceptor pair for FRET. Therefore, FRET experiments were carried out by exciting the donor DChol with 408 nm laser and observing the emission of acceptor DiD with 680/32 emission filter. When only donor DChol was present, no fluorescence signal was detected (Fig. 10.3E) because DChol did not emit at 680/32 region. When only DiD was present, no fluorescence emission signal was observed (Fig. 10.3F) because DiD did not absorb at 408 nm. When both DChol and DiD were present, the fluorescence emission of acceptor DiD was clearly observed (Fig. 10.3G), indicating DiD and DChol were in close proximity to permit energy transfer from DChol to DiD.

10.10.3 FRET Between DChol and DHE in Living Cells

DHE is a naturally occurring fluorescent cholesterol analog. To determine if DChol and DHE were codistributed in close proximity at the plasma membrane of living cells, FRET experiments between DChol (acceptor) and DHE (donor) were carried out using multiphoton laser scanning microscopy (MPLSM) with three photon excitation at 900 nm, and two separate emission filters and photomultipliers,

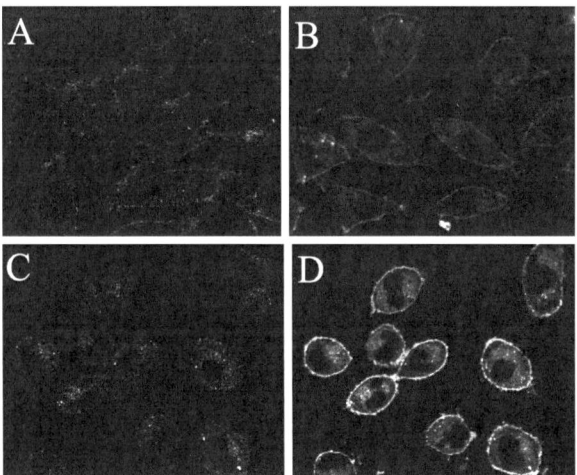

Fig. 10.4 Fluorescence images of FRET between DChol and DHE. FRET between DChol and DHE (*panels* A–D) was studies by multi-photon imaging with 900 nm laser three photon excitation, and D375/50 (350-400 nm) emission filter for DHE (*panels* A, C), BGG22 (410–490 nm) emission filter for DChol (*panels* B, D). *Panel* A, DHE emission when only donor DHE was present; *Panel* B, DChol emission when only accepter DChol was present; *Panel* C, DHE emission when both DHE and DChol were present; *Panel* D, DChol emission when both DHE and DChol were present. When both DHE and DChol were present, decreased DHE emission and increased DChol emission at plasma membrane indicated the DHE and DChol were closed enough for energy transfer from DHE to DChol

D375/50 (350–400 nm) filter for DHE and BGG22 (410–490 nm) filter for DChol (Huang et al., 2009). When cells were incubated with DHE only, DHE emission was detected mainly at the plasma membrane in a non-random pattern, and within the cells with less intensity (Fig. 10.4A). DChol emission exhibited a similar pattern when cells were incubated with DChol only (Fig. 10.4B). When L-cells were incubated with both DHE (FRET donor) and acceptor DChol (FRET acceptor), the emission of donor DHE became weaker (Fig. 10.4C), and the emission of accepter DChol became stronger (Fig. 10.4D), clearly demonstrated energy transfer from donor to accepter. Quantitative analysis of the FRET in the whole cell and at the plasma membrane revealed the DChol emission was significantly increased severalfold (Huang et al., 2009). The results indicated that in living L-cells DChol and DHE were codistributed in close proximity, and appeared clustered into bright and not so bright regions.

10.10.4 Colocalization and Fluorescence Resonance Energy Transfer (FRET) Between DChol and BCθ

BCθ, the non-lytic fragment of a bacterial cytolysin, binds cholesterol in cholesterol-rich domains with high affinity (Ohno-Iwashita et al., 2004) (*see also* Chapter 22).

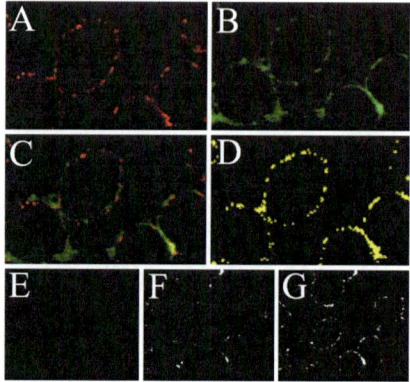

Fig. 10.5 Fluorescence images of Colocalization and FRET between DChol and BCθ. *Panels A–D* show the colocalization between DChol and BCθ. DChol was imaged with 408 nm laser excitation and HQ530/40 emission filter. Alexa Fluor 660 BCθ was imaged with 647 nm laser excitation and 680/32 emission filter. A, Alexa Fluor 660 BCθ emission; B, DChol emission; C, colocalization of DChol and Alexa Fluor 660 BCθ; D, colocalized pixels are shown as yellow. *Panels E–G* show FRET between DChol (donor) and Alexa Fluor 660 BCθ (acceptor). The images were taken with 408 nm laser donor excitation and 680/32 acceptor emission filter. E, image when only DChol was present; F, image when only Alexa Fluor 660 BCθ was present; G, image when both DChol and Alexa Fluor 660 BCθ were present. When both donor and acceptor were present, the enhanced acceptor emission when donor was excited showed the donor and acceptor were located in proximity to permit energy transfer

LSCM imaging showed Alexa Fluor 660 BCθ was distributed non-randomly at the cell surface plasma membrane (Fig. 10.5A) (Huang et al., 2009). DChol was highly colocalized with BCθ as shown in superposed image (Fig. 10.5C) of DChol (Fig. 10.5B) and BCθ (Fig. 10.5A). Display of only colocalized pixels showed that DChol/BCθ colocalized pixels were distributed in a clustered pattern at the plasma membrane (Fig. 10.5D). Confocal imaging colocalization and statistical analysis of DChol codistribution with BCθ in plasma membranes of living L-cells showed that DChol strongly codistributed with cholesterol-rich microdomain probe BCθ (Huang et al., 2009). To increase the resolution, FRET experiments were performed by excitation of donor DChol (408 nm laser) and detection of acceptor Alexa Fluor 660 BCθ emission (emission filter 680/32, 664–696 nm). When DChol and Alexa Fluor 660 BCθ were both present, the acceptor emission was significantly higher (Fig. 10.5G) as compared with when only the acceptor was present (Fig. 10.5F). The increase in fluorescence was not due to emission from donor DChol was confirmed by the absence of emission signal when only DChol was present (Fig. 10.5E). Therefore FRET experiments confirmed that DChol was distributed in close proximity with BCθ in the plasma membrane of L-cells.

10.10.5 Weak Colocalization and Absence of Fluorescence Resonance Energy Transfer (FRET) Between DChol and N-Rh-DOPE

It is important to note that DChol distribution, like cholesterol distribution, into cholesterol-rich versus -poor microdomains is not absolute, but rather subjective such that these microdomains are relatively rich or poor in sterol (Schroeder et al., 2005; Huang et al., 2009). Therefore, DChol colocalization and FRET was performed with N-Rh-DOPE [a marker for fluid, liquid-disordered, lipid-poor, cholesterol-poor phase in membranes (Samsonov et al., 2001; de Almeida et al., 2005)] in L-cell fibroblasts. Confocal imaging colocalization and statistical analysis of DChol codistribution with N-Rh-DOPE in the plasma membrane of living L-cells showed that DChol only weakly, partially distributed with cholesterol-poor microdomain probe N-Rh-DOPE (Huang et al., 2009). Fluorescence resonance energy transfer (FRET) experiments were carried out to investigate if N-Rh-DOPE and DChol were in close proximity to permit FRET (Huang et al., 2009). The data showed that DChol was not in sufficiently close proximity to N-Rh-DOPE in the plasma membrane to permit FRET (Huang et al., 2009).

In summary, colocalization, statistical analyses, and FRET experiments demonstrated that DChol was in close proximity with cholesterol-rich microdomain markers (DHE, DiD, BCθ), but not the cholesterol-poor microdomain marker (N-Rh-DOPE) (Huang et al., 2009).

10.11 Real-Time Imaging of Sterol Carrier Protein-2 Mediated Cholesterol Dynamics Through Cholesterol-Rich and -Poor Microdomains in Plasma Membranes of Cultured Cells

Sterol carrier protein-2 (SCP-2) is known to directly interact with caveolin-1 in L-cell plasma membrane caveolae (a subset of cholesterol-rich microdomains) (Parr et al., 2007; Zhou et al., 2004) and to facilitate cholesterol transfer from purified cholesterol-rich (but not cholesterol-poor) microdomains (Atshaves et al., 2003, 2007c; Gallegos et al., 2001a, 2004, 2006, 2008; Schroeder et al., 2005). Since SCP-2 binds DChol with high affinity (K_d =33.8 ± 2.7 nM), SCP-2 expressing and control (mock transfected) L-cell fibroblasts were used to examine the effects of SCP-2 on DChol uptake from DChol-MβCD and efflux to MβCD in real time by confocal microscopy. DChol fluorescence intensity in whole cells (WC), plasma membranes (PM), intracellular regions, cholesterol-rich microdomains (pixels colocalized with cholesterol-rich microdomain marker Alexa Fluor 594 CT-B at PM), and cholesterol-poor microdomain (pixels not colocalized with the cholesterol-rich microdomain marker Alexa Fluor 594 CT-B cholesterol-rich domain maker at PM) were determined.

For DChol uptake, cells were incubated with a cholesterol-rich microdomain maker Alexa Fluor 594 CT-B and DChol-MβCD complex. For control L-cells: (i) at early time points, such as 2 min, much more DChol were taken up into cholesterol-rich than into cholesterol-poor microdomains (Fig. 10.6D,E); (ii) at 15 min, much more DChol were taken up into whole cell, PM, intracellular sites, cholesterol-rich and -poor microdomains than at 2 min; (iii) at 15 min, DChol was more concentrated in PM than in intracellular sites. While the above observations were also true for SCP-2 expressing cells, expression of SCP-2 resulted in enhanced DChol uptake into whole cells compared to control L-cells (Fig. 10.6A, 2 min), enhanced DChol trafficking through cholesterol-rich microdomains to intracellular sites, and increased DChol in intracellular sites concomitant with decreased DChol concentration in cholesterol-rich microdomains at 15 min (Fig. 10.6C,D, 15 min). Even though SCP-2 expression significantly decreased DChol appearing in cholesterol-rich microdomains (Fig. 10.6D), it did not significantly alter the appearance of DChol in cholesterol-poor microdomains (Fig. 10.6E). These results were consistent with SCP-2 facilitating DChol trafficking through cholesterol-rich domains, but not cholesterol-poor domains, for targeting to intracellular sites.

For DChol efflux, after pre-incubating the cells with DChol-MβCD complex for 15 min, efflux was initiated by addition of MβCD (3 mM). Alexa Fluor 594 CT-B and DChol emission were monitored in real time by LSCM. Percentage of DChol fluorescence remaining in whole cells (WC), plasma membranes (PM), intracellular regions, cholesterol-rich and -poor microdomains was determined as described. In control L-cells, DChol efflux was primarily from the plasma membrane (Fig. 10.7B) and not from intracellular sites (Fig. 10.7C). While SCP-2 overexpression did not

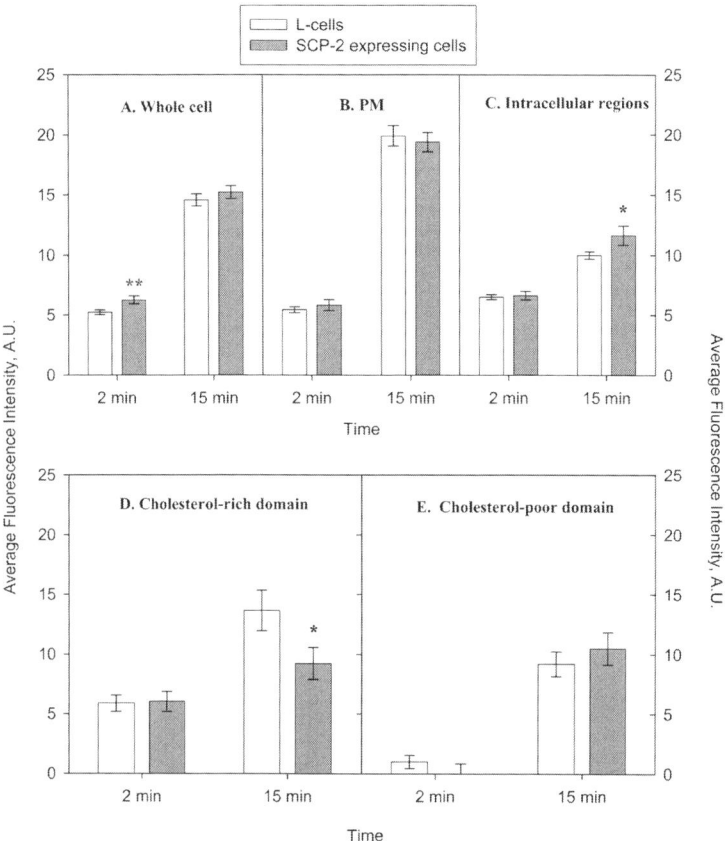

Fig. 10.6 Effects of SCP-2 expression on DChol uptake. SCP-2 expressing and control L-cells were labeled first with a cholesterol-rich domain marker – Alexa Fluor CT-B, and then with DChol-MβCD (DChol concentration 10 μg/ml). Fluorescence images of DChol and Alexa Fluor CT-B emission were acquired in two different channels at 2 and 15 min after addition of DChol-MβCD. DChol was excited with 408 nm laser, and imaged with HQ530/40 emission filter. Alexa Fluor CT-B was excited with 568 nm laser, and imaged with HQ598/40 emission filter. Average DChol fluorescence intensity ($n=15$–25 cells) was calculated in the following regions of interest, A, whole cell; B, PM; C, intracellular regions; D, cholesterol-rich (PM that colocalized with cholesterol-rich domain marker); E, cholesterol-poor domain (PM that were not colocalized with cholesterol-rich domain marker)

alter this overall qualitative pattern, SCP-2 expression inhibited DChol efflux from whole cells (Fig. 10.7A, 10 min), primarily due to decreased efflux from cholesterol-rich microdomains (Fig. 10.6D). The average % DChol fluorescence remaining in cholesterol-poor microdomains was lower for SCP-2 expressing cells than for control L-cells (Fig. 10.7E), but the difference was not statistically significant.

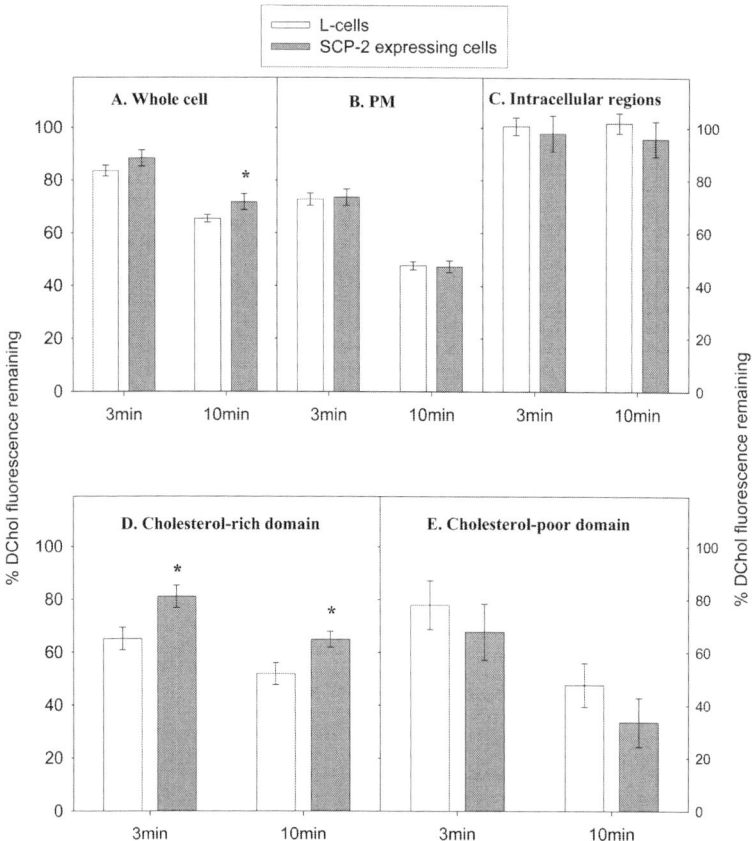

Fig. 10.7 Effects of SCP-2 expression on DChol efflux. SCP-2 expressing and control L-cells were labeled first with a cholesterol-rich domain marker – Alexa Fluor CT-B, and then with DChol-MβCD (DChol concentration 10 μg/ml) for 15 min. MβCD (3 mM) was added to the buffer to initiate DChol efflux. Fluorescence images of DChol and Alexa Fluor CT-B emission were acquired in two different channels at 3 and 10 min after addition of MβCD. DChol was excited with 408 nm laser, and imaged with HQ530/40 emission filter. Alexa Fluor CT-B was excited with 568 nm laser, and imaged with HQ598/40 emission filter. Percentage of DChol fluorescence remaining ($n=15-25$ cells) was calculated in the following regions of interest, A, whole cell; B, PM; C, intracellular regions; D, cholesterol-rich domain (PM that colocalized with cholesterol-rich domain marker); E, cholesterol-poor domain (PM that were not colocalized with cholesterol-rich domain marker)

In summary, taken together these data were consistent with SCP-2 enhancing DChol uptake from an extracellular acceptor by facilitating DChol trafficking through (but not retention in) cholesterol-rich plasma membrane microdomains for increased distribution to intracellular sites. Conversely, SCP-2 increased retention of DChol, by decreasing DChol efflux from plasma membrane cholesterol-rich microdomains to extracellular acceptors.

10.12 Physiological Studies of Effects of SCP-2 Overexpression and Gene Ablation on Cholesterol Dynamics

In order to examine the physiological impact of cholesterol binding proteins such as SCP-2, the expression of SCP-2 in mice has been increased (SCP-2 overexpression) or decreased (antisense cDNA treatment, SCP-2/SCP-x gene ablation). Although studies with L-FABP overexpressing mice have not yet been reported, the L-FABP gene has been independently ablated by several groups. The following sections summarize the effects of these manipulations on hepatic cholesterol phenotype.

10.12.1 Effects of SCP-2 Over-Expression and Antisense Treatment on Hepatic and Biliary Cholesterol

Studies with genetically engineered mice indicated that SCP-2 over-expression increases hepatic cholesterol accumulation and biliary excretion of cholesterol and bile acids (Zanlungo et al., 2000; Amigo et al., 2003; Atshaves et al., 2009). Hepatic cholesterol accumulation in SCP-2 over-expressing mice correlated with increased expression of select proteins involved in cholesterol uptake [LDL-receptor, SRB1 (HDL-receptor), SRB1 regulatory protein PDZK1, 3α-hydroxysteroid dehydrogenase]. Hepatic lipid accumulation was further exacerbated by a cholesterol-rich diet in both female (cholesterol/cholesteryl esters) and male (cholesterol/cholesteryl esters and triacylglycerol) SCP-2 over-expressing mice (Atshaves et al., 2009). In contrast, SCP-2 antisense cDNA treatment in rats decreases biliary cholesterol excretion (Puglielli et al., 1996). These findings were consistent with SCP-2 being an important contributor to hepatic cholesterol content and biliary cholesterol secretion.

10.12.2 Effects of SCP-2/SCP-x Gene Ablation on Hepatic and Biliary Cholesterol

Based on the preceding studies with SCP-2 overexpressing and antisense cDNA treated mice, it was expected that SCP-2 gene ablation would decrease hepatic cholesterol accumulation and biliary cholesterol secretion. While mice ablated only in SCP-2 expression do not yet to our knowledge exist, SCP-2/SCP-x gene ablation (loss of both SCP-2 and SCP-x) led to decreased hepatic lipid accumulation (cholesterol esters, triglycerides), as predicted (Fuchs et al., 2001; Seedorf et al., 1998). Further, the finding that decreased hepatic lipid accumulation (cholesterol, triacylglycerol) in SCP-x gene-ablated mice suggests that the loss of SCP-x alone at least contributes to decreased hepatic lipid accumulation (Atshaves et al., 2007b). In contrast to these findings, however biliary cholesterol excretion was not decreased, but unexpectedly increased dramatically in SCP-2/SCP-x gene-ablated mice, an effect attributed to concomitant upregulation of liver fatty acid binding protein L-FABP

(Fuchs et al., 2001). Thus, while these studies with SCP-2/SCP-x gene-ablated mice provided important insights, they were complicated by simultaneous disruption of two proteins (SCP-2, SCP-x) and massive (5-6 fold) upregulation of another protein (L-FABP) involved in fatty acid and cholesterol metabolism. Finally, studies with cultured fibroblasts and serum from humans with genetic variations in the SCP-2/SCP-x gene showed that this gene impacts both bile acid (cholesterol branched side chain oxidation) and branched-chain fatty acid metabolism (Ferdinandusse et al., 2006). Unfortunately, hepatic lipid data were not reported in the latter studies to allow direct comparison with the phenotype of SCP-2/SCP-x gene-ablated mice.

10.13 Potential Compensation by Other Cholesterol-binding Proteins

As indicated above studies with SCP-2/SCP-x null mice, hypersecretion of cholesterol in bile in these mice was attributed to upregulation of the liver fatty acid binding protein (L-FABP) (Fuchs et al., 2001; Seedorf et al., 1998). L-FABP also binds cholesterol and other sterols (NBD-cholesterol, DHE), but with lower affinity than SCP-2 (Fischer et al., 1985; Sams et al., 1991; Schroeder et al., 1998, 2000, 1998; Stolowich et al., 1999; Martin et al., 2008b, 2009a; Avdulov et al., 1999).

It is important to note that L-FABP is 40-fold more prevalent than SCP-2 in hepatocytes (McArthur et al., 1999; Gallegos et al., 2001b; Stolowich et al., 2002; Schroeder et al., 1996) and results obtained with purified plasma membranes (Nemecz and Schroeder, 1991; Schroeder et al., 1996; Woodford et al., 1993, 1995a) and transfected cells (Jefferson et al., 1991; Incerpi et al., 1991, 1992) suggest this protein may play a role not only in intracellular trafficking of fatty acids (McArthur et al., 1999; Weisiger, 2005; Schroeder et al., 2008), but also of cholesterol. However, there is disagreement as to whether L-FABP gene ablation elicits obesity and hepatic cholesterol accumulation. Our lab first detected obesity in older (>6 mo, but not younger) female (but not male) L-FABP null mice generated by complete L-FABP gene ablation (Martin et al., 2008a, 2009b). Furthermore, increased hepatic cholesterol was observed in female N2–N6 backcross generation L-FABP null mice generated by complete L-FABP gene ablation and fed either standard rodent chow or cholesterol-rich diet (Martin et al., 2003a, 2006, 2008a), a phenotype maintained by the current N10 backcross, as shown by 1.8-, 1.5-, and 1.7-fold increased hepatic total ($p<0.01$), free ($p<0.01$), and esterified ($p<0.05$) cholesterol in female L-FABP null versus wild-type age- and sex- matched littermates (Martin et al., 2009a). In contrast, while young (<5mo old) independently generated control chow-fed L-FABP null mice generated by a GFP knock-in strategy did not display obesity, the phenotype of older mice was not examined (Newberry et al., 2006). Likewise, while young (<5 mo old) independently generated control chow-fed L-FABP null mice generated by a GFP knock-in strategy did not display hepatic cholesterol accumulation when backcrossed to the N10 generation, the hepatic cholesterol phenotype of older (>6 mo) mice was not reported (Newberry et al., 2008b). Interestingly, hepatic cholesterol accumulation was observed under some (western, lithogenic) but not other (extended time on high cholesterol) dietary conditions (Newberry et al., 2006; Xie et al., 2009). Since backcross generation number did not account for the apparent discrepancy in hepatic cholesterol phenotype, several other possibilities will now be considered.

1) Significant differences in gene-construct strategies:

The small L-FABP gene is composed of four exons: exon 1 (encodes aa1–22), exon 2 (aa23–79), exon 3 (aa80–112), and exon 4 (aa 113–126) (Matarese et al., 1989). Since others have reported that different ablation construct strategies for proteins involved in lipid metabolism can yield mice sharing some but not other aspects of lipid phenotype (Fex et al., 2006; Tansey et al., 2001; Martinez-Botas et al., 2000), our complete L-FABP gene ablation construct was prepared to delete most of the 5′ non-coding (promoter) region, all 4 exons of the small L-FABP gene, and part of the 3′ noncoding region, thereby avoiding the potential for expressing L-FABP fragments (Martin et al., 2003a). In contrast, the other L-FABP null construct was generated by knock-in of GFP at the N-terminal protein translation start ATG that left intact the 5′ noncoding region (includes part of untranslated region of exon 1), a portion of intron 2, all of exons 3 and 4, and the 3′ non-coding region (Newberry et al., 2003). This strategy resulted in over-expression of GFP in liver as well as other tissues, but no controls were provided demonstrating that GFP over-expression did not affect lipid phenotype, especially in the liver of these mice. Furthermore, the presence of the 5′ non-coding region left unresolved the issue of whether L-FABP peptide fragments were expressed in these L-FABP null mice. Since peptide fragments of SCP-2 (e.g. N-terminal α-helical domain) do not directly bind/transfer cholesterol, but bind membranes and potentiate SCP-2's sterol transfer activity (Huang et al., 1999a, b, 2002), the possibility that the latter strategy may have resulted in peptide fragments that contribute biological activity was not precluded. These subtly different cholesterol phenotypes obtained with different gene ablation strategies may provide new insights regarding potential roles of L-FABP's N-terminal α-helical membrane interaction domain, whose full function has yet to be completely determined.

2) L-FABP null mice generated by different construct strategies were back-crossed to C57BL/6 mice obtained from different vendors:

The L-FABP null mice generated with the complete L-FABP gene ablation construct were back-crossed to C57BL/6NCr mice from Charles River (Wilmington, MA) obtained through the National Cancer Institute (Frederick Cancer Research and Development Center, Frederick, MD) (Martin et al., 2003b, 2005, 2008a, 2009a, b). In contrast, L-FABP null mice generated by the GFP knock-in strategy were backcrossed to C57BL/6 J mice from Jackson Labs (Bar Harbor, ME) (Newberry et al., 2006, 2008a, b; Xie et al., 2009). Whether C57BL/6 mice from these two sources might differ sufficiently to be different substrains is not known.

3) Hepatic lipid phenotypes were examined at dramatically different ages for the L-FABP null mice generated by different construct strategies:

While L-FABP null mice generated by complete L-FABP gene ablation strategy did not show hepatic cholesterol accumulation at young age (<6 mo), these mice demonstrated sex-dependent hepatic cholesterol accumulation (in females, but not males) at >6 months of age (Martin et al., 2003a, 2006, 2008a, 2009a). While L-FABP null mice generated with a GFP knock-in strategy and aged <6 mo also did not exhibit hepatic cholesterol accumulation, hepatic lipid phenotype was not reported for mice >6 mo of age (Newberry et al., 2006, 2008a, b; Xie et al., 2009). It is not known if L-FABP null mice generated with the GFP knock-in strategy exhibit a sex-dependent hepatic cholesterol phenotype at ages >6mo.

4) L-FABP null mice generated by different construct strategies were fed control chow diets obtained from different vendors:

L-FABP null mice generated by complete L-FABP gene ablation strategy were fed control chow from Harlan Teklad (Madison, WI) (Martin et al., 2003a; Martin et al., 2006; Martin et al., 2008a; Martin et al., 2009a), while L-FABP null mice generated with a GFP knock-in strategy were maintained on control chow diets (Picolab Rodent Diet, LabDiet, Richmond, IN) or MP Biomedicals (Solon, OH) (Newberry et al., 2006, 2008a, b; Xie et al., 2009). Standard control rodent chows from different vendors (or even from the same vendor using different sources of components) can vary significantly in their contents of phytol and phytoestrogens, both of which markedly influence hepatic lipid metabolism (Atshaves et al., 2004, 2005, 2007b; Mackie et al., 2009; Martin et al., 2005, 2006, 2008a, 2009b; Seedorf et al., 1998; Thigpen et al., 1999a, b).

5) L-FABP null mice generated by different construct strategies were fed experimental diets differed in source, caloric content, and controls:

L-FABP null mice generated by complete L-FABP gene ablation strategy were fed a defined control diet (phytol-free, phytoestrogen-free) or isocaloric cholesterol-rich control diet based on the control diet—both from Research Diets (New Brunswick, NJ) (Martin et al., 2006, 2008a). In contrast, L-FABP null mice generated with a GFP knock-in strategy were fed control diets (Picolab Rodent Diet, LabDiet, Richmond, IN) or MP Biomedicals (Solon, OH) while non-isocaloric experimental diets (cholesterol-rich, high fat, western, lithogenic) were from completely different sources including Harlan Teklad (Madison, WI), Research Diets (New Brunswick, NJ), or MP Biomedicals (Solon, OH) (Newberry et al., 2006, 2008a, b; Xie et al., 2009). Finally, the time on experimental diets (e.g. high cholesterol) differed markedly—5 weeks on 1.25% cholesterol-diet for L-FABP null mice generated by complete L-FABP gene ablation (Martin et al., 2006) versus 12 weeks on 1.25 and 2% cholesterol diet for L-FABP null mice generated by GFP knock-in strategy (Newberry et al., 2008b). Hepatic cholesterol content of control-chow fed female mice is known to be higher than that of male mice (Atshaves et al., 2004; Turley et al., 1998; Martin et al., 2005, 2006) and this phenotype is maintained when fed for short times (3–4 weeks) on a 1.0–1.25% cholesterol-diet (Turley et al., 1998; Martin et al., 2005, 2006). Since hepatic cholesterol accumulation increases with increasing time on high cholesterol diet much more so for female than male mice (Turley et al., 1998; Schwarz et al., 2001; Yu et al., 2000), it is possible that

L-FABP null mice generated by GFP knock-in strategy fed cholesterol-rich diet for 3mo (Newberry et al., 2008b) may have overcome differences noted with L-FABP null mice generated by complete L-FABP gene ablation and fed for only 5 weeks, a much shorter duration (Martin et al., 2006). It remains to be shown whether L-FABP null mice generated by GFP knock-in strategy will exhibit elevated hepatic cholesterol accumulation when fed 1.25% cholesterol-rich diet for shorter time (i.e. 5 weeks) rather than the much longer time (3 months) used previously.

6) L-FABP null mice generated by different construct strategies were fasted for markedly different times prior to sacrifice:

In all studies with L-FABP null mice generated by complete L-FABP gene ablation strategy, the mice were fasted 12 h prior to sacrifice (Martin et al., 2003a, b, 2005, 2006, 2008a, 2009a, b; Atshaves et al., 2004, 2005; Mackie et al., 2009). In contrast, in studies with L-FABP null mice generated by GFP knock-in strategy the mice were fasted for variable times, 4 to 48 h (Newberry et al., 2003, 2006, 2008a, b; Spann et al., 2006; Xie et al., 2009). Fasting for 12 h (but not 4 h) is sufficient to clear most chylomicrons from the serum without inducing starvation while fasting for 48 h represents starvation, a condition that significantly alters hepatic lipid metabolism.

7) L-FABP null mice generated by different construct strategies, backcrossed to C57Bl/6 mice from different vendors, fed control chow from different vendors, and maintained in different laboratory animal facilities are likely to exhibit different intestinal microflora:

In summary, while L-FABP null mice generated by different construct strategies share significant features [e.g. reduced hepatic fatty acid uptake, reduced hepatic fatty acid oxidation, and reduced hepatic triglyceride (males only)], only those generated by complete gene ablation exhibit an age- and sex-dependent obese and hepatic cholesterol phenotype. While it has been suggested that the L-FABP null mice generated by GFP knock-in strategy do not exhibit the latter phenotype, careful comparison of the respective work indicates studies with the L-FABP null mice generated by GFP knock-in strategy were not performed under the same conditions, in particular with a different vendor source of C57Bl/6 mice for backcrossing, use of significantly younger mice, comparisons to control diets not isocaloric and/or from different vendors, length of time on experimental diet and comparison to appropriate control diet, as well as other factors more difficult to control. Thus, the hepatic cholesterol phenotype of L-FABP null mice generated by a GFP knock-in strategy and sex-matched, aged at least 6 months, fed the same diets (from the same vendor) and for experimental diets compared to the same control diets (from the same vendor), and fasted for the same length of time as the L-FABP null mice generated by complete L-FABP gene ablation remains to be determined. The physiological significance of studies with L-FABP null mice to humans is underscored by recent findings that human genetic variations in the L-FABP gene impact blood lipoprotein/lipid levels and the response to lipid-lowering therapy with fenofibrate, a cholesterol synthesis inhibitor (Brouillette et al., 2004; Fisher et al., 2007).

10.14 Conclusions and Future Perspectives

Real-time fluorescence imaging has now established the existence of cholesterol-rich microdomains in the plasma membranes of mammalian cells. Likewise, studies performed in vitro and in transfected cells show that intracellular cholesterol binding proteins such as SCP-2 and L-FABP are both involved in cholesterol uptake, intracellular trafficking, and targeting. While studies with genetically engineered mice have proven valuable in elucidating physiological roles of SCP-2 and L-FABP in cholesterol metabolism, the phenotypes of such mice are complicated by concomitant upregulation of L-FABP in SCP-2/SCP-x null mice and upregulation of SCP-2 in L-FABP null mice. Furthermore, studies with two independently generated L-FABP null mice demonstrate the sensitivity of the hepatic cholesterol phenotype to differences in construct strategy, source of mouse strain for backcrossing, age and sex of the mice, experimental diet (source, caloric content), conditions for experimental diet (concentration of cholesterol added, time on diet), use of appropriate control diets used for comparison to experimental diets, duration of fasting (decrease chylomicrons, starvations), intestinal microflora, and other nuanced variables difficult to control. Thus, it is important that conclusions regarding the role of individual cholesterol-binding proteins be based not only on studies with gene-targeted mice but that these conclusions should also be supported by experiments performed in vitro (cholesterol binding, cholesterol transfer between purified membranes, stimulation of cholesterol metabolism, etc.) with the respective purified proteins and in transfected cells (cholesterol uptake, efflux, intracellular diffusion/transport) over-expressing the respective proteins where variables are easier to control.

Acknowledgements This work was supported in part by the USPHS National Institutes of Health GM31651 (FS, ABK) and DK70965 (BPA). The helpful technical assistance of Ms. Kerstin Landrock was much appreciated.

References

Amigo, L., Zanlungo, S., Miquel, J.F., Glick, J.M., Hyogo, H., Cohen, D.E., Rigotti, A., and Nervi, F., 2003, Hepatic overexpression of sterol carrier protein-2 inhibits VLDL production and reciprocally enhances biliary lipid secretion. *J. Lipid Res.* **44:** 399–407.

Anderson, R.G.W. and Jacobson, K., 2002, A role for lipid shells in targeting proteins to caveolae, rafts, and other lipid domains. *Science* **296:** 1821–1825.

Anniss, A.M., Apostolopoulos, J., Dworkin, S., Purton, L.E., and Sparrow, R.L., 2002, An oxysterol binding protein family identified in the mouse. *DNA Cell Biol.* **21:** 571–580.

Assanasen, C., Mineo, C., Seetharam, D., Yuhanna, I.S., Marcel, Y.L., Connelly, M.A., Williams, D.L., de la Llera-Moya, M., Shaul, P.W., and Silver, D.L., 2005, Cholesterol binding, efflux, and PDZ-interacting domain of scavenger receptor-B1 mediate HDL-initiated signaling. *J. Clin. Invest.* **115:** 969–977.

Atshaves, B.P., Gallegos, A., McIntosh, A.L., Kier, A.B., and Schroeder, F., 2003, Sterol carrier protein-2 selectively alters lipid composition and cholesterol dynamics of caveolae/lipid raft vs non-raft domains in L-cell fibroblast plasma membranes. *Biochemistry* **42:** 14583–14598.

Atshaves, B.P., Jefferson, J.R., McIntosh, A.L., McCann, B.M., Landrock, K., Kier, A.B., and Schroeder, F., 2007a, Effect of sterol carrier protein-2 expression on sphingolipid distribution in plasma membrane lipid rafts/caveolae. *Lipids* **42**: 871–884.

Atshaves, B.P., McIntosh, A.L., Landrock, D., Payne, H.R., Mackie, J., Maeda, N., Ball, J.M., Schroeder, F., and Kier, A.B., 2007b, Effect of SCP-x gene ablation on branched-chain fatty acid metabolism. *Am. J. Physiol.* **292**: 939–951.

Atshaves, B.P., McIntosh, A.L., Martin, G.G., Landrock, D., Payne, H.R., Bhuvanendran, S., Landrock, K.K., Lyuksyutova, O.I., Johnson, J.D., Macfarlane, R.D., Kier, A.B., and Schroeder, F., 2009, Overexpression of sterol carrier protein-2 differentiallly alters hepatic cholesterol accumulation in cholesterol-fed mice. *J. Lipid Res.* **50**: 1429–47.

Atshaves, B.P., McIntosh, A.L., Payne, H.R., Gallegos, A.M., Landrock, K., Maeda, N., Kier, A.B., and Schroeder, F., 2007c, Sterol carrier protein-2/sterol carrier protein-x gene ablation alters lipid raft domains in primary cultured mouse hepatocytes. *J. Lipid Res.* **48**: 2193–2211.

Atshaves, B.P., McIntosh, A.L., Payne, H.R., Mackie, J., Kier, A.B., and Schroeder, F., 2005, Effect of branched-chain fatty acid on lipid dynamics in mice lacking liver fatty acid binding protein gene. *Am. J. Physiol.* **288**: C543–C558.

Atshaves, B.P., Payne, H.R., McIntosh, A.L., Tichy, S.E., Russell, D., Kier, A.B., and Schroeder, F., 2004, Sexually dimorphic metabolism of branched chain lipids in C57BL/6 J mice. *J. Lipid Res.* **45**: 812–830.

Atshaves, B.P., Starodub, O., McIntosh, A.L., Roths, J.B., Kier, A.B., and Schroeder, F., 2000, Sterol carrier protein-2 alters HDL-mediated cholesterol efflux. *J. Biol. Chem.* **275**: 36852–36861.

Avdulov, N.A., Chochina, S.V., Igbavboa, U., Warden, C.H., Schroeder, F., and Wood, W.G., 1999, Lipid binding to sterol carrier protein-2 is inhibited by ethanol. *Biochim. Biophys. Acta* **1437**: 37–45.

Babiychuk, E.B. and Draeger, A., 2006, Biochemical characterization of detergent-resistant membranes: a systematic approach. *Biochem. J.* **397**: 407–416.

Balbis, A.B.G., Mounier, C., and Posner, B.I., 2004, Effect of insulin on caveolin-enriched membrane domains in rat liver. *J. Biol. Chem.* **279**: 39348–39357.

Baldan, A., Tarr, P., Lee, R., and Edwards, P.A., 2006, ATP-binding cassette transporter G1 and lipid homeostasis. *Curr. Opin. Lipidol.* **17**: 227–232.

Beltroy, E.P., Richardson, J.A., Horton, J.D., Turley, S.D., and Dietschy, J.M., 2005, Cholesterol accumulation and liver cell death in mice with NPC disease. *Hepatology* **42**: 886–893.

Bourret, G., Brodeur, M.R., Luangrath, V., Lapointe, J., Falstrault, L., and Brissette, L., 2006, *In vivo* cholesteryl ester selective uptake of mildly and standardly oxidized LDL occurs by both parenchymal and nonparenchymal mouse hepatic cells but SR-BI is only responsible for standardly oxidized LDL selective uptake by nonparenchymal cells. *Int. J. Biochem. Cell Biol.* **38**: 1160–1170.

Bretscher, M.S. and Munro, S., 1993, Cholesterol and the Golgi apparatus. *Science* **261**: 1280–1281.

Brouillette, C., Bose, Y., Perusse, L., Gaudet, D., and Vohl, M.-C., 2004, Effect of liver fatty acid binding protein (FABP) T94A missense mutation on plasma lipoprotein responsiveness to treatment with fenofibrate. *J. Hum. Genet.* **49**: 424–432.

Chang, T.Y., Chang, C.C., Lin, S., Yu, C., Li, B.L., and Miyazaki, A., 2001, Roles of acyl CoA:cholesterol acyltrasnferase-1 and -2. *Curr. Opin. Lipidol.* **12**: 289–296.

Chang, T.-Y., Chang, C.C.Y., and Cheng, D., 1997, Acyl-coenzyme A:cholesterol acyltransferase. *Annu. Rev. Biochem.* **66**: 613–638.

Chang, T.-Y., Reid, P.C., Sugii, S., Ohgami, N., Cruz, J.C., and Chang, C.C.Y., 2005, Niemann-Pick Type C disease and intracellular cholesterol trafficking. *J. Biol. Chem.* **280**: 20917–20920.

Cohen, D.E., 1999, Hepatocellular transport and secretion of biliary lipids. *Cur. Opin. Lipidology* **10**: 295–302.

Connelly, M.A. and Williams, D.L., 2004, Scavenger receptor B1: A scavenger receptor with a mission to transport high density lipoprotein lipids. *Cur. Opin. Lipidol.* **15**: 287–295.

de Almeida, R.F.M., Loura, L.M.S., Fedorov, A., and Prieto, M., 2005, Lipid rafts have different sizes depending on membrane composition: a time resolved fluorescence resonance energy transfer study. *J. Mol. Biol.* **346:** 1109–1120.

de la Llera-Moya, M., Rothblat, G.H., Connelly, M.A., Kellner-Weibel, G., Sakr, S.W., Phillips, M.C., and Williams, D.L., 1999, Scavenger receptor B1 (SRB1) mediates free cholesterol flux independently of HDL tethering to the surface. *J. Lipid Res.* **40:** 575–580.

Everson, W.V. and Smart, E.J., 2005, Caveolae and the regulation of cellular cholesterol homeostasis. *In* Caveolae and Lipid Rafts: Roles in Signal Transduction and the Pathogenesis of Human Disease. M. P. Lisanti and P. G. Frank, editors. Elsevier Academic Press, San Diego, pp. 37–55.

Ferdinandusse, S., Kostopoulos, P., Denis, S., Rusch, R., Overmars, H., Dillman, U., Reith, W., Haas, D., Wanders, R.J.A., Duran, M., and Marziniak, M., 2006, Mutations in the gene encoding peroxisomal sterol carrier protein-x (SCP-x) cause leukencephalopathy with dystonia and motor neuropathy. *Am. J. Hum. Genet.* **78:** 1046–1052.

Feul-Lagerstedt, E., Movitz, C., Pelime, S., Dahlgren, C., and Karlsson, A., 2007, Lipid raft proteome of the human neutrophil azurophil granule. *Proteomics* **7:** 194–205.

Fex, M., Lucas, S., Winzell, M.S., Ahren, B., Holm, C., and Mulder, H., 2006, β-Cell lipases and insulin secretion. *Diabetes* **55:** S24–S31.

Fischer, R.T., Cowlen, M.S., Dempsey, M.E., and Schroeder, F., 1985, Fluorescence of delta 5,7,9(11),22-ergostatetraen-3 beta-ol in micelles, sterol carrier protein complexes, and plasma membranes. *Biochemistry* **24:** 3322–3331.

Fisher, E., Weikert, C., Klapper, M., Lindner, I., Mohlig, M., Spranger, J., Boeing, H., Schrezenmeir, J., and Doring, F., 2007, L-FABP T94A is associated with fasting triglycerides and LDL-cholesterol in women. *Mol. Gen. Metab.* **91:** 278–284.

Foster, L.J., de Hoog, C.L., and Mann, M., 2003, Unbiased quantitative proteomics of lipid rafts reveals high specificity for signaling factors. *Proc. Natl. Acad. Sci. U. S. A.* **100:** 5813–5818.

Frolov, A., Petrescu, A., Atshaves, B.P., So, P.T.C., Gratton, E., Serrero, G., and Schroeder, F., 2000, High density lipoprotein mediated cholesterol uptake and targeting to lipid droplets in intact L-cell fibroblasts. *J. Biol. Chem.* **275:** 12769–12780.

Frolov, A., Woodford, J.K., Murphy, E.J., Billheimer, J.T., and Schroeder, F., 1996a, Spontaneous and protein-mediated sterol transfer between intracellular membranes. *J. Biol. Chem.* **271:** 16075–16083.

Frolov, A.A., Woodford, J.K., Murphy, E.J., Billheimer, J.T., and Schroeder, F., 1996b, Fibroblast membrane sterol kinetic domains: modulation by sterol carrier protein 2 and liver fatty acid binding protein. *J. Lipid Res.* **37:** 1862–1874.

Fuchs, M., Hafer, A., Muench, C., Kannenberg, F., Teichmann, S., Scheibner, J., Stange, E.F., and Seedorf, U., 2001, Disruption of the sterol carrier protein 2 gene in mice impairs biliary lipid and hepatic cholesterol metabolism. *J. Biol. Chem.* **276:** 48058–48065.

Gallegos, A.M., Atshaves, B.P., Storey, S., McIntosh, A., Petrescu, A.D., and Schroeder, F., 2001a, Sterol carrier protein-2 expression alters plasma membrane lipid distribution and cholesterol dynamics. *Biochemistry* **40:** 6493–6506.

Gallegos, A.M., Atshaves, B.P., Storey, S.M., Starodub, O., Petrescu, A.D., Huang, H., McIntosh, A., Martin, G., Chao, H., Kier, A.B., and Schroeder, F., 2001b, Gene structure, intracellular localization, and functional roles of sterol carrier protein-2. *Prog. Lipid Res.* **40:** 498–563.

Gallegos, A.M., McIntosh, A.L., Atshaves, B.P., and Schroeder, F., 2004, Structure and cholesterol domain dynamics of an enriched caveolae/raft isolate. *Biochem. J.* **382:** 451–461.

Gallegos, A.M., McIntosh, A.L., Kier, A.B., and Schroeder, F., 2008, Membrane domain distributions: Analysis of fluorescent sterol exchange kinetics. *Curr. Anal. Chem.* **4:** 1–7.

Gallegos, A.M., Storey, S.M., Kier, A.B., Schroeder, F., and Ball, J.M., 2006, Structure and cholesterol dynamics of caveolae/raft and nonraft plasma membrane domains. *Biochemistry* **45:** 12100–12116.

Gulbins, E. and Li, P.L., 2006, Physiological and pathophysiological aspects of ceramide. *Am. J. Physiol. Regul. Integr. Comp. Physiol.* **290:** 11–26.

Hale, J.E. and Schroeder, F., 1982, Asymmetric transbilayer distribution of sterol across plasma membranes determined by fluorescence quenching of dehydroergosterol. *Eur. J. Biochem.* **122:** 649–661.

Hao, M., Mukherjee, S., and Maxfield, F.R., 2001, Cholesterol depletion induces large scale domain segregation in living cell membranes. *Proc. Natl. Acad. Sci. U. S. A.* **98:** 13072–13077.

Harder, C.J., Meng, A., Rippstein, P., McBride, H.M., and McPherson, R., 2007, SR-BI undergoes cholesterol-stimulated transcytosis to the bile canaliculus in polarized WIF-B cells. *J. Biol. Chem.* **282:** 1445–1455.

Harris, W.S., Mozaffarian, D., Rimm, E., Kris-Etherton, P., Rudel, L.L., Appel, L.J., Engler, M.M., Engler, M.B., and Sacks, F., 2009, Omega-6 fatty acids and the risk for cardiovascular disease. *Circulation* **119:** 902–907.

Harroun, T.A., Katsaras, J., and Wassall, S.R., 2008, Cholesterol is found to reside in the center of a polyunsaturated lipid membrane. *Biochemistry* **47:** 7090–7096.

Hayakawa, E., Naganuma, M., Mukasa, K., Shimozawa, T., and Araiso, T., 1998, Change of Motion and Localization of Cholesterol Molecule during L_α-H_{II} Transition. *Biophys. J.* **74:** 892–898.

Haynes, M.P., Phillips, M.C., and Rothblat, G.H., 2000, Efflux of cholesterol from different cellular pools. *Biochemistry* **39:** 4508–4517.

Holthuis, J.C.M. and Levine, T.P., 2005, Lipid traffic: floppy drives and superhighway. *Nat. Rev. Mol. Cell Biol.* **6:** 209–220.

Holtta-Vuori, M., Uronen, R.-L., Repakova, J., Salonen, E., Vattulainen, L., Panula, P., Li, Z., Bittman, R., and Ikonen, E., 2008, BODIPY-cholesterol: a new tool to visualize sterol trafficking in living cells and organisms. *Traffic* **9:** 1839–1849.

Huang, H., Ball, J.A., Billheimer, J.T., and Schroeder, F., 1999a, Interaction of the N-terminus of sterol carrier protein-2 with membranes: role of membrane curvature. *Biochem. J.* **344:** 593–603.

Huang, H., Ball, J.A., Billheimer, J.T., and Schroeder, F., 1999b, The sterol carrier protein-2 amino terminus: a membrane interaction domain. *Biochemistry* **38:** 13231–13243.

Huang, H., Gallegos, A., Zhou, M., Ball, J.M., and Schroeder, F., 2002, Role of sterol carrier protein-2 N-terminal membrane binding domain in sterol transfer. *Biochemistry* **41:** 12149–12162.

Huang, H., McIntosh, A.L., Atshaves, B.P., Ohno-Iwashita, Y., Kier, A.B., and Schroeder, F., 2009, Use of dansyl-cholestanol as a probe of cholesterol behavior in membranes of living cells. *J. Lipid Res.* published December 11, 2009, doi:10.1194/jlr.M003244.

Hui, D.Y., Labonte, E.D., and Howles, P.N., 2008, Development and physiological regulation of intestinal absorption. III. Intestinal transporters and cholesterol absorption. *Am. J. Physiol. Gastrointest. Liver Phys.* **294:** G839–G843.

Igbavboa, U., Avdulov, N.A., Schroeder, F., and Wood, W.G., 1996, Increasing age alters transbilayer fluidity and cholesterol asymmetry in synaptic plasma membranes of mice. *J. Neurochem.* **66:** 1717–1725.

Ikonen, E. and Vainio, S., 2005, Lipid microdomains and insulin resistance: is there a connection? *Sci. STKE* **pe3:** 1–3.

Incerpi, S., Jefferson, J.R., Wood, W.G., Ball, W.J., and Schroeder, F., 1992, Na pump and plasma membrane structure in L-cell fibroblasts expressing rat liver fatty acid binding protein. *Arch. Biochem. Biophys.* **298:** 35–42.

Incerpi, S., Vito, P.D.. Luly, P., Jefferson, J.R., and Schroeder, F., 1991. The sodium pump and plasma membrane structure: effect of insulin and aging. *In* Biotechnology of Cell Regulation, Serono Symposia Series Adv. in Exptl. Med. R. Verna and Y. Nishizuka, editors. Raven Press, NY, pp. 409–423.

Ismair, M.G., Hausler, S., Stuermer, C.A., Guyot, C., Meier, P.J., Roth, J., and Stieger, B., 2009, ABC-transporters are localized in caveolin-1-positive and reggie-1-negative and reggie-2-negative microdomains of the canalicular membrane in rat hepatoctyes. *Hepatology* **49**: 1673–1682.

Jefferson, J.R., Powell, D.M., Rymaszewski, Z., Kukowska-Latallo, J., and Schroeder, F., 1990, Altered membrane structure in transfected mouse L-Cell fibroblasts expressing rat liver fatty acid-binding protein. *J. Biol. Chem.* **265**: 11062–11068.

Jefferson, J.R., Slotte, J.P., Nemecz, G., Pastuszyn, A., Scallen, T.J., and Schroeder, F., 1991, Intracellular sterol distribution in transfected mouse L-cell fibroblasts expressing rat liver fatty acid binding protein. *J. Biol. Chem.* **266**: 5486–5496.

Jessup, W., Gelissen, I., Gaus, K., and Kritharides, L., 2006, Roles of ATP binding cassette transporters A1 and G1, scavenger receptor BI and membrane lipid domains in cholesterol export from macrophages. *Curr. Opin. Lipidol.* **17**: 247–267.

Ji, Y., Wang, N., Ramakrishnan, R., Sehayek, E., Huszar, D., Breslow, J.L., and Tall, A.R., 1999, Hepatic scavenger receptor B1 promotes rapid clearance of high density lipoprotein free cholesterol and its transport into bile. *J. Biol. Chem.* **274**: 33398–33402.

Kamau, S.W., Kramer, S.D., Gunthert, M., and Wunderlich-Allenspach, H., 2005, Effect of the modulation of the membrane lipid composition on the localization and function of P-glycoprotein in MDR1-MDCK cells. *In Vitro Cell Dev. Biol.* **41**: 207–216.

Kanno, K., Wu, M.K., Scapa, E.F., Roderick, S.L., and Cohen, D.E., 2007, Structure and function of phosphatidylcholine transfer protein (PC-TP)/StarD2. *Biochim. Biophys. Acta* **1771**: 654–662.

Karten, B. et al., 2006, *J. Biol. Chem.* **281**: 4049–4057.

Kellner-Weibel, G., de la Llera-Moya, M., Connelly, M.A., Stoudt, G., Christian, A.E., Haynes, M.P., Williams, D.L., and Rothblat, G.H., 2000, *Biochemistry* **43**: 544–549.

Kier, A.B., Sweet, W.D., Cowlen, M.S., and Schroeder, F., 1986, Regulation of transbilayer distribution of a fluorescent sterol in tumor cell plasma membranes. *Biochim. Biophys. Acta* **861**: 287–301.

Lagace, T.A., Byers, D.M., Cook, H.W., and Ridgway, N.D., 1997, Altered regulation of cholesterol and cholesteryl ester synthesis in Chinese hamster ovary cells overexpressing the oxysterol binding protein is dependent on the pleckstrin homology domain. *Biochem. J.* **326**: 205–213.

Laitinen, S., Lehto, M., Lehtonen, S., Hyvarinen, K., Heino, S., Lehtonen, E., Ehnholm, C., Ikonen, E., and Olkkonen, V.M., 2002, ORP2, a homolog of oxysterol binding protein, regulates cellular cholesterol metabolism. *J. Lipid Res.* **43**: 245–255.

Lakowicz, J. R. 2006. Energy transfer. *In* Principles of Fluorescence Spectroscopy. J. R. Lakowicz, editor. Springer Science, New York, NY, pp. 443–475.

Lange, A.J., Dolde, J., and Steck, T.L., 2007, The rate of transmembrane movement of cholesterol in the human erythrocyte. *J. Biol. Chem.* **256**: 5321–5323.

Li, Q., Zhang, Q., Wang, M., Zhao, S., Ma, J., Luo, N., Li, N., Li, Y., Xu, G., and Li, J., 2007, Eicosapentaenoic acid modifies lipid composition in caveolae and induces translocation of endothelial nitric oxide synthase. *Biochimie* **89**: 169–177.

Lichtenberg, D., Gonnelli, M., and Heerklotz, H., 2005, Detergent-resistant membranes should not be identified with membrane rafts. *Trends Biochem. Sci.* **30**: 430–436.

Ma, D.W.L., Seo, J., Davidson, L.A., Callaway, E.S., Fan, Y.-Y., Lupton, J.R., and Chapkin, R.S., 2004, n-3 PUFA alter caveolae lipid composition and resident protein localization in mouse colon. *FASEB J.* **18**: 1040–1042.

Mackie, J.T., Atshaves, B.P., Payne, H.R., McIntosh, A.L., Schroeder, F., and Kier, A.B., 2009, Phytol-induced hepatotoxicity in mice. *Toxicol. Pathol.* **37**: 201–208.

Martin, G.G., Atshaves, B.P., Huang, H., McIntosh, A.L., Williams, B.W., Russell, D.H., Kier, A.B., and Schroeder, F., 2009a, Hepatic phenotype of liver fatty acid binding protein (L-FABP) gene ablated mice. *Am. J. Physiol. Gastrointest. Liver Physiol.* **297**: G1053–G1065.

Martin, G.G., Atshaves, B.P., McIntosh, A.L., Mackie, J.T., Kier, A.B., and Schroeder, F., 2005, Liver fatty acid binding protein (L-FABP) gene ablation alters liver bile acid metabolism in male mice. *Biochem. J.* **391**: 549–560.

Martin, G.G., Atshaves, B.P., McIntosh, A.L., Mackie, J.T., Kier, A.B., and Schroeder, F., 2006, Liver fatty acid binding protein (L-FABP) gene ablation potentiates hepatic cholesterol accumulation in cholesterol-fed female mice. *Am. J. Physiol.* **290:** G36–G48.

Martin, G.G., Atshaves, B.P., McIntosh, A.L., Mackie, J.T., Kier, A.B., and Schroeder, F., 2008a, Liver fatty acid binding protein gene ablated female mice exhibit increased age dependent obesity. *J. Nutr.* **138:** 1859–1865.

Martin, G.G., Atshaves, B.P., McIntosh, A.L., Mackie, J.T., Kier, A.B., and Schroeder, F., 2009b, Liver fatty acid binding protein gene ablation enhances age-dependent weight gain in male mice. *Mol. Cell. Biochem.* **324:** 101–115.

Martin, G.G., Danneberg, H., Kumar, L.S., Atshaves, B.P., Erol, E., Bader, M., Schroeder, F., and Binas, B., 2003a, Decreased liver fatty acid binding capacity and altered liver lipid distribution in mice lacking the liver fatty acid binding protein (L-FABP) gene. *J. Biol. Chem.* **278:** 21429–21438.

Martin, G.G., Hostetler, H.A., Tichy, S.E., Russell, D.H., Berg, J.M., Woldegiorgis, G., Spencer, T.A., Ball, J.A., Kier, A.B., and Schroeder, F., 2008b, Structure and function of the sterol carrier protein-2 (SCP-2) N-terminal pre-sequence. *Biochemistry* **47:** 5915–5934.

Martin, G.G., Huang, H., Atshaves, B.P., Binas, B., and Schroeder, F., 2003b, Ablation of the liver fatty acid binding protein gene decreases fatty acyl CoA binding capacity and alters fatty acyl CoA pool distribution in mouse liver. *Biochemistry* **42:** 11520–11532.

Martin, S. and Parton, R.G., 2005, Caveolin, cholesterol, and lipid bodies. *Semin. Cell Dev. Biol.* **16:** 163–174.

Martinez-Botas, J., Anderson, J.B., Tessier, D., Lapillonne, A., Chang, B.H.J., Quast, M.J., Gorenstein, D., Chen, K.-H., and Chan, L., 2000, Absence of perilipin results in leanness and reverses obesity in Lepr$^{db/db}$ mice. *Nat. Genet.* **26:** 474–479.

Matarese, V., Stone, R.L., Waggoner, D.W., and Bernlohr, D.A., 1989, Intracellular fatty acid trafficking and the role of cytosolic lipid binding proteins. *Prog. Lipid Res.* **28:** 245–272.

Mazzone, A., Tietz, P., Jefferson, J.R., Pagano, R., and LaRusso, N.F., 2006, Isolation and characterization of lipid microdomains from apical and basolateral plasma membranes of rat hepatocytes. *Hepatology* **43:** 287–296.

McArthur, M.J., Atshaves, B.P., Frolov, A., Foxworth, W.D., Kier, A.B., and Schroeder, F., 1999, Cellular uptake and intracellular trafficking of long chain fatty acids. *J. Lipid Res.* **40:** 1371–1383.

McIntosh, A., Atshaves, B.P., Huang, H., Gallegos, A.M., Kier, A.B., Schroeder, F., Xu, H., Zhang, W., and Liu, S., 2007. Multiphoton laser scanning microscopy and spatial analysis of dehydroergosterol distributions on plasma membranes of living cells. *In* Lipid Rafts. T. McIntosh, editor. Humana Press Inc., Totowa, NJ, pp. 85–105.

McIntosh, A., Gallegos, A., Atshaves, B.P., Storey, S., Kannoju, D., and Schroeder, F., 2003, Fluorescence and multiphoton imaging resolve unique structural forms of sterol in membranes of living cells. *J. Biol. Chem.* **278:** 6384–6403.

McIntosh, A.L., Atshaves, B.P., Gallegos, A.M., Storey, S.M., Reibenspies, J.H., Kier, A.B., Meyer, E., and Schroeder, F., 2008c, Structure of dehydroergosterol monohydrate and interaction with sterol carrier protein-2. *Lipids* **43:** 1165–1184.

McIntosh, A.L., Atshaves, B.P., Huang, H., Gallegos, A.M., Kier, A.B., and Schroeder, F., 2008a, Fluorescence techniques using dehydroergosterol to study cholesterol trafficking. *Lipids* **43:** 1208.

McIntosh, A.L., Huang, H., Atshaves, B.P., Storey, S.M., Gallegos, A., Spencer, T.A., Bittman, R., Ohno-Iwashita, Y., Kier, A.B., and Schroeder, F., 2008b. Fluorescent sterols for the study of cholesterol trafficking in living cells. *In* Probes and Tags to Study Biomolecular Function for Proteins, RNA, and Membranes. L. W. Miller, editor. Wiley VCH Verlag GmbH & Co., Weinheim, pp. 1–33.

Miller, W.L., 2007, Steroidogenic acute regulatory protein (StAR), a novel mitochondrial cholesterol transporter. *Biochim. Biophys. Acta* **1771:** 663–676.

Moncecchi, D.M., Murphy, E.J., Prows, D.R., and Schroeder, F., 1996, Sterol carrier protein-2 expression in mouse L-cell fibroblasts alters cholesterol uptake. *Biochim. Biophys. Acta* **1302**: 110–116.

Mondal, M., Mesmin, B., Mukherjee, S., and Maxfield, F.R., 2009, Sterols are mainly in the cytoplasmic leaflet of the plasma membrane and the endocytic recycling compartment in CHO cells. *Mol. Biol. Cell* **20**: 581–588.

Morris, D. H. 2007a. Background on omega-3 fatty acids. *In* FLAX- A Health and Nutrition Primer. D. H. Morris, editor. Flax Council of Canada, pp. 22–33.

Morris, D. H. 2007b. Importance of omega-3 fatty acids for adults and infants. *In* FLAX-A Health and Nutrition Primer. D. H. Morris, editor. Flax Council of Canada, pp. 34–43.

Mukherjee, S., Zha, X., Tabas, I., and Maxfield, F.R., 1998, Cholesterol distribution in living cells: fluorescence imaging using dehydroergosterol as a fluorescent cholesterol analog. *Biophys. J.* **75**: 1915–1925.

Muller, P. and Herrmann, A., 2002, Rapid transbilayer movement of spin-labeled steroids in human erythrocytes and in liposomes. *Biophys. J.* **82**: 1418–1428.

Murcia, M., Faraldo-Gomez, J.D., Maxfield, F.R., and Roux, B., 2006, Modeling the structure of the StART domains of MLN64 and StAR proteins in complex with cholesterol. *J. Lipid Res.* **47**: 2614–2630.

Murphy, E.J., 1998, Sterol carrier protein-2 expression increases fatty acid uptake and cytoplasmic diffusion in L-cell fibroblasts. *Am. J. Physiol.* **275**: G237–G243.

Murphy, E.J. and Schroeder, F., 1997, Sterol carrier protein-2 mediated cholesterol esterification in transfected L-cell fibroblasts. *Biochim. Biophys. Acta* **1345**: 283–292.

Nemecz, G. and Schroeder, F., 1988, Time-resolved fluorescence investigation of membrane cholesterol heterogeneity and exchange. *Biochemistry* **27**: 7740–7749.

Nemecz, G. and Schroeder, F., 1991, Selective binding of cholesterol by recombinant fatty acid-binding proteins. *J. Biol. Chem.* **266**: 17180–17186.

Newberry, E.P., Kennedy, S.M., Xie, Y., Luo, J., Stanley, S.E., Semenkovich, C.F., Crooke, R.M., Graham, M.J., and Davidson, N.O., 2008a, Altered hepatic triglyceride content after partial hepatectomy without impaired liver regeneration in multiple muring genetic models. *Hepatology* **48**: 1097–1105.

Newberry, E.P., Kennedy, S.M., Xie, Y., Sternard, B.T., Luo, J., and Davidson, N.O., 2008b, Diet-induced obesity and hepatic steatosis in L-FABP-/- mice is abrogated with SF, but not PUFA, feeding and attenuated after cholesterol supplementation. *Am. J. Physiol. Gastrointest. Liver Physiol.* **294**: G307–G314.

Newberry, E.P., Xie, Y., Kennedy, S., Buhman, K.K., Luo, J., Gross, R.W., and Davidson, N.O., 2003, Decreased hepatic triglyceride accumulation and altered fatty acid uptake in mice with deletion of the liver fatty acid binding protein gene. *J. Biol. Chem.* **278**: 51664–51672.

Newberry, E.P., Xie, Y., Kennedy, S.M., Luo, J., and Davidson, N.O., 2006, Protection against western diet-induced obesity and hepatic steatosis in liver fatty acid binding protein knockout mice. *Hepatology* **44**: 1191–1205.

Ohgami, N., Ko, D.C., Thomas, M., Scott, M.P., Chang, C.C.Y., and Chang, T.-Y., 2004, Binding between NPC1 protein and a photoactivatable cholesterol analog requires a functional sterol sensing domain. *Proc. Natl. Acad. Sci. U. S. A.* **34**: 12473–12478.

Ohno-Iwashita, Y., Shimada, Y., Waheed, A.A., Hayashi, M., Inomata, M., Nakamura, M., Maruya, M., and Iwashita, S., 2004, Perfringolysin O, a cholesterol-binding cytolysin, as a probe for lipid rafts. *Anaerobe* **10**: 125–134.

Orlowski, S., Martin, S., and Escargueil, A., 2006, P-glycoprotein and 'lipid rafts': some ambiguous mutual relationships (floating on them, building them or meeting them by chance?). *Cell. Mol. Life Sci.* **63**: 1038–1059.

Ortegren, U., Karlsson, M., Blazic, N., Blomqvist, M., Nysrom, F.H., Gustavsson, J., Fredman, P., and Stralfors, P., 2007, Lipids and glycosphingolipids in caveolae and surrounding plasma membrane of primary rat adipocytes. *Eur. J. Biochem.* **271**: 2028–2036.

Parathath, S., Connelly, M.A., Rieger, R.A., Klein, S.M., Abumrad, N.A., de la Llera-Moya, M., Iden, C.R., Rothblat, G.H., and Williams, D.L., 2004, Changes in plasma membrane properties and phosphatidylcholine subspecies of insect Sf9 cells due to expression of scavenger receptor class B, type 1, and CD36. *J. Biol. Chem.* **279:** 41310–41318.

Parr, R.D., Martin, G.G., Hostetler, H.A., Schroeder, M.E., Mir, K.D., Kier, A.B., Ball, J.M., and Schroeder, F., 2007, A new N-terminal recognition domain in caveolin-1 interacts with sterol carrier protein-2 (SCP-2). *Biochemistry* **46:** 8301–8314.

Parton, R.G., Hanzal-Bayer, M., and Handcock, J.F., 2006, Biogenesis of caveolae: a structural model for caveolin-induced domain formation. *J. Cell Sci.* **119:** 787–796.

Parton, R.G. and Simons, K., 2007, The multiple faces of caveolae. *Nat. Rev. Mol. Cell Biol.* **8:** 185–194.

Perry, R.J. and Ridgway, N.D., 2006, Oxysterol binding protein and vesicle associated membrane protein associated protein are required for sterol dependent activation of the ceramide transport protein. *Mol. Biol. Cell* **17:** 2604–2616.

Petersen, N.H., Faergeman, N.J., Yu, L., and Wustner, D., 2008, Kinetic imaging of NPC1L1 and sterol trafficking between plasma membrane and recycling endosomes in hepatoma cells. *J. Lipid Res.* **49:** 2023–2037.

Petrescu, A.D., Gallegos, A.M., Okamura, Y., Strauss, I.J.F., and Schroeder, F., 2001, Steroidogenic acute regulatory protein binds cholesterol and modulaes mitochondrial membrane sterol domain dynamics. *J. Biol. Chem.* **276:** 36970–36982.

Pike, L.J., 2004, Lipid rafts: heterogeneity on the high seas. *Biochem. J.* **378:** 281–292.

Pike, L.J., 2006, Rafts defined : A report on the Keystone symposium on lipid rafts and cell function. *J. Lipid Res.* **47:** 1597–1598.

Pike, L.J., Han, X., Chung, K.-N., and Gross, R.W., 2002, Lipid rafts are enriched in arachidonic acid and plasmenylethanolamine and their composition is independent of caveolin-1 expression: a quantitative electrospray ionization/mass spectrometric analysis. *Biochemistry* **41:** 2075–2088.

Pitto, M., Brunner, J., Ferraretto, A., Ravasi, D., Palestini, P., and Masserini, M., 2000, Use of a photoactivatable GM1 ganglioside analogue to assess lipid distribution in caveolae bilayer. *Glycoconj. J.* **17:** 215–222.

Puglielli, L., Rigotti, A., Amigo, L., Nunez, L., Greco, A.V., Santos, M.J., and Nervi, F., 1996, Modulation on intrahepatic cholesterol trafficking: Evidence by *in vivo* antisense treatment for the involvement of sterol carrier protein-2 in newly synthesized cholesterol transfer into bile. *Biochem. J.* **317:** 681–687.

Robins, S.J. and Fasulo, J.M., 1999, Delineation of a novel hepatic route for the selective transfer of unesterified sterols from high density lipoproteins to bile: studies using the perfused rat liver. *Hepatology* **29:** 1541–1548.

Rodriguez-Agudo, D., Ren, S., Hylemon, P.B., Redford, K., Natarajan, R., Del Castillo, A., Gil, G., and Pandak, W.M., 2005, Human StarD5, a cytosolic StAR-related lipid binding protein. *J. Lipid Res.* **46:** 1615–1623.

Roff, C.F., Pastuszyn, A., Strauss, J.F.I., Billheimer, J.T., Vanier, M.T., Brady, R.O., Scallen, T.J., and Pentchev, P.G., 1992, Deficiencies in sex-regulated expression and levels of two hepatic sterol carrier proteins in a murine model of Niemann-Pick type C disease. *J. Biol. Chem.* **267:** 15902–15908.

Roper, K., Corbeil, D., and Huttner, W.B., 2000, Retention of prominin in microvilli reveals distinct cholesterol-based lipid microdomains in the apical plasma membrane. *Nat. Cell Biol.* **2:** 582–592.

Saltiel, A.R. and Pessin, J.E., 2003, Insulin signaling in microdomains of the plasma membrane. *Traffic* **4:** 711–716.

Sams, G.H., Hargis, B.M., and Hargis, P.S., 1991, Identification of two lipid binding proteins from liver of *Gallus domesticus*. *Comp. Biochem. Physiol.* **99B:** 213–219.

Samsonov, A.V., Mihalyov, I., and Cohen, F.S., 2001, Characterization of cholesterol-sphingomyelin domains and their dynamics in bilayer membranes. *Biophys. J.* **81:** 1486–1500.

Schoer, J., Gallegos, A., Starodub, O., Petrescu, A., Roths, J.B., Kier, A.B., and Schroeder, F., 2000, Lysosomal membrane cholesterol dynamics: role of sterol carrier protein-2 gene products. *Biochemistry* **39**: 7662–7677.

Schram, V. and Thompson, T.E., 1997, Influence of the intrinsic membrane protein bacteriorhodopsin on gel-phase domain topology in two-component phase-separated bilayers. *Biophys. J.* **72**: 2217–2225.

Schroeder, F. 1988. Use of fluorescence spectroscopy in the assessment of biological membrane properties. *In* Advances in Membrane Fluidity: Methods for Studying Membrane Fluidity. R. C. Aloia, C. C. Cirtain, and L. M. Gordon, editors. Alan R. Liss, Inc., New York, NY, pp. 193–217.

Schroeder, F., Atshaves, B.P., Gallegos, A.M., McIntosh, A.L., Liu, J.C., Kier, A.B., Huang, H., and Ball, J.M., 2005, Lipid rafts and caveolae organization. *In* Advances in Molecular and Cell Biology. P. G. Frank and M. P. Lisanti, editors. Elsevier, Amsterdam, pp. 3–36.

Schroeder, F., Butko, P., Nemecz, G., Jefferson, J.R., Powell, D., Rymaszewski, Z., Dempsey, M.E., Kukowska-Latallo, J., and Lowe, J.B., 1989a, Sterol carrier protein: a ubiquitous protein in search of a function. *In* Bioengineered Molecules: Basic and Clinical Aspects. R. Verna, R. Blumenthal, and L. Frati, editors. Raven Press, New York, NY, pp. 29–45.

Schroeder, F., Fontaine, R.N., and Kinden, D.A., 1982, LM fibroblast plasma membrane subfractionation by affinity chromatography on ConA-sepharose. *Biochim. Biophys. Acta* **690**: 231–242.

Schroeder, F., Frolov, A., Schoer, J., Gallegos, A., Atshaves, B.P., Stolowich, N.J., Scott, A.I., and Kier, A.B., 1998, Intracellular sterol binding proteins, cholesterol transport and membrane domains. *In* Intracellular Cholesterol Trafficking. T. Y. Chang and D. A. Freeman, editors. Kluwer Academic Publishers, Boston, pp. 213–234.

Schroeder, F., Frolov, A., Starodub, O., Russell, W., Atshaves, B.P., Petrescu, A.D., Huang, H., Gallegos, A., McIntosh, A., Tahotna, D., Russell, D., Billheimer, J.T., Baum, C.L., and Kier, A.B., 2000, Pro-sterol carrier protein-2: role of the N-terminal presequence in structure, function, and peroxisomal targeting. *J. Biol. Chem.* **275**: 25547–25555.

Schroeder, F., Frolov, A.A., Murphy, E.J., Atshaves, B.P., Jefferson, J.R., Pu, L., Wood, W.G., Foxworth, W.B., and Kier, A.B., 1996, Recent advances in membrane cholesterol domain dynamics and intracellular cholesterol trafficking. *Proc. Soc. Exp. Biol. Med.* **213**: 150–177.

Schroeder, F., Gallegos, A.M., Atshaves, B.P., McIntosh, A., Petrescu, A.D., Huang, H., Chao, H., Yang, H., Frolov, A., and Kier, A.B., 2001a, Recent advances in membrane microdomains: rafts, caveolae and intracellular cholesterol trafficking. *Exp. Biol. Med.* **226**: 873–890.

Schroeder, F., Gallegos, A.M., Atshaves, B.P., Storey, S.M., McIntosh, A., Petrescu, A.D., Huang, H., Starodub, O., Chao, H., Yang, H., Frolov, A., and Kier, A.B., 2001b, Recent advances in membrane cholesterol microdomains: rafts, caveolae, and intracellular cholesterol trafficking. *Exp. Biol. Med.* **226**: 873–890.

Schroeder, F., Jefferson, J.R., Kier, A.B., Knittell, J., Scallen, T.J., Wood, W.G., and Hapala, I., 1991a, Membrane cholesterol dynamics: cholesterol domains and kinetic pools. *Proc. Soc. Exp. Biol. Med.* **196**: 235–252.

Schroeder, F. and Nemecz, G., 1989, Interaction of sphingomyelins and phosphatidylcholines with fluorescent dehydroergosterol. *Biochemistry* **28**: 5992–6000.

Schroeder, F. and Nemecz, G., 1990, Transmembrane Cholesterol Distribution. *In* Advances in Cholesterol Research. M. Esfahami and J. Swaney, editors. Telford Press, Caldwell, NJ, pp. 47–87.

Schroeder, F., Nemecz, G., Barenholz, Y., Gratton, E., and Thompson, T.E., 1988a, Cholestatrienol Time Resolved Fluorescence in Phosphatidylcholine Bilayers. *In* Time Resolved Laser Spectroscopy in Biochemistry. J.R. Lakowicz, M. Eftink, and J. Wampler, editors. SPIE Press, pp. 457–465.

Schroeder, F., Nemecz, G., Gratton, E., Barenholz, Y., and Thompson, T.E., 1988b, Fluorescence properties of cholestatrienol in phosphatidylcholine bilayer vesicles. *Biophys. Chem.* **32:** 57–72.

Schroeder, F., Nemecz, G., Wood, W.G., Joiner, C., Morrot, G., Ayraut-Jarrier, M., and Devaux, P.F., 1991b, Transmembrane distribution of sterol in the human erythrocyte. *Biochim. Biophys. Acta* **1066:** 183–192.

Schroeder, F., Petrescu, A.D., Huang, H., Atshaves, B.P., McIntosh, A.L., Martin, G.G., Hostetler, H.A., Vespa, A., Landrock, K., Landrock, D., Payne, H.R., and Kier, A.B., 2008, Role of fatty acid binding proteins and long chain fatty acids in modulating nuclear receptors and gene transcription. *Lipids* **43:** 1–17.

Schroeder, F. and Sweet, W.D., 1988. The role of membrane lipid and structure asymmetry on transport systems. *In* Advances in Biotechnology of Membrane Ion Transport. P. L. Jorgensen and R. Verna, editors. Serono Symposia, New York, NY, pp. 183–195.

Schroeder, F., Wood, W.G., Morrison, W.J., Fontaine, R.N., and Kier, A.B., 1989b, Synaptosomal plasma membrane lipid and structural asymmetry. *In* Neurochemical Aspects of Phospholipid Metabolism. L. Freysz, J. N. Hawthorne, and G. Toffano, editors. Liviana Press, Padova, Italy, pp. 17–33.

Schwarz, M., Russell, D.W., Dietschy, J.M., and Turley, S.D., 2001, Alternate pathways of bile acid synthesis in the cholesterol 7 alpha hydroxylase knockout mouse are not upregulated by either cholesterol or cholestyramine feeding. *J. Lipid Res.* **42:** 1594–1603.

Seedorf, U., Raabe, M., Ellinghaus, P., Kannenberg, F., Fobker, M., Engel, T., Denis, S., Wouters, F., Wirtz, K.W.A., Wanders, R.J.A., Maeda, N., and Assmann, G., 1998, Defective peroxisomal catabolism of branched fatty acyl coenzyme A in mice lacking the sterol carrier protein-2/sterol carrier protein-x gene function. *Genes Dev.* **12:** 1189–1201.

Shahedi, V., Oradd, G., and Lindblom, G., 2006, Domain-formation in DOPC/SM bilayers studied by pfg-NMR: Effect of sterol structure. *Biophys. J.* **91:** 2501–2507.

Shaw, J.E., Epand, R.F., Epand, R.M., Li, Z., Bittman, R., and Yip, C.M., 2006, Correlated fluorescence-atomic force microscopy of membrane domains: Structure of fluorescence probes determines lipid localization. *Biophys. J.* **90:** 2170–2178.

Shimada, Y., Maruya, M., Iwashita, S., and Ohno-Iwashita, Y., 2002, The C-terminal domain of perfringolysin O is an essential cholesterol-binding unit targeting to cholesterol-rich domains. *Eur. J. Biochem.* **269:** 6195–6203.

Silver, D.L., 2004, SRB1 and protein-protein interactions in hepatic high density lipoprotein metabolism. *Rev. Endocr. Metab. Dis.* **5:** 327–333.

Simopoulos, A.P., 2006, Evolutional aspects of diet, the omega-6/omega-3 ratio and genetic variation: nutritional implications for chronic diseases. *Biomed. Pharmacother.* **60:** 502–507.

Smart, E.J. and van der Westhuyzen, D.R., 1998, Scavenger receptors, caveolae, caveolin, and cholesterol trafficking. *In* Intracellular Cholesterol Trafficking. T. Y. Chang and D. A. Freeman, editors. Kluwer Academic Publishers, Boston, pp. 253–272.

Smart, E.J., Ying, Y., Mineo, C., and Anderson, R.G.W., 1995, A detergent-free method for purifying caveolae membrane from tissue culture cells. *Proc. Natl. Acad. Sci. U. S. A.* **92:** 10404–10408.

Soccio, R. and Breslow, J.L., 2003, StAR-related lipid transfer (START) proteins: mediators of intracellular lipid metabolism. *J. Biol. Chem.* **278:** 22183–22186.

Spann, N.J., Kang, S., Li, A.C., Chen, A.Z., Newberry, E.P., Davidson, N.O., Hui, S.T.Y., and Davis, R.A., 2006, Coordinate transcriptional repression of liver fatty acid binding protein and microsomal triglyceride transfer protein blocks hepatic VLDL secretion without hepatosteatosis. *J. Biol. Chem.* **281:** 33066–33077.

Spink, C.H., Yeager, M., and Feigenson, G.W., 1990, Partitioning behavior of indocarbocyanine probes between coexisting gel and fluid phases in model membranes. *Biochim. Biophys. Acta* **1023:** 25–33.

Stillwell, W. and Wassall, S.R., 2003, Docosahexaenoic acid: membrane properties of a unique fatty acid. *Chem. Phys. Lipids* **126**: 1–27.

Stolowich, N.J., Frolov, A., Petrescu, A.D., Scott, A.I., Billheimer, J.T., and Schroeder, F., 1999, Holo-sterol carrier protein-2: ^{13}C-NMR investigation of cholesterol and fatty acid binding sites. *J. Biol. Chem.* **274**: 35425–35433.

Stolowich, N.J., Petrescu, A.D., Huang, H., Martin, G., Scott, A.I., and Schroeder, F., 2002, Sterol carrier protein-2: structure reveals function. *Cell. Mol. Life Sci.* **59**: 193–212.

Storey, S.M., Gallegos, A.M., Atshaves, B.P., McIntosh, A.L., Martin, G.G., Landrock, K., Kier, A.B., Ball, J.A., and Schroeder, F., 2007a, Selective cholesterol dynamics between lipoproteins and caveolae/lipid rafts. *Biochemistry* **46**: 13891–13906.

Storey, S.M., Gibbons, T.F., Williams, C.V., Parr, R.D., Schroeder, F., and Ball, J.M., 2007b, Full-length, glycosylated NSP4 is localized to plasma membrane caveolae by a novel raft isolation technique. *J. Virol.* **81**: 5472–5483.

Stulnig, T.M., Huber, J., Leitinger, N., Imre, E.M., Angelisova, P., Nowotny, P., and Waldhaus, W., 2001, Polyunsaturated eicosapentaenoic acid displaces proteins from membrane lipid rafts by altering raft lipid composition. *J. Biol. Chem.* **276**: 37335–37340.

Subramanian, K. and Balch, W.E., 2008, NPC1/NPC2 function as a tag team duo to mobilize cholesterol. *Proc. Natl. Acad. Sci. U. S. A.* **105**: 15223–15224.

Suchanek, M., Hynynen, R., Wohlfahrt, G., Lehto, M., Johansson, M., Saarinen, H., Radzikowska, A., Thiele, C., and Olkkonen, V.M., 2007, The mammalian oxysterol-binding protein related proteins (ORPs) bind 25-hydroxycholesterol in an evolutionarily conserved pocket. *Biochem. J.* **405**: 473–480.

Sugii, S., Reid, P.C., Ohgami, N., Shimada, Y., Maue, R.A., Ninomiya, H., Ohno-Iwashita, Y., and Chang, T.-Y., 2003, Biotinylated theta-toxin derivative as a probe to examine intracellular cholesterol-rich domains in normal and Niemann-Pick type C1 cells. *J. Lipid Res.* **44**: 1033–1041.

Sweet, W.D. and Schroeder, F., 1988a. Lipid domains and enzyme activity. *In* Advances in Membrane Fluidity: Lipid Domains and the Relationship to Membrane Function. R. C. Aloia, C. C. Cirtain, and L. M. Gordon, editors. Alan R. Liss, Inc., New York, NY, pp. 17–42.

Sweet, W.D. and Schroeder, F., 1988b, Polyunsaturated fatty acids alter sterol transbilayer domains in LM fibroblast plasma membrane. *FEBS Lett.* **229**: 188–192.

Tansey, J.T., Sztalryd, C., Gruia-Gray, J., Roush, D.L., Zee, J.V., Gavrilova, O., Reitman, M.L., Deng, C.-X., Li, C., Kimmel, A.R., and Londos, C., 2001, Perilipin ablation results in a lean mouse with aberrant adipocyte lipolysis, enhanced leptin production, and resistance to diet-induced obesity. *Proc. Natl. Acad. Sci. U. S. A.* **98**: 6494–6499.

Thiele, C., Hannah, M.J., Fahrenholz, F., and Huttner, W.B., 2000, Cholesterol binds to synaptophysin and is required for biogenesis of synaptic vesicles. *Nat. Cell Biol.* **2**: 42–49.

Thigpen, J.E., Setchell, K.D., Ahlmark, K.B., Kocklear, J., Spahr, T., Caviness, G.F., Goelz, M.F., Haseman, J.K., Newbold, R.R., and Forsythe, D.B., 1999a, Phytoestrogen content of purified, open- and closed-formula laboratory animal diets. *Lab. An. Sci.* **49**: 530–536.

Thigpen, J.E., Setchell, K.D., Goelz, M.F., and Forsythe, D.B., 1999b, The phytoestrogen content of rodent diets. *Environ. Health Perspect.* **107**: A182–A183.

Troup, G.M., Tulenko, T.N., Lee, S.P., and Wrenn, S.P., 2003, Detection and characterization of laterally phase separated cholesterol domains in model lipid membranes. *Colloids Surf. B: Biointerfaces* **29**: 217–231.

Tsujishita, Y. and Hurley, J.H., 2000, Structure and lipid transport mechanism of a StAR-related transport domain. *Nat. Struct. Biol.* **7**: 408–411.

Tulenko, T.N., Chen, M., Mason, P.E., and Mason, R.P., 1998, Physical effects of cholesterol on arterial smooth muscle membranes: evidence of immiscible cholesterol domains and alterations in bilayer width during atherogenesis. *J. Lipid Res.* **39**: 947–956.

Turley, S.D., Schwarz, M., Spady, D.K., and Dietschy, J.M., 1998, Gender related differences in bile acid and sterol metabolism in outbred CD-1 mice fed low and high cholesterol diets. *Hepatology* **28**: 1088–1094.

Vainio, S., Bykov, I., Hermansson, M., Jokitalo, E., Somerharju, P., and Ikonen, E., 2005, Defective insulin receptor activation and altered lipid rafts in Niemann-Pick type C disease hepatocytes. *Biochem. J.* **391:** 465–472.

Vainio, S., Heino, S., Mansson, J.E., Fredman, P., Kuismanen, E., Vaarala, O., and Ikonen, E., 2002, Dynamic association of human insulin receptor with lipid rafts in cells lacking caveolae. *EMBO Rep.* **3:** 95–100.

Vaughan, A.M. and Oram, J.F., 2005, *J. Biol. Chem.* **280:** 30150–30157.

Wanaski, S., Ng, B.K., and Glaser, M., 2003, Caveolin scaffolding region and the membrane binding region of Src form lateral membrane domains. *Biochemistry* **42:** 42–46.

Wang, C., JeBailey, L., and Ridgway, N.D., 2002, Oxysterol binding protein (OSBP)-related protein 4 binds 25-hydroxycholesterol and interacts with vimentin intermediate filaments. *Biochem. J.* **361:** 461–472.

Wang, P.-Y., Weng, J., and Anderson, R.G.W., 2005, OSBP is a cholesterol-regulated scaffolding protein in control of ERK1/2 activation. *Science* **307:** 1472–1476.

Weisiger, R.A., 2005, Cytosolic fatty acid binding proteins catalyze two distinct steps in intracellular transport of their ligands. *Mol. Cell. Biochem.* **239:** 35–42.

Wood, W.G. and Schroeder, F., 1992. Membrane Exofacial and Cytofacial Leaflets: A New Approach to Understanding How Ethanol Alters Brain Membranes. *In* Alcohol and Neurobiology: Receptors, Membranes, and Channels. R. R. Watson, editor. CRC Press, Boca Raton, FL, pp. 161–184.

Wood, W.G., Schroeder, F., Hogy, L., Rao, A.M., and Nemecz, G., 1990, Asymmetric distribution of a fluorescent sterol in synaptic plasma membranes: effects of chronic ethanol consumption. *Biochim. Biophys. Acta* **1025:** 243–246.

Woodford, J.K., Behnke, W.D., and Schroeder, F., 1995a, Liver fatty acid binding protein enhances sterol transfer by membrane interaction. *Mol. Cell. Biochem.* **152:** 51–62.

Woodford, J.K., Colles, S.M., Myers-Payne, S., Billheimer, J.T., and Schroeder, F., 1995b, Sterol carrier protein-2 stimulates intermembrane sterol transfer by direct membrane interaction. *Chem. Phys. Lipids* **76:** 73–84.

Woodford, J.K., Jefferson, J.R., Wood, W.G., Hubbell, T., and Schroeder, F., 1993, Expression of liver fatty acid binding protein alters membrane lipid composition and structure in transfected L-cell fibroblasts. *Biochim. Biophys. Acta* **1145:** 257–265.

Wustner, D., Herrmann, A., Hao, M., and Maxfield, F.R., 2002, Rapid nonvesicular transport of sterol between the plasma membrane domains of polarized hepatic cells. *J. Biol. Chem.* **277:** 30325–30336.

Wustner, D., Mondal, M., Huang, A., and Maxfield, F.R., 2004, Different routes of transport for high density lipoprotein and its associated free sterol in polarized hepatic cells. *J. Lipid Res.* **45:** 427–437.

Wustner, D., Mondal, M., Tabas, I., and Maxfield, F.R., 2005, Direct observation of rapid internalization and intracellular transport of sterol by macrophage foam cells. *Traffic* **6:** 396–412.

Xie, Y., Newberry, E.P., Kennedy, S.M., Luo, J., and Davidson, N.O., 2009, Increased susceptibility to diet-induced gallstones in liver fatty acid binding protein knockout mice. *J. Lipid Res.* **50:** 977–987.

Yan, D., Lehto, M., Rasilainen, L., Metso, J., Ehnholm, C., Yla-Herttuala, S., Jauhiainen, M., and Olkkonen, V.M., 2007, Oxysterol binding protein induces upregulation of SREBP-1c and enhances hepatic lipogenesis. *Arterioscler. Thromb. Vasc. Biol.* **27:** 1108–1114.

Yancey, P.G., Bortnick, A.E., Kellner-Weibel, G., de la Llera-Moya, M., Phillips, M.C., and Rothblat, G.H., 2003, Importance of Different Pathways of Cellular Cholesterol Efflux. *Arterioscler. Thromb. Vasc. Biol.* **23:** 712–719.

Yu, C., Alterman, M., and Dobrowsky, R.T., 2005, Ceramide displaces cholesterol from lipid rafts and decreases the association of the cholesterol binding protein caveolin-1. *J. Lipid Res.* **46:** 1678–1691.

Yu, C., Wang, F., Kan, M., Jin, C., Jones, R.B., Weintein, M., Deng, C.-X., and McKeehan, W.L., 2000, Elevated cholesterol metabolism and bile acid synthesis in mice lacking membrane tyrosine kinase receptor FGFR4. *J. Biol. Chem.* **275:** 15482–15489.

Zanlungo, S., Amigo, L., Mendoza, H., Glick, J., Rodriguez, A., Kozarsky, K., Miquel, J.F., Rigotti, A., and Nervi, F., 2000, Overexpression of sterol carrier protein-2 in mice leads to increased hepatic cholesterol content and enterohepatic circulation of bile acids. *Gastroenterology* **118:** 135 A1165.

Zanlungo, S., Rigotti, A., and Nervi, F., 2004, Hepatic cholesterol transport from plasma into bile: implications for gallstone disease. *Cur. Opin. Lipidol.* **15:** 279–286.

Zhang, W., McIntosh, A., Xu, H., Wu, D., Gruninger, T., Atshaves, B.P., Liu, J.C.S., and Schroeder, F., 2005, Structural analysis of sterol distribution in the plasma membrane of living cells. *Biochemistry* **44:** 2864–2984.

Zhou, M., Parr, R.D., Petrescu, A.D., Payne, H.R., Atshaves, B.P., Kier, A.B., Ball, J.A., and Schroeder, F., 2004, Sterol carrier protein-2 directly interacts with caveolin-1 in vitro and in vivo. *Biochemistry* **43:** 7288–7306.

Chapter 11
Cholesterol in Niemann–Pick Type C disease

Xiaoning Bi and Guanghong Liao

Abstract Niemann-Pick Type C (NPC) disease is associated with accumulation of cholesterol and other lipids in late endosomes/lysosomes in virtually every organ; however, neurodegeneration represents the fatal cause for the disease. Genetic analysis has identified loss-of-function mutations in NPC1 and NPC2 genes as the molecular triggers for the disease. Although the precise function of these proteins has not yet been clarified, recent research suggests that they orchestrate cholesterol efflux from late endosomes/lysosomes. NPC protein deficits result in impairment in intracellular cholesterol trafficking and dysregulation of cholesterol biosynthesis. Disruption of cholesterol homeostasis is also associated with deregulation of autophagic activity and early-onset neuroinflammation, which may contribute to the pathogenesis of NPC disease. This chapter reviews recent achievements in the investigation of disruption of cholesterol homeostasis-induced neurodegeneration in NPC disease, and provides new insight for developing a potential therapeutic strategy for this disorder.

Keywords Autophagy · Cholesterol · Cyclodextrin · Endosome · Inflammation · Lysosome · Neuronal death

Abbreviations

GABA	gamma-Aminobutyric acid
GD2	disialoganglioside
LDL	low density lipoprotein
NPC	Niemann-Pick Type C

X. Bi (✉)
Department of Basic Medical Sciences, COMP, Western University of Health Sciences, Pomona, CA 91766, USA
e-mail: xbi@westernu.edu

11.1 Introduction

Niemann–Pick disease type C (NPC) is a severe neurovisceral lysosomal lipid storage disorder first described by Niemann in 1914 (Niemann, 1914), and further characterized by Pick in 1933 (Pick, 1933). NPC disease is rare with a prevalence of 1:150,000 in the general population. The associated loss-of-function mutations in NPC1 (accounting for 95% of the cases) or NPC2 (accounting for the remaining 5%) genes were identified as the genetic cause of this disease in 1997 and 2000, respectively (Carstea et al., 1997; Naureckiene et al., 2000). Clinical manifestations of NPC include vertical gaze palsy, ataxia, dystonia, dementia, cognitive impairment, and seizures with hepatosplenomegaly in early childhood; progressive neurological function defects are considered the cause of death that often occurs in teenage years (Fink et al., 1989). Pathologic features in NPC brain include neuronal loss, especially in the cerebellum, axonal spheroids, meganeurite formation (Higashi et al., 1993), and ectopic neurites (*reviewed by* Walkley & Suzuki, 2004). The hallmark of NPC at the cellular level is accumulation of cholesterol and other lipids in late endosome/lysosomes.

How disruption of cholesterol metabolism contributes to NPC neuropathology remains largely unknown and currently no effective therapy is available for this disease. Extensive investigations have focused on characterization of NPC protein functions and the links between NPC loss-of-function and cholesterol storage, however the mechanism underlying NPC pathogenesis still remains to be further elucidated. Interest in studying NPC disease was markedly increased after a link between NPC disease and Alzheimer's disease was discovered (Love et al., 1995; Ohm et al., 2003). NPC and Alzheimer's disease exhibit several similarities, including endosomal/lysosomal abnormalities, cholesterol imbalance, neurofibrillary tangle formation, deregulation of the phosphatidylinositol-3 kinase signalling cascade, and glial-mediated neuroinflammation (Auer et al., 1995; Baudry et al., 2003; Bi & Liao, 2007; Bi et al., 2005; Distl et al., 2003; German et al., 2002; Liao et al., 2007; Lynch & Bi, 2003; Suzuki et al., 1995). In addition, amyloid-beta peptide deposition was also evident in brains of NPC patients with ApoE epsilon4 homozygosity (Saito et al., 2002) (*see also* Chapter 2). Similarly, accumulation of beta-C-terminal fragments of amyloid precursor proteins was found in brains of a mouse model for NPC (Burns et al., 2003). Recently, it has been reported that brains of some NPC patients also contain aberrant alpha-synuclein accumulation and Lewy bodies (Saito et al., 2004), which inspires the proposal to include NPC as a subclass of "Lewy body diseases" (Hardy et al., 2009). New methodological and technological developments have also greatly improved our understanding of the functions of cholesterol and lipoproteins in brain. This review focuses on disruption of cholesterol homeostasis, especially cholesterol intracellular trafficking-induced neurodegeneration in NPC disease.

11.2 NPC Proteins and Intracellular Cholesterol Transport

Human NPC1 protein contains 1278 amino acids and 13 putative transmembrane domains (Davies & Ioannou, 2000). To-date more than 200 mutations that induce

NPC phenotype have been identified in the *NPC1* gene (Runz et al., 2008). NPC1 proteins have been localized in late endosomes of various cell types using different methods (Berger et al., 2007; Chikh et al., 2004; Higgins et al., 1999; Neufeld et al., 1999; Urano et al., 2008; Zhang et al., 2003). Biochemical and structural analyses have indicated that the protein contains a sterol sensing domain, homologous to the sterol sensing domain found in other key proteins in cholesterol homeostasis such as morphogen receptor Patched, 3-hydroxy-3-methylglutaryl coenzyme A reductase, SREBP cleavage activating protein, and Niemann–Pick C1-like 1 (Carstea et al., 1997; Davies et al., 2000a, b; Loftus et al., 1997; Scott et al., 2004), which is located between the third and seventh transmembrane domains (Davies & Ioannou, 2000; Millard et al., 2005); the sterol sensing domain is essential for NPC1 binding of cholesterol as demonstrated by using a photoactivatable cholesterol analog (Ohgami et al., 2004). The sterol-sensing domain is also critically involved in regulation of NPC1 protein stability (Ohsaki et al., 2006), as well as its late endosomal targeting (Scott et al., 2004). Besides the sterol-sensing domain, the N-terminal domain (amino acids 25-264) also exhibits high affinity binding for cholesterol and side-chain oxysterols in vitro (Infante et al., 2008a); however, NPC1 proteins with mutations in this region affecting sterol binding still rescue NPC1-deficient cells (Infante et al., 2008b), suggesting that the binding function of this domain is not essential. Recently, it has been shown that this region may interact with NPC2 to facilitate cholesterol efflux from late endosome/lysosomes (Infante et al., 2008c).

Human NPC2 protein (also termed HE1) is a small soluble protein, which contains 132 amino acids (Kirchhoff et al., 1996; Okamura et al., 1999). Eighteen mutations in NPC2 gene have been identified (Runz et al., 2008). Structural and biochemical studies have shown that NPC2 has a hydrophobic ligand binding pocket (Friedland et al., 2003) and binds cholesterol with a 1:1 stoichiometry (Xu et al., 2007) and a high affinity ($K_d = 30\text{–}50$ nM) (Ko et al., 2003). Cholesterol binding is essential for NPC2 function since mutant NPC2 proteins that lack high affinity cholesterol binding also fail to rescue NPC2-null cells (Ko et al., 2003). Currently, it is generally accepted that NPC2 is mainly present in the lysosomal lumen (Naureckiene et al., 2000; Willenborg et al., 2005). Although the two proteins are very different structurally, recessive inheritance of either one leads to NPC disease with almost indistinguishable phenotypes (Vance, 2006; Vanier & Millat, 2004), suggesting that the two proteins must function in a closely related fashion. This notion has been further confirmed by a direct comparison study of mice with Npc1, Npc2, or Npc1/Npc2 double deficiency (Sleat et al., 2004). However, exactly how the two proteins participate in cholesterol efflux from late endosomes/lysosomes remains an open question. Using fluorescence-labelled NPC2 a recent study showed that NPC2 was able to transfer cholesterol to vesicular membranes (Cheruku et al., 2006), possibly by direct NPC2-membrane interaction (Xu et al., 2008). Using in vitro assays, Infante et al. (2008c) showed that a bidirectional transfer of cholesterol occurs between liposomes and either NPC1 or NPC2, although that mediated by NPC1 is much slower compared to NPC2. However, in the presence of NPC2, the bidirectional transfer is enhanced over 100 fold. These data suggest a model in which NPC1 and NPC2 may bind cholesterol sequentially and promote its egress from late endosomes/lysosomes.

11.3 Cholesterol Accumulation in Niemann-Pick Type C Disease

The essential role of cholesterol in maintaining functional integrity of virtually all types of cell has gained tremendous attention. Cholesterol homeostasis is critical for normal function of the central nervous system (CNS), which is particularly rich in cholesterol. Although the human brain comprises only 2% of the body mass, it contains about 25% of the total body unesterified cholesterol (Dietschy & Turley, 2001). In contrast to other peripheral tissues that obtain cholesterol from both de novo synthesis within the cells and uptake of cholesterol-containing lipoprotein particles from serum, nearly all cholesterol supply in the CNS comes from in situ synthesis (Turley et al., 1996). Previous studies have shown that during early development neurons rely heavily on de novo cholesterol synthesis (Jurevics et al., 1997; Turley et al., 1996), whereas uptake of exogenous cholesterol provided by glia may be critical for mature neurons later (Cruz & Chang, 2000; Pitas et al., 1987; Weisgraber et al., 1994). Dysfunction of either de novo synthesis or uptake of exogenous cholesterol can lead to disruption of cholesterol homeostasis in neurons.

Cholesterol esterification impairment in NPC disease was first revealed in fibroblasts cultured from NPC patients, which distinguished NPC disease from other lysosomal storage diseases (Pentchev et al., 1985; Vanier et al., 1988). Subsequent research found that not only unesterified cholesterol, but also gangliosides GM2 and GM3, and bis-monoacylglycerol phosphate accumulated in late endosomes/lysosomes (Kobayashi et al., 1999; Liscum & Munn, 1999; Sokol et al., 1988; te Vruchte et al., 2004; Watanabe et al., 1998; Zervas et al., 2001b). The discovery that lipids other than cholesterol also accumulated in late endosomes/lysosomes has led to the debate over whether aberrant trafficking of cholesterol or of other lipids is the primary cause of the NPC phenotype (te Vruchte et al., 2004; Zervas et al., 2001b). It was suggested that cholesterol accumulation was ganglioside-dependent since depletion of the ganglioside-related enzyme GM2/GD2 synthase in NPC-deficient neurons diminished cholesterol accumulation (Gondre-Lewis et al., 2003). However, a more recent study failed to reproduce these results; Li et al reported that deprivation of either GM2/GD2 or GM3 did not reduce cholesterol accumulation or pathology in Npc1-/- mice (Li et al., 2008). In fact, the lifespan was shortened by these manipulations (Li et al., 2008). Another recent study suggested that sphingosine storage was an initiating factor that caused altered calcium homeostasis in lysosomes, leading to the secondary accumulation of sphingolipids and cholesterol (Lloyd-Evans et al., 2008). The caveat for this hypothesis is that both NPC1 and NPC2 have been shown to directly bind cholesterol, and not sphingosine. Therefore, to argue that sphingosine storage is the initiating event, some additional mechanism is needed. In this regard, one recent study reported that mutation in the sterol-sensing domain of a yeast NPC-related protein led to subcellular sphingolipid redistribution (Malathi et al., 2004). Whether this holds true in mammals remains to be determined.

Besides abnormal late endosomes/lysosomes (Zervas et al., 2001a), early endosomes were also reported to be substantially enlarged and to contain high levels of the lysosomal hydrolase cathepsin D in Purkinje cells and microglia in brain tissues of NPC patients (Jin et al., 2004). Furthermore, our previous research has shown that cathepsin D immunoreactivity was increased not only in microglia, but also in neurons in Npc1-/- mice (Liao et al., 2007, 2009), a murine model of NPC disease. The phenotype in these mice is almost identical to that in humans except that only hyperphosphorylation of tau, but not neurofibrillary tangles, has been observed in the mutant mouse brain (German et al., 2001). These observations suggest that mutations in NPC1 gene impair functions in both early and late endocytic pathways; whether disruption of early endosomes is induced by accumulation of cholesterol in late endosomes/lysosomes or is an independent deficit needs further study.

Cholesterol accumulation was detected as early as postnatal day 9 in various brain regions in Npc1-/- mice (Reid et al., 2004). In the cerebellum, although the morphology of Purkinje cells was normal at this age, cholesterol accumulation was already evident in cell bodies and dendritic arbors. In other brain areas, cholesterol accumulation was first observed in neuronal perikarya and at the base of axonal hillocks, especially in the thalamus (Reid et al., 2004). In later stages, cholesterol accumulation was also found in astrocytes (Mutka et al., 2004; Reid et al., 2004) and active microglia (Liao et al., 2009). Cholesterol accumulation in cell bodies, and to a smaller degree in axons, was observed in sympathetic neurons cultured from Npc1-/- mice and maintained in serum-free medium for only one day (Karten et al., 2002). However, whether cholesterol accumulation occurs in embryonic brain tissues is still under debate. Interestingly, the percentage of Npc1-/-pups bred from heterozygous parents is about 12% instead of the predicted 25% (Karten et al., 2002), implicating possible embryonic lethality in Npc1-/- mice. At embryonic day 16, the percentage of homozygous embryos is still 25%, which indicates that death takes place after E16 (Henderson et al., 2000). Additional research is needed to define the potential links between disruption of cholesterol homeostasis and embryonic death. Nevertheless, studies reviewed in this section have clearly shown that cholesterol accumulation occurs early in life in Npc1-/- mice, although the mechanism for this early event remains obscure.

11.4 Suppression of Brain Cholesterol Synthesis in NPC Disease

A paradox in brain cholesterol metabolism in Npc1-/- mice is that although cholesterol accumulation in neurons and glia is clearly evident, the total amount of brain cholesterol is not significantly increased, which is in contrary to what is found in other organs. A direct measurement study showed that the total amount of cholesterol in brain of newborn Npc1-/- mice was more than that of wild-type mice, but gradually reduced with age (Xie et al., 2000). By 7-week postnatal, cholesterol levels were significantly reduced in midbrain, brainstem and spinal cord in

Npc1-/- mice and the reduction was paralleled with an increase in net cholesterol turnover (Xie et al., 2000). Further study from the same research group demonstrated that the synthesis rate of cholesterol was reduced while its excretion from brain was enhanced (Xie et al., 2003). Excretion was independent of the 24-hydroxycholesterol pathway that the brain normally uses to transfer excess cholesterol to plasma. Research from other groups supported the notion that cholesterol synthesis in Npc1-deficient mice was decreased. For instance, an in vitro study showed that cholesterol synthesis in Npc1-deficient astrocytes was reduced (Reid et al., 2003). Furthermore, the synthesis of neurosteroids, such as allopregnanolone (Griffin et al., 2004) and testosterone (Roff et al., 1993), was also decreased in Npc1-/- mice. Using microarray analysis we found that mRNAs for several key proteins in the sterol biosynthesis pathway were significantly reduced (Liao et al., unpublished data). However, other studies indicated that there were no significant changes in cholesterol synthesis in Npc1-/- mice (Karten et al., 2005; Reid et al., 2008). These controversial findings suggest that the impairment in cholesterol synthesis requires further investigation.

11.5 Impairment of Cholesterol Transport in NPC Disease

The cloning of NPC1 protein, and later of NPC2 protein, sped up the investigation of the mechanisms underlying pathogenesis in the disease. NPC1 protein is generally located in late endosomes (Higgins et al., 1999; Kobayashi et al., 1999). In situ hybridization study showed that in mouse brain, Npc1 mRNA was detected in the majority of neurons in nearly all regions, but at significantly higher levels in cerebellum and in specific pontine nuclei; this regional specificity was established by postnatal day 7 (Prasad et al., 2000). The earliest neuronal expression of Npc1 mRNA was detected at embryonic day 15 (Prasad et al., 2000). As discussed above, while the structure of the NPC1 protein is well characterized, little is known regarding its function in vivo. Several lines of evidence indicate that NPC1 may be involved in the trafficking of both LDL-derived and endogenously synthesized cholesterol from the endoplasmic reticulum to the trans-Golgi network (Higgins et al., 1999; Reid et al., 2003; *see* Scott & Ioannou, 2004 for a recent review).

Brain cholesterol homeostasis is achieved through different mechanisms from those in other organs. In vivo, direct measurement of the uptake of low density lipoproteins (LDL) in different brain regions has indicated that cholesterol carried in LDL circulating in serum plays little or no role in the process of sterol acquisition during brain development or in cholesterol turnover in the mature central nervous system (Turley et al., 1996). In contrast, lipoproteins in brain transport exogenous cholesterol generated in glia to neurons. Several members of the LDL receptor family, including apolipoprotein E, A1, D, and J, are expressed in brains with apolipoprotein E and apolipoprotein J being the major apolipoproteins in CNS (Gong et al., 2002). Apolipoprotein E is mainly synthesized by astrocytes and microglia and to a small extent by neurons (Brecht et al., 2004).

Extracellular cholesterol in the brain is transported mostly by apolipoprotein E (Boyles et al., 1985), and a small amount by apolipoprotein A1, apolipoprotein D, and apolipoprotein J (Patel et al., 1995). Expression of apolipoprotein D is increased in Npc1-deficient mice (Li et al., 2005; Ong et al., 2002; Suresh et al., 1998), although the exact function of apolipoprotein D in brain is not clear. Levels of apolipoprotein E mRNA (Li et al., 2005) and protein (Liao et al., unpublished data) are also increased in Npc1-/- mice. However, using a functional assay, Karten and colleagues have shown that apolipoprotein E-containing lipoproteins generated by Npc1-/- and Npc1+/+ glia were equally capable of stimulating axonal elongation (Karten et al., 2005). Furthermore, degeneration of neurons and glia in double Npc1-/-/LDLR-/- deficient mice was similar to that in Npc1-/- mice, which indicates an LDLR-independent pathogenic process (German et al., 2001). On the other hand, neurons cultured from Npc1-/- mice exhibited cholesterol accumulation in cell bodies, while distal axons had reduced cholesterol (Karten et al., 2002), suggesting impairment in intracellular cholesterol trafficking. Overall, although the precise role of lipoproteins in NPC disease needs to be further defined, these studies suggest that disruption of cholesterol transport, especially inside neurons, may play a critical role in NPC pathogenesis.

11.6 Cholesterol Accumulation-Associated Autophagy in NPC Disease

Although NPC1 gene is expressed in all tissues, the nervous system manifestations of the disease are predominant and lethal. The reason why neurons are most vulnerable to NPC1 deficiency remains unknown. Apoptosis was found in cortical neurons treated with a blocker of cholesterol transport, U18666A (Koh et al., 2006, 2007), in liver cells of Npc1-/- mice (Beltroy et al., 2005), and in brains of NPC patients and Npc1-/- mice (Wu et al., 2005). However, additional results support the notion that another type of programmed cell death, autophagic cell death, plays a critical role in neuronal death in NPC disease.

Autophagy or "self-eating" is an adaptation process conserved in cells from yeasts to mice and humans (Klionsky & Emr, 2000). As a house-keeping mechanism, autophagy engulfs fragments of damaged organelles and long-lived membrane proteins and transfers packaged cargos to lysosomes for degradation (Xie & Klionsky, 2007). Recent evidence indicates that autophagy is associated with neurodegeneration in Alzheimer's disease (Nixon, 2007), Parkinson's disease (Pan et al., 2008), and Huntington disease (Ravikumar et al., 2004). Research from our laboratory and others have also shown that autophagy activity is increased in Npc1-/- mice (Ko et al., 2005; Liao et al., 2007; Pacheco et al., 2007). Levels of LC3 (microtubule-associated protein 1 light chain 3 protein)-II, a marker of autophagic activation (Kabeya et al., 2000; Klionsky et al., 2008; Tanida et al., 2005), are increased in brain of Npc1-/- mice (Liao et al., 2007) and in fibroblasts with NPC1 deficiency (Pacheco et al., 2007). LC3-immunopositive granules were

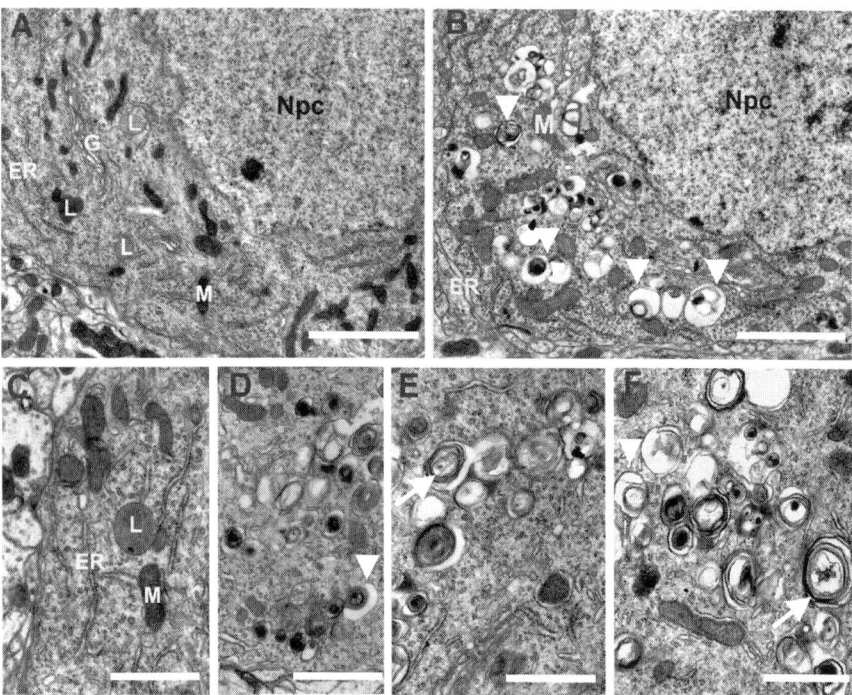

Fig. 11.1 Ultrastructure of Purkinje cells in Npc1-/- and Npc1+/+ mice. **A** A Purkinje cell of a 6-week-old Npc1+/+ mice. Npc, nucleus of Purkinje cell. ER, endoplasmic reticulum; G, Golgi apparatus; L, lysosome; M, mitochondria. **B** A Purkinje cell of a 6-week-old Npc1-/- mice. Numerous vacuoles (arrowheads) of different sizes with various levels of electron-dense materials are present in the cytoplasm. **C** Lysosome-like structures exist in Purkinje cells in Npc1+/+ mice. **D–F** Morphology of various membranous vacuoles. Some of them are with double membranes (arrowheads), whereas others have multilamellated structures (arrows). Scale bars = 2 μm (A and B), 1 μm (C–F). (Adapted from Liao et al., 2007)

also labelled with filipin-stained cholesterol, suggesting that autophagy in NPC is closely associated with cholesterol accumulation (Liao et al., 2007). This notion is further supported by our recent finding that suppression of autophagy by treatment of mice with allopregnanolone, a neurosteroid that is deficient in brain of Npc1-/- mice, was associated with reduction in cholesterol accumulation (Liao et al., 2009). Ultrastructural analysis with electron microscopy revealed the existence of classic double membrane vacuole-like structures in 6-weeks old Npc1-/- mice (Fig. 11.1) (Liao et al., 2007). These results suggest an increase in autophagosomes in NPC. However, as the volume of autophagosomes depends on the dynamics of influx and efflux, whether this increase represents a net increase in autophagic activity or a efflux jam because of lysosomal dysfunction remains an open question (Bi & Liao, 2007).

The mechanism by which autophagic activity is elevated is largely unknown. It is generally agreed that amino acid starvation induces autophagic activity; whether lipid/cholesterol starvation also results in enhanced autophagic activity is not as certain. Depletion of cholesterol in human fibroblasts, by either acute chemical treatment or metabolic suppression of cholesterol synthesis, increased levels of LC3-II and LC3-II-immunopositive granules suggesting an increase in autophagic activity (Cheng et al., 2006). Electron microscopy examination revealed that autophagic vacuoles induced by cholesterol depletion were indistinguishable from that induced by amino acid starvation, which further supports the idea that cholesterol starvation can also initiate autophagy. More convincing evidence suggesting an increase in autophagic induction in NPC was obtained by Ishibashi and colleagues, who recently reported that cholesterol depletion by U18666A inhibited the formation of filipin-labeled LC3-immunopositive granules but promoted the formation of ring-shaped filipin-negative LC3-immunopositive structures (Ishibashi et al., 2009). However, the molecular basis for cholesterol depletion-induced autophagy remains elusive. Blocking intracellular cholesterol trafficking by U18666A in wild-type fibroblasts increased the expression of LC3 and the conversion of LC3-I to LC3-II, a process that was dependent on the Beclin-1 rather than the mTOR (mammalian target of rapamycin) signalling pathway (Pacheco et al., 2007), which may imply that cholesterol depletion-induced autophagy uses different molecular mechanisms from those induced by amino acid starvation. In reviewing the literature, it is clear that an increase in autophagosomes, possibly by enhanced initiation rather than decreased efflux, is associated with cholesterol accumulation in NPC disease. However, the underlying mechanism is not as clear.

11.7 Treatment Development for NPC Disease

Currently there is no effective treatment for NPC disease. Clinically, NPC patients are often placed on a cholesterol-lowering treatment, although the results are not very convincing. An animal study showed that introduction of functional *npc1* gene in Npc1-/- mouse brain with a prion promoter prevented neurodegeneration, normalized lifespan, and corrected sterility (Loftus et al., 2002). Results from this study further emphasize the importance of neurodegeneration in NPC disease. Another recent study has shown that restoring Npc1 function only in astrocytes triples Npc1-/- mice lifespan, indicating that astrocytes play a critical role in NPC disease (Zhang et al., 2008). Substrate-reduction therapy, by using N-butyldeoxynojirimycin (Miglustat), an inhibitor of glycosphingolipid biosynthesis, has also shown promising results; it extended the average lifespan from 67 days to 89 days in the NPC mouse model (Zervas et al., 2001b). Supplementing the neurosteroid, allopregnanolone, by a single injection at postnatal day 7 has been shown to double Npc1-/- mice lifespan (Griffin et al., 2004). Regarding the potential mechanisms of allopregnanolone treatment, in vitro experiments indicated that allopregnanolone-mediated Purkinje cell survival was blocked by the $GABA_A$ receptor antagonist, bicuculline, suggesting that the effect of the drug might be mediated by $GABA_A$ receptors (Griffin et al., 2004). However, this hypothesis has been challenged by the finding that ent-allopregnanolone, an allopregnanolone stereoisomer without $GABA_A$ receptor agonist function, has identical effects as natural allopregnanolone, which strongly suggests the existence of $GABA_A$-independent mechanisms (Langmade et al., 2006). On the other hand, T0901317, a synthetic oxysterol ligand, acts in concert with allopregnanolone to promote survival and to delay the onset of neurological symptoms (Langmade et al., 2006). The effects of allopregnanolone and T0901317 correlate with their ability to activate the pregnane X receptor, suggesting a role for this receptor. However, other researchers have reported that there is no detectable pregnane X receptor activity in mouse cerebellum (Bookout et al., 2006; Gofflot et al., 2007; Repa et al., 2007). Liu et al. (2008) recently reported that administration of β-cyclodextrin, the vehicle used in the allopregnanolone studies, also rescued Npc1-/- mice. This study has inspired a "compassionate use" of β-cyclodextrin to twin NPC patients, which was approved by the FDA (http://www.addiandcassi.com). However, results from our recent study showed that while combined allopregnanolone and cyclodextrin treatment markedly reduced cholesterol accumulation, autophagic/lysosomal dysfunction, microglia- and astrocyte-mediated inflammation, and increased myelination in brain of Npc1-/- mice at one month (Ahmad et al., 2005; Liao et al., 2009), cyclodextrin treatment alone only slightly reduced cholesterol accumulation and had little effect on other pathological features (Liao et al., 2009). These results raise caution regarding the clinical use of cyclodextrin in NPC.

11.8 Conclusions

Recent studies have indicated that disruption in cholesterol homeostasis plays an important role in several neurodegenerative diseases, including Alzheimer's disease (*see* Chapter 2) and NPC disease. In NPC, cholesterol accumulation occurs early and is closely associated with neurodegeneration. Although loss-of-function mutations in NPC1 and NPC2 genes have been identified as the genetic cause of this disorder, the precise mechanism by which NPC deficit leads to neuronal death remains elusive. Recent research has led to a better understanding of the roles of NPC1 and NPC2 in cholesterol flux through late endosomes/lysosomes, which may reveal new therapeutic strategies. Other pathological features such as neuroinflammation and autophagy are also linked to the development of the disease. Therefore, we speculate that multiple therapeutic strategies, including lipid transport improvement, inflammation suppression, and autophagy manipulation should be considered along with gene therapy to provide a comprehensive treatment of this disease.

Acknowledgements This work was supported by a grant from NINDS (NS048423 to XB) and by funds from Western University of Health Sciences (Pomona, CA) to X.B. Xiaoning Bi was also supported by funds from the Daljit and Elaine Sarkaria Chair.

References

Ahmad, I., Lope-Piedrafita, S., Bi, X., Hicks, C., Yao, Y., Yu, C., Chaitkin, E., Howison, C. M., Weberg, L., Trouard, T. P., Erickson, R. P., 2005, Allopregnanolone treatment, both as a single injection or repetitively, delays demyelination and enhances survival of Niemann-Pick C mice. *J. Neurosci. Res.* **82:** 811–821

Auer, I. A., Schmidt, M. L., Lee, V. M., Curry, B., Suzuki, K., Shin, R. W., Pentchev, P. G., Carstea, E. D., Trojanowski, J. Q., 1995, Paired helical filament tau (PHFtau) in Niemann-Pick type C disease is similar to PHFtau in Alzheimer's disease. *Acta Neuropathol.* **90:** 547–551

Baudry, M., Yao, Y., Simmons, D., Liu, J., Bi, X., 2003, Postnatal development of inflammation in a murine model of Niemann-Pick type C disease: immunohistochemical observations of microglia and astroglia. *Exp. Neurol.* **184:** 887–903

Beltroy, E. P., Richardson, J. A., Horton, J. D., Turley, S. D., Dietschy, J. M., 2005, Cholesterol accumulation and liver cell death in mice with Niemann-Pick type C disease. *Hepatology* **42:** 886–893

Berger, A. C., Salazar, G., Styers, M. L., Newell-Litwa, K. A., Werner, E., Maue, R. A., Corbett, A. H., Faundez, V., 2007, The subcellular localization of the Niemann-Pick Type C proteins depends on the adaptor complex AP-3. *J. Cell Sci.* **120:** 3640–3652

Bi, X., Liao, G., 2007, Autophagic-lysosomal dysfunction and neurodegeneration in Niemann-Pick Type C mice: lipid starvation or indigestion? *Autophagy* **3:** 646–648

Bi, X., Liu, J., Yao, Y., Baudry, M., Lynch, G., 2005, Deregulation of the phosphatidylinositol-3 kinase signaling cascade is associated with neurodegeneration in Npc1-/- mouse brain. *Am. J. Pathol.* **167:** 1081–1092

Bookout, A. L., Jeong, Y., Downes, M., Yu, R. T., Evans, R. M., Mangelsdorf, D. J., 2006, Anatomical profiling of nuclear receptor expression reveals a hierarchical transcriptional network. *Cell* **126:** 789–799

Boyles, J. K., Pitas, R. E., Wilson, E., Mahley, R. W., Taylor, J. M., 1985, Apolipoprotein E associated with astrocytic glia of the central nervous system and with nonmyelinating glia of the peripheral nervous system. *J. Clin. Invest.* **76:** 1501–1513

Brecht, W. J., Harris, F. M., Chang, S., Tesseur, I., Yu, G. Q., Xu, Q., Dee Fish, J., Wyss-Coray, T., Buttini, M., Mucke, L., Mahley, R. W., Huang, Y., 2004, Neuron-specific apolipoprotein e4 proteolysis is associated with increased tau phosphorylation in brains of transgenic mice. *J. Neurosci.* **24:** 2527–2534

Burns, M., Gaynor, K., Olm, V., Mercken, M., LaFrancois, J., Wang, L., Mathews, P. M., Noble, W., Matsuoka, Y., Duff, K., 2003, Presenilin redistribution associated with aberrant cholesterol transport enhances beta-amyloid production in vivo. *J. Neurosci.* **23:** 5645–5649

Carstea, E. D., Morris, J. A., Coleman, K. G., Loftus, S. K., Zhang, D., Cummings, C., Gu, J., Rosenfeld, M. A., Pavan, W. J., Krizman, D. B., Nagle, J., Polymeropoulos, M. H., Sturley, S. L., Ioannou, Y. A., Higgins, M. E., Comly, M., Cooney, A., Brown, A., Kaneski, C. R., Blanchette-Mackie, E. J., Dwyer, N. K., Neufeld, E. B., Chang, T. Y., Liscum, L., Strauss, J. F., 3rd, Ohno, K., Zeigler, M., Carmi, R., Sokol, J., Markie, D., O'Neill, R. R., van Diggelen, O. P., Elleder, M., Patterson, M. C., Brady, R. O., Vanier, M. T., Pentchev, P. G., Tagle, D. A., 1997, Niemann-Pick C1 disease gene: homology to mediators of cholesterol homeostasis. *Science* **277:** 228–231

Cheng, J., Ohsaki, Y., Tauchi-Sato, K., Fujita, A., Fujimoto, T., 2006, Cholesterol depletion induces autophagy. *Biochem. Biophys. Res. Commun.* **351:** 246–252

Cheruku, S. R., Xu, Z., Dutia, R., Lobel, P., Storch, J., 2006, Mechanism of cholesterol transfer from the Niemann-Pick type C2 protein to model membranes supports a role in lysosomal cholesterol transport. *J. Biol. Chem.* **281:** 31594–31604

Chikh, K., Vey, S., Simonot, C., Vanier, M. T., Millat, G., 2004, Niemann-Pick type C disease: importance of N-glycosylation sites for function and cellular location of the NPC2 protein. *Mol. Genet. Metab.* **83:** 220–230

Cruz, J. C., Chang, T. Y., 2000, Fate of endogenously synthesized cholesterol in Niemann-Pick type C1 cells. *J. Biol. Chem.* **275:** 41309–41316

Davies, J. P., Chen, F. W., Ioannou, Y. A., 2000a, Transmembrane molecular pump activity of Niemann-Pick C1 protein. *Science* **290:** 2295–2298

Davies, J. P., Ioannou, Y. A., 2000, Topological analysis of Niemann-Pick C1 protein reveals that the membrane orientation of the putative sterol-sensing domain is identical to those of 3-hydroxy-3-methylglutaryl-CoA reductase and sterol regulatory element binding protein cleavage-activating protein. *J. Biol. Chem.* **275**: 24367–24374

Davies, J. P., Levy, B., Ioannou, Y. A., 2000b, Evidence for a Niemann-pick C (NPC) gene family: identification and characterization of NPC1L1. *Genomics* **65**: 137–145

Dietschy, J. M., Turley, S. D., 2001, Cholesterol metabolism in the brain. *Curr. Opin. Lipidol.* **12**: 105–112

Distl, R., Treiber-Held, S., Albert, F., Meske, V., Harzer, K., Ohm, T. G., 2003, Cholesterol storage and tau pathology in Niemann-Pick type C disease in the brain. *J. Pathol.* **200**: 104–111

Fink, J. K., Filling-Katz, M. R., Sokol, J., Cogan, D. G., Pikus, A., Sonies, B., Soong, B., Pentchev, P. G., Comly, M. E., Brady, R. O., et al., 1989, Clinical spectrum of Niemann-Pick disease type C. *Neurology* **39**: 1040–1049

Friedland, N., Liou, H. L., Lobel, P., Stock, A. M., 2003, Structure of a cholesterol-binding protein deficient in Niemann-Pick type C2 disease. *Proc. Natl. Acad. Sci. USA* **100**: 2512–2517

German, D. C., Liang, C. L., Song, T., Yazdani, U., Xie, C., Dietschy, J. M., 2002, Neurodegeneration in the Niemann-Pick C mouse: glial involvement. *Neuroscience* **109**: 437–450

German, D. C., Quintero, E. M., Liang, C., Xie, C., Dietschy, J. M., 2001, Degeneration of neurons and glia in the Niemann-Pick C mouse is unrelated to the low-density lipoprotein receptor. *Neuroscience* **105**: 999–1005

Gofflot, F., Chartoire, N., Vasseur, L., Heikkinen, S., Dembele, D., Le Merrer, J., Auwerx, J., 2007, Systematic gene expression mapping clusters nuclear receptors according to their function in the brain. *Cell* **131**: 405–418

Gondre-Lewis, M. C., McGlynn, R., Walkley, S. U., 2003, Cholesterol accumulation in NPC1-deficient neurons is ganglioside dependent. *Curr. Biol.* **13**: 1324–1329

Gong, J. S., Kobayashi, M., Hayashi, H., Zou, K., Sawamura, N., Fujita, S. C., Yanagisawa, K., Michikawa, M., 2002, Apolipoprotein E (ApoE) isoform-dependent lipid release from astrocytes prepared from human ApoE3 and ApoE4 knock-in mice. *J. Biol. Chem.* **277**: 29919–29926

Griffin, L. D., Gong, W., Verot, L., Mellon, S. H., 2004, Niemann-Pick type C disease involves disrupted neurosteroidogenesis and responds to allopregnanolone. *Nat. Med.* **10**: 704–711

Hardy, J., Lewis, P., Revesz, T., Lees, A., Paisan-Ruiz, C., 2009, The genetics of Parkinson's syndromes: a critical review. *Curr. Opin. Genet. Dev.* **19**: 254–265

Henderson, L. P., Lin, L., Prasad, A., Paul, C. A., Chang, T. Y., Maue, R. A., 2000, Embryonic striatal neurons from Niemann-Pick type C mice exhibit defects in cholesterol metabolism and neurotrophin responsiveness. *J. Biol. Chem.* **275**: 20179–20187

Higashi, Y., Murayama, S., Pentchev, P. G., Suzuki, K., 1993, Cerebellar degeneration in the Niemann-Pick type C mouse. *Acta Neuropathol.* **85**: 175–184

Higgins, M. E., Davies, J. P., Chen, F. W., Ioannou, Y. A., 1999, Niemann-Pick C1 is a late endosome-resident protein that transiently associates with lysosomes and the trans-Golgi network. *Mol. Genet. Metab.* **68**: 1–13

Infante, R. E., Abi-Mosleh, L., Radhakrishnan, A., Dale, J. D., Brown, M. S., Goldstein, J. L., 2008a, Purified NPC1 protein. I. Binding of cholesterol and oxysterols to a 1278-amino acid membrane protein. *J. Biol. Chem.* **283**: 1052–1063

Infante, R. E., Radhakrishnan, A., Abi-Mosleh, L., Kinch, L. N., Wang, M. L., Grishin, N. V., Goldstein, J. L., Brown, M. S., 2008b, Purified NPC1 protein: II. Localization of sterol binding to a 240-amino acid soluble luminal loop. *J. Biol. Chem.* **283**: 1064–1075

Infante, R. E., Wang, M. L., Radhakrishnan, A., Kwon, H. J., Brown, M. S., Goldstein, J. L., 2008c, NPC2 facilitates bidirectional transfer of cholesterol between NPC1 and lipid bilayers, a step in cholesterol egress from lysosomes. *Proc. Natl. Acad. Sci. USA* **105**: 15287–15292

Ishibashi, S., Yamazaki, T., Okamoto, K., 2009, Association of autophagy with cholesterol-accumulated compartments in Niemann-Pick disease type C cells. *J. Clin. Neurosci.* **16**: 954–959

Jin, L. W., Shie, F. S., Maezawa, I., Vincent, I., Bird, T., 2004, Intracellular accumulation of amyloidogenic fragments of amyloid-beta precursor protein in neurons with Niemann-Pick type C defects is associated with endosomal abnormalities. *Am. J. Pathol.* **164:** 975–985

Jurevics, H. A., Kidwai, F. Z., Morell, P., 1997, Sources of cholesterol during development of the rat fetus and fetal organs. *J. Lipid Res.* **38:** 723–733

Kabeya, Y., Mizushima, N., Ueno, T., Yamamoto, A., Kirisako, T., Noda, T., Kominami, E., Ohsumi, Y., Yoshimori, T., 2000, LC3, a mammalian homologue of yeast Apg8p, is localized in autophagosome membranes after processing. *EMBO J.* **19:** 5720–5728

Karten, B., Hayashi, H., Francis, G. A., Campenot, R. B., Vance, D. E., Vance, J. E., 2005, Generation and function of astroglial lipoproteins from Niemann-Pick type C1-deficient mice. *Biochem. J.* **387:** 779–788

Karten, B., Vance, D. E., Campenot, R. B., Vance, J. E., 2002, Cholesterol accumulates in cell bodies, but is decreased in distal axons, of Niemann-Pick C1-deficient neurons. *J. Neurochem.* **83:** 1154–1163

Kirchhoff, C., Osterhoff, C., Young, L., 1996, Molecular cloning and characterization of HE1, a major secretory protein of the human epididymis. *Biol. Reprod.* **54:** 847–856

Klionsky, D. J., Abeliovich, H., Agostinis, P., Agrawal., D. K., Aliev, G., Askew, D. S., Baba, M., Baehrecke, E. H., Bahr, B. A., Ballabio, A., Bamber, B. A., Bassham, D. C., Bergamini, E., Bi, X., Biard-Piechaczyk, M., Blum, J. S., Bredesen, D. E., Brodsky, J. L., Brumell, J. H., Brunk, U. T., Bursch, W., Camougrand, N., Cebollero, E., Cecconi, F., Chen, Y., Chin, L. S., Choi, A., Chu, C. T., Chung, J., Clarke, P. G., Clark, R. S., Clarke, S. G., Clave, C., Cleveland, J. L., Codogno, P., Colombo, M. I., Coto-Montes, A., Cregg, J. M., Cuervo, A. M., Debnath, J., Demarchi, F., Dennis, P. B., Dennis, P. A., Deretic, V., Devenish, R. J., Di Sano, F., Dice, J. F., Difiglia, M., Dinesh-Kumar, S., Distelhorst, C. W., Djavaheri-Mergny, M., Dorsey, F. C., Droge, W., Dron, M., Dunn, W. A., Jr., Duszenko, M., Eissa, N. T., Elazar, Z., Esclatine, A., Eskelinen, E. L., Fesus, L., Finley, K. D., Fuentes, J. M., Fueyo, J., Fujisaki, K., Galliot, B., Gao, F. B., Gewirtz, D. A., Gibson, S. B., Gohla, A., Goldberg, A. L., Gonzalez, R., Gonzalez-Estevez, C., Gorski, S., Gottlieb, R. A., Haussinger, D., He, Y. W., Heidenreich, K., Hill, J. A., Hoyer-Hansen, M., Hu, X., Huang, W. P., Iwasaki, A., Jaattela, M., Jackson, W. T., Jiang, X., Jin, S., Johansen, T., Jung, J. U., Kadowaki, M., Kang, C., Kelekar, A., Kessel, D. H., Kiel, J. A., Kim, H. P., Kimchi, A., Kinsella, T. J., Kiselyov, K., Kitamoto, K., Knecht, E., Komatsu, M., Kominami, E., Kondo, S., Kovacs, A. L., Kroemer, G., Kuan, C. Y., Kumar, R., Kundu, M., Landry, J., Laporte, M., Le, W., Lei, H. Y., Lenardo, M. J., Levine, B., Lieberman, A., Lim, K. L., Lin, F. C., Liou, W., Liu, L. F., Lopez-Berestein, G., Lopez-Otin, C., Lu, B., Macleod, K. F., Malorni, W., Martinet, W., Matsuoka, K., Mautner, J., Meijer, A. J., Melendez, A., Michels, P., Miotto, G., Mistiaen, W. P., Mizushima, N., Mograbi, B., Monastyrska, I., Moore, M. N., Moreira, P. I., Moriyasu, Y., Motyl, T., Munz, C., Murphy, L. O., Naqvi, N. I., Neufeld, T. P., Nishino, I., Nixon, R. A., Noda, T., Nurnberg, B., Ogawa, M., Oleinick, N. L., Olsen, L. J., Ozpolat, B., Paglin, S., Palmer, G. E., Papassideri, I., Parkes, M., Perlmutter, D. H., Perry, G., Piacentini, M., Pinkas-Kramarski, R., Prescott, M., Proikas-Cezanne, T., Raben, N., Rami, A., Reggiori, F., Rohrer, B., Rubinsztein, D. C., Ryan, K. M., Sadoshima, J., Sakagami, H., Sakai, Y., Sandri, M., Sasakawa, C., Sass, M., Schneider, C., Seglen, P. O., Seleverstov, O., Settleman, J., Shacka, J. J., Shapiro, I. M., Sibirny, A., Silva-Zacarin, E. C., Simon, H. U., Simone, C., Simonsen, A., Smith, M. A., Spanel-Borowski, K., Srinivas, V., Steeves, M., Stenmark, H., Stromhaug, P. E., Subauste, C. S., Sugimoto, S., Sulzer, D., Suzuki, T., Swanson, M. S., Tabas, I., Takeshita, F., Talbot, N. J., Talloczy, Z., Tanaka, K., Tanida, I., Taylor, G. S., Taylor, J. P., Terman, A., Tettamanti, G., Thompson, C. B., Thumm, M., Tolkovsky, A. M., Tooze, S. A., Truant, R., Tumanovska, L. V., Uchiyama, Y., Ueno, T., Uzcategui, N. L., van der Klei, I., Vaquero, E. C., Vellai, T., Vogel, M. W., Wang, H. G., Webster, P., Wiley, J. W., Xi, Z., Xiao, G., Yahalom, J., Yang, J. M., Yap, G., Yin, X. M., Yoshimori, T., Yu, L., Yue, Z., Yuzaki, M., Zabirnyk, O., Zheng, X., Zhu, X., Deter, R. L., 2008, Guidelines for the use and interpretation of assays for monitoring autophagy in higher eukaryotes. *Autophagy* **4:** 151–175

Klionsky, D. J., Emr, S. D., 2000, Autophagy as a regulated pathway of cellular degradation. *Science* **290:** 1717–1721

Ko, D. C., Binkley, J., Sidow, A., Scott, M. P., 2003, The integrity of a cholesterol-binding pocket in Niemann-Pick C2 protein is necessary to control lysosome cholesterol levels. *Proc. Natl. Acad. Sci. USA* **100:** 2518–2525

Ko, D. C., Milenkovic, L., Beier, S. M., Manuel, H., Buchanan, J., Scott, M. P., 2005, Cell-autonomous death of cerebellar purkinje neurons with autophagy in Niemann-Pick type C disease. *PLoS Genet.* **1:** 81–95

Kobayashi, T., Beuchat, M. H., Lindsay, M., Frias, S., Palmiter, R. D., Sakuraba, H., Parton, R. G., Gruenberg, J., 1999, Late endosomal membranes rich in lysobisphosphatidic acid regulate cholesterol transport. *Nat. Cell Biol.* **1:** 113–118

Koh, C. H., Peng, Z. F., Ou, K., Melendez, A., Manikandan, J., Qi, R. Z., Cheung, N. S., 2007, Neuronal apoptosis mediated by inhibition of intracellular cholesterol transport: microarray and proteomics analyses in cultured murine cortical neurons. *J. Cell Physiol.* **211:** 63–87

Koh, C. H., Qi, R. Z., Qu, D., Melendez, A., Manikandan, J., Bay, B. H., Duan, W., Cheung, N. S., 2006, U18666A-mediated apoptosis in cultured murine cortical neurons: role of caspases, calpains and kinases. *Cell Signal.* **18:** 1572–1583

Langmade, S. J., Gale, S. E., Frolov, A., Mohri, I., Suzuki, K., Mellon, S. H., Walkley, S. U., Covey, D. F., Schaffer, J. E., Ory, D. S., 2006, Pregnane X receptor (PXR) activation: a mechanism for neuroprotection in a mouse model of Niemann-Pick C disease. *Proc. Natl. Acad. Sci. USA* **103:** 13807–13812

Li, H., Repa, J. J., Valasek, M. A., Beltroy, E. P., Turley, S. D., German, D. C., Dietschy, J. M., 2005, Molecular, anatomical., and biochemical events associated with neurodegeneration in mice with Niemann-Pick type C disease. *J. Neuropathol. Exp. Neurol.* **64:** 323–333

Li, H., Turley, S. D., Liu, B., Repa, J. J., Dietschy, J. M., 2008, GM2/GD2 and GM3 gangliosides have no effect on cellular cholesterol pools or turnover in normal or NPC1 mice. *J. Lipid Res.* **49:** 1816–1828

Liao, G., Cheung, S., Galeano, J., Ji, A. X., Qin, Q., Bi, X., 2009, Allopregnanolone treatment delays cholesterol accumulation and reduces autophagic/lysosomal dysfunction and inflammation in Npc1-/- mouse brain. *Brain Res.* **1270:** 140–151

Liao, G., Yao, Y., Liu, J., Yu, Z., Cheung, S., Xie, A., Liang, X., Bi, X., 2007, Cholesterol accumulation is associated with lysosomal dysfunction and autophagic stress in Npc1 -/- mouse brain. *Am. J. Pathol.* **171:** 962–975

Liscum, L., Munn, N. J., 1999, Intracellular cholesterol transport. *Biochim. Biophys. Acta* **1438:** 19–37

Liu, B., Li, H., Repa, J. J., Turley, S. D., Dietschy, J. M., 2008, Genetic variations and treatments that affect the lifespan of the NPC1 mouse. *J. Lipid Res.* **49:** 663–669

Lloyd-Evans, E., Morgan, A. J., He, X., Smith, D. A., Elliot-Smith, E., Sillence, D. J., Churchill, G. C., Schuchman, E. H., Galione, A., Platt, F. M., 2008, Niemann-Pick disease type C1 is a sphingosine storage disease that causes deregulation of lysosomal calcium. *Nat. Med.* **14:** 1247–1255

Loftus, S. K., Erickson, R. P., Walkley, S. U., Bryant, M. A., Incao, A., Heidenreich, R. A., Pavan, W. J., 2002, Rescue of neurodegeneration in Niemann-Pick C mice by a prion-promoter-driven Npc1 cDNA transgene. *Hum. Mol. Genet.* **11:** 3107–3114

Loftus, S. K., Morris, J. A., Carstea, E. D., Gu, J. Z., Cummings, C., Brown, A., Ellison, J., Ohno, K., Rosenfeld, M. A., Tagle, D. A., Pentchev, P. G., Pavan, W. J., 1997, Murine model of Niemann-Pick C disease: mutation in a cholesterol homeostasis gene. *Science* **277:** 232–235

Love, S., Bridges, L. R., Case, C. P., 1995, Neurofibrillary tangles in Niemann-Pick disease type C. *Brain* **118 (Pt 1):** 119–129

Lynch, G., Bi, X., 2003, Lysosomes and brain aging in mammals. *Neurochem. Res.* **28:** 1725–1734

Malathi, K., Higaki, K., Tinkelenberg, A. H., Balderes, D. A., Almanzar-Paramio, D., Wilcox, L. J., Erdeniz, N., Redican, F., Padamsee, M., Liu, Y., Khan, S., Alcantara, F., Carstea, E. D., Morris, J. A., Sturley, S. L., 2004, Mutagenesis of the putative sterol-sensing domain of yeast Niemann Pick C-related protein reveals a primordial role in subcellular sphingolipid distribution. *J. Cell Biol.* **164:** 547–556

Millard, E. E., Gale, S. E., Dudley, N., Zhang, J., Schaffer, J. E., Ory, D. S., 2005, The sterol-sensing domain of the Niemann-Pick C1 (NPC1) protein regulates trafficking of low density lipoprotein cholesterol. *J. Biol. Chem.* **280:** 28581–28590

Mutka, A. L., Lusa, S., Linder, M. D., Jokitalo, E., Kopra, O., Jauhiainen, M., Ikonen, E., 2004, Secretion of sterols and the NPC2 protein from primary astrocytes. *J. Biol. Chem.* **279:** 48654–48662

Naureckiene, S., Sleat, D. E., Lackland, H., Fensom, A., Vanier, M. T., Wattiaux, R., Jadot, M., Lobel, P., 2000, Identification of HE1 as the second gene of Niemann-Pick C disease. *Science* **290:** 2298–2301

Neufeld, E. B., Wastney, M., Patel, S., Suresh, S., Cooney, A. M., Dwyer, N. K., Roff, C. F., Ohno, K., Morris, J. A., Carstea, E. D., Incardona, J. P., Strauss, J. F., 3rd, Vanier, M. T., Patterson, M. C., Brady, R. O., Pentchev, P. G., Blanchette-Mackie, E. J., 1999, The Niemann-Pick C1 protein resides in a vesicular compartment linked to retrograde transport of multiple lysosomal cargo. *J. Biol. Chem.* **274:** 9627–9635

Niemann, A. (1914) *Ein unbekanntes Krankheitsbild*. Jahrbuch der Kinderheikunde 1.

Nixon, R. A., 2007, Autophagy, amyloidogenesis and Alzheimer disease. *J. Cell Sci.* **120:** 4081–4091

Ohgami, N., Ko, D. C., Thomas, M., Scott, M. P., Chang, C. C., Chang, T. Y., 2004, Binding between the Niemann-Pick C1 protein and a photoactivatable cholesterol analog requires a functional sterol-sensing domain. *Proc. Natl. Acad. Sci. USA* **101:** 12473–12478

Ohm, T. G., Treiber-Held, S., Distl, R., Glockner, F., Schonheit, B., Tamanai, M., Meske, V., 2003, Cholesterol and tau protein–findings in Alzheimer's and Niemann Pick C's disease. *Pharmacopsychiatry* **36 Suppl 2:** S120–S126

Ohsaki, Y., Sugimoto, Y., Suzuki, M., Hosokawa, H., Yoshimori, T., Davies, J. P., Ioannou, Y. A., Vanier, M. T., Ohno, K., Ninomiya, H., 2006, Cholesterol depletion facilitates ubiquitylation of NPC1 and its association with SKD1/Vps4. *J. Cell Sci.* **119:** 2643–2653

Okamura, N., Kiuchi, S., Tamba, M., Kashima, T., Hiramoto, S., Baba, T., Dacheux, F., Dacheux, J. L., Sugita, Y., Jin, Y. Z., 1999, A porcine homolog of the major secretory protein of human epididymis, HE1, specifically binds cholesterol. *Biochim. Biophys. Acta* **1438:** 377–387

Ong, W. Y., Hu, C. Y., Patel, S. C., 2002, Apolipoprotein D in the Niemann-Pick type C disease mouse brain: an ultrastructural immunocytochemical analysis. *J. Neurocytol.* **31:** 121–129

Pacheco, C. D., Kunkel, R., Lieberman, A. P., 2007, Autophagy in Niemann-Pick C disease is dependent upon Beclin-1 and responsive to lipid trafficking defects. *Hum. Mol. Genet.* **16:** 1495–1503

Pan, T., Kondo, S., Le, W., Jankovic, J., 2008, The role of autophagy-lysosome pathway in neurodegeneration associated with Parkinson's disease. *Brain* **131:** 1969–1978

Patel, S. C., Asotra, K., Patel, Y. C., McConathy, W. J., Patel, R. C., Suresh, S., 1995, Astrocytes synthesize and secrete the lipophilic ligand carrier apolipoprotein D. *Neuroreport* **6:** 653–657

Pentchev, P. G., Comly, M. E., Kruth, H. S., Vanier, M. T., Wenger, D. A., Patel, S., Brady, R. O., 1985, A defect in cholesterol esterification in Niemann-Pick disease (type C) patients. *Proc. Natl. Acad. Sci. USA* **82:** 8247–8251

Pick, L., 1933, Niemann–Pick's disease and other forms of so called xanthomatosis. *Am. J. Med. Sci.* **185:** 601–616

Pitas, R. E., Boyles, J. K., Lee, S. H., Foss, D., Mahley, R. W., 1987, Astrocytes synthesize apolipoprotein E and metabolize apolipoprotein E-containing lipoproteins. *Biochim. Biophys. Acta* **917:** 148–161

Prasad, A., Fischer, W. A., Maue, R. A., Henderson, L. P., 2000, Regional and developmental expression of the Npc1 mRNA in the mouse brain. *J. Neurochem.* **75:** 1250–1257

Ravikumar, B., Vacher, C., Berger, Z., Davies, J. E., Luo, S., Oroz, L. G., Scaravilli, F., Easton, D. F., Duden, R., O'Kane, C. J., Rubinsztein, D. C., 2004, Inhibition of mTOR induces autophagy and reduces toxicity of polyglutamine expansions in fly and mouse models of Huntington disease. *Nat. Genet.* **36:** 585–595

Reid, P. C., Lin, S., Vanier, M. T., Ohno-Iwashita, Y., Harwood, H. J., Jr., Hickey, W. F., Chang, C. C., Chang, T. Y., 2008, Partial blockage of sterol biosynthesis with a squalene synthase inhibitor in early postnatal Niemann-Pick type C npcnih null mice brains reduces neuronal cholesterol accumulation, abrogates astrogliosis, but may inhibit myelin maturation. *J. Neurosci. Meth.* **168:** 15–25

Reid, P. C., Sakashita, N., Sugii, S., Ohno-Iwashita, Y., Shimada, Y., Hickey, W. F., Chang, T. Y., 2004, A novel cholesterol stain reveals early neuronal cholesterol accumulation in the Niemann-Pick type C1 mouse brain. *J. Lipid Res.* **45:** 582–591

Reid, P. C., Sugii, S., Chang, T. Y., 2003, Trafficking defects in endogenously synthesized cholesterol in fibroblasts, macrophages, hepatocytes, and glial cells from Niemann-Pick type C1 mice. *J. Lipid Res.* **44:** 1010–1019

Repa, J. J., Li, H., Frank-Cannon, T. C., Valasek, M. A., Turley, S. D., Tansey, M. G., Dietschy, J. M., 2007, Liver X receptor activation enhances cholesterol loss from the brain, decreases neuroinflammation, and increases survival of the NPC1 mouse. *J. Neurosci.* **27:** 14470–14480

Roff, C. F., Strauss, J. F., 3rd, Goldin, E., Jaffe, H., Patterson, M. C., Agritellis, G. C., Hibbs, A. M., Garfield, M., Brady, R. O., Pentchev, P. G., 1993, The murine Niemann-Pick type C lesion affects testosterone production. *Endocrinology* **133:** 2913–2923

Runz, H., Dolle, D., Schlitter, A. M., Zschocke, J., 2008, NPC-db, a Niemann-Pick type C disease gene variation database. *Hum. Mutat* .**29:** 345–350

Saito, Y., Suzuki, K., Hulette, C. M., Murayama, S., 2004, Aberrant phosphorylation of alpha-synuclein in human Niemann-Pick type C1 disease. *J. Neuropathol. Exp. Neurol.* **63:** 323–328

Saito, Y., Suzuki, K., Nanba, E., Yamamoto, T., Ohno, K., Murayama, S., 2002, Niemann-Pick type C disease: accelerated neurofibrillary tangle formation and amyloid beta deposition associated with apolipoprotein E epsilon 4 homozygosity. *Ann. Neurol.* **52:** 351–355

Scott, C., Higgins, M. E., Davies, J. P., Ioannou, Y. A., 2004, Targeting of NPC1 to late endosomes involves multiple signals, including one residing within the putative sterol-sensing domain. *J. Biol. Chem.* **279:** 48214–48223

Scott, C., Ioannou, Y. A., 2004, The NPC1 protein: structure implies function. *Biochim. Biophys. Acta* **1685:** 8–13

Sleat, D. E., Wiseman, J. A., El-Banna, M., Price, S. M., Verot, L., Shen, M. M., Tint, G. S., Vanier, M. T., Walkley, S. U., Lobel, P., 2004, Genetic evidence for nonredundant functional cooperativity between NPC1 and NPC2 in lipid transport. *Proc. Natl. Acad. Sci. USA* **101:** 5886–5891

Sokol, J., Blanchette-Mackie, J., Kruth, H. S., Dwyer, N. K., Amende, L. M., Butler, J. D., Robinson, E., Patel, S., Brady, R. O., Comly, M. E., et al., 1988, Type C Niemann-Pick disease. Lysosomal accumulation and defective intracellular mobilization of low density lipoprotein cholesterol. *J. Biol. Chem.* **263:** 3411–3417

Suresh, S., Yan, Z., Patel, R. C., Patel, Y. C., Patel, S. C., 1998, Cellular cholesterol storage in the Niemann-Pick disease type C mouse is associated with increased expression and defective processing of apolipoprotein D. *J. Neurochem.* **70:** 242–251

Suzuki, K., Parker, C. C., Pentchev, P. G., Katz, D., Ghetti, B., D'Agostino, A. N., Carstea, E. D., 1995, Neurofibrillary tangles in Niemann-Pick disease type C. *Acta Neuropathol.* **89:** 227–238

Tanida, I., Minematsu-Ikeguchi, N., Ueno, T., Kominami, E., 2005, Lysosomal turnover, but not a cellular level, of endogenous LC3 is a marker for autophagy. *Autophagy* **1:** 84–91

te Vruchte, D., Lloyd-Evans, E., Veldman, R. J., Neville, D. C., Dwek, R. A., Platt, F. M., van Blitterswijk, W. J., Sillence, D. J., 2004, Accumulation of glycosphingolipids in Niemann-Pick C disease disrupts endosomal transport. *J. Biol. Chem.* **279:** 26167–26175

Turley, S. D., Burns, D. K., Rosenfeld, C. R., Dietschy, J. M., 1996, Brain does not utilize low density lipoprotein-cholesterol during fetal and neonatal development in the sheep. *J. Lipid Res.* **37:** 1953–1961

Urano, Y., Watanabe, H., Murphy, S. R., Shibuya, Y., Geng, Y., Peden, A. A., Chang, C. C., Chang, T. Y., 2008, Transport of LDL-derived cholesterol from the NPC1 compartment to the ER involves the trans-Golgi network and the SNARE protein complex. *Proc. Natl. Acad. Sci. USA* **105:** 16513–16518

Vance, J. E., 2006, Lipid imbalance in the neurological disorder, Niemann-Pick C disease. *FEBS Lett.* **580:** 5518–5524

Vanier, M. T., Millat, G., 2004, Structure and function of the NPC2 protein. *Biochim. Biophys. Acta* **1685:** 14–21

Vanier, M. T., Wenger, D. A., Comly, M. E., Rousson, R., Brady, R. O., Pentchev, P. G., 1988, Niemann-Pick disease group C: clinical variability and diagnosis based on defective cholesterol esterification. A collaborative study on 70 patients. *Clin. Genet.* **33:** 331–348

Walkley, S. U., Suzuki, K., 2004, Consequences of NPC1 and NPC2 loss of function in mammalian neurons. *Biochim. Biophys. Acta* **1685:** 48–62

Watanabe, Y., Akaboshi, S., Ishida, G., Takeshima, T., Yano, T., Taniguchi, M., Ohno, K., Nakashima, K., 1998, Increased levels of GM2 ganglioside in fibroblasts from a patient with juvenile Niemann-Pick disease type C. *Brain Dev.* **20:** 95–97

Weisgraber, K. H., Roses, A. D., Strittmatter, W. J., 1994, The role of apolipoprotein E in the nervous system. *Curr. Opin. Lipidol.* **5:** 110–116

Willenborg, M., Schmidt, C. K., Braun, P., Landgrebe, J., von Figura, K., Saftig, P., Eskelinen, E. L., 2005, Mannose 6-phosphate receptors, Niemann-Pick C2 protein, and lysosomal cholesterol accumulation. *J. Lipid Res.* **46:** 2559–2569

Wu, Y. P., Mizukami, H., Matsuda, J., Saito, Y., Proia, R. L., Suzuki, K., 2005, Apoptosis accompanied by up-regulation of TNF-alpha death pathway genes in the brain of Niemann-Pick type C disease. *Mol. Genet. Metab.* **84:** 9–17

Xie, C., Burns, D. K., Turley, S. D., Dietschy, J. M., 2000, Cholesterol is sequestered in the brains of mice with Niemann-Pick type C disease but turnover is increased. *J. Neuropathol. Exp. Neurol.* **59:** 1106–1117

Xie, C., Lund, E. G., Turley, S. D., Russell, D. W., Dietschy, J. M., 2003, Quantitation of two pathways for cholesterol excretion from the brain in normal mice and mice with neurodegeneration. *J. Lipid Res.* **44:** 1780–1789

Xie, Z., Klionsky, D. J., 2007, Autophagosome formation: core machinery and adaptations. *Nat. Cell Biol.* **9:** 1102–1109

Xu, S., Benoff, B., Liou, H. L., Lobel, P., Stock, A. M., 2007, Structural basis of sterol binding by NPC2, a lysosomal protein deficient in Niemann-Pick type C2 disease. *J. Biol. Chem.* **282:** 23525–23531

Xu, Z., Farver, W., Kodukula, S., Storch, J., 2008, Regulation of sterol transport between membranes and NPC2. *Biochemistry* **47:** 11134–11143

Zervas, M., Dobrenis, K., Walkley, S. U., 2001a, Neurons in Niemann-Pick disease type C accumulate gangliosides as well as unesterified cholesterol and undergo dendritic and axonal alterations. *J. Neuropathol. Exp. Neurol.* **60:** 49–64

Zervas, M., Somers, K. L., Thrall, M. A., Walkley, S. U., 2001b, Critical role for glycosphingolipids in Niemann-Pick disease type C. *Curr. Biol.* **11:** 1283–1287

Zhang, M., Strnatka, D., Donohue, C., Hallows, J. L., Vincent, I., Erickson, R. P., 2008, Astrocyte-only Npc1 reduces neuronal cholesterol and triples life span of Npc1(-/-) mice. *J. Neurosci. Res.* **86:** 2848–2456

Zhang, M., Sun, M., Dwyer, N. K., Comly, M. E., Patel, S. C., Sundaram, R., Hanover, J. A., Blanchette-Mackie, E. J., 2003, Differential trafficking of the Niemann-Pick C1 and 2 proteins highlights distinct roles in late endocytic lipid trafficking. *Acta Paediatr. Suppl.* **92:** 63–73

Chapter 12
Protein Mediators of Sterol Transport Across Intestinal Brush Border Membrane

J. Mark Brown and Liqing Yu

Abstract Dysregulation of cholesterol balance contributes significantly to atherosclerotic cardiovascular disease (ASCVD), the leading cause of death in the United States. The intestine has the unique capability to act as a gatekeeper for entry of cholesterol into the body, and inhibition of intestinal cholesterol absorption is now widely regarded as an attractive non-statin therapeutic strategy for ASCVD prevention. In this chapter we discuss the current state of knowledge regarding sterol transport across the intestinal brush border membrane. The purpose of this work is to summarize substantial progress made in the last decade in regards to protein-mediated sterol trafficking, and to discuss this in the context of human disease.

Keywords Intestinal cholesterol absorption · Niemann-Pick C1-Like 1 · Sterol-sensing domain · ATP-binding cassette transporters G5 and G8 · Scavenger receptor class B type I · Animal model · Cell model · Ezetimibe

Abbreviations

ABC	ATP-binding cassette transporter
Apo	Apolipoprotein
ASCVD	Atherosclerotic cardiovascular disease
CD36	Cluster determinant 36
ER	Endoplasmic reticulum
HDL	High density lipoprotein
HNF4α	Hepatocyte nuclear factor 4 alpha
LDL	Low density lipoprotein
LDL-C	Low density lipoprotein-cholesterol
LXR	Liver X receptor
NPC1	Niemann-Pick C1

L. Yu (✉)
Department of Pathology Section on Lipid Sciences, Wake Forest University School of Medicine, Winston-Salem, Medical Center Blvd, Winston-Salem, NC 27157-1040, USA
e-mail: lyu@wfubmc.edu

NPC1L1	Niemann-Pick C1-Like 1
PPAR	Peroxisome proliferators-activated receptor
SCAP	Sterol regulatory element-binding protein-cleavage activating protein
SR-BI	Scavenger receptor class B type I
SREBP	Sterol regulatory element-binding protein

12.1 Introduction

As discussed in detail in several previous chapters, cholesterol is essential for the growth and function of all mammalian cells. However, elevated low-density lipoprotein (LDL) cholesterol (LDL-C) represents a major risk factor for the development of atherosclerotic cardiovascular disease (ASCVD). For many years now, statin-mediated inhibition of endogenous cholesterol biosynthesis has been the major therapeutic means to lower LDL-C, yet ASCVD still persists in most of the world (Rosamond et al., 2007). Therefore, additional LDL-C lowering is now recommended, and the search for therapeutic strategies that work in synergy with statins has now begun. As a result of this search, drugs that inhibit intestinal cholesterol absorption have become an attractive therapeutic strategy to use in combination with statins. However, only within the last decade have we begun to understand how intestinal sterol absorption occurs at the molecular level. Recently we have learned that intestinal sterol absorption is tightly regulated by key proteins located at the brush border membrane. One of these proteins, Niemann-Pick C1-Like 1 (NPC1L1), was recently identified to be essential for intestinal cholesterol absorption, and will be discussed in detail. In direct opposition of NPC1L1, the heterodimer of ATP-binding cassette transporters G5 and G8 (ABCG5/G8) has been shown to be critical for intestinal disposal of sterols. In addition, the scavenger receptor class B type I (SR-BI) has been implicated in modulating intestinal cholesterol absorption. The purpose of this chapter is to summarize the current state of knowledge regarding the structure and function of these apically-localized cholesterol transporters, and to provide a detailed review as to how these proteins and others interact to regulate the complex process of intestinal sterol absorption.

12.2 Intestinal Cholesterol Absorption and Ezetimibe

Cholesterol in the intestinal lumen is mainly derived from bile and diet. A physiological process by which cholesterol enters intestinal or thoracic duct lymph across the small intestine is called intestinal cholesterol absorption (Wang, 2007a) (Fig. 12.1). Cholesterol absorption involves at least the following three phases: (1) intralumenal solubilization; (2) movement across the apical membrane of absorptive enterocytes; and (3) intracellular metabolism for incorporation into chylomicrons destined for lymph. Intestinal cholesterol absorption rates range widely from 29 to 80%

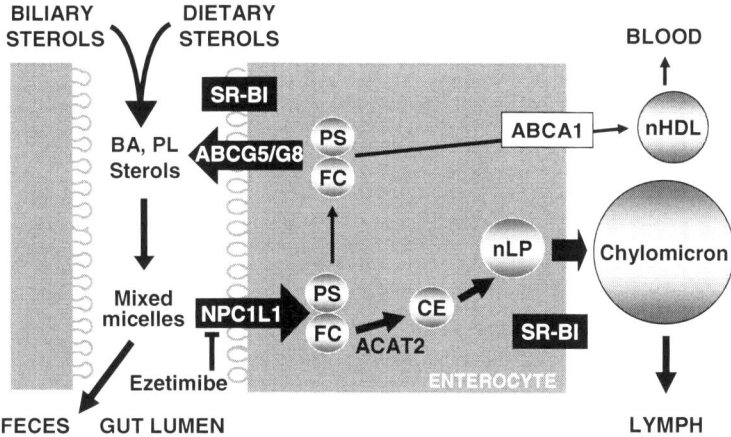

Fig. 12.1 Intestinal sterol absorption and secretion. Sterols including free cholesterol (FC) and free plant sterols (PS) from diet and bile are mixed with phospholipids (PL) and bile acids (BA) to form micelles. FC and PS solubilized in mixed micelles are transported into absorptive enterocytes via an NPC1L1-dependent and ezetimibe-inhibitable mechanism. FC is delivered to the ER for esterification by acyl-CoA:cholesterol acyltransferase-2 (ACAT2) to form cholesterol esters (CE) that is then packaged into nascent lipoprotein particles (nLP) and secreted as a constituent of chylomicron. PS and FC that escapes ACAT2 esterification may be directly transported to nascent HDL (nHDL) through basolateral ABCA1, or back to the gut lumen via ABCG5/G8

in normal men and women consuming a moderately low cholesterol diet (Bosner et al., 1999). Molecular mechanisms underlying this large inter-individual variation remain to be elucidated.

Since a detailed review of intestinal cholesterol absorption and its potential protein mediators is beyond the focus of this chapter, readers interested in this topic are referred to many excellent reviews available (Wilson and Rudel, 1994; Dawson and Rudel, 1999; Davis and Veltri, 2007; Levy et al., 2007; Wang, 2007a, Turley and Dietschy, 2003; Iqbal and Hussain, 2009; Hui and Howles, 2005).

Intestinal cholesterol absorption represents an attractive target for developing cholesterol-lowering drugs because it is a major pathway governing whole-body cholesterol homeostasis. The Schering-Plough Research Institute successfully identified ezetimibe as a potent and specific inhibitor of intestinal cholesterol absorption using in vivo models of cholesterol absorption (Clader, 2004). The drug is now widely used in monotherapy or in combination with statins (inhibitors of cholesterol biosynthesis) to efficiently treat hypercholesterolemia in the general population (Davis and Veltri, 2007).

Intestinal cholesterol absorption was once thought to be a passive process. The fact that ezetimibe potently inhibits intestinal cholesterol absorption at very low doses (Van Heek et al., 1997, 2001a,2001b; Sudhop et al., 2002) suggests that specific protein(s) must be involved in cholesterol absorption (Turley and Dietschy, 2003). In a search for the ezetimibe-inhibitory protein(s), NPC1L1, a previously-identified protein of unknown function (Davies et al., 2000a), was discovered in the ezetimibe-sensitive pathway because disruption of NPC1L1 in mice reduces intestinal cholesterol absorption to the level seen in ezetimibe-treated animals (Altmann et al., 2004, Davis et al., 2004). Whether NPC1L1 is the molecular target of ezetimibe has been under considerable debate (Smart et al., 2004; Kramer et al., 2005; Labonte et al., 2007; Knopfel et al., 2007). Ezetimibe can bind to intestinal brush border membrane vesicles from wild-type mice but not mice lacking NPC1L1 (Garcia-Calvo et al., 2005). Recently, Weinglass and associates from Merck Research Laboratories purified the NPC1L1-ezetimibe complex from NPC1L1-expressing and ezetimibe-treated cells and found that NPC1L1 is the only protein to account for ezetimibe binding (Weinglass et al., 2008). These pieces of biochemical evidence, together with findings from animal, genetic and cell biology studies (Yu, 2008) (*see below*), strongly supports that NPC1L1 is the molecular target of ezetimibe.

12.3 NPC1L1

12.3.1 Structure: Gene, mRNA, and Protein Domains

In the human genome, NPC1L1 gene spans about 29 kb in chromosome 7p13 and contains 20 exons. It produces a predominant mRNA transcript that skips exon 15. This transcript, like that from rodents, encodes a 1332-amino acid protein, and has been used in most, if not all, NPC1L1 studies (Davies et al., 2000a; Yu et al., 2006; Altmann et al., 2004). The human NPC1L1 gene also produces two alternatively spliced transcripts (Davies et al., 2000a). One contains a 27-amino acid insertion transcribed from the in-frame exon 15. The other skips exon 7 and terminates within intron 8, encoding a truncated protein of 724 amino acids. The physiological significance of these two alternatively spliced variants has yet to be defined.

The human NPC1L1 protein is a homolog of Niemann-Pick C1 (NPC1), having ~50% amino acid homology to NPC1 protein (Davies et al., 2000a; Davies and Ioannou, 2006). Deficiency of human NPC1 causes an autosomal recessive lipid storage disorder, Niemann-Pick disease type C1 that is characterized by defective trafficking of intracellular cholesterol and lysosomal accumulation of free cholesterol, gangliosides and other lipids (Carstea et al., 1997; Loftus et al., 1997) (*see* Chapter 11). NPC1L1, like its homolog NPC1, also shares similarity with the resistance-nodulation-division family of prokaryotic permeases that can pump out lipophilic drugs, detergents, fatty acids, bile acids, metal ions, and dyes from the cytosol of bacteria (Davies et al., 2000b; Davies and Ioannou, 2006).

Based on the amino acid sequences, human NPC1L1 is predicted to have a typical signal peptide of 21 amino acids, 13 putative transmembrane domains (Fig. 12.2A), and extensive potential *N*-linked glycosylation sites located within the extracellular loops of the protein facing intestinal lumen (Altmann et al., 2004; Davies et al., 2000a) or within the luminal loops of the protein if endocytosed from plasma

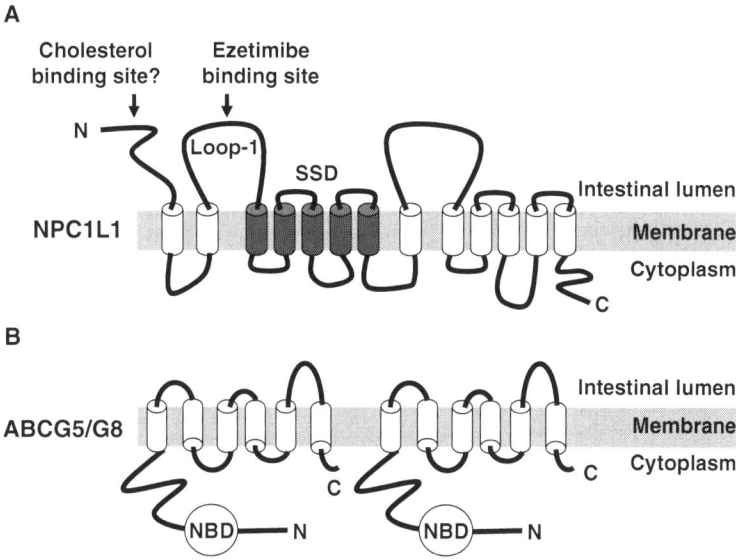

Fig. 12.2 Proposed membrane topologies and domains of NPC1L1 (**A**) and ABCG5/G8 (**B**) proteins. SSD, sterol-sensing domain; N, N-terminus; C, C-terminus; NBD, nucleotide-binding domain

membrane into intracellular vesicles (Yu et al., 2006; Ge et al., 2008; Wang et al., 2009). The membrane topology and *N*-glycosylation of NPC1L1 have been examined experimentally and the findings are consistent with sequence-based prediction (Iyer et al., 2005; Temel et al., 2007; Altmann et al., 2004; Davies et al., 2000a; Wang et al., 2009).

A sterol-sensing domain (SSD) is another signature of NPC1L1 protein (Davies et al., 2000a; Altmann et al., 2004) (Fig. 12.2A). This domain consists of ~180 amino acids that form five predicted membrane-spanning helices with short intervening loops (Radhakrishnan et al., 2004). The SSD is conserved in at least 7 other membrane proteins, all of which have relations to cholesterol (Kuwabara and Labouesse, 2002), including the aforementioned NPC1 (Davies et al., 2000a; Carstea et al., 1997; Loftus et al., 1997); 3-hydroxy-3-methylglutaryl (HMG) CoA reductase, the rate-limiting enzyme of cholesterol biosynthesis (Brown and Goldstein, 1999; Goldstein and Brown, 1990); sterol regulatory element-binding protein (SREBP)-cleavage activating protein (SCAP), a protein that controls the endoplasmic reticulum (ER)-to-Golgi transport and proteolytic activation of membrane-bound transcription factors SREBPs (Brown and Goldstein, 1999; Horton et al., 2002a; Goldstein and Brown, 1990); and Patched, a membrane receptor for the cholesterol-linked signalling peptide Hedgehog (Cooper et al., 2003). The function of NPC1L1 SSD remains unknown and may be implicated in cholesterol regulation of NPC1L1 protein trafficking (Yu et al., 2006).

12.3.2 Function: Lessons Learned from Animal Models and Human Genetics

In mammals, NPC1L1 mRNA and protein are highly expressed in small intestine (Altmann et al., 2004; Davies et al., 2005). In the small intestine, NPC1L1 protein localizes at the apical surface of absorptive enterocyte (Altmann et al., 2004; Davis et al., 2004). Gene knockout studies in mice have unambiguously established an essential role of NPC1L1 in intestinal cholesterol absorption (Altmann et al., 2004; Davis et al., 2004; Tang et al., 2008a, b; Temel et al., 2009). The cholesterol absorption inhibitor ezetimibe cannot further reduce intestinal cholesterol absorption in NPC1L1-deficient mice, demonstrating that NPC1L1 is in the ezetimibe-sensitive pathway (Altmann et al., 2004; Davis et al., 2004).

Interestingly, genetic ablation of NPC1L1 in mice also protects against obesity, insulin resistance and fatty liver induced by a nutrient-rich diet (Labonte et al., 2008; Davies et al., 2005) (Yu, L., unpublished observation). Ezetimibe treatment also improves dyslipidemia, hepatic steatosis, non-alcoholic fatty liver disease, and insulin resistance in several animal models (van Heek et al., 2001a; Assy et al., 2006; Deushi et al., 2007; Zheng et al., 2008). Another interesting observation is that NPC1L1 ablation in mice greatly attenuates dyslipidemia, lipogenic gene overexpression, and hepatic steatosis induced by activation of nuclear receptor liver X receptors (LXRs) (Tang et al., 2008b). LXR forms a heterodimer with retinoid X receptor to regulate expression of their target genes in response to fluctuations of cellular cholesterol content (Repa and Mangelsdorf, 2002). Given that rodent NPC1L1 is almost exclusively expressed in the small intestine and the primary defect of NPC1L1-null mice is the blockade of intestinal cholesterol absorption (Altmann et al., 2004), the observed phenotypes are likely attributable to reduced intestinal cholesterol absorption. Efficient intestinal cholesterol absorption may be essential for maximizing LXR activities in the liver and small intestine, and perhaps other tissues where cholesterol and its derivatives are likely used as endogenous LXR ligands. LXR target genes involve the regulation of many metabolic pathways, including metabolism of cholesterol, fatty acids, and carbohydrates (Kalaany and Mangelsdorf, 2006). Inhibition of NPC1L1-dependent intestinal cholesterol absorption may improve some metabolic disorders by down-regulating tissue LXR activities. Consistent with this notion, LXR knockout mice are protected against obesity induced by a high fat diet (Kalaany et al., 2005).

Although NPC1L1 mRNA and protein are very abundant in the small intestine of all mammals examined, their levels in the liver differ remarkably among species. Rodents express only a negligible amount of NPC1L1 in the liver (Altmann et al., 2004; Tang et al., 2006). In contrast, human livers have readily detectable levels of NPC1L1 mRNA and proteins (Altmann et al., 2004; Davies et al., 2005; Temel et al., 2007). The reason for the different expression pattern of NPC1L1 among species is unknown, but may result from their differences in cholesterol metabolism (Dietschy and Turley, 2002; Yu, 2007). In the liver of nonhuman primates and humans, NPC1L1 localizes to the canalicular membrane of hepatocyte (Yu et al., 2006; Temel et al., 2007). When over-expressed in the mouse liver by transgenic technology, human NPC1L1 also concentrates to the canalicular membrane of hepatocyte (Temel et al., 2007). Whereas the function of NPC1L1 in human livers remains to be elucidated, transgenic over-expression of human NPC1L1 in the mouse liver dramatically reduces biliary cholesterol concentrations without altering hepatic expression levels of the cholesterol efflux transporter ABCG5/G8 (Temel et al., 2007). This finding implies that hepatic NPC1L1 may inhibit biliary cholesterol excretion by transporting cholesterol from the canalicular bile back into hepatoctyes (Yu, 2008). The inhibition of NPC1L1 overexpression on biliary cholesterol excretion can be rescued by ezetimibe treatment, suggesting that hepatic NPC1L1 is another target for ezetimibe, at least in mice (Temel et al., 2007).

Increased cholesterol concentrations in the bile can result in gallstone formation. If ezetimibe treatment results in an increase in biliary cholesterol concentration in humans by inhibiting hepatic NPC1L1, it may have a potential to promote gallstone formation, particularly in subjects in whom NPC1L1 is more abundantly expressed in the liver than in the intestine. Currently, there is no evidence that ezetimibe increases the incidence of gallstone disease and it remains unknown as to whether hepatic NPC1L1 regulates biliary cholesterol excretion in humans. In a small human study, inhibition of NPC1L1 for 30 days by ezetimibe treatment at 20 mg/day did not alter biliary cholesterol molar percentage, cholesterol to phospholipid ratio, and cholesterol saturation index in 5 overweight subjects without gallstones, but did reduce these parameters significantly in 7 patients with gallstones (Wang et al., 2008a). Ezetimibe may have its predominant effect at the intestinal level, thereby reducing cholesterol that is transported from the gut lumen to the liver for biliary secretion. In animals that express NPC1L1 predominantly in the small intestine, inhibition of NPC1L1 by ezetimibe may reduce biliary cholesterol levels and prevent gallstone disease. This notion is consistent with the following observations: (1) NPC1L1 knockout mice have lower biliary cholesterol concentrations, even after being challenged with a high cholesterol diet (Davis et al., 2004); (2) In Golden Syrian hamsters, ezetimibe prevents high cholesterol diet-induced increase in biliary cholesterol (Valasek et al., 2008); and (3) In wild-type mice, ezetimibe treatment protects against lithogenic diet-induced increase in biliary cholesterol concentration and gallstone formation (Zuniga et al., 2008).

Human genetic studies have shown that sequence variations in NPC1L1 are associated with sterol absorption efficiency, LDL-C levels, and LDL-C response to ezetimibe therapy (Cohen et al., 2006; Wang et al., 2005; Hegele et al., 2005; Simon et al., 2005; Fahmi et al., 2008). These findings strongly support a key role of NPC1L1 and NPC1L1-depedent cholesterol transport in whole-body cholesterol homeostasis in humans.

12.3.3 Function: Lessons Learned from Cell Model Systems

In whole animals, NPC1L1 protein is asymmetrically enriched in the intestinal brush border membrane (Altmann et al., 2004; Garcia-Calvo et al., 2005; Iyer et al., 2005; Labonte et al., 2007; Sane et al., 2006) or hepatic canalicular membrane (Yu et al.,

2006; Temel et al., 2007). In cultured cells, NPC1L1 proteins can localize to both the plasma membrane and intracellular compartments (Yu et al., 2006; Davies et al., 2005; Yamanashi et al., 2007; Iyer et al., 2005). Human NPC1L1 with a C-terminal green fluorescent protein tag predominantly localizes at the endocytic recycling compartment in actively growing McArdle RH7777 rat hepatoma cells (Yu et al., 2006). Intriguingly, the intracellular itinerary of NPC1L1 protein in these cells and in HepG2 hepatic carcinoma cells is under control of cellular cholesterol availability (Yu et al., 2006). Cholesterol depletion results in a redistribution of NPC1L1 from the intracellular endocytic recycling compartment to the cell surface likely via a mechanism involved in microfilament-associated myosin Vb/Rab11a/Rab11-FIP2 complex, and conversely, cholesterol reloading causes the proteins to transit from the cell surface back into the cell interior likely through clathrin-mediated enocytosis, which is coupled to NPC1L1-facilitated and ezetimibe-inhibitable cholesterol uptake (Yu et al., 2006; Ge et al., 2008; Brown et al., 2007; Chu et al., 2009; Petersen et al., 2008). This cholesterol regulated trafficking may explain why both intracellular and cell surface locations have been observed for NPC1L1 protein (Altmann et al., 2004; Garcia-Calvo et al., 2005; Iyer et al., 2005; Labonte et al., 2007; Sane et al., 2006; Knopfel et al., 2007; Yamanashi et al., 2007; Temel et al., 2007; Yu et al., 2006).

The establishment of NPC1L1-dependent and ezetimibe-sensitive cholesterol-uptake assay in cell models allows an opportunity to examine if NPC1L1 differentiates plant sterols from cholesterol (Fig. 12.3). Each day, a large amount of plant-derived sterol (mainly sitosterol and campesterol) are consumed. Although these phytosterols are structurally similar to cholesterol, in normal individuals phytosterols are poorly absorbed. The rank order of fractional intestinal sterol absorption is cholesterol (~45%) > campesterol (~20%) > sitosterol (~5%)

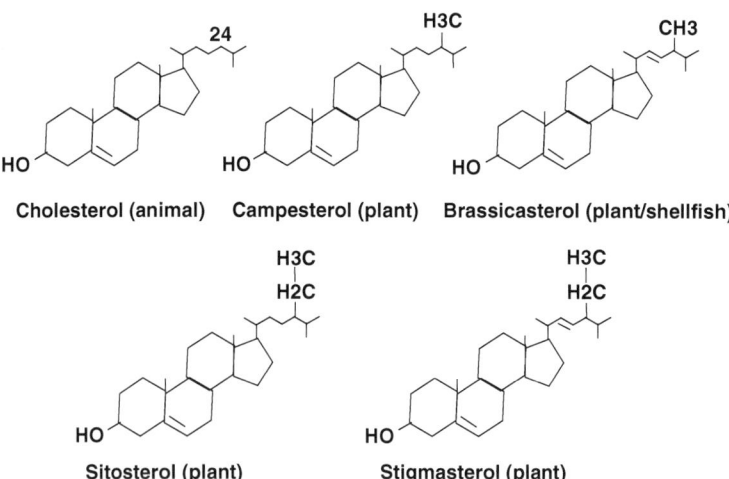

Fig. 12.3 Structures of sterols derived from animals and plants

(Lutjohann et al., 1995). In mammals, this discrimination for phytosterols may be protective because phytosterols can displace cholesterol in the cell membrane and interfere with cell functions (Wang et al., 1981; Su et al., 2006; Kruit et al., 2008; Kim et al., 2008). Mechanisms underlying the defense against phytosterols remain elusive. Although mutations in ABCG5 and/or ABCG8 cause accumulation of phytosterols in the body (Berge et al., 2000; Lee et al., 2001), the rank order of intestinal sterol absorption rates is maintained in mice lacking ABCG5 and ABCG8 (Yu et al., 2002a), implying that ABCG5/G8 is a gatekeeper rather than a discriminator for phytosterols. Is NPC1L1 a discriminator of phytosterols? Studies in mice and humans suggest that NPC1L1 is a common transporter for all sterols because NPC1L1 and ezetimibe-sensitive pathway are also essential for phytosterol absorption (Davis et al., 2004; Salen et al., 2004, 2006; Yu et al., 2005; Tang et al., 2008a). However, cell culture studies suggest that NPC1L1 may not mediate cellular uptake of all sterols equally. NPC1L1-mediated uptake is 60% lower for sitosterol than cholesterol in intestine-derived CaCo-2 cells over-expressing NPC1L1 (Yamanashi et al., 2007). Overexpression of NPC1L1 in McArdle RH7777 rat hepatoma cells facilitates cholesterol, but not sitosterol uptake (Yu et al., 2006; Brown et al., 2007; Yamanashi et al., 2007; Ge et al., 2008). These cell-based assays imply that NPC1L1 has lower affinity to sitosterol than cholesterol. The lower affinity of NPC1L1 to phytosterols has the potential to determine the rank order of intestinal sterol absorption rates. NPC1L1 might be the first genetic defense against phytosterol absorption. But this defense is not complete; otherwise, mutations in ABCG5/G8 would not cause sitosterolemia in the presence of NPC1L1.

12.3.4 Regulation of Expression

Regulation of NPC1L1 gene expression is largely unknown and inconsistent. Since activation of many nuclear receptors, including peroxisome proliferators-activated receptor alpha (PPARα), PPARδ, liver X receptor (LXR), and retinoid X receptor (RXR), reduces intestinal cholesterol absorption (Knight et al., 2003; Repa et al., 2000; McNamara et al., 1980; Umeda et al., 2001; Vanhanen and Miettinen, 1995; Yu et al., 2003; Oliver et al., 2001; van der Veen et al., 2005), effects of these nuclear receptors on intestinal NPC1L1 expression have been examined in CaCo-2 cells and in mice. In 2005, van der Veen and colleagues showed that PPARδ activation reduces intestinal cholesterol absorption by 43% in DBA mice after 8 days of treatment, coinciding with a significant reduction in intestinal NPC1L1 mRNA levels by 40% (van der Veen et al., 2005). PPARδ activation may explain why fish oil and docosahexaenoic acid treatments reduces NPC1L1 expression in CaCo-2 cells and in the proximal small intestine of hamsters (Mathur et al., 2007). In 2007, Valasek and colleagues made the interesting observation that fenofibrate significantly reduces intestinal NPC1L1 mRNA and protein levels via a PPARα-dependent mechanism in mice, but NPC1L1 gene does not seem to be the direct target of PPARα, because short-term treatment with fenofibrate does not reduce NPC1L1 mRNA expression and yet nuclear receptors generally enhance expression of their direct targets within several hours (Valasek et al., 2007). In the same study, these workers also found that the effect of PPARα on intestinal NPC1L1 expression is LXR-independent because the similar effect was observed in the LXRα/LXRβ double knockout mice.

Whether LXR directly modulates NPC1L1 expression is also uncertain. In a cell culture study, LXR activation by 1 μM of a synthetic agonist T0901317 or GW3965 for 24 h reduces NPC1L1 mRNA levels by ∼30 or ∼60%, respectively, in CaCo-2/TC7 cells (Duval et al., 2006). In studies with whole animals, LXR activation reduces NPC1L1 mRNA levels by ∼35% in the duodena of apolipoprotein (apo) E2-KI mice (lacking endogenous apoE but expressing human apoE2; Sullivan et al., 1998) fed a Western diet (Duval et al., 2006), and ∼50% in the small intestine of ABCB4 knockout mice (Kruit et al., 2005). However, LXR activation by T0901317 or GW3965 does not alter intestinal mRNA levels significantly in chow-fed wild-type mice (Yu L, unpublished data) (Kruit et al., 2005). NPC1L1 mRNA levels remains unaffected in the proximal small intestine of chow-fed LXRα/LXRβ double knockout mice (Valasek et al., 2007). Thus, it is unlikely that intestinal NPC1L1 is a direct target of LXR.

The human NPC1L1 promoter has a putative sterol regulatory element (Davies et al., 2000a). In 2007, Alrefai and colleagues reported that SREBP-2 (Horton et al., 2002b) can bind to the two putative sterol regulatory elements in the human NPC1L1 promoter (Alrefai et al., 2007) that differ from the one predicted previously (Davies et al., 2000a). They also showed that the NPC1L1 mRNA levels and promoter activities in CaCo-2 cells are decreased by 25-hydroxycholesterol that suppresses SREBP activation, and are increased by mevinolin that induces SREBP activation (Alrefai et al., 2007). This study is consistent with that NPC1L1 mRNA levels are ∼3-fold and ∼2.4-fold higher in cholesterol-depleted CaCo-2 cells induced by cyclodextrin and by taurocholate/phosphatidylcholine for 24 h, respectively, and are significantly lower in CaCo-2 cells treated with cholesterol and 25-hydroxycholesterol (Field et al., 2007). Alrefai et al. further showed that co-expression of the human NPC1L1 promoter-luciferase reporter and the active form of SREBP-2 dramatically increases human NPC1L1 promoter activity (Alrefai et al., 2007). Interestingly, SREBP-2 and the hepatocyte nuclear factor 4α (HNF4α) can synergistically increase human NPC1L1 promoter activity despite HNF4α alone having no effect, and the presence of HNF4α appears to be essential for cholesterol-dependent regulation of NPC1L1 expression (Iwayanagi et al., 2008).

Taken together, these studies suggest that NPC1L1 expression may be regulated by cellular cholesterol availability in a SREBP-dependent manner. Consistent with this regulation, we found that the hepatic NPC1L1 mRNA level is drastically increased in ABCG5/G8 transgenic mice (Yu et al., 2002b) treated with lovastatin, a condition that causes a compensatory increase in hepatic mRNA levels of all cholesterol biosynthetic genes (Tang et al., 2006). Additionally, the intestinal NPC1L1 mRNA level is reduced by ∼35% in mice lacking acyl-CoA:cholesterol acyltransferase-2 versus wild-type mice fed a synthetic diet containing 20% energy from palm oil and 0.17% cholesterol (Temel et al., 2005), and by ∼45% in phospholipid transport protein-deficient versus wild-type mice fed a chow diet (Liu et al., 2007), which are two conditions under which free cholesterol is accumulated in the intestine. Further, in ezetimibe-treated miniature pigs fed a Western-type diet, intestinal and hepatic NPC1L1 mRNA levels are significantly increased (Telford et al., 2007).

Despite several pieces of evidence supporting regulation of NPC1L1 expression by cellular cholesterol availability, discrepancies exist in animal and cell culture studies. For example, high cholesterol diet feeding does not suppress intestinal NPC1L1 expression in mice (Valasek et al., 2007; Plosch et al., 2006), unless 0.5% cholate is added to the diet (Davis et al., 2004). Ezetimibe treatment that reduces cholesterol entry into enterocytes does not increase intestinal NPC1L1 expression in the chow-fed mice (Valasek et al., 2007). In one cell culture study, ezetimibe treatment for 16 h reduces rather than increases NPC1L1 mRNA by 65% in CaCo-2 cells (During et al., 2005). In another study, 24 h treatment of CaCo-2 cells with ezetimibe causes no alterations in NPC1L1 mRNA levels or protein mass (Field et al., 2007). Currently, it is unclear if these obvious discrepancies in cholesterol regulation of NPC1L1 expression are related to differences in animal species, diet compositions, duration of drug treatment, or experimental systems used or other factors.

Other factors influencing NPC1L1 expression include estrogen receptors and diabetic state. Administration of high-doses of 17β-estradiol (6 μg/day) to ovarectomized AKR or C57L mice increases NPC1L1 mRNA expression in duodena and jejuna, but not ilea (Duan et al., 2006). Intestinal and hepatic NPC1L1 mRNA levels are significantly higher in streptozotocin-induced diabetic versus non-diabetic rats (Lally et al., 2007a), and in Zucker diabetic fatty versus lean rats (Lally et al., 2007b). NPC1L1 mRNA levels are also ~2-fold higher in the intestinal biopsy samples from type 2 diabetic patients than non-diabetic patients (Lally et al., 2006, 2007b). Given the diabetes epidemic and clinical use of ezetimibe (NPC1L1 inhibitor), exploring mechanisms underlying the relationship between diabetes and NPC1L1 expression levels represents an important future direction and has enormous translation potential.

12.3.5 Cholesterol and Ezetimibe Binding Studies

It is currently unknown if NPC1L1 protein binds cholesterol and other sterols. The purified SSD-containing membrane region of SCAP can directly bind cholesterol through receptor-ligand interaction and SCAP is thus considered as an ER receptor for cholesterol (Radhakrishnan et al., 2004). The binding between NPC1 (a homolog of NPC1L1) and photoactivatable cholesterol analog appears to be SSD-dependent (Ohgami et al., 2004). Unexpectedly, detailed cholesterol binding assays using purified NPC1 protein and its truncated versions localize the binding site of cholesterol and oxysterols to the luminal N-terminus (a 240-amino acid domain with 18 cysteines), instead of SSD (Infante et al., 2008a, b). The N-terminal extracellular domain of NPC1L1 protein consists of 263-amino acids (the signal peptide of 21 amino acids excluded), 18 of which are also cysteines. NPC1L1, like its homolog NPC1, may also bind cholesterol via this N-terminal cysteine-rich domain (Fig. 12.2A).

The NPC1L1 protein has two large extracellular loops facing intestinal lumen (Fig. 12.2A). A 61-amino acid region in the extracellular loop-1 has been recently shown to be critical for ezetimibe binding by Weinglass and associates (Weinglass et al., 2008). In this study, these workers purified a NPC1L1-ezetimibe complex from cultured cells and analyzed its constituents by mass spectrometry. They found that NPC1L1 is the only protein to account for ezetimibe binding. Taking advantage of the large difference in affinity between dog and mouse NPC1L1 for ezetimibe, they further identified two residues in this loop of NPC1L1 that are mostly responsible for the large differences in affinity between the two species. These residues reside adjacent to a hotspot of human NPC1L1 polymorphisms that are associated with reduced intestinal cholesterol absorption (Cohen et al., 2006). These findings indicate that the extracellular loop-1 of NPC1L1 plays an important role in mediating ezetimibe action and intestinal cholesterol absorption.

12.3.6 Potential Mechanisms for NPC1L1 to Mediate Sterol Uptake

Several lines of evidence suggest that NPC1L1 may mediate cholesterol uptake via the clathrin-mediated endocytic pathway (Yu, 2008). These include: (1) NPC1L1 cycles in a cholesterol-regulated manner to and from the cell surface (Yu et al., 2006); (2) NPC1L1 physically resides at both plasma membrane and intracellular compartments in cultured cells and in absorptive enterocytes of small intestine (Altmann et al., 2004; Garcia-Calvo et al., 2005; Iyer et al., 2005; Labonte et al., 2007; Sane et al., 2006; Knopfel et al., 2007; Yamanashi et al., 2007; Temel et al., 2007; Yu et al., 2006); (3) After extraction from cultured McArdle rat hepatoma cells, NPC1L1 co-immunoprecipitates with the $\mu 2$ (mu2) subunit of an adaptor protein complex AP2 and the clathrin heavy chain (Ge et al., 2008), two proteins that are involved in the clathrin endocytic pathway; (4) Potassium depletion, a condition known to arrest clathrin-mediated endocytosis (Larkin et al., 1983), inhibits NPC1L1-dependent cholesterol uptake (Brown et al., 2007); and (5) Mice lacking caveolin-1, a structural molecule of caveolae, display normal intestinal cholesterol absorption (Valasek et al., 2005), demonstrating that caveolin-mediated endocytosis (another important endocytic pathway) is not the cellular basis for NPC1L1-dependent cholesterol uptake. Clathrin-mediated endocytosis appears to be the cellular basis for intestinal fat absorption (Hansen et al., 2003, 2007). A similar mechanism may operate for intestinal cholesterol absorption and perhaps hepatic cholesterol retrieval from canalicular bile (Temel et al., 2007). Ezetimibe appears to inhibit the sterol-induced internalization of NPC1L1 via clathrin-mediated endocytosis in cultured hepatoma cells (Ge et al., 2008; Chang and Chang, 2008; Petersen et al., 2008).

To definitively identify molecular mechanisms underlying NPC1L1-mediated cholesterol transport, many important questions have to be answered. For example, how does cholesterol regulate NPC1L1 trafficking? What sorting signals in the protein and sorting platforms in cells are involved in this regulation? Does NPC1L1 bind cholesterol? If it does, what is the physiological significance of cholesterol binding to NPC1L1? Do NPC1L1-cholesterol interactions function as a signal for inducing clathrin-mediated endocytosis of NPC1L1 or as a process to recruit and transfer free cholesterol to NPC1L1-containing membrane microdomains? Does NPC1L1 bind cholesterol that resides in the plasma membrane or cholesterol present in extracellular spaces such as the gut lumen and hepatic bile canaliculus (Temel et al., 2007; Yu, 2008)? How does ezetimibe inhibit NPC1L1 internalization from the plasma membrane? Does ezetimibe interfere with NPC1L1-cholesterol binding? In addition, due to the dense and rigid microvillar cytoskeleton (Yamamoto, 1982), the base of the microvilli of small intestine is likely the only site for endocytic membrane traffic to occur (Hansen et al., 2003). Given that NPC1L1 abundantly localizes at the microvilli, a dilemma is how the microvillus-localized NPC1L1 and its cargos are endocytosed. If the microvillus-localized NPC1L1 has to move to the base of microvilli to mediate cholesterol uptake, what are signals triggering this translocation or movement? Elucidation of all these questions will greatly enhance our understanding of how cells such as enterocytes and hepatocytes handle extracellular unesterfied cholesterol.

12.3.7 *Therapeutic Perspectives*

NPC1L1 is undoubtedly a gatekeeper of intestinal cholesterol absorption, thus positioning itself as an attractive drug target for prevention of cholesterol-driven diseases such as ASCVD and gallstone disease. Additionally, several studies have shown that NPC1L1 inhibition may provide benefits for important metabolic diseases such as hepatic steatosis, dyslipidemia, high fat diet-induced obesity and insulin resistance (Labonte et al., 2008; Davies et al., 2005; van Heek et al., 2001a; Assy et al., 2006; Deushi et al., 2007; Zheng et al., 2008) (Yu, L., unpublished data). Further studies are needed to define molecular mechanisms underlying these intriguing observations. Intestinal cholesterol has been reported to regulate fat storage (Kalaany et al., 2005). Perhaps, simply by controlling how much cholesterol enters the body via the small intestine, NPC1L1 can regulate responses of other metabolic pathways and physiological processes to over-nutrition, thereby improving metabolic diseases induced by metabolic overload.

Although the NPC1L1 inhibitor ezetimibe can efficiently lower blood cholesterol, it is important to point out that inter-individual variations exist in response to ezetimibe treatment (Hegele et al., 2005). Comparison of multiple species NPC1L1 orthologs have shown that the in vivo responsiveness to ezetimibe correlates with NPC1L1 binding affinity (Hawes et al., 2007). Ezetimibe non-responders may have distinct NPC1L1 sequence variations, and these individuals may be responsive to other NPC1L1 inhibitors that work via distinct mechanisms. In addition, small molecule NPC1L1 inhibitors, after entering the body, have the potential to cause adverse effects. NPC1L1 localizes at the intestinal brush border membrane and its extracellular domains are exposed to contents in the gut lumen. This unique location and membrane topology makes NPC1L1 an ideal target for large molecule nonabsorbable NPC1L1 inhibitors (Davidson, 2009). Thus, NPC1L1 will remain an attractive drug target in the future.

12.4 ATP-Binding Cassette Transporters G5 and G8 (ABCG5/G8)

12.4.1 Discovery of ABCG5/G8: The Power of Human Genetics

The discovery of the heterodimeric transporters ABCG5 and ABCG8 represents a powerful example of human genetics leading to mechanistic understanding of the underlying disease process. In this case, it has long been appreciated that cholesterol is structurally very similar to plant sterols, with only minor differences in side chain configurations (Schoenheimer, 1929) (Fig. 12.3). However, it has been known for nearly a century that mammals consume large amounts of dietary plant sterols, yet these phytosterols are excluded from the body, primarily at the level of the intestine (Sudhop et al., 2005; Schoenheimer and Breusch, 1933; Borgstrom, 1968; Gould et al., 1969; Hernandez et al., 1954, Huang and Kuksis, 1965). A major breakthrough regarding this issue came when Bhattacharayya and Connor described a novel disease in which two sisters presented to the clinic with tendon xanthomas, and were surprisingly shown to have extremely elevated levels of plasma plant sterols (primarily β-sitosterol) (Bhattacharyya and Connor, 1974). Hence, the disease was named β-sitosterolemia, and the authors proposed that this disease was caused by a single genetic defect, which prevented the intestine from excluding plant sterols from being absorbed. It was later discovered that sitosterolemic patients have elevated fractional absorption of dietary sterols and diminished ability to secrete multiple sterols into bile (Miettinen, 1980; Lutjohann et al., 1995; Gregg et al., 1986). As a result, sitosterolemia manifests as the accumulation of both plant and animal sterols in the plasma, skin, tendons, coronary arteries and other tissues, and most affected individuals suffer from premature coronary heart disease (Salen et al., 1985; Kolovou et al., 1996; Mymin et al., 2003; Bjorkhem et al., 2001). The sitosterolemia locus was subsequently mapped to a single site on human chromosome 2p21 that contains two genes, ABCG5 and ABCG8 (Patel et al., 1998), and mutations in either of these genes is causative of sitosterolemia (Berge et al., 2000; Lee et al., 2001).

12.4.2 Structure: Gene, mRNA, and Protein Domains

ABCG5 and ABCG8 genes are arranged in a head-to-head orientation with less than 400 base pairs between their respective start codons, and each has 13 exons and twelve introns (Berge et al., 2000). The two genes encode two distinct proteins known as sterolin-1 (ABCG5) and sterolin-2 (ABCG8), which are mammalian homologues of the *Drosophila* gene *White*. In addition to being in the same genetic neighborhood, ABCG5 and ABCG8 are also in close proximity at the protein level. Both of these proteins contain an ATP-binding cassette near the N-terminus followed by six putative transmembrane domains (Fig. 12.2B), and serve only as non-functional half-transporters when expressed alone (Graf et al., 2002, 2003, 2004). ABCG5 and ABCG8 must heterodimerize to transport sterols across membranes (Graf et al., 2002, 2003, 2004). This idea is supported by data demonstrating that, when expressed together, the proteins colocalize in the ER and the plasma membrane, they can be co-immunoprecipitated, and the exit of these proteins from the ER to the plasma membrane requires coexpression of both proteins (Graf et al., 2002, 2003, 2004). Even stronger evidence for obligate heterodimerization comes from the fact that single mutations in either of these genes alone causes sitosterolemia (Berge et al., 2000; Lee et al., 2001). The current data support a model where ABCG5 and ABCG8 heterodimerize in the ER, traffic together through the Golgi apparatus, and subsequently target to apical subdomains in the plasma membrane. Both ABCG5 and ABCG8 undergo *N*-linked glycosylation, and glycosylation at Asn-619 in ABCG8 is critical for efficient trafficking of the heterodimer (Graf et al., 2004). The ABCG5/G8 heterodimer requires the molecular chaperones, calnexin and calreticulin for proper folding and trafficking out of the ER (Graf et al., 2004; Okiyoneda et al., 2006). Subsequent site-directed mutagenesis experiments demonstrated that the majority of mutants causative of sitosterolemia exhibit impaired transport of the heterodimer from the ER to the plasma membrane (Graf et al., 2004).

12.4.3 Function: Lessons Learned from Animal Models

Since the discovery of the genetic basis for sitosterolemia, there has been intensive effort put forth to understand the function of the ABCG5/G8 heterodimer. Like NPC1L1, ABCG5 and ABCG8 are expressed almost exclusively on the apical membrane of enterocytes in the intestine and hepatocytes in the liver (Patel et al., 1998; Berge et al., 2000; Lee et al., 2001; Graf et al., 2003; Klett et al., 2004a). It is generally accepted that the ABCG5/G8 heterodimeric complex serves as an efflux pump to remove sterols (cholesterol and phytosterols) from hepatocytes and enterocytes. In the intestinal brush border membrane, this function would allow for transport of intracellular sterols back into the lumen of small intestine for fecal excretion. If this were true, ABCG5/G8 would likely play an important role in intestinal sterol absorption by opposing the action of NPC1L1.

A potential role for ABCG5/G8 in intestinal cholesterol absorption was first discovered in sitosterolemic patients, who have elevated intestinal absorption of both plant sterols and cholesterol (Miettinen, 1980; Lutjohann et al., 1995; Gregg et al., 1986; Salen et al., 1989, 1992; Bhattacharyya et al., 1991). In fact, sitosterolemic patients absorb roughly 20–30% of dietary sitosterol, compared to <5% absorption in unaffected individuals (Bhattacharyya and Connor, 1974; Miettinen, 1980; Salen et al., 1989, 1992). Importantly, mice lacking either ABCG5 alone (Plosch et al., 2004), ABCG8 alone (Klett et al., 2004b), or both transporters (Yu et al.,

2002a) exhibit sitosterolemia, which can probably in part be explained by increased intestinal absorption of phytosterols. In support of this concept, in mice lacking both ABCG5 and ABCG8 fed a chow diet, intestinal absorption of cholesterol is not significantly altered, yet the absorption of sitosterol, cholestanol, campesterol is significantly increased (Yu et al., 2002a). In a recent study, Wang and colleagues demonstrated that the lymphatic transport rate of cholesterol and sitostanol is increased by ~40 and 500%, respectively in mice lacking ABCG8 (Wang et al., 2007b). Furthermore, mice transgenically overexpressing ABCG5 and ABCG8 in the intestine and liver exhibit a 50% reduction in fraction cholesterol absorption, and fecal neutral sterol levels increase 3–6 fold (Yu et al., 2002b). Collectively, studies in both sitosterolemic humans and animal models support an important role for ABCG5/G8 in intestinal sterol absorption. Although ABCG5 and ABCG8 heterodimer is thought to influence intestinal sterol absorption by serving as an efflux pump to deliver sterols from enterocytes back into the gut lumen for fecal disposal, direct experiment evidence of the concept is still lacking.

12.4.4 Function: Lessons Learned from Cell Model Systems

There is a strong body of work examining the subcellular trafficking and dimerization of ABCG5 and ABCG8, but characterization of the cellular sterol transport properties of the heterodimer is still incomplete. For unknown reasons, although ABCG5 and ABCG8 can be readily expressed and properly localized in mammalian cells, a standard reproducible functional assay for ABCG5/G8-dependent sterol transport has not been established. Recently, one group was able to show that overexpression of ABCG5 and ABCG8 in human kidney and gallbladder epithelial cells promotes the efflux of cholesterol and plant sterols (Vrins et al., 2007; Tachibana et al., 2007). In these studies it was shown that ABCG5/G8-dependent efflux is dependent on the presence of mixed bile salt micelles as an acceptor, whereas other cholesterol acceptors such as apoAI, high-density lipoprotein (HDL), or methyl-β-cyclodextrin do not efficiently promote ABCG5/G8-dependent efflux.

Given that mutations in either ABCG5 or ABCG8 cause abnormal accumulation of plant sterols in the body (Berge et al., 2000; Lee et al., 2001), most have assumed that the ABCG5/G8 heterodimer is the primary, if not the sole protein complex, responsible for sterol discrimination in the intestine. However, the current knowledge base does not firmly support this conclusion. For example, genetically sitosterolemic patients (Lutjohann et al., 1995), rats (Hamada et al., 2007), and mice (Yu et al., 2002a) still possess the ability to discriminate between cholesterol, campesterol, and sitosterol at the level of intestinal absorption, indicating that ABCG5/G8 is a gatekeeper rather than the intestinal sterol discriminator. Given this information, caution should be taken when assuming that the ABCG5/G8 heterodimer is the sole mediator of sterol selectivity in the intestine. Much more work is needed in this area, and with recent progress using cell-based systems for ABCG5/G8-dependent sterol transport, we may gain further mechanistic insights into the critical question of substrate specificity.

12.4.5 Regulation of Expression

Both ABCG5 and ABCG8 seem to be primarily controlled at the transcriptional level. The sterol-sensing transcription factors LXRα and LXRβ appear to be the primary regulator of ABCG5 and ABCG8 mRNA expression. In support of this, both dietary cholesterol and synthetic LXR agonists upregulate ABCG5 and ABCG8 mRNA expression in the small intestine and liver of wild type mice, but not LXR knockout mice (Berge et al., 2000; Repa et al., 2002; Plosch et al., 2002; Kaneko et al., 2003). This is indicative of a direct effect, but the presence of a bona fide LXR response element in the ABCG5/G8 intergenic promoter or surrounding areas has not been identified. The physiological relevance of LXR-driven upregulation of ABCG5 and ABCG8 mRNA has recently been highlighted in two separate studies. It was first shown that LXR-driven increases in hepatobiliary and fecal cholesterol excretion rely on functional ABCG5 and ABCG8, since ABCG5/G8 knockout mice could not elevate biliary and fecal sterol secretion in response to a synthetic LXR agonist (Yu et al., 2003). In agreement, using ABCG5/G8 knockout mice, Calpe-Berdiel and colleagues demonstrated that LXR-mediated induction of macrophage to feces reverse cholesterol transport requires functional ABCG5/G8 (Calpe-Berdiel et al., 2008).

Another transcriptional activator of the sitosterolemia locus is the liver receptor homolog-1 (LRH-1) (Freeman et al., 2004). However, it is important to point out that a functional LRH-1 binding site is only present in the human gene, and rodent orthologs do not possess LRH-1-sensitive transcriptional activation (Freeman et al., 2004). More recently, it was shown that three additional transcription factors known as HNF-4α, GATA binding protein 4, and GATA binding protein 6 act in a cooperative fashion to transactivate the human intergenic promoter (Sumi et al., 2007). There is also evidence that bacterial endotoxin can downregulate ABCG5 and ABCG8 mRNA levels (Khovidhunkit et al., 2003), yet the transcription factors involved in this response have yet to be clearly elucidated. Collectively, it is quite clear that ABCG5 and ABCG8 are coordinately regulated at the transcriptional level. More work in this area may prove to be critical for future ABCG5/G8-centered therapies.

More recently, there has been evidence for hormonal control of ABCG5/G8 expression at both transcriptional and post-transcriptional levels. In support of this, hepatic insulin resistance promotes cholesterol gallstone formation, which is driven in part by transcriptional regulation of ABCG5/G8 (Biddinger et al., 2008). In this study it was shown that hepatocyte-specific deletion of the insulin receptor resulted in reduced insulin-driven inhibition of the transcription factor forkhead box O1 (FOXO1), and that FOXO1 was a direct transcriptional activator of ABCG5/G8. In addition to insulin-mediated regulation, ABCG5/G8 expression has also been linked to the leptin axis in the liver (Sabeva et al., 2007). In this case, the hepatic protein (not mRNA) abundance of ABCG5/G8 is reduced in mice lacking leptin signalling. Furthermore, hypophesectomized rats have dramatic reductions in hepatic ABCG5/G8 mRNA levels that are coupled to decreased fecal sterol loss, and these alterations can be normalized by thyroid hormone replacement (Galman et al.,

2008). The mechanism by which thyroid hormone and leptin regulates ABCG5/G8 expression requires further exploration.

12.4.6 Biochemical Studies on ABCG5/G8-dependent Sterol Transport

Although cellular functional assays have been somewhat elusive, ABCG5/G8-dependent sterol transport has been successfully reconstituted in vitro with either a recombinant or purified native ABCG5/G8, and the heterodimer can directly transport sterols, sterol esters, and phospholipids from donor vesicles to proteoliposomes in an ATP-dependent fashion and vanadate-sensitive manner (Wang et al., 2006, 2008b). With these established in vitro assays and complimentary cell-based models (Vrins et al., 2007; Tachibana et al., 2007) for ABCG5/G8-dependent sterol transport we now have the tool to address important unanswered questions. For instance, does the ABCG5/G8-transported pool of sterols originate from a cytosolic pool, or does the heterodimer act primarily on a membrane-associated pool as a flippase? Are other proteins required for ABCG5/G8-dependent sterol transport? Additionally, what is the substrate specificity for ABCG5/G8?

12.4.7 Therapeutic Perspectives for ABCG5/G8

ABCG5/G8 plays a crucial role in cholesterol excretion from the body and therefore it seems logical to assume that mice lacking the heterodimer should have increased atherosclerotic burden, while mice overexpressing ABCG5/G8 should be protected against atherosclerosis. Consistent with these hypotheses, transgenic overexpression of ABCG5/G8 in both the intestine and liver of LDL receptor knockout mice results in reduced LDL-C and significantly less atherosclerosis (Wilund et al., 2004a). In contrast, hepatocyte-specific ABCG5/G8 overexpression, which elevates biliary sterol secretion by $\sim$2-fold, does not protect against atherosclerosis in the LDL receptor or apoE knockout backgrounds (Wu et al., 2004). This result is very confusing, since it was anticipated that a 2-fold increase in biliary cholesterol output would result in atheroprotection. In a subsequent study, these same authors clarified this confusing outcome, and showed that ezetimibe treatment in mice overexpressing ABCG5/G8 specifically in the liver produces profound protection against atherosclerosis (Basso et al., 2007). These data strongly support the opposing roles of ABCG5/G8 and NPC1L1 (target of ezetimibe), and demonstrate that therapies that simply increase biliary sterol output may not be effective at altering sterol balance and atherosclerosis, since these sterols can re-enter the body via the action of intestinal NPC1L1 (Altmann et al., 2004). However, dual therapies that increase biliary sterol output and block re-uptake of biliary sterols by the intestine continue to hold promise.

It is well known that many sitosterolemic patients suffer from premature ASCVD (Salen et al., 1985; Kolovou et al., 1996; Mymin et al., 2003; Bjorkhem et al., 2001).

It is however not known whether this ASCVD is caused by the massive accumulation of plant sterols in the blood or the associated hypercholesterolemia. Without a doubt hypercholesterolemia is a major driving force in ASCVD progression, but little is known about the consequences of elevated blood plant sterols because in organisms with intact ABCG5/G8 plant sterols are efficiently excluded from the body. Sitosterolemic patients and ABCG5/G8 knockout mice provide a unique research opportunity to ask the age-old question: How are phytosterols metabolized, and do phytosterols like their animal counterparts contribute to atherogenesis? Importantly, the plant sterol stigmasterol has been shown to regulate cholesterol metabolism by directly inhibiting SREBP-2 processing and activating LXR-driven gene expression (Yang et al., 2004). In addition sitosterol has recently been shown to drive macrophage cell death (Bao et al., 2006), an important feature of the late stages of ASCVD. These and many other findings have promoted a potential role for plant sterols in the progression of ASCVD in sitosterolemic patients. Recently, a study examined the relationship between blood sitosterol levels and atherosclerosis in both ABCG5/G8 knockout mice and in sitosterolemic patients (Wilund et al., 2004b). It was found that plasma levels of plant sterols do not correlate with atherosclerosis extent in either of these sitosterolemic models, leading the authors to conclude that perhaps elevated plasma cholesterol levels are responsible for the development of premature ASCVD in sitosterolemic patients.

Inter-individual variations exist in responses to statins (Kajinami et al., 2004), and it was thought to be attributable to differences in the baseline of intestinal cholesterol absorption and/or endogenous cholesterol synthesis among individuals (Miettinen et al., 2000; Gylling and Miettinen, 2002). Stimulation of ABCG5/G8-mediated cholesterol excretion not only promotes fecal disposal of cholesterol, but also increases endogenous cholesterol synthesis (Yu et al., 2002b, 2004a). Co-administration of a statin and a yet-to-be developed ABCG5/G8 activator is expected to have a synergistic cholesterol-lowering effect. Consistent with this notion is that transgenic overexpression of ABCG5/G8 in the liver and intestine confers mice hypersensitivity to lovastatin in reducing plasma cholesterol (Tang et al., 2006).

Unlike the successful story for NPC1L1 and its inhibitor ezetimibe, a specific pharmacological activator for ABCG5/G8-dependent sterol transport has yet to be developed. Most current attempts at promoting ABCG5/G8-dependent sterol transport involve the use of synthetic LXR agonists, which robustly and transcriptionally upregulates heterodimer expression in the liver and intestine (Berge et al., 2000; Repa et al., 2002; Plosch et al., 2002; Kaneko et al., 2003). LXR activation, however, elicits the unwanted side effect of increased de novo lipogenesis, resulting in pronounced hepatic steatosis (Rader, 2007; Cao et al., 2004), which rules out synthetic LXR agonists as a safe way to promote ABCG5/G8 function. Alternative strategies for ABCG5/G8 modulation need to be pursued. ABCG5/G8 heterodimer may be directly targeted by small molecule activators. With in vitro and cell-based functional assays for ABCG5/G8-dependent sterol transport in place, it is now possible to screen for such compounds. Based on the limited data generated in mice transgenically overexpressing ABCG5 and ABCG8 (Yu et al., 2002b; Tang et al.,

2006), pharmacologic activators of this pathway could serve as powerful promoters of cholesterol removal from the body, and thus providing additional protection against atherosclerosis when given in combination with a statin.

12.5 Scavenger Receptor Class B Type l (SR-Bl)

12.5.1 Discovery of SR-BI

SR-BI was originally discovered based on its sequence homology to another scavenger receptor known as cluster determinant 36 (CD36) (Calvo and Vega, 1993; Acton et al., 1994). SR-BI is expressed in a wide variety of tissues, with the highest level of expression in tissues regulating cholesterol metabolism such as the liver, intestine, adrenal gland, testes, and ovary (Acton et al., 1994, 1996; Landschulz et al., 1996). SR-BI is widely accepted as an HDL receptor, and can bi-directionally transport sterols across biological membranes (Landschulz et al., 1996; Acton et al., 1996). SR-BI-driven selective uptake into the liver promotes biliary and fecal excretion of cholesterol (Kozarsky et al., 1997; Ji et al., 1999; Zhang et al., 2005a). In addition, SR-BI-driven selective uptake into the adrenal gland is linked to steroid hormone production (Rigotti et al., 1996, 1997; Temel et al., 1997). Since the role of SR-BI in these processes has been the focus of previous reviews (Trigatti, 2005; Connelly, 2009), these concepts will not be discussed in detail here. The purpose of this section is to specifically discuss the structure and function of SR-BI in regards to intestinal sterol transport.

12.5.2 Structure: Gene, mRNA, and Protein Domains

SR-BI belongs to the class B scavenger receptor superfamily, which includes other proteins such as lysosomal integral membrane protein II (LIMP II), *Drosophila melanogaster* croquemort membrane protein, and Snmp-1 (Franc et al., 1996, 1999; Hart and Wilcox, 1993; Karakesisoglou et al., 1999; Vega et al., 1991). The human SR-BI gene contains 13 exons, and resides on chromosome 12q24.2-qter (Cao et al., 1997). The pattern of SR-BI full-length mRNA expression is similar in humans and rodents, with the highest level of mRNA in the adrenal gland, ovary, liver, intestine, and placenta (Acton et al., 1996; Landschulz et al., 1996; Cao et al., 1997). The full-length SR-BI mRNA encodes a 509 amino acid protein containing a short N-terminal cytoplasmic domain of 9 amino acids, a transmembrane spanning domain of 22 amino acids, a predominant extracellular domain of 408 amino acids, a second transmembrane domain of 23 amino acids, and a cytoplasmic C-terminus of 44 amino acids (Krieger, 1999; Williams et al., 1999). The predicted molecular weight of SR-BI is $\sim$57 kDa, yet extensive *N*-linked glycosylation yields a protein that runs $\sim$82–84 kDa on a Western blot (Krieger, 1999; Williams et al., 1999). SR-BI is also myristoylated and palmitoylated (Babitt et al., 1997), yet these post-translational modifications do not seem to impact SR-BI function. The extracellular domain of SR-BI is likely to form disulfide bridges since it contains six cysteine residues (Krieger, 1999). An alternatively spliced variant of SR-BI mRNA was identified, and this protein is known as SR-BII (Webb et al., 1998; Eckhardt et al., 2004, 2006). This alternatively spliced form has a slightly modified C-terminal tail, yet can also function to transport sterols across biological membranes (Webb et al., 1998; Eckhardt et al., 2004, 2006). Most recently, the PDZ domain-containing adaptor protein PDZK1 has been shown to bind to the C-terminus of SR-BI and this protein-protein interaction controls SR-BI activity via a tissue-specific post-transcriptional mechanism (Silver, 2002; Yesilaltay et al., 2006; Kocher et al., 2003). In fact, mice deficient in PDZK1 have elevated plasma cholesterol levels due to liver-specific downregulation of SR-BI, implicating PDZK1 as a critical regulator of hepatic SR-BI function (Kocher et al., 2003).

12.5.3 Function: Lessons Learned from Animal Models

Animal models of SR-BI deficiency or overexpression have shed light on the critical role of this protein in whole body sterol balance. Transgenic overexpression of SR-BI results in diminished very low density lipoprotein cholesterol, LDL-C, and HDL-cholesterol levels, increased biliary cholesterol secretion, and marked protection against atherosclerosis (Wang et al., 1998; Ueda et al., 1999, 2000; Arai et al., 1999; Kozarsky et al., 2000; Ji et al., 1999). Conversely, mice lacking SR-BI accumulate cholesterol in the plasma, have decreased biliary sterol excretion and steroid hormone insufficiency, and develop severe atherosclerosis (Rigotti et al., 1997; Varban et al., 1998; Trigatti et al., 1999). Mice lacking SR-BI in the apoE-deficient background develop occlusive coronary artery atherosclerosis, myocardial infarction, and die at a very early age (Braun et al., 2002; Zhang et al., 2005b). The hyperlipidemic and proatherogenic effects seen with SR-BI deficiency are thought to be primarily due to impairment of SR-BI's critical role in selective uptake of HDL cholesteryl esters into the liver. However, SR-BI is also abundantly expressed in the intestine, and has been proposed to play a role in intestinal cholesterol absorption. The primary focus of this chapter is to discuss the current state of knowledge regarding SR-BI's role in the intestine, which is still a matter of debate.

Using immunohistochemical and biochemical approaches SR-BI protein expression has been documented to be expressed in enterocytes, and can be found on both apical and basolateral membranes (Cai et al., 2001; Voshol et al., 2001; Hauser et al., 1998; Hatzopoulos et al., 1998). The intestinal gradient of SR-BI expression is found most prominently along the proximal regions of the gastrocolic axis, including the duodenum and jejunum where cholesterol absorption is thought to occur (Cai et al., 2001; Voshol et al., 2001; Hauser et al., 1998; Hatzopoulos et al., 1998),. The concept of SR-BI being involved in intestinal cholesterol absorption was first supported by in vitro studies demonstrating that cholesterol uptake in intestinal brush border membranes is reduced by anti-SR-BI blocking antibodies (Hauser et al., 1998). Further, early evidence showed that intestinal cholesterol absorption inhibitor ezetimibe could physically interact with SR-BI (Hatzopoulos et al., 1998). Also, it was shown that SR-BI is a high-affinity cholesterol-binding protein in the intestinal brush border membrane (Altmann et al., 2002). However, mice genetically lacking SR-BI have either slightly elevated or normal intestinal cholesterol absorption (Altmann et al., 2002; Mardones et al., 2001; Nguyen et al., 2009). SR-BI-null mice still exhibit reduced fractional cholesterol absorption when treated with ezetimibe, ruling out SR-BI as the ezetimibe-sensitive transport protein (Altmann et al., 2002). In contrast to the results from SR-BI-deficient mice, intestine-specific transgenic overexpression of SR-BI promotes the appearance of radiolabelled cholesterol into the plasma following an oil-based gavage administration (Bietrix et al., 2006). This was interpreted as an increase in intestinal cholesterol absorption, yet this method may not provide an accurate quantitative measure of fractional cholesterol absorption. Collectively, there is evidence to support the ability of SR-BI to facilitate the binding of cholesterol from bile salt micelle donors into brush border membrane vesicles (Nguyen et al., 2009; Labonte et al., 2007; Knopfel et al., 2007), but data from knockout mice confirm that this phenomenon is not rate limiting in net cholesterol absorption (Altmann et al., 2002; Mardones et al., 2001; Nguyen et al., 2009).

12.5.4 Function: Lessons Learned from Cell Model Systems

Although SR-BI may have been implicated in the apical to basolateral transport process of intestinal cholesterol absorption, most cell models of SR-BI-dependent sterol transport suggest a quite different trafficking pattern of SR-BI-delivered sterols. Despite a matter of debate, in the steady state SR-BI seems to localize to both apical and basolateral membranes in polarized cells (Cai et al., 2001; Voshol et al., 2001; Hauser et al., 1998; Hatzopoulos et al., 1998; Silver et al., 2001; Burgos et al., 2004; Harder et al., 2007; Wustner et al., 2004; Sehayek et al., 2003). In polarized cells SR-BI undergoes sterol-dependent directional basolateral to apical transcytosis (Silver et al., 2001; Burgos et al., 2004; Harder et al., 2007; Wustner et al., 2004; Sehayek et al., 2003). This directional pattern of sterol transport is consistent with SR-BI's role as an HDL receptor promoting selective uptake of cholesteryl esters from circulation to liver for biliary disposal (Silver et al., 2001; Burgos et al., 2004; Harder et al., 2007; Wustner et al., 2004; Sehayek et al., 2003). This selective uptake involves the internalization of HDL-associated cholesteryl esters into the cell, without the net internalization and degradation of the lipoprotein itself. Additionally, SR-BI can also promote selective uptake of free sterols, phospholipids, triglycerides, and cholesteryl ethers (Stangl et al., 1999; Urban et al., 2000; Greene et al., 2001). Selective uptake may involve the "retro-endocytosis" pathway, which involves holoparticle uptake and the subsequent re-secretion of the cholesteryl-ester poor HDL (Silver et al., 2001; Rhainds et al., 2004; de la Llera-Moya et al., 1999). The precise molecular mechanism by which SR-BI facilitates selective sterol uptake remains incompletely understood, but purified SR-BI reconstituted into liposomes can recapitulate high-affinity HDL binding and selective uptake, indicating that these processes are intrinsic to the receptor itself (Liu and Krieger, 2002).

Although SR-BI is generally thought to be an HDL receptor, it can bind a wide variety of ligands. In fact, the first SR-BI ligands described were modified apoB-containing lipoproteins including acetylated-LDL and oxidized-LDL (Acton et al., 1994; Gillotte-Taylor et al., 2001). In addition to binding native and modified lipoproteins, SR-BI binds maleylated BSA (Acton et al., 1994), advanced glycated proteins (Ohgami et al., 2001), anionic phospholipids (Rigotti et al., 1995; Fukasawa et al., 1996), and β-amyloid (Paresce et al., 1996; Husemann et al., 2001; Husemann and Silverstein, 2001). The majority of SR-BI ligand binding studies have been carried out using native lipoproteins. Lipoprotein binding to SR-BI exhibits intrinsic nonreciprocal cross competition (Ashkenas et al., 1993; Gu et al., 2000a). In this case, HDL binding efficiently blocks subsequent LDL binding, yet the ability of LDL to block HDL binding is relatively weak (Ashkenas et al., 1993; Gu et al., 2000b). Therefore, it is generally accepted that LDL does not typically interfere with HDL binding in vivo, further supporting the claim that SR-BI is the HDL receptor (Acton et al., 1996; Landschulz et al., 1996). Importantly, not all HDL particles bind to SR-BI with the same affinity. Lipid-free apoA-I and pre-β HDL particles are poor substrates for SR-BI, whereas spherical large α-HDL particles bind with much higher affinity (Xu et al., 1997; Liadaki et al., 2000; de Beer et al., 2001; Williams et al., 2000). The interaction of HDL particles with SR-BI relies on interactions with intrinsic HDL apolipoproteins including apoA-I, apoA-II, and apoE (Li et al., 2002; Pilon et al., 2000; Bultel-Brienne et al., 2002; Liu et al., 2002).

In addition to facilitating cholesterol uptake from lipoproteins, SR-BI can serve as a bi-directional sterol transporter, facilitating cholesterol efflux from cells (Kozarsky et al., 1997; Ji et al., 1997; Jian et al., 1998; Yancey et al., 2000; Gu et al., 2000a). This activity was originally established when SR-BI expression levels in a variety of different cell lines are highly correlated to HDL-mediated sterol efflux (Ji et al., 1997). Later it was shown in SR-BI gain of function experiments in cells that SR-BI promotes cholesterol efflux to HDL and phosphatidylcholine-containing liposomes (Jian et al., 1998). Studies done with mutated forms of apoA-I suggest that a direct interaction between apoA-I and SR-BI must occur for efficient cholesterol efflux (Liu et al., 2002), yet other studies have challenged the concept that physical binding of the HDL particle to SR-BI is required for sterol efflux (Kellner-Weibel et al., 2000; de la Llera-Moya et al., 1999). In the context of atherosclerosis, the process of cholesterol efflux is thought to be most important in immune cells present in the artery wall such as macrophages, but macrophage SR-BI does not seem to play a major role in macrophage cholesterol efflux or in vivo reverse cholesterol transport (Yu et al., 2004b; Yvan-Charvet et al., 2008; Wang et al., 2007c). The physiological implication of SR-BI-mediated sterol efflux has yet to be understood.

12.5.5 Regulation of Expression

Since SR-BI is so critical to sterol balance and steroid hormone production, its expression is highly controlled at the transcriptional level. In fact SR-BI expression is regulated by trophic hormones, cholesterol, and fatty acids in a tissue specific manner (Rigotti et al., 1996; Wang et al., 1996; Cao et al., 1999; Towns et al., 2005; Fluiter et al., 1998; Spady et al., 1999; Li et al., 1998; McLean and Sandhoff, 1998; Mizutani et al., 1997, 2000; Reaven et al., 1998, 1999; Azhar et al., 1998; Li et al., 2001; Svensson et al., 1999). In the case of hormone regulation, SR-BI expression is upregulated by adrenocorticotrophic hormone (Kozarsky et al., 1997; Rigotti et al., 1996; Wang et al., 1996; Cao et al., 1999), luteinizing hormone (Landschulz et al., 1996), human chorionic gonadotropin (Kozarsky et al., 1997; Li et al., 1998; McLean and Sandhoff, 1998), pregnant mare serum gonadotropin (Li et al., 1998; Mizutani et al., 1997; McLean and Sandhoff, 1998), cyclic AMP analogs (Azhar et al., 1998), and insulin (Li et al., 2001). It is generally accepted that trophic hormone-driven stimulation of SR-BI expression in steroidogenic cells relies on cAMP-dependent activation of protein kinase A, and subsequent promoter transactivation that depends on both steroidogenic factor-1 (SF-1) (Cao et al., 1997, 1999; Parker, 1998; Lopez and McLean, 1999) and CCAAT/enhancer binding proteins (Lekstrom-Himes and Xanthopoulos, 1998). SF-1 is critical for both basal and trophic hormone-stimulated transactivation of the SR-BI promoter, and directly promotes the expression of SR-BI during the development of several steroidogenic tissues (Cao et al., 1997, 1999; Parker, 1998; Lopez and McLean, 1999).

In addition to hormonal regulation, SR-BI expression is also sensitive to alteration in cellular cholesterol levels (Wang et al., 1996; Cao et al., 1999; Sun et al., 1999). In this case, a bona fide sterol response element has been identified in the SR-BI promoter (Lopez and McLean, 1999). This element confers SREBP-1a-dependent transactivation of the gene. Interestingly, in the ovary, gonadotropin treatment induces the expression of SREBP-1a and subsequent enhancement of SR-BI expression (McConihay et al., 2001). In line with this, depletion of plasma cholesterol levels results in induction of SR-BI expression in the adrenal gland, an effect that does not depend on plasma levels of adrenocorticotrophic hormone (Sun et al., 1999). In further support of sterol-dependent transcriptional regulation, SR-BI expression levels are dramatically altered in mouse models of altered cholesterol metabolism (Wang et al., 1996; Ng et al., 1997). For example, mice genetically lacking the steroidogenic acute regulatory (StAR) protein have elevated levels of adrenal SR-BI (Cao et al., 1999). Mice lacking either apoA-I, lecithin:cholesterol acyltransferase, or hepatic lipase all exhibit increased levels of adrenal SR-BI (Cao et al., 1999; Wang et al., 1996; Ng et al., 1997). Furthermore, the human SR-BI promoter can be regulated by LXR in hepatocytes and adipocyte (Malerod et al., 2002).

Two transcriptional pathways are known to repress SR-BI expression, including YY1 (Yin Yang 1) zinc transcription factor and DAX-1 (dosage-sensitive sex reversal, adrenal hypoplasia congenital, critical region on the X chromosome, gene 1). YY1-mediated repression occurs in the basal state due to direct binding of the SR-BI promoter, and can alternatively bind directly to SREBP-1 functionally preventing its transactivation of the SR-BI promoter (Shea-Eaton et al., 2001). In contrast, DAX-1 represses SR-BI expression by directly interacting with the known transcriptional activators such as SF-1 and SREBP-1a (Lopez et al., 2001). In addition to these defined pathways, numerous additional modes of transcriptional regulation of the SR-BI promoter are being discovered including PPARγ, HNF4α, and the prolactin regulatory element-binding (PREB) transcription factor (Malerod et al., 2003; Murao et al., 2008). Collectively, the SR-BI promoter is under complex transcriptional control, and additional work is needed to more completely define physiological implications of these regulatory pathways.

12.5.6 Cholesterol Binding Studies

Although it is well accepted that SR-BI can bi-directionally transfer cholesterol across biological membranes, the precise molecular mechanism by which this is carried out is a matter of debate. The vast majority of work has focused on the direct interactions between SR-BI and apolipoproteins resident on HDL. However, there has been some recent evidence that SR-BI can directly bind cholesterol (Assanasen et al., 2005). In this work the authors tested whether SR-BI could bind a photoactivatable cholesterol analog, and indeed found direct binding between the C-terminal transmembrane domain of SR-BI and this cholesterol analog. The molecular base for SR-BI to bind cholesterol has yet to be determined. The C-terminal transmembrane region of SR-BI does not contain a cholesterol recognition/interaction amino acid consensus (CRAC) or steroidogenic acute regulatory protein-related lipid transfer (START) domain found in other cholesterol-binding proteins (Li and Papadopoulos, 1998; Ponting and Aravind, 1999; Epand, 2006). However, SR-BI is not alone in its lacking of a canonical cholesterol-binding motif. Proteins in the tetraspanin and synaptophysin families also bind cholesterol through a CRAC-, or START-domain independent fashion (Thiele et al., 2000; Charrin et al., 2003). Recent evidence suggests that SR-BI may dimerize or form higher order oligomers, providing the potential for multiple C-termini to serve as a functional cholesterol binding pocket (Reaven et al., 2004; Sahoo et al., 2007a, b). Additional work is needed to define this possibility.

12.5.7 Therapeutic Perspectives for SR-BI

The hyperlipidemic and proatherogenic effects seen with SR-BI deficiency are thought to be primarily due to loss of SR-BI's critical role in selective uptake of HDL cholesteryl esters into the liver, thereby reducing biliary and fecal sterol excretion. However, effects of SR-BI deficiency in the intestine and other tissues on whole-body cholesterol homeostasis should be considered as well. Several investigators have proposed that SR-BI facilitates intestinal cholesterol absorption (Altmann et al., 2002; Mardones et al., 2001; Nguyen et al., 2009; Bietrix et al., 2006; Labonte et al., 2007; Knopfel et al., 2007). The main argument for this is that SR-BI is present in intestinal brush border membrane vesicles, and an SR-BI antibody reduces cholesterol binding to brush border membrane vesicles. However these in vitro findings have not been substantiated in mice with targeted disruption of SR-BI. In fact, mice lacking SR-BI have normal or enhanced intestinal cholesterol absorption (Altmann et al., 2002; Mardones et al., 2001; Nguyen et al., 2009). The most thorough of these studies revealed that SR-BI deficiency significantly increases fractional cholesterol absorption, under several conditions of dietary cholesterol challenge (Mardones et al., 2001). Collectively, these studies have revealed that SR-BI is highly expressed in the intestine, but its role in intestinal cholesterol absorption has not been definitively established. In contrast to a role of SR-BI in the apical to basolateral transport process of cholesterol absorption, we propose that intestinal SR-BI may play a role in basolateral to apical transport of cholesterol in enterocytes. This directional process of transintestinal cholesterol excretion or non-biliary fecal sterol loss is a critical pathway to rid the body of excess cholesterol directly through intestinal secretion (Kruit et al., 2006; Brown et al., 2008; van der Velde et al., 2008, 2007). Based on a number of cellular trafficking studies, SR-BI-mediated selective uptake results in directional basolateral to apical trafficking of sterol cargo (Silver et al., 2001; Burgos et al., 2004; Harder et al., 2007; Wustner et al., 2004; Sehayek et al., 2003). This directional trafficking pattern for SR-BI cargo has been described in vivo in the liver (Silver et al., 2001), and investigations are now underway to examine whether a similar pathway exists in the proximal small intestine. At this point, given our incomplete and opposite understanding of SR-BI's role in intestinal sterol trafficking, it is difficult to know whether intestinal SR-BI remains a viable therapeutic target for ASCVD. More work is needed before this possibility can be realized.

12.6 Other Proteins Influencing Intestinal Cholesterol Absorption

Other proteins thought to influence intestinal cholesterol absorption at the intestinal brush border membrane level include CD36 and aminopeptidase N (CD13). CD36 is a scavenger receptor. Deletion of CD36 in mice appear to reduce cholesterol absorption from the proximal small intestine, but increase cholesterol absorption from the distal small intestine, and therefore does not affect net cholesterol absorption (Nguyen et al., 2009; Nauli et al., 2006). CD13 localizes to the apical membrane of enterocyte and was reported to interact with ezetimibe (Kramer et al., 2005). Physiological evidence for a role of CD13 in intestinal cholesterol absorption is currently unavailable.

Cholesterol enters absorptive enterocytes as a free form. Cholesterol delivered to intestinal and thoracic duct lymph mainly exists as an esterified form. The enzyme that catalyzes cholesterol esterification in absorptive enterocytes is acyl-CoA:cholesterol acyltransferase-2 (Anderson et al., 1998) (Fig. 12.1). Thus, genetic inactivation of acyl-CoA:cholesterol acyltransferase-2 in mice significantly reduces intestinal cholesterol absorption (Buhman et al., 2000; Repa et al., 2004; Temel et al., 2005).

Unesterified cholesterol, if not transported out back to the gut lumen by the apically-localized heterodimer of ABCG5/G8, can enter the circulation via the basolaterally-localized ABCA1 (Temel et al., 2005) (Fig. 12.1) and this ABCA1-dependent pathway plays a significant role in HDL biogenesis because mice lacking ABCA1 display a reduced plasma HDL-cholesterol concentration (Brunham et al., 2006).

In addition to proteins discussed in this article, many other cellular proteins may also regulate intestinal sterol absorption, such as serine palmitoytransferase (Li et al., 2009) and those listed in a classic review on regulation of intestinal cholesterol absorption (Wang, 2007a), to which interested readers are referred.

12.7 Concluding Comments

Within the last decade, our understanding of protein-mediated intestinal cholesterol transport has seen immense progress. We have gained important insights into the structure, function, and regulation of a dedicated apical sterol transporter NPC1L1, and its opposing heterodimeric protein complex ABCG5/G8. Furthermore, the development of small molecule inhibitors of intestinal cholesterol absorption has provided important tools to dissect the pathway. Such rapid progress in our understanding of this complex physiological process has been a powerful integration of many scientific disciplines, including human genetics, pharmacology, cell biology, biochemistry, genetically altered mouse models, and physiological approaches. Although we have made substantial progress in identifying several protein-mediators of sterol transport across intestinal brush border membrane, it is unlikely this list is complete in its current form. Importantly, mechanistic understanding of protein-mediated sterol transport across plasma membrane and subsequent intracellular trafficking of sterols in the absorptive enterocyte is still in its infancy. By taking advantage of this same multidisciplinary approach, we now have the tools in place to further our understanding of intestinal cholesterol absorption and its protein mediators.

Acknowledgements Dr. Mark Brown is supported by a career transition award (1K99HL096166-01) generously provided by the National Heart, Lung, and Blood Institute. Dr. Liqing Yu is supported by a Scientist Development Grant (#0635261 N) from the American Heart Association, and by funds generously provided by the Department of Pathology of Wake Forest University Health Sciences.

References

2002, Third Report of the National Cholesterol Education Program (NCEP) Expert Panel on Detection, Evaluation, and Treatment of High Blood Cholesterol in Adults (Adult Treatment Panel III) final report. *Circulation* **106**: 3143–3421. (http://circ.ahajournals.org/cgi/content/full/106/25/3143)

Acton, S., Rigotti, A., Landschulz, K.T., Xu, S., Hobbs, H.H., and Krieger, M., 1996, Identification of scavenger receptor SR-BI as a high density lipoprotein receptor. *Science* **271**: 518–520.

Acton, S.L., Scherer, P.E., Lodish, H.F., and Krieger, M., 1994, Expression cloning of SR-BI, a CD36-related class B scavenger receptor. *J. Biol. Chem.* **269**: 21003–21009.

Alrefai, W.A., Annaba, F., Sarwar, Z., Dwivedi, A., Saksena, S., Singla, A., Dudeja, P.K., and Gill, R.K, 2007, Modulation of human Niemann-Pick C1-like 1 gene expression by sterol: role of sterol regulatory element binding protein 2. *Am. J. Physiol. Gastrointest. Liver Physiol.* **292:** G369–376.

Altmann, S.W., Davis, H.R., Jr., Yao, X., Laverty, M., Compton, D.S., Zhu, L.J., Crona, J.H., Caplen, M.A., Hoos, L.M., Tetzloff, G., Priestley, T., Burnett, D.A., Strader, C.D., and Graziano, M.P., 2002, The identification of intestinal scavenger receptor class B, type I (SR-BI) by expression cloning and its role in cholesterol absorption. *Biochim. Biophys. Acta* **1580:** 77–93.

Altmann, S.W., Davis, H.R., Jr., Zhu, L.J., Yao, X., Hoos, L.M., Tetzloff, G., Iyer, S.P., Maguire, M., Golovko, A., Zeng, M., Wang, L., Murgolo, N., and Graziano, M.P., 2004, Niemann-Pick C1 Like 1 protein is critical for intestinal cholesterol absorption. *Science* **303:** 1201–1204.

Anderson, R.A., Joyce, C., Davis, M., Reagan, J.W., Clark, M., Shelness, G.S., and Rudel, L.L., 1998, Identification of a form of acyl-CoA:cholesterol acyltransferase specific to liver and intestine in nonhuman primates. *J. Biol. Chem.* **273:** 26747–26754.

Arai, T., Wang, N., Bezouevski, M., Welch, C., and Tall, A.R., 1999, Decreased atherosclerosis in heterozygous low density lipoprotein receptor-deficient mice expressing the scavenger receptor BI transgene. *J. Biol. Chem.* **274:** 2366–2371.

Ashkenas, J., Penman, M., Vasile, E., Acton, S., Freeman, M., and Krieger, M., 1993, Structures and high and low affinity ligand binding properties of murine type I and type II macrophage scavenger receptors. *J. Lipid Res.* **34:** 983–1000.

Assanasen, C., Mineo, C., Seetharam, D., Yuhanna, I.S., Marcel, Y.L., Connelly, M.A., Williams, D.L., de la Llera-Moya, M., Shaul, P.W., and Silver, D.L., 2005, Cholesterol binding, efflux, and a PDZ-interacting domain of scavenger receptor-BI mediate HDL-initiated signaling. *J. Clin. Invest.* **115:** 969–977.

Assy, N., Grozovski, M., Bersudsky, I., Szvalb, S., and Hussein, O., 2006, Effect of insulin-sensitizing agents in combination with ezetimibe, and valsartan in rats with non-alcoholic fatty liver disease. *World J. Gastroenterol.* **12:** 4369–4376.

Azhar, S., Nomoto, A., Leers-Sucheta, S., and Reaven, E., 1998, Simultaneous induction of an HDL receptor protein (SR-BI) and the selective uptake of HDL-cholesteryl esters in a physiologically relevant steroidogenic cell model. *J. Lipid Res.* **39:** 1616–1628.

Babitt, J., Trigatti, B., Rigotti, A., Smart, E.J., Anderson, R.G., Xu, S., and Krieger, M., 1997, Murine SR-BI, a high density lipoprotein receptor that mediates selective lipid uptake, is N-glycosylated and fatty acylated and colocalizes with plasma membrane caveolae. *J. Biol. Chem.* **272:** 13242–13249.

Bao, L., Li, Y., Deng, S.X., Landry, D., and Tabas, I., 2006, Sitosterol-containing lipoproteins trigger free sterol-induced caspase-independent death in ACAT-competent macrophages. *J. Biol. Chem.* **281:** 33635–33649.

Basso, F., Freeman, L.A., Ko, C., Joyce, C., Amar, M.J., Shamburek, R.D., Tansey, T., Thomas, F., Wu, J., Paigen, B., Remaley, A.T., Santamarina-Fojo, S., and Brewer, H.B., Jr., 2007, Hepatic ABCG5/G8 overexpression reduces apoB-lipoproteins and atherosclerosis when cholesterol absorption is inhibited. *J. Lipid Res.* **48:** 114–126.

Berge, K.E., Tian, H., Graf, G.A., Yu, L., Grishin, N.V., Schultz, J., Kwiterovich, P., Shan, B., Barnes, R., and Hobbs, H.H., 2000, Accumulation of dietary cholesterol in sitosterolemia caused by mutations in adjacent ABC transporters. *Science* **290:** 1771–1775.

Bhattacharyya, A.K., and Connor, W.E., 1974, Beta-sitosterolemia and xanthomatosis. A newly described lipid storage disease in two sisters. *J. Clin. Invest.* **53:** 1033–1043.

Bhattacharyya, A.K., Connor, W.E., Lin, D.S., McMurry, M.M., and Shulman, R.S., 1991, Sluggish sitosterol turnover and hepatic failure to excrete sitosterol into bile cause expansion of body pool of sitosterol in patients with sitosterolemia and xanthomatosis. *Arterioscler. Thromb.* **11:** 1287–1294.

Biddinger, S.B., Haas, J.T., Yu, B.B., Bezy, O., Jing, E., Zhang, W., Unterman, T.G., Carey, M.C., and Kahn, C.R., 2008, Hepatic insulin resistance directly promotes formation of cholesterol gallstones. *Nat. Med.* **14:** 778–782.

Bietrix, F., Yan, D., Nauze, M., Rolland, C., Bertrand-Michel, J., Comera, C., Schaak, S., Barbaras, R., Groen, A.K., Perret, B., Terce, F., and Collet, X., 2006, Accelerated lipid absorption in mice overexpressing intestinal SR-BI. *J. Biol. Chem.* **281**: 7214–7219.

Bjorkhem, I., Boberg, K., and Leitersdorf, E., 2001, *The Metabolic and Molecular Basis of Inherited Diseases*. McGraw-Hill, New York.

Borgstrom, B., 1968, Quantitative aspects of the intestinal absorption and metabolism of cholesterol and beta-sitosterol in the rat. *J. Lipid Res.* **9**: 473–481.

Bosner, M.S., Lange, L.G., Stenson, W.F., and Ostlund, R.E., Jr., 1999, Percent cholesterol absorption in normal women and men quantified with dual stable isotopic tracers and negative ion mass spectrometry. *J. Lipid Res.* **40**: 302–308.

Braun, A., Trigatti, B.L., Post, M.J., Sato, K., Simons, M., Edelberg, J.M., Rosenberg, R.D., Schrenzel, M., and Krieger, M., 2002, Loss of SR-BI expression leads to the early onset of occlusive atherosclerotic coronary artery disease, spontaneous myocardial infarctions, severe cardiac dysfunction, and premature death in apolipoprotein E-deficient mice. *Circ. Res.* **90**: 270–276.

Brown, J.M., Bell, T.A., 3rd, Alger, H.M., Sawyer, J.K., Smith, T.L., Kelley, K., Shah, R., Wilson, M.D., Davis, M.A., Lee, R.G., Graham, M.J., Crooke, R.M., and Rudel, L.L., 2008, Targeted depletion of hepatic ACAT2-driven cholesterol esterification reveals a non-biliary route for fecal neutral sterol loss. *J. Biol. Chem.* **283**: 10522–10534.

Brown, J.M., Rudel, L.L., and Yu, L., 2007, NPC1L1 (Niemann-Pick C1-like 1) mediates sterol-specific unidirectional transport of non-esterified cholesterol in McArdle-RH7777 hepatoma cells. *Biochem. J.* **406**: 273–283.

Brown, M.S., and Goldstein, J.L., 1999, A proteolytic pathway that controls the cholesterol content of membranes, cells, and blood. *Proc Natl Acad Sci USA* **96**: 11041–11048.

Brunham, L.R., Kruit, J.K., Iqbal, J., Fievet, C., Timmins, J.M., Pape, T.D., Coburn, B.A., Bissada, N., Staels, B., Groen, A.K., Hussain, M.M., Parks, J.S., Kuipers, F., and Hayden, M.R., 2006, Intestinal ABCA1 directly contributes to HDL biogenesis in vivo. *J. Clin. Invest.* **116**: 1052–1062.

Buhman, K.K., Accad, M., Novak, S., Choi, R.S., Wong, J.S., Hamilton, R.L., Turley, S., and Farese, R.V., Jr., 2000, Resistance to diet-induced hypercholesterolemia and gallstone formation in ACAT2-deficient mice. *Nat. Med.* **6**: 1341–1347.

Bultel-Brienne, S., Lestavel, S., Pilon, A., Laffont, I., Tailleux, A., Fruchart, J.C., Siest, G., and Clavey, V., 2002, Lipid free apolipoprotein E binds to the class B Type I scavenger receptor I (SR-BI) and enhances cholesteryl ester uptake from lipoproteins. *J. Biol. Chem.* **277**: 36092–36099.

Burgos, P.V., Klattenhoff, C., de la Fuente, E., Rigotti, A., and Gonzalez, A., 2004, Cholesterol depletion induces PKA-mediated basolateral-to-apical transcytosis of the scavenger receptor class B type I in MDCK cells. *Proc. Natl. Acad. Sci. USA* **101**: 3845–3850.

Cai, S.F., Kirby, R.J., Howles, P.N, and Hui, D.Y., 2001, Differentiation-dependent expression and localization of the class B type I scavenger receptor in intestine. *J. Lipid Res.* **42**: 902–909.

Calpe-Berdiel, L., Rotllan, N., Fievet, C., Roig, R., Blanco-Vaca, F., and Escola-Gil, J.C., 2008, Liver X receptor-mediated activation of reverse cholesterol transport from macrophages to feces in vivo requires ABCG5/G8. *J. Lipid Res.* **49**: 1904–1911.

Calvo, D., and Vega, M.A., 1993, Identification, primary structure, and distribution of CLA-1, a novel member of the CD36/LIMPII gene family. *J. Biol. Chem.* **268**: 18929–18935.

Cao, G., Garcia, C.K., Wyne, K.L., Schultz, R.A., Parker, K.L., and Hobbs, H.H., 1997, Structure and localization of the human gene encoding SR-BI/CLA-1. Evidence for transcriptional control by steroidogenic factor 1. *J. Biol. Chem.* **272**: 33068–33076.

Cao, G., Liang, Y., Jiang, X.C., and Eacho, P.I., 2004, Liver X receptors as potential therapeutic targets for multiple diseases. *Drug News Perspect.* **17**: 35–41.

Cao, G., Zhao, L., Stangl, H., Hasegawa, T., Richardson, J.A., Parker, K.L., and Hobbs, H.H., 1999, Developmental and hormonal regulation of murine scavenger receptor, class B, type 1. *Mol. Endocrinol.* **13**: 1460–1473.

Carstea, E.D., Morris, J.A., Coleman, K.G., Loftus, S.K., Zhang, D., Cummings, C., Gu, J., Rosenfeld, M.A., Pavan, W.J., Krizman, D.B., Nagle, J., Polymeropoulos, M.H., Sturley, S.L., Ioannou, Y.A., Higgins, M.E., Comly, M., Cooney, A., Brown, A., Kaneski, C.R., Blanchette-Mackie, E.J., Dwyer, N.K., Neufeld, E.B., Chang, T.Y., Liscum, L., Tagle, D.A., and et al., 1997, Niemann-Pick C1 disease gene: homology to mediators of cholesterol homeostasis. *Science* **277**: 228–231.

Chang, T.Y., and Chang, C., 2008, Ezetimibe blocks internalization of the NPC1L1/cholesterol complex. *Cell Metab.* **7**: 469–471.

Charrin, S., Manie, S., Thiele, C., Billard, M., Gerlier, D., Boucheix, C., and Rubinstein, E., 2003, A physical and functional link between cholesterol and tetraspanins. *Eur. J. Immunol.* **33**: 2479–2489.

Chu, B.B., Ge, L., Xie, C., Zhao, Y., Miao, H.H., Wang, J., Li, B.L., and Song, B.L., 2009, Requirement of myosin Vb/Rab11a/Rab11-FIP2 complex in cholesterol-regulated translocation of Niemann-Pick C1 like 1 protein to the cell surface. *J. Biol. Chem.* **284**: 22481–22490.

Clader, J.W., 2004, The discovery of ezetimibe: a view from outside the receptor. *J. Med. Chem.* **47**: 1–9.

Cohen, J.C., Pertsemlidis, A., Fahmi, S., Esmail, S., Vega, G.L., Grundy, S.M., and Hobbs, H.H., 2006, Multiple rare variants in NPC1L1 associated with reduced sterol absorption and plasma low-density lipoprotein levels. *Proc. Natl. Acad. Sci. USA* **103**: 1810–1815.

Connelly, M.A., 2009, SR-BI-mediated HDL cholesteryl ester delivery in the adrenal gland. *Mol. Cell Endocrinol.* **300**: 83–88.

Cooper, M.K., Wassif, C.A., Krakowiak, P.A., Taipale, J., Gong, R., Kelley, R.I., Porter, F.D., and Beachy, P.A., 2003, A defective response to Hedgehog signaling in disorders of cholesterol biosynthesis. *Nat. Genet.* **33**: 508–513.

Davidson, M.H., 2009, Novel nonstatin strategies to lower low-density lipoprotein cholesterol. *Curr. Atheroscler. Rep.* **11**: 67–70.

Davies, J.P., Levy, B., and Ioannou, Y.A., 2000a, Evidence for a Niemann-pick C (NPC) gene family: identification and characterization of NPC1L1. *Genomics* **65**: 137–145.

Davies, J.P., and Ioannou, Y.A., 2006, The role of the Niemann-Pick C1-like 1 protein in the subcellular transport of multiple lipids and their homeostasis. *Curr. Opin. Lipidol.* **17**: 221–226.

Davies, J.P., Chen, F.W., and Ioannou, Y.A., 2000b, Transmembrane molecular pump activity of Niemann-Pick C1 protein. *Science* **290**: 2295–2298.

Davies, J.P., Scott, C., Oishi, K., Liapis, A., and Ioannou, Y.A., 2005, Inactivation of NPC1L1 causes multiple lipid transport defects and protects against diet-induced hypercholesterolemia. *J. Biol. Chem.* **280**: 12710–12720.

Davis, H.R., and Veltri, E.P., 2007, Zetia: inhibition of Niemann-Pick C1 Like 1 (NPC1L1) to Reduce Intestinal Cholesterol Absorption and Treat Hyperlipidemia. *J. Atheroscler. Thromb.* **14**: 99–108.

Davis, H.R., Jr., Zhu, L.J., Hoos, L.M., Tetzloff, G., Maguire, M., Liu, J., Yao, X., Iyer, S.P., Lam, M.H., Lund, E.G., Detmers, P.A., Graziano, M.P., and Altmann, S.W., 2004, Niemann-Pick C1 Like 1 (NPC1L1) is the intestinal phytosterol and cholesterol transporter and a key modulator of whole-body cholesterol homeostasis. *J. Biol. Chem.* **279**: 33586–33592.

Dawson, P.A., and Rudel, L.L., 1999, Intestinal cholesterol absorption. *Curr. Opin. Lipidol.* **10**: 315–320.

de Beer, M.C., Durbin, D.M., Cai, L., Jonas, A., de Beer, F.C., and van der Westhuyzen, D.R., 2001, Apolipoprotein A-I conformation markedly influences HDL interaction with scavenger receptor BI. *J. Lipid Res.* **42**: 309–313.

de la Llera-Moya, M., Rothblat, G.H., Connelly, M.A., Kellner-Weibel, G., Sakr, S.W., Phillips, M.C., and Williams, D.L., 1999, Scavenger receptor BI (SR-BI) mediates free cholesterol flux independently of HDL tethering to the cell surface. *J. Lipid Res.* **40**: 575–580.

Deushi, M., Nomura, M., Kawakami, A., Haraguchi, M., Ito, M., Okazaki, M., Ishii, H., and Yoshida, M., 2007, Ezetimibe improves liver steatosis and insulin resistance in obese rat model of metabolic syndrome. *FEBS Lett.* **581:** 5664–5670.

Dietschy, J.M., and Turley, S.D., 2002, Control of cholesterol turnover in the mouse. *J. Biol. Chem.* **277:** 3801–3804.

Duan, L.P., Wang, H.H., Ohashi, A., and Wang, D.Q., 2006, Role of intestinal sterol transporters Abcg5, Abcg8, and Npc1l1 in cholesterol absorption in mice: gender and age effects. *Am. J. Physiol. Gastrointest. Liver Physiol.* **290:** G269–276.

During, A., Dawson, H.D., and Harrison, E.H., 2005, Carotenoid transport is decreased and expression of the lipid transporters SR-BI, NPC1L1, and ABCA1 is downregulated in CaCo-2 cells treated with ezetimibe. *J. Nutr.* **135:** 2305–2312.

Duval, C., Touche, V., Tailleux, A., Fruchart, J.C., Fievet, C., Clavey, V., Staels, B., and Lestavel, S., 2006, Niemann-Pick C1 like 1 gene expression is down-regulated by LXR activators in the intestine. *Biochem. Biophys. Res. Commun.* **340:** 1259–1263.

Eckhardt, E.R., Cai, L., Shetty, S., Zhao, Z., Szanto, A., Webb, N.R., and Van der Westhuyzen, D.R., 2006, High density lipoprotein endocytosis by scavenger receptor SR-BII is clathrin-dependent and requires a carboxyl-terminal dileucine motif. *J. Biol. Chem.* **281:** 4348–4353.

Eckhardt, E.R., Cai, L., Sun, B., Webb, N.R., and van der Westhuyzen, D.R., 2004, High density lipoprotein uptake by scavenger receptor SR-BII. *J. Biol. Chem.* **279:** 14372–14381.

Epand, R.M., 2006, Cholesterol and the interaction of proteins with membrane domains. *Prog. Lipid Res.* **45:** 279–294.

Fahmi, S., Yang, C., Esmail, S., Hobbs, H.H., and Cohen, J.C., 2008, Functional characterization of genetic variants in NPC1L1 supports the sequencing extremes strategy to identify complex trait genes. *Hum. Mol. Genet.* **17:** 2101–2107.

Field, F.J., Watt, K., and Mathur, S.N., 2007, Ezetimibe interferes with cholesterol trafficking from the plasma membrane to the endoplasmic reticulum in CaCo-2 cells. *J. Lipid Res.* **48:** 1735–1745.

Fluiter, K., van der Westhuijzen, D.R., and van Berkel, T.J., 1998, In vivo regulation of scavenger receptor BI and the selective uptake of high density lipoprotein cholesteryl esters in rat liver parenchymal and Kupffer cells. *J. Biol. Chem.* **273:** 8434–8438.

Franc, N.C., Dimarcq, J.L., Lagueux, M., Hoffmann, J., and Ezekowitz, R.A., 1996, Croquemort, a novel Drosophila hemocyte/macrophage receptor that recognizes apoptotic cells. *Immunity* **4:** 431–443.

Franc, N.C., Heitzler, P., Ezekowitz, R.A., and White, K., 1999, Requirement for croquemort in phagocytosis of apoptotic cells in Drosophila. *Science* **284:** 1991–1994.

Freeman, L.A., Kennedy, A., Wu, J., Bark, S., Remaley, A.T., Santamarina-Fojo, S., and Brewer, H.B., Jr., 2004, The orphan nuclear receptor LRH-1 activates the ABCG5/ABCG8 intergenic promoter. *J. Lipid Res.* **45:** 1197–1206.

Fukasawa, M., Adachi, H., Hirota, K., Tsujimoto, M., Arai, H., and Inoue, K., 1996, SRB1, a class B scavenger receptor, recognizes both negatively charged liposomes and apoptotic cells. *Exp. Cell Res.* **222:** 246–250.

Galman, C., Bonde, Y., Matasconi, M., Angelin, B., and Rudling, M., 2008, Dramatically increased intestinal absorption of cholesterol following hypophysectomy is normalized by thyroid hormone. *Gastroenterology* **134:** 1127–1136.

Garcia-Calvo, M., Lisnock, J., Bull, H.G., Hawes, B.E., Burnett, D.A., Braun, M.P., Crona, J.H., Davis, H.R., Jr., Dean, D.C., Detmers, P.A., Graziano, M.P., Hughes, M., Macintyre, D.E., Ogawa, A., O'Neill K, A., Iyer, S.P., Shevell, D.E., Smith, M.M., Tang, Y.S., Makarewicz, A.M., Ujjainwalla, F., Altmann, S.W., Chapman, K.T., and Thornberry, N.A., 2005, The target of ezetimibe is Niemann-Pick C1-Like 1 (NPC1L1). *Proc. Natl. Acad. Sci. USA* **102:** 8132–8137.

Ge, L., Wang, J., Qi, W., Miao, H.H., Cao, J., Qu, Y.X., Li, B.L., and Song, B.L., 2008, The cholesterol absorption inhibitor ezetimibe acts by blocking the sterol-induced internalization of NPC1L1. *Cell Metab.* **7:** 508–519.

Gillotte-Taylor, K., Boullier, A., Witztum, J.L., Steinberg, D., and Quehenberger, O., 2001, Scavenger receptor class B type I as a receptor for oxidized low density lipoprotein. *J. Lipid Res.* **42:** 1474–1482.

Goldstein, J.L., and Brown, M.S., 1990, Regulation of the mevalonate pathway. *Nature* **343:** 425–430.

Gould, R.G., Jones, R.J., LeRoy, G.V., Wissler, R.W., and Taylor, C.B., 1969, Absorbability of beta-sitosterol in humans. *Metabolism*, **18,** 652–662.

Graf, G.A., Cohen, J.C. and Hobbs, H.H. (2004) Missense mutations in ABCG5 and ABCG8 disrupt heterodimerization and trafficking. *J. Biol. Chem.* **279:** 24881–24888.

Graf, G.A., Li, W.P., Gerard, R.D., Gelissen, I., White, A., Cohen, J.C., and Hobbs, H.H., 2002, Coexpression of ATP-binding cassette proteins ABCG5 and ABCG8 permits their transport to the apical surface. *J. Clin. Invest.* **110:** 659–669.

Graf, G.A., Yu, L., Li, W.P., Gerard, R., Tuma, P.L., Cohen, J.C., and Hobbs, H.H., 2003, ABCG5 and ABCG8 are obligate heterodimers for protein trafficking and biliary cholesterol excretion. *J. Biol. Chem.* **278:** 48275–48282.

Greene, D.J., Skeggs, J.W., and Morton, R.E., 2001, Elevated triglyceride content diminishes the capacity of high density lipoprotein to deliver cholesteryl esters via the scavenger receptor class B type I (SR-BI). *J. Biol. Chem.* **276:** 4804–4811.

Gregg, R.E., Connor, W.E., Lin, D.S., and Brewer, H.B., Jr., 1986, Abnormal metabolism of shellfish sterols in a patient with sitosterolemia and xanthomatosis. *J. Clin. Invest.* **77:** 1864–1872.

Gu, X., Kozarsky, K., and Krieger, M., 2000a, Scavenger receptor class B, type I-mediated [3H]cholesterol efflux to high and low density lipoproteins is dependent on lipoprotein binding to the receptor. *J. Biol. Chem.* **275:** 29993–30001.

Gu, X., Lawrence, R., and Krieger, M., 2000b, Dissociation of the high density lipoprotein and low density lipoprotein binding activities of murine scavenger receptor class B type I (mSR-BI) using retrovirus library-based activity dissection. *J. Biol. Chem.* **275:** 9120–9130.

Gylling, H., and Miettinen, T.A., 2002, Baseline intestinal absorption and synthesis of cholesterol regulate its response to hypolipidaemic treatments in coronary patients. *Atherosclerosis* **160:** 477–481.

Hamada, T., Egashira, N., Nishizono, S., Tomoyori, H., Nakagiri, H., Imaizumi, K., and Ikeda, I., 2007, Lymphatic absorption and deposition of various plant sterols in stroke-prone spontaneously hypertensive rats, a strain having a mutation in ATP binding cassette transporter G5. *Lipids* **42:** 241–248.

Hansen, G.H., Niels-Christiansen, L.L., Immerdal, L., and Danielsen, E.M., 2003, Scavenger receptor class B type I (SR-BI) in pig enterocytes: trafficking from the brush border to lipid droplets during fat absorption. *Gut* **52:** 1424–1431.

Hansen, G.H., Niels-Christiansen, L.L., Immerdal, L., Nystrom, B.T., and Danielsen, E.M., 2007, Intestinal alkaline phosphatase: selective endocytosis from the enterocyte brush border during fat absorption. *Am. J. Physiol. Gastrointest. Liver Physiol.* **293:** G1325–1332.

Harder, C.J., Meng, A., Rippstein, P., McBride, H.M., and McPherson, R., 2007, SR-BI undergoes cholesterol-stimulated transcytosis to the bile canaliculus in polarized WIF-B cells. *J. Biol. Chem.*, **282,** 1445–1455.

Hart, K. and Wilcox, M. (1993) A Drosophila gene encoding an epithelial membrane protein with homology to CD36/LIMP II. *J. Mol. Biol.* **234:** 249–253.

Hatzopoulos, A.K., Rigotti, A., Rosenberg, R.D., and Krieger, M., 1998, Temporal and spatial pattern of expression of the HDL receptor SR-BI during murine embryogenesis. *J. Lipid Res.* **39:** 495–508.

Hauser, H., Dyer, J.H., Nandy, A., Vega, M.A., Werder, M., Bieliauskaite, E., Weber, F.E., Compassi, S., Gemperli, A., Boffelli, D., Wehrli, E., Schulthess, G., and Phillips, M.C., 1998, Identification of a receptor mediating absorption of dietary cholesterol in the intestine. *Biochemistry* **37:** 17843–17850.

Hawes, B.E., O'Neill K, A., Yao, X., Crona, J.H., Davis, H.R., Jr., Graziano, M.P., and Altmann, S.W., 2007, In vivo responsiveness to ezetimibe correlates with Niemann-Pick C1 like-1 (NPC1L1) binding affinity: comparison of multiple species NPC1L1 orthologs. *Mol. Pharmacol.* **71**: 19–29.

Hegele, R.A., Guy, J., Ban, M.R., and Wang, J., 2005, NPC1L1 haplotype is associated with interindividual variation in plasma low-density lipoprotein response to ezetimibe. *Lipids Health Dis.* **4**: 16.

Hernandez, H.H., Chaikoff, I.L., Dauben, W.G., and Abraham, S., 1954, The absorption of C14-labeled epicholesterol in the rat. *J. Biol. Chem.* **206**: 757–765.

Horton, J.D., Goldstein, J.L., and Brown, M.S., 2002a, SREBPs: activators of the complete program of cholesterol and fatty acid synthesis in the liver. *J. Clin. Invest.* **109**: 1125–1131.

Horton, J.D., Goldstein, J.L., and Brown, M.S., 2002b, SREBPs: transcriptional mediators of lipid homeostasis. *Cold Spring Harb. Symp. Quant. Biol.* **67**: 491–498.

Huang, T.C., and Kuksis, A., 1965, Lymphatic absorption of cholesterol in the dog following corn oil and butterfat feeding. *Can. J. Physiol. Pharmacol.* **43**: 299–311.

Hui, D.Y., and Howles, P.N., 2005, Molecular mechanisms of cholesterol absorption and transport in the intestine. *Semin. Cell Dev. Biol.* **16**: 183–192.

Husemann, J., Loike, J.D., Kodama, T., and Silverstein, S.C., 2001, Scavenger receptor class B type I (SR-BI) mediates adhesion of neonatal murine microglia to fibrillar beta-amyloid. *J. Neuroimmunol.* **114**: 142–150.

Husemann, J., and Silverstein, S.C., 2001, Expression of scavenger receptor class B, type I, by astrocytes and vascular smooth muscle cells in normal adult mouse and human brain and in Alzheimer's disease brain. *Am. J. Pathol.* **158**: 825–832.

Infante, R.E., Abi-Mosleh, L., Radhakrishnan, A., Dale, J.D., Brown, M.S., and Goldstein, J.L., 2008a, Purified NPC1 protein. I. Binding of cholesterol and oxysterols to a 1278-amino acid membrane protein. *J. Biol. Chem.* **283**: 1052–1063.

Infante, R.E., Radhakrishnan, A., Abi-Mosleh, L., Kinch, L.N., Wang, M.L., Grishin, N.V., Goldstein, J.L., and Brown, M.S., 2008b, Purified NPC1 protein: II. Localization of sterol binding to a 240-amino acid soluble luminal loop. *J. Biol. Chem.* **283**: 1064–1075.

Iqbal, J., and Hussain, M.M., 2009, Intestinal lipid absorption. *Am. J. Physiol. Endocrinol. Metab.* **296**: E1183–1194.

Iwayanagi, Y., Takada, T., and Suzuki, H., 2008, HNF4alpha is a crucial modulator of the cholesterol-dependent regulation of NPC1L1. *Pharm. Res.* **25**: 1134–1141.

Iyer, S.P., Yao, X., Crona, J.H., Hoos, L.M., Tetzloff, G., Davis, H.R., Jr., Graziano, M.P., and Altmann, S.W., 2005, Characterization of the putative native and recombinant rat sterol transporter Niemann-Pick C1 Like 1 (NPC1L1) protein. *Biochim. Biophys. Acta* **1722**: 282–292.

Ji, Y., Jian, B., Wang, N., Sun, Y., Moya, M.L., Phillips, M.C., Rothblat, G.H., Swaney, J.B., and Tall, A.R., 1997, Scavenger receptor BI promotes high density lipoprotein-mediated cellular cholesterol efflux. *J. Biol. Chem.* **272**: 20982–20985.

Ji, Y., Wang, N., Ramakrishnan, R., Sehayek, E., Huszar, D., Breslow, J.L., and Tall, A.R., 1999, Hepatic scavenger receptor BI promotes rapid clearance of high density lipoprotein free cholesterol and its transport into bile. *J. Biol. Chem.* **274**: 33398–33402.

Jian, B., de la Llera-Moya, M., Ji, Y., Wang, N., Phillips, M.C., Swaney, J.B., Tall, A.R., and Rothblat, G.H., 1998, Scavenger receptor class B type I as a mediator of cellular cholesterol efflux to lipoproteins and phospholipid acceptors. *J. Biol. Chem.* **273**: 5599–5606.

Kajinami, K., Takekoshi, N., Brousseau, M.E., and Schaefer, E.J., 2004, Pharmacogenetics of HMG-CoA reductase inhibitors: exploring the potential for genotype-based individualization of coronary heart disease management. *Atherosclerosis* **177**: 219–234.

Kalaany, N.Y., Gauthier, K.C., Zavacki, A.M., Mammen, P.P., Kitazume, T., Peterson, J.A., Horton, J.D., Garry, D.J., Bianco, A.C., and Mangelsdorf, D.J., 2005, LXRs regulate the balance between fat storage and oxidation. *Cell Metab.*, **1**: 231–244.

Kalaany, N.Y., and Mangelsdorf, D.J., 2006, LXRS and FXR: the yin and yang of cholesterol and fat metabolism. *Annu. Rev. Physiol.* **68**: 159–191.

Kaneko, E., Matsuda, M., Yamada, Y., Tachibana, Y., Shimomura, I., and Makishima, M., 2003, Induction of intestinal ATP-binding cassette transporters by a phytosterol-derived liver X receptor agonist. *J. Biol. Chem.* **278**: 36091–36098.

Karakesisoglou, I., Janssen, K.P., Eichinger, L., Noegel, A.A., and Schleicher, M., 1999, Identification of a suppressor of the Dictyostelium profilin-minus phenotype as a CD36/LIMP-II homologue. *J. Cell Biol.* **145**: 167–181.

Kellner-Weibel, G., de La Llera-Moya, M., Connelly, M.A., Stoudt, G., Christian, A.E., Haynes, M.P., Williams, D.L., and Rothblat, G.H., 2000, Expression of scavenger receptor BI in COS-7 cells alters cholesterol content and distribution. *Biochemistry* **39**: 221–229.

Khovidhunkit, W., Moser, A.H., Shigenaga, J.K., Grunfeld, C., and Feingold, K.R., 2003, Endotoxin down-regulates ABCG5 and ABCG8 in mouse liver and ABCA1 and ABCG1 in J774 murine macrophages: differential role of LXR. *J. Lipid Res.* **44**: 1728–1736.

Kim, H.J., Fan, X., Gabbi, C., Yakimchuk, K., Parini, P., Warner, M., and Gustafsson, J.A., 2008, Liver X receptor beta (LXRbeta): a link between beta-sitosterol and amyotrophic lateral sclerosis-Parkinson's dementia. *Proc. Natl. Acad. Sci. USA* **105**: 2094–2099.

Klett, E.L., Lee, M.H., Adams, D.B., Chavin, K.D., and Patel, S.B., 2004a, Localization of ABCG5 and ABCG8 proteins in human liver, gall bladder and intestine. *BMC Gastroenterol.* **4**: 21.

Klett, E.L., Lu, K., Kosters, A., Vink, E., Lee, M.H., Altenburg, M., Shefer, S., Batta, A.K., Yu, H., Chen, J., Klein, R., Looije, N., Oude-Elferink, R., Groen, A.K., Maeda, N., Salen, G., and Patel, S.B., 2004b, A mouse model of sitosterolemia: absence of Abcg8/sterolin-2 results in failure to secrete biliary cholesterol. *BMC Med.* **2**: 5.

Knight, B.L., Patel, D.D., Humphreys, S.M., Wiggins, D., and Gibbons, G.F., 2003, Inhibition of cholesterol absorption associated with a PPAR alpha-dependent increase in ABC binding cassette transporter A1 in mice. *J. Lipid Res.* **44**: 2049–2058.

Knopfel, M., Davies, J.P., Duong, P.T., Kvaerno, L., Carreira, E.M., Phillips, M.C., Ioannou, Y.A., and Hauser, H., 2007, Multiple plasma membrane receptors but not NPC1L1 mediate high-affinity, ezetimibe-sensitive cholesterol uptake into the intestinal brush border membrane. *Biochim. Biophys. Acta* **1771**: 1140–1147.

Kocher, O., Yesilaltay, A., Cirovic, C., Pal, R., Rigotti, A., and Krieger, M., 2003, Targeted disruption of the PDZK1 gene in mice causes tissue-specific depletion of the high density lipoprotein receptor scavenger receptor class B type I and altered lipoprotein metabolism. *J. Biol. Chem.* **278**: 52820–52825.

Kolovou, G., Voudris, V., Drogari, E., Palatianos, G., and Cokkinos, D.V., 1996, Coronary bypass grafts in a young girl with sitosterolemia. *Eur. Heart J.* **17**: 965–966.

Kozarsky, K.F., Donahee, M.H., Glick, J.M., Krieger, M., and Rader, D.J., 2000, Gene transfer and hepatic overexpression of the HDL receptor SR-BI reduces atherosclerosis in the cholesterol-fed LDL receptor-deficient mouse. *Arterioscler. Thromb. Vasc. Biol.* **20**: 721–727.

Kozarsky, K.F., Donahee, M.H., Rigotti, A., Iqbal, S.N., Edelman, E.R., and Krieger, M., 1997, Overexpression of the HDL receptor SR-BI alters plasma HDL and bile cholesterol levels. *Nature* **387**: 414–417.

Kramer, W., Girbig, F., Corsiero, D., Pfenninger, A., Frick, W., Jahne, G., Rhein, M., Wendler, W., Lottspeich, F., Hochleitner, E.O., Orso, E., and Schmitz, G., 2005, Aminopeptidase N (CD13) is a molecular target of the cholesterol absorption inhibitor ezetimibe in the enterocyte brush border membrane. *J. Biol. Chem.* **280**: 1306–1320.

Krieger, M., 1999, Charting the fate of the "good cholesterol": identification and characterization of the high-density lipoprotein receptor SR-BI. *Annu. Rev. Biochem.* **68**: 523–558.

Kruit, J.K., Drayer, A.L., Bloks, V.W., Blom, N., Olthof, S.G., Sauer, P.J., de Haan, G., Kema, I.P., Vellenga, E., and Kuipers, F., 2008, Plant sterols cause macrothrombocytopenia in a mouse model of sitosterolemia. *J. Biol. Chem.* **283**: 6281–6287.

Kruit, J.K., Groen, A.K., van Berkel, T.J., and Kuipers, F., 2006, Emerging roles of the intestine in control of cholesterol metabolism. *World J. Gastroenterol.* **12**: 6429–6439.

Kruit, J.K., Plosch, T., Havinga, R., Boverhof, R., Groot, P.H., Groen, A.K., and Kuipers, F., 2005, Increased fecal neutral sterol loss upon liver X receptor activation is independent of biliary sterol secretion in mice. *Gastroenterology* **128**: 147–156.

Kuwabara, P.E., and Labouesse, M., 2002, The sterol-sensing domain: multiple families, a unique role? *Trends Genet.* **18**: 193–201.

Labonte, E.D., Camarota, L.M., Rojas, J.C., Jandacek, R.J., Gilham, D.E., Davies, J.P., Ioannou, Y.A., Tso, P., Hui, D.Y., and Howles, P.N., 2008, Reduced absorption of saturated fatty acids and resistance to diet-induced obesity and diabetes by ezetimibe-treated and Npc1l1-/- mice. *Am. J. Physiol. Gastrointest. Liver Physiol.* **295**: G776–G783.

Labonte, E.D., Howles, P.N., Granholm, N.A., Rojas, J.C., Davies, J.P., Ioannou, Y.A., and Hui, D.Y., 2007, Class B type I scavenger receptor is responsible for the high affinity cholesterol binding activity of intestinal brush border membrane vesicles. *Biochim. Biophys. Acta* **1771**: 1132–1139.

Lally, S., Owens, D., and Tomkin, G.H., 2007a, Genes that affect cholesterol synthesis, cholesterol absorption, and chylomicron assembly: the relationship between the liver and intestine in control and streptozotosin diabetic rats. *Metabolism* **56**: 430–438.

Lally, S., Owens, D., and Tomkin, G.H., 2007b, The different effect of pioglitazone as compared to insulin on expression of hepatic and intestinal genes regulating post-prandial lipoproteins in diabetes. *Atherosclerosis* **193**: 343–351.

Lally, S., Tan, C.Y., Owens, D., and Tomkin, G.H., 2006, Messenger RNA levels of genes involved in dysregulation of postprandial lipoproteins in type 2 diabetes: the role of Niemann-Pick C1-like 1, ATP-binding cassette, transporters G5 and G8, and of microsomal triglyceride transfer protein. *Diabetologia* **49**: 1008–1016.

Lally, S.E., Owens, D., and Tomkin, G.H., 2007, Sitosterol and cholesterol in chylomicrons of type 2 diabetic and non-diabetic subjects: the relationship with ATP binding cassette proteins G5 and G8 and Niemann-Pick C1-like 1 mRNA. *Diabetologia* **50**: 217–219.

Landschulz, K.T., Pathak, R.K., Rigotti, A., Krieger, M., and Hobbs, H.H., 1996, Regulation of scavenger receptor, class B, type I, a high density lipoprotein receptor, in liver and steroidogenic tissues of the rat. *J. Clin. Invest.* **98**: 984–995.

Larkin, J.M., Brown, M.S., Goldstein, J.L., and Anderson, R.G., 1983, Depletion of intracellular potassium arrests coated pit formation and receptor-mediated endocytosis in fibroblasts. *Cell* **33**: 273–285.

Lee, M.H., Lu, K., Hazard, S., Yu, H., Shulenin, S., Hidaka, H., Kojima, H., Allikmets, R., Sakuma, N., Pegoraro, R., Srivastava, A.K., Salen, G., Dean, M., and Patel, S.B., 2001, Identification of a gene, ABCG5, important in the regulation of dietary cholesterol absorption. *Nat. Genet.* **27**: 79–83.

Lekstrom-Himes, J., and Xanthopoulos, K.G., 1998, Biological role of the CCAAT/enhancer-binding protein family of transcription factors. *J. Biol. Chem.* **273**: 28545–28548.

Levy, E., Spahis, S., Sinnett, D., Peretti, N., Maupas-Schwalm, F., Delvin, E., Lambert, M., and Lavoie, M.A., 2007, Intestinal cholesterol transport proteins: an update and beyond. *Curr. Opin. Lipidol.* **18**: 310–318.

Li, H., and Papadopoulos, V., 1998, Peripheral-type benzodiazepine receptor function in cholesterol transport. Identification of a putative cholesterol recognition/interaction amino acid sequence and consensus pattern. *Endocrinology* **139**: 4991–4997.

Li, X., Kan, H.Y., Lavrentiadou, S., Krieger, M., and Zannis, V., 2002, Reconstituted discoidal ApoE-phospholipid particles are ligands for the scavenger receptor BI. The amino-terminal 1-165 domain of ApoE suffices for receptor binding. *J. Biol. Chem.* **277**: 21149–21157.

Li, X., Peegel, H., and Menon, K.M., 1998, In situ hybridization of high density lipoprotein (scavenger, type 1) receptor messenger ribonucleic acid (mRNA) during folliculogenesis and luteinization: evidence for mRNA expression and induction by human chorionic gonadotropin specifically in cell types that use cholesterol for steroidogenesis. *Endocrinology* **139**: 3043–3049.

Li, X., Peegel, H., and Menon, K.M., 2001, Regulation of high density lipoprotein receptor messenger ribonucleic acid expression and cholesterol transport in theca-interstitial cells by insulin and human chorionic gonadotropin. *Endocrinology* **142:** 174–181.

Li, Z., Park, T.S., Li, Y., Pan, X., Iqbal, J., Lu, D., Tang, W., Yu, L., Goldberg, I.J., Hussain, M.M., and Jiang, X.C., 2009, Serine palmitoyltransferase (SPT) deficient mice absorb less cholesterol. *Biochim. Biophys. Acta* **1791:** 297–306.

Liadaki, K.N., Liu, T., Xu, S., Ishida, B.Y., Duchateaux, P.N., Krieger, J.P., Kane, J., Krieger, M., and Zannis, V.I., 2000, Binding of high density lipoprotein (HDL) and discoidal reconstituted HDL to the HDL receptor scavenger receptor class B type I. Effect of lipid association and APOA-I mutations on receptor binding. *J. Biol. Chem.* **275:** 21262–21271.

Liu, B., and Krieger, M., 2002, Highly purified scavenger receptor class B, type I reconstituted into phosphatidylcholine/cholesterol liposomes mediates high affinity high density lipoprotein binding and selective lipid uptake. *J. Biol. Chem.* **277:** 34125–34135.

Liu, R., Iqbal, J., Yeang, C., Wang, D.Q., Hussain, M.M., and Jiang, X.C., 2007, Phospholipid transfer protein-deficient mice absorb less cholesterol. *Arterioscler. Thromb. Vasc. Biol.* **27:** 2014–2021.

Liu, T., Krieger, M., Kan, H.Y., and Zannis, V.I., 2002, The effects of mutations in helices 4 and 6 of ApoA-I on scavenger receptor class B type I (SR-BI)-mediated cholesterol efflux suggest that formation of a productive complex between reconstituted high density lipoprotein and SR-BI is required for efficient lipid transport. *J. Biol. Chem.* **277:** 21576–21584.

Loftus, S.K., Morris, J.A., Carstea, E.D., Gu, J.Z., Cummings, C., Brown, A., Ellison, J., Ohno, K., Rosenfeld, M.A., Tagle, D.A., Pentchev, P.G., and Pavan, W.J., 1997, Murine model of Niemann-Pick C disease: mutation in a cholesterol homeostasis gene. *Science* **277:** 232–235.

Lopez, D., and McLean, M.P., 1999, Sterol regulatory element-binding protein-1a binds to cis elements in the promoter of the rat high density lipoprotein receptor SR-BI gene. *Endocrinology* **140:** 5669–5681.

Lopez, D., Shea-Eaton, W., Sanchez, M.D., and McLean, M.P., 2001, DAX-1 represses the high-density lipoprotein receptor through interaction with positive regulators sterol regulatory element-binding protein-1a and steroidogenic factor-1. *Endocrinology* **142:** 5097–5106.

Lutjohann, D., Bjorkhem, I., Beil, U.F., and von Bergmann, K., 1995, Sterol absorption and sterol balance in phytosterolemia evaluated by deuterium-labeled sterols: effect of sitostanol treatment. *J. Lipid Res.* **36:** 1763–1773.

Malerod, L., Juvet, L.K., Hanssen-Bauer, A., Eskild, W., and Berg, T., 2002, Oxysterol-activated LXRalpha/RXR induces hSR-BI-promoter activity in hepatoma cells and preadipocytes. *Biochem. Biophys. Res. Commun.* **299:** 916–923.

Malerod, L., Sporstol, M., Juvet, L.K., Mousavi, A., Gjoen, T., and Berg, T., 2003, Hepatic scavenger receptor class B, type I is stimulated by peroxisome proliferator-activated receptor gamma and hepatocyte nuclear factor 4alpha. *Biochem. Biophys. Res. Commun.* **305:** 557–565.

Mardones, P., Quinones, V., Amigo, L., Moreno, M., Miquel, J.F., Schwarz, M., Miettinen, H.E., Trigatti, B., Krieger, M., VanPatten, S., Cohen, D.E., and Rigotti, A., 2001, Hepatic cholesterol and bile acid metabolism and intestinal cholesterol absorption in scavenger receptor class B type I-deficient mice. *J. Lipid Res.* **42:** 170–180.

Mathur, S.N., Watt, K.R., and Field, F.J., 2007, Regulation of intestinal NPC1L1 expression by dietary fish oil and docosahexaenoic acid. *J. Lipid Res.* **48:** 395–404.

McConihay, J.A., Horn, P.S., and Woollett, L.A., 2001, Effect of maternal hypercholesterolemia on fetal sterol metabolism in the Golden Syrian hamster. *J. Lipid Res.* **42:** 1111–1119.

McLean, M.P., and Sandhoff, T.W., 1998, Expression and hormonal regulation of the high-density lipoprotein (HDL) receptor scavenger receptor class B type I messenger ribonucleic acid in the rat ovary. *Endocrine* **9:** 243–252.

McNamara, D.J., Davidson, N.O., Samuel, P., and Ahrens, E.H., Jr., 1980, Cholesterol absorption in man: effect of administration of clofibrate and/or cholestyramine. *J. Lipid Res.* **21:** 1058–1064.

Miettinen, T.A., 1980, Phytosterolaemia, xanthomatosis and premature atherosclerotic arterial disease: a case with high plant sterol absorption, impaired sterol elimination and low cholesterol synthesis. *Eur. J. Clin. Invest.* **10:** 27–35.

Miettinen, T.A., Strandberg, T.E., and Gylling, H., 2000, Noncholesterol sterols and cholesterol lowering by long-term simvastatin treatment in coronary patients: relation to basal serum cholestanol. *Arterioscler. Thromb. Vasc. Biol.* **20:** 1340–1346.

Mizutani, T., Sonoda, Y., Minegishi, T., Wakabayashi, K., and Miyamoto, K., 1997, Cloning, characterization, and cellular distribution of rat scavenger receptor class B type I (SRBI) in the ovary. *Biochem. Biophys. Res. Commun.* **234:** 499–505.

Mizutani, T., Yamada, K., Minegishi, T., and Miyamoto, K., 2000, Transcriptional regulation of rat scavenger receptor class B type I gene. *J. Biol. Chem.* **275:** 22512–22519.

Murao, K., Imachi, H., Yu, X., Cao, W.M., Muraoka, T., Dobashi, H., Hosomi, N., Haba, R., Iwama, H., and Ishida, T., 2008, The transcriptional factor prolactin regulatory element-binding protein mediates the gene transcription of adrenal scavenger receptor class B type I via 3',5'-cyclic adenosine 5'-monophosphate. *Endocrinology* **149:** 6103–6112.

Mymin, D., Wang, J., Frohlich, J., and Hegele, R.A., 2003, Image in cardiovascular medicine. Aortic xanthomatosis with coronary ostial occlusion in a child homozygous for a nonsense mutation in ABCG8. *Circulation* **107:** 791.

Nauli, A.M., Nassir, F., Zheng, S., Yang, Q., Lo, C.M., Vonlehmden, S.B., Lee, D., Jandacek, R.J., Abumrad, N.A., and Tso, P., 2006, CD36 is important for chylomicron formation and secretion and may mediate cholesterol uptake in the proximal intestine. *Gastroenterology* **131:** 1197–1207.

Ng, D.S., Francone, O.L., Forte, T.M., Zhang, J., Haghpassand, M., and Rubin, E.M., 1997, Disruption of the murine lecithin:cholesterol acyltransferase gene causes impairment of adrenal lipid delivery and up-regulation of scavenger receptor class B type I. *J. Biol. Chem.* **272:** 15777–15781.

Nguyen, D.V., Drover, V.A., Knopfel, M., Dhanasekaran, P., Hauser, H., and Phillips, M.C., 2009, Influence of class B scavenger receptors on cholesterol flux across the brush border membrane and intestinal absorption. *J. Lipid Res.* **50:** 2235–2244.

Ohgami, N., Ko, D.C., Thomas, M., Scott, M.P., Chang, C.C., and Chang, T.Y., 2004, Binding between the Niemann-Pick C1 protein and a photoactivatable cholesterol analog requires a functional sterol-sensing domain. *Proc. Natl. Acad. Sci. USA* **101:** 12473–12478.

Ohgami, N., Nagai, R., Miyazaki, A., Ikemoto, M., Arai, H., Horiuchi, S., and Nakayama, H., 2001, Scavenger receptor class B type I-mediated reverse cholesterol transport is inhibited by advanced glycation end products. *J. Biol. Chem.* **276:** 13348–13355.

Okiyoneda, T., Kono, T., Niibori, A., Harada, K., Kusuhara, H., Takada, T., Shuto, T., Suico, M.A., Sugiyama, Y., and Kai, H., 2006, Calreticulin facilitates the cell surface expression of ABCG5/G8. *Biochem. Biophys. Res. Commun.* **347:** 67–75.

Oliver, W.R., Jr., Shenk, J.L., Snaith, M.R., Russell, C.S., Plunket, K.D., Bodkin, N.L., Lewis, M.C., Winegar, D.A., Sznaidman, M.L., Lambert, M.H., Xu, H.E., Sternbach, D.D., Kliewer, S.A., Hansen, B.C., and Willson, T.M., 2001, A selective peroxisome proliferator-activated receptor delta agonist promotes reverse cholesterol transport. *Proc. Natl. Acad. Sci. USA* **98:** 5306–5311.

Paresce, D.M., Ghosh, R.N., and Maxfield, F.R., 1996, Microglial cells internalize aggregates of the Alzheimer's disease amyloid beta-protein via a scavenger receptor. *Neuron* **17:** 553–565.

Parker, K.L., 1998, The roles of steroidogenic factor 1 in endocrine development and function. *Mol. Cell Endocrinol.* **140:** 59–63.

Patel, S.B., Salen, G., Hidaka, H., Kwiterovich, P.O., Stalenhoef, A.F., Miettinen, T.A., Grundy, S.M., Lee, M.H., Rubenstein, J.S., Polymeropoulos, M.H., and Brownstein, M.J., 1998, Mapping a gene involved in regulating dietary cholesterol absorption. The sitosterolemia locus is found at chromosome 2p21. *J. Clin. Invest.* **102:** 1041–1044.

Petersen, N.H., Faergeman, N.J., Yu, L., and Wustner, D., 2008, Kinetic imaging of NPC1L1 and sterol trafficking between plasma membrane and recycling endosomes in hepatoma cells. *J. Lipid Res.* **49**: 2023–2037.

Pilon, A., Briand, O., Lestavel, S., Copin, C., Majd, Z., Fruchart, J.C., Castro, G., and Clavey, V., 2000, Apolipoprotein AII enrichment of HDL enhances their affinity for class B type I scavenger receptor but inhibits specific cholesteryl ester uptake. *Arterioscler. Thromb. Vasc. Biol.* **20**: 1074–1081.

Plosch, T., Bloks, V.W., Terasawa, Y., Berdy, S., Siegler, K., Van Der Sluijs, F., Kema, I.P., Groen, A.K., Shan, B., Kuipers, F., and Schwarz, M., 2004, Sitosterolemia in ABC-transporter G5-deficient mice is aggravated on activation of the liver-X receptor. *Gastroenterology* **126**: 290–300.

Plosch, T., Kok, T., Bloks, V.W., Smit, M.J., Havinga, R., Chimini, G., Groen, A.K., and Kuipers, F., 2002, Increased hepatobiliary and fecal cholesterol excretion upon activation of the liver X receptor is independent of ABCA1. *J. Biol. Chem.* **277**: 33870–33877.

Plosch, T., Kruit, J.K., Bloks, V.W., Huijkman, N.C., Havinga, R., Duchateau, G.S., Lin, Y., and Kuipers, F., 2006, Reduction of cholesterol absorption by dietary plant sterols and stanols in mice is independent of the Abcg5/8 transporter. *J. Nutr.* **136**: 2135–2140.

Ponting, C.P., and Aravind, L., 1999, START: a lipid-binding domain in StAR, HD-ZIP and signalling proteins. *Trends Biochem. Sci.* **24**: 130–132.

Rader, D.J., 2007, Liver X receptor and farnesoid X receptor as therapeutic targets. *Am. J. Cardiol.* **100**: n15–n19.

Radhakrishnan, A., Sun, L.P., Kwon, H.J., Brown, M.S., and Goldstein, J.L., 2004, Direct binding of cholesterol to the purified membrane region of SCAP: mechanism for a sterol-sensing domain. *Mol. Cell* **15**: 259–268.

Reaven, E., Cortez, Y., Leers-Sucheta, S., Nomoto, A., and Azhar, S., 2004, Dimerization of the scavenger receptor class B type I: formation, function, and localization in diverse cells and tissues. *J. Lipid Res.* **45**: 513–528.

Reaven, E., Lua, Y., Nomoto, A., Temel, R., Williams, D.L., van der Westhuyzen, D.R., and Azhar, S., 1999, The selective pathway and a high-density lipoprotein receptor (SR-BI) in ovarian granulosa cells of the mouse. *Biochim. Biophys. Acta* **1436**: 565–576.

Reaven, E., Nomoto, A., Leers-Sucheta, S., Temel, R., Williams, D.L., and Azhar, S., 1998, Expression and microvillar localization of scavenger receptor, class B, type I (a high density lipoprotein receptor) in luteinized and hormone-desensitized rat ovarian models. *Endocrinology* **139**: 2847–2856.

Repa, J.J., Berge, K.E., Pomajzl, C., Richardson, J.A., Hobbs, H., and Mangelsdorf, D.J., 2002, Regulation of ATP-binding cassette sterol transporters ABCG5 and ABCG8 by the liver X receptors alpha and beta. *J. Biol. Chem.* **277**: 18793–18800.

Repa, J.J., Buhman, K.K., Farese, R.V., Jr., Dietschy, J.M., and Turley, S.D., 2004, ACAT2 deficiency limits cholesterol absorption in the cholesterol-fed mouse: impact on hepatic cholesterol homeostasis. *Hepatology* **40**: 1088–1097.

Repa, J.J., and Mangelsdorf, D.J., 2002, The liver X receptor gene team: potential new players in atherosclerosis. *Nat. Med.* **8**: 1243–1248.

Repa, J.J., Turley, S.D., Lobaccaro, J.A., Medina, J., Li, L., Lustig, K., Shan, B., Heyman, R.A., Dietschy, J.M., and Mangelsdorf, D.J., 2000, Regulation of absorption and ABC1-mediated efflux of cholesterol by RXR heterodimers. *Science* **289**: 1524–1529.

Rhainds, D., Bourgeois, P., Bourret, G., Huard, K., Falstrault, L., and Brissette, L., 2004, Localization and regulation of SR-BI in membrane rafts of HepG2 cells. *J. Cell Sci.* **117**: 3095–3105.

Rigotti, A., Acton, S.L., and Krieger, M., 1995, The class B scavenger receptors SR-BI and CD36 are receptors for anionic phospholipids. *J. Biol. Chem.* **270**: 16221–16224.

Rigotti, A., Edelman, E.R., Seifert, P., Iqbal, S.N., DeMattos, R.B., Temel, R.E., Krieger, M., and Williams, D.L., 1996, Regulation by adrenocorticotropic hormone of the in vivo expression of scavenger receptor class B type I (SR-BI), a high density lipoprotein receptor, in steroidogenic cells of the murine adrenal gland. *J. Biol. Chem.* **271**: 33545–33549.

Rigotti, A., Trigatti, B.L., Penman, M., Rayburn, H., Herz, J., and Krieger, M., 1997, A targeted mutation in the murine gene encoding the high density lipoprotein (HDL) receptor scavenger receptor class B type I reveals its key role in HDL metabolism. *Proc. Natl. Acad. Sci. USA* **94**: 12610–12615.

Rosamond, W., Flegal, K., Friday, G., Furie, K., Go, A., Greenlund, K., Haase, N., Ho, M., Howard, V., Kissela, B., Kittner, S., Lloyd-Jones, D., McDermott, M., Meigs, J., Moy, C., Nichol, G., O'Donnell, C.J., Roger, V., Rumsfeld, J., Sorlie, P., Steinberger, J., Thom, T., Wasserthiel-Smoller, S., and Hong, Y., 2007, Heart disease and stroke statistics–2007 update: a report from the American Heart Association Statistics Committee and Stroke Statistics Subcommittee. *Circulation* **115**: e69–e171.

Sabeva, N.S., Rouse, E.J., and Graf, G.A., 2007, Defects in the leptin axis reduce abundance of the ABCG5-ABCG8 sterol transporter in liver. *J. Biol. Chem.* **282**: 22397–22405.

Sahoo, D., Darlington, Y.F., Pop, D., Williams, D.L., and Connelly, M.A., 2007a, Scavenger receptor class B Type I (SR-BI) assembles into detergent-sensitive dimers and tetramers. *Biochim. Biophys. Acta* **1771**: 807–817.

Sahoo, D., Peng, Y., Smith, J.R., Darlington, Y.F., and Connelly, M.A., 2007b, Scavenger receptor class B, type I (SR-BI) homo-dimerizes via its C-terminal region: fluorescence resonance energy transfer analysis. *Biochim. Biophys. Acta* **1771**: 818–829.

Salen, G., Horak, I., Rothkopf, M., Cohen, J.L., Speck, J., Tint, G.S., Shore, V., Dayal, B., Chen, T., and Shefer, S., 1985, Lethal atherosclerosis associated with abnormal plasma and tissue sterol composition in sitosterolemia with xanthomatosis. *J. Lipid Res.* **26**: 1126–1133.

Salen, G., Shore, V., Tint, G.S., Forte, T., Shefer, S., Horak, I., Horak, E., Dayal, B., Nguyen, L., Batta, A.K., and et al., 1989, Increased sitosterol absorption, decreased removal, and expanded body pools compensate for reduced cholesterol synthesis in sitosterolemia with xanthomatosis. *J. Lipid Res.* **30**: 1319–1330.

Salen, G., Starc, T., Sisk, C.M., and Patel, S.B., 2006, Intestinal cholesterol absorption inhibitor ezetimibe added to cholestyramine for sitosterolemia and xanthomatosis. *Gastroenterology* **130**: 1853–1857.

Salen, G., Tint, G.S., Shefer, S., Shore, V., and Nguyen, L., 1992, Increased sitosterol absorption is offset by rapid elimination to prevent accumulation in heterozygotes with sitosterolemia. *Arterioscler. Thromb.* **12**: 563–568.

Salen, G., von Bergmann, K., Lutjohann, D., Kwiterovich, P., Kane, J., Patel, S.B., Musliner, T., Stein, P., and Musser, B., 2004, Ezetimibe effectively reduces plasma plant sterols in patients with sitosterolemia. *Circulation* **109**: 966–971.

Sane, A.T., Sinnett, D., Delvin, E., Bendayan, M., Marcil, V., Menard, D., Beaulieu, J.F., and Levy, E., 2006, Localization and role of NPC1L1 in cholesterol absorption in human intestine. *J. Lipid Res.* **47**: 2112–2120.

Schoenheimer, R.,1929, Uber die bedeutung der pflanzensterine fur den tierschen organismus. *Hoppe-Seyler's Fur Physiol. Chem.* **180**: 1–5.

Schoenheimer, R., and Breusch, F., 1933, Synthesis and destruction of cholesterol in the organism. *J. Biol. Chem.* **103**: 439–448.

Sehayek, E., Wang, R., Ono, J.G., Zinchuk, V.S., Duncan, E.M., Shefer, S., Vance, D.E., Ananthanarayanan, M., Chait, B.T., and Breslow, J.L., 2003, Localization of the PE methylation pathway and SR-BI to the canalicular membrane: evidence for apical PC biosynthesis that may promote biliary excretion of phospholipid and cholesterol. *J. Lipid Res.* **44**: 1605–1613.

Shea-Eaton, W., Lopez, D., and McLean, M.P., 2001, Yin yang 1 protein negatively regulates high-density lipoprotein receptor gene transcription by disrupting binding of sterol regulatory element binding protein to the sterol regulatory element. *Endocrinology* **142**: 49–58.

Silver, D.L., 2002, A carboxyl-terminal PDZ-interacting domain of scavenger receptor B, type I is essential for cell surface expression in liver. *J. Biol. Chem.* **277**: 34042–34047.

Silver, D.L., Wang, N., Xiao, X., and Tall, A.R., 2001, High density lipoprotein (HDL) particle uptake mediated by scavenger receptor class B type 1 results in selective sorting of HDL cholesterol from protein and polarized cholesterol secretion. *J. Biol. Chem.* **276**: 25287–25293.

Simon, J.S., Karnoub, M.C., Devlin, D.J., Arreaza, M.G., Qiu, P., Monks, S.A., Severino, M.E., Deutsch, P., Palmisano, J., Sachs, A.B., Bayne, M.L., Plump, A.S., and Schadt, E.E., 2005, Sequence variation in NPC1L1 and association with improved LDL-cholesterol lowering in response to ezetimibe treatment. *Genomics* **86**: 648–656.

Smart, E.J., De Rose, R.A., and Farber, S.A., 2004, Annexin 2-caveolin 1 complex is a target of ezetimibe and regulates intestinal cholesterol transport. *Proc. Natl. Acad. Sci. USA* **101**: 3450–3455.

Spady, D.K., Kearney, D.M., and Hobbs, H.H., 1999, Polyunsaturated fatty acids up-regulate hepatic scavenger receptor B1 (SR-BI) expression and HDL cholesteryl ester uptake in the hamster. *J. Lipid Res.* **40**: 1384–1394.

Stangl, H., Hyatt, M., and Hobbs, H.H., 1999, Transport of lipids from high and low density lipoproteins via scavenger receptor-BI. *J. Biol. Chem.* **274**: 32692–32698.

Su, Y., Wang, Z., Yang, H., Cao, L., Liu, F., Bai, X., and Ruan, C., 2006, Clinical and molecular genetic analysis of a family with sitosterolemia and co-existing erythrocyte and platelet abnormalities. *Haematologica* **91**: 1392–1395.

Sudhop, T., Lutjohann, D., Kodal, A., Igel, M., Tribble, D.L., Shah, S., Perevozskaya, I., and von Bergmann, K., 2002, Inhibition of intestinal cholesterol absorption by ezetimibe in humans. *Circulation* **106**: 1943–1948.

Sudhop, T., Lutjohann, D., and von Bergmann, K., 2005, Sterol transporters: targets of natural sterols and new lipid lowering drugs. *Pharmacol. Ther.* **105**: 333–341.

Sullivan, P.M., Mezdour, H., Quarfordt, S.H., and Maeda, N., 1998, Type III hyperlipoproteinemia and spontaneous atherosclerosis in mice resulting from gene replacement of mouse Apoe with human Apoe*2. *J. Clin. Invest.* **102**: 130–135.

Sumi, K., Tanaka, T., Uchida, A., Magoori, K., Urashima, Y., Ohashi, R., Ohguchi, H., Okamura, M., Kudo, H., Daigo, K., Maejima, T., Kojima, N., Sakakibara, I., Jiang, S., Hasegawa, G., Kim, I., Osborne, T.F., Naito, M., Gonzalez, F.J., Hamakubo, T., Kodama, T., and Sakai, J., 2007, Cooperative interaction between hepatocyte nuclear factor 4 alpha and GATA transcription factors regulates ATP-binding cassette sterol transporters ABCG5 and ABCG8. *Mol. Cell Biol.* **27**: 4248–4260.

Sun, Y., Wang, N., and Tall, A.R., 1999, Regulation of adrenal scavenger receptor-BI expression by ACTH and cellular cholesterol pools. *J. Lipid Res.* **40**: 1799–1805.

Svensson, P.A., Johnson, M.S., Ling, C., Carlsson, L.M., Billig, H., and Carlsson, B., 1999, Scavenger receptor class B type I in the rat ovary: possible role in high density lipoprotein cholesterol uptake and in the recognition of apoptotic granulosa cells. *Endocrinology* **140**: 2494–2500.

Tachibana, S., Hirano, M., Hirata, T., Matsuo, M., Ikeda, I., Ueda, K., and Sato, R., 2007, Cholesterol and plant sterol efflux from cultured intestinal epithelial cells is mediated by ATP-binding cassette transporters. *Biosci. Biotechnol. Biochem.* **71**: 1886–1895.

Tang, W., Ma, Y., Jia, L., Ioannou, Y.A., Davies, J.P., and Yu, L., 2008a, Genetic inactivation of NPC1L1 protects against sitosterolemia in mice lacking ABCG5/ABCG8. *J. Lipid. Res.* **50**: 293–300.

Tang, W., Ma, Y., Jia, L., Ioannou, Y.A., Davies, J.P., and Yu, L., 2008b, Niemann-Pick C1-Like 1 Is Required for an LXR Agonist to Raise Plasma HDL Cholesterol in Mice. *Arterioscler. Thromb. Vasc. Biol.* **28**: 448–454.

Tang, W., Ma, Y., and Yu, L., 2006, Plasma cholesterol is hyperresponsive to statin in ABCG5/ABCG8 transgenic mice. *Hepatology* **44**: 1259–1266.

Telford, D.E., Sutherland, B.G., Edwards, J.Y., Andrews, J.D., Barrett, P.H., and Huff, M.W., 2007, The molecular mechanisms underlying the reduction of LDL apoB-100 by ezetimibe plus simvastatin. *J. Lipid Res.* **48**: 699–708.

Temel, R.E., Brown, J.M., Ma, Y., Tang, W., Rudel, L.L., Ioannou, Y.A., Davies, J.P., and Yu, L., 2009, Diosgenin stimulation of fecal cholesterol excretion in mice is not NPC1L1-dependent. *J. Lipid Res.* **50:** 915–923.

Temel, R.E., Lee, R.G., Kelley, K.L., Davis, M.A., Shah, R., Sawyer, J.K., Wilson, M.D., and Rudel, L.L., 2005, Intestinal cholesterol absorption is substantially reduced in mice deficient in both ATP-binding cassette transporter A1 (ABCA1) and Acyl-CoA:Cholesterol O-acyltransferase 2 (ACAT2). *J. Lipid Res.* **46:** 2423–2431.

Temel, R.E., Tang, W., Ma, Y., Rudel, L.L., Willingham, M.C., Ioannou, Y.A., Davies, J.P., Nilsson, L.M., and Yu, L., 2007, Hepatic Niemann-Pick C1-like 1 regulates biliary cholesterol concentration and is a target of ezetimibe. *J. Clin. Invest.* **117:** 1968–1978.

Temel, R.E., Trigatti, B., DeMattos, R.B., Azhar, S., Krieger, M., and Williams, D.L., 1997, Scavenger receptor class B, type I (SR-BI) is the major route for the delivery of high density lipoprotein cholesterol to the steroidogenic pathway in cultured mouse adrenocortical cells. *Proc. Natl. Acad. Sci. USA* **94:** 13600–13605.

Thiele, C., Hannah, M.J., Fahrenholz, F., and Huttner, W.B., 2000, Cholesterol binds to synaptophysin and is required for biogenesis of synaptic vesicles. *Nat. Cell Biol.*, **2:** 42–49.

Towns, R., Azhar, S., Peegel, H., and Menon, K.M., 2005, LH/hCG-stimulated androgen production and selective HDL-cholesterol transport are inhibited by a dominant-negative CREB construct in primary cultures of rat theca-interstitial cells. *Endocrine* **27:** 269–277.

Trigatti, B., Rayburn, H., Vinals, M., Braun, A., Miettinen, H., Penman, M., Hertz, M., Schrenzel, M., Amigo, L., Rigotti, A., and Krieger, M., 1999, Influence of the high density lipoprotein receptor SR-BI on reproductive and cardiovascular pathophysiology. *Proc. Natl. Acad. Sci. USA* **96:** 9322–9327.

Trigatti, B.L, 2005, Hepatic high-density lipoprotein receptors: roles in lipoprotein metabolism and potential for therapeutic modulation. *Curr. Atheroscler. Rep.* **7:** 344–350.

Turley, S.D., and Dietschy, J.M., 2003, Sterol absorption by the small intestine. *Curr. Opin. Lipidol.* **14:** 233–240.

Ueda, Y., Gong, E., Royer, L., Cooper, P.N., Francone, O.L., and Rubin, E.M., 2000, Relationship between expression levels and atherogenesis in scavenger receptor class B, type I transgenics. *J. Biol. Chem.* **275:** 20368–20373.

Ueda, Y., Royer, L., Gong, E., Zhang, J., Cooper, P.N., Francone, O., and Rubin, E.M., 1999, Lower plasma levels and accelerated clearance of high density lipoprotein (HDL) and non-HDL cholesterol in scavenger receptor class B type I transgenic mice. *J. Biol. Chem.* **274:** 7165–7171.

Umeda, Y., Kako, Y., Mizutani, K., Iikura, Y., Kawamura, M., Seishima, M., and Hayashi, H., 2001, Inhibitory action of gemfibrozil on cholesterol absorption in rat intestine. *J. Lipid Res.* **42:** 1214–1219.

Urban, S., Zieseniss, S., Werder, M., Hauser, H., Budzinski, R., and Engelmann, B., 2000, Scavenger receptor BI transfers major lipoprotein-associated phospholipids into the cells. *J. Biol. Chem.* **275:** 33409–33415.

Valasek, M.A., Clarke, S.L., and Repa, J.J., 2007, Fenofibrate reduces intestinal cholesterol absorption via PPAR{alpha}-dependent modulation of NPC1L1 expression in mouse. *J. Lipid Res.* **48:** 2725–2735.

Valasek, M.A., Repa, J.J., Quan, G., Dietschy, J.M., and Turley, S.D., 2008, Inhibiting intestinal NPC1L1 activity prevents diet-induced increase in biliary cholesterol in Golden Syrian hamsters. *Am. J. Physiol. Gastrointest. Liver Physiol.* **295:** G813–G822.

Valasek, M.A., Weng, J., Shaul, P.W., Anderson, R.G., and Repa, J.J., 2005, Caveolin-1 is not required for murine intestinal cholesterol transport. *J. Biol. Chem.* **280:** 28103–28109.

van der Veen, J.N., Kruit, J.K., Havinga, R., Baller, J.F., Chimini, G., Lestavel, S., Staels, B., Groot, P.H., Groen, A.K., and Kuipers, F., 2005, Reduced cholesterol absorption upon PPARdelta activation coincides with decreased intestinal expression of NPC1L1. *J. Lipid Res.* **46:** 526–534.

van der Velde, A.E., Vrins, C.L., van den Oever, K., Kunne, C., Oude Elferink, R.P., Kuipers, F., and Groen, A.K., 2007, Direct intestinal cholesterol secretion contributes significantly to total fecal neutral sterol excretion in mice. *Gastroenterology* **133**: 967–975.

van der Velde, A.E., Vrins, C.L., van den Oever, K., Seemann, I., Oude Elferink, R.P., van Eck, M., Kuipers, F., and Groen, A.K., 2008, Regulation of direct transintestinal cholesterol excretion in mice. *Am. J. Physiol. Gastrointest. Liver Physiol.* **295**: G203–G208.

van Heek, M., Austin, T.M., Farley, C., Cook, J.A., Tetzloff, G.G., and Davis, H.R., 2001a, Ezetimibe, a potent cholesterol absorption inhibitor, normalizes combined dyslipidemia in obese hyperinsulinemic hamsters. *Diabetes* **50**: 1330–1335.

van Heek, M., Farley, C., Compton, D.S., Hoos, L., and Davis, H.R., 2001b, Ezetimibe selectively inhibits intestinal cholesterol absorption in rodents in the presence and absence of exocrine pancreatic function. *Br. J. Pharmacol.* **134**: 409–417.

Van Heek, M., France, C.F., Compton, D.S., McLeod, R.L., Yumibe, N.P., Alton, K.B., Sybertz, E.J., and Davis, H.R., Jr., 1997, In vivo metabolism-based discovery of a potent cholesterol absorption inhibitor, SCH58235, in the rat and rhesus monkey through the identification of the active metabolites of SCH48461. *J. Pharmacol. Exp. Ther.* **283**: 157–163.

Vanhanen, H.T., and Miettinen, T.A., 1995, Cholesterol absorption and synthesis during pravastatin, gemfibrozil and their combination. *Atherosclerosis* **115**: 135–146.

Varban, M.L., Rinninger, F., Wang, N., Fairchild-Huntress, V., Dunmore, J.H., Fang, Q., Gosselin, M.L., Dixon, K.L., Deeds, J.D., Acton, S.L., Tall, A.R., and Huszar, D., 1998, Targeted mutation reveals a central role for SR-BI in hepatic selective uptake of high density lipoprotein cholesterol. *Proc. Natl. Acad. Sci. USA* **95**: 4619–4624.

Vega, M.A., Rodriguez, F., Segui, B., Cales, C., Alcalde, J., and Sandoval, I.V., 1991, Targeting of lysosomal integral membrane protein LIMP II. The tyrosine-lacking carboxyl cytoplasmic tail of LIMP II is sufficient for direct targeting to lysosomes. *J. Biol. Chem.* **266**: 16269–16272.

Voshol, P.J., Schwarz, M., Rigotti, A., Krieger, M., Groen, A.K., and Kuipers, F., 2001, Downregulation of intestinal scavenger receptor class B, type I (SR-BI) expression in rodents under conditions of deficient bile delivery to the intestine. *Biochem. J.* **356**: 317–325.

Vrins, C., Vink, E., Vandenberghe, K.E., Frijters, R., Seppen, J., and Groen, A.K., 2007, The sterol transporting heterodimer ABCG5/ABCG8 requires bile salts to mediate cholesterol efflux. *FEBS Lett.* **581**: 4616–4620.

Wang, C., Lin, H.J., Chan, T.K., Salen, G., Chan, W.C., and Tse, T.F., 1981, A unique patient with coexisting cerebrotendinous xanthomatosis and beta-sitosterolemia. *Am. J. Med.* **71**: 313–319.

Wang, D.Q., 2007a, Regulation of intestinal cholesterol absorption. *Annu. Rev. Physio.* **69**: 221–248.

Wang, H.H., Patel, S.B., Carey, M.C., and Wang, D.Q., 2007b, Quantifying anomalous intestinal sterol uptake, lymphatic transport, and biliary secretion in Abcg8(-/-) mice. *Hepatology* **45**: 998–1006.

Wang, H.H., Portincasa, P., Mendez-Sanchez, N., Uribe, M., and Wang, D.Q., 2008a, Effect of ezetimibe on the prevention and dissolution of cholesterol gallstones. *Gastroenterology* **134**: 2101–2110.

Wang, J., Chu, B.B., Ge, L., Li, B.L., Yan, Y., and Song, B.L., 2009, Membrane topology of human NPC1L1, a key protein in enterohepatic cholesterol absorption. *J. Lipid Res.* [epub ahead of print].

Wang, J., Sun, F., Zhang, D.W., Ma, Y., Xu, F., Belani, J.D., Cohen, J.C., Hobbs, H.H., and Xie, X.S., 2006, Sterol transfer by ABCG5 and ABCG8: in vitro assay and reconstitution. *J. Biol. Chem.* **281**: 27894–27904.

Wang, J., Williams, C., and Hegele, R., 2005, Compound heterozygosity for two non-synonymous polymorphisms in NPC1L1 in a non-responder to ezetimibe. *Clin. Genet.* **67**: 175–177.

Wang, J., Zhang, D.W., Lei, Y., Xu, F., Cohen, J.C., Hobbs, H.H., and Xie, X.S., 2008b, Purification and reconstitution of sterol transfer by native mouse ABCG5 and ABCG8. *Biochemistry* **47**: 5194–5204.

Wang, N., Arai, T., Ji, Y., Rinninger, F., and Tall, A.R., 1998, Liver-specific overexpression of scavenger receptor BI decreases levels of very low density lipoprotein ApoB, low density lipoprotein ApoB, and high density lipoprotein in transgenic mice. *J. Biol. Chem.* **273**: 32920–32926.

Wang, N., Weng, W., Breslow, J.L., and Tall, A.R., 1996, Scavenger receptor BI (SR-BI) is upregulated in adrenal gland in apolipoprotein A-I and hepatic lipase knock-out mice as a response to depletion of cholesterol stores. In vivo evidence that SR-BI is a functional high density lipoprotein receptor under feedback control. *J. Biol. Chem.* **271**: 21001–21004.

Wang, X., Collins, H.L., Ranalletta, M., Fuki, I.V., Billheimer, J.T., Rothblat, G.H., Tall, A.R., and Rader, D.J., 2007c, Macrophage ABCA1 and ABCG1, but not SR-BI, promote macrophage reverse cholesterol transport in vivo. *J. Clin. Invest.* **117**: 2216–2224.

Webb, N.R., Connell, P.M., Graf, G.A., Smart, E.J., de Villiers, W.J., de Beer, F.C., and van der Westhuyzen, D.R., 1998, SR-BII, an isoform of the scavenger receptor BI containing an alternate cytoplasmic tail, mediates lipid transfer between high density lipoprotein and cells. *J. Biol. Chem.* **273**: 15241–15248.

Weinglass, A.B., Kohler, M., Schulte, U., Liu, J., Nketiah, E.O., Thomas, A., Schmalhofer, W., Williams, B., Bildl, W., McMasters, D.R., Dai, K., Beers, L., McCann, M.E., Kaczorowski, G.J., and Garcia, M.L., 2008, Extracellular loop C of NPC1L1 is important for binding to ezetimibe. *Proc. Natl. Acad. Sci. USA* **105**: 11140–11145.

Williams, D.L., Connelly, M.A., Temel, R.E., Swarnakar, S., Phillips, M.C., de la Llera-Moya, M., and Rothblat, G.H., 1999, Scavenger receptor BI and cholesterol trafficking. *Curr. Opin. Lipidol.* **10**: 329–339.

Williams, D.L., de La Llera-Moya, M., Thuahnai, S.T., Lund-Katz, S., Connelly, M.A., Azhar, S., Anantharamaiah, G.M., and Phillips, M.C., 2000, Binding and cross-linking studies show that scavenger receptor BI interacts with multiple sites in apolipoprotein A-I and identify the class A amphipathic alpha-helix as a recognition motif. *J. Biol. Chem.* **275**: 18897–18904.

Wilson, M.D., and Rudel, L.L., 1994, Review of cholesterol absorption with emphasis on dietary and biliary cholesterol. *J. Lipid Res.* **35**: 943–955.

Wilund, K.R., Yu, L., Xu, F., Hobbs, H.H., and Cohen, J.C., 2004a, High-level expression of ABCG5 and ABCG8 attenuates diet-induced hypercholesterolemia and atherosclerosis in Ldlr-/- mice. *J. Lipid Res.* **45**: 1429–1436.

Wilund, K.R., Yu, L., Xu, F., Vega, G.L., Grundy, S.M., Cohen, J.C., and Hobbs, H.H., 2004b, No association between plasma levels of plant sterols and atherosclerosis in mice and men. *Arterioscler. Thromb. Vasc. Biol.* **24**: 2326–2332.

Wu, J.E., Basso, F., Shamburek, R.D., Amar, M.J., Vaisman, B., Szakacs, G., Joyce, C., Tansey, T., Freeman, L., Paigen, B.J., Thomas, F., Brewer, H.B., Jr., and Santamarina-Fojo, S., 2004, Hepatic ABCG5 and ABCG8 overexpression increases hepatobiliary sterol transport but does not alter aortic atherosclerosis in transgenic mice. *J. Biol. Chem.* **279**: 22913–22925.

Wustner, D., Mondal, M., Huang, A., and Maxfield, F.R., 2004, Different transport routes for high density lipoprotein and its associated free sterol in polarized hepatic cells. *J. Lipid Res.* **45**: 427–437.

Xu, S., Laccotripe, M., Huang, X., Rigotti, A., Zannis, V.I., and Krieger, M., 1997, Apolipoproteins of HDL can directly mediate binding to the scavenger receptor SR-BI, an HDL receptor that mediates selective lipid uptake. *J. Lipid Res.* **38**: 1289–1298.

Yamamoto, T., 1982, Ultrastructural basis of intestinal absorption. *Arch. Histol. Jpn.* **45**: 1–22.

Yamanashi, Y., Takada, T., and Suzuki, H., 2007, Niemann-Pick C1-like 1 overexpression facilitates ezetimibe-sensitive cholesterol and beta-sitosterol uptake in CaCo-2 cells. *J. Pharmacol. Exp. Ther.* **320**: 559–564.

Yancey, P.G., de la Llera-Moya, M., Swarnakar, S., Monzo, P., Klein, S.M., Connelly, M.A., Johnson, W.J., Williams, D.L., and Rothblat, G.H., 2000, High density lipoprotein phospholipid composition is a major determinant of the bi-directional flux and net movement of cellular free cholesterol mediated by scavenger receptor BI. *J. Biol. Chem.* **275**: 36596–36604.

Yang, C., Yu, L., Li, W., Xu, F., Cohen, J.C., and Hobbs, H.H., 2004, Disruption of cholesterol homeostasis by plant sterols. *J. Clin. Invest.* **114:** 813–822.

Yesilaltay, A., Kocher, O., Pal, R., Leiva, A., Quinones, V., Rigotti, A., and Krieger, M., 2006, PDZK1 is required for maintaining hepatic scavenger receptor, class B, type I (SR-BI) steady state levels but not its surface localization or function. *J. Biol. Chem.* **281:** 28975–28980.

Yu, L., 2007, Dual action of the ezetimibe target Niemann-Pick C1-Like 1. *Future Lipidol.* **2:** 379–382.

Yu, L., 2008, The structure and function of Niemann-Pick C1-like 1 protein. *Curr. Opin. Lipidol.* **19:** 263–269.

Yu, L., Bharadwaj, S., Brown, J.M., Ma, Y., Du, W., Davis, M.A., Michaely, P., Liu, P., Willingham, M.C., and Rudel, L.L., 2006, Cholesterol-regulated translocation of NPC1L1 to the cell surface facilitates free cholesterol uptake. *J. Biol. Chem.* **281:** 6616–6624.

Yu, L., von Bergmann, K., Lutjohann, D., Hobbs, H.H., and Cohen, J.C., 2004a, Selective sterol accumulation in ABCG5/ABCG8-deficient mice. *J. Lipid Res.* **45:** 301–307.

Yu, L., Hammer, R.E., Li-Hawkins, J., Von Bergmann, K., Lutjohann, D., Cohen, J.C., and Hobbs, H.H., 2002a, Disruption of Abcg5 and Abcg8 in mice reveals their crucial role in biliary cholesterol secretion. *Proc. Natl. Acad. Sci. USA* **99:** 16237–16242.

Yu, L., Li-Hawkins, J., Hammer, R.E., Berge, K.E., Horton, J.D., Cohen, J.C., and Hobbs, H.H., 2002b, Overexpression of ABCG5 and ABCG8 promotes biliary cholesterol secretion and reduces fractional absorption of dietary cholesterol. *J. Clin. Invest.* **110:** 671–680.

Yu, L., Cao, G., Repa, J., and Stangl, H., 2004b, Sterol regulation of scavenger receptor class B type I in macrophages. *J. Lipid Res.* **45:** 889–899.

Yu, L., von Bergmann, K., Lutjohann, D., Hobbs, H.H., and Cohen, J.C., 2005, Ezetimibe normalizes metabolic defects in mice lacking ABCG5 and ABCG8. *J. Lipid Res* **46:** 1739–1744.

Yu, L., York, J., von Bergmann, K., Lutjohann, D., Cohen, J.C., and Hobbs, H.H., 2003, Stimulation of cholesterol excretion by the liver X receptor agonist requires ATP-binding cassette transporters G5 and G8. *J. Biol. Chem.* **278:** 15565–15570.

Yvan-Charvet, L., Pagler, T.A., Wang, N., Senokuchi, T., Brundert, M., Li, H., Rinninger, F., and Tall, A.R., 2008, SR-BI inhibits ABCG1-stimulated net cholesterol efflux from cells to plasma HDL. *J. Lipid Res.* **49:** 107–114.

Zhang, Y., Da Silva, J.R., Reilly, M., Billheimer, J.T., Rothblat, G.H., and Rader, D.J., 2005a, Hepatic expression of scavenger receptor class B type I (SR-BI) is a positive regulator of macrophage reverse cholesterol transport in vivo. *J. Clin. Invest.* **115:** 2870–2874.

Zhang, S., Picard, M.H., Vasile, E., Zhu, Y., Raffai, R.L., Weisgraber, K.H., and Krieger, M., 2005b, Diet-induced occlusive coronary atherosclerosis, myocardial infarction, cardiac dysfunction, and premature death in scavenger receptor class B type I-deficient, hypomorphic apolipoprotein ER61 mice. *Circulation* **111:** 3457–3464.

Zheng, S., Hoos, L., Cook, J., Tetzloff, G., Davis, H., Jr., van Heek, M., and Hwa, J.J., 2008, Ezetimibe improves high fat and cholesterol diet-induced non-alcoholic fatty liver disease in mice. *Eur. J. Pharmacol.* **584:** 118–124.

Zuniga, S., Molina, H., Azocar, L., Amigo, L., Nervi, F., Pimentel, F., Jarufe, N., Arrese, M., Lammert, F, and Miquel, J.F., 2008, Ezetimibe prevents cholesterol gallstone formation in mice. *Liver Int.* **28:** 935–947.

Chapter 13
Cholesterol at the Endoplasmic Reticulum: Roles of the Sigma-1 Receptor Chaperone and Implications thereof in Human Diseases

Teruo Hayashi and Tsung-Ping Su

Abstract Despite substantial data elucidating the roles of cholesterol in lipid rafts at the plasma membrane, the roles of cholesterol and related lipids in lipid raft microdomains at the level of subcellular membrane, such as the endoplasmic reticulum (ER) membrane, remain less understood. Growing evidence, however, begins to unveil the importance of cholesterol and lipids on the lipid raft at the ER membrane. A few ER proteins including the sigma-1 receptor chaperone were identified at lipid raft-like microdomains of the ER membrane. The sigma-1 receptor, which is highly expressed at a subdomain of ER membrane directly apposing mitochondria and known as the mitochondria-associated ER membrane or MAM, has been shown to associate with steroids as well as cholesterol. The sigma-1 receptor has been implicated in ER lipid metabolisms/transports, lipid raft reconstitution at the plasma membrane, trophic factor signalling, cellular differentiation, and cellular protection against β-amyloid-induced neurotoxicity. Recent studies on sigma-1 receptor chaperones and other ER proteins clearly suggest that cholesterol, in concert with those ER proteins, may regulate several important functions of the ER including folding, degradation, compartmentalization, and segregation of ER proteins, and the biosynthesis of sphingolipids.

Keywords Cholesterol · Steroid · Sigma-1 receptor chaperone · Endoplasmic reticulum · Mitochondria-associated ER membrane · Lipid raft · Detergent-resistant microdomain · Trophic factor

Abbreviations

ER	endoplasmic reticulum
MAM	mitochondria-associated ER membrane
SREBP	sterol regulatory element binding protein
PHB	prohibitin domain-containing

T. Hayashi and T.-P. Su (✉)
Cellular Pathobiology Section, Cellular Neurobiology Research Branch, Intramural Research Program, National Institute on Drug Abuse, Department of Health and Human Services, National Institutes of Health, 333 Cassell Drive, Baltimore, MD 21224, USA
e-mail: TSU@intra.nida.nih.gov; thayashi@intra.nida.nih.gov

PrP	prion protein
erlin	ER lipid raft protein
IP3 receptors	inositol 1,4,5-trisphosphate receptors
PtSer	phosphatidylserine
PtEt	phosphatidylethanolamine
SBDL	sterol-binding domain-like
IAF	iodo-azido fenpropimorph
NGF	nerve growth factor
EGF	epidermal growth factor
BDNF	brain-derived neurotrophic factor
MAPK	mitogen-activated protein kinase
NMDA	N-methyl-D-aspartate
Hsp	heat shock protein
BiP	immunoglobulin binding protein

13.1 Introduction

Cholesterol is an important molecule not only for constituting the biological membrane but also for regulating a spectrum of cellular processes including gene transcription and signal transduction. Cholesterol serving as a universal precursor of steroidgenesis can regulate gene transcriptions indirectly via steroid receptors. Cholesterol can also directly activates cholesterol–sensing transcription factor, the sterol regulatory element binding proteins (SREBPs), thus regulating transcriptions of a spectrum of lipid enzymes (Brown and Goldstein, 1997; Bengoechea-Alonso and Ericsson, 2007; Lavoie and King, 2009). Further, cholesterol, by forming lipid microdomains with sphingolipids (also called as detergent-insoluble microdomains or lipid rafts), can regulate membrane curvature, protein sorting, vesicle transport, endocytosis, and trophic factor-induced signal transduction at the plasma membrane (Simons and Ikonen, 1997; Simons and Toomre, 2000). Lipid rafts composed of cholesterol and sphingolipids compartmentalize signalling molecules (e.g., receptors, kinases) at the phospholipids bilayer, thus promoting selective and efficient signal transduction (Pike, 2003).

The discovery of the lipid raft microdomain shed lights on the importance of cholesterol as a direct modulator of signal transduction. Roles of cholesterol, particularly those in plasma membrane lipid rafts, have been thus extensively examined by employing relevant models such as raft-residing trophic factor receptors and lipid raft-dependent endocytosis (particularly that via caveolae, a subtype of rafts containing caveolins) (Ikonen and Vainio, 2005; Lajoie and Nabi, 2007). On the other hand, roles of cholesterol-rich lipid microdomains at subcellular membranes, except for Golgi apparatus, have been less well-known.

Growing number of studies, however, has begun to elucidate the existence and importance of lipid microdomains at subcellular organelles, which include mitochondria and the endoplasmic reticulum (ER). Although lipid rafts are considered as ubiquitous constituents of plasma membrane or Golgi, which is in agreement with the enrichment of lipid rafts in these loci, recent studies indicated that lipid rafts are similarly formed at the subcellular membranes with relatively low concentrations of cholesterol and glycosphingolipids (van Meer and van Genderen, 1994; van Meer, 2000). For example, it is known that mitochondrial membranes contain lipid rafts that are remodeled and redistributed during apoptotic and immune reaction, thus initiating cell death signals derived from mitochondria (Malorni et al., 2007).

Most recent studies also demonstrated that the ER membrane, although contains considerably low levels of cholesterol and glycosphingolipids compared to other organelles, can form lipid raft-like microdomains. A few ER proteins were identified as being present in lipid raft-like microdomains, that include ER lipid raft protein (erlin)-1 and -2, PrP, and the sigma-1 receptor chaperone (Hayashi and Su, 2003a, , 2004b; Sarnataro et al., 2004; Hayashi and Su, 2005; Browman et al., 2006; Campana et al., 2006; Hoegg et al., 2009). Erlin, the family of prohibitin domain-containing (PHB) proteins including the prohibitins, the stomatins and the flotillins, was found exclusively at detergent-resistant microdomains of the ER membrane (Browman et al., 2006). The biological function of erlin, however, remains elusive. A few recent studies implicated the potential role of ER rafts in protein folding of prion protein (PrP). Campana et al. (2006) found that processing of the PrP^{sc}, the pathogenic glycoprotein causing neurodegenerative Creutzfeldt-Jakob disease, depends on the association of PrP with lipid microdomains at the ER. The pathogenic mutant PrP T182A mainly retains at lipid raft-like domains at the ER. Importantly, the association of the protein with lipid rafts promotes the folding of the protein, thus inhibiting scrapie-like conversion of PrP mutants and protecting cells (Campana et al., 2006).

A line of recent study from our laboratory and others has demonstrated that cholesterol plays important roles at the ER together with the ER protein sigma-1 receptor chaperone. We reported that sigma-1 receptor chaperones targeting ER rafts regulate protein degradation and lipid transports (Hayashi and Su, 2003b, 2007). The sigma receptor is a protein originally proposed as a subtype of opioid receptors, but later confirmed not to be an opioid receptor, rather being a novel protein mainly localized at the ER (Su and Hayashi, 2003). Based on the ligand-binding profile, it has been postulated that the sigma receptor consists of at least two subtypes: sigma-1 and sigma-2. The sigma-1 receptor possesses high affinity for the (+)-isomer of prototypic ligands such as (+)-SKF10047, whereas the sigma-2 receptor does it for the (−)-isomers (Su and Hayashi, 2003). Recent studies demonstrated that both type-1 and -2 sigma receptors are present at lipid rafts (Hayashi and Su, 2003b, 2004b; Gebreselassie and Bowen, 2004). Especially, the sigma-1 receptor, which was cloned in 1996 (Hanner et al., 1996), has been implicated in lipid metabolisms, transports, and cellular survival (Hayashi and Su, 2005; Hayashi et al., 2009). Further, recent studies begin to unveil the cholesterol-binding characteristic of the sigma-1 receptors (Palmer et al., 2007). These findings given from the sigma receptor research indeed provide several interesting aspects in terms of roles of cholesterol at the ER. Thus, we seek in this chapter to revisit recent findings on the sigma-1 receptor that may ultimately help understanding roles of cholesterol and lipid microdomains at the ER as well as those in pathophysiology of certain human diseases.

13.2 Structure and Subcellular Localization of the Sigma-1 Receptor

In agreement with a fact that the N-terminus of the sigma-1 receptor has a double-arginine ER retention signal (Hanner et al., 1996), several immunocytochemical or biochemical studies have demonstrated the ER localization of sigma-1 receptors (Alonso et al., 2000; Hayashi and Su, 2003a; Jiang et al., 2006; Dun et al., 2007). Although a number of studies found that sigma-1 receptors modulate functions of proteins at the plasma membrane (e.g., K+ channel) (Aydar et al., 2002; Fontanilla et al., 2009; Johannessen et al., 2009), the possibility of sigma-1 receptors in localizing at the plasma membrane has not been thoroughly clarified at present. Importantly, sigma-1 receptors at the ER show typically a punctate staining pattern in immunocytochemistry (Hayashi and Su, 2003a, 2004b), that is distinctive from the pattern seen with other ER proteins such as cytochrome P450 reductase that typically show the reticular or diffuse cytoplasmic pattern in immunostaining. A recent study identified the ER subdomains enriched with sigma-1 receptors as the mitochondria-associated ER membrane (MAM) (Hayashi and Su, 2007) (Fig. 13.1), the ER membrane physically associating with the outer membrane of mitochondria (Rusinol et al., 1994). The MAM plays critical roles in supplying Ca2+ directly from the MAM via inositol 1,4,5-trisphosphate receptors (IP3 receptors) to mitochondria, thus regulating metabolism and bioenergetics in mitochondria (Rizzuto et al., 1999; Hajnoczky and Hoek, 2007; Duchen et al., 2008). The MAM is also important for lipid synthesis. The MAM express high levels of enzymes involved in syntheses of phospholipids, cholesterol, neutral lipids, and ceramides (Vance, 1990; Rusinol et al., 1993, 1994; Bionda et al., 2004). Phosphatidylserine (PtSer) synthesized at the MAM is transported to mitochondria via intermembrane lipid transport (without a demand of energy and transport vesicles) for the synthesis of phosphatidylethanolamine (PtEt) (Voelker, 2000). On the other hand, the role of the MAM in cholesterol transport is poorly understood. Nevertheless, since the MAM accommodates enzymes involved in cholesterol biosynthesis and since cholesterol transported from ER to mitochondria is the primary step in steroidgenesis, the MAM is proposed to serve as one of loci operating the cholesterol transport from the ER to mitochondria (Hayashi et al., 2009). A study demonstrated that the MAM, which was visualized by expressing sigma-1 receptors fused to the yellow fluorescent protein, accommodates the much higher level of free cholesterol when compared to other ER membranes (Hayashi and Su, 2003b).

The sigma-1 receptor is a 24-kDa protein possessing two transmembrane domains at the N-terminus and the center of the protein. The sigma-1 receptor also possesses one membrane-anchoring domain at the C-terminus. A study analyzing membrane topology of the sigma-1 receptor found the sigma-1 receptor possessing one cytosolic loop flanked by two transmembrane domains with a long ER lumenal domain at the C-terminus (Hayashi and Su, 2007) (Fig. 13.2). A recent study identified the sigma-1 receptor as a new class of molecular chaperone forming a complex with another ER chaperone BiP/GRP78 (Hayashi and Su, 2007). The long ER lumenal domain at the C-terminus is demonstrated to possess chaperone activity,

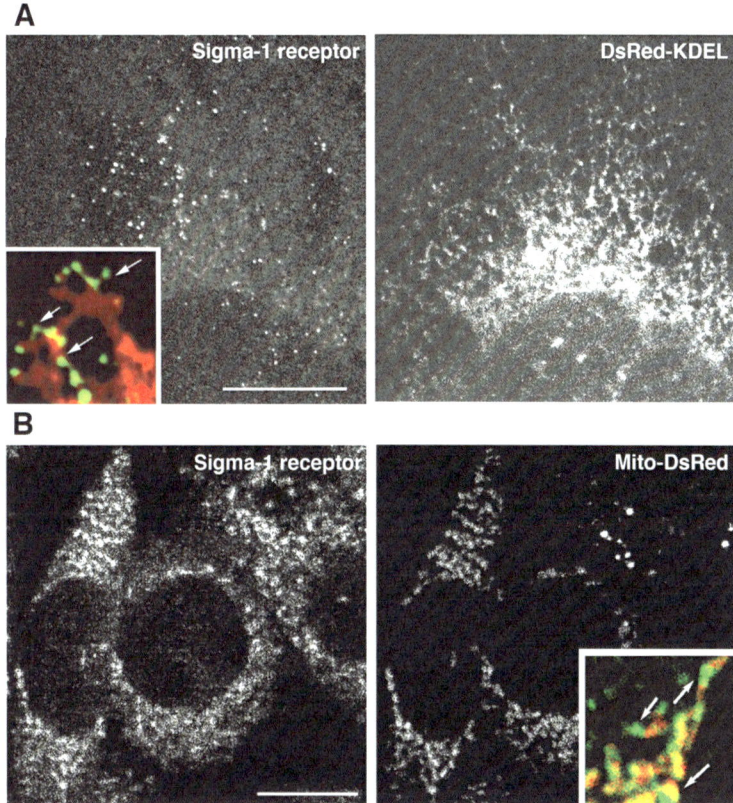

Fig. 13.1 Mitochondria-associated ER membrane (MAM). **A** Spatial localization of MAM and bulk ER membranes. MAM is visualized by immunofluorescence using anti-sigma-1 receptor antibodies. ER membranes are visualized by expressing DsRed-tagged KDEL in the CHO cell. Bar=10 μm. Note that sigma-1 receptors accumulated at punctate substructures of ER membranes (arrows in the inset at higher magnification). **B** Spatial localization of MAM and mitochondria. MAM is visualized by immunofluorescence using anti-sigma-1 receptor antibodies. Mitochondria were visualized by expressing mitochondria-targeting red fluorescent proteins (Mito-DsRed) in the CHO cell. Bar=10 μm. MAM expressing sigma-1 receptors apposes to mitochondria (arrows in the inset at higher magnification)

which is negatively regulated by the association with BiP (Hayashi and Su, 2007). The molecular function of the sigma-1 receptor will be discussed later with more details.

Several studies have sought to identify the ligand-binding site(s) of sigma-1 receptors including that for steroids. A study showed that the amino acids in the second transmembrane domain are responsible for binding of the ligands with a slight difference in responsible amino acids for binding to agonists and antagonists, respectively (Yamamoto et al., 1999). On the other hands, a few charged amino acids in the ER lumenal domain at the C-terminus are shown to contribute to the formation of the ligand-binding site as well (Ganapathy et al., 1999; Seth et al.,

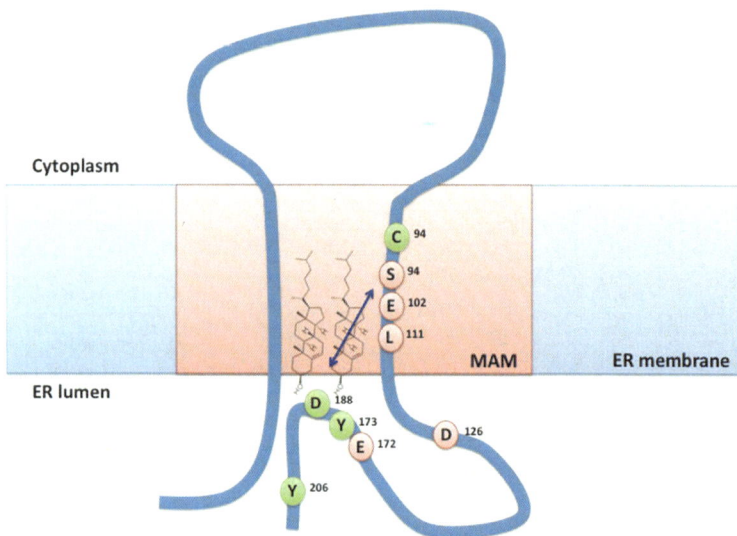

Fig. 13.2 ER lipid rafts composed of cholesterol and sigma-1 receptors. The mitochondria-associated ER membrane (MAM) contains the relatively high level of cholesterol, when compared to other bulk ER membranes, thus forming lipid rafts as reported by recent studies. The sphingolipid counterpart(s) constituting lipid rafts with cholesterol at the ER is not well defined. In oligodendrocytes, galactosylceramide appears to represent a major sphingolipid component of ER rafts. The sigma-1 receptor, which possesses two transmembrane domains, resides preferentially at lipid rafts of the MAM. Amino acids marked with pink are shown to be involved in ligand-binding of the sigma-1 receptor. Those in green are proposed to constitute sterol or cholesterol-binding domains. Sigma-1 receptors at ER rafts have been shown to regulate lipid transport/metabolism, Ca2+ signalling, reconstitution of plasma membrane lipid rafts, ganglioside synthesis, cellular survival and differentiation. The arrow indicates the domains in the juxtaposed position as demonstrated by sulfhydryl-reactive, radioiodinated photo-crosslinking

2001). The amino acids in the membrane-anchoring domain seem to be responsible for binding of cocaine or progesterone (Chen et al., 2007; Pal et al., 2007, 2008; Fontanilla et al., 2008). Therefore multiple domains, including the transmembrane domains or amino acids in vicinity to the ER membrane, may constitute the ligand/sterol-binding pocket of the sigma-1 receptor (Pal et al., 2008). Whether the pocket is composed within a single molecule of the sigma-1 receptor or by homo oligomerization of the protein is not examined yet.

13.3 The Potential Link Between the Sigma-1 Receptor and Sterols

Sigma-1 receptors ubiquitously express in several organs of mammals including brain, liver, pancreas, testis, overlay, placenta, and adrenal gland, as well as in malignant tumors (Vilner et al., 1995; Hanner et al., 1996; Spruce et al., 2004).

A number of studies using selective sigma receptor ligands have demonstrated that sigma-1 receptors are involved in regulations of morphogenesis of neuronal cells (e.g., synaptogenesis, neuronal differentiation, myelination), neuroprotection, pain, pathophysiology of certain human diseases including depression, drug abuse, Alzheimer's disease and cancer (Nakazawa et al., 1998; Goyagi et al., 2001; Maurice, 2004; Spruce et al., 2004; Liu et al., 2005; Marrazzo et al., 2005; Achison et al., 2007; Bermack and Debonnel, 2007; Dun et al., 2007; Martin-Fardon et al., 2007; Mei and Pasternak, 2007; Renaudo et al., 2007; Hayashi and Su, 2008; Smith et al., 2008; Tchedre and Yorio, 2008). Sigma-1 receptors bind a variety of psychotropic drugs and progesterone with submicromolar K_i (Su and Hayashi, 2003); particularly the finding on progesterone raised first time a possibility that sigma-1 receptors may interact with sterols (Su et al., 1988).

The success of cloning of the sigma-1 receptor provided another striking evidence supporting the link between the sigma-1 receptor and sterols. The sequence of the sigma-1 receptor, although having no homology to any mammalian proteins, shares a similarity to that of the yeast sterol C8-C7 isomerase (Hanner et al., 1996). In spite of the high homology between these proteins, following studies however negated the sigma-1 receptor serving as a mammalian C8-C7 sterol isomerase; the sigma-1 receptor lacks the enzymatic activity and the cloned mammalian sterol isomerase shares homology with neither the yeast sterol isomerase nor with the sigma-1 receptor (Hanner et al., 1996; Moebius et al., 1997; Bae et al., 2001). Nevertheless, the structural similarity, particularly that between the membrane-spanning domain of the sigma-1 receptor and the sterol-binding pocket of the yeast sterol isomerase (Hanner et al., 1996), argues the possibility that the sigma-1 receptor possesses the sterol-binding domain.

13.4 Sigma-1 Receptors Interact with Cholesterol

Although sigma-1 receptors were shown to bind some steroids in early in vitro binding assays, the possibility of sigma-1 receptors associating with cholesterol has just begun to be examined recently. Immunocytochemical studies have revealed sigma-1 receptors highly compartmentalized at the cholesterol-enriched subdomains of the ER membranes (i.e., MAM) (Hayashi and Su, 2003b; Hayashi and Su, 2004b) (Fig. 13.2). The cholesterol content in the ER membrane is generally kept at a considerably low level when compared to that in the plasma or Golgi membrane. However, filipin staining visualizing subcellular distributions of free cholesterol found that cholesterol is accumulated at the MAM and co-localized with sigma-1 receptors (Hayashi and Su, 2003b). The MAM was originally found as a specialized membrane domain for transports of lipids between ER and mitochondria (Vance, 1990). Thus, it is intriguing that the cholesterol is exceptionally enriched at the MAM. Transfection of dominant-negative sigma-1 receptors, which fail to target the MAM, is shown to disrupt the compartmentalization of cholesterol at MAM, leading to the diffuse distribution of cholesterol over the ER membrane (Hayashi and Su,

2003b). Seemingly, sigma-1 receptors regulate, at least in part, the accumulation of free cholesterol at the MAM.

Existence of lipid rafts at the ER has been under a debate for several years. However, growing evidence supports the existence of lipid rafts at the ER membrane (Hayashi and Su, 2003a, b, 2004b, 2005; Sarnataro et al., 2004; Browman et al., 2006; Campana et al., 2006; Hoegg et al., 2009). Sigma-1 receptors were found to reside in lipid rafts in NG108 neuroblastoma x glioma hybrid cells (Hayashi and Su, 2003b). In oligodendrocytes, sigma-1 receptors also accumulate at lipids rafts by forming the complex with cholesterol and galactosylceramides (Hayashi and Su, 2004b). In light of exceptional enrichment of cholesterol and sphingolipids (e.g., ceramides) at MAM among bulk ER membranes (Vance, 1990; Rusinol et al., 1993, 1994; Bionda et al., 2004), it is conceivable for the cell to form lipid rafts at the specialized subdomains. The identity of sphingolipid(s) forming rafts at the MAM is however not clarified, except for that in oligodendrocytes (Hayashi and Su, 2004b). In the sucrose-floatation centrifugation with Triton X-100, the ER-lipid rafts generally show lower buoyancy than that of plasma membrane rafts that mostly contain gangliosides such as GM1 (Hayashi and Su, 2003b). Interestingly, a certain class of caveolin shows similar buoyancy to ER rafts. Calveolin-2, but not caveolin-1 which is enriched at the plasma membrane and Golgi, was indeed shown to co-localize with sigma-1 receptors at the MAM (Hayashi and Su, 2003b).

Lipid rafts at the plasma membrane was classically proposed as platforms of the signal transduction, which compartmentalize receptors and signalling molecules for promoting protein-protein interactions (Simons and Ikonen, 1997; Simons and Toomre, 2000). Several trophic factor receptors, upon stimulation, dimerize and translocate into or out from lipid rafts to pursue activation of downstream signallings (Simons and Ikonen, 1997; Simons and Toomre, 2000). Similarly, sigma-1 receptors at ER rafts, upon stimulation with agonists, translocate from rafts to non-raft ER membranes (Hayashi and Su, 2003b; Palmer et al., 2007). Treatment with selective sigma-1 receptor agonist (+)pentazocine or (+)SKF10047 has been shown to shift sigma-1 receptors from Triton X-100-insoluble to the soluble fractions in a sucrose gradient centrifugation (Hayashi and Su, 2003b; Palmer et al., 2007). As discussed below, sigma-1 receptor agonists replace cholesterol binding to sigma-1 receptors (Palmer et al., 2007) and promote translocation of the receptor at the ER (Hayashi and Su, 2003b; Hayashi and Su, 2003a). Thus, the ligand-free, cholesterol-associating form of sigma-1 receptors seems to associate with ER rafts.

Although the above-mentioned evidence supports the possibility of sigma-1 receptors to bind cholesterol, the direct demonstration of cholesterol binding to sigma-1 receptors has begun to be examined recently. Ruoho and colleagures identified two potential sterol-binding domains (SBDL-1, SDBL-II) of the sigma-1 receptors by using a series of photoaffinity labelings (Chen et al., 2007; Pal et al., 2007, 2008; Fontanilla et al., 2008). They found that derivatives of fenpropimorph (e.g., [^{125}I]IAF), the inhibitor of yeast sterol isomerase that has also a high affinity for sigma-1 receptors, can selectively photolabel amino acids 91-109 and 176-194 with showing a single population of binding sites for [^{125}I]IAF to interact with the sigma-1 receptor (Pal et al., 2007). These data propose a model in which the

SBDL-I and SBDL-II are juxtaposed to form a sterol-binding site of the sigma-1 receptor (Pal et al., 2008). Palmer et al. (2007) by using the sequence matching analysis found that the sigma-1 receptor possesses amino acid sequences similar to the cholesterol-binding motif of the benzodiazepine receptor (i.e., L/V-X_{1-2}-Y-X_{1-5}-K/R). They postulated two potential cholesterol-binding domains on the sigma-1 receptors: residues 171–175 and 199–208. They demonstrated that the synthesized peptides containing these putative cholesterol-binding motifs (a.a. 161–180, a.a. 191–210) indeed bind cholesterol on immobilized nitrocellulose membranes (Palmer et al., 2007). The binding of cholesterol in the same system is reduced by co-incubation with sigma-1 receptor agonist (+)-SKF10047 (Palmer et al., 2007). Following energy minimization by Universal Force Field prediction, they proposed that tyrosine173 and 206 at the surface of the lipid bilayer are critical for the cholesterol-binding property of the sigma-1 receptors, since the substitutions of tyrosine to serine at the sites abolished the cholesterol-binding property of the two peptides (Palmer et al., 2007). These findings support the notion that sigma-1 receptors may directly interacts with cholesterol at lipid rafts of the MAM (Fig. 13.2). Further, the ability of the ligands in altering the sigma-1 receptor-cholesterol association may partly explain the underlying mechanism in ligand-induced translocation of sigma-1 receptors from rafts to non-raft ER membranes (Hayashi and Su, 2003b; Hayashi and Su, 2003a).

13.5 Molecular Function of the Sigma-1 Receptor

Since the 1970 s, a numerous number of studies have demonstrated a variety of pharmacological and physiological effects of sigma-1 receptors and their ligands in both in vitro and in vivo systems; that include neuroprotection, anti/pro-apoptotic action, cellular differentiation, potentiation of Ca2+ signalling via IP3 receptors, regulation of ion channels such as K+ channel and NMDA receptor, potentiation of trophic factor signalling (NGF, EGF, BDNF and MAPKs), cellular proliferation, protein secretion, carcinogenesis, long-term potentiation of hippocampal neurons, learning and memory, mood and cognition, and drug-dependence and craving (Maurice et al., 2002; Takebayashi et al., 2002, 2004a, b; Matsumoto et al., 2003; Su and Hayashi, 2003; Chen et al., 2006; Yagasaki et al., 2006; Martina et al., 2007; Hayashi and Su, 2008; Sabeti and Gruol, 2008; Fontanilla et al., 2009; Hayashi et al., 2009). However, the basic molecular function of the sigma-1 receptor has been elusive until the recent discovery of the innate activity. The sigma-1 receptor is a ligand-operated molecular chaperone at MAM (Hayashi and Su, 2007). Under the physiological concentration of Ca2+ in the ER lumen (0.5 mM), the lumenal domain of the sigma-1 receptor forms a complex with BiP, an ER homologue of heat-shock protein 70 (Hsp 70), in a Ca2+/Mn2+-dependent manner (Hayashi and Su, 2007). Sigma-1 receptors forming the complex with BiP are basically at the dormant state, thus minimizing the chaperone activity (Hayashi and Su, 2007). The depletion of ER Ca2+ by activation of IP3 receptors or inhibition of the ER Ca2+ pump by thapsigargin triggers the dissociation of sigma-1 receptors from BiP, which in turn fully activates sigma-1 receptor chaperones (Hayashi and Su, 2007; Hayashi et al., 2009).

Because of their specific localization at the MAM, sigma-1 receptor chaperones are assumed to stabilize mostly MAM-residing proteins. So far, the type-3 IP3 receptor is only protein identified as a substrate of the sigma-1 receptor chaperone. Knockdown of sigma-1 receptors causes rapid ubiquitination and degradation of type-3 IP3 receptors residing at MAM, thus causing decrease of direct Ca2+ influx from MAM to mitochondria (Hayashi and Su, 2007). Ca2+ uptaken into mitochondria in turn activates dehydrogenases in the TCA cycle, leading to potentiation of the ATP production in mitochondria (Rizzuto et al., 1999; Hajnoczky and Hoek, 2007; Duchen et al., 2008), the stabilization of IP3 receptors by sigma-1 receptor chaperones is thus postulated to contribute to bioenergetics and cellular survival. In this context, the role of cholesterol potentially specifying the MAM localization of sigma-1 receptors seems vast.

13.6 Ligand-Binding Profile of the Sigma-1 Receptor

The prominent uniqueness of the sigma-1 receptor is that the innate chaperone activity can be activated/inactivated pharmacologically by synthetic compounds or by sterols. Sigma-1 receptor agonists promote the dissociation of sigma-1 receptors from BiP, which in turn activates chaperone activity of sigma-1 receptors in an ER Ca2+-independent manner (Hayashi and Su, 2007). The action of the agonists is blocked by antagonists. Structurally diverse compounds including steroids and some clinically used psychotropic drugs and immunesuppressants exert high affinities for sigma-1 receptors (Table 13.1). A recent study found that endogenous hallucinogen dimethyltryptamine binds to sigma-1 receptors (Fontanilla et al., 2009). Selective

Table 13.1 Ligands of the sigma-1 receptor chaperone

Synthetic compounds
  (+)Pentazocine
  (+)Dextromethorphan
  (+)SKF10047
  Haloperidol
  Fluvoxamine
  Imipramine
  Donepezil
  SA31747

Natural/endogenous compounds
  Progesterone
  Dehydroepiandrosterone sulfate
  Pregnenolone sulfate
  Hyperforin/hypericin
  Dimethyltryltamine

sigma-1 receptor agonists have been demonstrated to possess therapeutic actions in animal models of depression, schizophrenia, and strokes (Hayashi and Su, 2008). Some steroids that have affinities to sigma-1 receptors indeed exert antiamnesic or antidepressant-like actions via sigma-1 receptors (Hayashi and Su, 2008). Whether cholesterol at the MAM may affect the chaperone activity of the sigma-1 receptor as a potential endogenous ligand is an interesting unsolved question. The recent finding showing ER rafts regulating folding of PrP (Campana et al., 2006), however, raises a tempting speculation that ER cholesterol may gain the folding capability of the ER by either mimicking a molecular chaperone by itself or by regulating activity of ER chaperone proteins.

13.7 Roles of Sigma-1 Receptors in Subcellular Distribution of Lipids and Reconstitution of Lipid Rafts

Studies exploring roles of sigma-1 receptors in lipid biology has just emerged. Recent data suggest that sigma-1 receptors play important roles in the subcellular distribution of cholesterol as well as its partitioning in lipid rafts at the plasma membrane (Hayashi and Su, 2005). For example, overexpression of sigma-1 receptors, although they are mainly at the ER (e.g., MAM), can increase cholesterol at the plasma membrane rafts (Takebayashi et al., 2004b). Further, expression of dominantly negative sigma-1 receptors decreases plasma membrane cholesterol with concomitantly increasing cholesterol at the ER (Hayashi and Su, 2003b). Thus, sigma-1 receptors at the ER seem to be involved in export of cholesterol from ER to the plasma membrane.

Stable overexpression of sigma-1 receptors in PC12 cells promotes the alteration of glycosphingolipids in plasma membrane lipid rafts (Takebayashi et al., 2004b). Overexpression of sigma-1 receptors increases ganglioside GD1a in plasma membrane rafts, the endproduct of the ganglioside synthesis, but decreases less-glycosylated precursors of GD1a such as GM1 and GM2 (Takebayashi et al., 2004b). Because the biosynthesis of glycopshingolipids downstream of ceramides is entirely processed at Golgi (van Meer, 2000), this finding suggests the sigma-1 receptor accelerating glycosylation of gangliosides at the Golgi apparatus. The reconstitution of plasma membrane rafts caused by sigma-1 receptors therefore promotes a significant alteration in signal transduction triggered by receptors of trophic factors as mentioned below. How sigma-1 receptors associating with cholesterol at the MAM regulate glycosylation of sphingolipids at Golgi is, however, an open question at present.

Several trophic factor receptors, upon stimulation, dimerize and translocate into lipid rafts for activation of downstream signallings. However, some trophic factor receptors such as the EGF receptor conversely translocate from rafts to non-raft membranes upon the activation; thus lipid rafts serve as both positive and negative regulators of trophic factor-related signals in the receptor-specific manner. Importantly, it is known that the association of receptors with lipid rafts depends on sugar moieties of residing gangliposides; for example, EGF receptors have the highest affinity for GM1 or GM2 ganglioside, but the lowest affinity for highly glycosylated GD1a (Miljan et al., 2002). Thus, reconstitution of gangliosides in lipid rafts caused by sigma-1 receptors promotes the relocation of EGF receptors from rafts to non-rafts membranes, leading to potentiation of the EGF receptor activity (Takebayashi et al., 2004b). In such, sigma-1 receptors by modulating lipid components of the plasma membrane may regulate diverse signal transductions mediated by trophic factor receptors as is indeed seen in other systems such as NGF or BFND (Takebayashi et al., 2002; Yagasaki et al., 2006).

13.8 The Sigma-1 Receptor in Human Diseases

13.8.1 Neuropsychiatric Disorders

Since the first prototypic sigma receptor ligand SKF-10047 induces psychotomimeric actions and some antipsychotics such as haloperidol have low-nM affinities for sigma receptors (Su and Hayashi, 2003), the sigma-1 receptor ligands had been expected to serve as a new class of psychotherapeutic drugs. However, newly synthesized selective sigma-1 receptor ligands do not necessarily have psychotimimetic effects (Hayashi and Su, 2004a). Further clinical studies failed to show significant effect of sigma-1 receptor ligands on positive symptoms of schizophrenia (Volz and Stoll, 2004). On the other hand, recent preclinical studies have accumulated substantial data showing that sigma-1 receptor ligands possess robust neuroprotective action against, for example, β-amyloid- or ischemia (Hayashi and Su, 2008). Sigma-1 receptor agonists also show anti-amnesic and anti-depressant-like actions in animal models (Maurice et al., 2002).

On the analogy between psychostimulant-induced psychosis and schizophrenia, sigma receptor ligands have been extensively examined in animal models of drug-dependence. The use of psychostimulants such as methamphetamine and cocaine acutely causes psychosis in humans similar to positive symptoms of schizophrenia. Further, the long-term use of psychostimulants promotes neuroadaptative/neuroplastic changes, often those are irreversible and promote hard-to-cure symptoms such as withdrawal symptoms, craving, anxiety, aberrant stress-responses and depression (Hyman et al., 2006). Recent studies have found that sigma-1 receptors are upregulated in particular brains regions relating to dopaminergic reward and motor systems following chronic treatments with psychosimulants (Stefanski et al., 2004; Liu and Matsumoto, 2008).

It is not totally clarified how sigma-1 receptor chaperones at the MAM promote neuronal plasticity that likely involves morphogenesis of cells in the central nervous system. In in vitro studies, however, upregulation of sigma-1 receptors is shown to promote neuronal differentiation and potentiation of BDNF signallings (Hayashi and Su, 2004b; Yagasaki et al., 2006); both have been proven to be implicated in the pathophysiology of depression and drug dependence (Hyman et al., 2006). Indeed, certain antidepressants promote neuritegenesis by activating sigma-1 receptors in PC12 cells (Takebayashi et al., 2002). The action in activating neuronal differentiation may involve at least in part the action of sigma-1 receptors to promote reconstitution of plasma membrane lipid rafts that alters signal transduction of the trophic factor receptors (Hayashi and Su, 2005). Recent studies demonstrate that cholesterol is de novo synthesized in the brain, which is independent of cholesterol supplied by the blood circulation. Further, cholesterol provided by astrocytes is shown to serve as a potent inducer of neuronal morphogenesis including synaptogenesis (Barres and Smith, 2001; Suzuki et al., 2007). In contrast, the aberrant metabolism of cholesterols is known to cause neuronal dysfunction and degeneration (Pregelj, 2008). Thus, sigma-1 receptors regulating cholesterol transport and raft formation may play important roles in neuroprotection and neuroplasticity in the brain.

13.8.2 Cancer

Early binding studies found that sigma-1 receptors highly express in many cancer cells (Vilner et al., 1995). Recent studies demonstrated that some sigma-1 receptor ligands inhibit unstrained proliferation of carcinoma both in vitro and in vivo (Spruce et al., 2004). A recent study provides an intriguing mechanism of sigma-1 receptors in regulating proliferation in cancers. The study found that sigma-1 receptor ligands cause the dissociation of cholesterol from sigma-1 receptors, which subsequently leads to reconstitution of plasma membrane rafts (Palmer et al., 2007). The reconstitution of lipid rafts promotes dramatic inhibition of cell adhesion mediated by β-integrin, the strength is indicative of invasiveness of cancers (Palmer et al., 2007). Knockdown of sigma-1 receptors by siRNA shows the similar phenotype. The effect of sigma-1 receptor ligands was abolished if the composition of lipid rafts is altered by using drugs that affect lipid raft formation such as methyl- β -cyclodextrin (2% for 30 min) (Palmer et al., 2007). These findings further suggest that cholesterol trafficking and raft formation regulated by sigma-1 receptors may be involved in promoting pathogenic processes involved in human diseases.

13.9 Conclusions

In this book chapter, we introduced a novel sterol-binding ER protein, the sigma-1 receptor chaperone. Sigma-1 receptor chaperones localize at the MAM, an ER subcompartment apposing to mitochondria (Hayashi and Su, 2007), where sigma-1 receptors associate with lipid rafts (Hayashi and Su, 2005). The association of sigma-1 receptors with cholesterol is altered by the ligand treatment, leading the relocation of sigma-1 receptors at the ER membrane (Hayashi and Su, 2005; Palmer et al., 2007). The sigma-1 receptor regulates the ER distribution of cholesterol as well as the formation of lipid rafts at the plasma membrane (Hayashi and Su, 2005). Further, sigma-1 receptors accelerate glycosylation of gangliosides at Golgi (Hayashi and Su, 2005). It is unclear at present how exactly the innate activity of the sigma-1 receptors (i.e., molecular chaperone activity) contributes to the regulations of cholesterol transport, ganglioside syntheses, and reconstitution of lipid rafts at the plasma membrane. Nevertheless, the regulation of lipid redistribution and metabolisms certainly affect a variety of signal transductions mediated by receptors of several trophic factors (e.g., EGF, NGF, BDNF), thus regulating cellular survival, differentiation, and cell adhesion.

The role of ER cholesterol in regulation of SREBP has been exhaustively examined, and elucidated evidence constitutes a clear picture wherein cholesterol regulates one of fundamental functions of the ER: the lipid biosyntheses. However, information for depicting roles of cholesterol in other basic ER functions (e.g., protein synthesis, protein folding, protein/lipid secretion, post-translational protein modifications, Ca2+ signalling), if any, is considerably scarce. Thus, one ultimate goal of this chapter is to cast fresh light on potential functions of cholesterol at the ER by reviewing the sigma-1 receptor chaperone, and to provide a scope for future investigations. The data from sigma-1 receptor chaperones and others clearly suggest the potentials of ER cholesterol in regulating 1) protein folding and degradation at the ER, 2) compartmentalization/segregation of ER resident proteins, 3) the function of mitochondria via the ER-mitochondria contact, and 4) sphingolipid biosyntheses.

Acknowledgement This work is supported by Intramural Research Program, NIDA, NIH, DHHS.

References

Achison, M, Boylan, MT., Hupp, TR. and Spruce, BA., 2007, HIF-1alpha contributes to tumour-selective killing by the sigma receptor antagonist rimcazole. *Oncogene* **26**: 1137–1146.
Alonso, G, Phan, V, Guillemain, I, Saunier, M, Legrand, A, Anoal, M and Maurice, T, 2000, Immunocytochemical localization of the sigma(1) receptor in the adult rat central nervous system. *Neuroscience* **97**: 155–170.
Aydar, E, Palmer, CP., Klyachko, VA. and Jackson. M.B., 2002, The sigma receptor as a ligand-regulated auxiliary potassium channel subunit. *Neuron* **34**: 399–410.
Bae, S, Seong, J and Paik, Y, 2001, Cholesterol biosynthesis from lanosterol: molecular cloning, chromosomal localization, functional expression and liver-specific gene regulation of rat sterol delta8-isomerase, a cholesterogenic enzyme with multiple functions. *Biochem J* **353**: 689–699.
Barres, BA. and Smith, SJ., 2001, Neurobiology. Cholesterol–making or breaking the synapse. *Science* **294**: 1296–1297.
Bengoechea-Alonso, MT. and Ericsson, J, 2007, SREBP in signal transduction: cholesterol metabolism and beyond. *Curr Opin Cell Biol* **19**: 215–222.
Bermack, JE. and Debonnel, G, 2007, Effects of OPC-14523, a combined sigma and 5-HT1a ligand, on pre- and post-synaptic 5-HT1a receptors. *J Psychopharmacol* **21**: 85–92.
Bionda, C, Portoukalian, J, Schmitt, D, Rodriguez-Lafrasse, C and Ardail, D, 2004, Subcellular compartmentalization of ceramide metabolism: MAM (mitochondria-associated membrane) and/or mitochondria? *Biochem J* **382**: 527–533.

Browman, DT., Resek, ME., Zajchowski, LD. and Robbins, SM., 2006, Erlin-1 and erlin-2 are novel members of the prohibitin family of proteins that define lipid-raft-like domains of the ER. *J Cell Sci* **119**: 3149–3160.

Brown, MS. and Goldstein, JL., 1997, The SREBP pathway: regulation of cholesterol metabolism by proteolysis of a membrane-bound transcription factor. *Cell* **89**: 331–340.

Campana, V, Sarnataro, D, Fasano, C, Casanova, P, Paladino, S and Zurzolo, C, 2006, Detergent-resistant membrane domains but not the proteasome are involved in the misfolding of a PrP mutant retained in the endoplasmic reticulum. *J Cell Sci* **119**: 433–442.

Chen, L, Dai, XN. and Sokabe, M, 2006, Chronic administration of dehydroepiandrosterone sulfate (DHEAS) primes for facilitated induction of long-term potentiation via sigma 1 (sigma1) receptor: optical imaging study in rat hippocampal slices. *Neuropharmacology* **50**: 380–392.

Chen, Y, Hajipour, AR., Sievert, MK., Arbabian, M and Ruoho, AE., 2007, Characterization of the cocaine binding site on the sigma-1 receptor. *Biochemistry* **46**: 3532–3542.

Duchen, MR., Verkhratsky, A and Muallem, S, 2008, Mitochondria and calcium in health and disease. *Cell Calcium* **44**: 1–5.

Dun, Y, Thangaraju, M, Prasad, P, Ganapathy, V and Smith, SB., 2007, Prevention of excitotoxicity in primary retinal ganglion cells by (+)-pentazocine, a sigma receptor-1 specific ligand. *Invest Ophthalmol Vis Sci* **48**: 4785–4794.

Fontanilla, D, Hajipour, AR., Pal, A, Chu, UB., Arbabian, M and Ruoho, AE., 2008, Probing the steroid binding domain-like I (SBDLI) of the sigma-1 receptor binding site using N-substituted photoaffinity labels. *Biochemistry* **47**: 7205–7217.

Fontanilla, D, Johannessen, M, Hajipour, AR., Cozzi, NV., Jackson, MB. and Ruoho, AE., 2009, The hallucinogen N,N-dimethyltryptamine (DMT) is an endogenous sigma-1 receptor regulator. *Science* **323**: 934–937.

Ganapathy, ME., Prasad, PD., Huang, W, Seth, P, Leibach, FH. and Ganapathy, V, 1999, Molecular and ligand-binding characterization of the sigma-receptor in the Jurkat human T lymphocyte cell line. *J Pharmacol Exp Ther* **289**: 251–260.

Gebreselassie, D and Bowen, WD. 2004, Sigma-2 receptors are specifically localized to lipid rafts in rat liver membranes. *Eur J Pharmacol* **493**:19–28.

Goyagi, T, Goto, S, Bhardwaj, A, Dawson, VL., Hurn, PD. and Kirsch, JR., 2001, Neuroprotective effect of sigma(1)-receptor ligand 4-phenyl-1-(4-phenylbutyl) piperidine (PPBP) is linked to reduced neuronal nitric oxide production. *Stroke* **32**: 1613–1620.

Hajnoczky, G and Hoek, JB., 2007, Cell signalling. Mitochondrial longevity pathways. *Science* **315**: 607–609.

Hanner, M, Moebius, FF., Flandorfer, A, Knaus, HG., Striessnig, J, Kempner, E and Glossmann, H, 1996, Purification, molecular cloning, and expression of the mammalian sigma1-binding site. *Proc Natl Acad Sci USA* **93**: 8072–8077.

Hayashi, T, Rizzuto, R, Hajnoczky, G and Su, TP., 2009, MAM: more than just a housekeeper. *Trends Cell Biol* **19**: 81–88.

Hayashi, T and Su, TP., 2003a, Intracellular dynamics of sigma-1 receptors (sigma(1) binding sites) in NG108-15 cells. *J Pharmacol Exp Ther* **306**: 726–733.

Hayashi, T and Su, TP., 2003b, Sigma-1 receptors (sigma(1) binding sites) form raft-like microdomains and target lipid droplets on the endoplasmic reticulum: roles in endoplasmic reticulum lipid compartmentalization and export. *J Pharmacol Exp Ther* **306**: 718–725.

Hayashi, T and Su, TP., 2004a, Sigma-1 receptor ligands: potential in the treatment of neuropsychiatric disorders. *CNS Drugs* **18**: 269–284.

Hayashi, T and Su, TP., 2004b, Sigma-1 receptors at galactosylceramide-enriched lipid microdomains regulate oligodendrocyte differentiation. *Proc Natl Acad Sci U S A* **101**: 14949–14954.

Hayashi, T and Su, TP., 2005, The potential role of sigma-1 receptors in lipid transport and lipid raft reconstitution in the brain: implication for drug abuse. *Life Sci* **77**: 612–1624.

Hayashi, T and Su, TP., 2007, Sigma-1 Receptor Chaperones at the ER- Mitochondrion Interface Regulate Ca(2+) Signalling and Cell Survival. *Cell* **131**: 596–610.

Hayashi, T and Su, TP., 2008, An update on the development of drugs for neuropsychiatric disorders: focusing on the sigma 1 receptor ligand. *Expert Opin Ther Targets* **12**: 5–58.

Hoegg, MB., Browman, DT., Resek, ME. and Robbins, SM., 2009, Distinct regions within the erlins are required for oligomerization and association with high molecular weight complexes. *J Biol Chem* **284**: 7766–7776.

Hyman, SE., Malenka, RC.and Nestler, EJ., 2006, Neural mechanisms of addiction: the role of reward-related learning and memory. *Annu Rev Neurosci* **29**: 565–598.

Ikonen, E and Vainio, S, 2005, Lipid microdomains and insulin resistance: is there a connection? *Sci STKE* **2005**:pe3.

Jiang, G, Mysona, B, Dun, Y, Gnana-Prakasam, JP., Pabla, N, Li, W, Dong, Z, Ganapathy, V and Smith, SB., 2006, Expression, subcellular localization, and regulation of sigma receptor in retinal muller cells. *Invest Ophthalmol Vis Sci* **47**: 5576–5582.

Johannessen, MA., Ramachandran, S, Riemer, L, Ramos-Serrano, A, Ruoho, AE. and Jackson, MB., 2009, Voltage-Gated Sodium Channel Modulation by Sigma Receptors in Cardiac Myocytes and Heterologous Systems. *Am J Physiol Cell Physiol* **296**: C1049–C1057.

Lajoie, P and Nabi, IR., 2007, Regulation of raft-dependent endocytosis. *J Cell Mol Med* **11**: 644–653.

Lavoie, HA. and King, SR., 2009, Transcriptional regulation of steroidogenic genes: STARD1, CYP11A1 and HSD3B. *Exp Biol Med (Maywood)* **234**: 880–907.

Liu, Y, Chen, GD., Lerner, MR., Brackett, DJ. and Matsumoto, RR., 2005, Cocaine up-regulates Fra-2 and sigma-1 receptor gene and protein expression in brain regions involved in addiction and reward. *J Pharmacol Exp Ther* **314**: 770–779.

Liu, Y and Matsumoto, RR., 2008, Alterations in fos-related antigen 2 and sigma1 receptor gene and protein expression are associated with the development of cocaine-induced behavioral sensitization: time course and regional distribution studies. *J Pharmacol Exp Ther* **327**: 87–195.

Malorni, W, Giammarioli, AM., Garofalo, T and Sorice, M, 2007, Dynamics of lipid raft components during lymphocyte apoptosis: the paradigmatic role of GD3. *Apoptosis* **12**: 941–949.

Marrazzo, A, Caraci, F, Salinaro, ET., Su, TP., Copani, A and Ronsisvalle, G, 2005, Neuroprotective effects of sigma-1 receptor agonists against beta-amyloid-induced toxicity. *Neuroreport* **16**: 1223–1226.

Martin-Fardon, R, Maurice, T, Aujla, H, Bowen, WD. and Weiss, F, 2007, Differential effects of sigma1 receptor blockade on self-administration and conditioned reinstatement motivated by cocaine vs natural reward. *Neuropsychopharmacology* **32**: 1967–1973.

Martina, M, Turcotte, ME., Halman, S and Bergeron, R, 2007, The sigma-1 receptor modulates NMDA receptor synaptic transmission and plasticity via SK channels in rat hippocampus. *J Physiol* **578**: 143–157.

Matsumoto, RR., Liu, Y, Lerner, M, Howard, EW. and Brackett, DJ., 2003, Sigma receptors: potential medications development target for anti-cocaine agents. *Eur J Pharmacol* **469**: 1–12.

Maurice, T, 2004, Neurosteroids and sigma1 receptors, biochemical and behavioral relevance. *Pharmacopsychiatry* **37 Suppl 3**: S171–S182.

Maurice, T, Martin-Fardon, R, Romieu, P and Matsumoto, RR., 2002, Sigma(1) (sigma(1))-receptor antagonists represent a new strategy against cocaine addiction and toxicity. *Neurosci Biobehav Rev* **26**: 499–527.

Mei, J and Pasternak, GW., 2007, Modulation of brainstem opiate analgesia in the rat by sigma 1 receptors: a microinjection study. *J Pharmacol Exp Ther* **322**: 1278–1285.

Miljan, EA., Meuillet, EJ., Mania-Farnell, B, George, D, Yamamoto, H, Simon, HG. and Bremer, EG., 2002, Interaction of the extracellular domain of the epidermal growth factor receptor with gangliosides. *J Biol Chem* **277**: 10108–10113.

Moebius, FF., Reiter, RJ., Hanner, M and Glossmann, H, 1997, High affinity of sigma 1-binding sites for sterol isomerization inhibitors: evidence for a pharmacological relationship with the yeast sterol C8-C7 isomerase. *Br J Pharmacol* **121**: 1–6.

Nakazawa, M, Matsuno, K and Mita, S, 1998, Activation of sigma1 receptor subtype leads to neuroprotection in the rat primary neuronal cultures. *Neurochem Int* **32**: 337–343.

Pal, A, Chu, UB., Ramachandran, S, Grawoig, D, Guo, LW., Hajipour, AR. and Ruoho, AE., 2008, Juxtaposition of the steroid binding domain-like I and II regions constitutes a ligand binding site in the sigma-1 receptor. *J Biol Chem.* **283**: 19646–19656.

Pal, A, Hajipour, AR., Fontanilla, D, Ramachandran, S, Chu, UB., Mavlyutov, T and Ruoho, AE., 2007, Identification of regions of the sigma-1 receptor ligand binding site using a novel photoprobe. *Mol Pharmacol* **72**: 921–933.

Palmer, CP., Mahen, R, Schnell, E, Djamgoz, MB. and Aydar, E, 2007, Sigma-1 receptors bind cholesterol and remodel lipid rafts in breast cancer cell lines. *Cancer Res* **67**: 11166–11175.

Pike, LJ., 2003, Lipid rafts: bringing order to chaos. *J Lipid Res* **44**: 655–667.

Pregelj, P, 2008, Involvement of cholesterol in the pathogenesis of Alzheimer's disease: role of statins. *Psychiatr Danub* **20**:162–167.

Renaudo, A, L'Hoste, S, Guizouarn, H, Borgese, F and Soriani, O 2007, Cancer cell cycle modulated by a functional coupling between sigma-1 receptors and Cl- channels. *J Biol Chem* **282**: 2259–2267.

Rizzuto, R, Pinton, P, Brini, M, Chiesa, A, Filippin, L and Pozzan, T, 1999, Mitochondria as biosensors of calcium microdomains. *Cell Calcium* **26**: 193–199.

Rusinol, AE., Chan, EY. and Vance, JE., 1993, Movement of apolipoprotein B into the lumen of microsomes from hepatocytes is disrupted in membranes enriched in phosphatidyl-monomethylethanolamine. *J Biol Chem* **268**: 25168–25175.

Rusinol, AE., Cui, Z, Chen, MH. and Vance, JE., 1994, A unique mitochondria-associated membrane fraction from rat liver has a high capacity for lipid synthesis and contains pre-Golgi secretory proteins including nascent lipoproteins. *J Biol Chem* **269**: 27494–27502.

Sabeti, J and Gruol, DL., 2008, Emergence of NMDAR-independent long-term potentiation at hippocampal CA1 synapses following early adolescent exposure to chronic intermittent ethanol: role for sigma-receptors. *Hippocampus* **18**: 148–168.

Sarnataro, D, Campana, V, Paladino, S, Stornaiuolo, M, Nitsch, L and Zurzolo, C, 2004, PrP(C) association with lipid rafts in the early secretory pathway stabilizes its cellular conformation. *Mol Biol Cell* **15**: 4031–4042.

Seth, P, Ganapathy, ME., Conway, SJ., Bridges, CD., Smith, SB., Casellas, P and Ganapathy, V, 2001, Expression pattern of the type 1 sigma receptor in the brain and identity of critical anionic amino acid residues in the ligand-binding domain of the receptor. *Biochim Biophys Acta* **1540**: 59–67.

Simons, K and Ikonen, E, 1997, Functional rafts in cell membranes. *Nature* **387**: 569–572.

Simons, K and Toomre, D, 2000, Lipid rafts and signal transduction. *Nat Rev Mol Cell Biol* **1**: 1–39.

Smith, SB., Duplantier, J, Dun, Y, Mysona, B, Roon, P, Martin, PM. and Ganapathy, V, 2008, In vivo protection against retinal neurodegeneration by sigma receptor 1 ligand (+)-pentazocine. *Invest Ophthalmol Vis Sci* **49**: 4154–4161.

Spruce, BA., Campbell, LA., McTavish, N, Cooper, MA., Appleyard, MV., O'Neill, M, Howie, J, Samson, J, Watt, S, Murray, K, McLean, D, Leslie, NR., Safrany, ST., Ferguson, MJ., Peters, JA., Prescott, AR., Box, G, Hayes, A, Nutley, B, Raynaud, F, Downes, CP., Lambert, JJ., Thompson, AM. and Eccles, S, 2004, Small molecule antagonists of the sigma-1 receptor cause selective release of the death program in tumor and self-reliant cells and inhibit tumor growth in vitro and in vivo. *Cancer Res* **64**: 4875–4886.

Stefanski, R, Justinova, Z, Hayashi, T, Takebayashi, M, Goldberg, SR. and Su, TP., 2004, Sigma1 receptor upregulation after chronic methamphetamine self-administration in rats: a study with yoked controls. *Psychopharmacology (Berl)* **175**: 68–75.

Su, TP. and Hayashi, T 2003, Understanding the molecular mechanism of sigma-1 receptors: towards a hypothesis that sigma-1 receptors are intracellular amplifiers for signal transduction. *Curr Med Chem* **10**:2073–2080.

Su, TP., London, ED. and Jaffe, JH., 1988, Steroid binding at sigma receptors suggests a link between endocrine, nervous, and immune systems. *Science* **240**: 219–221.

Suzuki, S, Kiyosue, K, Hazama, S, Ogura, A, Kashihara, M, Hara, T, Koshimizu, H and Kojima, M, 2007, Brain-derived neurotrophic factor regulates cholesterol metabolism for synapse development. *J Neurosci* **27**: 6417–6427.

Takebayashi, M, Hayashi, T and Su, TP., 2002, Nerve growth factor-induced neurite sprouting in PC12 cells involves sigma-1 receptors: implications for antidepressants. *J Pharmacol Exp Ther* **303**: 1227–1237.

Takebayashi, M, Hayashi, T and Su, TP., 2004a, A perspective on the new mechanism of antidepressants: neuritogenesis through sigma-1 receptors. *Pharmacopsychiatry* **37 Suppl 3**: S208–213.

Takebayashi, M, Hayashi, T and Su, TP., 2004b, Sigma-1 receptors potentiate epidermal growth factor signalling towards neuritogenesis in PC12 cells: potential relation to lipid raft reconstitution. *Synapse* **53**: 90–103.

Tchedre, KT. and Yorio, T, 2008, Sigma-1 receptors protect RGC-5 cells from apoptosis by regulating intracellular calcium, Bax levels, and caspase-3 activation. *Invest Ophthalmol Vis Sci.* **49**: 2577–2588.

van Meer, G, 2000, Cellular organelles: how lipids get there, and back. *Trends Cell Biol* **10**: 550–552.

van Meer, G and van Genderen, IL., 1994, Intracellular lipid distribution, transport, and sorting. A cell biologist's need for physicochemical information. *Subcell Biochem* **23**: 1–24.

Vance, JE., 1990, Phospholipid synthesis in a membrane fraction associated with mitochondria. *J Biol Chem* **265**: 7248–7256.

Vilner, BJ., John, CS. and Bowen, WD., 1995, Sigma-1 and sigma-2 receptors are expressed in a wide variety of human and rodent tumor cell lines. *Cancer Res* **55**: 408–413.

Voelker, DR., 2000, Interorganelle transport of aminoglycerophospholipids. *Biochim Biophys Acta* **1486**: 97–107.

Volz, HP. and Stoll, KD., 2004, Clinical trials with sigma ligands. *Pharmacopsychiatry* **37 Suppl 3**: S214–220.

Yagasaki, Y, Numakawa, T, Kumamaru, E, Hayashi, T, Su, TP. and Kunugi, H, 2006, Chronic antidepressants potentiate via sigma-1 receptors the brain-derived neurotrophic factor-induced signalling for glutamate release. *J Biol Chem* **281**: 12941–12949.

Yamamoto, H, Miura, R, Yamamoto, T, Shinohara, K, Watanabe, M, Okuyama, S, Nakazato, A and Nukada, T, 1999, Amino acid residues in the transmembrane domain of the type 1 sigma receptor critical for ligand binding. *FEBS Lett* **445**: 19–22.

Chapter 14
Prominin-1: A Distinct Cholesterol-Binding Membrane Protein and the Organisation of the Apical Plasma Membrane of Epithelial Cells

Denis Corbeil, Anne-Marie Marzesco, Christine A. Fargeas, and Wieland B. Huttner

Abstract The apical plasma membrane of polarized epithelial cells is composed of distinct subdomains, that is, planar regions and protrusions (microvilli, primary cilium), each of which are constructed from specific membrane microdomains. Assemblies containing the pentaspan glycoprotein prominin-1 and certain membrane lipids, notably cholesterol, are characteristic features of these microdomains in apical membrane protrusions. Here we highlight the recent findings concerning the molecular architecture of the apical plasma membrane of epithelial cells and its dynamics. The latter is illustrated by the budding and fission of prominin-1-containing membrane vesicles from apical plasma membrane protrusions, which is controlled, at least in part, by the level of membrane cholesterol and the cholesterol-dependent organization of membrane microdomains.

Keywords Apical membrane · Cilium · Cholesterol · Ganglioside · Microvillus · Prominin

14.1 Introduction

The apical plasma membrane of polarized epithelial cells is composed of at least two principal types of subdomains, (i) planar, non-protruding membrane regions and (ii) membrane protrusions such as microvilli and primary cilia. In certain cells, deep-apical membrane invaginations have also been described (Hansen et al., 2003), yet these are not the topic of the present review. In resorptive epithelia such as those found in kidney proximal tubules and small intestine, microvilli forming the brush border membrane (BBM) play an important role in the absorption and processing of nutrients, by hosting the appropriate hydrolytic ectoenzymes and membrane-bound transporters (Semenza, 1986; Corbeil et al., 1992; Daniel and Rubio-Aliaga,

---

D. Corbeil (✉)
Tissue Engineering Laboratories, BIOTEC, Technische Universität Dresden, Tatzberg 47-49, 01307, Dresden, Germany
e-mail: denis.corbeil@biotec.tu-dresden.de

2003). BBM also acts as a semi-permeable barrier preventing the entry of luminal pathogens. The particular organization of microvilli results in an increase of the cellular apical membrane surface by ~40–100-fold. Each microvillus is about 1-μm long with a diameter of 100-nm, and contains an actin-based cytoskeleton. In other types of epithelial cells (e.g. neuroepithelial cells) where microvilli are less abundant, their biological relevance is poorly understood. However, like the primary cilium, a tubulin-based structure common to most eukaryotic cells and acting as a sensory organelle (for exhaustive reviews *see* Pazour and Bloodgood, 2008; Berbari et al., 2009), microvilli may play an underestimated role in membrane trafficking, as recently illustrated by the release of small extracellular membrane vesicles (Marzesco et al., 2005; Dubreuil et al., 2007). In the recent years, a great deal of attention has been directed to the plasma membrane of microvilli and primary cilia, as any alteration in their organization may lead to severe pathology in several organs of the mammalian body (Reinshagen et al., 2006; Gerdes et al., 2009). Nevertheless, a comprehensive picture of the molecular architecture of the apical plasma membrane of epithelial cells is still missing.

Both types of apical plasma membrane protrusions contain specific sets of membrane glycoproteins, some of which are organized with particular lipids in complex assemblies known as membrane microdomains (Danielsen and Hansen, 2003). The lipidic components of these microdomains are commonly named lipid rafts (for definition *see* Pike, 2006) and are viewed as liquid-ordered domains that are more tightly packed than the surrounding phase of the membrane bilayer (Simons and Ikonen, 1997). They are enriched in sterols and sphingolipids present in the exoplasmic membrane leaflet. Therein, membrane cholesterol appears to be an essential structural player. Certain membrane microdomain-associated proteins may interact as well with components of the submembraneous cytoskeleton (Föger et al., 2001; Viola and Gupta, 2007), as recently demonstrated for CD317/tetherin (Rollason et al., 2009).

Membrane microdomains have been suggested to play a role in signal transduction and in various membrane trafficking events, including membrane budding and fission (Ikonen and Simons, 1998; Simons and Toomre, 2000; Huttner and Zimmerberg, 2001; Ikonen, 2001; Sharma et al., 2002; Rajendran and Simons, 2005). The partitioning of particular proteins in and out of membrane microdomains could also play a role in certain human pathologies (Simons and Ehehalt, 2002). Technically, the classical biochemical method used to determine the association of a given membrane protein with such membrane microdomains is based on its resistance to extraction with certain non-ionic detergents (e.g. Triton X-100) in the cold (Brown and Rose, 1992; reviewed in Chamberlain, 2004; Waugh and Hsuan, 2009). Proteins associated with detergent–resistant membranes will float in a cholesterol-dependent manner to low-buoyant density fractions upon density gradient centrifugation (Brown and London, 2000; Lingwood and Simons, 2007). It is important to mention that the results obtained from detergent-based analyses do not necessary reflect the native state of membrane microdomains (Munro, 2003). For instance, some detergents such as Triton X-100 may create ordered domains in a homogeneous fluid membrane (Heerklotz, 2002). Consequently, these starting-point observations should always be complemented by morphological investigations based on light and/or transmission electron microscopy.

By studying the polarity of neuroepithelial cells, we previously identified prominin-1 (Weigmann et al., 1997), a cholesterol-interacting membrane glycoprotein with a distinct membrane topology (*see* Section 14.2). Using prominin-1 as a specific marker of plasma membrane protrusions, we have obtained new insights into the molecular organization of the apical plasma membrane of polarized epithelial cells, notably the co-existence of distinct cholesterol-based membrane microdomains (*see* Section 14.3). Studying the intracellular trafficking of prominin-1, we have observed that both microvilli and primary cilia constitute donor membranes of small extracellular membrane vesicles containing prominin-1. The latter vesicles, which are found into various body fluids, may become valuable diagnostic tools for certain human diseases (*see* Section 14.4). These recent findings concerning the architecture and dynamics of the apical plasma membrane will be presented in this review.

14.2 Prominin Molecules and Plasma Membrane Protrusions

14.2.1 Prominin – Basic Facts

Prominin-1 (also known as CD133, for nomenclature *see* Fargeas et al., 2007) is the first member of a novel evolutionarily conserved membrane glycoprotein family, exhibiting a unique membrane topology (Weigmann et al., 1997; Miraglia et al., 1997; Fargeas et al.,2003a). It contains five transmembrane segments with two large extracellular loops (Fig. 14.1A). Prominin-1 was identified using the rat monoclonal antibody 13A4 generated against mouse telencephalic neuroepithelium and its cDNA cloned from adult mouse kidney (Weigmann et al., 1997). Several splice variants have been identified and their expression characterized (Corbeil et al., 1998; Yu et al., 2002; Fargeas et al., 2003b, 2004, 2007). Prominin-1 is expressed at the apical plasma membrane of embryonic and adult epithelia, present in numerous mammalian organs (Weigmann et al., 1997; Corbeil et al., 2000, 2001b; Uchida et al., 2000; Florek et al., 2005; Jászai et al., 2007a, 2008; Lardon et al., 2008; Karbanová et al., 2008). Non-epithelial cells, notably photoreceptor cells (*see* below) and glial cells (i.e. oligodendrocytes and astrocytes), express it as well (Maw et al. 2000; Zacchigna et al. 2009; Corbeil et al. 2009). Although broadly expressed among various tissues, several organ-specific somatic stem and cancer stem cells were identified and isolated based on prominin-1 expression (reviewed in Fargeas et al., 2006; Bauer et al., 2008).

In mammals, a second prominin molecule has been identified and cloned (Fargeas et al., 2003a). The overall amino acid identity between prominin-1 and prominin-2 is low (<30%) (Corbeil et al., 2001a; Fargeas et al., 2003a), however their genomic organization is almost identical, indicating that both *prominin* genes originated from a common ancestral gene. Neither prominin exhibits obvious sequence similarity to other proteins and no motif that could provide clues

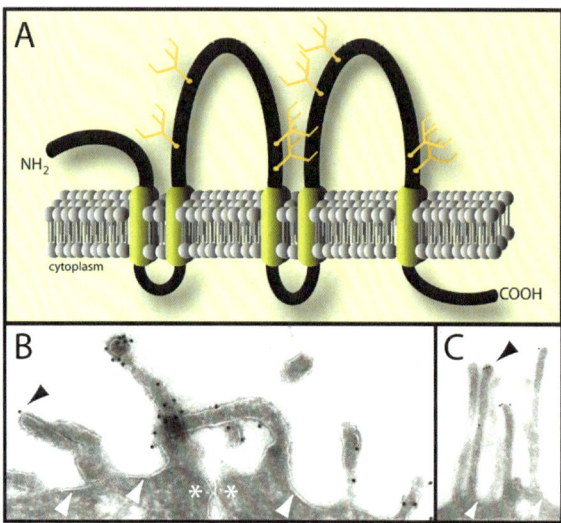

Fig. 14.1 Membrane topology of prominin-1 and its microvillar localization. (A) Topological model of human prominin-1. This molecule contains five transmembrane segments (*yellow* cylinders) separating two small intracellular domains and two large extracellular loops. The latters contain all eight potential *N*-glycosylation sites (forks). Note that their position within the protein sequence varies between species. Upon cleavage of the signal peptide the N-terminal domain is located in the extracellular space whereas the C-terminal domain is facing the cytoplasm. Its cytoplasmic location has been confirmed by antibody accessibility and epitope insertion analyses (Weigmann et al. 1997; Giebel et al. 2004). (B, C) Prominin-1 is specifically concentrated in microvilli of polarized epithelial cells. Ultrathin cryosections of either prominin-1–transfected MDCK cells (B) or Caco-2 cells (C) were staining with either rat mAb 13A4 (B) or mouse mAb AC133 (C) followed by rabbit anti-rat IgG/IgM and 9-nm Protein A-gold or goat-anti mouse IgG coupled to 15-nm gold particles (C), respectively. In both cell lines, prominin-1 immunoreactivity is confined to microvilli present at the apical cell surface and often concentrated at their tips (black arrowheads) mimicking the situation observed in native epithelia (Weigmann et al. 1997; Marzesco et al. 2005). Note the absence of labelling in the intermicrovillar regions of the plasma membrane (white arrowheads). Junctional complexes between two adjacent cells are indicated with asterisks

as to their physiological function has been identified in either sequence. Unlike prominin-1, which shows a widespread expression profile, its paralogue prominin-2 appears to be limited to epithelial cells (Fargeas et al., 2003a; Jászai et al., 2007a, 2008), where its distribution is either non-polarized or enriched at the basolateral plasma membrane, depending on the epithelium investigated (Florek et al., 2007; József Jászai, Lilla M. Farkas, C.A.F., Peggy Janich, Michael Haase, W.B.H., and D.C., manuscript submitted).

A remarkable characteristic of prominin-1 is its selective occurrence in specific subdomains of the plasma membrane that, although distinct in structure in various cell types, have one characteristic in common, i.e. to protrude from the planar regions of the plasmalemma (reviewed in Corbeil et al., 2001a). Within the apical plasma membrane of epithelial cells, prominin-1 is restricted to microvilli and the primary cilium, where it often appears to be concentrated at the tips of these membrane protrusions (Weigmann et al., 1997; Dubreuil et al., 2007). In particular epithelia such as epididymis and endometrium, it is also concentrated in stereocilia and motile cilia, respectively (Fargeas et al., 2004; Karbanová et al., 2008). In contrast, prominin-1 is very rarely detected in the inter-microvillar plasma membrane regions (Weigmann et al., 1997; Fargeas et al., 2004). The absence of prominin-1 in the planar subdomain of the apical plasma membrane is particularly evident in neural progenitors, i.e. neuroepithelial cells (Weigmann et al., 1997; Marzesco et al., 2005), which contain fewer microvilli compared to the kidney BBM. Interestingly, the plasma membrane protrusion-specific localization of prominin-1 is not restricted to epithelial cell, but is also observed in non-epithelial cells such as hematopoietic stem cells or prominin-1–transfected fibroblasts (Weigmann et al., 1997; Corbeil et al., 2000), suggesting that the molecular mechanism underlying such retention is conserved between epithelial and non-epithelial cells and may reflect a cell type-specific adaptation of a process common to all eukaryotic cells. Similarly, although not restricted to the apical plasma membrane, prominin-2 exhibits a preference for plasma membrane protrusions, and does so in all three plasmalemmal domains of polarized epithelial cells, i.e. the apical, lateral and basal plasma membrane (Fargeas et al., 2003a; Florek et al., 2007). Given their characteristic subcellular localization in plasma membrane protrusions, we named the first-discovered molecule and the entire family prominin (from the Latin word *prominere* – to stand out -; Weigmann et al., 1997; Fargeas et al., 2003b).

14.2.2 Prominin-1 and Photoreceptors

Insights into the significance of the concentration of prominin-1 in plasma membrane protrusions has come from the visual system, as the most conspicuous effect of its loss of function is retinal degeneration. These results specifically concern retinal photoreceptors, which are not the subject of the present review and will therefore only be briefly summarized here, although the conclusions from these data may be of general relevance.

In mammalian photoreceptors, the plasma membrane evaginations at the base of the rod outer segment, which represent the early stages in rod photoreceptor disc biogenesis (Steinberg et al., 1980), are enriched with prominin-1, as are the membrane disc evaginations of cones (Maw et al., 2000; Jászai et al., 2007b; Yang et al., 2008). Is this specific subcellular localization of prominin-1 related to disc formation, which includes complex transformations in membrane curvature? Nascent discs are generated from the connecting cilium as flattened plasma membrane evaginations enriched in cholesterol (Albert and Boesze-Battaglia, 2005). Disc formation occurs during the entire life of the photoreceptor cell, suggesting that any alteration may lead to visual impairment. The importance of prominin-1 in the maintenance of photoreceptor architecture is well illustrated by human pathology and further substantiated by data from prominin-1-deficient mice. Thus, mutations in the human *PROM-1* gene were reported to cause autosomal-recessive (Maw et al., 2000) and dominant macular (Zhang et al., 2007; Yang et al., 2008) photoreceptor degeneration. We have recently observed using a prominin-1-deficient mouse line that the lack of this molecule leads to progressive retinal photoreceptor degeneration (Zacchigna et al., 2009). Specifically, in young prominin-$1^{-/-}$ animals, the rod outer segments show a complete disorganization of membranes, whereas in older animals the entire photoreceptor layer is absent (Zacchigna et al., 2009). Together, these observations indicate that prominin-1 exerts an essential role in disc morphogenesis, the specific details of which remain to be elucidated.

14.3 The Apical Plasma Membrane Contains Distinct Cholesterol-Based Membrane Microdomains

A growing field of cell biology concerns the molecular mechanisms underlying the biogenesis and maintenance of the apical plasma membrane domain of polarized epithelial cells where distinct plasma membrane subdomains, e.g. microvilli, primary cilium and non-protruding regions, host-specific membrane proteins which are essential to their function. It becomes more and more evident that players other than the tight junctions or cytoskeletal elements are involved in this functional organization. By studying the biochemical and morphological properties of the membrane glycoprotein prominin-1, a selective marker of apical plasma membrane protrusions, we have gained insight into this organization.

14.3.1 Prominin-1 – A Cholesterol-Interacting Protein Associated with a Distinct Membrane Microdomain Subtype

How is prominin-1 selectively retained within apical plasma membrane protrusions of polarized epithelial cells? We have observed that prominin-1 maintains its protrusion-specific localization in neuroepithelial cells (Weigmann et al., 1997; Kosodo et al., 2004) even when the cells have lost functional tight junctions (Aaku-Saraste et al.,1996) and the polarized delivery of canonical apical and basal plasma membrane proteins (Aaku-Saraste et al., 1997), which is the case after the transition from the neural plate to the neural tube stage (*reviewed in* Götz and Huttner, 2005). These observations indicate that other mechanisms are operational in the protrusion-specific localization of prominin-1 (Corbeil et al., 1999).

The microvillar retention of prominin-1 could result from direct protein-protein interactions involving either a selective anchoring to sub-plasmalemmal cytoskeletal elements, e.g. members of the ezrin-radixin-moesin (ERM) family of proteins (Mangeat et al., 1999; Yonemura et al., 1999), as reported for other integral membrane proteins (Rotin et al., 1994; Chow et al., 1999; Rollason et al., 2009), or a binding to extracellular components (Stechly et al., 2009). Prominin-1 may also directly interact with other integral membrane proteins or membrane lipids. These types of partner(s), i.e. cytoplasmic, membraneous or peripheral, may not be mutually exclusive. We have addressed this issue in two epithelial cell models, prominin-1-transfected MDCK cells (Corbeil et al., 1999) and the human colon-carcinoma-derived Caco-2 cells, which express prominin-1 endogenously (Corbeil et al., 2000), as these cell lines are easy to manipulate and, importantly, reproduce the morphological characteristics of native epithelia expressing prominin-1. For instance, in both cell lines prominin-1 is restricted to apical plasma membrane protrusions (microvilli and primary cilium; Fig. 14.1B, C; Corbeil et al., 1999, 2000; Florek et al., 2005, 2007). Likewise, the microvillar retention of prominin-1 is maintained in prominin-1–transfected MDCK cells cultured in low-calcium medium, i.e. in the absence of functional tight junctions (Corbeil et al., 1999), as observed physiologically in neuroepithelial cells during the process of differentiation (Weigmann et al., 1997; Kosodo et al., 2004).

The cytoplasmic tail of prominin-1 contains a PDZ-binding domain (Fargeas et al., 2007) and could possibly interact with the cytoskeleton through interaction with PDZ-domain-containing proteins (e.g. NHERF proteins, reviewed in Seidler et al., 2009) that act as cytoplasmic adaptor/linker proteins between integral membrane proteins and the actin cytoskeleton (Muth et al., 1998). Yet, the specific localization of prominin-1 in plasma membrane protrusions does not depend on the cytoplasmic COOH-terminal domain (Fig. 14.1A), since it is not prevented by deletion of this segment (Corbeil et al., 1999). Moreover, the preferential localization of prominin-1 in microvilli is not abolished following a profound alteration of their cytoskeletal architecture, as observed in Caco-2 cells displaying rarefied and considerably altered microvilli (Röper et al., 2000) upon ectopic expression of antisense RNA for the actin-bundling protein villin (Costa de Beauregard et al., 1995). Villin is a key player in microvillar morphology (Friederich et al., 1989). Thus, a "vertical" interaction with cytoskeletal elements appears dispensable for the microvillar retention of prominin-1. Nielsen and colleagues (2007) have recently drawn a similar conclusion for another microvillar membrane glycoprotein, podocalyxin, suggesting that alternative determinants within the transmembrane segments and/or extracellular domains of these molecules are necessary for their microvillar localization.

Remarkably, using new tools to preserve and detect protein–lipid interactions, we could demonstrate that the specific localization of prominin-1 in microvillar membranes reflects its association with a novel cholesterol-based membrane microdomain in which prominin-1 directly interacts with plasma membrane cholesterol (Röper et al., 2000). The novel nature of these membrane microdomains is revealed by our observation that the prominin-1 molecules associated with them are solubilized in Triton X-100 but, importantly, are recovered in detergent-resistant membrane complexes upon extraction of microvillar membranes in the cold using another non-ionic detergent, Lubrol WX (Röper et al., 2000). Newly synthesized prominin-1 located in early intracellular compartments, i.e. the endoplasmic reticulum, remains Lubrol WX-soluble under these conditions, indicating that the formation of such detergent-resistant membrane complexes occurs during intracellular transport. Indeed, their formation was found to take place at the level of the trans-Golgi network (TGN) (Röper et al., 2000). Other mild non-ionic detergents such as Brij 58 and Triton X-102 give similar data, indicating that a detergent containing a large polar headgroup may preserve these distinct subtypes of membrane microdomains (Röper et al., 2000). As reported for several membrane microdomain-associated proteins recovered in Triton X-100–resistant membrane complexes (*see* Introduction), Lubrol WX-resistant membrane complexes containing prominin-1 float to lower-density fractions upon centrifugation in a sucrose density gradient (*see* Legend to Fig. 14.2A) and have a high content of cholesterol and glycolipids (Röper et al., 2000).

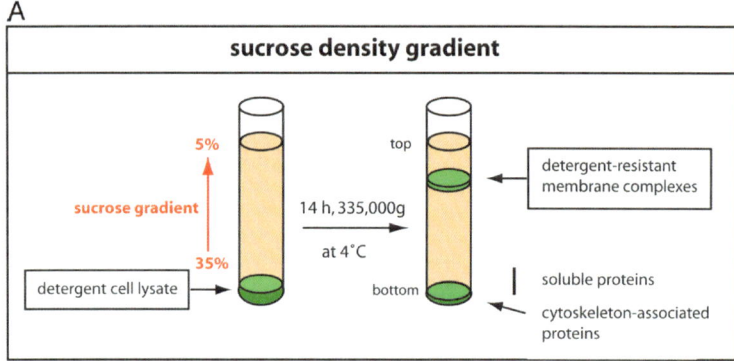

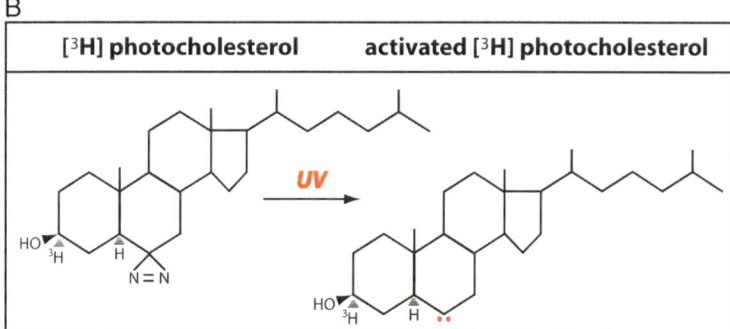

Fig. 14.2 Biochemical analysis of detergent-resistant membrane complexes and the photoactivatable analogue of cholesterol. (**A**) Equilibrium density gradient centrifugation. Classically, upon cell solubilization in non-ionic detergents such as Triton X-100 or Lubrol WX in the cold, detergent cell lysates are analysed by equilibrium density gradient centrifugation. The lower density of detergent-resistant membrane complexes allows their separation from the remaining solubilized membrane components and/or cytoskeleton-associated proteins (pellet). Either a linear (5–35%) or stepwise sucrose gradient approach can be used (Röper et al., 2000). The inclusion of detergents within the gradient solution may influence the outcome (Röper et al., 2000). For more details, appropriate controls and technical limitations, see Lingwood and Simons (2007). (**B**) Photoactivatable analogue of cholesterol, a tool to investigate protein-cholesterol interactions. Photoactivatable analogue of cholesterol (often referred to as photocholesterol; *left*) lacks the Δ5 double bond and the hydrogen at C-6 found in cholesterol, which are replaced by a photoactivatable diazirine ring (a carbene precursor). To monitor the protein-photocholesterol interaction by fluorography, a tritiated hydrogen is added. Upon irradiation by ultraviolet light (*uv*), activated [^{3}H] photocholesterol is formed (*right*). For more details concerning its synthesis and application, *see* Thiele et al. (2000)

Do prominin-1–containing membrane complexes defined solely on a biochemical basis (i.e. detergent extraction) have any physiological relevance? In other words, do they mirror an existing functional entity? These questions were addressed in vivo using a mild cholesterol depletion of the cell surface by adding methyl-β-cyclodextrin (mβCD) in the cold to prominin-1–transfected MDCK cells. It is important to note that under such experimental conditions the microvillar structures remain intact (Röper et al., 2000), which is not necessarily the case when a harsh mβCD–treatment is performed at physiological temperature (*see below*). The mβCD has a high affinity for sterols as compared to other lipids, which renders this methylated cyclic oligosaccharide compound useful for manipulating cellular cholesterol content (Klein et al., 1995; Christian et al., 1997). Biochemically, cholesterol depletion was found to lead to the fragmentation of the Lubrol WX-resistant membrane complexes (as observed by differential centrifugation) and to their loss of buoyancy. Morphologically, the mβCD–treatment caused a striking redistribution of prominin-1 from an exclusively microvillar localization to a more homogeneous distribution over the entire apical plasma membrane, as observed by confocal laser scanning microscopy. The effect of cholesterol depletion is reversible since the re-feeding of cells with cholesterol-loaded-mβCD restores the proper localization of prominin-1 (Röper et al., 2000). Interestingly, the cell surface-associated prominin-1 interacts directly and specifically with membrane cholesterol as shown by photoaffinity labelling using a photoactivatable radioactive derivative of cholesterol (Fig. 14.2B, Thiele et al., 2000). This cholesterol analogue mimics the biological characteristics of native membrane cholesterol (Thiele et al., 2000; Mintzer et al., 2002). Consistent with the absence of prominin-1-containing membrane microdomains in the endoplasmic reticulum, no interaction of cholesterol with newly synthesized prominin-1 was observed, suggesting that such interaction requires a particular configuration of prominin-1-membrane lipid assemblies (Röper et al., 2000). Taken together, these observations suggest that membrane cholesterol is an essential component of a particular subtype of membrane microdomains which play a role in the retention of prominin-1 within plasma membrane protrusions (Röper et al., 2000).

14.3.2 *Distinct Cholesterol-Based Membrane Microdomain Subtypes as Building Units of the Apical Plasma Membrane*

The complete solubility of prominin-1 in Triton X-100 is not unique to this molecule since it was previously demonstrated for numerous brush border enzymes, notably lactase-phlorizin hydrolase (Danielsen, 1995; Zheng et al., 1999; Jacob et al., 1997). This enzyme, like prominin-1, is specifically associated with microvillar membranes (Jacob et al., 1997). By contrast, certain apical components such as the glycosylphosphatidylinositol (GPI)-linked transferrin-like iron-binding protein and placental alkaline phosphatase, another GPI-anchored protein (Fiedler et al., 1993) are insoluble in Triton X-100 (Danielsen and van Deurs, 1995; Röper et al., 2000), suggesting that distinct apical membrane microdomains could be preserved in this detergent. Interestingly, Danielsen and van Deurs (1995) observed that the GPI-linked transferrin-like iron-binding protein is primarily localized in patches of flat or invaginated apical membrane subdomains rather than microvilli. Similarly, in MDCK cells expressing both prominin-1 and placental alkaline phosphatase, their segregation from each other was observed, with the latter protein being excluded from the prominin-1-containing microvilli (Röper et al., 2000). Thus, the differential biochemical behaviour of these membrane proteins has its morphological counterpart in segregation within the apical plasma membrane.

These morphological and biochemical features highlight a new structural facet of the apical plasma membrane architecture with the existence of distinct cholesterol-based membrane microdomain subtypes that are characteristic of either the protruding plasmalemma (biochemically defined to as Triton X-100–soluble but Lubrol WX–resistant membranes; operationally referred to as Lubrol-resistant membrane complexes; Fig. 14.3A, red) or the planar, intermicrovillar plasmalemma (Triton X-100 – plus Lubrol WX–resistant membranes; referred to as Triton-resistant membrane complexes; Fig. 14.3A, green) (Röper et al., 2000). Physiologically, the affinity of certain plasma membrane proteins for particular subtypes of membrane

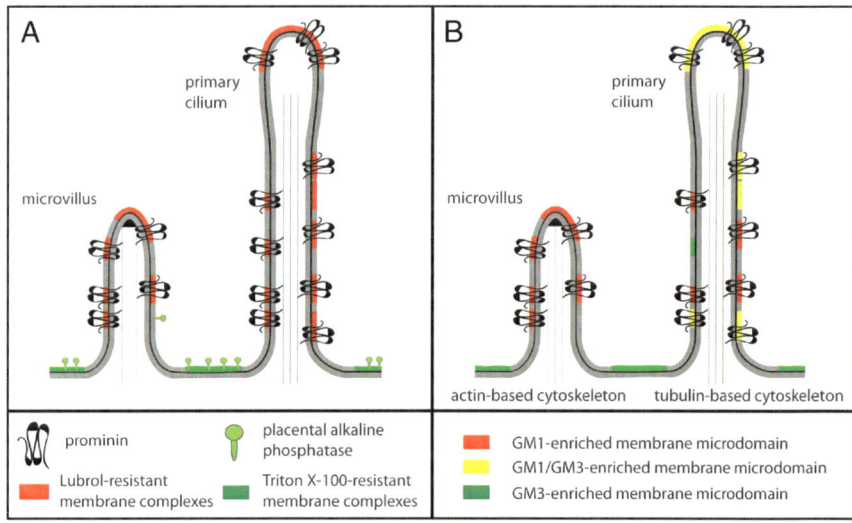

Fig. 14.3 Distinct subtypes of membrane microdomains as building units of the apical plasma membrane of polarized epithelial cells. (**A, B**) Illustration of two plasma membrane protrusions (microvillus and primary cilium) and the planar region of the apical domain with the membrane microdomain subtypes found therein. These are defined by either the segregation of particular membrane proteins (e.g. prominin-1 and placental alkaline phosphatase) that have differential detergent extraction properties (Lubrol WX versus Triton X-100) (**A**) or the segregation of two membrane microdomain-associated gangliosides, i.e. GM_1 and GM_3 (**B**). It remains to be determined whether GM_1 and GM_3 are located in the same membrane microdomain (*yellow*) or are part of two distinct (*red, green*) but close entities within the primary cilium

microdomains could maintain them in spatially distinct subdomains (protruding versus flat regions) of the plasma membrane.

Prominin-2 also binds to plasma membrane cholesterol, as demonstrated by photoaffinity labelling, and associates with a cholesterol-based membrane microdomain subtype similarly to prominin-1, suggesting that also for the lateral and basal domains of polarized epithelial cells the organization of plasma membrane protrusions may differ from the planar area (Florek et al., 2007). On a more general note, it is important to keep in mind that cholesterol-based membrane microdomain subtypes, which are preserved upon Lubrol WX, but not Triton X-100 extraction, are not restricted to epithelial cells (Röper et al., 2000; Slimane et al., 2003; Delaunay et al., 2007) but also found in non-epithelial cells (Chamberlain et al., 2001; Drobnik et al., 2002; Buechler et al., 2002; Kalus et al., 2002; Vetrivel et al., 2005; Won et al., 2008). Interestingly, the analysis of the corresponding detergent-resistant membrane fractions indicates that they differ considerably in the lipid ratio from those derived from Triton X-100 insoluble fractions (Drobnik et al., 2002). Moreover, certain membrane proteins are specifically incorporated into membrane microdomains that are preserved in Triton X-100, but not Lubrol WX, thus highlighting the complex organization of the plasma membrane (Drobnik et al., 2002; reviewed in Pike, 2004). It is of interest that the selective solubilization of different membrane fractions with Lubrol WX and Triton X-100 is not a unique characteristic of the plasma membrane, but may also occur in certain intracellular compartments (Moosic et al., 1982). A differential solubilization of inner plasma membrane leaflet components by these two detergents was also recently observed (Delaunay et al., 2008).

14.3.3 Distinct Ganglioside-Associated Membrane Microdomain Within the Apical Plasma Membrane

How many subtypes of membrane microdomains co-exist within the plasma membrane, and how can we define them in terms of specific lipids? Concerns about the ability of mild non-ionic detergents, e.g. Lubrol WX, to selectively solubilize membrane proteins and thus discriminate between those associated or not with membrane microdomains have been raised (Shogomori and Brown, 2003). Yet, our hypothesis that plasma membrane of protrusions contain distinct subtypes of cholesterol-based membrane microdomains compared to the planar portion of the plasma membrane was substantiated by the analysis of the co-distribution of prominin-1 with two membrane microdomain-associated gangliosides, GM_1 and GM_3, which are selectively enriched in distinct membrane microdomain subtypes in non-epithelial cells (Gómez-Móuton et al., 2001; Giebel et al., 2004; Fujita et al., 2009; reviewed in Bauer et al., 2008). Using confocal laser scanning microscopy of prominin-1–transfected MDCK cells we observed that GM_1 (as detected by cholera-toxin B subunit) co-localized with prominin-1 on microvilli (Fig. 14.3B, red) whereas GM_3 was segregated from there, suggesting its localization in the planar areas of the plasma membrane (Fig. 14.3B, green) (Janich and Corbeil, 2007). This is in agreement with the previous report of Chigorno and colleagues demonstrating by immunolabelling electron microscopy that GM_3 was found in the planar regions of the MDCK plasma membrane (Chigorno et al., 2000). Analysis of the primary cilium, where prominin-1 is also concentrated (Florek et al., 2007; Dubrcuil et al., 2007), revealed that both, GM_1 and GM_3, were detected there (Fig. 14.3B, yellow; Janich and Corbeil, 2007). Thus, GM_3 appears to be enriched in the primary cilium, but not in microvilli, whereas both membrane protrusions contain prominin-1 (Fig. 14.3B). These observations indicated that two plasma membrane protrusions based on different structural bases (actin for the microvillus and tubulin for the primary cilium) appear to be composed of distinct subtypes of membrane microdomains.

The co-localization of prominin-1 and GM_1 in microvilli and the primary cilium raises the possibility that these molecules may physically interact. Prominin's structure with its two large glycosylated extracellular loops (Fig. 14.1A) may well reflect its preference for plasma membrane protrusions, which exhibit positive membrane curvature (Huttner and Zimmerberg, 2001). Similar considerations may apply to the conically shaped glycolipid GM_1, which may also have a preference for positive-curvature plasma membrane protrusions (Iglic et al., 2006). In this context, it may be more than a coincidence that the first extracellular domain of prominin-1 contains a potential ganglioside-binding motif (Taïeb et al., 2009).

14.3.4 How Are Prominins Incorporated into the Protrusion-Specific Subtype of Membrane Microdomains? – Facts and Hypotheses

How are prominins incorporated into the protrusion-specific cholesterol-based membrane microdomains? At least three molecular mechanisms can be proposed. First, the covalent attachment of lipophilic moieties to proteins, such as palmitoylation, frequently promotes the interaction of proteins with membrane microdomains (Resh, 2004; Charollais and Van Der Goot, 2009). This lipid modification requires no strict consensus sequence other than the presence of cysteine residues and often occurs at cytoplasmic cysteine residues arranged in a cluster (McHaffie et al., 2007). Notably, both prominin molecules contain a conserved cluster of cysteine residues located at the boundary of the first transmembrane segment and the beginning of the first intracytoplasmic loop. Our preliminary observations indicate that prominin-1 can be labelled with [^{3}H]palmitic acid, suggesting that it undergoes palmitoylation (Katja Röper, Peggy Janich, D.C., W.B.H., unpublished data). It will be important to investigate whether mutations of these cysteine residues interfere with prominin's membrane microdomain association and it's correct subcellular localization.

Second, the length of transmembrane segments was shown to play a role in their retention in, versus their exit from, the Golgi complex (Munro, 1995a, b). Thus, in a similar way, a certain length of prominin's transmembrane segments may be required for its lateral diffusion into, and concentration in, the protrusion-specific membrane microdomains. In this context, it is of note that three of the five transmembrane segments of prominin-1 and prominin-2 contain over 25 amino acid residues and thus exceed the average length of transmembrane segments of plasma membrane proteins (Munro, 1995b). This may be significant for prominin's interaction with the surrounding membrane lipids, notably cholesterol - an interaction partner of prominins, as well as prominin's preference for curved rather than planar membrane domains.

Third, prominin's glycan moieties may play a role in membrane microdomain incorporation. Self-aggregation by homophilic interaction may cluster prominins and thereby form and/or stabilize a microdomain within the microvillar plasma membrane. A growing number of studies have presented evidence that some mammalian lectins such as galectin-4 and interlectin may also play a role in cross-linking lipid and protein glycoconjugates and consequently contribute to formation of stable membrane microdomains (Braccia et al., 2003; Wrackmeyer et al., 2006; Stechly et al., 2009). In this context, we have observed that the addition of the monoclonal antibody 13A4, which is directed against the extracellular domain of prominin-1, to prominin-1–transfected MDCK cells causes an increase in the amount of Lubrol-resistant membrane complexes observed upon differential centrifugation (Röper et al., 2000), consistent with a clustering of cell surface membrane microdomains containing prominin-1.

14.4 Dynamics of Apical Plasma Membrane Protrusions

An important aspect of the apical plasma membrane is its dynamics. Here, studies on the trafficking of prominin-1 have revealed novel forms of extracellular membrane traffic.

14.4.1 Prominin-Containing Extracellular Membrane Vesicles

The intracellular traffic of cytoplasmic vesicles has been comprehensively investigated for the past four decades, however, less is known about membrane particles released from cells into the extracellular milieu. The most commonly studied extracellular membrane vesicles are exosomes, the internal 50–90-nm vesicles of multivesicular bodies that are released into the extracellular space by exocytosis (*reviewed in* Lakkaraju and Rodriguez-Boulan, 2008). Many recent reports have revealed an increasing diversity of extracellular membrane particles, not only with regard to the kind but also their origin (*reviewed in* Doeuvre et al., 2009). Thus, extracellular membrane particles are not only derived from multivesicular bodies, but can also be released directly from the plasma membrane (Tanaka et al., 2005; Marzesco et al., 2005; Bachy et al., 2008), a route previously known to be used by enveloped viruses (Morita and Sundquist, 2004; Nayak et al., 2004). We have identified two novel classes of extracellular membrane particles that were initially observed in the ventricular fluid of the embryonic mouse brain, i.e. relatively large (0.5–1 μm) electron-dense particles and small (50–100 nm) electron-translucent vesicles (Marzesco et al., 2005). The large particles have been shown to derive from the apical midbody of dividing neuroepithelial cells (Dubreuil et al., 2007) and are not the topic of this review. The small membrane vesicles appear to have a widespread distribution, being found not only in the embryonic (Marzesco et al., 2005) and adult (Huttner et al., 2008) cerebrospinal fluid, but also in various external body fluids including saliva, seminal fluid and urine (Marzesco et al., 2005; Florek et al., 2007). Prominin-1 has been the marker in the characterization of these small extracellular membrane vesicles.

The sites of origin of the small extracellular prominin-1–containing vesicles appear to be microvilli and primary cilia – where the formation of budding-structures containing prominin-1 at their tips was observed (Marzesco et al., 2005; Florek et al., 2007; Marzesco et al., 2009; Dubreuil et al., 2007). Prominin-2 is also released into these body fluids in association with small membrane vesicles (Florek et al., 2007; Jászai et al., 2008).

Little is known about the molecular mechanism underlying the release of these membrane vesicles (*see* below). However, obtaining insight in this regard is not only of importance for basic cell biology, but may also be relevant for developmental biology and medicine, as several lines of evidence suggest a link between the release of prominin-1-containing membrane vesicles and cell differentiation or dedifferentiation. Thus, the number of membrane vesicles increases in the ventricular fluid of the embryonic mouse brain during neurogenesis (Marzesco et al., 2005), and decreases in human patients with glioblastoma during the final phase of the disease (Huttner et al., 2008). Given that certain glycosylated forms and splice variants of prominin-1 appear to be characteristic of various stem and progenitor cells as well as certain cancer stem cells (Weigmann et al., 1997; Yin et al., 1997; Uchida et al., 2000; Lee et al., 2005; Corbeil et al., 2009), an intriguing possibility arises that the release of the prominin-1–containing vesicles may contribute to cell differentiation by reducing a stem and progenitor cell-characteristic membrane microdomains within the plasma membrane (*reviewed in* Fargeas et al., 2006; Bauer et al., 2008). Moreover, it cannot be excluded that these vesicles may play a role in intercellular communication as well (*reviewed in* Bauer et al., 2008).

In a cell culture model, i.e. Caco-2 cells, we previously demonstrated that the release of small prominin-1-containing membrane vesicles increases with their enterocytic differentiation (Marzesco et al., 2005), which occurs about seven days after the culture reaches confluency (Pinto et al., 1982; Louvard et al., 1992). Differentiated Caco-2 cells thus appear as a suitable in vitro system to investigate in greater details the mechanism underlying the release of prominin-1–containing membrane vesicles from the donor membranes, the microvilli.

14.4.2 Role of Membrane Microdomains in the Release of Small Extracellular Membrane Vesicles

In intracellular membrane traffic, the formation of cytoplasmic vesicles from the TGN and plasma membrane is controlled not only by the respective protein machinery, but also by lipids, notably membrane cholesterol (Wang et al., 2000; Rodal et al., 1999; Subtil et al., 1999; Thiele et al., 2000; *reviewed in* van Meer and Sprong, 2004). Prominin-1 becomes incorporated into cholesterol-dependent membrane microdomains (biochemically defined as Triton X-100–soluble but Lubrol WX–resistant membrane complexes; *see above*) concomitant with the formation of post-TGN membrane vesicles (Röper et al., 2000). The latter vesicles are directly targeted to the apical plasma membrane (Corbeil et al. 1999; for a review concerning epithelial trafficking *see* Rodriguez-Boulan et al., 2004). At the level of the plasma membrane, our recent investigations suggest that changes in membrane microdomain organization of microvilli influences the release of extracellular vesicles (Marzesco et al., 2009). Specifically, cholesterol reduction (performed at physiological temperature using lovastatin and mβCD; Marzesco et al., 2009), which was previously shown to reduce the size of the Lubrol WX-resistant membrane complexes containing prominin-1 (Röper et al., 2000), was found to significantly enhance the release of prominin-1-containing membrane vesicles from microvilli of differentiated Caco-2 cells (Marzesco et al., 2009). Similarly, the appearance of small membrane vesicles in the immediate vicinity of microvilli has previously been observed upon cholesterol depletion of enterocytes (Hansen et al., 2000). The morphological correlate of the increased vesicle release from microvilli upon cholesterol reduction was a transition in the structure of the microvilli from a tubular shape to a pearling state – where membrane constrictions along their entire length were observed (Fig. 14.4). These membrane constrictions were found at an equal distance from one another ($\approx$ 50–100 nm) that matched the size of the resulting membrane vesicles (Fig. 14.4). When a microvillus showed only a single membrane constriction, it was typically found near its tip, indicating that this was the site where pearling was initiated (Fig. 14.4; Marzesco et al., 2009).

Pearling of tubular cell membranes and lipid bilayer tubes has previously been demonstrated to reflect the balance of two competing parameters, i.e. tension and curvature (Bar-Ziv and Moses, 1994; Roux et al., 2005). In the case of tubular plasma membrane protrusions (e.g. microvilli), tension is exerted, at least in part, by the actin-based cytoskeleton, the depolymerization of which (and hence the lowering of tension) leads to pearling (Bar-Ziv et al., 1999; McConnell and Tyska, 2007; Nambiar et al., 2009). Conversely, the transition from a tubular shape to a pearling state upon cholesterol reduction reflects an increase in membrane curvature, as tubular membranes exhibit curvature only in two-dimensions whereas pearling membranes does so in three-dimensions. Our demonstration that a reduction in membrane cholesterol, which is known to affect curvature (Evans and Rawicz, 1990; Needham and Nunn, 1990; Chen and Rand, 1997; Huttner and Zimmerberg, 2001; Wang et al., 2007), causes pearling of microvilli is consistent with early reports on artificial lipid bilayers (Baumgart et al., 2003; Yanagisawa et al., 2008), extends these studies to physiologically occurring plasma membrane protrusions and implies that lowering cholesterol levels in these tubular membranes increases curvature.

The influence of membrane cholesterol levels on the incidence of pearling and on the extent of extracellular membrane vesicle release appears to be physiologically pertinent. The pearling of microvilli and the presence of small membrane vesicles between duodenal microvilli of chicks fed on a low cholesterol diet have previously

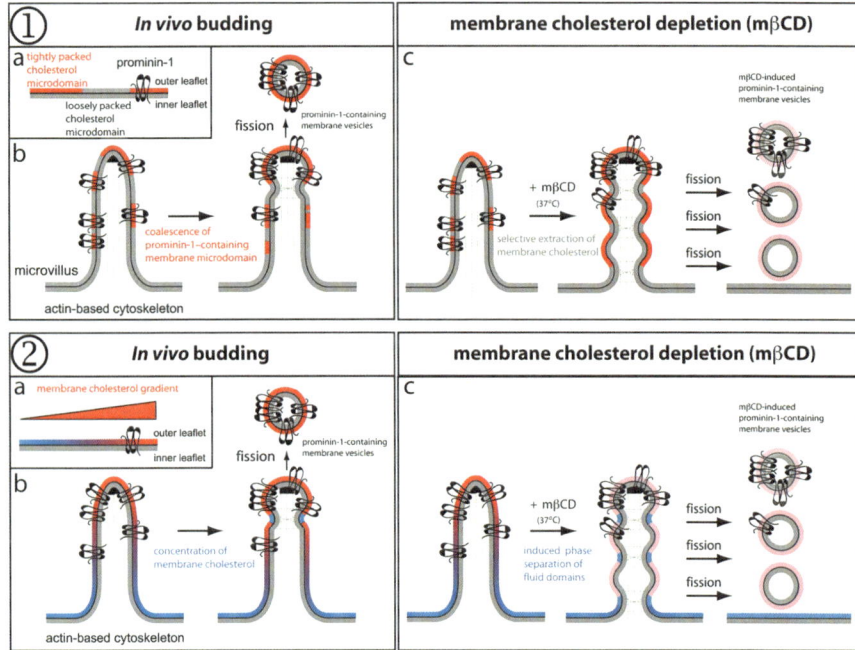

Fig. 14.4 Two hypothetical mechanisms underlying the release of prominin-1-containing membrane vesicles from microvilli. In model 1, small prominin-1-containing membrane microdomains (**a**, *red*) within the plasma membrane are converted into a larger one at the microvillar tip leading to phase separation (**b**) from the surrounding – more fluid – lipid microdomains where the membrane cholesterol is loosely packed (**a, b,** *grey*) and more sensitive to the depletion by mβCD (**c**). Prominin-1 within the released membrane vesicles displays the same biochemical properties (i.e. binding to cholesterol, differential solubility in distinct detergents) as the microvillus-associated one. In model 2, an increased concentration of membrane cholesterol (**a**, *red*) toward the microvillar tip together with the local curvature may lead to fission by promoting a restricted area of fluid phase separation (**b**, *blue*). This latter phenomenon may be induced by depleting membrane cholesterol (**c**). In both models, one cannot exclude that the cholesterol-binding protein prominin-1 itself plays a role either in the coalescence of particular membrane microdomain subtypes (model 1) or the concentration of membrane cholesterol (model 2). An active interplay between membrane constituents and membrane-associated actomyosin system should be considered as well. It remains to be determined whether mβCD-induced membrane vesicles (**c**, *pink*) have the same biochemical characteristics as those released under physiological conditions (**b**, *red*)

been reported (Hobbs, 1980), indicating that pearling is a physiological intermediate state in the release of membrane vesicles from microvilli. Furthermore, the differentiation of Caco-2 cells, which is associated with an increase in the release of prominin-1-containing membrane vesicles into the culture medium (Marzesco et al., 2005), is also accompanied by a decrease in membrane cholesterol levels (Jindrichova et al., 2003), and an induction of vesicle fission from artificial lipid bilayer tubes was observed upon cholesterol removal (Roux et al., 2005). Altogether, these observations are consistent with the hypothesis that the budding and fission of small membrane vesicles from microvilli, and perhaps also from the primary cilium, into extracellular fluids is controlled by the level of membrane cholesterol and the cholesterol-dependent organization of membrane microdomains (Fig. 14.4). Moreover, as proposed by Roux et al. (2005) for artificial lipid bilayer tubes, the microvillar pearling and increased membrane vesicle formation observed upon cholesterol removal may similarly reflect phase separation (Fig. 14.4). Two (alternative) models can be proposed. First, coalescence of small membrane microdomains containing the cholesterol-binding protein prominin-1 into a larger microdomain, particularly at the tip of microvilli, may induce a phase separation with regard to the surrounding fluid microenvironment – where the cholesterol is loosely packed –, leading to the fission of membrane vesicles (Fig. 14.4, model 1; Marzesco et al., 2009). In agreement with this model, the biochemical properties of prominin-1 within these vesicles are identical to that in microvilli, i.e. showing the same differential solubility/insolubility in Triton X-100 versus Lubrol WX and specific interaction with membrane cholesterol (Marzesco et al., 2009). Likewise, the pearling of microvilli induced upon interference with the microvillar actomyosin system (Tyska et al., 2005) may reflect phase separation resulting from the clustering of small membrane microdomains into larger ones. Second, the concentration of membrane cholesterol toward the microvillar tip together with the local curvature may lead to fission by promoting a restricted area of fluid phase separation (Fig. 14.4, model 2). Such concentration of membrane cholesterol has been previously demonstrated on filopodium-like processes (Möbius et al., 2002). Irrespective of the specific scenario, prominin molecules may constitute multivalent modules via their transmembrane segments and/or extracellular glycosylated loops, which could potentially change the membrane curvature and thus affect the architecture and dynamics of plasma membrane protrusions.

14.5 Perspectives

Our work on prominins has shown that these cholesterol-interacting pentaspan membrane proteins and the cholesterol-based membrane microdomains they are associated with exert a significant influence on the organization of plasma membrane protrusions. These findings set the stage for a more detailed dissection of the cell biology of prominin/membrane lipid assemblies. A functional analysis of their role in the biogenesis and maintenance of apical plasma membrane protrusions can now be undertaken using mouse models lacking prominins. Determination of the proteome and lipidome of prominin–containing membrane microdomains should reveal new players in the architecture of plasma membrane protrusions. Finally, prominin-1-containing membrane vesicles may offer a valuable tool for diagnostic purposes in clinical medicine, including cancer (Florek et al., 2005) and central nervous system disease (Huttner et al., 2008).

Acknowledgements We thank A. Hellwig for the excellent immunoelectron microscopy. D.C. was supported by the Deutsche Forschungsgemeinschaft (SFB/TR13-04 B1; SFB 655 A13/B3) and W.B.H. by the Deutsche Forschungsgemeinschaft (SPP 1111, Hu275/8-3; SFB/TR13-04 B1; SFB 655 A2).

References

Aaku-Saraste, E., Hellwig, A., and Huttner, W.B., 1996, Loss of occludin and functional tight junctions, but not ZO-1, during neural tube closure-remodeling of the neuroepithelium prior to neurogenesis. *Dev. Biol.* **180:** 664–679.

Aaku-Saraste, E., Oback, B., Hellwig, A., and Huttner, W.B., 1997, Neuroepithelial cells downregulate their plasma membrane polarity prior to neural tube closure and neurogenesis. *Mech. Dev.* **69:** 71–81.

Albert, A.D., and Boesze-Battaglia, K., 2005, The role of cholesterol in rod outer segment membranes. *Prog. Lipid. Res.* **44:** 99–124.

Bachy, I., Kozyraki, R., and Wassef, M., 2008, The particles of the embryonic cerebrospinal fluid: how could they influence brain development? *Brain Res. Bull.* **75:** 289–294.

Bar-Ziv, R., and Moses, E., 1994, Instability and "pearling" states produced in tubular membranes by competition of curvature and tension. *Phys. Rev. Lett.* **73:** 1392–1395.

Bar-Ziv, R., Tlusty, T., Moses, E., Safran, S.A. and Bershadsky, A., 1999, Pearling in cells: a clue to understanding cell shape. *Proc. Natl. Acad. Sci. USA* **96:** 10140–10145.

Bauer, N., Fonseca, A.-V., Florek, M., Freund, D., Jászai, J., Bornhäuser, M., Fargeas, C.A., and Corbeil D., 2008, New insights into the cell biology of hematopoietic progenitors by studying prominin-1 (CD133). *Cells Tissues Organs.* **188:** 127–138.

Baumgart, T., Hess, S.T. and Webb, W.W., 2003, Imaging coexisting fluid domains in biomembrane models coupling curvature and line tension. *Nature.* **425:** 821–824.

Berbari, N.F., O'Connor, A.K., Haycraft, C.J., Yoder, B.K., 2009, The primary cilium as a complex signaling center. *Curr. Biol.* 219: R526–R535.

Braccia, A., Villani, M., Immerdal, L., Niels-Christiansen, L.L., Nystrøm, B.T., Hansen, G.H., and Danielsen, E.M., 2003, Microvillar membrane microdomains exist at physiological temperature. Role of galectin-4 as lipid raft stabilizer revealed by "superrafts". *J. Biol. Chem.* **278:** 15679–15684.

Brown, D.A., and Rose, J.K., 1992, Sorting of GPI-anchored proteins to glycolipid-enriched membrane subdomains during transport to the apical cell surface. *Cell* **68:** 533–544.

Brown, D.A., and London, E., 2000, Structure and function of sphingolipid- and cholesterol-rich membrane rafts. *J. Biol. Chem.* **275:** 17221–17224.

Buechler, C., Boettcher, A., Bared, S.M., Probst, M.C., and Schmitz, G., 2002, The carboxyterminus of the ATP-binding cassette transporter A1 interacts with a beta2-syntrophin/utrophin complex. *Biochem. Biophys. Res. Commun.* **293:** 759–765.

Chamberlain, L.H., Burgoyne, R.D., and Gould, G.W., 2001, SNARE proteins are highly enriched in lipid rafts in PC12 cells: implications for the spatial control of exocytosis. *Proc. Natl. Acad. Sci. USA* **98:** 5619–5624.

Chamberlain, L.H., 2004, Detergents as tools for the purification and classification of lipid rafts. *FEBS Lett.* **559:** 1–5.

Charollais, J., and Van Der Goot, F.G., 2009, Palmitoylation of membrane proteins. *Mol. Membr. Biol.* **26:** 55–66.

Chen, Z. and Rand, R., 1997, The influence of cholesterol on phospholipid membrane curvature and bending elasticity. *Biophys. J.* **73:** 267–276.

Chigorno, V., Palestini, P., Sciannamblo, M., Dolo, V., Pavan, A., Tettamanti, G., and Sonnino, S., 2000, Evidence that ganglioside enriched domains are distinct from caveolae in MDCK II and human fibroblast cells in culture. *Eur. J. Biochem.* **267:** 4187–4197.

Chow, C.W., Woodside, M., Demaurex, N., Yu, F.H., Plant, P., Rotin, D., Grinstein, S., and Orlowski, J., 1999,Proline-rich motifs of the Na+/H+ exchanger 2 isoform. Binding of Src homology domain 3 and role in apical targeting in epithelia. *J. Biol. Chem.* **274:** 10481–10488.

Christian, A.E., Haynes, M.P., Phillips, M.C., and Rothblat, G.H., 1997, Use of cyclodextrins for manipulating cellular cholesterol content. *J. Lipid Res.* **38:** 2264–2272.

Corbeil, D., Gaudoux, F., Wainwright, S., Ingram, J., Kenny, A.J., Boileau, G., and Crine, P., 1992, Molecular cloning of the alpha-subunit of rat endopeptidase-24.18 (endopeptidase-2) and co-localization with endopeptidase-24.11 in rat kidney by in situ hybridization. *FEBS Lett.* **309:** 203–208.

Corbeil, D., Röper, K., Weigmann, A., and Huttner, WB., 1998, AC133 hematopoietic stem cell antigen: human homologue of mouse kidney prominin or distinct member of a novel protein family? *Blood.* **91:** 2625–2626.

Corbeil, D., Röper, K., Hannah, M.J., Hellwig, A., and Huttner, WB., 1999, Selective localization of the polytopic membrane protein prominin in microvilli of epithelial cells - a combination of apical sorting and retention in plasma membrane protrusions. *J. Cell Sci.* **112:** 1023–1033.

Corbeil, D., Röper, K., Hellwig, A., Tavian, M., Miraglia, S., Watt, S.M., Simmons, P.J., Peault, B., Buck, D.W., and Huttner, W.B., 2000, The human AC133 hematopoietic stem cell antigen is also expressed in epithelial cells and targeted to plasma membrane protrusions. *J. Biol. Chem.* **275:** 5512–5520.

Corbeil, D., Röper, K., Fargeas, C.A., Joester, A., and Huttner, W.B., 2001a, Prominin: a story of cholesterol, plasma membrane protrusions and human pathology. *Traffic.* **2:** 82–91.

Corbeil, D., Fargeas, C.A., and Huttner, W.B., 2001b, Rat prominin, like its mouse and human orthologues, is a pentaspan membrane glycoprotein. *Biochem. Biophys. Res. Commun.* **285:** 939–944.

Corbeil, D., Joester, A., Fargeas, C.A., Jászai, J., Garwood, J., Hellwig, A., Werner, H.B., and Huttner, W.B., 2009, Expression of distinct splice variants of the stem cell marker prominin-1 (CD133) in glial cells. *Glia.* **57:** 860–874.

Costa de Beauregard, M.A., Pringault, E., Robine, S., and Louvard, D., 1995, Suppression of villin expression by antisense RNA impairs brush border assembly in polarized epithelial intestinal cells. *EMBO. J.* **14:** 409–421.

Daniel, H., and Rubio-Aliaga, I., 2003, An update on renal peptide transporters. *Am J Physiol Renal Physiol.* **284:** F885–F892.

Danielsen, E.M., 1995, Involvement of detergent-insoluble complexes in the intracellular transport of intestinal brush border enzymes. *Biochemistry.* **34:** 1596–1605.

Danielsen, E.M., and van Deurs, B., 1995, A transferrin-like GPI-linked iron-binding protein in detergent-insoluble noncaveolar microdomains at the apical surface of fetal intestinal epithelial cells. *J. Cell Biol.* **131:** 939–950.

Danielsen, E.M., and Hansen, G.H., 2003, Lipid rafts in epithelial brush borders: atypical membrane microdomains with specialized functions. *Biochim. Biophys. Acta.* **1617:** 1–9.

Delaunay, J.L., Breton, M., Goding, J.W., Trugnan, G., and Maurice, M., 2007, Differential detergent resistance of the apical and basolateral NPPases: relationship with polarized targeting. *J. Cell Sci.* **120:** 1009–1016.

Delaunay, J.L., Breton, M., Trugnan, G., and Maurice, M., 2008, Differential solubilization of inner plasma membrane leaflet components by Lubrol WX and Triton X-100. *Biochim. Biophys. Acta.* **1778:** 105–112.

Doeuvre, L., Plawinski, L., Toti, F., and Anglés-Cano, E., 2009, Cell-derived microparticles: a new challenge in neuroscience. *J. Neurochem.* **110:** 457–468.

Drobnik, W., Borsukova, H., Böttcher, A., Pfeiffer, A., Liebisch, G., Schütz, G.J., Schindler, H., and Schmitz, G., 2002, Apo AI/ABCA1-dependent and HDL3-mediated lipid efflux from compositionally distinct cholesterol-based microdomains. *Traffic.* **3:** 268–278.

Dubreuil, V., Marzesco, A.-M., Corbeil, D., Huttner, W.B., and Wilsch-Bräuninger, M., 2007, Midbody and primary cilium of neural progenitors release extracellular membrane particles enriched in the stem cell marker prominin-1. *J. Cell Biol.* **176:** 483–495.

Evans, E. and Rawicz, W., 1990, Entropy-driven tension and bending elasticity in condensed-fluid membranes. *Physical. Review Letters.* **64:** 2094–2097.

Fargeas, C.A., Florek, M., Huttner, W.B., and Corbeil, D., 2003a, Characterization of prominin-2, a new member of the prominin family of pentaspan membrane glycoproteins. *J. Biol. Chem.* **278:** 8586–8596.

Fargeas, C.A., Corbeil, D., and Huttner, W.B., 2003b, AC133 antigen, CD133, prominin-1, prominin-2, etc.: prominin family gene products in need of a rational nomenclature. *Stem Cells.* **21:** 506–508.

Fargeas, C.A., Joester, A., Missol-Kolka, E., Hellwig, A., Huttner, W.B., and Corbeil, D., 2004, Identification of novel Prominin-1/CD133 splice variants with alternative C-termini and their expression in epididymis and testis. *J. Cell Sci.* **117:** 4301–4311.

Fargeas, C.A., Fonseca, A.-V., Huttner, W.B., and Corbeil, D., 2006, Prominin-1 (CD133) – from progenitor cells to human diseases. *Future Lipidol.* **1:** 213–225.

Fargeas, C.A., Huttner, W.B., and Corbeil, D., 2007, Nomenclature of prominin-1 (CD133) splice variants - an update. *Tissue Antig.* **69:** 602–606.

Fiedler, K., Kobayashi, T., Kurzchalia, T.V., and Simons, K., 1993, Glycosphingolipid-enriched, detergent-insoluble complexes in protein sorting in epithelial cells. *Biochemistry.* **32:** 6365–6373.

Florek, M., Haase, M., Marzesco, A.-M., Freund, D., Ehninger, G., Huttner, W.B., and Corbeil, D., 2005, Prominin-1/CD133, a neural and hematopoietic stem cell marker, is expressed in adult human differentiated cells and certain types of kidney cancer. *Cell Tissue Res.* **319:** 15–26.

Florek, M., Bauer, N., Janich, P., Wilsch-Bräuninger, M., Fargeas, C.A., Marzesco, A.-M., Ehninger, G., Thiele, C., Huttner ,W.B., and Corbeil, D., 2007, Prominin-2 is a cholesterol-binding protein associated with apical and basolateral plasmalemmal protrusions in polarized epithelial cells and released into urine. *Cell Tissue Res.* **328:** 31–47.

Föger, N., Marhaba, R., and Zöller, M., 2001, Involvement of CD44 in cytoskeleton rearrangement and raft reorganization in T cells. *J. Cell Sci.* **114:** 1169–1178.

Friederich, E., Huet, C., Arpin, M., and Louvard, D., 1989, Villin induces microvilli growth and actin redistribution in transfected fibroblasts. *Cell.* **59:** 461–475.

Fujita, A., Cheng, J., and Fujimoto, T., 2009, Segretation of GM1 and GM3 clusters in the cell membrane depends on the intact actin cytoskeleton. *Biochim. Biophys. Acta.* **1791:** 388–396.

Gerdes, J.M., Davis, E.E., and Katsanis, N., 2009, The vertebrate primary cilium in development, homeostasis, and disease. *Cell.* **137:** 32–45.

Giebel, B., Corbeil, D., Beckmann, J., Höhn, J., Freund, D., Giesen, K., Fischer, J., Kögler, G., and Wernet, P., 2004, Segregation of lipid raft markers including CD133 in polarized human hematopoietic stem and progenitor cells. *Blood.* **104:** 2332–2338.

Gómez-Móuton, C., Abad, J.L., Mira, E., Lacalle, R.A., Gallardo, E., Jiménez-Baranda, S., Illa, I., Bernad, A., Mañes, S., and Martínez-A. C., 2001, Segregation of leading-edge and uropod components into specific lipid rafts during T cell polarization. *Proc. Natl. Acad. Sci. USA* **98:** 9642–9647.

Götz, M., and Huttner, W.B., 2005, The cell biology of neurogenesis. *Nat. Rev. Mol. Cell Biol.* **6:** 777–788.

Hansen, G.H., Niels-Christiansen, L.L., Thorsen, E., Immerdal, L., and Danielsen, E.M., 2000, Cholesterol depletion of enterocytes. Effect on the Golgi complex and apical membrane trafficking. *J. Biol. Chem.* **275:** 5136–5142.

Hansen, G.H., Pedersen, J., Niels-Christiansen, L.L., Immerdal, L., and Danielsen, E.M., 2003, Deep-apical tubules: dynamic lipid-raft microdomains in the brush-border region of enterocytes. *Biochem. J.* **373:** 125–132.

Heerklotz, H., 2002, Triton promotes domain formation in lipid raft mixtures. *Biophys. J.* **83:** 2693–2701.

Hobbs, D.G., 1980, The origin and distribution of membrane-bound vesicles associated with the brush border of chick intestinal mucosa. *J. Anat.* **131:** 635–642.

Huttner, H.B., Janich, P., Köhrmann, M., Jászai, J., Siebzehnrubl, F., Blümcke, I., Suttorp, M., Gahr, M., Kuhnt, D., Nimsky, C., Krex, D., Schackert, G., Löwenbrück, K., Reichmann, H., Jüttler, E., Hacke, W., Schellinger, P.D., Schwab, S., Wilsch-Bräuninger, M., Marzesco,

A.-M., and Corbeil, D., 2008, The stem cell marker prominin-1/CD133 on membrane particles in human cerebrospinal fluid offers novel approaches for studying central nervous system disease. *Stem Cells.* **26:** 698–705.

Huttner, W.B., and Zimmerberg, J., 2001, Implications of lipid microdomains for membrane curvature, budding and fission. *Curr. Opin. Cell Biol.* **13:** 478–484.

Iglic, A., Hägerstrand, H., Veranic, P., Plemenitas, A., and Kralj-Iglic, V., 2006, Curvature-induced accumulation of anisotropic membrane components and raft formation in cylindrical membrane protrusions. *J. Theor. Biol.* **240:** 368–373.

Ikonen, E., and Simons, K., 1998, Protein and lipid sorting from the trans-Golgi network to the plasma membrane in polarized cells. *Semin. Cell Dev. Biol.* **9:** 503–509.

Ikonen E., 2001, Roles of lipid rafts in membrane transport. *Curr. Opin. Cell Biol.* **13:** 470–477.

Jacob, R., Zimmer, K.P., Naim, H., and Naim, H.Y., 1997, The apical sorting of lactase-phlorizin hydrolase implicates sorting sequences found in the mature domain. *Eur. J. Cell Biol.* **72:** 54–60.

Janich, P., and Corbeil, D., 2007, GM1 and GM3 gangliosides highlight distinct lipid microdomains within the apical domain of epithelial cells. *FEBS Lett.* **581:** 1783–1787.

Jászai, J., Janich, P., Farkas, L.M., Fargeas, C.A., Huttner, W.B., and Corbeil, D., 2007a, Differential expression of Prominin-1 (CD133) and Prominin-2 in major cephalic exocrine glands of adult mice. *Histochem. Cell Biol.* **128:** 409–419.

Jászai, J., Fargeas, C.A., Florek, M., Huttner, W.B., and Corbeil, D., 2007b, Focus on molecules: prominin-1 (CD133). *Exp. Eye Res.* **85:** 585–586.

Jászai, J., Fargeas, C.A., Haase, M., Farkas, L.M., Huttner, W.B., and Corbeil, D., 2008, Robust expression of Prominin-2 all along the adult male reproductive system and urinary bladder. *Histochem. Cell Biol.* **130:** 749–759.

Jindrichova, S., Novakova, O., Bryndova, J., Tvrzicka, E., Lisa, V., Novak, F. and Pacha, J., 2003, Corticosteroid effect on Caco-2 cell lipids depends on cell differentiation. *J. Steroid. Biochem. Mol. Biol.* **87:** 157–165.

Kalus, I., Hodel, A., Koch, A., Kleene, R., Edwardson, J.M., and Schrader, M., 2002, Interaction of syncollin with GP-2, the major membrane protein of pancreatic zymogen granules, and association with lipid microdomains. *Biochem. J.* **362:** 433–442.

Karbanová, J., Missol-Kolka, E., Fonseca, A.-V., Lorra, C., Janich, P., Hollerová, H., Jászai, J., Ehrmann, J., Kolár, Z., Liebers, C., Arl, S., Subrtová, D., Freund, D., Mokry, J., Huttner, W.B., and Corbeil, D., 2008, The stem cell marker CD133 (Prominin-1) is expressed in various human glandular epithelia. *J. Histochem. Cytochem.* **56:** 977–993.

Klein, U., Gimpl, G. and Fahrenholz, F., 1995, Alteration of the myometrial plasma membrane cholesterol content with beta cyclodextrin modulates the binding affinity of the oxytocin receptor. *Biochemistry* **34:** 13784–13793.

Kosodo, Y., Röper, K., Haubensak, W., Marzesco, A.-M., Corbeil, D., and Huttner, W.B., 2004, Asymmetric distribution of the apical plasma membrane during neurogenic divisions of mammalian neuroepithelial cells. *EMBO J.* **23:** 2314–2324.

Lakkaraju, A., and Rodriguez-Boulan, R., 2008, Itinerant exosomes: emerging roles in cell and tissue polarity. *Trends. Cell Biol.* **18:** 199–209.

Lardon, J., Corbeil, D., Huttner, W.B., Ling, Z., and Bouwens, L., 2008, Stem cell marker prominin-1/AC133 is expressed in duct cells of the adult human pancreas. *Pancreas.* **36:** e1–e6.

Lee, A., Kessler, J.D., Read, T.A., Kaiser, C., Corbeil, D., Huttner, W.B., Johnson, J.E., and Wechsler-Reya, R.J., 2005, Isolation of neural stem cells from the postnatal cerebellum. *Nat Neurosci.* **8:** 723–729.

Lingwood, D., and Simons, K., 2007, Detergent resistance as a tool in membrane research. *Nat. Protoc.* **2:** 2159–2165.

Louvard, D., Kedinger, M. and Hauri, H.P., 1992, The differentiating intestinal epithelial cell: Establishment and maintenance of functions through interactions between cellular structures. *Annu. Rev. Cell Biol.* **8:** 157–195.

Mangeat, P., Roy, C., and Martin, M., 1999, ERM proteins in cell adhesion and membrane dynamics. *Trends Cell Biol.* **9:** 187–192.

Marzesco, A.-M, Janich, P., Wilsch-Bräuninger, M., Dubreuil, V., Langenfeld, K., Corbeil, D., and Huttner, W.B., 2005, Release of extracellular membrane particles carrying the stem cell marker prominin-1 (CD133) from neural progenitors and other epithelial cells. *J. Cell Sci.* **118:** 2849–2858.

Marzesco, A.-M., Wilsch-Bräuninger, M., Dubreuil, V., Janich, P., Langenfeld, K., Thiele, C., Huttner, W.B., and Corbeil, D., 2009, Release of extracellular membrane vesicles from microvilli of epithelial cells is enhanced by depleting membrane cholesterol. *FEBS Lett.* **583:** 897–902.

Maw, M.A., Corbeil, D., Koch, J., Hellwig, A., Wilson-Wheeler, J.C., Bridges, R.J., Kumaramanickavel, G., John, S., Nancarrow, D., Röper, K., Weigmann, A., Huttner, W.B., and Denton, M.J., 2000, A frameshift mutation in prominin (mouse)-like 1 causes human retinal degeneration. *Hum. Mol. Genet.* **9:** 27–34.

McConnell, R.E., and Tyska, M.J., 2007, Myosin-1a powers the sliding of apical membrane along microvillar actin bundles. *J. Cell Biol.* **177:** 671–681.

McHaffie, G.S., Graham, C., Kohl, B., Strunck-Warnecke, U., and Werner, A., 2007, The role of an intracellular cysteine stretch in the sorting of the type II Na/phosphate cotransporter. *Biochim. Biophys. Acta.* **1768:** 2099–2106.

Mintzer, E.A., Waarts, B.L., Wilschut, J., and Bittman, R., 2002, Behavior of a photoactivatable analog of cholesterol, 6-photocholesterol, in model membranes. *FEBS Lett.* **510:** 181–184.

Miraglia, S., Godfrey, W., Yin, A.H., Atkins, K., Warnke, R., Holden, J.T., Bray, R.A., Waller, E.K., and Buck, D.W., 1997, A novel five-transmembrane hematopoietic stem cell antigen: isolation, characterization, and molecular cloning. *Blood.* **90:** 5013–5021.

Möbius, W., Ohno-Iwashita, Y., van Donselaar, E.G., Oorschot, V.M., Shimada, Y., Fujimoto, T., Heijnen, H.F., Geuze, H.J., and Slot, J.W., 2002, Immunoelectron microscopic localization of cholesterol using biotinylated and non-cytolytic perfringolysin O. *J. Histochem. Cytochem.* **50:** 43–55.

Moosic, J.P., Sung, E., Nilson, A., Jones, P.P., and McKean, D.J., 1982, The selective solubilization of different murine splenocyte membrane fractions with lubrol WX and triton X-100 distinguishes two forms of Ia antigens. A cell surface (alpha, beta) and an intracellular (alpha, Ii, beta). *J. Biol. Chem.* **257:** 9684–9691.

Morita, E., and Sundquist, W.I., 2004, Retrovirus budding. *Annu. Rev. Cell Dev. Biol.* **20:** 395–425.

Munro, S., 1995a, An investigation of the role of transmembrane domains in Golgi protein retention. *EMBO. J.* **14:** 4695–4704.

Munro, S., 1995b, A comparison of the transmembrane domains of Golgi and plasma membrane proteins. *Biochem. Soc. Trans.* **23:** 527–530.

Munro, S., 2003, Lipid rafts: elusive or illusive? *Cell.* **115:** 377–388.

Muth, T.R., Ahn, J., and Caplan, M.J., 1998, Identification of sorting determinants in the C-terminal cytoplasmic tails of the gamma-aminobutyric acid transporters GAT-2 and GAT-3. *J. Biol. Chem.* **273:** 25616–25627.

Nambiar, R., McConnell, R.E., and Tyska, M.J., 2009, Control of cell membrane tension by myosin-I. *Proc. Natl. Acad. Sci. USA* **106:** 11972–11977.

Nayak, D.P., Hui, E.K., and Barman, S., 2004, Assembly and budding of influenza virus. *Virus Res.* **106:** 147–165.

Needham, D. and Nunn, R.S., 1990, Elastic-deformation and failure of lipid bilayer-membranes containing cholesterol. *Biophys. J.* **58:** 997–1009.

Nielsen, J.S., Graves, M.L., Chelliah, S., Vogl, A.W., Roskelley, C.D., and McNagny, K.M., 2007, The CD34-related molecule podocalyxin is a potent inducer of microvillus formation. *PLoS One* **2:** e237.

Pazour, G.J., and Bloodgood, R.A., 2008, Targeting proteins to the ciliary membrane. *Curr. Top. Dev. Biol.* **85:** 115–149.

Pike, L.J., 2004, Lipid rafts: heterogeneity on the high seas. *Biochem. J.* **378:** 281–292.

Pike, L.J., 2006, Rafts defined: a report on the Keystone Symposium on Lipid Rafts and Cell Function. *J. Lipid. Res.* **47:** 1597–1598.

Pinto, M., Appay, M.-D., Simon-Assmann, P., Chevalier, G., Dracopoli, J., Fagh, J. and Zweibaum, A., 1982, Enterocyte differentiation of cultured human colon cancer cells by replacement of glucose by galactose in the medium. *Biol. Cell.* **44:** 193–196.

Rajendran, L., and Simons, K., 2005, Lipid rafts and membrane dynamics. *J. Cell Sci.* **118:** 1099–1102.

Reinshagen, K., Naim, H., Heusipp, G., and Zimmer, K.P., 2006, Pathophysiology in Microvillus inclusion disease. *Z. Gastroenterol.* **44:** 667–671.

Resh, M.D., 2004, Membrane targeting of lipid modified signal transduction proteins. *Subcell. Biochem.* **37:** 217–232.

Rodal, S.K., Skretting, G., Garred, O., Vilhardt, F., van Deurs, B. and Sandvig, K., 1999, Extraction of cholesterol with methyl-β-cyclodextrin perturbs formation of clathrin-coated endocytic vesicles. *Mol. Biol. Cell.* **10:** 961–974.

Rodriguez-Boulan, E., Müsch, A., and Le Bivic, A., 2004, Epithelial trafficking: new routes to familiar places. *Curr. Opin. Cell Biol.* **16:** 436–442.

Rollason, R., Korolchuk, V., Hamilton, C., Jepson, M., and Banting, G., 2009, A CD317/tetherin-RICH2 complex plays a critical role in the organization of the subapical actin cytoskeleton in polarized epithelial cells. *J. Cell Biol.* **184:** 721–736.

Röper, K., Corbeil, D., and Huttner, W.B., 2000, Retention of prominin in microvilli reveals distinct cholesterol-based lipid micro-domains in the apical plasma membrane. *Nat. Cell Biol.* **2:** 582–592.

Rotin, D., Bar-Sagi, D., O'Brodovich, H., Merilainen, J., Lehto, V.P., Canessa, C.M., Rossier, B.C., and Downey, G.P., 1994, An SH3 binding region in the epithelial Na+ channel (alpha rENaC) mediates its localization at the apical membrane. *EMBO. J.* **13:** 4440–4450.

Roux, A., Cuvelier, D., Nassoy, P., Prost, J., Bassereau, P. and Goud, B., 2005, Role of curvature and phase transition in lipid sorting and fission of membrane tubules. *EMBO. J.* **24:** 1537–1545.

Seidler, U., Singh, A.K., Cinar, A., Chen, M., Hillesheim, J., Hogema, B., and Riederer, B., 2009, The role of the NHERF family of PDZ scaffolding proteins in the regulation of salt and water transport. *Ann. N.Y. Acad. Sci.* **1165:** 249–260.

Semenza, G., 1986, Anchoring and biosynthesis of stalked brush border membrane proteins: glycosidases and peptidases of enterocytes and renal tubuli. *Annu. Rev. Cell Biol.* **2:** 255–313.

Sharma, P., Sabharanjak, S., and Mayor, S., 2002, Endocytosis of lipid rafts: an identity crisis. *Semin. Cell Dev. Biol.* **13:** 205–214.

Shogomori, H., and Brown, D.A., 2003, Use of detergents to study membrane rafts: the good, the bad, and the ugly. *Biol. Chem.* **384:** 1259–1263.

Simons, K., and Ikonen, E., 1997, Functional rafts in cell membranes. *Nature.* **387:** 569–572.

Simons, K., and Toomre, D., 2000, Lipid rafts and signal transduction. *Nat. Rev. Mol. Cell Biol.* **1:** 31–39.

Simons, K., and Ehehalt, R., 2002, Cholesterol, lipid rafts, and disease. *J. Clin. Invest.* **110:** 597–603.

Slimane, T.A., Trugnan, G., Van IJzendoorn, S.C., and Hoekstra, D., 2003, Raft-mediated trafficking of apical resident proteins occurs in both direct and transcytotic pathways in polarized hepatic cells: role of distinct lipid microdomains. *Mol. Biol. Cell.* **14:** 611–624.

Stechly, L., Morelle, W., Dessein, A.F., André, S., Grard, G., Trinel, D., Dejonghe, M.J., Leteurtre, E., Drobecq, H., Trugnan, G., Gabius, H.J., and Huet, G., 2009, Galectin-4-regulated delivery of glycoproteins to the brush border membrane of enterocyte-like cells. *Traffic.* **10:** 438–450.

Steinberg, R.H., Fisher, S.K., and Anderson, D.H., 1980, Disc morphogenesis in vertebrate photoreceptors. *J. Comp. Neurol.* **190:** 501–508.

Subtil, A., I., G., Kobylarz, K., Lampson, M.A., Keen, J.H. and McGraw, T.E., 1999, Acute cholesterol depletion inhibits clathrin-coated pit budding. *Proc. Natl. Acad. Sci. USA* **96:** 6775–6780.

Taïeb, N., Maresca, M., Guo, X.J., Garmy, N., Fantini, J., and Yahi, N., 2009, The first extracellular domain of the tumour stem cell marker CD133 contains an antigenic ganglioside-binding motif. *Cancer Lett.* **278:** 164–173.

Tanaka, Y., Okada, Y., and Hirokawa, N., 2005, FGF-induced vesicular release of Sonic hedgehog and retinoic acid in leftward nodal flow is critical for left-right determination. *Nature.* **435:** 172–177.

Thiele, C., Hannah, M.J., Fahrenholz, F. and Huttner, W.B., 2000, Cholesterol binds to synaptophysin and is required for biogenesis of synaptic vesicles. *Nat. Cell Biol.* **2:** 42–49.

Tyska, M.J., Mackey, A.T., Huang, J.D., Copeland, N.G., Jenkins, N.A., and Mooseker, M.S., 2005, Myosin-1a is critical for normal brush border structure and composition. *Mol. Biol. Cell.* **16:** 2443–2457.

Uchida, N., Buck, D.W., He, D., Reitsma, M.J., Masek, M., Phan, T.V., Tsukamoto, A.S., Gage, F.H., and Weissman, I.L., 2000, Direct isolation of human central nervous system stem cells. *Proc. Natl. Acad. Sci. USA.* **97:** 14720–14725.

van Meer, G., and Sprong, H., 2004, Membrane lipids and vesicular traffic. *Curr. Opin. Cell Biol.* **16:** 373–378.

Vetrivel, K.S., Cheng, H., Kim, S.H., Chen, Y., Barnes, N.Y., Parent, A.T., Sisodia, S.S., and Thinakaran, G., 2005, Spatial segregation of gamma-secretase and substrates in distinct membrane domains. *J. Biol. Chem.* **280:** 25892–25900.

Viola, A., and Gupta, N., 2007, Tether and trap: regulation of membrane-raft dynamics by actin-binding proteins. *Nat. Rev. Immunol.* **7:** 889–896.

Wang, W., Yang, L. and Huang, H.W., 2007, Evidence of cholesterol accumulated in high curvature regions: implication to the curvature elastic energy for lipid mixtures. *Biophys. J.* **92:** 2819–2830.

Wang, Y., Thiele, C. and Huttner, W.B., 2000, Cholesterol is required for the formation of regulated and constitutive secretory vesicles from the trans-Golgi network. *Traffic.* **1:** 952–962.

Waugh, M.G., and Hsuan, J.J., 2009, Preparation of membrane rafts. *Methods Mol. Biol.* **462:** 403–414.

Weigmann, A., Corbeil, D., Hellwig, A., and Huttner, W.B., 1997, Prominin, a novel microvilli-specific polytopic membrane protein of the apical surface of epithelial cells, is targeted to plasmalemmal protrusions of non-epithelial cells. *Proc. Natl. Acad. Sci. USA* **94:** 12425–12430.

Won, J.S., Im, Y.B., Khan, M., Contreras, M., Singh, A.K., and Singh, I., 2008, Lovastatin inhibits amyloid precursor protein (APP) beta-cleavage through reduction of APP distribution in Lubrol WX extractable low density lipid rafts. *J. Neurochem.* **105:** 1536–1549.

Wrackmeyer, U., Hansen, G.H., Seya, T., and Danielsen, E.M., 2006, Intelectin: a novel lipid raft-associated protein in the enterocyte brush border. *Biochemistry.* **45:** 9188–9197.

Yanagisawa, M., Imai, M. and Taniguchi, T., 2008, Shape deformation of ternary vesicles coupled with phase separation. *Phys. Rev. Lett.* **100:** 148102.1–148102.4.

Yang, Z., Chen, Y., Lillo, C., Chien, J., Yu, Z., Michaelides, M., Klein, M., Howes, K.A., Li, Y., Kaminoh, Y., Chen, H., Zhao, C., Chen, Y., Al-Sheikh, Y.T., Karan, G., Corbeil, D., Escher, P., Kamaya, S., Li, C., Johnson, S., Frederick, J.M., Zhao, Y., Wang, C., Cameron, D.J., Huttner, W.B., Schorderet, D.F., Munier, F.L., Moore, A.T., Birch, D.G., Baehr, W., Hunt, D.M., Williams, D.S., and Zhang, K., 2008, Mutant prominin 1 found in patients with macular degeneration disrupts photoreceptor disk morphogenesis in mice. *J. Clin. Invest.* **118:** 2908–2916.

Yin, A.H., Miraglia, S., Zanjani, E.D., Almeida-Porada, G., Ogawa, M., Leary, A.G., Olweus, J., Kearney, J., and Buck, D.W., 1997, AC133, a novel marker for human hematopoietic stem and progenitor cells. *Blood.* **90:** 5002–5012.

Yonemura, S., Tsukita, S., and Tsukita, S., 1999, Direct involvement of ezrin/radixin/moesin (ERM)-binding membrane proteins in the organization of microvilli in collaboration with activated ERM proteins. *J. Cell Biol.* **145:** 1497–1509.

Yu, Y., Flint, A., Dvorin, E.L., and Bischoff, J., 2002, AC133-2, a novel isoform of human AC133 stem cell antigen. *J. Biol. Chem.* **277:** 20711–20716.

Zacchigna, S., Oh, H., Wilsch-Bräuninger, M., Missol-Kolka, E., Jászai, J., Jansen, S., Tanimoto, N., Tonagel, F., Seeliger, M., Huttner, W.B., Corbeil, D., Dewerchin, M., Vinckier, S., Moons, L., and Carmeliet, P., 2009, Loss of the cholesterol-binding protein prominin-1/CD133 causes disk dysmorphogenesis and photoreceptor degeneration. *J. Neurosci.* **29:** 2297–2308.

Zhang, Q., Zulfiqar, F., Xiao, X., Riazuddin, S.A., Ahmad, Z., Caruso, R., MacDonald, I., Sieving, P., Riazuddin, S., and Hejtmancik, J.F., 2007, Severe retinitis pigmentosa mapped to 4p15 and associated with a novel mutation in the PROM1 gene. *Hum. Genet.* **122:** 293–299.

Zheng, X., Lu, D., and Sadler, J.E., 1999, Apical sorting of bovine enteropeptidase does not involve detergent-resistant association with sphingolipid-cholesterol rafts. *J. Biol. Chem.* **274:** 1596–1605.

Chapter 15
Mammalian StAR-Related Lipid Transfer (START) Domains with Specificity for Cholesterol: Structural Conservation and Mechanism of Reversible Binding

Pierre Lavigne, Rafael Najmanivich, and Jean-Guy LeHoux

Abstract The StAR-related lipid transfer (START) domain is an evolutionary conserved protein module of approximately 210 amino acids. There are 15 mammalian proteins that possess a START domain. Whereas the functions and specific ligands are being elucidated, 5 of them have already been shown to bind specifically cholesterol. The most intensively studied member of this subclass is the steroidogenic acute regulatory protein (StAR) or STARD1. While its role in steroid hormone production has been demonstrated, much less is understood about how its START domain specifically recognizes cholesterol and how it releases it to be transferred inside the mitochondria of steroidogenic cell of the gonads and adrenal cortex. A major obstacle that is slowing down progress in this area is the lack of knowledge of the 3D structures of the START domain of StAR in both its free and complexed forms. However, 3D models of the START domain of StAR and mechanisms of binding have been proposed. In addition biophysical studies aimed at validating the models and mechanism have been published. What's more, the crystal structures of the free forms of 3 START domains (STARD3, STARD4 and STARD5) known to specifically bind cholesterol have been elucidated so far. In this chapter, we will review and critically summarize existing data in order to provide the most current view and status of our understanding of the structure and reversible cholesterol binding mechanism of START domains.

Keywords StAR, START domains · Cholesterol · Steroidogenesis · LCAH

15.1 Introduction

The steroidogenic acute regulatory protein (StAR)-related lipid transfer (START) domain is a protein module of approximately 210 amino acids. This module

P. Lavigne (✉)
Département de Pharmacologie, Institut de Pharmacologie, Faculté de médecine et des sciences de la santé, Université de Sherbrooke, 3001 12e Avenue Nord, J1H 5N4, Sherbrooke, QC, Canada
e-mail: Pierre.lavigne@usherbrooke.ca

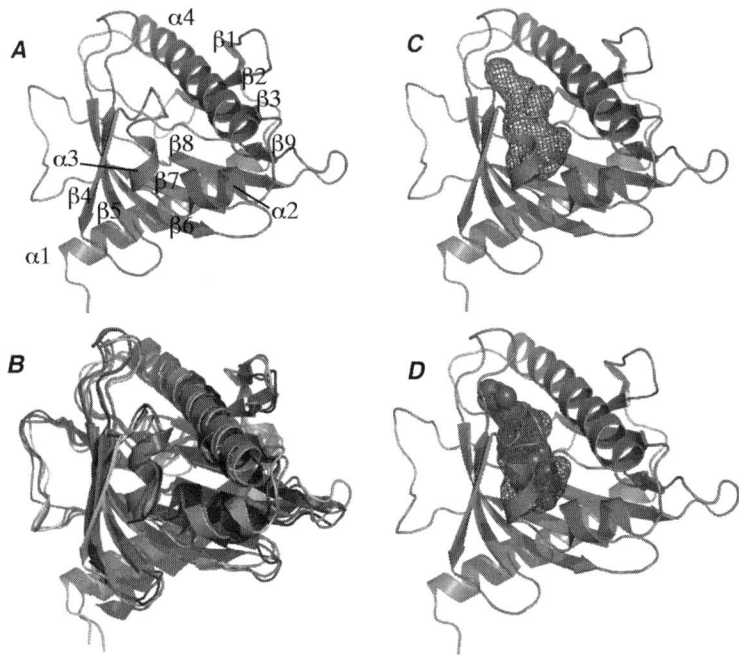

Fig. 15.1 A. The α/β helix grip fold. Note the presence of an N- and a C-Terminal helix gripping a twisted 9–10 stranded b-sheet. Depending on the algorithms used for secondary structure assignments, 9 or 10 β-strands can be identified. **B.** Superimposition of the backbone of the model of StAR (1IMG.pdb; Mathieu et al., 2002) and the crystal structure of MLN64 (1EM2.pdb, Tsujishita & Hurley, 2000), STARD4 (1JSS.pdb, Romonowski et al., 2002) and STARD5 (2R55.pdb, to be published). **C.** Molecular surface of the STARD1 internal cavity. **D.** One cholesterol molecule fits in the cavity. Figure made with Pymol (Delano, 2002)

has a α/β helix grip fold (Tsujishita and Hurley, 2000; Iyer et al., 2001 and Fig. 15.1A) and binds a wide variety of lipids, i.e. cholesterol, ceramides, Phosphatidylenthanolamine (PE), Phosphatidylcholine (PC) (Schrick et al., 2004). Though not present in yeast, START domains are conserved through evolution and are found in bacteria, plants, flies, nematodes and mammalians. In humans, 15 proteins possess a START domain (Alpy and Tomasetto, 2005). These proteins are involved in lipid metabolism, lipid transfer and cell signalling and have diverse expression patterns and cellular localizations. The diverse cellular functions and known ligands of the 15 mammalians START containing proteins were recently reviewed elsewhere (Alpy and Tomasetto, 2005). Here we will focus of the START domains that are known to bind cholesterol, i.e. StAR (STARD1), MLN-64 (STARD3), STARD4, STARD5 and STARD6 (Alpy and Tomasetto, 2005; Bose et al., 2008).

The discovery that StAR was responsible for the acute steroidogenesis has boosted much intense interest in an attempt to understand the molecular and structural biology of START domain containing proteins (Clark et al., 1994). In fact, StAR regulates the cholesterol mobilization to the mitochondria of steroidogenic cells (gonads and adrenal cortex), following a ACTH stimulus. This mobilization is the rate-limiting step of steroidogenesis and the production of pregnenolone by the p450 cholesterol side chain cleavage complex located in the inner mitochondrial membrane (Arakane et al., 1998; Bose et al., 2002). StAR has a mitochondrial target sequence N-Terminal to its START domain (Arakane et al., 1998; Wang et al., 1998). However, this signal sequence is dispensable for cholesterol transport across the mitochondrial membrane. Indeed, a truncated mutant of StAR lacking its first 62 residue (N62-StAR) retains full activity in the stimulation of pregnenolone production by steroidogeneic mitochondrial preparations (Bose et al. 2002). Mutations in the StAR gene cause lipoid congenital adrenal hyperplasia (LCAH), a severe autosomal recessive form of congenital adrenal hyperplasia (Bose et al., 1996; 2000). Most of these mutations are located at or near the C-terminal α-helix.

There are currently two active areas of research in the field of StAR. The first, which was recently reviewed in depth by Rone et al. (2009), focuses on the clarification of the elusive mechanism by which cholesterol enters the mitochondrial matrix. The other addresses the mechanism by which StAR reversibly binds and dissociates from cholesterol. This area will be updated and survey in detail in the present chapter.

MLN-64 is a member of the StAR group (Alpy and Tomasetto, 2005). Like StAR, MLN-64 possesses a sub-cellular localization domain that targets it to the late endosomes (Clark et al., 1994; Alpy et al., 2001) where it is thought to be involved in the mobilization of lysosomal cholesterol and its transfer to other organelles and membranes. Like StAR and in isolation, the START domain of MLN-64 can also stimulate steroidogenesis in steroidogenic mitochondrial preparations (Watari et al., 1997). In fact MLN-64 is thought to be involved in steroidogenesis in tissues that lack StAR expression such as the placenta (Watari et al., 1997). Conversely to StAR, the crystal structure of the START domain (without cholesterol) of MLN64 has been solved (Tsujishita and Hurley, 2000).

STARD4, STARD5 and STARD6 are members of the STARD4 group (Alpy and Tomasetto, 2005). They are the only START proteins with no subcellular localization domain and as such are thought to be cytosolic. Reports suggest that STARD4 and STARD5 are able to stimulate steroidogenesis. However, a recent study has shown that while STARD4, STARD5 and STARD6 can bind cholesterol, only STARD4 and STARD6 can stimulate pregnenolone production in vitro (Bose et al., 2008). By being able to diffuse through the cytoplasm, members of this family are though to be able to provide multiple sources (membranes, lipid droplets and organelles such as the ER, endosomes and lysosomes) with cholesterol for steroidogenesis (Rone et al., 2009).

Despite all of the data discussed here and reviewed extensively elsewhere (Rone et al., 2009), we still do not know the molecular determinants responsible for the recognition of cholesterol by these START domains and the mechanism by which the START domains bind and release cholesterol. Although mundane a first glance, the mechanism of reversible binding is complicated by the fact that the binding site of cholesterol is buried inside the START domain and hence necessitates an opening reaction of some sort. Consequently, in this Chapter, we will review with in some detail structural and functional data from the literature in order to present an up-to-date picture of the recent progress toward our understanding of the molecular recognition of cholesterol by START domains and how these protein modules reversibly bind cholesterol.

15.2 The START Domains That Specifically Bind Cholesterol Have a Highly Conserved α/β Helix Grip Fold

Our recent survey of the protein data bank has revealed that the crystal structures of 6 mammalian START domains have been solved so far. These START domains are those of STARD2 (Roderick et al., 2002), D3 (MLN64, Tsujishita and Hurley, 2000), D4 (Romanowski et al., 2002), D5 (2R55.pdb, to be published), D11 (Kudo et al., 2008) and D13 (2PSO.pdb, to be published). STARD3, D4 and D5 have specificity for cholesterol. STARD2 (PCPT) binds phosphatidylcholine (PC) and STARD11 (CERT) binds ceramides, (sphingolipid precursors). The lipid specificity of STAR13 (DCL-2) is not known as of yet.

Interestingly, apart from deletion and insertions in loops, the structures of all the START domains are very similar. Indeed, on can see in Fig. 15.1 that the α/β helix grip fold (Fig. 15.1A) of the START domains with affinity for cholesterol (Fig. 15.1B) and throughout the groups (not shown) is conserved. Another striking and common feature of these structures is the presence of an internal cavity large enough to fit one lipid molecule, i.e. cholesterol (Fig. 15.1C, D), ceramide (Kudo et al., 2008) or PC (Roderick et al., 2002).

While the structures of the START domains of STARD2 and STARD11 have been solved with their respective ligands, no structure of the START domain-cholesterol complex is currently available. Hence, the actual mode of binding and the determinants of the specificity of STARD1, 3, 4, 5 and 6 towards cholesterol remain to be understood and unravelled. In this regard, models for specific cholesterol complexes of MLN-64 (Murcia et al., 2006) and StAR (Mathieu et al., 2002; Yaworsky et al., 2005) have been proposed. In these models, the presence of conserved (between StAR and MLN-64) salt bridge between an acidic side-chain in β-strand 5 and an Arg in β-strand 6 at the bottom of the cavity was proposed to be a key determinant. More precisely, it was proposed that the cholesterol OH group forms a specific interaction with the guanidinium group of the conserved Arg (Fig. 15.2). On the other hand, it appears unlikely that molecular recognition and ligand selection rely only on one specific interaction (H-Bond). However, and as noted by others (Romanowski et al., 2002), it is quite possible that the actual shape of the cavity may play an important role in molecular recognition. To illustrate this, we present the molecular surface of the cavity of our 3D model of StAR with one cholesterol molecule located inside. As can be observed, the shape of the cavity matches almost perfectly that of the cholesterol molecule. Furthermore, by fitting the cholesterol molecule in this cavity, the OH group is perfectly positioned to interact with the guanidino group of the Arg side-chain of the salt-bridge (Fig. 15.2).

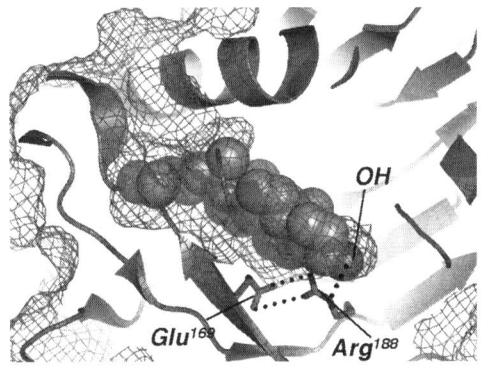

Fig. 15.2 Depiction of a putative specific interaction between cholesterol and a conserved salt-bridge in the model of StAR (1IMG.pdb, Mathieu et al., 2002). This salt-bridge is conserved in MLN64 and has been proposed to be involved in the specificity of interaction between its START domain and cholesterol (Murcia et al., 2006)

While surface complementarity is a hallmark of molecular recognition, it has to be emphasized that molecular recognition of small molecules by proteins usually involves more than one H-bond. In addition, it also has to be noted that the cholesterol binding site is totally buried inside the protein, which also a rather unusual feature. Hence, it is quite possible that the small number of H-bond may be balanced by perfect complementarity of non-polar and non-specific van der Waals interactions between cholesterol and the rest of the cavity. On the other hand, it appears as though these determinants are most likely different between the StAR (StAR, MLN64) and STARD4 (STARD4, 5 and 6) groups. Indeed, the salt bridge conserved in StAR and MLN-64 is not conserved in the STARTD4 group. However, another salt bridge Asp-Arg in helix C is present in the structures of STARD4 and STARD5. This salt bridge is conserved in the primary structure of STARD6 (not shown) and could be involved, as proposed for MLN-64 and StAR, in the recognition of cholesterol. As one can appreciate, while we have a fair amount of structural data on the structure of START domains that specifically bind cholesterol, we are still awaiting the validation of our models in order to understand exactly what makes a START domain recognize cholesterol. This will come only with the crystal or NMR structures of START-cholesterol complexes from the StAR and STARD4 groups.

15.3 "To Be or Not to Be" a Molten Globule to Bind and Release Cholesterol Reversibly?

As one can appreciate, besides having a highly conserved fold, one peculiar feature of the structure of the START domains that bind cholesterol is the fact that their binding sites are shielded from the solvent and in the very core of the protein modules. Even though we are still trying to decipher the determinants of molecular recognition, this fact has prompted and stimulated many laboratories to search for the mechanism by which cholesterol and other lipids can reach the interior of

START domains (the binding site) in order to be transported to and then transferred inside the mitochondria.

In the late 1990's, the group of Walter Miller discovered that the N62-START domain behaved like a molten globule at acidic pH (from pH 4.5 to 3). Indeed, while the amount of secondary structure did not significantly changes from pH 4.5 to 3, the thermodynamic stability at 25°C (i.e. $\Delta G°_u(25°)$) of the N62-START decreased significantly (Bose et al., 1999). Note that the stability of N62-StAR as measured by urea denaturation was observed to be constant from pH 8.3 to 4.5. This demonstrated that the while the secondary structure was maintained, the amount of stable and tertiary interaction was diminished at lower pH values. This is in complete agreement with the existence of a molten globule, i.e. a protein with native like secondary structure but loosely a packed native tertiary structure (Ptitsyn, 1995). In the same study, the Miller group found by proteolysis-MS analyses that the C-terminal region (193–285) was less folded at acidic pH than the N-terminus (63–188). Coupled to their hypothesis that the pH near the mitochondria OMM is acidic (~4.5), the Miller group put forth that a low pH induced molten globular state of StAR with the N-terminus more tightly folded than the C-terminus plays an important role in the cholesterol binding and transfer.

In 2002, we proposed an alternative mechanism for the reversible binding of cholesterol by N62-StAR (Mathieu et al., 2002). This mechanism is proposed to occur at neutral pH and was derived largely from the 3D model we had proposed for the N62-StAR, which was based on the crystal structure of STARD3. As initially uncovered in the crystal structures of STARD3 (Tsujishita and Hurley, 2000) and STARD4 (Romanowski et al., 2002)) our model depicted a large internal cavity assigned to the cholesterol-binding site. In fact the volume of the cavity corresponded exactly to that of cholesterol (Mathieu et al., 2002 and Fig. 15.1C,D).

Based on the fact that such internal cavities destabilizes tertiary structures and that the C-Terminal helix could move independently from the rest of the molecule, we proposed the existence of an intermediate state with the N-terminus intact and the C-terminal helix undergoing a microscopic (independent of global unfolding) and reversible local unfolding (Mathieu et al., 2002; Roostaee et al., 2008; 2009). In this intermediate state (Fig. 15.3), the cholesterol-binding site would become accessible and explains cholesterol binding. In this mechanism, the C-terminal helix acts as a gate. Indeed, the refolding of the C-terminal helix when a cholesterol molecule is in the binding site it will lead to a more stable complex, with a lifetime long enough to carry and deliver cholesterol to its target organelle and/or transporter (e.g. TSPO (*see* Rone et al., 2009)).

15.4 Experimental Validation of the Two-State Model

We reasoned that if this mechanism is correct, the helical content of free N62-StAR should be less than optimal and that upon addition of cholesterol the helical content as well as the thermodynamic stability of N62-StAR should be increased at neutral

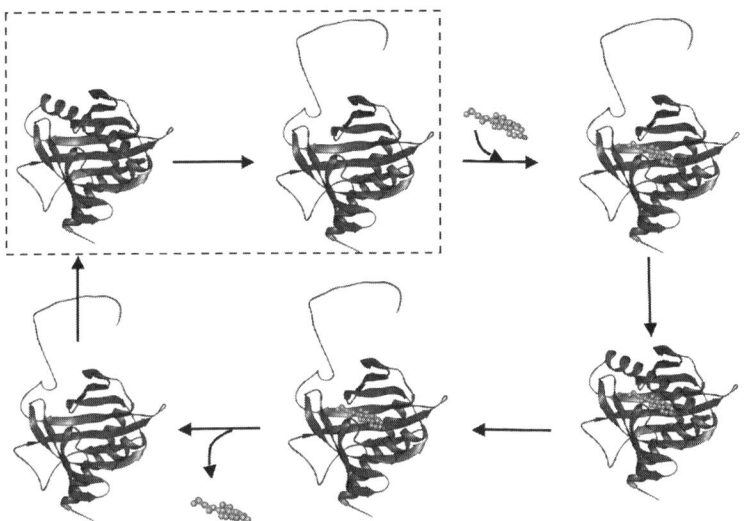

Fig. 15.3 The two-state model. Because of the presence of a cavity in the absence of cholesterol, the folded state of the closed and apo form of the START domain of StAR undergoes a local and reversible unfolding of its C-Terminal helix. *Boxed reaction.* The intermediate and partially unfolded state is proposed to initiate specific binding and allow for the dissociation of cholesterol

pH. As shown in Fig. 15.4, this is exactly what we have observed (Roostaee et al., 2008; 2009). Indeed, as one can see, the far-UV CD spectrum (Fig. 15.4A) of N62-StAR with an equimolar concentration of cholesterol depicts more negative molar ellipticities than the free construct (of course the contribution of cholesterol has been subtracted). This is indicative of an increase in secondary structure (α and/or β). While there are many computational routines to determine the percentage of secondary structure from CD spectra, these have sizable uncertainties. On the other hand and more reliably, if the addition of cholesterol is accompanied by a transition from a mostly random coil C-terminus to the stabilization of this region into an α-helix, an isosbestic point is expected at 203 nm. Random coil and α-helix have the same molar ellipticity at 203 nm. However, the isosbestic point between the β structure and random coil is close to 208 nm. Therefore, if cholesterol had induced a random coil to β transition, an isosbestic point at ~208 nm would have been observed. As shown in Fig. 15.4A, an isosbestic point at 203 nm observed, hence demonstrating that the addition of cholesterol stabilizes the C-terminus into an α-helix from an otherwise mostly random coil configuration. Moreover, as expected, the melting temperature $T°$ of the cholesterol-N62-StAR (1:1) complex is increase by almost $4°$ compared to the free construct (Fig. 15.4B). As described in detail elsewhere, the thermodynamic stability of the complex is also increased at all temperatures (Roostaee et al., 2008). Finally, by monitoring the increase in α-helical content at 222 nm, we have titrated N62-StAR with cholesterol and confirmed, like MLN-64 (Tsujishita and Hurley, 2000) that the START domain of StAR binds

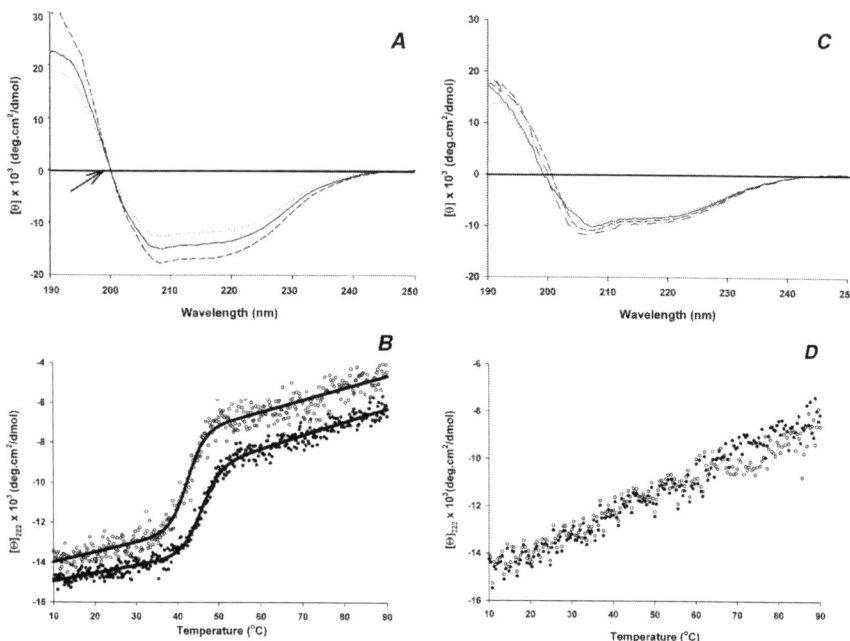

Fig. 15.4 The presence of stable tertiary structure in the START domain of StAR is a prerequisite for cholesterol binding. **A.** Far-UV CD spectra of the N62-StAR construct in absence (*solid* line), immediately (*dotted* line) and 20 minutes after addition of cholesterol at neutral pH (*dashed* line). Note the presence of an isosbestic point at 203 nm (*arrow*), which is indicative of the stabilization of α-helical structure from a random coil content upon addition of cholesterol and once equilibrium has been attained. **B.** Temperature-induced denaturation of the free (*open* circles) and 1:1 cholesterol complex (*solid* circles) of the N62-StAR construct at neutral pH. Note the cooperativity of the curves. This signifies that stable tertiary structure is present. **C.** Far-UV CD spectra of the N62-StAR construct in absence (*solid line*), immediately (*dotted line*), 20 min (*dashed line*) and an hour after addition of cholesterol at pH 3.5 (dashed and dotted line). **D.** Temperature-induced denaturation of the free N62-StAR construct (*open circles*) and a 1:1 N62-StAR:cholesterol mixture after 1 h of incubation (solid circles) at pH 3.5. Note that under these conditions, no cooperativity is observed, denoting the absence of stable tertiary structure even in presence of cholesterol

cholesterol in a 1:1 stoichiometry with apparent affinity of $\sim 3 \cdot 10^{-8}$ (Roostaee et al., 2008).

For the sake of comparison and in order to evaluate the ability of the molten globular state of N62-StAR to bind cholesterol, we have repeated the same experiment at pH 3.5. As can be seen on Fig. 15.4C, the far-UV CD spectra of N62-StAR depicts a sizable content of secondary structure at acidic pH. However, N62-STAR is devoid of stable tertiary structure and does not bind cholesterol. Indeed, as one can see, the temperature denaturation curves monitored by CD of N62-StAR does not show any cooperativity and the presence of cholesterol does not alter the curve (Fig. 15.4D).

This data indicates that the molten globule state cannot provide the minimal tertiary structure necessary to allow for the specific binding of cholesterol. Hence, while a putative pH induction of a molten globular state of StAR near the mitochondrial membrane could be an effective way to release cholesterol, however it is not sure how such a state could reversibly refold into a structure with a stable tertiary structure and undergo such a global transition cyclically. In fact, once acidified N62-STARs solutions irreversibly lose their capacity to bind cholesterol and induce steroidogenesis (Roostaee et al., 2008). Moreover, it is known that one StAR molecule can be responsible for the transfer of over 400 cholesterol molecules per minute (Artemenko et al., 2001). It appears to us that the local unfolding of the C-terminal is a much more efficient way to reversibly expose a well defined binding site to allow for a repetitive binding and release of many cholesterol molecule per START.

Furthermore, there are clear experimental evidences which show that restricting the movement of the C-terminal of N62-StAR hampers it's cholesterol-binding affinity and it's ability to induce pregnenolone production in vitro assays. By covalently attaching helix 4 to the loop between β-strands 1 and 2 with disulfide bridges S100C/S261C or D106C/A268C (Fig. 15.5), caused StAR to lose half or completely its binding and steroidogenic activity, respectively (Baker et al., 2005). However,

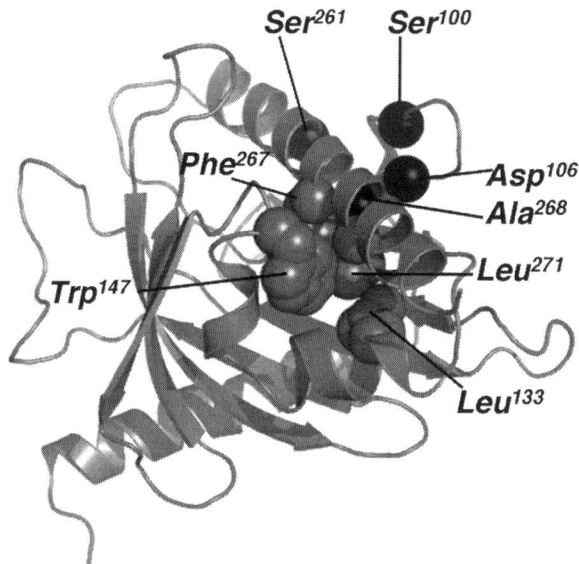

Fig. 15.5 A conserved hydrophobic core is present at the interface of helices 2, 3 and 4 (*grey* side chains). This hydrophobic cluster in present in the 3D model of StAR and the crystal structures of MLN64, STARD4 and STARD5 (not shown) and is proposed to stabilize the C-terminal helix in its folded state. Localization of the mutations made by Baker et al. (2005) in order to prevent movement of the C-terminal helix (*black spheres*)

adding a reducing agent restored the binding and activity. In addition, weakening hydrophobic and tertiary interactions at the interface of helices 2,3 and 4 (Fig. 15.5) in the fully folded state of N62-StAR in the bound and free forms, reduced the thermodynamic stability of both forms and abridged the steroidogenic activity of the mutants. This conserved hydrophobic cluster, involving Leu^{133}, Trp^{147}, Phe^{267} and Leu^{271}, is proposed to stabilize helix-4 in its closed form and provide the necessary stabilization free energy to generate a stable and functional complex (Roostaee et al., 200). While, mutating the conserved Phe^{267}, Leu^{271} to polar residues with similar respective volumes (i.e. Gln and Asn, respestively) promoted the opened form but prevented the formation of a stable and functional complex (Roostaee et al., 2008; 2009).

15.5 Towards a Consensual Model for the Reversible and Specific Binding of Cholesterol by START Domain

As discussed, the START domains with affinity for cholesterol, have or can be predicted to have an internal cavity with a volume equal to that of cholesterol. Hence, in absence of cholesterol, the tertiary α/β helix grip fold will be destabilized. Since, helix-4 is free to move from the rest of the molecule, this excess in free energy should promote its unfolding (Fig. 15.3). In other words, that the lack of stabilization free energy created by the absence of a ligand should be naturally compensated by an increase in conformational entropy following the unfolding of the C-terminal helix, and hence promote the population of an intermediate state. This local unfolding (and intermediate state) can be seen as serving two purposes: 1 – stabilization of an otherwise unstable state and 2 – unveiling of the buried binding site. We propose that this is the state that recognizes cholesterol and from which cholesterol will dissociate in order to be delivered to organelles by processes still not fully understood (Rone et al., 2009). As discussed here, the discovery of the specific determinants or interactions responsible for the molecular recognition await the resolution of the 3D structure(s) of START-cholesterol complex(es). Nonetheless, it is increasingly evident that the species of START domains responsible for the formation of an initial complex should possess a minimal tertiary structure content capable to present (at least in part) the required structural determinants to allow for the formation of a specific complex. As discussed by others (Tsujishita and Hurley, 2000) and shown experimentally, a molten globular state with non-defined (specific) tertiary structure is a somewhat unsatisfactory purported entity to carry out such a function.

Interestingly, many of the experimental results that originally led authors to propose or validate the molten globule model, agree with and support the two-state model. Namely, the fact that the C-terminal helix is the most susceptible region of N62-StAR and that restricting the movement of the C-terminal helix by the engineering of disulfide bridges prevent binding and impede on the steroidogenic activity of N62-StAR.

At this stage, we believe that ambiguity or divergence in the interpretation of the results available in the literature stems from the hypothesis of a role of low pH. While, it is true that N62-StAR behaves like a bona fide molten globule at low pH, the existence of pH values in the 3.5–4.5 range near the mitochondrial membrane still needs to be demonstrated and the ability of such a state to specifically recognize any ligand or protein is contrary to the established understanding of biomolecular recognition.

15.6 Conclusions and Perspectives

Structural data on START domains that bind cholesterol and biophysical analysis of START domain-cholesterol interactions reviewed in here point to a common mechanism of molecular recognition and reversible binding. Indeed, many lines of evidence suggest the occurrence of a local and microscopic unfolding event of the α/β helix grip fold that leads to the population of key intermediate state with the C-terminal helix unfolded (Fig. 15.3). It is this intermediate state that is proposed to initiate the specific binding and release the cholesterol to its end point. The formation and the stability of the fully folded complex is also consideredto be crucial. Indeed, the formation of an unstable complex would lead to a complex of short lifetime which would be unfit to deliver cholesterol to its final destination.

Finally, the definite validation of the two-state model awaits the characterization of the proposed movements or molecular unfolding events in the presence and absence of cholesterol. In this regards, we have recently published preliminary NMR data with and without cholesterol, which show that N62-StAR undergoes a slow exchange between two states with stable tertiary structure. The characterization of the structure and dynamics of both states is likely to shed the necessary light to definitely validate the two-state model.

References

Alpy, F., Stoeckel, M.E., Dierich, A., Escola, J.M., Wendling, C., Chenard, M.P., Vanier, M.T., Gruenberg, J., Tomasetto, C. and Rio, M.C., 2001, The steroidogenic acute regulatory protein homolog MLN64, a late endosomal cholesterol-binding protein. *J. Biol. Chem.* **276**: 4261–4269.

Alpy, F., and Tomasetto, C., 2005, Give lipids a START: the StAR-related lipid transfer (START) domain in mammals. *J. Cell Sci.* **118**: 2791–2801.

Arakane, F., Kallen, C.B., Watari, H., Foster, J.A., Sepuri, N.B., Pain, D., Stayrook, S.E., Lewis, M., Gerton, G.L., and Strauss, J.F., 3rd., 1998, The mechanism of action of steroidogenic acute regulatory protein (StAR). StAR acts on the outside of mitochondria to stimulate steroidogenesis. *J. Biol. Chem.* **273**: 16339–16345.

Artemenko, I.P., Zhao, D., Hales, D.B., Hales, K.H., and Jefcoate, C.R., 2001, Mitochondrial processing of newly synthesized steroidogenic acute regulatory protein (StAR), but not total StAR, mediates cholesterol transfer to cytochrome P450 side chain cleavage enzyme in adrenal cells. *J. Biol. Chem.* **276**: 46583–46596.

Baker, B.Y., Yaworsky, D.C., and Miller, W.L., 2005, A pH-dependent molten globule transition is required for activity of the steroidogenic acute regulatory protein, StAR. *J. Biol. Chem.* **280**: 41753–41760.

Bose, H.S., Sugawara, T., Strauss, J.F., 3rd, and Miller, W.L., 1996, The pathophysiology and genetics of congenital lipoid adrenal hyperplasia. International Congenital Lipoid Adrenal Hyperplasia Consortium. *N. Engl. J. Med.* **335**: 1870–1878.

Bose, H.S., Whittal, R.M., Baldwin, M.A., and Miller, W.L., 1999, The active form of the steroidogenic acute regulatory protein, StAR, appears to be a molten globule. *Proc. Natl. Acad. Sci. USA* **96**: 7250–7255.

Bose, H.S., Sato, S., Aisenberg, J., Shalev, S.A., Matsuo, N., and Miller, W.L., 2000, Mutations in the steroidogenic acute regulatory protein (StAR) in six patients with congenital lipoid adrenal hyperplasia. *J. Clin. Endocrinol. Metab.* **85**: 3636–3639.

Bose, H.S., Lingappa, V.R., and Miller, W.L., 2002, The steroidogenic acute regulatory protein, StAR, works only at the outer mitochondrial membrane. *Endocr. Res.* **28**: 295–308.

Bose, H.S., Whittal, R.M., Ran, Y., Bose, M., Baker, B.Y., and Miller W.L., 2008, StAR-like activity and molten globule behavior of StARD6, a male germ-lineprotein. *Biochemistry* **47**: 2277–2288.

Clark, B.J., Wells, J., King, S.R., and Stocco, D.M., 1994, The purification, cloning, and expression of a novel luteinizing hormone-induced mitochondrial protein in MA-10 mouse Leydig tumor cells. Characterization of the steroidogenic acute regulatory protein (StAR). *J. Biol. Chem.* **269**: 28314–28322.

DeLano, W.L., 2002, The PyMOL Molecular Graphics System. DeLano Scientific, Palo Alto, CA, USA.

Iyer, L.M., Koonin, E.V., and Aravind, L., 2001, Adaptations of the helix-grip fold for ligand binding and catalysis in the START domain superfamily. *Proteins* **43**: 134–144.

Kudo, N., Kumagai, K., Tomishige, N., Yamaji, T., Wakatsuki, S., Nishijima, M., Hanada, K., and Kato R., 2008, Structural basis for specific lipid recognition by CERT responsible for nonvesicular trafficking of ceramide. *Proc. Natl. Acad. Sci. USA* **105**: 488–493.

Mathieu, A.P., Fleury, A., Ducharme, L., Lavigne, P., and LeHoux, J.G., 2002, Insights into steroidogenic acute regulatory protein (StAR)-dependent cholesterol transfer in mitochondria: evidence from molecular modeling and structure-based thermodynamics supporting the existence of partially unfolded states of StAR. *J. Mol. Endocrinol.* **29**: 327–345.

Murcia, M., Faráldo-Gómez, J.D., Maxfield, F.R., and Roux, B., 2006, Modeling the structure of the StART domains of MLN64 and StAR proteins in complex with cholesterol. J. Lipid Res. **47**: 2614–2630.

Ptitsyn, O.B., 1995, Molten globule and protein folding. *Adv. Protein Chem.* **47**: 83–229.

Roderick, S.L., Chan, W.W., Agate, D.S., Olsen, L.R., Vetting, M.W., Rajashankar, K.R., and Cohen, D.E., 2002, Structure of human phosphatidylcholine transfer protein in complex with its ligand. *Nat. Struct. Biol.* **9**: 507–511.

Romanowski, M.J., Soccio, R.E., Breslow, J.L., and Burley, S.K., 2002, Crystal structure of the Mus musculus cholesterol-regulated START protein 4 (StarD4) containing a StAR-related lipid transfer domain. *Proc. Natl. Acad. Sci. USA* **99**: 6949–6954.

Rone, M. B., Fan, J., and Papadopoulos V., 2009, Cholesterol transport in steroid biosynthesis: role of protein-protein interactions and implications in disease states. *Biochim. Biophys. Acta* **1791**: 646–658.

Roostaee, A., Barbar, E., LeHoux, J.G., and Lavigne, P., 2008, Cholesterol binding is a prerequisite for the activity of the steroidogenic acute regulatory protein (StAR). *Biochem. J.* **412**: 553–562.

Roostaee, A., Barbar, E., Lavigne, P., and LeHoux, J.G., 2009, The mechanism of specific binding of free cholesterol by the Steroidogenic Acute Regulatory Protein : Evidence for a role of the C-terminal helix in the gating of the binding site. *Biosci. Rep.* **29**: 89–101.

Schrick, K., Nguyen, D., Karlowski, W.M., and Mayer, K.F., 2004, START lipid/sterol-binding domains are amplified in plants and are predominantly associated with homeodomain transcription factors. *Genome Biol.* **5**: R41.

Tsujishita, Y., and Hurley, J.H., 2000, Structure and lipid transport mechanism of a StAR-related domain. *Nat. Struct. Biol.* **7**: 408–414.

Wang, X., Liu, Z., Eimerl, S., Timberg, R., Weiss, A.M., Orly, J., and Stocco, D.M., 1998, Effect of truncated forms of the steroidogenic acute regulatory protein on intramitochondrial cholesterol transfer. *Endocrinology* **139**: 3903–3912.

Watari, H., Arakane, F., Moog-Lutz, C., Kallen, C.B., Tomasetto, C., Gerton, G.L., Rio, M.C., Baker, M.E., and Strauss, J.F., 3rd, 1997, MLN64 contains a domain with homology to the steroidogenic acute regulatory protein (StAR) that stimulates steroidogenesis. *Proc. Natl. Acad. Sci. USA* **94**: 8462–8467.

Yaworsky, D.C., Baker, B.Y., Bose, H.S., Best, K.B., Jensen, L.B., Bell, J.D., Baldwin, M.A., and Miller, W.L., 2005, pH-dependent Interactions of the carboxyl-terminal helix of steroidogenic acute regulatory protein with synthetic membranes. *J. Biol. Chem.* **280**: 2045–2054.

Chapter 16
Membrane Cholesterol in the Function and Organization of G-Protein Coupled Receptors

Yamuna Devi Paila and Amitabha Chattopadhyay

Abstract Cholesterol is an essential component of higher eukaryotic membranes and plays a crucial role in membrane organization, dynamics and function. The G-protein coupled receptors (GPCRs) are the largest class of molecules involved in signal transduction across membranes, and represent major targets in the development of novel drug candidates in all clinical areas. Membrane cholesterol has been reported to have a modulatory role in the function of a number of GPCRs. Two possible mechanisms have been previously suggested by which membrane cholesterol could influence the structure and function of GPCRs (i) through a direct/specific interaction with GPCRs, or (ii) through an indirect way by altering membrane physical properties in which the receptor is embedded, or due to a combination of both. Recently reported crystal structures of GPCRs have shown structural evidence of cholesterol binding sites. Against this backdrop, we recently proposed a novel mechanism by which membrane cholesterol could affect structure and function of GPCRs. According to our hypothesis, cholesterol binding sites in GPCRs could represent 'nonannular' binding sites. Interestingly, previous work from our laboratory has demonstrated that membrane cholesterol is required for the function of the serotonin$_{1A}$ receptor (a representative GPCR), which could be due to specific interaction of the receptor with cholesterol. Based on these results, we envisage that there could be specific/nonannular cholesterol binding site(s) in the serotonin$_{1A}$ receptor. We have analyzed putative cholesterol binding sites from protein databases in the serotonin$_{1A}$ receptor. Our analysis shows that cholesterol binding sites are inherent characteristic features of serotonin$_{1A}$ receptors and are conserved through natural evolution. Progress in deciphering molecular details of the GPCR-cholesterol interaction in the membrane would lead to better insight into our overall understanding of GPCR function in health and disease, thereby enhancing our ability to design better therapeutic strategies to combat diseases related to malfunctioning of GPCRs.

A. Chattopadhyay (✉)
Centre for Cellular and Molecular Biology, Council of Scientific and Industrial Research, Uppal Road, Hyderabad 500 007, India
e-mail: amit@ccmb.res.in

Keywords G-protein coupled receptor · Membrane cholesterol · Nonannular binding site · Serotonin$_{1A}$ receptor · Specific interaction · Specific cholesterol binding site

Abbreviations

5-HT$_{1A}$ receptor	5-hydroxytryptamine-1A receptor
7-DHC	7-dehydrocholesterol
8-OH-DPAT	8-hydroxy-2(di-*N*-propylamino)tetralin
CCK	cholecystokinin
CCM	cholesterol consensus motif
DPH	1,6-diphenyl-1,3,5-hexatriene
FRET	fluorescence resonance energy transfer
GPCR	G-protein coupled receptor
MβCD	methyl-β-cyclodextrin
SLOS	Smith-Lemli-Opitz syndrome

16.1 Introduction

Biological membranes are complex two-dimensional, non-covalent assemblies of a diverse variety of lipids and proteins. They impart an identity to the cell and its organelles and represent an ideal *milieu* for the proper function of a diverse set of membrane proteins. Membrane proteins mediate a wide range of essential cellular processes such as signaling across the membrane, cell-cell recognition, and membrane transport. About 30% of all open reading frames (ORFs) are predicted to encode membrane proteins and almost 50% of all proteins encoded by eukaryotic genomes are membrane proteins (Liu et al., 2002; Granseth et al., 2007). Importantly, they represent prime candidates for the generation of novel drugs in all clinical areas (Drews, 2000; Dailey et al., 2009) owing to their involvement in a wide variety of cellular processes. Since a significant portion of integral membrane proteins remains in contact with the membrane (Lee, 2003), and reaction centers in them are often buried within the membrane, the function of membrane proteins depends on the surrounding membrane lipid environment. Work spanning several years from a number of groups has contributed to our understanding of the requirement of specific lipids and/or the membrane environment for maintaining the proper topology, structure and function of membrane proteins (Opekarová and Tanner, 2003; Lee, 2004; Palsdottir and Hunte 2004; Nyholm et al., 2007). These effects have been attributed either to specific interactions of lipids with amino acids in proteins or to bulk properties of membranes. Considering the diverse array of lipids in natural membranes, it is believed that physiologically relevant processes occurring in membranes involve a precise coordination of multiple lipid-protein interactions. Such lipid-protein interactions are of particular importance because cells possess the ability to vary the lipid composition of their membranes in response to a variety of stresses and stimuli, thereby changing the environment and the activity of the proteins in their membranes. Insights into the structure of membrane proteins and specific lipid-protein interactions required for their function are therefore of considerable interest and physiological relevance.

16.2 Cholesterol in Biological Membranes: A Tale of Two Faces

Cholesterol is an essential and representative lipid in higher eukaryotic cellular membranes and is crucial in membrane organization, dynamics, function, and sorting (Liscum and Underwood, 1995; Simons and Ikonen, 2000; Mouritsen and Zuckermann, 2004). Cholesterol is a predominantly hydrophobic molecule comprising a near planar tetracyclic fused steroid ring and a flexible isooctyl hydrocarbon tail (*see* Fig. 16.1a). The basic hydrocarbon skeleton of cholesterol (and other sterols found in eukaryotes) is sterane (Fig. 16.1b). Since sterane resists microbial attack and is stable over long periods of time, sterols have emerged as important fossil markers for paleontologists (Kodner et al., 2008). The 3β-hydroxyl moiety provides cholesterol its amphiphilic character and helps cholesterol to orient and anchor in the membrane (Villalaín, 1996). The tetracyclic nucleus and isooctyl side chain create the bulky wedge-type shape of the cholesterol molecule. Interestingly, the planar tetracyclic ring system of cholesterol is asymmetric about the ring plane. The sterol ring has a flat and smooth side with no substituents (the α face) and a rough side with methyl substitutions (the β face; *see* Fig. 16.1c). The smooth α face of the sterol nucleus helps in favorable van der Waals interaction with the saturated fatty acyl chains of phospholipids (Lange and Steck, 2008). The α face of cholesterol contains only axial hydrogen atoms. The absence of any bulky group in this face facilitates close contact between the sterol nucleus and phospholipid chains. The bumpiness of the β face of cholesterol molecule is due to the protruding methyl groups at positions C_{18}, C_{19} and C_{21}. The molecular structure of cholesterol has been exceedingly fine-tuned over a very long time scale of natural evolution. This is exemplified by the recent report that removal of methyl groups from cholesterol results in altered tilt angle which affects ordering and condensing effects, as shown by atomic scale molecular dynamics simulations (Róg et al., 2007; Pöyry et al., 2008). Molecular simulation approaches have earlier shown that the α face of cholesterol promoted a stronger ordering effect on saturated alkyl chains compared to the β face (Róg and Pasenkiewicz-Gierula, 2001). In addition, molecular dynamics simulation has shown that cholesterol orients its smooth α face toward saturated chains and its uneven β face toward unsaturated chains of phospholipids (Pandit et al., 2004), or with a bumpy transmembrane domain of an integral membrane protein (*see* Fig. 16.1d).

Cholesterol is oriented in the membrane bilayer with its long axis perpendicular to the plane of the membrane (Fig. 16.1d), so that it's polar hydroxyl group encounters the aqueous environment and the hydrophobic steroid ring is oriented parallel to and immersed in the hydrophobic fatty acyl chains of the phospholipids (Yeagle, 1985). It has been previously shown using X-ray and neutron diffraction that cholesterol is aligned in bilayers with its 3β-hydroxyl group in the proximity

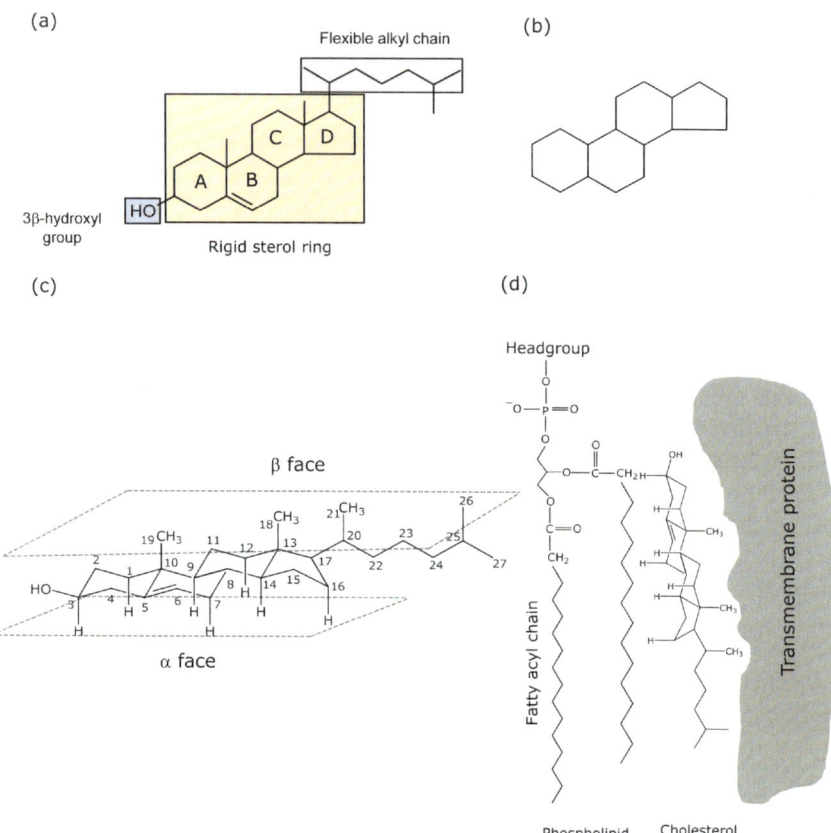

Fig. 16.1 Chemical structure and membrane orientation of cholesterol: (**a**) *Structure of cholesterol showing the individual rings (A–D)*. Three structurally distinct regions are shown as shaded boxes: the 3β-hydroxyl group, the rigid steroid ring, and the flexible alkyl chain. The 3β-hydroxyl moiety is the only polar group in cholesterol thereby contributing to its amphiphilic character and it helps cholesterol to orient and anchor in the membrane. Reproduced from Paila et al. (2009). (**b**) *Chemical structure of sterane*. Sterane is the basic hydrocarbon skeleton of cholesterol and other sterols found in eukaryotes. (**c**) *Two faces of cholesterol*. Cholesterol is characterized by a flat and smooth α face, and a rough β face. The α face of cholesterol contains only axial hydrogen atoms. The roughness of the β face is due to the protruding bulky methyl groups. (**d**) *Schematic orientation of cholesterol in relation to a phospholipid molecule in a lipid bilayer*. The smooth α face of the sterol nucleus helps in favorable van der Waals interaction with the saturated fatty acyl chains of phospholipids. α and β faces of cholesterol can simultaneously interact with a saturated fatty acyl chain of phospholipids and uneven transmembrane domain of an integral membrane protein, respectively. Cholesterol is shown to align in bilayers with its 3β-hydroxyl group in the vicinity of the ester carbonyls of phospholipids and its tetracyclic ring immersed in the bilayer interior, in close contact with a part of the phospholipid fatty acyl chain. Since the length of the cholesterol molecule including the isooctyl tail in all-*trans* energy minimum conformation is ~20 Å, a single cholesterol molecule can traverse each leaflet of a bilayer composed of phospholipids typically found in eukaryotic plasma membranes. It should be noted that the effective length of cholesterol molecule in membranes could vary, depending on the nature of the phospholipids. See text for other details

of the ester bonds of phospholipids and its tetracyclic ring buried in the bilayer interior, in close contact with a part of the phospholipid fatty acyl chains (Villalaín, 1996; Bittman, 1997). It should be mentioned here that although the hydroxyl group of cholesterol is shown to be aligned at the level of *sn*-2 ester carbonyl group of the phospholipid in Fig. 16.1d, unambiguous experimental evidence supporting the interaction (hydrogen bonding) between the hydroxyl group of cholesterol and the lipid carbonyl group is lacking. Since the length of the cholesterol molecule including the isooctyl tail in all-trans energy minimum conformation is ~ 20 Å, a single cholesterol molecule can traverse one leaflet of a bilayer composed of phospholipids (Bittman, 1997), typically found in eukaryotic plasma membranes (Fig. 16.1d). In fact, cholesterol has previously shown to exist as transbilayer ('tail-to-tail') dimers spanning the two leaflets of the membrane bilayer at low concentrations (Harris et al., 1995; Mukherjee and Chattopadhyay, 1996; Loura and Prieto, 1997). Interestingly, the transbilayer dimer arrangement of cholesterol was shown to be sensitive to the membrane surface curvature and is stringently controlled by a narrow window of membrane thickness (Rukmini et al., 2001). The environment around the cholesterol dimers appears to be more rigid (Mukherjee and Chattopadhyay, 2005) and the dimer population exhibits relatively slow lateral diffusion (Pucadyil et al., 2007). Importantly, such transbilayer tail-to-tail cholesterol dimers have been implicated in atherogenesis (Tulenko et al., 1998) and in human ocular lens fiber cell plasma membranes, especially in cataractous condition (Jacob et al., 1999; 2001; Mason et al., 2003).

Cholesterol is often found distributed non-randomly within domains found in biological and model membranes (Liscum and Underwood, 1995; Schroeder et al., 1995; Simons and Ikonen, 1997, 2000; Xu and London, 2000; Mukherjee and Maxfield, 2004). Many of these domains (sometimes termed as 'lipid rafts') are believed to be important for the maintenance of membrane structure and function. The idea of such specialized membrane domains assumes significance in cell biology since physiologically important functions such as membrane sorting and trafficking (Simons and van Meer, 1988), signal transduction processes (Simons and Toomre, 2000), and the entry of pathogens (Simons and Ehehalt, 2002; Riethmüller et al., 2006; Pucadyil and Chattopadhyay, 2007) have been attributed to these domains. Importantly, cholesterol plays a vital role in the function and organization of membrane proteins and receptors (Burger et al., 2000; Pucadyil and Chattopadhyay, 2006).

16.3 Role of Membrane Cholesterol in the Function of G-Protein Coupled Receptors

The G-protein coupled receptor (GPCR) superfamily is the largest and most diverse protein family in mammals, involved in signal transduction across membranes (Pierce et al., 2002; Perez, 2003; Rosenbaum et al., 2009). Cellular signaling by GPCRs involves their activation by ligands present in the extracellular environment and the subsequent transduction of signals to the interior of the cell through concerted changes in their transmembrane domain structure (Gether, 2000). GPCRs are prototypical members of the family of seven transmembrane domain proteins and include >800 members which together constitute ~1–2% of the human genome (Fredriksson and Schiöth, 2005). GPCRs dictate physiological responses to a diverse array of stimuli that include endogenous ligands such as biogenic amines, peptides, glycoproteins, lipids, nucleotides, Ca^{2+} ions and various exogenous ligands for sensory perception such as odorants, pheromones, and even photons. As a consequence, these receptors mediate multiple physiological processes such as neurotransmission, cellular metabolism, secretion, cellular differentiation, growth, inflammatory and immune responses. It is therefore only natural that GPCRs have emerged as major targets for the development of novel drug candidates in all clinical areas (Nature reviews drug discovery GPCR questionnaire participants 2004; Jacoby et al., 2006; Schlyer and Horuk, 2006; Insel et al., 2007; Heilker et al., 2009). Interestingly, although GPCRs represent 30–50% of current drug targets, only a small fraction of all GPCRs are presently targeted by drugs (Lin and Civelli, 2004). This points out the exciting possibility that the receptors which are not yet recognized could be potential drug targets for diseases that are difficult to treat by currently available drugs.

GPCRs are integral membrane proteins with a significant portion of the protein embedded in the membrane. In the case of rhodopsin, molecular dynamics simulation studies have estimated that the lipid-protein interface corresponds to ~38% of the total surface area of the receptor (Huber et al., 2004). This raises the obvious possibility that the membrane lipid environment could be an important modulator of receptor structure and function (Lee, 2004). The importance of a membrane lipid environment for optimal function of membrane proteins in general, and GPCRs in particular, is evident from the adverse effects of delipidation on receptor function (Kirilovsky and Schramm, 1983; Jones et al., 1988). Importantly, membrane cholesterol has been shown to modulate the function of a number of GPCRs. From the available data on the role of cholesterol on GPCR function (*see* Table 16.1), it appears that there is a lack of consensus on the manner in which cholesterol modulates receptor function. For example, while cholesterol is found to be essential for the proper function of several GPCRs, the function of rhodopsin has been shown to be inhibited in the presence of cholesterol. This calls for a detailed mechanistic analysis of the effect of cholesterol on any given receptor. What follows is a critical analysis of the available literature on the role of membrane cholesterol in GPCR function, with an overall objective to distinguish specific and general effects.

16.3.1 Effect of Membrane Cholesterol on the Function of GPCRs: General Effect or Specific Interaction ?

The mechanism underlying the effect of cholesterol on the structure and function of integral membrane proteins and receptors is complex and as yet no general consensus has emerged (Burger et al., 2000; Pucadyil and Chattopadhyay, 2006; Paila and Chattopadhyay, 2009). It has been proposed that cholesterol can modulate the

Table 16.1 Membrane cholesterol and GPCR function

GPCR	References
Rhodopsin	Straume and Litman (1988); Mitchell et al. (1990); Albert and Boesze-Battaglia (2005)
Cholecystokinin (CCK)	Gimpl et al. (1997); Burger et al. (2000); Harikumar et al. (2005)
Galanin (GAL2)	Pang et al. (1999)
Serotonin$_{1A}$ (5-HT$_{1A}$)	Pucadyil and Chattopadhyay (2004, 2005, 2006); Paila et al. (2005, 2008)
Serotonin$_7$ (5-HT$_7$)	Sjögren et al. (2006)
Metabotropic glutamate[a]	Eroglu et al. (2002, 2003)
δ Opioid	Huang et al. (2007); Levitt et al. (2009)
κ Opioid	Xu et al. (2006)
μ Opioid	Lagane et al. (2000); Levitt et al., (2009)
Oxytocin	Gimpl et al. (1995, 1997, 2002b); Fahrenholz et al. (1995); Klein et al. (1995)
β$_2$-adrenergic	Kirilovsky and Schramm (1983); Kirilovsky et al. (1987); Ben-Arie et al. (1988)
Chemokine (CXCR4, CCR5)	Nguyen and Taub (2002a, 2002b, 2003)
Neurokinin (NK1)	Monastyrskaya et al. (2005)
Cannabinoid (CB1)	Bari et al. (2005a, 2005b)
M$_2$ Muscarinic	Colozo et al. (2007)

[a]These studies were carried out in the *Drosophila* eye where the major sterol present is ergosterol

function of GPCRs in two ways: (i) by a direct/specific interaction with the GPCR, which could induce a conformational change in the receptor (Gimpl et al., 2002a,b), or (ii) through an indirect way by altering the membrane physical properties in which the receptor is embedded (Ohvo-Rekilä et al., 2002; Lee, 2004) or due to a combination of both. There could be yet another mechanism by which membrane cholesterol could affect structure and function of membrane proteins. This mechanism invokes the concept of 'nonannular' binding sites of membrane lipids (Lee et al., 1982; Simmonds et al., 1982). We recently proposed that cholesterol binding sites in GPCRs could represent nonannular binding sites (Paila et al., 2009; *see below*). A comprehensive discussion on the representative GPCRs, for which the mechanism of cholesterol-dependence of function has been addressed, is provided below.

16.3.1.1 Rhodopsin

Rhodopsin, the photoreceptor of retinal rod cells, undergoes a series of conformational changes upon exposure to light. The light-activated receptor exists in equilibrium with various intermediates, collectively termed metarhodopsins. The state of equilibrium is sensitive to the presence of cholesterol in the membrane (Straume and Litman, 1988; Mitchell et al., 1990; Bennet and Mitchell, 2008). An increase in the amount of cholesterol in the membrane shifts this equilibrium toward the inactive conformation of the protein. The inhibitory effect of cholesterol on rhodopsin function has been explained by direct (*see below*) as well as indirect modes of action. The indirect mode of action has been rationalized on the basis of the free-volume theory of membranes, which relates the alteration in membrane physical properties due to the presence of cholesterol, to receptor function (Mitchell et al., 1990). The conversion of the photointermediates, metarhodopsin I to metarhodopsin II, upon exposure to light involves an expansion of the protein in the plane of the bilayer (Attwood and Gutfreund, 1980), which occupies the available partial free volume from the surrounding bilayer. The presence of cholesterol in the membrane has been reported to inhibit the formation of metarhodopsin II, due to its role in reducing the partial free volume in the membrane (Niu et al., 2002). Importantly, fluorescence resonance energy transfer (FRET) measurements have indicated an inherent property of rhodopsin to partition out of cholesterol-rich regions of the membrane (Polozova and Litman, 2000). These results have been reinforced by molecular dynamics simulation with rhodopsin in membranes containing a mixture of cholesterol and polyunsaturated phospholipids (Pitman et al., 2005).

16.3.1.2 Oxytocin and Cholecystokinin Receptors

Oxytocin and cholecystokinin (CCK) receptors have been shown to require membrane cholesterol for their function (Fahrenholz et al., 1995; Klein et al., 1995; Gimpl et al., 1995, 1997, 2002b; Harikumar et al., 2005). Interestingly, while the interaction between the oxytocin receptor and cholesterol is believed to be specific, the function of the CCK receptor appears to be dependent on the physical properties of membranes, which are a function of cholesterol content. This is demonstrated by the fact that these receptors exhibited different types of correlation, when fluorescence anisotropy of the membrane probe DPH was correlated with ligand binding activity. In case of the CCK receptor, ligand binding showed linear increase with measured anisotropy values (Gimpl et al., 1997). On the other hand, the ligand binding activity of the oxytocin receptor showed a slight reduction upon cholesterol depletion followed by a sharp decline, when the membrane cholesterol content reached a certain critical level ($\sim$57% of the original cholesterol content). This shows that membrane cholesterol could affect the ligand binding activity of the oxytocin receptor by a cooperative mechanism. Hill analysis of cholesterol content versus ligand binding revealed that the oxytocin receptor binds several molecules of cholesterol ($n \geq 6$) in a positive cooperative manner (Burger et al., 2000; Gimpl et al., 2002b). These conclusions were reinforced by structure-activity analysis of the oxytocin and CCK receptor using a variety of cholesterol analogues (Gimpl et al., 1997). In order to examine the specific molecular features of cholesterol required to maintain the high-affinity state of the oxytocin receptor, MβCD was used to replenish cholesterol-depleted membranes with a broad range of cholesterol analogues that are subtly different from cholesterol in the headgroup, the steroid ring, or in the hydrocarbon tail. Interestingly, ligand binding of the oxytocin receptor could be restored only with certain analogues, thereby indicating a specific molecular feature in cholesterol to support receptor function. Although cholesterol depletion reduces ligand binding to the CCK receptor, this effect could be reversed with most analogues of cholesterol that could restore membrane order. The ligand binding of the CCK receptor therefore was supported by each of the cholesterol analogues and was well correlated with the corresponding fluorescence anisotropy values. However, similar effects on the oxytocin receptor could be demonstrated only with certain analogues that structurally resembled cholesterol in some critical features. Taken together, these data provide support for a specific molecular interaction between the oxytocin receptor and cholesterol. In addition, molecular modeling studies have indicated a putative docking site (involving residues on the surface of transmembrane segments 5 and 6) for cholesterol in the oxytocin receptor that is absent in the CCK receptor (Politowska et al., 2001). Further, it has been reported that cholesterol stabilizes oxytocin receptor against thermal inactivation and protects the receptor from proteolytic degradation (Gimpl and Fahrenholz, 2002).

16.3.1.3 Galanin Receptors

Membrane cholesterol has been shown to be required for ligand binding and intracellular signaling of the subtype 2 galanin receptor (GalR2) (Pang et al., 1999). The role of membrane cholesterol in modulating ligand binding to the galanin receptor was monitored by treating membranes with MβCD or by culturing cells expressing the receptor in lipoprotein-deficient serum. These studies revealed a marked reduction in galanin binding to the receptor in cholesterol-deficient membranes. Importantly, replenishment of cholesterol to cholesterol-depleted membranes restored galanin binding to normal levels. This interaction appears to be specific, as only a limited number of cholesterol analogues were able to rescue galanin binding. In addition, treatment of membranes either with filipin (which binds cholesterol) or with cholesterol oxidase markedly reduced galanin binding. Hill analysis suggested that several molecules of cholesterol ($n \geq 3$) could bind in a positively cooperative manner to GalR2 (Pang et al., 1999).

16.4 Nonannular Lipids in the Function of Membrane Proteins

It has been proposed for the nicotinic acetylcholine receptor (which requires cholesterol for its function) that cholesterol could be present at the 'nonannular' sites of the receptor (Jones and McNamee, 1988). Early evidence for the presence of nonannular lipids was obtained from experiments monitoring effects of cholesterol and fatty acids on the Ca^{2+}/Mg^{2+}-ATPase (Lee et al., 1982; Simmonds et al., 1982). Integral membrane proteins are surrounded by a shell or annulus of lipid molecules, which mimics the immediate layer of solvent surrounding soluble proteins (Jost et al., 1973; Lee, 2003). These are termed 'annular' lipids surrounding the membrane protein. After several years of moderate controversy surrounding the interpretation of spectroscopic data, it later became clear that the annular lipids are exchangeable with bulk lipids (Devaux and Seigneuret, 1985). The rate of exchange of lipids between the annular lipid shell and the bulk lipid phase was shown to be approximately an order of magnitude slower than the rate of exchange of bulk lipids, resulting from translational diffusion of lipids in the plane of the membrane. It therefore appears that exchange between annular and bulk lipids is relatively slow, since lipid-protein interaction is favorable compared to lipid-lipid interaction. However, the difference in interaction energy is modest, consistent with the observation that lipid-protein binding constants (affinity) depend weakly on lipid structure (Lee, 2003). Interestingly, the two different types of lipid environments (annular and bulk) can be readily detected using electron spin resonance (ESR) spectroscopy (Marsh, 1990). In addition to the annular lipids, there is evidence for other lipid molecules in the immediate vicinity of integral membrane proteins. These are termed as 'nonannular' lipids. Nonannular sites are characterized by lack of accessibility to the annular lipids, *i.e.*, these sites cannot be displaced by competition with annular lipids. This is evident from analysis of fluorescence quenching of intrinsic tryptophans of membrane proteins by phospholipids or cholesterol covalently labelled with bromine (Simmonds et al., 1982; Jones and McNamee, 1988), which acts as a quencher due to the presence of the heavy bromine atom (Chattopadhyay, 1992). These results indicate that nonannular lipid binding sites remain vacant even in the presence of annular lipids around the protein (Marius et al., 2008). The exchange of lipid molecules between nonannular sites and bulk lipids would be relatively slow compared to the exchange between annular sites and bulk lipids (although this has not yet been shown experimentally), and binding to the nonannular sites is considered to be more specific compared to annular binding sites (Lee, 2003).

The location of the postulated nonannular sites merits comment. It has been suggested that the possible locations for the nonannular sites could be either inter or intramolecular (interhelical) protein interfaces, characterized as deep clefts (or cavities) on the protein surface (Simmonds et al., 1982; Marius et al., 2008). For example, in the crystal structure of the potassium channel KcsA from *S. lividans*, a negatively charged lipid molecule was found to be bound as 'anionic nonannular' lipid at each of the protein-protein interface in the homotetrameric structure (Marius et al., 2005). These nonannular sites show high selectivity for anionic lipids over zwitterionic lipids, and it has been proposed that the change in the nature of the nonannular lipid leads to a change in packing at the protein-protein interface which modulates the open channel probability and conductance. Interestingly, the relationship between open channel probability of KcsA and negative phospholipid content exhibits cooperativity. This is consistent with a model in which the nonannular sites in the KcsA homotetramer have to be occupied by anionic lipids for the channel to remain open (Marius et al., 2008). This example demonstrates the crucial requirement of nonannular lipids in the function of membrane proteins and the stringency associated with regard to specificity of nonannular lipids.

In the context of GPCRs, it is interesting to note that many GPCRs are believed to function as oligomers (Park et al., 2004). More importantly, cholesterol has been shown to improve stability of GPCRs such as the β_2-adrenergic receptor (Yao and

Kobilka, 2005), and appears to be a necessary component for crystallization of the receptor since it facilitates receptor-receptor interaction and consequent oligomerization (Cherezov et al., 2007). Since a possible location of the nonannular sites is inter-protein interfaces (Simmonds et al., 1982; Jones and McNamee, 1988), it is possible that cholesterol molecules located between individual receptor molecules (*see below*, Fig. 16.2b) occupy nonannular sites and modulate receptor structure and function.

16.4.1 Presence of Specific (Nonannular?) Cholesterol binding Sites in the Crystal Structures of GPCRs

16.4.1.1 Rhodopsin

Specific interaction between rhodopsin and cholesterol has been monitored utilizing FRET between tryptophan residues of rhodopsin (donor) and cholestatrienol (acceptor) (Albert et al., 1996). Cholestatrienol is a naturally occurring fluorescent cholesterol analogue and has been reported to be a faithful mimic of cholesterol (Gimpl and Gehrig-Burger, 2007; Wüstner, 2007). In the aforementioned work (Albert et al., 1996), replenishment of cholesterol or ergosterol into cholesterol-depleted rod outer segment disk membranes was carried out and their ability to inhibit the quenching of donor tryptophan fluorescence was monitored. Interestingly, cholesterol was able to inhibit tryptophan quenching, whereas in presence of ergosterol, quenching was observed due to energy transfer between tryptophan residues of rhodopsin and cholestatrienol, indicating a specific interaction between rhodopsin and cholesterol. In addition, it was postulated that one cholesterol molecule per rhodopsin monomer would be present at the lipid-protein interface (Albert et al., 1996). This has been supported by the crystal structure of a photo-stationary state, highly enriched in metarhodopsin I, which shows a cholesterol molecule between two rhodopsin monomers, which could possibly represent a nonannular site for cholesterol binding (Ruprecht et al., 2004; see Fig. 16.2a). These authors also reported that cholesterol could improve the reliability and yield of crystallization. In this structure, cholesterol is shown to be oriented with its tetracyclic ring aligned normal to the membrane bilayer. Interestingly, these authors proposed that some of the tryptophans in transmembrane helices would be able to interact with the cholesterol tetracyclic ring. Recent crystallographic structures of the β_2-adrenergic receptor have shown similar interactions (*see below*).

16.4.1.2 β_2-Adrenergic Receptor

Lipid molecules that are resolved in crystal structures of membrane proteins are believed to be tightly bound. These lipid molecules, which are preserved even in the crystal structure, are often localized at protein-protein interfaces in multimeric proteins and belong to the class of nonannular (sometimes termed as 'co-factor') lipids (Lee, 2003, 2005). Cholesterol has been shown to improve stability of the

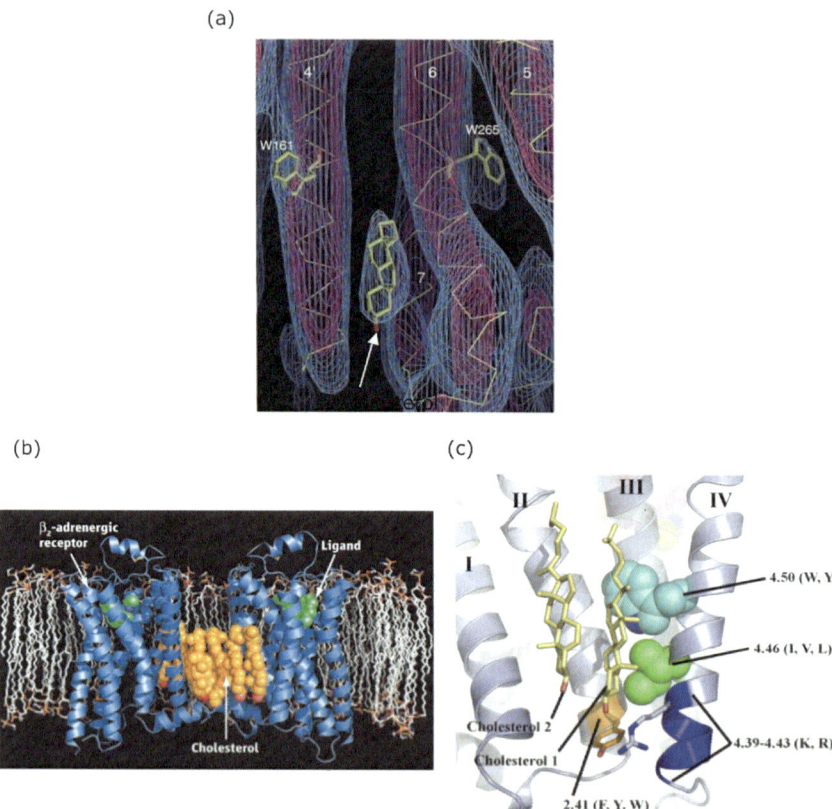

Fig. 16.2 Presence of tightly bound cholesterol molecules in the transmembrane regions in the crystal structures of metarhodopsin I (*panel* a) and human β_2-adrenergic receptor (*panels* b and c). *Panel* (**a**) shows side view of metarhodopsin I showing cholesterol between transmembrane helices. Notice the close proximity of tryptophan residues (W161 and W265) to cholesterol, independently confirmed by FRET studies (see text for more details). *Panel* (**b**) depicts the structure of the human β_2-adrenergic receptor (shown in *blue*) bound to the partial inverse agonist carazolol (in *green*) embedded in a lipid bilayer. Cholesterol molecules between two receptor molecules are shown in *orange*. *Panel* (**c**) shows the Cholesterol Consensus Motif (CCM) in the β_2-adrenergic receptor (bound to the partial inverse agonist timolol) crystal structure. The side chain positions of the β_2-adrenergic receptor and two bound cholesterol molecules are shown. Residues at positions 4.39–4.43 fulfill the CCM requirement (if one or more of these positions contains an arginine or lysine residue) and constitute site 1 (shown in *blue*) toward the cytoplasmic end of transmembrane helix IV. Site 2 (in *cyan*) represents the most important site at position 4.50 on transmembrane helix IV since it is the most conserved site with tryptophan occupying this position in 94% of class A GPCRs. The other choice of amino acid for this site is tyrosine. Site 3 (in *green*) at position 4.46 on transmembrane helix IV satisfies the CCM requirement if isoleucine, valine, or leucine occupy the position. Site 4 (in *orange*) on transmembrane helix II is at position 2.41 and can be either phenylalanine or tyrosine. Reproduced from Paila et al. (2009)

β_2-adrenergic receptor (Yao and Kobilka, 2005), and appears to be necessary for crystallization of the receptor (Cherezov et al., 2007). The cholesterol analogue, cholesterol hemisuccinate, has recently been shown to stabilize the β_2-adrenergic receptor against thermal inactivation (Hanson et al., 2008). Since a possible location of the nonannular sites is at inter-protein interfaces (Simmonds et al., 1982; Jones and McNamee, 1988), it is possible that cholesterol molecules located between individual receptor molecules (see Fig. 16.2b) occupy nonannular sites and modulate receptor structure and function. Importantly, the recent crystal structure of the β_2-adrenergic receptor has revealed structural evidence of a specific cholesterol binding site (Fig. 16.2c, Hanson et al., 2008). The crystal structure shows a cholesterol binding site between transmembrane helices I, II, III and IV with two cholesterol molecules bound per receptor monomer. The cholesterol binding site appears to be characterized by the presence of a cleft located at the membrane interfacial region. Both cholesterol molecules bind in a shallow surface groove formed by segments of the above mentioned helices (I–IV), thereby providing an increase in the intramolecular occluded surface area, a parameter often correlated to the enhanced thermal stability of proteins (DeDecker et al., 1996). Calculation of packing values of various helices in the β_2-adrenergic receptor which are involved in the cholesterol interacting site showed that the packing of transmembrane helices II and IV increases upon cholesterol binding, which would restrict their mobility rendering greater thermal stability to the protein (Hanson et al., 2008).

Earlier literature suggests that there are several structural features of proteins that are believed to result in preferential association with cholesterol (Epand, 2006). In many cases, proteins interacting with cholesterol have a characteristic stretch of amino acids, termed the cholesterol recognition/interaction amino acid consensus (CRAC) motif (Li and Papadopoulos, 1998). Another important cholesterol interacting domain is the sterol-sensing domain (SSD). The SSD is relatively large and consists of five transmembrane segments and is involved in cholesterol biosynthesis and homeostasis (Kuwabara and Labouesse, 2002; Brown and Goldstein, 1999). It has been recently proposed that cholesterol binding sequence or motif should contain at least one aromatic amino acid, which could interact with ring D of cholesterol (Hanson et al., 2008) and a positively charged residue (Epand et al., 2006; Jamin et al., 2005), capable of participating in electrostatic interactions with the 3β-hydroxyl group. In the crystal structure of the β_2-adrenergic receptor, three amino acids in transmembrane helix IV, along with an amino acid in transmembrane helix II, have been suggested to constitute a cholesterol consensus motif (CCM, see Fig. 16.2c). The aromatic Trp $158^{4.50}$ [according to the Ballesteros-Weinstein numbering system (Ballesteros and Weinstein, 1995)] is conserved to a high degree ($\sim$94%) among rhodopsin-like GPCRs and appears to contribute the most significant interaction with ring D of cholesterol (see Fig. 16.1a; Hanson et al., 2008). In this structure, the hydrophobic residue Ile$154^{4.46}$ would interact with rings A and B of cholesterol and is largely conserved ($\sim$60%) in rhodopsin family GPCRs. The aromatic residue Tyr$70^{2.41}$ in transmembrane helix II could interact with ring A of cholesterol and with Arg$151^{4.43}$ of transmembrane helix IV through hydrogen bonding. The criterion of specific residues in CCM (as described above) could be somewhat broadened by conservative substitutions of amino acids (see legend to Fig. 16.2c).

The above description of CCM in the recently reported crystal structure of the β_2-adrenergic receptor raises the interesting possibility of the presence of putative nonannular binding sites in transmembrane inter-helical locations in GPCRs in general. It was previously proposed, from quenching analysis of intrinsic tryptophan fluorescence in the nicotinic acetylcholine receptor by brominated phospholipids and cholesterol analogues, that there could be 5–10 nonannular sites per ~250 kDa monomer of the receptor (Jones and McNamee, 1988). This is consistent with the above proposal of two putative nonannular sites per ~50 kDa monomer of the β_2-adrenergic receptor.

16.5 The Serotonin$_{1A}$ Receptor: A Representative Member of the GPCR Superfamily in the Context of Membrane Cholesterol Dependence for Receptor Function

The serotonin$_{1A}$ (5-HT$_{1A}$) receptor is an important neurotransmitter receptor and is the most extensively studied of the serotonin receptors for a number of reasons (Pucadyil et al., 2005a; Kalipatnapu and Chattopadhyay, 2007). Serotonin receptors have been classified into at least 14 subtypes on the basis of their pharmacological responses to specific ligands, sequence similarities at the gene and amino acid levels, gene organization, and second messenger coupling pathways (Hoyer et al., 2002). The serotonin$_{1A}$ receptor was the first among all types of serotonin receptors to be cloned as an intronless genomic clone (G-21) of the human genome, which cross-hybridized with a full length β-adrenergic receptor probe at reduced stringency (Kobilka et al., 1987; Pucadyil et al., 2005a). Sequence analysis of this genomic clone (later identified as the serotonin$_{1A}$ receptor gene) showed ~43% amino acid similarity with the β_2-adrenergic receptor in the transmembrane domain. The serotonin$_{1A}$ receptor was therefore initially discovered as an 'orphan' receptor and was identified ('deorphanized') later (Fargin et al., 1988). The human gene for the receptor encodes a protein of 422 amino acids (*see* Fig. 16.3). Considering the presence of three consensus sequences for N-linked glycosylation in the amino terminus, and the homology of the receptor with β-adrenergic receptor, it is predicted that the receptor is oriented in the plasma membrane with the amino terminus facing the extracellular region and the carboxy terminus facing the intracellular cytoplasmic region (Raymond et al., 1999; Pucadyil et al., 2005a; Kalipatnapu and Chattopadhyay, 2007; *see* Fig. 16.3). The transmembrane domains (TM1–TM7) of the receptor are connected by hydrophilic sequences of three extracellular loops (EC1, EC2, EC3) and three intracellular loops (IC1, IC2, IC3). Such an arrangement is typical of the G-protein coupled receptor superfamily (Gether and Kobilka, 1998). Although the structure of the serotonin$_{1A}$ receptor has not yet been experimentally determined, mutagenesis studies have helped in identifying amino acid residues important for ligand binding and G-protein coupling of the serotonin$_{1A}$ receptor (reviewed in Pucadyil et al., 2005a). Among the predicted structural features of the serotonin$_{1A}$ receptor, palmitoylation status of the receptor has been confirmed in a recent report (Papoucheva et al., 2004). An interesting aspect of this study is that palmitoylation of the serotonin$_{1A}$ receptor was found to be stable and independent of stimulation by the agonist. This is unusual for GPCRs, which undergo repeated cycles of palmitoylation and depalmitoylation (Milligan et al., 1995). It

16 Cholesterol and G-Protein Coupled Receptors

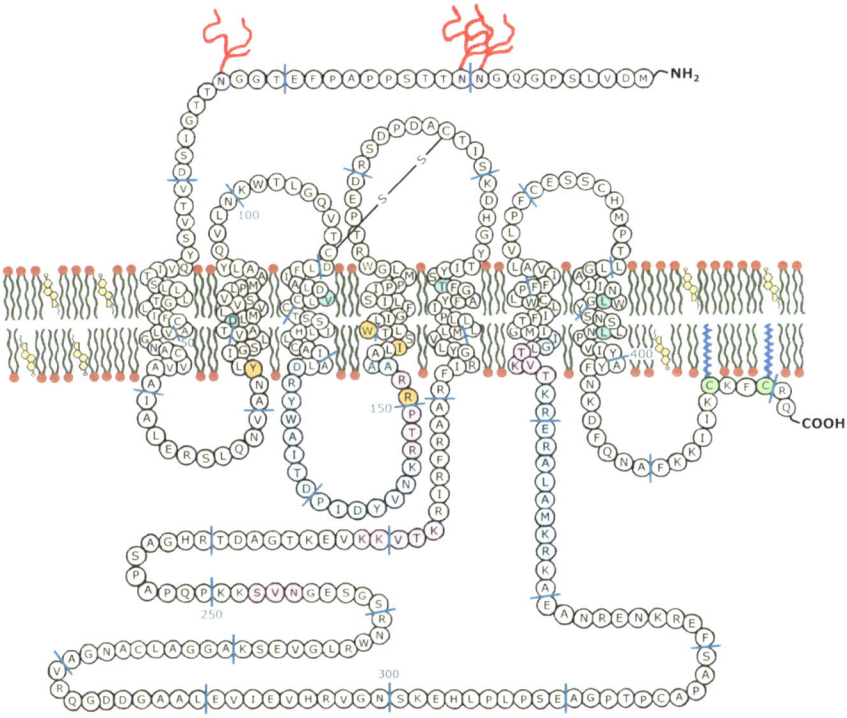

Fig. 16.3 A schematic representation of the membrane embedded human serotonin$_{1A}$ receptor showing its topological and other structural features. The membrane is shown as a bilayer of phospholipids and cholesterol, representative of typical eukaryotic membranes. The transmembrane helices of the receptor were predicted using TMHMM2. Seven transmembrane stretches, each composed of ~22 amino acids, are depicted as putative α-helices. The exact boundary between the membrane and the aqueous phase is not known and therefore the location of the residues relative to the membrane bilayer is putative. The amino acids in the receptor sequence are shown as circles and are marked for convenience. The potential sites (shown in *lavender*) for N-linked glycosylation (depicted as branching trees in *red*) on the amino terminus are shown. A putative disulfide bond between Cys^{109} and Cys^{187} is shown. The transmembrane domains contain residues (shown in *cyan*) that are important for ligand binding. The putative cholesterol binding site (see text) is highlighted (in *orange*). The receptor is stably palmitoylated (shown in *blue*) at residues Cys^{417} and/or Cys^{420} (shown in *green*). Light *blue* circles represent contact sites for G-proteins. Light *pink* circles represent sites for protein kinase mediated phosphorylation. Further structural details of the receptor are available in (Pucadyil et al., 2005a; Pucadyil and Chattopadhyay, 2006). Reproduced from Paila et al. (2009). It is probable, based on comparison with known crystal structures of similar GPCRs such as rhodopsin and β$_2$-adrenergic receptor, that there are motionally restricted water molecules that could be important in inducing conformational transitions in the transmembrane portion of the receptor (*see*, for example, Angel et al., 2009)

has therefore been proposed that stable palmitoylation of the receptor could play an important role in maintaining the receptor structure (Papoucheva et al., 2004).

Serotonergic signaling plays a key role in the generation and modulation of various cognitive, developmental and behavioral functions. The serotonin$_{1A}$ receptor agonists and antagonists have been shown to possess potential therapeutic effects in anxiety-or stress-related disorders (Pucadyil et al., 2005a). As a result, the serotonin$_{1A}$ receptor serves as an important target in the development of therapeutic agents for neuropsychiatric disorders such as anxiety and depression. Interestingly, mutant (knockout) mice lacking the serotonin$_{1A}$ receptor exhibit enhanced anxiety-related behavior, and represent an important animal model for genetic vulnerability to complex traits such as anxiety disorders and aggression in higher animals (Toth, 2003; Gardier, 2009). Taken together, the serotonin$_{1A}$ receptor is a central player in a multitude of physiological processes, and an important drug target.

Seminal work from our laboratory has comprehensively demonstrated the requirement of membrane cholesterol in the function of the serotonin$_{1A}$ receptor (Pucadyil and Chattopadhyay, 2004, 2006). We demonstrated the crucial modulatory role of membrane cholesterol on the ligand binding activity and G-protein coupling of the hippocampal serotonin$_{1A}$ receptor using a number of approaches such as treatment with (i) MβCD, which physically depletes cholesterol from membranes (Pucadyil and Chattopadhyay, 2004, 2005) (ii) the sterol-complexing detergent digitonin (Paila et al., 2005), and (iii) the sterol-binding antifungal polyene antibiotic nystatin (Pucadyil et al., 2004). Interestingly, while treatment with MβCD physically depletes membrane cholesterol, treatment with other agents merely modulates the availability of membrane cholesterol without physical depletion. The common message from these observations is that it is the *non-availability of membrane cholesterol*, rather than the manner in which its availability is modulated, is crucial for ligand binding of the serotonin$_{1A}$ receptor. Importantly, replenishment of membrane cholesterol using MβCD-cholesterol complex resulted in recovery of ligand binding activity to a considerable extent. However, it was not clear from these results whether the effect of membrane cholesterol on the function of the serotonin$_{1A}$ receptor is due to specific interaction of membrane cholesterol with the receptor, or general effect of cholesterol on the membrane bilayer, or a combination of both.

In order to further examine the mechanism of cholesterol-dependent function of the serotonin$_{1A}$ receptor and monitor the stringency of the process, membranes were treated with cholesterol oxidase, which catalyzes the oxidation of cholesterol to cholestenone. These results showed that oxidation of membrane cholesterol led to inhibition of the ligand binding activity of the serotonin$_{1A}$ receptor without altering overall membrane order (Figs. 16.4a and 16.5; Pucadyil et al., 2005b). Based on these results, it was proposed that there could be specific interaction between membrane cholesterol and the serotonin$_{1A}$ receptor. Toward this effect, we have recently generated a cellular model of the Smith-Lemli-Opitz Syndrome (SLOS), using cells stably expressing the human serotonin$_{1A}$ receptor (Paila et al., 2008). SLOS is a congenital and developmental malformation syndrome associated with defective cholesterol biosynthesis in which the immediate biosynthetic precursor

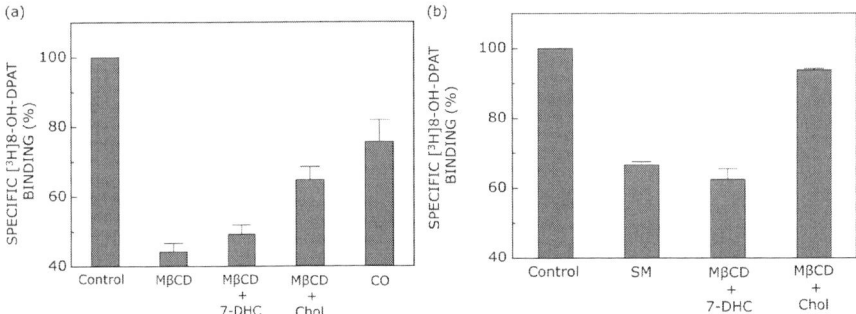

Fig. 16.4 (a) Effect of replenishment of 7-DHC and cholesterol into cholesterol-depleted membranes on the specific binding of [^{3}H]8-OH-DPAT to the hippocampal serotonin$_{1A}$ receptor. Cholesterol depletion in native hippocampal membranes was achieved using MβCD followed by replenishment with 7-DHC or cholesterol. In addition, this *panel* shows the effect of oxidation of membrane cholesterol on the specific binding of [^{3}H]8-OH-DPAT to the hippocampal serotonin$_{1A}$ receptor. Membrane cholesterol was oxidized using cholesterol oxidase (CO). Values (means ± standard error) are expressed as percentages of specific binding obtained in native membranes. (b) Effect of replenishment of 7-DHC or cholesterol into solubilized membranes (denoted as SM) on specific binding of [^{3}H]8-OH-DPAT to the hippocampal serotonin$_{1A}$ receptor. Solubilized membranes were replenished with 7-DHC or cholesterol, using the corresponding sterol:MβCD complex. Values (means ± standard error) are expressed as percentages of specific ligand binding obtained in native membranes. Adapted and modified from Pucadyil et al. (2005b) and Paila and Chattopadhyay (2009)

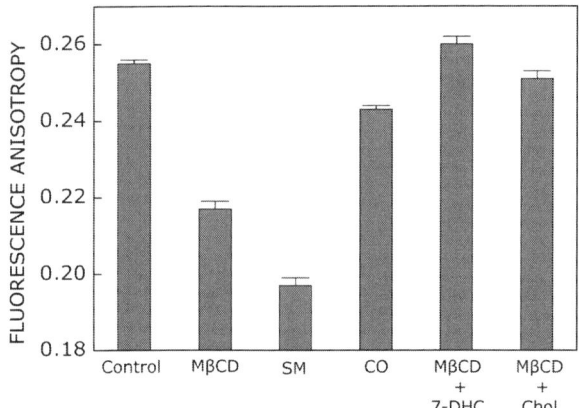

Fig. 16.5 Effect of replenishment of 7-DHC or cholesterol into cholesterol-depleted, solubilized membranes on fluorescence anisotropy (means ± standard error) of the membrane probe DPH. Cholesterol depletion was carried out using MβCD. Fluorescence anisotropy of cholesterol-oxidase treated membranes is also shown. Cholesterol was oxidized using cholesterol oxidase (CO). Membranes (cholesterol-depleted or solubilized) were replenished with 7-DHC or cholesterol, using the corresponding sterol:MβCD complex. Adapted and modified from Pucadyil et al. (2005b) and Paila and Chattopadhyay (2009)

of cholesterol (7-dehydrocholesterol or 7-DHC) is accumulated (Porter, 2008). We have recently shown that the effects of 7-DHC and cholesterol on membrane organization and dynamics are considerably different (Shrivastava et al., 2008). The cellular model of SLOS was generated by metabolically inhibiting the biosynthesis of cholesterol, utilizing a specific inhibitor (AY 9944) of the enzyme required in the final step of cholesterol biosynthesis. Importantly, AY 9944 treatment has previously been shown to generate animal (rat) models of SLOS (Wolf et al., 1996; Gaoua et al., 2000). SLOS serves as an appropriate condition to ensure the specific effect of membrane cholesterol in the function of the serotonin$_{1A}$ receptor, since the two aberrant sterols that are accumulated in SLOS, $i.e.,$ 7- and 8-DHC, differ with cholesterol only by a double bond. Our results showed a progressive and drastic reduction in specific ligand binding with increasing concentrations of AY 9944 (Paila et al., 2008). In addition, these results show that the G-protein coupling and downstream signaling of serotonin$_{1A}$ receptors are impaired in SLOS-like condition, although the membrane receptor level does not exhibit any reduction. Importantly, metabolic replenishment of cholesterol using serum partially restored the ligand binding activity of the serotonin$_{1A}$ receptor under these conditions.

Figure 16.4a shows that cholesterol depletion from native hippocampal membranes followed by replenishment with 7-DHC did not result in restoration of the ligand binding to the serotonin$_{1A}$ receptor, in spite of recovery of membrane order (Fig. 16.5) (Singh et al., 2007). In addition, solubilization of the hippocampal serotonin$_{1A}$ receptor is accompanied by loss of membrane cholesterol, which results in a reduction in specific ligand binding activity and overall membrane order (Chattopadhyay et al., 2005, 2007). It is important to note here that the loss in ligand binding of the serotonin$_{1A}$ receptor is not necessarily related to the reduction in overall membrane order. For example, solubilized membranes retained higher ligand binding compared to cholesterol-depleted membranes (Fig. 16.4), although overall membrane order appears to be lower in solubilized membranes (Fig. 16.5). This implies that general effects may not be an important factor.

Replenishment of cholesterol to solubilized membranes restores the cholesterol content of the membrane and significantly enhances specific ligand binding activity (Fig. 16.4b) and overall membrane order (Fig. 16.5). Importantly, replenishment of solubilized hippocampal membranes with 7-DHC does not result in restoration of ligand binding activity of the serotonin$_{1A}$ receptor (Fig. 16.4b), in spite of recovery of membrane order (Fig. 16.5). We therefore conclude that the requirement for maintaining ligand binding activity is more stringent than the requirement for maintaining membrane order. Taken together, these results indicate that the molecular basis for the requirement of membrane cholesterol in maintaining the ligand binding activity of serotonin$_{1A}$ receptors could be specific interaction, although global bilayer effects may not be ruled out (Prasad et al., 2009). In the light of these results, it is indeed interesting to note that there are reported cholesterol binding sites (possibly nonannular) in the crystal structure of a closely related receptor $i.e.,$ the β_2-adrenergic receptor, as discussed above.

16.5.1 Cholesterol binding Motif(s) in Serotonin$_{1A}$ Receptors?

In the overall context of the presence of CCM in the recently reported crystal structure of the β_2-adrenergic receptor (Hanson et al., 2008), it is tempting to consider whether there is a similar CCM(s) present in the serotonin$_{1A}$ receptor and if present, whether it is conserved during the natural evolution of the receptor. This is particularly relevant in view of the similarity between the serotonin$_{1A}$ and β_2-adrenergic receptors (~43% amino acid similarity in the transmembrane domain) (Kalipatnapu and Chattopadhyay, 2007), and the reported cholesterol dependence of serotonin$_{1A}$ receptor function (Pucadyil and Chattopadhyay, 2006). In order to examine the evolution of specific cholesterol binding site(s) of the serotonin$_{1A}$ receptor over various phyla, we analyzed amino acid sequences of the serotonin$_{1A}$ receptor from available databases (*see* Fig. 16.6). Partial, duplicate and other non-specific sequences were removed from the set of sequences obtained. The amino acid sequences used for the analysis belong to diverse taxons that include insects, fish and other marine species, amphibians and extending up to mammals. Initial alignment was carried out using ClustalW. It is apparent from this alignment that the cholesterol binding motif, which includes Tyr73 in the putative transmembrane helix II and Arg151, Ile157 and Trp161 in the putative transmembrane helix IV (Figs. 16.3 and 16.6), is conserved in most species. Realignment with ClustalW (after eliminating the relatively divergent parts of the receptor) resulted in conservation of the motif across all phyla analyzed, except in organisms such as *T. adhaerens* and *S. purpuratus*. Interestingly, pairwise alignment of the human serotonin$_{1A}$ receptor with the human β_2-adrenergic receptor and rhodopsin exhibited conservation of the motif in all sequences. Cholesterol binding sites may therefore represent an inherent characteristic feature of serotonin$_{1A}$ receptors, which are conserved during the course of evolution. It is interesting to note here that cholesterol binding sites appear to be present even in organisms which are not capable of biosynthesis of cholesterol. Organisms which lack cholesterol biosynthesis could, however, acquire cholesterol through diet (Bloch, 1983). Organisms such as insects possess sterols that are different from cholesterol, and which have diverged from cholesterol during the evolution of the sterol pathway (Clark and Bloch, 1959). The presence of CCM in these organisms could be due to binding of closely related sterols or dietary cholesterol to CCM.

16.6 Conclusion and Future Perspectives

Previous work from our laboratory has comprehensively demonstrated that membrane cholesterol is required for the function of the serotonin$_{1A}$ receptor, which could be due to specific interaction of the receptor with cholesterol. Based on these results, we envisage that there could be specific/nonannular cholesterol binding site(s) in the serotonin$_{1A}$ receptor. Mutation of the amino acid residues involved in the cholesterol binding site of the serotonin$_{1A}$ receptor, followed by functional

(a)

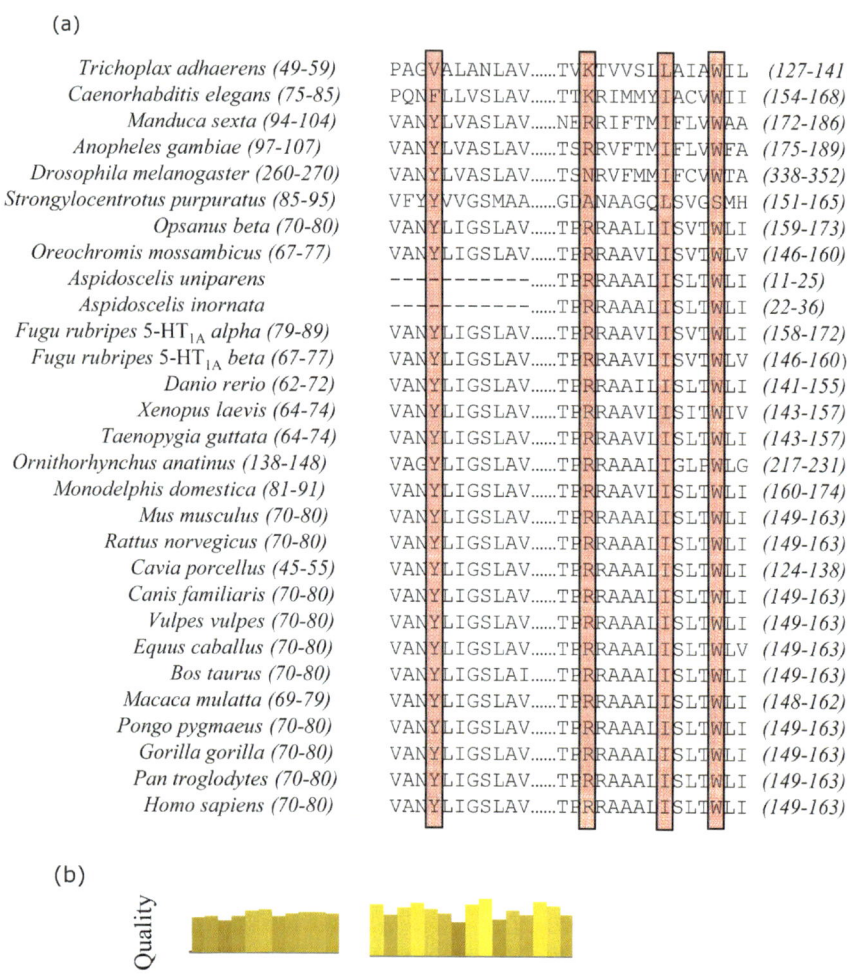

(b)

Fig. 16.6 Multiple alignment of the serotonin$_{1A}$ receptor around the CCM of interest with the conserved residues highlighted. As evident from *panel* (**a**), Trp161 is the most conserved residue, except in *S. purpuratus*. The sequences of *T. adhaerens*, *M. sexta* and *A. gambiae* are putative serotonin$_{1A}$ receptors whereas those of *S. pupuratus*, *B. taurus*, *O. anatinus*, *D. rerio*, *M. domestica*, *M. mulatta* and *T. guttata* are predicted by homology. The sequence of *C. elegans* belongs to the serotonin receptor family. The sequences of *C. porcellus*, *A. uniparens* and *A. inornata* are partial. The numbers of amino acid residues in respective sequences are mentioned in parentheses. Amino acid sequences of serotonin$_{1A}$ receptors are from NCBI and Expasy databases. *Panel* (**b**) is a graphical representation displaying the quality of alignment, with lighter shades representing higher quality. Adapted and modified from Paila et al. (2009)

and organizational analyses of the receptor, are likely to provide further insight into the membrane cholesterol-dependence of receptor function.

GPCRs are involved in a multitude of physiological functions and represent important drug targets. Although the pharmacological and signaling features of

GPCRs have been extensively studied, aspects related to their interaction with membrane lipids have been addressed in very few cases. In this context, the realization that lipids such as cholesterol could influence the function of GPCRs has remarkably transformed our ideas regarding the function of this important class of membrane proteins. With progress in deciphering molecular details on the nature of this interaction, our overall understanding of GPCR function in health and disease would improve significantly, thereby enhancing our ability to design better therapeutic strategies to combat diseases related to malfunctioning of these receptors. A comprehensive understanding of GPCR function in relation to the membrane lipid environment is important, in view of the enormous implications of GPCR function in human health (Jacoby et al., 2006; Schlyer and Horuk, 2006), and the observation that several diagnosed diseases are attributed to altered lipid-protein interactions (Pavlidis et al., 1994; Chattopadhyay and Paila, 2007).

Acknowledgements Work in A.C.'s laboratory was supported by the Council of Scientific and Industrial Research, Department of Biotechnology, Life Sciences Research Board, and the International Society for Neurochemistry. Y.D.P. thanks the Council of Scientific and Industrial Research for the award of a Senior Research Fellowship. A.C. is an Adjunct Professor at the Special Centre for Molecular Medicine of Jawaharlal Nehru University (New Delhi, India), and Honorary Professor of the Jawaharlal Nehru Centre for Advanced Scientific Research (Bangalore, India). A.C. gratefully acknowledges J.C. Bose Fellowship (Dept. Science and Technology, Govt. of India). We thank Sourav Haldar for helpful discussions, Roopali Saxena for help with the figures, Shrish Tiwari for his help in generating Fig. 16.6 and members of our laboratory for critically reading the manuscript

References

Albert, A.D., Young, J.E., Yeagle, P.L., 1996, Rhodopsin-cholesterol interactions in bovine rod outer segment disk membranes. *Biochim. Biophys. Acta* **1285**: 47–55.
Albert, A.D., Boesze-Battaglia, K., 2005, The role of cholesterol in rod outer segment membranes. *Prog. Lipid Res.* **44**: 99–124.
Angel, T.E., Chance, M.R., Palczewski, K., 2009, Conserved waters mediate structural and functional activation of family A (rhodopsin-like) G protein-coupled receptors. *Proc. Natl. Acad. Sci. USA* **106**: 8555–8560.
Attwood, P.V., Gutfreund, H., 1980, The application of pressure relaxation to the study of the equilibrium between metarhodopsin I and II from bovine retinas. *FEBS Lett.* **119**: 323–326.
Ballesteros, J.A., Weinstein, H., 1995, Integrated methods for the construction of three dimensional models and computational probing of structure-function relations in G-protein coupled receptors. *Methods Neurosci.* **25**: 366–428.
Bari, M., Battista, N., Fezza, F., Finazzi-Agrò, A., Maccarrone, M., 2005a, Lipid rafts control signaling of type-1 cannabinoid receptors in neuronal cells. Implications for anandamide-induced apoptosis. *J. Biol. Chem.* **280**: 12212–12220.
Bari, M., Paradisi, A., Pasquariello, N., Maccarrone, M., 2005b, Cholesterol-dependent modulation of type 1 cannabinoid receptors in nerve cells. *J. Neurosci. Res.* **81**: 275–283.
Ben-Arie, N., Gileadi, C., Schramm, M., 1988, Interaction of the β-adrenergic receptor with Gs following delipidation. Specific lipid requirements for Gs activation and GTPase function. *Eur. J. Biochem.* **176**: 649–654.
Bennet, M.P., Mitchell, D.C., 2008, Regulation of membrane proteins by dietary lipids: effects of cholesterol and docosahexanoic acid acyl chain-containing phospholipids on rhodopsin stability and function. *Biophys. J.* **95**: 1206–1216.

Bittman, R., 1997, Has nature designed the cholesterol side chain for optimal interaction with phospholipids? *Subcell. Biochem.* **28**: 145–171.

Bloch, K.E., 1983, Sterol structure and membrane function. *CRC Crit. Rev. Biochem.* **14**: 47–92.

Brown, M.S., Goldstein, J.L., 1999, A proteolytic pathway that controls the cholesterol content of membranes, cells, and blood. *Proc. Natl. Acad. Sci. USA* **96**: 11041–11048.

Burger, K., Gimpl, G., Fahrenholz, F., 2000, Regulation of receptor function by cholesterol. *Cell. Mol. Life Sci.* **57**: 1577–1592.

Chattopadhyay, A., 1992, Membrane penetration depth analysis using fluorescence quenching: a critical review. In: Biomembrane Structure and Function: The State of The Art (Gaber, B.P. and Easwaran, K.R.K., eds.), Adenine Press, Schenectady, NY, pp. 153–163.

Chattopadhyay, A., Jafurulla, Md., Kalipatnapu, S., Pucadyil, T.J., Harikumar, K.G., 2005, Role of cholesterol in ligand binding and G-protein coupling of serotonin$_{1A}$ receptors solubilized from bovine hippocampus. *Biochem. Biophys. Res. Commun.* **327**: 1036–1041.

Chattopadhyay, A., Paila, Y.D., 2007, Lipid-protein interactions, regulation and dysfunction of brain cholesterol. *Biochem. Biophys. Res. Commun.* **354**: 627–633.

Chattopadhyay, A., Paila, Y.D., Jafurulla, Md., Chaudhuri, A., Singh, P., Murty, M.R.V.S., Vairamani, M., 2007, Differential effects of cholesterol and 7-dehydrocholesterol on ligand binding of solubilized hippocampal serotonin$_{1A}$ receptors: Implications in SLOS. *Biochem. Biophys. Res. Commun.* **363**: 800–805.

Cherezov, V., Rosenbaum, D.M., Hanson, M.A., Rasmussen, S.G.F., Thian, F.S., Kobilka, T.S., Choi, H.-J., Kuhn, P., Weis, W.I., Kobilka, B.K., Stevens, R.C., 2007, High-resolution crystal structure of an engineered human β_2-adrenergic G protein-coupled receptor. *Science* **318**: 1258–1265.

Clark, A.J., Bloch, K., 1959, The absence of sterol synthesis in insects. *J. Biol. Chem.* **234**: 2578–2582.

Colozo, A.T., Park, P.S.-H., Sum, C.S., Pisterzi, L.F., Wells, J.W., 2007, Cholesterol as a determinant of cooperativity in the M_2 muscarinic cholinergic receptor. *Biochem. Pharmacol.* **74**: 236–255.

Dailey, M.M., Hait, C., Holt, P.A., Maguire, J.M., Meier, J.B., Miller, M.C., Petraccone, L., Trent, J.O., 2009, Structure-based drug design: From nucleic acid to membrane protein targets. *Exp. Mol. Pathol.* **86**:141–150.

DeDecker, B.S., O'Brien, R., Fleming, P.J., Geiger, J.H., Jackson, S.P., Sigler, P.B., 1996, The crystal structure of a hyperthermophilic archaeal TATA-box binding protein. *J. Mol. Biol.* **264**: 1072–1084.

Devaux, P.F., Seigneuret, M., 1985, Specificity of lipid-protein interactions as determined by spectroscopic techniques. *Biochim. Biophys. Acta* **822**: 63–125.

Drews, J., 2000, Drug discovery: a historical perspective. *Science* **287**: 1960–1964.

Epand, R.F., Thomas, A., Brasseur, R., Vishwanathan, S.A., Hunter, E., Epand, R.M., 2006, Juxtamembrane protein segments that contribute to recruitment of cholesterol into domains. *Biochemistry* **45**: 6105–6114.

Epand, R.M., 2006, Cholesterol and the interaction of proteins with membrane domains. *Prog. Lipid Res.* **45**: 279–294.

Eroglu, C., Cronet, P., Panneels, V., Beaufils, P., Sinning, I., 2002, Functional reconstitution of purified metabotropic glutamate receptor expressed in the fly eye. *EMBO Rep.* **3**: 491–496.

Eroglu, C., Brugger, B., Wieland, F., Sinning, I., 2003, Glutamate-binding affinity of *Drosophila* metabotropic glutamate receptor is modulated by association with lipid rafts. *Proc. Natl. Acad. Sci. USA* **100**: 10219–10224.

Fahrenholz, F., Klein, U., Gimpl, G., 1995, Conversion of the myometrial oxytocin receptor from low to high affinity state by cholesterol. *Adv. Exp. Med. Biol.* **395**: 311–319.

Fargin, A., Raymond, J.R., Lohse, M.J., Kobilka, B.K., Caron, M.G., Lefkowitz. R.J., 1988, The genomic clone G-21 which resembles a β-adrenergic receptor sequence encodes the 5-HT$_{1A}$ receptor. *Nature* **335**: 358–360.

Fredriksson, R., Schiöth, H.B., 2005, The repertoire of G-protein-coupled receptors in fully sequenced genomes. *Mol. Pharmacol.* **67**: 1414–1425.

Gaoua, W., Wolf, C., Chevy, F., Ilien, F., Roux, C., 2000, Cholesterol deficit but not accumulation of aberrant sterols is the major cause of the teratogenic activity in the Smith-Lemli-Opitz syndrome animal model. *J. Lipid Res.* **41**: 637–646.

Gardier, A.M., 2009, Mutant mouse models and antidepressant drug research: focus on serotonin and brain-derived neurotrophic factor. *Behav. Pharmacol.* **20**: 18–32.

Gether, U., 2000, Uncovering molecular mechanisms involved in activation of G protein-coupled receptors. *Endocr. Rev.* **21**: 90–113.

Gether, U., Kobilka, B.K., 1998, G protein-coupled receptors. II. Mechanism of agonist activation. *J. Biol. Chem.* **273**:17979–17982.

Gimpl, G., Klein, U., Reiländer, H., Fahrenholz, F., 1995, Expression of the human oxytocin receptor in baculovirus-infected insect cells: high-affinity binding is induced by a cholesterol-cyclodextrin complex. *Biochemistry* **34**: 13794–13801.

Gimpl, G., Burger, K., Fahrenholz, F., 1997, Cholesterol as modulator of receptor function. *Biochemistry* **36**: 10959–10974.

Gimpl, G., Fahrenholz, F., 2002, Cholesterol as stabilizer of the oxytocin receptor. *Biochim. Biophys. Acta* **1564**: 384–392.

Gimpl, G., Burger, K., Fahrenholz, F., 2002a, A closer look at the cholesterol sensor. *Trends Biochem. Sci.* **27**: 596–599.

Gimpl, G., Wiegand, V., Burger, K., Fahrenholz, F., 2002b, Cholesterol and steroid hormones: modulators of oxytocin receptor function. *Prog. Brain Res.* **139**: 43–55.

Gimpl, G., Gehrig-Burger, K., 2007, Cholesterol reporter molecules. *Biosci. Rep.* **27**: 335–358.

Granseth, E., Seppälä, S., Rappi, M., Daleyi, D.O., von Heijne, G., 2007, Membrane protein structural biology-How far can the bugs take us? *Mol. Membr. Biol.* **24**: 329–332.

Hanson, M.A., Cherezov, V., Griffith, M.T., Roth, C.B., Jaakola, V.-P., Chien, E.Y.T., Velasquez, J., Kuhn, P., Stevens, R.C., 2008, A specific cholesterol binding site is established by the 2.8 Å structure of the human β_2-adrenergic receptor. *Structure* **16**: 897–905.

Harikumar, K.G., Puri, V., Singh, R.D., Hanada, K., Pagano, R.E., Miller, L.J., 2005, Differential effects of modification of membrane cholesterol and sphingolipids on the conformation, function, and trafficking of the G protein-coupled cholecystokinin receptor. *J. Biol. Chem.* **280**: 2176–2185.

Harris, J.S., Epps, D.E., Davio, S.R., Kezdy, F.J., 1995, Evidence for transbilayer, tail-to-tail cholesterol dimers in dipalmitoylglycerophosphocholine liposomes. *Biochemistry* **34**: 3851–3857.

Heilker, R., Wolff, M., Tautermann, C.S., Bieler, M., 2009, G-protein-coupled receptor-focused drug discovery using a target class platform approach. *Drug Discov. Today* **14**: 231–240.

Hoyer, D., Hannon, J.P., Martin, G.R., 2002, Molecular, pharmacological and functional diversity of 5-HT receptors. *Pharmacol. Biochem. Behav.* **71**: 533–554.

Huang, P., Xu, W., Yoon, S.-I., Chen, C., Chong, P.L.-G., Liu-Chen, L.-Y., 2007, Cholesterol reduction by methyl-β-cyclodextrin attenuates the delta opioid receptor-mediated signaling in neuronal cells but enhances it in non-neuronal cells. *Biochem. Pharmacol.* **73**: 534–549.

Huber, T., Botelho, A.V., Beyer, K., Brown, M.F., 2004, Membrane model for the G-protein-coupled receptor rhodopsin: hydrophobic interface and dynamical structure. *Biophys. J.* **86**: 2078–2100.

Insel, P.A., Tang, C.M., Hahntow, I., Michel, M.C., 2007, Impact of GPCRs in clinical medicine: monogenic diseases, genetic variants and drug targets. *Biochim. Biophys. Acta* **1768**: 994–1005.

Jacob, R.F., Cenedella, R.J., Mason, R.P., 1999, Direct evidence for immiscible cholesterol domains in human ocular lens fiber cell plasma membranes. *J. Biol. Chem.* **274**: 31613–31618.

Jacob, R.F., Cenedella, R.J., Mason, R.P., 2001, Evidence for distinct cholesterol domains in fiber cell membranes from cataractous human lenses. *J. Biol. Chem.* **276**: 13573–13578.

Jacoby, E., Bouhelal, R., Gerspacher, M., Seuwen, K., 2006, The 7TM G-protein-coupled receptor target family. *Chem. Med. Chem.* **1**: 760–782.

Jamin, N., Neumann, J.M., Ostuni, M.A., Vu, T.K., Yao, Z.X., Murail, S., Robert, J.C., Giatzakis, C., Papadopoulos, V., Lacapere, J.J., 2005, Characterization of the cholesterol recognition amino acid consensus sequence of the peripheral-type benzodiazepine receptor. *Mol. Endocrinol.* **19**: 588–594.

Jones, O.T., Eubanks, J.H., Earnest, J.P., McNamee, M.G., 1988, A minimum number of lipids are required to support the functional properties of the nicotinic acetylcholine receptor. *Biochemistry* **27**: 3733–3742.

Jones, O.T., McNamee, M.G., 1988, Annular and nonannular binding sites for cholesterol associated with the nicotinic acetylcholine receptor. *Biochemistry* **27**: 2364–2374.

Jost, P.C., Griffith, O.H., Capaldi, R.A., Vanderkooi, G., 1973, Evidence for boundary lipids in membranes. *Proc. Natl. Acad. Sci. USA* **70**: 480–484.

Kalipatnapu, S., Chattopadhyay, A., 2007, Membrane organization and function of the serotonin$_{1A}$ receptor. *Cell. Mol. Neurobiol.* **27**: 1097–1116.

Kirilovsky, J., Schramm, M., 1983, Delipidation of a β-adrenergic receptor preparation and reconstitution by specific lipids. *J. Biol. Chem.* **258**: 6841–6849.

Kirilovsky, J., Eimerl, S., Steiner-Mordoch, S., Schramm, M., 1987, Function of the delipidated β-adrenergic receptor appears to require a fatty acid or a neutral lipid in addition to phospholipids. *Eur. J. Biochem.* **166**: 221–228.

Klein, U., Gimpl, G., Fahrenholz, F., 1995, Alteration of the myometrial plasma membrane cholesterol content with beta-cyclodextrin modulates the binding affinity of the oxytocin receptor. *Biochemistry* **34**: 13784–13793.

Kobilka, B.K., Frielle, T., Collins, S., Yang-Feng, T., Kobilka, T.S., Francke, U., Lefkowitz, R.J., Caron, M.G., 1987, An intronless gene encoding a potential member of the family of receptors coupled to guanine nucleotide regulatory proteins. *Nature* **329**: 75–79.

Kodner, R.B., Summons, R.E., Pearson, A., King, N., Knoll, A.H., 2008, Sterols in unicellular relative of the metazoans. *Proc. Natl. Acad. Sci. USA* **105**: 9897–9902.

Kuwabara, P.E., Labouesse, M., 2002, The sterol-sensing domain: multiple families, a unique role. *Trends Genet.* **18**: 193–201.

Lagane, B., Gaibelet, G., Meilhoc, E., Masson, J.-M., Cézanne, L., Lopez, A., 2000, Role of sterols in modulating the human μ-opioid receptor function in *Saccharomyces cerevisiae*. *J. Biol. Chem.* **275**: 33197–33200.

Lange, Y., Steck, T.L., 2008, Cholesterol homeostasis and the escape tendency (activity) of plasma membrane cholesterol. *Prog. Lipid Res.* **47**: 319–332.

Lee, A.G., East, J.M., Jones, O.T., McWhirter, J., Rooney, E.K., Simmonds, A.C., 1982, Interaction of fatty acids with the calcium-magnesium ion dependent adenosinetriphosphatase from sarcoplasmic reticulum. *Biochemistry* **21**: 6441–6446.

Lee, A.G., 2003, Lipid-protein interactions in biological membranes: a structural perspective. *Biochim. Biophys. Acta* **1612**: 1–40.

Lee, A.G., 2004, How lipids affect the activities of integral membrane proteins. *Biochim. Biophys. Acta* **1666**: 62–87.

Lee, A.G., 2005, How lipids and proteins interact in a membrane: a molecular approach. *Mol. BioSyst.* **1**: 203–212.

Levitt, E.S., Clark, M.J., Jenkins, P.M., Martens, J.R., Traynor, J.R., 2009, Differential effect of membrane cholesterol removal on μ and δ opioid receptors: a parallel comparison of acute and chronic signaling to adenylyl cyclase. *J. Biol. Chem.* **284**: 22108–22122.

Li, H., Papadopoulos, V., 1998, Peripheral-type benzodiazepine receptor function in cholesterol transport. Identification of a putative cholesterol recognition/interaction amino acid sequence and consensus pattern. *Endocrinology* **139**: 4991–4997.

Lin, S.H., Civelli, O., 2004, Orphan G protein-coupled receptors: targets for new therapeutic interventions. *Ann. Med.* **36**: 204–214.

Liscum, L., Underwood, K.W., 1995, Intracellular cholesterol transport and compartmentation. *J. Biol. Chem.* **270**: 15443–15446.

Liu, Y., Engelman, D.M., Gerstein, M., 2002, Genomic analysis of membrane protein families: abundance and conserved motifs. *Genome Biol.* **3**: 1–12.

Loura, L.M.S., Prieto, M., 1997, Dehydroergosterol structural organization in aqueous medium and in a model system of membranes. *Biophys. J.* **72**: 2226–2236.

Marius, P., Alvis, S.J., East, J.M., Lee, A.G., 2005, The interfacial lipid binding site on the potassium channel KcsA is specific for anionic phospholipids. *Biophys. J.* **89**: 4081–4089.

Marius, P., Zagnoni, M., Sandison, M.E., East, J.M., Morgan, H., Lee, A.G., 2008, Binding of anionic lipids to at least three nonannular sites on the potassium channel KcsA is required for channel opening. *Biophys. J.* **94**: 1689–1698.

Marsh, D., 1990, Lipid-protein interactions in membranes. *FEBS Lett.* **268**: 371–375.

Mason, R.P., Tulenko, T.N., Jacob, R.F., 2003, Direct evidence for cholesterol crystalline domains in biological membranes: role in human pathobiology. *Biochim. Biophys. Acta* **1610**: 198–207.

Milligan, G., Parenti, M., Magee, A.I., 1995, The dynamic role of palmitoylation in signal transduction. *Trends Biochem. Sci.* **20**: 181–187.

Mitchell, D.C., Straume, M., Miller, J.L., Litman, B.J., 1990, Modulation of metarhodopsin formation by cholesterol-induced ordering of bilayer lipids. *Biochemistry* **29**: 9143–9149.

Monastyrskaya, K., Hostettler, A., Buergi, S., Draeger, A., 2005, The NK1 receptor localizes to the plasma membrane microdomains, and its activation is dependent on lipid raft integrity. *J. Biol. Chem.* **280**: 7135–7146.

Mouritsen, O.G., Zuckermann, M.J., 2004, What's so special about cholesterol? *Lipids* **39**: 1101–1113.

Mukherjee, S., Chattopadhyay, A., 1996, Membrane organization at low cholesterol concentrations: a study using 7-nitrobenz-2-oxa-1,3-diazol-4-yl-labeled cholesterol. *Biochemistry* **35**: 1311–1322.

Mukherjee, S., Maxfield, F.R., 2004, Membrane domains. *Annu. Rev. Cell Dev. Biol.* **20**: 839–866.

Mukherjee, S., Chattopadhyay, A., 2005, Monitoring cholesterol organization in membranes at low concentrations utilizing the wavelength-selective fluorescence approach. *Chem. Phys. Lipids* **134**: 79–84.

Nature reviews drug discovery GPCR questionnaire participants, The state of GPCR research in 2004. *Nat. Rev. Drug Discov.* **3**: 577–626.

Nguyen, D.H., Taub, D., 2002a, CXCR4 function requires membrane cholesterol: implications for HIV infection. *J. Immunol.* **168**: 4121–4126.

Nguyen, D.H., Taub, D., 2002b, Cholesterol is essential for macrophage inflammatory protein 1 beta binding and conformational integrity of CC chemokine receptor 5. *Blood* **99**: 4298–4306.

Nguyen, D.H., Taub, D.D., 2003, Inhibition of chemokine receptor function by membrane cholesterol oxidation. *Exp. Cell Res.* **291**: 36–45.

Niu, S.L., Mitchell, D.C., Litman, B.J., 2002, Manipulation of cholesterol levels in rod disk membranes by methyl-β-cyclodextrin: effects on receptor activation. *J. Biol. Chem.* **277**: 20139–20145.

Nyholm, T.K.M., Özdirekcan, S., Killian, J.A., 2007, How transmembrane segments sense the lipid environment. *Biochemistry* **46**: 1457–1465.

Ohvo-Rekilä, H., Ramstedt, B., Leppimäki, P., Slotte, J.P., 2002, Cholesterol interactions with phospholipids in membranes. *Prog. Lipid Res.* **41**: 66–97.

Opekarová, M., Tanner, W., 2003, Specific lipid requirements of membrane proteins – a putative bottleneck in heterologous expression. *Biochim. Biophys. Acta* **1610**: 11–22.

Paila, Y.D., Pucadyil, T.J., Chattopadhyay, A., 2005, The cholesterol-complexing agent digitonin modulates ligand binding of the bovine hippocampal serotonin$_{1A}$ receptor. *Mol. Membr. Biol.* **22**: 241–249.

Paila, Y.D., Murty, M.R.V.S., Vairamani, M., Chattopadhyay, A., 2008, signaling by the human serotonin$_{1A}$ receptor is impaired in cellular model of Smith-Lemli-Opitz Syndrome. *Biochim. Biophys. Acta* **1778**: 1508–1516.

Paila, Y.D., Chattopadhyay, A., 2009, The function of G-protein coupled receptors and membrane cholesterol: specific or general interaction? *Glycoconj. J.* **26**: 711–720.

Paila, Y.D., Tiwari, S., Chattopadhyay, A., 2009, Are specific nonannular cholesterol binding sites present in G-protein coupled receptors ? *Biochim. Biophys. Acta* **1788**: 295–302.

Palsdottir, H., Hunte, C., 2004, Lipids in membrane protein structures. *Biochim. Biophys. Acta* **1666**: 2–18.

Pandit, S.A., Jakobsson, E., Scott, H.L., 2004, Simulation of the early stages of nano-domain formation in mixed bilayers of sphingomyelin, cholesterol, and dioleylphosphatidylcholine. *Biophys. J.* **87**: 3312–3322.

Pang, L., Graziano, M., Wang, S., 1999, Membrane cholesterol modulates galanin-GalR2 interaction. *Biochemistry* **38**: 12003–12011.

Papoucheva, E., Dumuis, A., Sebben, M., Richter, D.W., Ponimaskin, E.G., 2004, The 5-hydroxytryptamine$_{1A}$ receptor is stably palmitoylated, and acylation is critical for communication of receptor with Gi protein. *J. Biol. Chem.* **279**:3280–3291.

Park, P.S., Filipek, S., Wells, J.W., Palczewski, K., 2004, Oligomerization of G protein-coupled receptors: past, present, and future. *Biochemistry* **43**: 15643–15656.

Pavlidis, P., Ramaswami, M., Tanouye, M.A., 1994, The Drosophila easily shocked gene: a mutation in a phospholipid synthetic pathway causes seizure, neuronal failure, and paralysis. *Cell* **79**: 23–33.

Perez, D.M., 2003, The evolutionarily triumphant G-protein-coupled receptor. *Mol. Pharmacol.* **63**: 1202–1205.

Pierce, K.L., Premont, R.T., Lefkowitz, R.J., 2002, Seven-transmembrane receptors. *Nat. Rev. Mol. Cell Biol.* **3**: 639–650.

Pitman, M.C., Grossfield, A., Suits, F., Feller, S.E., 2005, Role of cholesterol and polyunsaturated chains in lipid-protein interactions: molecular dynamics simulation of rhodopsin in a realistic membrane environment. *J. Am. Chem. Soc.* **127**: 4576–4577.

Politowska, E., Kazmierkiewicz, R., Wiegand, V., Fahrenholz, F., Ciarkowski, J., 2001, Molecular modeling study of the role of cholesterol in the stimulation of the oxytocin receptor. *Acta Biochim. Pol.* **48**: 83–93.

Polozova, A., Litman, B.J., 2000, Cholesterol dependent recruitment of di22:6-PC by a G protein-coupled receptor into lateral domains. *Biophys. J.* **79**: 2632–2643.

Porter, F.D., 2008, Smith-Lemli-Opitz syndrome: pathogenesis, diagnosis and management. *Eur. J. Hum. Genet.* **16**: 535–541.

Pöyry, S., Róg, T., Karttunen, M., Vattulainen, I., 2008, Significance of cholesterol methyl groups. *J. Phys. Chem. B* **112**: 2922–2929.

Prasad, R., Singh, P., Chattopadhyay, A., 2009, Effect of capsaicin on ligand binding activity of the hippocampal serotonin$_{1A}$ receptor. *Glycoconj. J.* **26**: 733–738.

Pucadyil, T.J., Chattopadhyay, A., 2004, Cholesterol modulates the ligand binding and G-protein coupling to serotonin$_{1A}$ receptors from bovine hippocampus. *Biochim. Biophys. Acta* **1663**: 188–200.

Pucadyil, T.J., Shrivastava, S., Chattopadhyay, A., 2004, The sterol-binding antibiotic nystatin differentially modulates ligand binding of the bovine hippocampal serotonin$_{1A}$ receptor. *Biochem. Biophys. Res. Commun.* **320**: 557–562.

Pucadyil, T.J., Chattopadhyay, A., 2005, Cholesterol modulates the antagonist-binding function of bovine hippocampal serotonin$_{1A}$ receptors. *Biochim. Biophys. Acta* **1714**: 35–42.

Pucadyil, T.J., Kalipatnapu, S., Chattopadhyay, A., 2005a, The serotonin$_{1A}$ receptor: A representative member of the serotonin receptor family. *Cell. Mol. Neurobiol.* **25**: 553–580.

Pucadyil, T.J., Shrivastava, S., Chattopadhyay, A., 2005b, Membrane cholesterol oxidation inhibits ligand binding function of hippocampal serotonin$_{1A}$ receptors. *Biochem. Biophys. Res. Commun.* **331**: 422–427.

Pucadyil, T.J., Chattopadhyay, A., 2006, Role of cholesterol in the function and organization of G-protein coupled receptors. *Prog. Lipid Res.* **45**: 295–333.

Pucadyil, T.J., Chattopadhyay, A., 2007, Cholesterol: a potential therapeutic target in Leishmania infection? *Trends Parasitol.* **23**: 49–53.

Pucadyil, T.J., Mukherjee, S., Chattopadhyay, A., 2007, Organization and dynamics of NBD-labeled lipids in membranes analyzed by fluorescence recovery after photobleaching. *J. Phys. Chem. B* **111**: 1975–1983.

Raymond, J.R., Mukhin, Y.V., Gettys, T.W., Garnovskaya, M.N., 1999, The recombinant 5-HT$_{1A}$ receptor: G protein coupling and signaling pathways. *Br. J. Pharmacol.* **27**: 1751–1764.

Riethmüller, J., Riehle, A., Grassmé, H., Gulbins, E., 2006, Membrane rafts in host-pathogen interactions. *Biochim. Biophys. Acta* **1758**: 2139–2147.

Róg, T., Pasenkiewicz-Gierula, M., 2001, Cholesterol effects on the phosphatidylcholine bilayer nonpolar region: A molecular simulation study. *Biophys. J.* **81**: 2190–2202.

Róg, T., Pasenkiewicz-Gierula, M., Vattulainen, I., Karttunen, M., 2007, What happens if cholesterol is made smoother: importance of methyl substituents in cholesterol ring structure on phosphatidylcholine-sterol interaction. *Biophys. J.* **92**: 3346–3357.

Rosenbaum, D.M., Rasmussen, S.G.F., Kobilka, B.K., 2009, The structure and function of G-protein-coupled receptors. *Nature* **459**: 356–363.

Rukmini, R., Rawat, S.S., Biswas, S.C., Chattopadhyay, A., 2001, Cholesterol organization in membranes at low concentrations: effects of curvature stress and membrane thickness. *Biophys. J.* **81**: 2122–2134.

Ruprecht, J.J., Mielke, T., Vogel, R., Villa, C., Schertler, G.F., 2004, Electron crystallography reveals the structure of metarhodopsin I. *EMBO J.* **23**: 3609–3620.

Schlyer, S., Horuk, R., 2006, I want a new drug: G-protein-coupled receptors in drug development. *Drug Discov. Today* **11**: 481–493.

Schroeder, F., Woodford, J.K., Kavecansky, J., Wood, W.G., Joiner, C., 1995, Cholesterol domains in biological membranes. *Mol. Membr. Biol.* **12**: 113–119.

Shrivastava, S., Paila, Y.D., Dutta, A., Chattopadhyay, A., 2008, Differential effects of cholesterol and its immediate biosynthetic precursors on membrane organization. *Biochemistry* **47**: 5668–5677.

Simmonds, A.C., East, J.M., Jones, O.T., Rooney, E.K., McWhirter, J., Lee, A.G., 1982, Annular and nonannular binding sites on the ($Ca^{2+} + Mg^{2+}$)-ATPase. *Biochim. Biophys. Acta* **693**: 398–406.

Simons, K., van Meer, G., 1988, Lipid sorting in epithelial cells. *Biochemistry* **27**: 6197–6202.

Simons, K., Ikonen, E., 1997, Functional rafts in cell membranes. *Nature* **387**: 569–572.

Simons, K., Ikonen, E., 2000, How cells handle cholesterol. *Science* **290**: 1721–1725.

Simons, K., Toomre, D., 2000, Lipid rafts and signal transduction. *Nat. Rev. Mol. Cell Biol.* **1**: 31–39.

Simons, K., Ehehalt, R., 2002, Cholesterol, lipid rafts, and disease. *J. Clin. Invest.* **110**: 597–603.

Singh, P., Paila, Y.D., Chattopadhyay, A., 2007, Differential effects of cholesterol and 7-dehydrocholesterol on the ligand binding activity of the hippocampal serotonin$_{1A}$ receptors: Implications in SLOS. *Biochem. Biophys. Res. Commun.* **358**: 495–499.

Sjögren, B., Hamblin, M.W., Svenningsson, P., 2006, Cholesterol depletion reduces serotonin binding and signaling via human 5-HT$_{7(a)}$ receptors. *Eur. J. Pharmacol.* **552**: 1–10.

Straume, M., Litman, B.J., 1988, Equilibrium and dynamic bilayer structural properties of unsaturated acyl chain phosphatidylcholine-cholesterol-rhodopsin recombinant vesicles and rod outer segment disk membranes as determined from higher order analysis of fluorescence anisotropy decay. *Biochemistry* **27**: 7723–7733.

Toth, M., 2003, 5-HT$_{1A}$ receptor knockout mouse as a genetic model of anxiety. *Eur. J. Pharmacol.* **463**: 177–184.

Tulenko, T.N., Chen, M., Mason, P.E., Mason, R.P., 1998, Physical effects of cholesterol on arterial smooth muscle membranes: evidence of immiscible cholesterol domains and alterations in bilayer width during atherogenesis. *J. Lipid Res.* **39**: 947–956.

Villalaín, J., 1996, Location of cholesterol in model membranes by magic-angle-sample-spinning NMR. *Eur. J. Biochem.* **241**: 586–593.

Wolf, C., Chevy, F., Pham, J., Kolf-Clauw, M., Citadelle, D., Mulliez, N., Roux, C., 1996, Changes in serum sterols of rats treated with 7-dehydrocholesterol-Δ^7-reductase inhibitors: comparison to levels in humans with Smith-Lemli-Opitz syndrome. *J. Lipid Res.* **37**: 1325–1333.

Wüstner, D., 2007, Fluorescent sterols as tools in membrane biophysics and cell biology. *Chem. Phys. Lipids* **146**: 1–25.

Xu, W., Yoon, S.-I., Huang, P., Wang, Y., Chen, C., Chong, P.L.-G., Liu-Chen, L.-Y., 2006, Localization of the κ opioid receptor in lipid rafts. *J. Pharmacol. Exp. Ther.* **317**: 1295–1306.

Xu, X., London, E., 2000, The effect of sterol structure on membrane lipid domains reveals how cholesterol can induce lipid domain formation. *Biochemistry* **39**: 843–849.

Yao, Z., Kobilka, B., 2005, Using synthetic lipids to stabilize purified β_2 adrenoceptor in detergent micelles. *Anal. Biochem.* **343**: 344–346.

Yeagle, P.L., 1985, Cholesterol and the cell membrane. *Biochim. Biophys. Acta* **822**: 267–287.

Chapter 17
Cholesterol Effects on Nicotinic Acetylcholine Receptor: Cellular Aspects

Francisco J. Barrantes

Abstract Cholesterol is an essential partner of the nicotinic acetylcholine receptor (AChR). It is not only an abundant component of the postsynaptic membrane but also affects the stability of the receptor protein in the membrane, its supramolecular organization and function. In the absence of innervation, early on in ontogenetic development of the muscle cell, embryonic AChRs occur in the form of diffusely dispersed molecules. At embryonic day 13, receptors organize in the form of small aggregates. This organization can be mimicked in mammalian cells in culture.

Trafficking to the plasmalemma is a cholesterol-dependent process. Receptors acquire association with the sterol as early as the endoplasmic reticulum and the Golgi apparatus. Once AChRs reach the cell surface, their stability is also highly dependent on cholesterol levels. Acute cholesterol depletion reduces the number of receptor domains by accelerating the rate of endocytosis. In muscle cells, AChRs are internalized via a recently discovered dynamin- and clathrin-independent, cytoskeleton-dependent endocytic mechanism. Unlike other endocytic pathways, cholesterol depletion accelerates internalization and re-routes AChR endocytosis to an Arf6-dependent pathway. Cholesterol depletion also results in ion channel gain-of-function of the remaining cell-surface AChRs, whereas cholesterol enrichment has the opposite effect.

Wide-field microscopy shows AChR clusters as diffraction-limited puncta of ~200 nm diameter. Stimulated emission depletion (STED) fluorescence microscopy resolves these puncta into nanoclusters with an average diameter of ~55 nm. Exploiting the enhanced resolution, the effect of acute cholesterol depletion can be shown to alter the short- and long-range organization of AChR nanoclusters. In the short range, AChRs form bigger nanoclusters. On larger scales (0.5–3.5 µm) nanocluster distribution becomes non-random, attributable to the cholesterol-related abolition of cytoskeletal physical barriers normally preventing the lateral diffusion of AChR nanoclusters. The dependence of AChR numbers at the cell surface on membrane cholesterol raises the possibility that cholesterol depletion leads to AChR

F.J. Barrantes (✉)
UNESCO Chair of Biophysics and Molecular Neurobiology, Instituto de Investigaciones Bioquímicas de Bahía Blanca, C.C. 857, B8000FWB, Bahía Blanca, Argentina
e-mail: rtfjb1@criba.edu.ar

conformational changes that alter its stability and its long-range dynamic association with other AChR nanoclusters, accelerate its endocytosis, and transiently affect the channel kinetics of those receptors remaining at the surface. Cholesterol content at the plasmalemma may thus homeostatically modulate AChR dynamics, cell-surface organization and lifetime of receptor nanodomains, and fine tune the ion permeation process.

Keywords Cholesterol · Acetylcholine receptor · Lipid domains · Cyclodextrins · Fluorescence microscopy · Lipid–protein interactions · Cell-surface receptor · "Raft" lipid domains · Trafficking · Membrane · Nicotinic

Abbreviations

AChR	nicotinic acetylcholine receptor
αBTX	α-bungarotoxin
Chol	cholesterol
GP	generalized polarization
M-β-CDx	methyl-β-cyclodextrin
Chol- M-β-CDx	cholesterol-methyl-β-cyclodextrin
NMJ	neuromuscular junction
STED	stimulated emission depletion

17.1 Introduction

Cholesterol is an abundant component of the postsynaptic membrane where the nicotinic acetylcholine receptor (AChR) is located (*for early reviews, see* Barrantes 1979, 1989). As we learn more about the effects of this sterol on the properties of the AChR, the accumulated evidence indicates that cholesterol constitutes an essential partner in the life of the neurotransmitter receptor. The experimental evidence supporting this conclusion is extensive: cholesterol affects the structural and functional properties of the receptor protein, its trafficking from the site of synthesis to the cell surface, its spatiotemporal distribution and organization–including clustering- at the plasmalemma, its rate of endocytosis, and even single-channel behavior (*see reviews,* Barrantes, 2003, 2004, 2007). The most important message that emerges is that cholesterol effects on the AChR protein are multiple, are exerted at various levels of organization ranging from the molecular to the cellular level, and occur within time windows from milliseconds (single-channel properties) to hours (endocytic/exocytic trafficking), during ontogenetic development and adulthood.

Lipids in general, and cholesterol in particular, have been proposed to interact with the AChR protein either directly or indirectly (Barrantes, 2002). Direct interactions imply the binding of cholesterol to the AChR protein. Where this binding occurs and the precise nature of the interaction of this lipid with the AChR is still not known with precision, nor are the mechanisms by which these interactions are finally traduced into the observed epiphenomenological changes in the receptor's ligand binding affinity (Criado et al., 1982) or ion channel properties (Borroni et al., 2007). In this Chapter I discuss the influence of cholesterol on AChR properties, with special emphasis on the cellular aspects: cholesterol effects on the biosynthesis, trafficking and stability of the receptor at the cell membrane.

17.2 The Natural Scenario of AChR-Cholesterol Interactions

Postsynaptic receptor localization is crucial for synapse formation and function. In the postsynaptic membrane the AChR molecules are tightly packed at extraordinarily high concentrations – 10,000–20,000 particles per μm^2 (Barrantes, 1979, 1989) – in a lipid microenvironment that differs from the rest of the bulk membrane bilayer and has the biophysical properties of the liquid-ordered (l_o) lipid phase (Antollini et al., 1996). This biophysical description highlights one important feature of the medium in which AChRs occur: they are surrounded by a "lipid-belt" or "annular" lipid, that is the protein-vicinal lipid in the immediate perimeter of the protein. In the case of the AChR, this region was identified for the first time by means of electron spin resonance (ESR) techniques (Marsh and Barrantes, 1978). The main characteristic difference between the lipid immediately adjacent to the protein and the rest of the bilayer lipid is its mobility: the lipid mobility surrounding the AChR protein is reduced relative to that of the bulk membrane lipid, giving rise to a two-component ESR spectrum from which the number and selectivity of the lipids at the lipid-protein interface may be quantified. Spin-labelled sterols, phosphatidic acid, and fatty acids were also shown to associate preferentially with the AChR (Marsh et al., 1981; Ellena et al., 1983; Mantipragada et al., 2003).

Together with a few other neutral lipids, cholesterol has been found to influence the activity of various ion channels (Bolotina et al., 1989) and to be a key modulator of AChR function in particular (Criado et al., 1982; Jones and McNamee, 1988; *see reviews* in Barrantes 1983, 1989, 1993a,b, 2003, 2004). This modulation is exerted at different levels: molecular, cellular and metabolic. Since the two main functional abilities of the receptor protein are the recognition of the ligand and the subsequent opening and closure of its ion channel, over the course of recent decades, studies have addressed the functional modulation exerted by cholesterol on these separate but interconnected properties of the AChR. And indeed, modulatory roles have been found. Cholesterol and analogs are needed for the AChR to undergo agonist-induced affinity state transitions (Criado et al., 1982, 1984) and AChR-mediated ion influx increases as the membrane cholesterol content is raised to a certain concentration (Dalziel et al., 1980; Criado et al., 1982; McNamee et al., 1982). It is necessary to add cholesterol to AChR preparations reconstituted in pure phospholipid to increase the thermal stability of the protein induced by cholinergic ligands (Castresana et al., 1992; Fernández et al., 1993; Fernandez-Ballester et al., 1992, 1994; da Costa and Baezinger, 2009). When AChRs were reconstituted into lipid bilayers lacking cholesterol, agonists no longer stimulated cation flux (Rankin et al., 1997).

Cholesterol interacts with high affinity with the AChR, as demonstrated initially in reconstituted systems using planar lipid bilayers (Popot et al., 1978) and using ESR spectroscopy on reconstituted liposomes (Ellena et al., 1983) or native membranes (Marsh et al., 1981). Sterols in general were found to exhibit selectivity for the boundary lipid surrounding the AChR protein (Ellena et al. 1983; Marsh and Barrantes, 1978; Dreger et al., 1997) and cholesterol stabilizes AChR structure in reconstituted vesicles (Artigues et al., 1989; Fernandez-Ballester et al., 1992, 1994). I shall briefly discuss the potential sites for cholesterol binding and their implication on AChR structural stability, a subject that is currently a focus of attention.

17.3 Lipid-AChR Interactions at the Cellular Scale. Tentative Association of AChR Clusters with a Specific Subset of Lipid Domains, the Lipid "Rafts"

Proteins are seldom distributed uniformly in the plasmalemma; most often they are segregated into supramolecular aggregates that range from the nanometer scale, below the resolution of the light microscope (Jacobson and Dietrich, 1999) to the micrometer scale, well accessible to optical microscopy. The AChR is unique in this respect: not only does it occur at extraordinarily high densities in the postsynaptic membrane, as indicated in the preceding section, but such molecular aggregates occupy a relatively small proportion of the membrane surface, since these micron-scale clusters containing up to 10^6-10^7 molecules are highly concentrated in a restricted area in the fully developed neuromuscular junction in skeletal muscle or in the electromotor synapse in electric fish (Barrantes, 1979). Remarkably, just a few microns away from the synaptic region the AChR density drops sharply to values 100–500-fold lower. These low densities are also characteristically observed at early stages of embryonic muscle development, where the AChR protein undergoes changes from the monodisperse distribution at the surface of myoblasts in the embryo to the fully developed, several micron-sized clusters in the mature NMJ (Sanes and Lichtman, 2001). Analogously, the so-called lipid "raft" hypothesis postulates the existence of compositional inhomogeneities in the lipid content of the plasma membrane, in particular that sphingolipids and cholesterol are distributed non-homogeneously, occurring in laterally segregated domains or "rafts" with similar spatial organizations (Edidin, 1997; Brown and London, 2000). Raft lipid domains have been postulated to concentrate signalling molecules and various types of receptors in particular regions of the cell surface (Maxfield, 2002).

The association of AChR with lipid "rafts" was postulated on the basis of biochemical criteria: cold detergent extraction procedures combined with subcellular fractionation techniques resulting in detergent-resistant (DRM) and detergent-soluble fractions. The DRMs are thought to represent liquid order (l_o) domains, which coexist in the same membrane with liquid disorder (l_d) domains (Brown and London, 1998; 2000). The resistance of l_o domains to detergent solubilization is ascribed to the close packing of lipids in the l_o phase, which prevents detergent incorporation into the bilayer (Xu et al., 2001). The homomeric neuronal α7-type AChR was the first to be suggested to occur in lipid "rafts" at the surface of the somatic spines in chick ciliary ganglion sympathetic neurons (Bruses et al., 2001). Subsequently, muscle-type AChR was overexpressed in COS-7 cells (Marchand et al., 2002) and the receptor protein was found to be present in the insoluble fraction obtained by cold 1% Triton X-100 extraction followed by gradient centrifugation. The correlation between isolated detergent-resistant membranes in vitro and raft lipid domains in intact cells is still a contentious issue: the effects of detergents on biological membranes are much too complex and in many cases "raft" markers are spatially segregated in the membrane into physically distinct compartments, their association upon subcellular fractionation and detergent extraction being mantailored rather than reflecting the nanoscales organization in situ (*see reviews by* Kusumi and Suzuki, 2003; Lichtenberg et al., 2005).

More recently, morphological criteria such as diminution in the size or changes in the shape of large, micron-sized AChR clusters have been used as a measure of lipid raft-AChR association upon methyl-β-cyclodextrin (M-β-CDx)-mediated cholesterol extraction (Stetzkowski-Marden et al., 2006; Willmann et al., 2006). Independently of these caveats, these studies clearly indicate the sensitivity of AChR clusters, or the stability of such clusters (Willmann et al., 2006) to changes in cholesterol content.

Agrin is a protein that promotes NMJ maturation and maintenance by inducing, strengthening and sustaining AChR clustering via activation of the muscle-specific kinase, MuSK. In fact agrin activates an AChR "clustering cascade" by phosphorylating src family kinases in a rapsyn-dependent manner, downstream of MuSK (Mittaud et al., 2001). The clustering-promoting cascade is counteracted by another nerve-derived factor, the natural neurotransmitter acetylcholine, which disperses extrasynaptic AChR clusters (Lin et al., 2005; Chen et al., 2007). In a recent study Campagna and Fallon (2006) showed that a fragment of agrin induced AChR clustering in C2C12 myoblasts in culture, and that this phenomenon is sensitive to M-β-CDx-mediated cholesterol depletion and cholera toxin-triggered lipid "patching". C2C12 cells were treated with fluorescent-labeled cholera toxin B (which labels and aggregates ganglioside GM1 (Holmgren et al., 1973), and fluorescent αBTX, and circular-shaped 2–4 μm patches were measured. Overlap of the two signals was only partial. From this, the authors concluded that AChRs in C2C12 reside in lipid "rafts", and that agrin treatment increased by ∼3-fold the association of AChRs and the shape of the AChR clusters, within lipid rafts. Campagna and Fallon (2006) also showed that cholesterol depletion prior to agrin stimulation results in more sparse fluorescent αBTX-stained AChR clusters with atypical circular morphology instead of the usual narrow, elongated shape. The authors did not investigate the "raft" morphology in parallel. When M-β-CDx extraction was undertaken after agrin stimulation, AChR clusters were larger than normal and also atypical in shape.

Cholesterol has also been found to stabilize AChR clusters in denervated muscle in vivo and in nerve-muscle explants. In paralyzed muscles cholesterol triggered maturation of nerve sprout-induced AChR clusters into the adult-type, pretzel-shaped large clusters. A specific defect in AChR cluster stability in cultured double knockout myotubes carrying defective (src$^{-/-}$;fyn$^{-/-}$) kinases could be rescued, and clusters became stable upon addition of cholesterol (Willmann et al., 2006). When long-term cholesterol depletion is accomplished by metabolic inhibition of a key enzyme of cholesterol biosynthesis, cell-surface delivery of the AChR is disrupted in CHO cells (Pediconi et al., 2004). The latter results provide a possible explanation for the instability of the mature receptor clusters.

17.4 Cholesterol Sensitivity of AChR Exocytic Trafficking

Does the association of the AChR with cholesterol-sensitive regions occur exclusively at the plasma membrane? Marchand et al. (2002) suggested that the exocytic trafficking of the AChR could be mediated by cholesterol and sphingolipid-enriched microdomains, and found AChRs in Triton X-100 insoluble fractions from whole cells. Likewise, Zhu et al. (2006) and Stetzkowski-Marden et al. (2006) suggested the association of the AChR with "raft" domains in the Golgi complex, but their suggestion was based on experiments using DRMs from whole cells. In a recent study, we analyzed the distribution of the AChR in lipid domains resistant to detergent extraction (the so-called DRMs) prepared from intracellular membranes. Procedures resulting in depletion of cholesterol and sphingolipid levels were carried out to evaluate whether they affected the association of the receptor protein with intracellular lipid domains. Impairment of sphingolipid biosynthesis in CHO-K1/A5 cells, a clonal cell line expressing muscle-type AChR, resulted in a 40–50% decrease in the amount of AChR in DRMs from both Golgi- and endoplasmic reticulum-enriched membranes. Chronic metabolic cholesterol depletion by Mevinolin treatment produced similar changes. These results suggest that a pool of AChRs becomes associated with lipid domains early on in the endoplasmic reticulum, and that such association is sensitive to the sphingolipid and cholesterol content of the cell (Gallegos, Baier, Pediconi and Barrantes, in preparation). Disruption of these lipid domains by chronic cholesterol depletion could affect the insertion of the receptor in exocytic vesicles, impairing its correct delivery to the plasma membrane, with the concomitant accumulation of receptor molecules in the trans-Golgi/trans-Golgi network (Pediconi et al., 2004) and diminution in the number of AChRs at the cell surface.

In spite of all these recent efforts, no *direct* evidence has been produced to date unambiguously demonstrating the occurrence of the AChR in cholesterol or sphingolipid-enriched "raft" domains in situ. In the electrocyte, AChRs are located exclusively at the innervated, ventral cell surface, where they colocalize with e.g. some components of the so-called "raft" lipid domains, such as glycosphingolipids (Marcheselli et al., 1993), but the low resolution of this early morphological study precludes any firm conclusion. Higher resolution techniques will be needed to demonstrate the association of the AChR protein with the bona fide "raft" lipids, cholesterol and sphingolipids.

17.5 Cholesterol Sensitivity of AChR Endocytosis

Endocytosis of the AChR is clearly an important mechanism regulating the number of receptors at the cell surface, exerting neuromodulation at the neuromuscular junction and possibly playing a role in the synaptic plasticity and in the pathology of synapses in the central nervous system. We have recently characterized the endocytic mechanism operating on AChRs expressed heterologously in CHO cells or endogenously in C2C12 myocytes (Kumari et al., 2008). The endocytic internalization of the AChR is a rather slow process (Fig. 17.1). We have further shown that binding of αBTX or antibody-mediated crosslinking induces the internalization of cell-surface AChR to late endosomes (Kumari et al. 2008). Internalization occurs via sequestration of AChR-αBTX complexes in narrow, tubular, surface-connected compartments, indicated by differential surface-accessibility of fluorescently-tagged αBTX-AChR complexes to small and large molecules, and

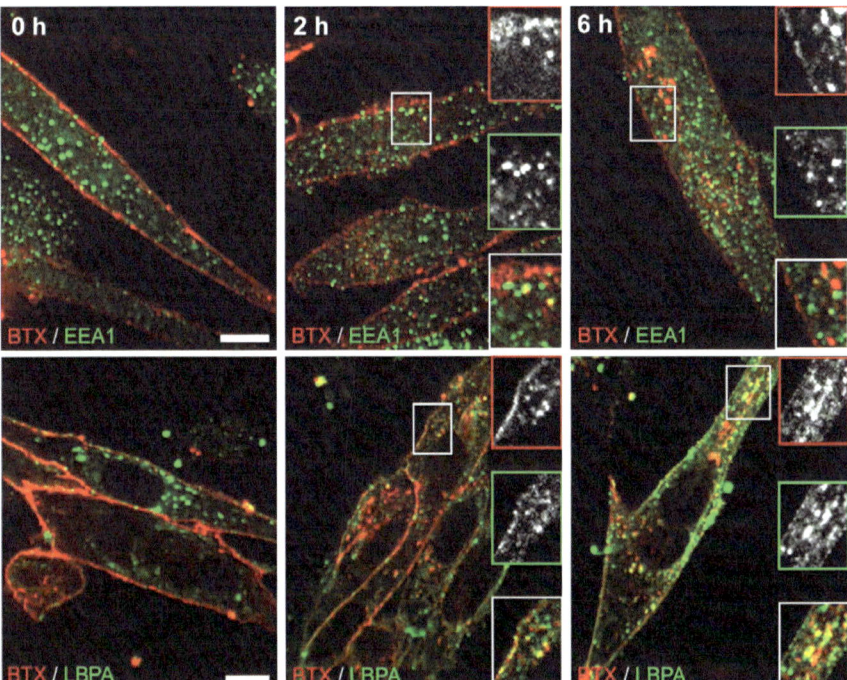

Fig. 17.1 Kinetics of AChR endocytosis in C2C12 myoblasts. Cells were labeled with Cy3αBTX and chased for 0, 2 or 6 h, washed, fixed, and processed for indirect immunofluorescence for the early endosomal marker EEA1 (*upper panels*) or the lysosomal marker LBPA (*lower panels*). Insets are magnified areas, marked by the rectangle on the image; box with red outline for αBTX, and with *green* outline for EEA1 or LBPA. Images of Cy3αBTX (*red*) and organelle specific markers (*green*) were collected from single slices using confocal microscopy and color-combined. Note that Cy3αBTX-bound to AChR is initially extensively colocalized with EEA1, and following a chase of 6 h it is located with the late endosomal marker. Scale bars, 10 μm. From Kumari et al. (2008)

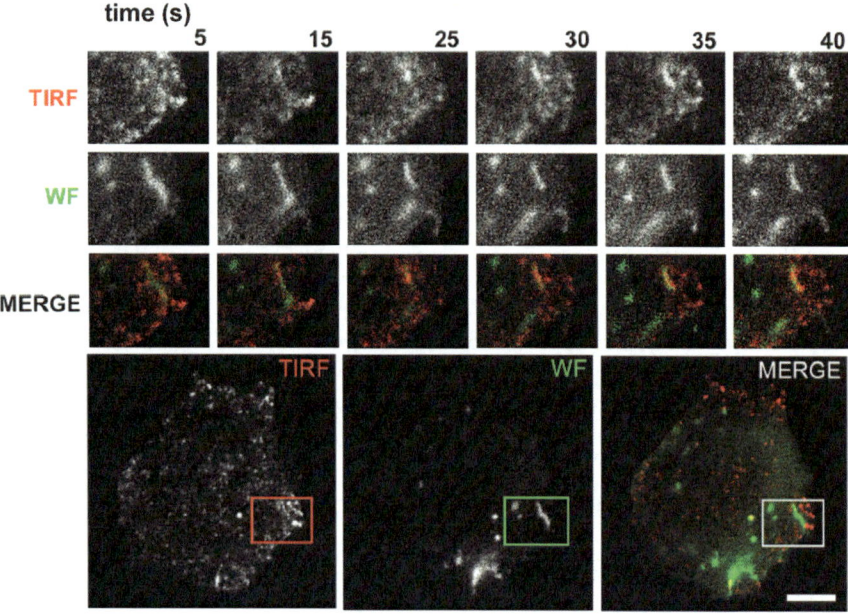

Fig. 17.2 Kinetics of AChR sequestration and internalization into surface-accessible tubular structures. Total internal reflection fluorescence (TIRF) imaging in combination with wide-field (WF) epifluorescence microscopy depict the presence of AChR in membrane proximal tubules prior to late endosomal delivery. CHO-K1/A5 cells were labeled with Alexa^{488}αBTX on ice and chased at 37°C for 2 h and imaged live, using TIRF microscopy. The figure shows snapshots at the indicated time in seconds from a movie [(TIRF images in *uppermost panels*; *middle panels* show their wide-field counterpart, and *lower panels* are pseudocolored merged images of the WF (*green*) and TIRF (*red*) images)], showing αBTX-labeled tubular structures close to the cell surface. The montage of insets is a magnified time-lapse sequence from a region outlined in the cell depicted in the *lowermost panels*, where *red* and *green* outlined boxes represent αBTX distribution in TIRF and WF, respectively. Scale, 10 μm. From Kumari et al. (2008)

real-time total intensity reflection fluorescence (TIRF) microscopy (Fig. 17.2). Internalization occurs in the absence of clathrin, caveolin or dynamin, but requires actin-polymerization. Furthermore, αBTX-binding triggers a cascade of reactions involving c-Src phosphorylation, and subsequent activation of the Rho GTPase Rac1. Consequently, inhibition of c-Src kinase activity, Rac1 activity or actin polymerization inhibits internalization via this unusual endocytic mechanism. This pathway may regulate AChR levels at ligand-gated synapses and in pathological conditions such as the autoimmune disease myasthenia gravis.

The plasma membrane is estimated to contain half of the total cellular cholesterol content in Chinese hamster ovary (CHO) cells (Warnock et al., 1993). Our laboratory has introduced a CHO-derived cell, CHO-K1/A5, that stably expresses adult-type muscle AChR (Roccamo et al., 1999). These cells are devoid of the AChR-anchoring proteins involved in AChR clustering, such as rapsyn and muscle-specific tyrosine kinases. Rapsyn is a scaffold protein that interacts with the cytoplasmic domain of the AChR and links AChRs to cytoskeletal proteins and also to other integral membrane proteins of the postsynaptic membrane, including tyrosine kinases that are receptors for nerve-derived factors that regulate AChR clustering. These latter kinases, and in particular *src* family kinases, are present in lipid "rafts" (Simons and Ikonen, 1997) and appear to be important in the assembly and stability of the adult NMJ. Thus, this cell line constitutes a minimalist mammalian cell expression system ideally suited to study the putative association of the AChR with lipid domains under conditions that mimic early stages of receptor development: absence of innervation and lack of scaffolding receptor-associated scaffolding proteins.

How does cholesterol affect the AChR internalization mechanism? We should first recall that in most cells, M-β-CDx-mediated cholesterol depletion *inhibits* clathrin and caveolar endocytic pathways, disrupts endosomal traffic (Le et al., 2002), perturbs the actin network (Kwik et al., 2003), and partially inhibits cholera toxin B uptake without affecting transferrin uptake (Kirkham et al., 2005). Thus, in general, cholesterol depletion severely hinders endocytic processes, slowing them down or bringing them to a complete standstill. The accelerated internalization of a transmembrane protein, as observed with the AChR (Borroni et al., 2007), appears to be an exception. Normally, AChRs submicron-sized puncta (240–280 nm) remain stable at the cell-surface membrane of CHO-K1/A5 cells over a period of hours. Concomitant with the decrease in cholesterol content, the fluorescent staining of AChRs sub-micron domains in CHO-K1/A5 cells stained with a fluorescent adduct of the competitive antagonist α-bungarotoxin (Alexa^{488}-αBTX) diminished by ~50%, in agreement with independent estimates from [^{125}I] αBTX binding and whole-cell patch-clamp recording experiments (Borroni et al., 2007). Surface Alexa^{488}-αBTX fluorescence, with a rate of disappearance $t_{1/2}$ of 1.5 h in control cells, diminished to 0.5 h in cholesterol-depleted cells. The accelerated internalization was mirrored by the appearance of vesicular structures inside the cells (Fig. 17.3). In addition, cholesterol depletion produced ion channel gain-of-function of the remaining cell-surface AChR, whereas cholesterol enrichment had the opposite effect. Fluorescence measurements under conditions of direct excitation of the probe Laurdan and of Förster-type resonance energy transfer (FRET) using the intrinsic protein fluorescence as donor both indicated an increase in membrane fluidity in the bulk membrane and in the immediate environment of the AChR protein upon cholesterol depletion. It is worth pointing out that cholesterol-depleted CHO cells do not, by themselves, replenish cholesterol within the time range (Vrljic et al., 2005). Other constitutive cell-surface proteins were not affected by cholesterol depletion; e.g. the cell-surface fluorescence intensity of the transferrin receptor did not decrease but in fact increased significantly, in agreement with literature reports (*reviewed by* Pichler and Riezman, 2004; Subtil et al.,1999).

Zhu et al. (2006) reported that cholesterol depletion by M-β-CDx (0–2 mM) did not affect AChR expression in C2C12 differentiated myoblasts subjected to agrin stimulation. Neural agrin produced a dramatic increase (30-fold) in the aggregation of AChR into micron-sized clusters displaying a longer lifetime (Phillips et al.,

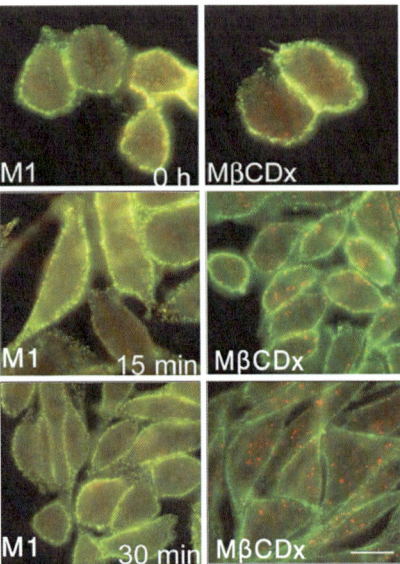

Fig. 17.3 Cholesterol depletion decreases the number of cell-surface AChRs and shifts their distribution in living cells. (**A**) CHO-K1/A5 cells, heterologously expressing adult-type AChR were first labeled with Alexa^{596}-αBTX (*red*) for 1 h at 4°C and incubated at the indicated times at 37°C with M1 buffer (control) or 15 mM methyl-β-cyclodextrin (MβCDx) and subsequently labeled with Alexa^{488}-αBTX for 1 h at 4°C and examined by fluorescence microscopy. The time-course of AChR internalization can be followed by simple inspection, as well as the acceleration of the process associated with cholesterol depletion: The AChR-containing endosomes can be seen as distinct (*red*) puncta inside the cells in cholesterol-depleted cells as early as 15 min after MβCDx treatment. Scale, 10 μm (from Borroni et al., 2007)

1997). According to Zhu et al. (2006), highly stable agrin-induced AChR clusters appear to be cholesterol insensitive, at variance with other studies on micron-sized (Bruses et al., 2001; Marchand et al., 2002; Stetzkowski-Marden et al., 2006; Willmann et al., 2006) or nanometer-sized, agrin- and rapsyn-less AChR clusters (Borroni et al., 2007; Kellner et al., 2007). One possible explanation for the fact that the results of Zhu et al. (2006) differ from those in the rest of the literature could be that the M-β-CDx concentrations they used were insufficient to achieve some critically low cholesterol level required for triggering AChR endocytosis in C2C12 cells. AChR endocytosis in response to CDx treatment is a dose-dependent phenomenon (Borroni and Barrantes, 2009, unpublished results).

We recently explored the possible pathway(s) involved in receptor loss in cholesterol-depleted cells (Borroni and Barrantes, in preparation). We found that AChRs maintain their clathrin- and dynamin-independence and utilize an endocytic mechanism that does not involve the presence of the AChR-associated protein rapsyn. The small GTPase Rac1 is also required: expression of a dominant negative form of Rac1, Rac1N17, abrogates receptor endocytosis. However, at variance with the default endocytic pathway in control CHO cells, the accelerated AChR internalization proceeds even upon disruption of the cortical actin cytoskeleton and does not depend on the cytoskeleton-associated inositol lipid $PI(4,5)P_2$; its sequestration by the PH domain of phospholipase C does not alter internalization. AChR endocytosis is, furthermore, found to require the activity of Arf6 and its effectors Rac1 and phospholipase D. Thus, this non-canonical cholesterol-sensitive mechanism constitutes a new alternative Arf6-dependent route for AChR endocytic internalization.

17.6 "Diffuse" AChRs Are in Fact Organized in the Form of "Nanoclusters" at the Cell Surface

Conventional far-field epifluorescence and confocal microscopies fall short of resolving the fine structure of the sub-micron sized AChR aggregates owing to their diffraction-limited resolving power. Thus, in conventional (wide-field) fluorescence microscopy, AChRs are observed as diffusely distributed submicron-sized puncta all over the surface of CHO-K1/A5 cells (Borroni et al., 2007). Reducing the dimensionality of the specimens by "unroofing" the cells and thus imaging only the coverslip-adhered ventral surface of the cells enabled the visualization of AChR fluorescent spots of ~0.25 μm, still beyond the resolution of conventional light microscopy. When cells were subjected to cholesterol depletion by acute CDx treatment and treatment with receptor-specific antibodies, although no changes could be observed in the mean diameter of the spots, the mean fluorescence intensity increased by ~50% with respect to the spots in control specimens (Borroni et al., 2007), suggesting the antibody-mediated recruitment of AChRs into the diffraction-limited puncta. The sub-micron sized AChR domains are much smaller than the several micron-sized (macro) clusters observed at later stages in developing muscle cells or in the adult NMJ (Sanes and Lichtman 2001). Plaque-shaped AChR clusters are stabilized and adopt a pretzel-shaped morphology, with AChRs located at the crests of the mature postjunctional folds.

The diffraction limit of far-field fluorescence microscopy can be overcome by applying new principles of physical optics (*reviewed by* Hell, 2009). There are various experimental modalities to accomplish super-resolution light microscopy. One such modality is termed STED (stimulated emission depletion microscopy). STED is considered a member of a new family of microscopy concepts that, despite using regular lenses, entails diffraction-unlimited resolution (Hell 1997; 2004). STED microscopy enabled the imaging of the supramolecular organization of the AChR below the diffraction limit (Kellner et al., 2007). Since the puncta represent a convolution of the particles with the finite effective point spread function, the actual protein aggregates have an estimated average diameter <55 nm, and are hereafter referred to as AChR "nanoclusters". This nomenclature takes into account the size of fully developed clusters in the plaque-shaped adult vertebrate NMJ (*reviewed in* Sanes and Lichtman, 2001; Willmann and Fuhrer, 2002) or

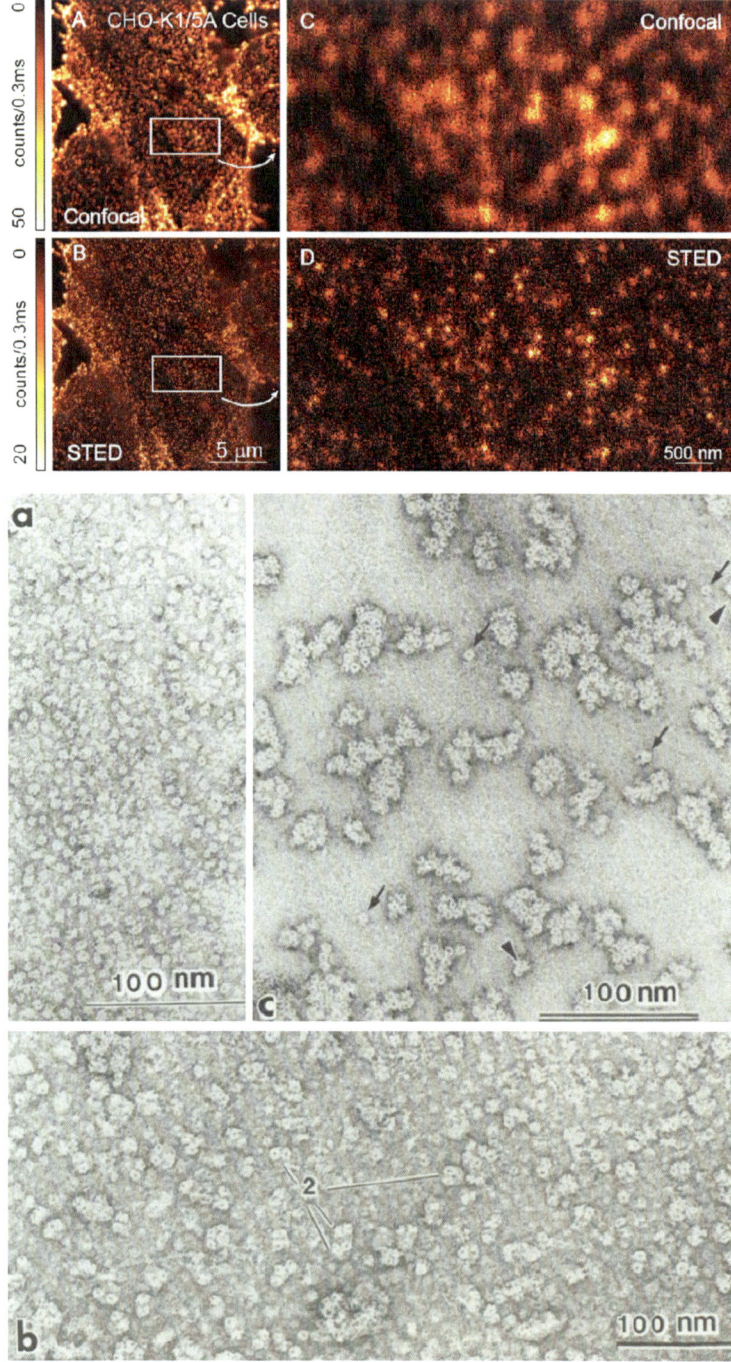

Fig. 17.4 (continued)

aneural C2 myotubes (23–94 μm in length, Kummer et al., 2004) and the smallest AChR "sub-micron aggregates" visualized by light microscopy in aneural myotubes (Kishi et al., 2005).

We analyzed the distribution of AChR clusters in the fluorescence images at larger scales by applying Ripley's K-function (Ripley, 1979). Representing a second-order analysis of spatial point patterns, the K-function searches for spatial randomness. As opposed to nearest neighbor methods, all inter-particle distances over the study area are incorporated into the analysis, thus providing a thorough topographical characterization which is compared to that of a complete spatial randomness pattern (Ripley, 1979). Cholesterol depletion was accompanied by an increase in long-range interactions (as compared to the nanometer scale of the AChR clusters themselves) and hence a change in AChR cluster distribution was revealed by STED microscopy (Kellner et al., 2007). The hypothesis of antibody-mediated AChR recruitment upon cholesterol depletion (Borroni et al., 2007) received strong support from the STED microscopy data (Kellner et al., 2007). Both control AChR nanoclusters and those disclosed by STED microscopy upon cholesterol depletion exhibit a size distribution of tens of nanometers (Fig. 17.4, upper panel). Nanoclusters observed in negatively stained electron micrographs of *Torpedo* AChR in Triton X-100 appear as small oligomers ($n = 4$–6), dimers and monomers of AChR molecules, the latter with a minimum diameter of about 8 nm ((Fig. 17.4, lower panel) and see Barrantes, 1982). In other words, STED microscopy, using standard microcopical lenses and visible light, gives access to the AChR supramolecular organization at a scale so far restricted to the realms of electron microscopy.

17.7 How Do Cholesterol Levels Modulate AChR Stability at the Cell Membrane?

The physical state of the AChR-vicinal lipid in *Torpedo* electrocytes (Antollini et al., 1996) and more relevantly in CHO-K1/A5 cells (Zanello et al., 1996) is in the liquid-ordered (L_o) state. Cholesterol can modulate the physical state of the bulk bilayer

Fig. 17.4 AChR nanoclusters imaged on the cell surface of CHO-K1/A5 cells using confocal or STED microscopy and AChR oligomers imaged by electron microscopy. *Upper panel*: CHO-K1/A5 cells were fixed and labeled with mAb210 monoclonal anti-AChR antibody and Atto 532-labeled secondary antibodies. High magnification views of the marked areas provide a side-to-side comparison of the resolution achieved in confocal (A, C) and STED (C, D) microscopy respectively (from Kellner et al., 2007). *Lower panel:* Transmission electron micrographs of (**a**) Reduced and alkylated AChR monomers solubilized in Triton X-100. (**b**) Membranes prepared throughout in *N*-ethyl-maleimide are predominantly made up of the dimeric, 13S AChR species. (**c**) Dithio-nitrobenzoic acid-oxidized AChR-rich membranes. Higher-order oligomers are observed. Negative contrast (1% uranyl acetate). From Barrantes (1982)

and the AChR-vicinal lipid belt region (Borroni et al., 2007). How can we rationalize these observations in the framework of AChR stability at the cell surface? L_o domain stability or lifetime is a function of size and protein–protein interactions of constituent proteins (Hancock, 2006). This is an active, energy-dependent process that confines lipid domain size. Recent views of lipid domain dynamics suggest that L_o domains form spontaneously, diffuse laterally in the plasma membrane, but have a limited lifetime (Turner et al., 2005). In this hypothesis, the more stable L_o lipid–protein complexes can be captured by endocytic pathways that disassemble the complexes and return lipid and protein constituents back to the plasma membrane. In our hypothesis, a similar fate may be followed by the cholesterol-rich L_o-AChR nanometer-sized clusters upon destabilization by cholesterol depletion. However, in contradistinction to the hypothesis of Turner et al. (2005) we envision cholesterol depletion as a perturbation that shifts the distribution of AChR from the surface to intracellular compartments by accelerating an endocytic process of the *larger-than-normal, less stable* AChR nanoclusters. This process would not normally operate in CHO-K1/A5 cells or C2C12 myoblasts within the time window of a few hours, unless triggered by external stimuli (e.g. anti-AChR antibodies, Kumari et al., 2008). Interestingly, the number of nanoclusters depends on cholesterol levels (in agreement with recent results of Willmann et al., 2006, in C2C12 myotubes), whereas the number of receptors within these clusters appears to be fairly independent of cholesterol levels.

17.8 Possible Relationship Between AChR Nanocluster Organization and the Membrane-Associated Cortical Cytoskeletal Network

Before establishment of the mature postsynaptic specializations, AChRs shift from a diffusely dispersed form to a submicron-sized cluster distribution during the early stages of embryonic development of the NMJ (Sanes and Lichtman, 2001; Willmann and Fuhrer, 2002). These changes in AChR supramolecular organization occur within a very narrow time window in ontogeny, between embryonic stages E13 and E14 (Sanes and Lichtman, 2001). Postsynaptic maturation and cluster formation can occur in the absence of nerve (Kummer et al., 2004), but it is not known how these supramolecular aggregates are constructed at the cell surface in the absence of innervation. We earlier entertained the hypothesis that AChR aggregation resembled a protein-protein "nucleation" process (Barrantes, 1979) and we currently surmise that cholesterol is involved in this process.

The influence of cholesterol levels on surface AChR organization is substantial: about half of the AChR nanoclusters in CHO cells are sensitive to membrane cholesterol content (Borroni et al., 2007), in agreement with the results of Bruses et al. (2001) on neuronal AChR; however, differences with the endogenous neuronal α7 AChR and muscle-type AChR are also apparent. The most likely reason for such differences is the presence of the receptor-anchoring machinery (neuronal agrin, MusK, rapsyn, etc.) in those cellular systems endogenously expressing AChR and their absence in heterologous expression systems such as the CHO cell line. Thus the cholesterol modulatory effect on AChR supramolecular organization in the latter expression system must be related to other factors.

Mature AChR clusters in the fully developed NMJ are believed to be tethered to the cytoskeleton via scaffolding connections (Hall et al., 1981; Prives et al., 1982; Wallace, 1992; Sanes and Lichtman, 2001). Campagna and Fallon (2006) discussed their findings on cholesterol sensitivity of AChR cluster organization in the light of an apparent mutually exclusive contribution of lipid "raft" recruitment of AChR versus cytoskeletal effects in stabilizing AChRs. Lipid raft aggregation, by for example cholera toxin-mediated patching, may be an artifactual phenomenon that does not necessarily reflect actual associations of lipid and protein (AChR) at that scale of organization. Furthermore, we employed a cell line, CHO-K1/A5, which lacks ganglioside GM1; hence no such ganglioside-cholera toxin mediated "raft" aggregation can occur in these cells. Patching can also cause a polymerization of F-actin and clustered lipid "rafts" may themselves be tethered to the cytoskeleton (Harder and Simons, 1999; Schutz et al., 2000). CDx-mediated cholesterol depletion can cause F-actin depolymerization (Bruses et al., 2001). Harder and Simons (1999) argued that lipid raft clustering can result in patch-associated tyrosine phosphorylation, and this phosphorylation is required for actin polymerization. According to Harder and Simons (1999) CDx treatment would disrupt the cytoskeleton by preventing src kinase association with rafts, and therefore prevent F-actin phosphorylation.

The results of the Ripley's spatial point pattern analysis revealed that the long-range AChR organization at the plasmalemma of CHO-K1/A5 cells depends on cholesterol-sensitive interactions that normally extend over the range of a few microns in untreated cells (Kellner et al., 2007). A likely candidate for the maintenance of such an influence is the cortical cytoskeleton and, particularly, the actin network (e.g. Kwik et al., 2003). The ability of AChR nanoclusters to aggregate upon cholesterol depletion in whole cells was apparently lost in single plasma membrane sheets devoid of the subcortical cytoskeleton. This provided additional, albeit indirect, evidence that the long-range AChR supramolecular organization is likely to be associated with the presence of an intact cytoskeletal network under physiological energy supply and normal cholesterol levels (Borroni et al., 2007). According to Kusumi and Suzuki (2003) membrane constituent molecules undergo short-term confined diffusion within a compartment and long-term hop diffusion between compartments. Compartment boundaries are made up of the actin-based, membrane-associated cytoskeleton mesh ("fence") and the transmembrane proteins ("pickets") anchored to and lined up along the membrane skeleton fence. In muscle, AChRs have been reported to be associated with actin via urotrophin and rapsyn (Willmann and Fuhrer, 2002). In response to agrin, the AChR-rapsyn association translates into binding to cytoskeletal proteins (Moransard et al., 2003). The short-range extent and composition of the AChR nanoclusters may also be maintained by protein-protein interactions (i.e. receptor-receptor and receptor-nonreceptor proteins) by both fences and pickets, as well as receptor-lipid interactions, of which AChR-cholesterol interactions may constitute a prevailing stabilizing force (Barrantes, 2004).

17.9 A Word on Cholesterol Binding Sites

Attempts to identify the cholesterol recognition site on the AChR protein have made preferential use of photoaffinity labeling techniques that rely on photo-activatable cholesterol analogs. Early experiments were targeted at the characterization of labeling to the intact subunit level (Middlemas and Raftery, 1987; Fernández et al., 1993) and/or employed photoactivatable sterols that were purported to be functional cholesterol substitutes (Corbin et al., 1998; Blanton et al., 1999). The most recent attempts using photoaffinity labeling confirmed earlier results and led to the identification of putative cholesterol-AChR interaction sites at the M4, M3, and M1 segments of each subunit, fully overlapping the lipid-protein interface of the AChR (Hamouda et al., 2006). The M4 segment showed the most extensive interaction with the cholesterol analog. For αM4, the labeling pattern was consistent with azicholesterol incorporation into αGlu-398, αAsp-407, and αCys-412, i.e. amino acid residues that lie in a rather shallow region in the NH-term of the M4 transmembrane segment. Hamouda et al. (2006) also point out that it is striking that the conserved Asp at the N-terminus of each M4 segment (αAsp-407, βAsp-436, γAsp-448, δAsp-454) is labeled, as well as βAsp-457, the only acidic side chain at the C-terminus of the M4 segments.

Recent molecular dynamics simulations of the AChR in the presence or absence of cholesterol led Brannigan et al. (2008) to conclude that the AChR possesses multiple cholesterol binding sites, most of which would be deeply buried in the protein, and that the AChR collapses in the absence of the sterol. Their argument is based on the observation of "holes" in the electron density maps of the AChR cryoelectron microscopy images of Unwin and colleagues, which could accommodate up to 15 cholesterol molecules. We had calculated such a number of cholesterol molecules from ESR experiments (Mantipragada et al., 2003), but at variance with Brannigan et al. (2008) all the cholesterol molecules readily exchange with bulk lipids in the ESR experiment, unlike the deeply buried cholesterols postulated from in silico calculations. In fact only five out of the 15 cholesterol molecules are localized at the protein–lipid interface, in agreement with the wealth of information gained from experimental approaches (Barrantes, 2007), whereas the remainder suggested by the molecular dynamics simulations occupy the deeply buried sites. Brannigan et al. (2008) further propose that each cholesterol molecule consistently interacts with at least 10 (mostly hydrophobic) residues in the AChR protein. Further studies are needed to resolve this matter.

Acknowledgements Experimental work described in this Chapter was supported in part by PICT 5-20155 from the Ministry of Science and Technology of Argentina; PIP No. 6367 from the Argentinian Scientific Research Council (CONICET); Philip Morris USA Inc. and Philip Morris International; and PGI No. 24/B135 from Universidad Nacional del Sur, Argentina, to F.J.B.

References

Antollini, S.S., Soto, M.A., Bonini de Romanelli, I.C., Gutierrez-Merino C., Sotomayor, P., Barrantes, F.J., 1996, Physical state of bulk and protein-associated lipid in nicotinic acetylcholine receptor-rich membrane studied by laurdan generalized polarization and fluorescence energy transfer. *Biophys. J.* **70**: 1275–1284.

Artigues, A., Villar, M.T., Fernández, A.M., Ferragut, J.A., Gonzalez-Ros, J.M., 1989, Cholesterol stabilizes the structure of the nicotinic acetylcholine receptor reconstituted in lipid vesicles. *Biochim. Biophys. Acta* **985**: 325–330.

Barrantes, F. J., 1979, Endogenous chemical receptors: some physical aspects. *Annu. Rev. Biophys. Bioeng.* **8**: 287–321.

Barrantes, F.J., 1982, Oligomeric forms of the membrane-bound acetylcholine receptor disclosed upon extraction of the Mr 43,000 nonreceptor peptide. *J. Cell Biol.* **92**: 60–68.

Barrantes, F.J., 1983, Recent developments in the structure and function of the acetylcholine receptor. *Int. Rev. Neurobiol.* **24**: 259–341.

Barrantes, F.J., 1989, The lipid environment of the nicotinic acetylcholine receptor in native and reconstituted membranes. *Crit. Rev. Biochem. Mol. Biol.* **24**: 437–478.

Barrantes, F.J., 1993a, The lipid annulus of the nicotinic acetylcholine receptor as a locus of structural-functional interactions. In: Watts, A. (ed.), New Comprehensive Biochemistry, Elsevier, Amsterdam, pp. 231–257.

Barrantes, F.J., 1993b, Structural-functional correlates of the nicotinic acetylcholine receptor and its lipid microenvironment. *FASEB J.* **7**: 1460–1467.

Barrantes, F.J., 2002, Lipid matters: nicotinic acetylcholine receptor-lipid interactions. *Molec. Membr. Biol.* **19**: 277–284.

Barrantes, F.J., 2003, Modulation of nicotinic acetylcholine receptor function through the outer and middle rings of transmembrane domains. *Current Opinion in Drug Discovery & Development* **6**: 620–632.

Barrantes, F.J., 2004, Structural basis for lipid modulation of nicotinic acetylcholine receptor function. *Brain Res. Rev. Brain Res. Rev.* **47**: 71–95.

Barrantes, F.J, 2007, Cholesterol effects on nicotinic acetylcholine receptor. *J. Neurochem.* **103 (Suppl. 1)**: 72–80.

Blanton, M.P., Xie, Y., Dangott, L.J., Cohen, J.B., 1999, The steroid promegestone is a non-competitive antagonist of the Torpedo nicotinic acetylcholine receptor that interacts with the lipid-protein interface. *Mol. Pharmacol.* **55**: 269–278.

Bolotina, V., Omelyanenko, V., Heyes, B., Ryan, U., Bregestovski, P., 1989, Variations of membrane cholesterol alter the kinetics of Ca2+- dependent K+ channels and membrane fluidity in vascular smooth muscle cells. *Pflügers Archiv.-Eur. J. Physiol.* **415**: 262–268.

Borroni, V., Baier, C.J., Lang, T., Bonini, I., White, M.M., Garbus I., Barrantes, F.J., 2007, Cholesterol depletion activates rapid internalization of submicron.sized acetylcholine receptor domains at the cell membrane. *Molec. Membr. Biol.* **24**: 1–15.

Brannigan, G., Hénin, J., Law, R., Eckenhoff, R., Klein, M.L., 2008, Embedded cholesterol in the nicotinic acetylcholine receptor. *Proc. Natl. Acad. Sci. USA* **105**: 14418–14423.

Brown, D.A., London, E., 1998, Functions of lipid rafts in biological membranes. *Annu. Rev. Cell Dev. Biol.* **14**: 111–136.

Brown, D.A., London, E., 2000, Structure and function of sphingolipid- and cholesterol-rich membrane rafts. *J. Biol. Chem.* **275**: 17221–17224.

Bruses, J., Chauvet, N., Rutishauser, U., 2001, Membrane lipid rafts are necessary for the maintenance of the (alpha)7 nicotinic acetylcholine receptor in somatic spines of ciliary neurons. *J. Neurosci.* **21**: 504–512.

Campagna, J.A., Fallon, J. (2006), Lipid rafts are involved in C95 (4,8) agrin fragment-induced acetylcholine receptor clustering. *Neuroscience* **138**: 123–132.

Castresana, J., Fernandez-Ballester, G., Fernández, A.M., Laynez, J.L., Arrondo, J.-L.R., Ferragut, J.A., González-Ros, J.M., 1992, Protein structural effects of agonist binding to the nicotinic acetylcholine receptor. *FEBS Lett.* **314**: 171–175.

Chen, F., Quian, L., Yang, Z.H., Huan, Y., Ngo, S.T., Ruan, N.J., Wang, J., Schneider, C., Noakes, P.G., Ding, Y.Q., Mei, L., Luo, Z.G., 2007, Rapsyn interaction with calpain stabilizes AChR clusters at the neuromuscular junction. *Neuron* **55:** 247–260.

Corbin, J., Wang, H.H., Blanton, M.P., 1998, Identifying the cholesterol binding domain in the nicotinic acetylcholine receptor with [125I]azido-cholesterol. *Biochim. Biophys. Acta* **1414:** 65–74.

Criado, M., Eibl, H., Barrantes, F.J., 1982, Effects of lipids on acetylcholine receptor. Essential need of cholesterol for maintenance of agonist-induced state transitions in lipid vesicles. *Biochemistry* **21:** 3622–3629.

Criado, M., Eibl, H., Barrantes, F.J., 1984, Functional properties of the acetylcholine receptor incorporated in model lipid membranes. Differential effects of chain length and head group of phospholipids on receptor affinity states and receptor-mediated ion translocation. *J. Biol. Chem.* **259:** 9188–9198.

da Costa, C.J., Baezinger, J.E., 2009, A lipid-dependent uncoupled conformation of the acetylcholine receptor. *J. Biol. Chem.* **284:** 17819–17825.

Dalziel, A.W., Rollins, E.S., McNamee, M.G., 1980, The effect of cholesterol on agonist-induced flux reconstituted acetylcholine receptor vesicles. *FEBS Lett.* **122:** 193–196.

Dreger, M., Krauss, M., Herrmann, A., Hucho, F., 1997, Interactions of the nicotinic acetylcholine receptor transmembrane segments with the lipid bilayer in native receptor-rich membranes. *Biochemistry* **36:** 839–847.

Edidin, M., 1997, Lipid microdomains in cell surface membranes. *Curr. Opin. Struct. Biol.* **7:** 528–532.

Ellena, J.F., Blazing, M.A., McNamee, M.G., 1983, Lipid-protein interactions in reconstituted membranes containing acetylcholine receptor. *Biochemistry* **22:** 5523–5535.

Fernández, A.M., Fernandez-Ballester, G., Ferragut, J.A., González-Ros, J.M. 1993, Labeling of the nicotinic acetylcholine receptor by a photoactivatable steroid probe: effects of cholesterol and cholinergic ligands. *Biochim. Biophys. Acta Bio-Membr.* **1149:** 135–144.

Fernandez-Ballester, G., Castresana, J., Fernandez, A.M., Arrondo, J.-L.R., Ferragut, J. A., Gonzalez-Ros, J.M., 1994, A role for cholesterol as a structural effector of the nicotinic acetylcholine receptor. *Biochemistry* **33:** 4065–4071.

Fernandez-Ballester, G., Castresana, J., Arrondo, J.-.L.R., Ferragut, J.A., Gonzalez-Ros, J.M., 1992, Protein stability and interaction of the nicotinic acetylcholine receptor with cholinergic ligands studied by Fourier-transform infrared spectroscopy. *Biochem. J.* **288:** 421–426.

Hall, Z.W., Lubit, B.W., Schwartz, J.H., 1981, Cytoplasmic actin in postsynaptic structures at the neuromuscular junction. *J. Cell Biol.* **90:** 789–792.

Hamouda, A.K., Chiara, D.C., Sauls, D., Cohen, J.B., Blanton, M.P. 2006, Cholesterol interacts with transmembrane alpha-helices M1, M3, and M4 of the Torpedo nicotinic acetylcholine receptor: photolabeling studies using [3H]Azicholesterol. *Biochemistry* **45:** 976–986.

Hancock, J. F., 2006, Lipid rafts: contentious only from simplistic standpoints. *Nature Rev. Molec. Cell Biol.* **7:** 456–462.

Harder, T., Simons, K., 1999, Clusters of glycolipid and glycosylphosphatidyl inositol-anchored proteins in lymphoid cells: accumulation of actin regulated by local tyrosine phosphorylation. *Eur. J. Immunol.* **29:** 556–562.

Hell, S.W., 1997, Increasing the resolution of far-field fluorescence light microscopy by point-spread-function engineering. In: Lakowicz J. R. (ed), Topics in fluorescence spectroscopy, Plenum Press, New York, pp. 361–422.

Hell, S.W., 2004, Strategy for far-field optical imaging and writing without diffraction limit. *Phys. Lett. A* **326:** 140–145.

Hell, S.W., 2009, Microscopy and its focal switch. *Nat. Meth.* **6:** 24–32.

Holmgren, J., Lonnroth, I., Svennerholm, L., 1973, Tissue receptor for cholera exotoxin: postulated structure from studies with GM1gangliosidea and related glycolipids. *Infect. Immunol.* **8:** 208–214.

Jacobson K., Dietrich, C., 1999, Looking at lipid rafts? *Trends Cell Biol.* **9**: 87–91.

Jones, O.T., McNamee, M.G., 1988, Annular and nonannular binding sites for cholesterol associated with the nicotinic acetylcholine receptor. *Biochemistry* **27**: 2364–2374.

Kellner, R., Baier, J., Willig, K.I., Hell, S.W., Barrantes, F.J., 2007, Nanoscale organization of nicotinic acetylcholine receptors revealed by STED microscopy. *Neuroscience* **144**: 135–143.

Kirkham, M., Fujita, A., Chadda, R., Nixon, S.J., Kurzchalia, T.V., Sharma, D.K., Pagano, R.E., Hancock, J.F., Mayor, S., Parton, R.G., 2005, Ultrastructural identification of uncoated caveolin-independent early endocytic vehicles. *J. Cell Biol.* **168**: 465–476.

Kishi, M., Kummer, T.T., Eglen, S.J., Sanes, J.R., 2005, LL5beta: a regulator of postsynaptic differentiation identified in a screen for synaptically enriched transcripts at the neuromuscular junction. *J. Cell Biol.* **169**: 355–366.

Kumari, S., Borroni, V., Chaudhry, A., Chanda, B., Massol, R., Mayor, S., Barrantes, F.J., 2008, Nicotinic acetylcholine receptor is internalized via a Rac-dependent dynamin-independent endocytic pathway. *J. Cell Biol.* **181**: 1179–1193.

Kummer, T.T., Misgeld, T., Lichtman, J.W., Sanes, J.R., 2004, Nerve independent formation of a topologically complex postsynaptic apparatus. *J.Cell Biol.* **164**: 1077–1087.

Kusumi, A., Suzuki, K., 2003, Toward understanding the dynamics of membrane-raft-based molecular interactions. *Biochim Biophys. Acta* **1746**: 234–251.

Kwik, J., Boyle, S., Fooksman, D., Margolis, L., Sheetz, M.P., Edidin, M., 2003, Membrane cholesterol, lateral mobility, and the phosphatidylinositol 4,5-bisphosphate-dependent organization of cell actin. *Proc. Natl Acad. Sci. U.S.A.* **100**: 13964–13969.

Le, P.U., Guay, G., Altschuler, Y., Nabi, I.R., 2002, Caveolin-1 is a negative regulator of caveolae-mediated endocytosis to the endoplasmic reticulum. *J. Biol. Chem.* **277**: 371–3379.

Lichtenberg, D., Goñi, F.M., Heerklotz, H., 2005, Detergent-resistant membranes should not be identified with membrane rafts. *Trends Biochem. Sci.* **30**: 430–436.

Lin, W., Dominguez, B., Yang, J., Aryal, P., Brandon, E.P., Gage, F.H., Lee, K.F., 2005, Neurotransmitter acetylcholine negatively regulates neuromuscular synapse formation by a Cdk5-dependent mechanism. *Neuron* **46**: 569–579.

Mantipragada, S.B., Horvath, L.I., Arias, H.R., Schwarzmann, G., Sandhoff, K., Barrantes, F.J., Marsh, D., 2003, Lipid-protein interactions and effect of local anesthetics in acetylcholine receptor-rich membranes from Torpedo marmorata electric organ. *Biochemistry* **42**: 9167–9175.

Marchand, S., Devillers-Thiery, A., Pons, S., Changeux, J. P., Cartaud, J., 2002, Rapsyn escorts the nicotinic acetylcholine receptor along the exocytic pathway via association with lipid rafts. *J. Neurosci.* **22**: 8891–8901.

Marcheselli, V., Daniotti, J.L., Vidal, A.C., Maccioni, H., Marsh, D., Barrantes F.J., 1993, Gangliosides in acetylcholine receptor-rich membranes from Torpedo marmorata and Discopyge tschudii. *Neurochem Res.* **18**: 599–603.

Marsh, D., Barrantes, F.J. (1978), Immobilized lipid in acetylcholine receptor-rich membranes from Torpedo marmorata. *Proc. Natl Acad. Sci. U.S.A.* **75**: 4329–4333.

Marsh, D., Watts, A., Barrantes, F.J., 1981, Phospholipid chain immobilization and steroid rotational immobilization in acetylcholine receptor-rich membranes from Torpedo marmorata. *Biochim. Biophys. Acta* **645**: 97–101.

Maxfield, F.R., 2002, Plasma membrane microdomains. *Curr. Opin. Cell Biol.* **14**: 483–487.

McNamee, M.G., Ellena, J. F., Dalziel, A.W., 1982, Lipid-protein interactions in membranes containing the acetylcholine receptor. *Biophys. J.* **37**: 103–104.

Middlemas, D.S., Raftery, M.A., 1987, Identification of subunits of acetylcholine receptor that interact with a cholesterol photoaffinity probe. *Biochemistry* **26**: 1219–1223.

Mittaud, P., Marangi, P.A., Erb-Vögtli, S., Fuhrer, C., 2001, Agrin-induced activation of acetylcholine receptor-bound src family kinases requires rapsyn and correlates with acetylcholine receptor clustering. *J. Biol. Chem.* **276**: 14505–14513.

Moransard, M., Borges, L.S., Willmann, R., Marangi, P.A., Brenner, H.R., Ferns, M.J., Fuhrer, C., 2003, Agrin regulates rapsyn interaction with surface acetylcholine receptors, and this underlies cytoskeletal anchoring and clustering. *J. Biol. Chem.* **278:** 7350–7359.

Pediconi, M.F., Gallegos, C.E., De los Santos, E.B., Barrantes, F.J., 2004, Metabolic cholesterol depletion hinders cell-surface trafficking of the nicotinic acetylcholine receptor. *Neuroscience* **128:** 239–249.

Phillips, D.W., Vladeta, D., Han, H., Noakes P.G., 1997, Rapsyn and agrin slow the metabolic degradation of the acetylcholine receptor. *Mol. Cel. Neurosci.* **10:** 16–26.

Pichler, H., Riezman, H., 2004, Where sterols are required for endocytosis. *Biochim. Biophys. Acta* **1666:** 51–61.

Popot, J.L., Demel, R.A., Sobel, A., van Deenen, L.L., Changeux, J.P., 1978, Interaction of the acetylcholine (nicotinic) receptor protein from Torpedo marmorata electric organ with monolayers of pure lipids. *Eur..J. Biochem.* **85:** 27–42.

Prives, J., Fulton, A.B., Penman, S., Daniels, M.P., Christian, C.N., 1982, Interaction of the cytoskeletal framework with acetylcholine receptor on the surface of embryonic muscle cells in culture. *J. Cell Biol.* **92:** 231–236.

Rankin, S.E., Addona, G.H., Kloczewiak, M.A., Bugge, B., Miller, K.W., 1997, The cholesterol dependence of activation and fast desensitization of the nicotinic acetylcholine receptor. *Biophys J.* **73:** 2446–2455.

Ripley, B.D., 1979, Test of randomness for spatial point patterns. *J. R. Stat. Soc. B* **41:** 368–374.

Roccamo, A.M., Pediconi, M.F., Aztiria, E., Zanello, L., Wolstenholme, A., Barrantes F.J., 1999, Cells defective in sphingolipids biosynthesis express low amounts of muscle nicotinic acetylcholine receptor. *Eur. J. Neurosci.* **11:** 1615–1623.

Sanes, J.R., Lichtman, J.W., 2001, Induction, assembly, maturation and maintenance of a postsynaptic apparatus. *Nat. Rev. Neurosci.* **2:** 791–805.

Schutz, G.J., Kada, G., Pastushenko, V.P., Schindler, H., 2000, Properties of lipid microdomains in a muscle cell membrane visualized by single molecule microscopy. *EMBO J* **19:** 892–901.

Simons, K., Ikonen, E., 1997, Functional rafts in cell membranes. *Nature* **387:** 569–572.

Stetzkowski-Marden, F., Gaus, K., Recouvreur, M., Cartaud, A., Cartaud, J., 2006, Agrin elicits membrane lipid condensation at sites of acetylcholine receptor clusters in C2C12 myotubes. *J. Lipid Res.* **47:** 2121–2133.

Subtil, A., Gaidarov, I., Kobylarz, K., Lampson, M.A., Keen, J.H., McGraw, T.E., 1999, Acute cholesterol depletion inhibits clathrin-coated pit budding. *Proc. Natl Acad. Sci. U.S.A.* **96:** 6775–6780.

Turner, M.S., Sens, P., Socci, N.D., 2005, Nonequilibrium raft-like domains under continuous recycling. *Phys. Rev. Lett.* **95:** 168301.

Vrljic, M., Nishimura, S.Y., Moerner, W. E., McConnell, H.M., 2005, Cholesterol depletion suppresses the translational diffusion of class II major histocompatibility complex proteins in the plasma membrane. *Biophys J.* **88:** 334–347.

Wallace, B.G., 1992, Mechanism of agrin-induced acetylcholine receptor aggregation. *J. Neurobiol.* **23:** 592–604.

Warnock, D.E., Roberts, C., Lutz, M.S., Blackburn, W.A., Young, W.W. Jr, Baenziger, J.U., 1993, Determination of plasma membrane lipid mass and composition in cultured Chinese hamster ovary cells using high gradient magnetic affinity chromatography. *J. Biol. Chem.* **268:** 10145–10153.

Willmann, R., Fuhrer, C., 2002, Neuromuscular synaptogenesis: clustering of acetylcholine receptors revisited. *Cell Mol Life Sci* **59:** 1296–1316.

Willmann, R., Pun, S., Stallmach, L., Sadasivam, G., Santos, A.F., Caroni, P., Fuhrer, C., 2006, Cholesterol and lipid microdomains stabilize the postsynapse at the neuromuscular junction. *EMBO. J.* **25:** 4050–4060.

Xu, X., Bittman. R., Duportail, G., Heissler, D., Vilcheze, C., London, E., 2001, Effect of the structure of natural sterols and sphingolipids on the formation of ordered sphingolipid/sterol domains (rafts). Comparison of cholesterol to plant, fungal, and disease-associated sterols and comparison of sphingomyelin, cerebrosides, and ceramide. *J. Biol. Chem.* **276:** 33540–33546.

Zanello, L.P., Aztiria, E., Antollini, S., Barrantes, F.J., 1996, Nicotinic acetylcholine receptor channels are influenced by the physical state of their membrane environment. *Biophys. J.* **70**, 2155–2164.

Zhu, D., Xiong, W.C., Mei, L., 2006, Lipid rafts serve as a signaling platform for nicotinic acetylcholine receptor clustering. *J. Neurosci.* **26:** 4841–4851.

Chapter 18
Cholesterol and Myelin Biogenesis

Gesine Saher and Mikael Simons

Abstract Myelin consists of several layers of tightly compacted membranes wrapped around axons in the nervous system. The main function of myelin is to provide electrical insulation around the axon to ensure the rapid propagation of nerve conduction. As the myelinating glia terminally differentiates, they begin to produce myelin membranes on a remarkable scale. This membrane is unique in its composition being highly enriched in lipids, in particular galactosylceramide and cholesterol. In this review we will summarize the role of cholesterol in myelin biogenesis in the central and peripheral nervous system.

Keywords Oligodendrocyte · Schwann cell · Myelin · Cholesterol

18.1 Introduction

The myelin sheath is formed by the spiral wrapping of glial plasma membrane extensions around axons, followed by the extrusion of cytoplasm and the compaction of the stacked membrane bilayers (Sherman and Brophy, 2005). These tightly packed membrane stacks provide the neuron with an insulating layer to prevent the exposure to extracellular fluids and ions. The insulation around the axon changes the electrical properties of the neurons. It dramatically increases the electrical resistance across the cell membrane and decreases the capacitance thereby speeding up propagation of electrical signals through the nervous system. To fulfil its function as an insulating barrier myelin requires a specific molecular composition. One essential structural component is cholesterol. The hydroxyl group of cholesterol interacts with the polar head group of phospholipids and sphingolipids, while the bulky hydrophobic portion closely associates with the fatty acid chains of the lipids within the membrane bilayer. Cholesterol might also directly interact with some myelin

M. Simons (✉)
Max-Planck-Institute for Experimental Medicine, Hermann-Rein-Str. 3, Göttingen, Germany;
Department of Neurology, University of Göttingen, Robert-Koch-Str. 40, Göttingen, Germany
e-mail: msimons@gwdg.de

proteins. The structure of cholesterol is involved in stabilizing and sealing the membrane, in particular to proton and sodium ions. In addition to its role in regulating the permeability of a membrane, cholesterol is a critical factor for membrane fluidity. Many of these established functions of cholesterol are particular relevant in myelin as its main function is to shield the axon from the extracellular environment. In this chapter, we will provide an overview of the role of cholesterol in myelin of the central and peripheral nervous system (CNS; PNS).

18.2 Myelin Structure and Composition

Myelin has several typical structural features (Fig. 18.1) such as the periodic structure with alternating electron-dense and light layers (Baumann and Pham-Dinh, 2001). The major dense line (electron-dense) represents the closely attached cytoplasmic myelin membranes, whereas the interperiod lines consist of the tightly apposed outer membranes. The compaction of these membranes is so tight that it results in a periodicity of about 12 nm.

The myelinated segments of the axons are around 150 μm in length and interrupted by spaces where myelin is lacking, the nodes of Ranvier. At the nodes of Ranvier axons contain the sodium channels at high density that are required for the saltatory propagation of the action potentials (Peles and Salzer, 2000; Salzer, 2003). The nodes are flanked on either side by lateral membrane loops formed by

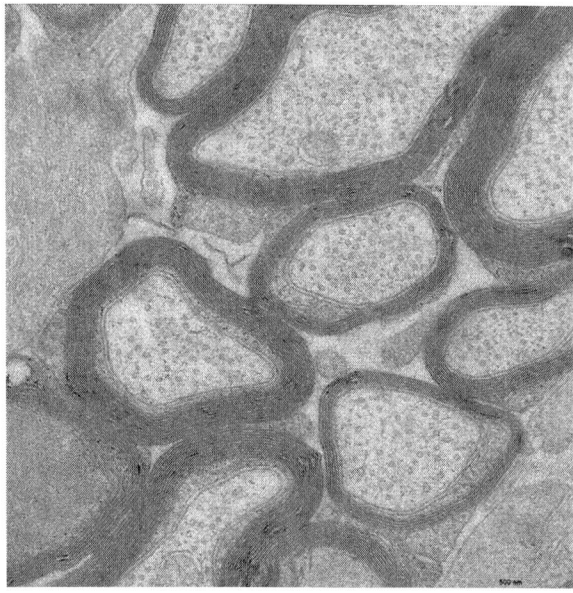

Fig. 18.1 Myelin structure. An electron micrograph of the optic nerve is shown. The apposition of the external faces of the membrane form the intraperiodic line whereas the cytoplasmic faces form the major dense lines

the myelinating glia. These paranodal loops form septate-like junctions with the axonal membrane. The juxtaparanodal domain is located underneath the compact myelin sheath adjacent to the paranodes.

Myelin has not only distinctive structural features, but also a unique molecular composition. In contrast to most plasma membranes, myelin contains more lipids than proteins. Lipids constitute 70–80% of the dry weight of myelin. Although myelin-specific lipids are lacking, some lipids are found in high abundance. Glycosphingolipids are particularly enriched in myelin. The major glycosphingolipids in myelin are galactosylceramide and its sulfated derivative, sulfatide (20% of lipid dry weight) (Baumann and Pham-Dinh, 2001). These glycosphingolipids are particularly rich in very long chain fatty acids containing 22–26 carbon atoms that are saturated or monounsaturated. There is also an unusually high proportion of ethanolamine phosphoglycerides in the plasmalogen form, which accounts for one-third of the phospholipids. More than 25% of the total lipid content is cholesterol, compared to less then 20% in most membranes. The molar ratio of cholesterol: phospholipids: glycosphingolipids in myelin is between 4:3:2 and 4:4:2. This high amount of cholesterol and glycosphingolipids might lead to an increase in membrane lipid order, which could be important for myelin to perform its insulating function.

A striking feature about myelin is not only the lipid, but also the protein composition. Myelin contains a relatively simple array of proteins, myelin basic protein (MBP) and the proteolipid proteins (PLP/DM20) being the two major CNS myelin proteins. PLP spans the myelin membrane bilayer four times, with two extracellular loops stabilized by disulfide bonds. PLP contains a large number of hydrophobic amino acids and several cysteine-bound acyl-chains (Weimbs and Stoffel, 1992). MBPs are cytoplasmic proteins with a high density of basic amino acids that interact with negatively charged lipid head groups such as phosphatidylinositol 4,5 – bisphosphate (Harauz et al., 2004; Musse et al., 2008; Nawaz et al., 2009). One important function of MBP is to promote the compaction between the two cytoplasmic membranes and to condense specific lipids in a lateral dimension (Fitzner et al., 2006).

MBP and the P0 protein constitute the majority of the proteins in PNS myelin. P0 is a single transmembrane protein with an extracellular domain containing one immunoglobulin (Ig)-like domain. The adhesive properties of the extracellular domain and the highly basic cytoplasmic domain are responsible for the compaction of the myelin membrane (Quarles, 2002).

In addition to compact myelin membranes, myelin contains regions of non-compact myelin. Some cytoplasm remains within the innermost and outermost tongues of myelin membranes, in paranodal loops bordering nodes of Ranvier, and in Schmidt-Lanterman incisures of the PNS. Non-compact myelin regions are believed to facilitate transport of metabolites and ions. The protein composition of non-compact myelin is distinct of compact myelin. While compact myelin proteins are absent, non-compact myelin is characterized by the presence of marker proteins, such as myelin associated glycoprotein (MAG) and the gap junction protein connexin 32.

18.3 Schwann Cells

18.3.1 The Origin and Differentiation of Schwann Cells

Schwann cells are the myelinating glial cells of the peripheral nervous system of vertebrates. They are derived from the neural crest cells, a transient cell population that emerge at the dorsal part of the neural tube. From there, neural crest cells migrate to various locations including the developing embryonic nerves (Le Douarin and Dupin, 2003; Woodhoo and Sommer, 2008). Glial specification already occurs at early stages of embryonic development and can be divided into three main phases (Jessen and Mirsky, 2005). In the spinal nerves of the mouse, Schwann cell precursors (SCPs) are born around embryonic day 12/13 (E12/13) from migrating neural crest (Fig. 18.2). SCPs differentiate into immature Schwann cells around E15/16 followed by the maturation into myelinating and non-myelinating Schwann cells (Dong et al., 1999). Finally, peripheral myelination in rodents begins around birth.

Each of the developmental stages is characterized by a distinct molecular profile and signal responsiveness (Jessen and Mirsky, 2005). In embryonic nerves, axons

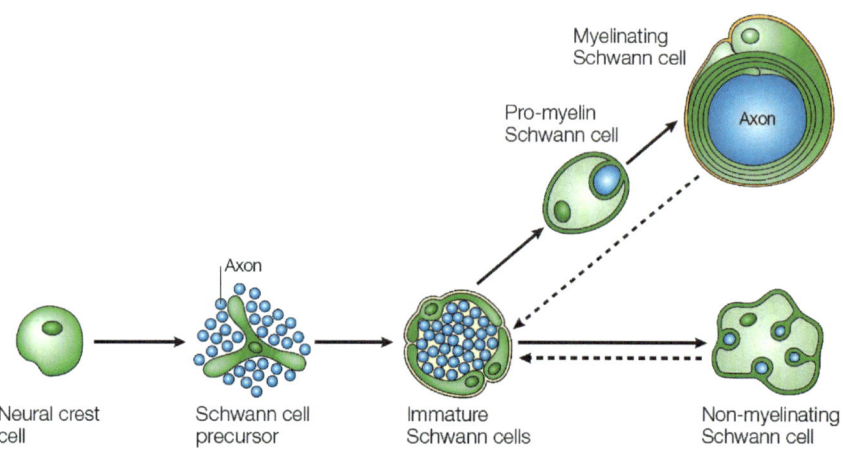

Fig. 18.2 The Schwann cell lineage. A schematic illustration of the main cell types and developmental transitions in Schwann cell development is shown. Stippled arrows indicate the reversibility of the final, and in rodents largely post-natal, transition that generates mature myelinating and non-myelinating cells. The embryonic phase of Schwann cell development involves three transient cell populations. First, migrating neural crest cells. Second, Schwann cell precursors (SCPs). These cells express a number of differentiation markers not found on migrating crest cells. Third, immature Schwann cells. All these cells are considered to have the same developmental potential, and their fate is dictated by the axons with which they associate. Only those Schwann cells that by chance envelop the large-diameter axons will be induced to myelinate, while those cells that ensheath small-diameter axons progress to become mature non-myelinating cells. Reproduced from Nature Reviews Neuroscience with minor modifications (Jessen and Mirsky, 2005), with permission

and SCPs are tightly packed lacking substantial connective tissue or blood vessels. SCPs associate with axon fascicles and extensively with each other (Woodhoo and Sommer, 2008). SCPs already express low levels of myelin proteins like P0, PMP22 and PLP. Nerves are profoundly remodelled when axons start to establish synaptic connections, and SCPs differentiate into immature Schwann cells. The tight structure of nerves is loosened, coinciding with the development of the endoneurial space containing blood vessels, connective tissue and endoneurial fibroblasts (Morell and Norton, 1980). At this stage, immature Schwann cells surround small bundles of axons forming "axon-Schwann cell-families" (Jessen and Mirsky, 2005). A basal lamina encloses the Schwann cell together with associated axons. Among other markers, immature Schwann cells are characterized by the expression of the transcription factor Oct6 (SCIP, Tst-1, POU3fl), and the appearance of sulfatide (sulfogalactosylceramide, O4 epitope) on their surface. In the process of radial sorting, axon bundles are defasciculated such that finally each Schwann cell forms a unit with a single axon. Myelinating Schwann cells downregulate Oct6 and start to express the pro-myelin transcription factor Krox-20 (Egr-2). Myelin proteins are strongly upregulated together with an enhanced lipid synthesis, including the biosynthesis of cholesterol. Myelin is then formed as a spiral extension of the plasma membrane with a unique composition of proteins and lipids. It has been estimated that the membrane surface area of Schwann cells expands by 6500-fold during myelination (Webster, 1971).

A number of factors that positively or negatively control the development of Schwann cells have been found (Jessen and Mirsky, 2008). A factor that accompanies the entire Schwann cell development is the epidermal growth factor Neuregulin 1 (NRG1) (Nave and Salzer, 2006). NRG1 binds to the family of ErbB receptor tyrosine kinases, influencing various cellular processes. NRG1 is essential for the survival of neural crest and SCPs and is by this means involved in the control of Schwann cell numbers in peripheral nerves (Grinspan et al., 1996). At this stage of development, NRG1 is bound to the axonal membrane (NRG1 type III isoform). Moreover, axonal NRG1 is involved in the regulation of myelin formation by Schwann cells. Unmyelinated axons normally express low levels of NRG1, while the myelinated axons express high levels. Neurons lacking NRG1 expression fail to be myelinated completely (Taveggia et al., 2005). Schwann cells "measure" the axon calibre via the amount of NRG1 bound to ErbB receptors and adjust the myelin thickness accordingly (Michailov et al., 2004). Mice heterozygous for NRG1 type III show a reduced myelin thickness, while NRG1 type III transgenic mice are hypermyelinated.

In the context of cholesterol metabolism, NRG1 is also of interest. In cell culture, NRG1 increases the transcription of the HMG-CoA-reductase (3-hydroxy-3-methylglutaryl-coenzyme-A-reductase) gene, the rate-limiting enzyme of cholesterol biosynthesis. It is hypothesized that axonal signals stimulate glial lipid synthesis to accomplish the increased lipid demand (Fig. 18.3) that Schwann cells encounter during myelination (Pertusa et al., 2007). The biosynthesis of myelin proteins and lipids relevant for myelin formation might be coupled (Leblanc et al., 2005). In vitro the pro-myelin transcription factor Krox-20 acts synergistically with

Fig. 18.3 A model for the axonal control of Schwann cell cholesterol biosynthesis. Membrane-attached axonal neuregulin binds and activates ErbB receptors exposed on the membrane surface of Schwann cells. ErbB activation induces Schwann cells to up-regulate cholesterol biosynthesis by activating the PI3K pathway and a CREB/ATF transcription factor. Question mark indicates a putative CRE-independent pathway involved in HMG-CoA reductase promoter activation by neuregulin. Reproduced from Glia with minor modifications (Pertusa et al., 2007), with permission

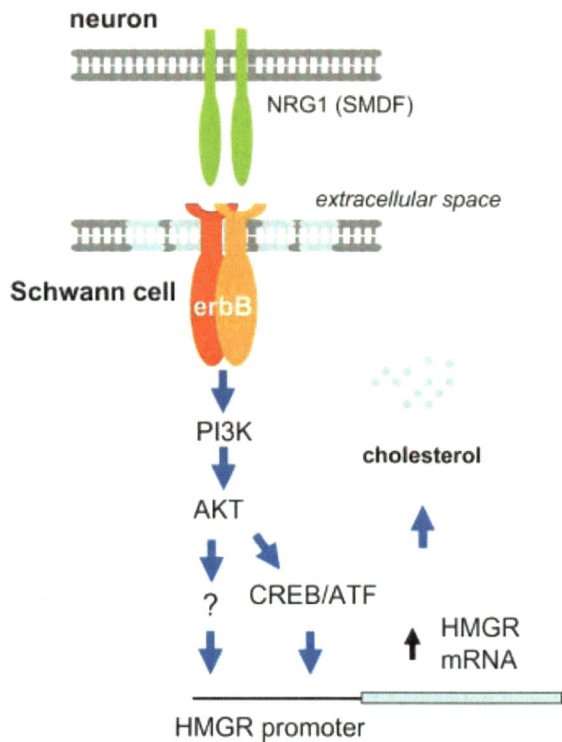

SREBP (sterol-responsive element binding protein) on the transcription of SREBP target genes enhancing lipid and sterol biosynthesis.

18.3.2 Origin and Differentiation of Oligodendrocytes

Myelination of axons in the CNS by oligodendrocytes takes place early postnatally (Miller, 2002; Rowitch, 2004). Oligodendrocyte precursor cells (OPCs) derive from the neuroepithelium of the ventricular/subventricular zone from where they migrate into the developing white matter. OPCs are highly dynamic and continuously extend and retract numerous processes as they move towards their final position (Kirby et al., 2006). These dynamic processes sense nearby OPCs, thereby regulating the uniform spacing of oligodendrocytes that is required to ensure complete myelination with evenly spaced nodes of Ranvier.

Upon arrival at their final position, OPCs stop to proliferate and differentiate into myelin-forming oligodendrocytes (Baumann and Pham-Dinh, 2001). In parallel, oligodendrocytes become highly metabolically active, and many of the lipid synthesising pathways become upregulated (Cahoy et al., 2008). Oligodendrocytes not only start to synthesize large amounts of myelin components, but also undergo a dramatic change of their morphology with the formation of a large network of branching processes (Sherman and Brophy, 2005; Simons and Trotter, 2007). Since oligodendrocytes, unlike Schwann cells, myelinate multiple axons and can establish as many as 50 myelin segments, these processes have to reach many different axons. Analogous to the dynamic behaviour of OPC processes, the process activity of mature oligodendrocytes might be important to fill unoccupied space and facilitate complete myelination.

To ensure full myelination of all axonal tracts, the timing of OPC differentiation needs to be tightly controlled by neurons (Barres and Raff, 1999; Simons and Trajkovic, 2006). It was therefore surprising when cell culture experiments demonstrated that the differentiation of oligodendrocytes occurs normally in the absence of neurons (Dubois-Dalcq et al., 1986; Mirsky et al., 1980). In cell culture, the differentiation of oligodendrocytes seems to follow a default pathway in which intrinsic signals define the number of cell divisions before the cells exit the cell cycle (Temple and Raff, 1986). In vivo, the situation is likely to be more complex with a number of negative and positive regulators that operate in highly regulated fashion to ensure timely oligodendrocyte differentiation and myelination.

Many studies now point to the importance of neuron-derived signalling molecules at different stages of oligodendrocyte development (Barres and Raff, 1999). These signals help to control the proper timing of OPC differentiation and they control and match the number of oligodendrocytes to the axonal surface area requiring myelination. Several growth factors and trophic factors, such as PDGF-A, FGF-2, IGF-1, NT-3 and CNTF, have been shown to regulate oligodendrocyte development (Baron et al., 2005; Barres et al., 1994a, 1994b; Miller, 2002). PDGF-A is produced by both astrocytes and neurons and regulates the proliferation and survival of OPCs (Noble et al., 1988; Raff et al., 1988; Richardson et al., 1988). It is likely that many of these growth factors directly influence signalling pathways involved in the control of myelin membrane synthesis. Recently, the myelin gene regulatory factor (MRF) was identified from a microarray analysis of highly expressed genes in oligodendrocytes (Emery et al., 2009). Deletion of this factor prevented the expression of several myelin genes, whereas overexpression of MRF promoted the appearance of myelin proteins within cultured oligodendrocyte progenitors. The upregulation of myelin protein content is accompanied by increased synthesis of lipids within oligodendrocytes. In fact, oligodendrocytes keep a fixed stoichiometry of cholesterol to proteins in the myelin membrane, and this ratio is likely to be important for the function of myelin (Saher et al., 2005). One important open issue is the mechanism by which the expression of myelin protein gene products is coordinated with the synthesis of cholesterol and other lipids in oligodendrocytes.

18.4 Source of Cholesterol in Myelin

The brain contains the highest concentration of cholesterol with ~15–20 mg/g of cholesterol, compared to ten times lower levels in the whole animal (2.2 mg/g) (Dietschy and Turley, 2004; Dietschy and Wilson, 1968). In the rat and mouse, around 70% of brain cholesterol is found in myelin (Snipes and Suter, 1997; Muse et al., 2001). Many endocrine tissues also contain high amounts of cholesterol (~13 mg/g of cholesterol in the adrenal gland). Most of the cholesterol in these tissues is found in the cytosol in an esterified form within lipid droplets. In contrast, cholesteryl esters are virtually absent from the brain (Dietschy and Turley, 2004). While the cellular concentration of cholesteryl esters can fluctuate dramatically, free cholesterol is kept constant in a membrane. By the means of complex regulatory mechanisms the cholesterol level in a membrane is adjusted (Espenshade and Hughes, 2007). One important mechanism controls the activity of acyl-CoA:cholesterol acyltransferase (ACAT) in the endoplasmic reticulum (ER), which rapidly converts excess free cholesterol to cholesteryl esters. In the brain, however, the role of ACAT seems to be less important. The reason for this difference is based on the way cholesterol is supplied to the cells in the brain. Whereas most cells in the body acquire cholesterol from the blood by uptake of lipoprotein particles, cholesterol seems to be almost entirely produced by de novo synthesis in the brain (Dietschy and Turley, 2004). Endothelial cells of the blood-brain barrier express functional lipoprotein transporters such as the low density lipoprotein receptor, the scavenger receptor class B type 1 and the ABCA1 (Rubin and Staddon, 1999). However, there is little evidence for unselective transport of cholesterol across the blood-brain barrier as the capillaries of the nervous system are particular tight lacking fenestrae and little bulk flow transcytosis. Indeed, LDL particles might not cross the blood-brain barrier in significant amounts. When ^{125}I-labeled LDL was applied into the blood circulation, the uptake into brain was undetectable, while the liver and adrenal gland incorporated high amounts of radioactivity (Spady et al., 1987; Osono et al., 1995). Similar results were obtained using HDL labelled [^{14}C]cholesteryl esters. Isotopically labelled sterols or precursors have also been used to measure the permeability of the blood-brain barrier and the incorporation into myelin. After injection of [^{14}C]cholesterol into rats or feeding deuterium-labelled milk, less then 8% of brain sterol could be accounted for circulating cholesterol (Dobbing, 1963; Edmond et al., 1991). When ^{3}H$_2$O was applied to rats to determine the uptake of cholesterol in the brain, none of the radioactive cholesterol came from the circulation (Morell and Jurevics, 1996). All of these studies have failed to show transport of cholesterol across the blood-brain barrier and thus established the concept that cholesterol is produced locally in the brain.

The sole dependence of brain biosynthesis as a source of cholesterol in myelin may have important therapeutic consequences. The treatment with cholesterol lowering drugs such as statins that target the HMG-CoA reductase might reduce cholesterol biosynthesis in oligodendrocytes below a critical level and thereby block myelination. There is indeed evidence that statins inhibit remyelination after chemically induced demyelination using cuprizone in the central nervous system in mice (Klopfleisch et al., 2008; Miron et al., 2009).

In addition, in diseases that affect cholesterol biosynthesis, treatment with dietary cholesterol might not be effective in the brain due to the dependence on de novo synthesis. For example in the Smith-Lemli-Opitz syndrome, an autosomal recessive disorder in which cholesterol biosynthesis is inhibited due to an enzymatic defect of the 7-dehydrocholesterol reductase, raising circulatory cholesterol levels does not alleviate the neurological deficits.

Although it has been clearly shown that cholesterol is produced de novo in the brain, little is known about the transfer of cholesterol between the individual cells within the brain. One important cholesterol transferring apolipoprotein in the brain is apoE (Bu, 2009). Astrocytes are believed to be the major source of apoE in the brain (Boyles et al., 1985). In culture, these cells secrete cholesterol together with apoE into the medium from where it can be taken up by neurons (Shanmugaratnam et al., 1997). There is evidence that neurons depend on cholesterol derived from astrocytes for efficient synapse formation (Mauch et al., 2001). Whether astrocytes deliver cholesterol to oligodendrocytes for the generation of myelin has not been shown. However, genetic inactivation of cholesterol biosynthesis in oligodendrocytes does not abolish myelin formation completely, as these cells are able to take up cholesterol from a currently unknown external source (Saher et al., 2005).

In the brain, the rate of cholesterol turnover is extremely low. In the mouse around 0.4% of brain cholesterol is metabolised per day as compared to a whole-body turnover rate of 8% (Dietschy and Turley, 2004). The reason for the low turnover of myelin is the slow metabolism of cholesterol in myelin. The half-life of cholesterol in myelin has been calculated to be more than 8 months (Smith, 1968). The fate of the cholesterol that is turned over in myelin is not known. One major excretory pathway from the CNS into the periphery involves the hydroxylation of cholesterol at the 24 position by the cholesterol 24-hydroxylase (CYP 46A1) (Lutjohann et al., 1996). Because 24(S)-hydroxycholesterol is able to cross the blood-brain barrier, the generation of hydroxylated cholesterol represents the major pathway to remove cholesterol from the brain. However, CYP46A1 seems to be primarily expressed in a subtype of neurons and not in glia (Lund et al., 2003). It is therefore possible that oligodendrocytes use CYP46A1-independent mechanisms to metabolise cholesterol. Whether this involves the formation of lipoprotein particles is not known. Recently it has been shown that oligodendrocytes secrete cholesterol-rich exosomes into the extracellular space (Trajkovic et al., 2008). Exosomes are vesicles with a diameter of 40–100 nm that are secreted upon fusion of multivesicular endosomes with the cell surface (Simons and Raposo, 2009; Thery et al., 2009). These vesicles contain relatively large amounts of cholesterol and may thus represent one mechanism to remove free cholesterol from oligodendrocytes.

18.5 Cholesterol-Binding Proteins in Myelin

Despite the fact that myelin ranks among the cellular membranes richest in cholesterol in the vertebrate organism, there are limited definitive data about myelin proteins that directly associate with cholesterol. Plenty of candidates are found in myelin: members of the tetraspan family of proteins (e.g. PLP, PMP22), proteins with a single transmembrane domain that associate with detergent-resistant membranes (DRMs) (e.g. P0), and proteins lacking a transmembrane domain that are

Table 18.1 Myelin proteins that harbour a CRAC consensus sequence. The location of the CRAC motif relative to the predicted transmembrane domains (TM) is given on the basis of the human sequence (amino acids) and the Uniprot software

Protein	CRAC motif (aa)	Location relative to putative TM regions
Caveolin1	94–101	Before TM 1 of 1 TM
CD82	105–112	After TM 3 of 4
CNP	Multiple	No TM, but lipid anchor
MAG	526–537	After TM 1 of 1
MOG	174–183	After TM 2 of 3
P0	170–180	After TM 1 of 1
Plasmolipin	156–166	After TM 4 of 4
PLP	86–98	Between TM 2 and 3 of 4
PMP22	147–157	After TM 4 of 4

tightly associated with myelin by lipid anchors (e.g. CNP). In addition, many myelin proteins contain putative cholesterol-binding domains such as the CRAC (cholesterol recognition/interaction amino acid consensus) sequence (Li et al., 2001), the SSD (sterol sensing domain) (Espenshade and Hughes, 2007), or the CCM (cholesterol consensus motif) (Hanson et al., 2008) (Table 18.1). However, the tertiary structure of most myelin proteins is unknown. Without this knowledge, a meaningful interpretation of the occurrence of the respective consensus sequences or even a predictive value is not feasible. To date, only few interactions with cholesterol have been proven directly. PLP (Simons et al., 2000) and P0 (Saher et al., 2009), were found to directly associate with cholesterol by binding to photo-activatable cholesterol.

The feature of cholesterol-binding could be directly linked to the partitioning into lipid raft microdomains. PLP is found in detergent-resistant membranes in cultured oligodendrocytes and in myelin (Simons et al., 2000). Moreover, missense mutations in the PLP gene that lead to leukodystrophies in patients result in the expression of mutated PLP proteins with lowered affinity to cholesterol and reduced association with DRMs (Kramer-Albers et al., 2006). Not all of these mutations target the putative cholesterol interaction sites. Cholesterol-binding could influence the trafficking properties of proteins. In the case of PLP, there are additional functional implications. When oligodendrocyte-like cell lines are transfected with PLP, the protein is endocytosed in a cholesterol-dependent manner and stored in late endosomes/lysosomes. Both, PLP and cholesterol accumulate in late endosomes/lysosomes implying a co-transport (Trajkovic et al., 2006; Simons et al., 2002). These PLP-enriched late endosomes/lysosomes disappear when oligodendrocyte-like cells are cultured in the presence of neurons, suggesting that neuronal signals regulate oligodendroglial cholesterol trafficking.

Cholesterol-binding also plays a role in the PNS in P0 protein trafficking. The integrity of the CRAC motif of P0 was found to be required for ER-exit and trafficking to the myelin sheath (Saher et al., 2009). In contrast, the CRAC motif present in MAG might not be functional. MAG was not recovered from DRMs (Erne et al., 2002; Bosse et al., 2003) and its trafficking did not depend on cholesterol.

18.6 Cholesterol Depletion in Oligodendrocytes and Schwann cells In Vivo

The metabolism of cholesterol in the nervous system is presumed to be largely independent of the periphery (Jurevics et al., 1998; Morell and Jurevics, 1996). Oligodendrocytes and Schwann cells, respectively, synthesize the cholesterol needed for the formation of myelin membranes essentially cell autonomously (Saher et al, 2005; Fu et al., 1998; Jurevics and Morell, 1995). What is the role of cholesterol and cholesterol-binding proteins in myelin in vivo? Several studies have addressed this issue by gene targeting or by modifying the cholesterol level. In vivo cholesterol supplementation studies are hampered by the fact that the blood brain and blood nerve barriers efficiently shield the nervous system from cholesterol in the circulation (Rechthand and Rapoport, 1987). However, ex vivo studies using cultures of dorsal root ganglia (DRG) comprising DRG neurons and endogenous Schwann cells have revealed that supplementation with cholesterol advanced myelination at early stages (Saher et al., 2009). This finding implied that cholesterol could be rate limiting at the beginning of peripheral myelin formation.

Genetic inactivation of cholesterol biosynthesis in oligodendrocytes revealed that cholesterol is rate-limiting for CNS myelination, as myelin formation was severely delayed in conditional mutants (Fig. 18.4) (Saher et al., 2005). Most likely, cholesterol that was synthesized by other cells of the brain is provided to mutant oligodendrocytes for myelination. Cholesterol uptake appeared to be less efficient than endogenous cholesterol synthesis, as the process of myelin formation by mutant oligodendrocytes was severely slowed and extended into adulthood. A yet unidentified quality control ensures that myelin synthesized by cholesterol mutant oligodendrocytes appears rather normal. It showed a basically unaltered morphology regarding myelin periodicity and compaction correlating with the basically unchanged biochemical composition. Cholesterol and other lipids were still enriched in myelin to about 95% of control levels when normalized to phospholipid content. This implied that only minimal changes in cholesterol content of myelin are tolerated by oligodendrocytes. A high amount of cholesterol in myelin membranes might be required for physiological membrane fluidity and curving (Huttner and Zimmerberg, 2001). As cholesterol limits ion leakage through membranes it presumably contributes to the insulator function of myelin membranes and thus facilitates the rapid impulse conduction of myelinated axons (Salzer, 2003).

In the many genetic null mutant mice targeting genes for myelin structural proteins of the PNS, each individual myelin protein was dispensable for Schwann cells to synthesize myelin. In the PNS, cholesterol emerged as the first membrane component that is essential for myelination by Schwann cells (Fig. 18.5) (Saher et al., 2009). Cholesterol synthesis in Schwann cells has been targeted in two different

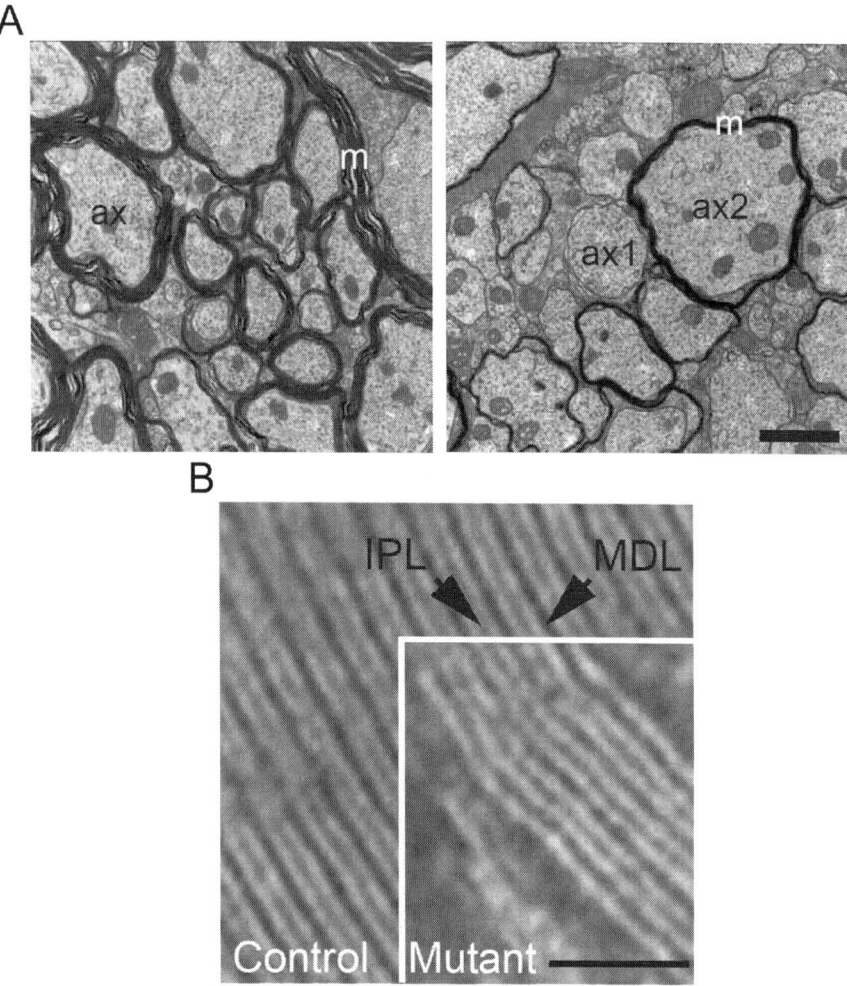

Fig. 18.4 Hypomyelination in the CNS of conditional cholesterol mutants. (**A**) Ultrastructural analysis of spinal cord at P20. In white matter of control animals (*left*), virtually all axons (ax) are myelinated, whereas fibers in mutants (*right*) possess a thin myelin sheath or lack myelin. This dysmyelinating phenotype is less pronounced in gray matter, where axons have myelin of comparable thickness in both control and mutant mice (ax1, unmyelinated axon; ax2, axon with thinner myelin; m, myelin; scale bar, 1 mm). (**B**) Ultrastructure of the lumbar spinal cord from mutant and control mice at P20. Compact myelin shows the same periodicity with normal major dense line (MDL) and intraperiod lines (IPL) in mutant (inset at *lower right*) and control mice. Photomicrographs are aligned to show similarity in Q35 periodicity. Scale bar, 50 nm. Reproduced from *Nature Neuroscience* with minor modifications (Saher et al., 2005), with permission

ways: One approach conditionally inactivated squalene synthase, the first enzyme which is strictly committed to cholesterol biosynthesis (Saher et al., 2009). A second approach targeted cholesterol homeostasis by conditionally inactivating SCAP (SREBP cleavage activating protein) in Schwann cells (Verheijen et al., 2009). It

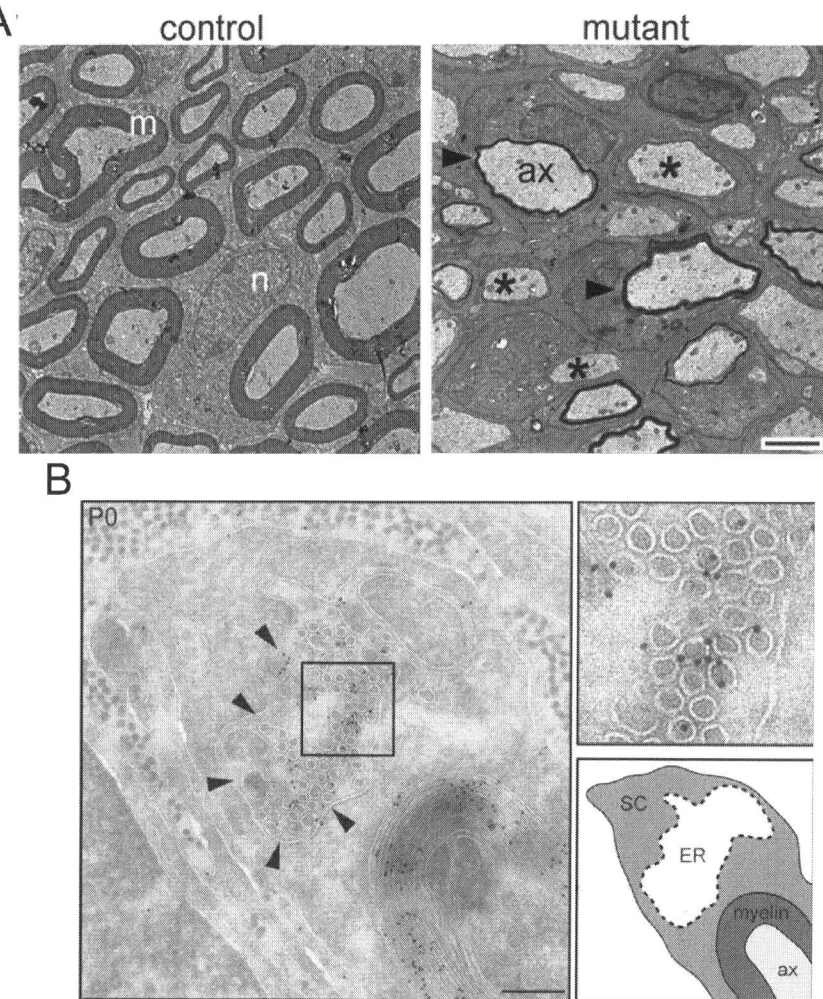

Fig. 18.5 Hypomyelination and a P0 trafficking defect in the PNS of conditional cholesterol mutants. (**A**) Ultrastructure of sciatic nerves (P20) reveals normal myelination of control animals but strong hypomyelination of mutants with thinly myelinated axons (*arrowheads*) and axons that were devoid of myelin (*asterisk*). Scale bar, 2 μm. (**B**) P0 detection by immunoelectron microscopy on mutant P14 sciatic nerve (10 nm gold). In addition to compact myelin, P0 is found in vesicular/tubular profiles within the Schwann cell ER (arrows and detail of boxed area in *top right*). The diagram clarifies structures of the picture. Scale bar, 250 nm (ax, Axon; ER, endoplasmic reticulum; m, myelin; n, Schwann cell nucleus; Schwann cell, Schwann cell). Reproduced from *Journal of Neurosciences* with minor modifications (Saher et al., 2009), with permission

has been shown in liver cells that SCAP senses cholesterol through its membranous sterol-sensing domain (Horton, 2002). In the case of excess cholesterol, SCAP binds to immature SREBP (sterol responsive element binding protein) transcription factors and the ER resident protein Insig (Insulin induced gene) (Sato, 2009). When cholesterol content falls below a critical level, the cholesterol-dependent binding of SCAP to Insig is lost, releasing SCAP-SREBP into ER transport vesicles. In the Golgi, the SREBP transcription factors are then activated by proteolytic cleavage. Finally, SREBPs induce target genes that are involved in the biosynthesis and uptake of cholesterol and fatty acids. Hence, inactivating SCAP affects cholesterol as well as fatty acid homeostasis. This regulatory system may also be functional in Schwann cells, as all components are expressed by these cells (Pertusa et al., 2007). Moreover during Schwann cell development, the expression of SREBP-2 follows the same time course not only as the expression of genes encoding enzymes of the cholesterol and lipid biosynthesis pathways, but also of myelin proteins (Lemke and Axel, 1985; D'Antonio et al. 2006; Nagarajan et al., 2002; Verheijen et al., 2003). This implied a close connection between lipid biosynthesis and myelin formation in the sciatic nerve.

By both approaches, the lack of cholesterol synthesis as well as the ablation of cholesterol and lipid regulation, peripheral myelin formation is severely affected (Saher et al., 2009; Verheijen et al., 2009). The severe delay of myelin formation in mutant mice causes typical neuropathic signs, including tremors and ataxic gait. Genes encoding several myelin proteins and relevant proteins involved in cholesterol and lipid synthesis are profoundly down-regulated. Cholesterol and lipids provided by other cells of the sciatic nerve (that are wild-type for the respective mutation) are most probably taken up by mutant Schwann cells. While cholesterol mutants remain severely hypomyelinated, conditional SCAP mutant mice overcome the pathology unexpectedly fast in adulthood. This is surprising, as inactivated SCAP targets cholesterol as well as lipid biosynthesis. Moreover, uptake of lipoprotein particles via the LDLR (low density lipoprotein receptor) should be also affected in these mutants. Hence, the (amazingly efficient) lipid uptake mechanism by Schwann cells remains unsolved. In line with this is the finding that the LDLR is dispensable for remyelination (Goodrum et al., 2000).

Inactivating major proteins of peripheral myelin that are candidates to associate with cholesterol causes completely different pathologies. P0 deficient mice are dysmyelinated, but still produce significant amounts of myelin (Giese et al., 1992). In P0-deficient myelin, there is a nearly complete lack of extracellular membrane compaction, whereas intracellular adhesion appears normal, probably because of the presence of MBP (Martini et al., 1995). Inactivation of the PMP22 gene also results in dysmyelination, however, in combination with focal hypermyelination (Adlkofer et al., 1995; Amici et al., 2006). These myelin outfoldings (tomacula) represent instabilities of myelin that eventually lead to the degeneration of myelin and finally of axons.

The comparison of the phenotypes of cholesterol mutants in PNS and CNS revealed that in both systems cholesterol appeared to be rate limiting for the formation of myelin membranes (Saher et al., 2005; Saher et al., 2009). In CNS myelin, the cholesterol content normalized to phosphatidyl-choline (PC) was about 95% in mutants compared to controls, implying that cholesterol levels were regulated in concert with other lipids. In contrast to the CNS, however, the composition of PNS myelin was strikingly altered. Here, cholesterol levels were only about 60% of

18 Cholesterol and Myelin Biogenesis

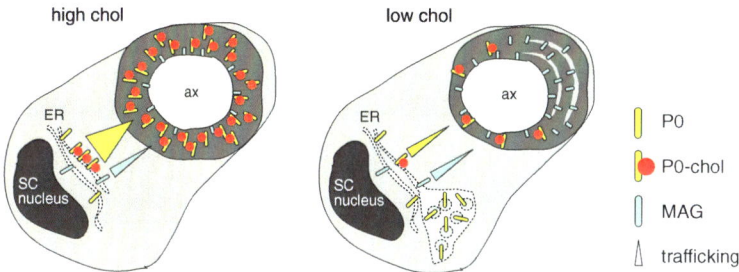

Fig. 18.6 Model of cholesterol mediated trafficking in the PNS. In the ER of Schwann cells, P0 normally associates with cholesterol enabling its transport to the myelin sheath (high chol). When cholesterol is limiting (low chol), i.e., during early myelination, P0 transport is adjusted to cholesterol availability and less P0 is shuttled from the ER to the myelin sheath. In case of severe cholesterol depletion, P0 is stored within endoplasmic vesicular/tubular structures. Note that MAG trafficking is independent of cholesterol. Consequently, the ratio of compact myelin (with P0-chol) to noncompact myelin (with MAG) is reduced (SC, Schwann cell; ax, axon). Reproduced from Journal of Neurosciences with minor modifications (Saher et al., 2009), with permission

control values (Saher et al., 2009). This implied a different quality control mechanism during PNS versus CNS myelin formation. The histological analysis of sciatic nerves showed that the compaction of myelin membranes of mutants was affected; frequently showing stretches of uncompact myelin. In addition, the composition of myelin proteins was altered showing higher levels of proteins of non-compact myelin and lower levels of proteins of compact myelin. Apparently the ratio of compact to non-compact myelin was shifted. It is possible that the changed composition of myelin also reflects this shift of compact to non-compact myelin. These data argue in favour for a yet unproven but tempting hypothesis that compact myelin may contain higher cholesterol levels then non-compact myelin.

The reason for non-compact myelin stretches may rest in the trafficking defect of P0 in cholesterol mutant mice. A fraction of P0 appeared to be retained in the ER of mutant Schwann cells in vivo. It is hypothesized that these P0 proteins failed to interact with cholesterol precluding the export towards the myelin sheath (Fig. 18.6). The mechanism by which P0-cholesterol controls myelin compaction remains to be shown. Taken together, in the PNS the level of cholesterol in Schwann cells may determine the rate of P0 trafficking to the myelin sheath enabling myelin compaction with proper cholesterol content.

18.7 Conclusions

As an integral component of most membranes, cholesterol plays an essential role in regulating its fluidity and permeability. Cholesterol is of particular importance for the insulating function of myelin that is required to ensure the rapid propagation of nerve conduction. In contrast to other organs, cholesterol is mainly synthesized locally within the brain and only minor amounts seem to be taken up by lipoproteins from the circulation. Nevertheless, mice that lack the ability to synthesize cholesterol in myelin-forming cells are still able to produce myelin, however, at a much lower rate. Mutant glia compensate for their loss of cholesterol biosynthesis by using cholesterol from neighboring cells. How cholesterol is delivered between the cells of the nervous system and how cells regulate the amount of cholesterol in myelin are important questions that need to be addressed in the future.

References

Adlkofer, K., Martini, R., Aguzzi, A., Zielasek, J., Toyka, K. V., and Suter, U., 1995, Hypermyelination and demyelinating peripheral neuropathy in Pmp22-deficient mice. *Nat Genet* **11**:274–80.

Amici, S. A., Dunn, W. A., Jr., Murphy, A. J., Adams, N. C., Gale, N. W., Valenzuela, D. M., Yancopoulos, G. D., and Notterpek, L., 2006, Peripheral myelin protein 22 is in complex with alpha6beta4 integrin, and its absence alters the Schwann cell basal lamina. *J Neurosci* **26**: 1179–89.

Baron, W., Colognato, H., and ffrench-Constant, C., 2005, Integrin-growth factor interactions as regulators of oligodendroglial development and function. *Glia* **49**:467–79.

Barres, B. A., Lazar, M. A., and Raff, M. C., 1994a, A novel role for thyroid hormone, glucocorticoids and retinoic acid in timing oligodendrocyte development. *Development* **120**: 1097–108.

Barres, B. A., and Raff, M. C., 1999, Axonal control of oligodendrocyte development, *J Cell Biol* **147**(6):1123–8.

Barres, B. A., Raff, M. C., Gaese, F., Bartke, I., Dechant, G., and Barde, Y. A., 1994b, A crucial role for neurotrophin-3 in oligodendrocyte development. *Nature* **367**:371–5.

Baumann, N., and Pham-Dinh, D., 2001, Biology of oligodendrocyte and myelin in the mammalian central nervous system. *Physiol Rev* **81**:871–927.

Bosse, F., Hasse, B., Pippirs, U., Greiner-Petter, R., and Müller, H.W., 2003, Proteolipid plasmolipin: Localization in polarized cells, regulated expression and lipid raft association in CNS and PNS myelin. *J Neurochem* **86**:508–18.

Boyles, J. K., Pitas, R. E., Wilson, E., Mahley, R. W., and Taylor, J. M., 1985, Apolipoprotein E associated with astrocytic glia of the central nervous system and with nonmyelinating glia of the peripheral nervous system. *J Clin Invest* **76**:1501–13.

Bu, G., 2009, Apolipoprotein E and its receptors in Alzheimer's disease: pathways, pathogenesis and therapy. *Nat Rev Neurosci* **10**:333–44.

Cahoy, J. D., Emery, B., Kaushal, A., Foo, L. C., Zamanian, J. L., Christopherson, K. S., Xing, Y., Lubischer, J. L., Krieg, P. A., Krupenko, S. A., Thompson, W. J., and Barres, B. A., 2008, A transcriptome database for astrocytes, neurons, and oligodendrocytes: a new resource for understanding brain development and function. *J Neurosci* **28**:264–78.

D'Antonio, M., Michalovich, D., Paterson, M., Droggiti, A., Woodhoo, A., Mirsky, R., and Jessen, K. R., 2006, Gene profiling and bioinformatic analysis of Schwann cell embryonic development and myelination. *Glia* **53**:501–15.

Dietschy, J. M., and Turley, S. D., 2004, Thematic review series: brain Lipids. Cholesterol metabolism in the central nervous system during early development and in the mature animal. *J Lipid Res* **45**:1375–97.

Dietschy, J. M., and Wilson, J. D., 1968, Cholesterol synthesis in the squirrel monkey: relative rates of synthesis in various tissues and mechanisms of control. *J Clin Invest* **47**:166–74.

Dobbing, J., 1963, The Blood-Brain Barrier: Some Recent Developments. *Guys Hosp Rep* **112**:267–86.

Dong, Z., Sinanan, A., Parkinson, D., Parmantier, E., Mirsky, R., and Jessen, K. R., 1999, Schwann cell development in embryonic mouse nerves. *J Neurosci Res* **56**:334–48.

Dubois-Dalcq, M., Behar, T., Hudson, L., and Lazzarini, R. A., 1986, Emergence of three myelin proteins in oligodendrocytes cultured without neurons. *J Cell Biol* **102**:384–92.

Edmond, J., Korsak, R. A., Morrow, J. W., Torok-Both, G., and Catlin, D. H., 1991, Dietary cholesterol and the origin of cholesterol in the brain of developing rats. *J Nutr* **121**:1323–30.

Emery, B., Agalliu, D., Cahoy, J. D., Watkins, T. A., Dugas, J. C., Mulinyawe, S. B., Ibrahim, A., Ligon, K. L., Rowitch, D. H., and Barres, B. A., 2009, Myelin gene regulatory factor is a critical transcriptional regulator required for CNS myelination. *Cell* **138**:172–85.

Erne, B., Sansano, S., Frank, M., and Schaeren-Wiemers, N., 2002, Rafts in adult peripheral nerve myelin contain major structural myelin proteins and myelin and lymphocyte protein (MAL) and CD59 as specific markers. *J Neurochem* **82**:550–62.

Espenshade, P. J., and Hughes, A. L., 2007, Regulation of sterol synthesis in eukaryotes. *Annu Rev Genet* **41**:401–27.

Fitzner, D., Schneider, A., Kippert, A., Mobius, W., Willig, K. I., Hell, S. W., Bunt, G., Gaus, K., and Simons, M., 2006, Myelin basic protein-dependent plasma membrane reorganization in the formation of myelin. *Embo J* **25**:5037–48.

Fu, Q., Goodrum, J. F., Hayes, C., Hostettler, J. D., Toews, A. D., and Morell, P., 1998, Control of cholesterol biosynthesis in Schwann cells. *J Neurochem* **71**:549–55.

Giese, K. P., Martini, R., Lemke, G., Soriano, P., and Schachner, M., 1992, Mouse P0 gene disruption leads to hypomyelination, abnormal expression of recognition molecules, and degeneration of myelin and axons. *Cell* **71**:565–76.

Goodrum, J. F., Fowler, K. A., Hostettler, J. D., and Toews, A. D., 2000, Peripheral nerve regeneration and cholesterol reutilization are normal in the low-density lipoprotein receptor knockout mouse. *J Neurosci Res* **59**:581–6.

Grinspan, J. B., Marchionni, M. A., Reeves, M., Coulaloglou, M., and Scherer, S. S., 1996, Axonal interactions regulate Schwann cell apoptosis in developing peripheral nerve: neuregulin receptors and the role of neuregulins. *J Neurosci* **16**:6107–18.

Hanson, M. A., Cherezov, V., Griffith, M. T., Roth, C. B., Jaakola, V. P., Chien, E. Y., Velasquez, J., Kuhn, P., and Stevens, R. C., 2008, A specific cholesterol binding site is established by the 2.8 A structure of the human beta2-adrenergic receptor. *Structure* **16**:897–905.

Harauz, G., Ishiyama, N., Hill, C. M., Bates, I. R., Libich, D. S., and Fares, C., 2004, Myelin basic protein-diverse conformational states of an intrinsically unstructured protein and its roles in myelin assembly and multiple sclerosis. *Micron* **35**:503–42.

Horton, J. D., 2002, Sterol regulatory element-binding proteins: transcriptional activators of lipid synthesis. *Biochem Soc Trans* **30**:1091–5.

Huttner, W. B., and Zimmerberg, J., 2001, Implications of lipid microdomains for membrane curvature, budding and fission. *Curr Opin Cell Biol* **13**:478–84.

Jessen, K. R., and Mirsky, R., 2005, The origin and development of glial cells in peripheral nerves. *Nat Rev Neurosci* **6**:671–82.

Jessen, K. R., and Mirsky, R., 2008, Negative regulation of myelination: relevance for development, injury, and demyelinating disease. *Glia* **56**:1552–65.

Jurevics, H., Bouldin, T. W., Toews, A. D., and Morell, P., 1998, Regenerating sciatic nerve does not utilize circulating cholesterol. *Neurochem Res* **23**:401–6.

Jurevics, H., and Morell, P., 1995, Cholesterol for synthesis of myelin is made locally, not imported into brain. *J Neurochem* **64**:895–901.

Kirby, B. B., Takada, N., Latimer, A. J., Shin, J., Carney, T. J., Kelsh, R. N., and Appel, B., 2006, In vivo time-lapse imaging shows dynamic oligodendrocyte progenitor behavior during zebrafish development. *Nat Neurosci* **9**:1506–11.

Klopfleisch, S., Merkler, D., Schmitz, M., Kloppner, S., Schedensack, M., Jeserich, G., Althaus, H. H., and Bruck, W., 2008, Negative impact of statins on oligodendrocytes and myelin formation in vitro and in vivo. *J Neurosci* **28**:13609–14.

Kramer-Albers, E. M., Gehrig-Burger, K., Thiele, C., Trotter, J., and Nave, K. A., 2006, Perturbed interactions of mutant proteolipid protein/DM20 with cholesterol and lipid rafts in oligodendroglia: implications for dysmyelination in spastic paraplegia. *J Neurosci* **26**:11743–52.

Le Douarin, N. M., and Dupin, E., 2003, Multipotentiality of the neural crest. *Curr Opin Genet Dev* **13**:529–36.
Leblanc, S. E., Srinivasan, R., Ferri, C., Mager, G. M., Gillian-Daniel, A. L., Wrabetz, L., and Svaren, J., 2005, Regulation of cholesterol/lipid biosynthetic genes by Egr2/Krox20 during peripheral nerve myelination. *J Neurochem* **93**:737–48.
Lemke, G., and Axel, R., 1985, Isolation and sequence of a cDNA encoding the major structural protein of peripheral myelin. *Cell* **40**:501–8.
Li, H., Yao, Z., Degenhardt, B., Teper, G., and Papadopoulos, V., 2001, Cholesterol binding at the cholesterol recognition/ interaction amino acid consensus (CRAC) of the peripheral-type benzodiazepine receptor and inhibition of steroidogenesis by an HIV TAT-CRAC peptide. *Proc Natl Acad Sci U S A* **98**:1267–72.
Lund, E. G., Xie, C., Kotti, T., Turley, S. D., Dietschy, J. M., and Russell, D. W., 2003, Knockout of the cholesterol 24-hydroxylase gene in mice reveals a brain-specific mechanism of cholesterol turnover. *J Biol Chem* **278**:22980–8.
Lutjohann, D., Breuer, O., Ahlborg, G., Nennesmo, I., Siden, A., Diczfalusy, U., and Bjorkhem, I., 1996, Cholesterol homeostasis in human brain: evidence for an age-dependent flux of 24S-hydroxycholesterol from the brain into the circulation. *Proc Natl Acad Sci U S A* **93**: 9799–804.
Martini, R., Mohajeri, M. H., Kasper, S., Giese, K. P., and Schachner, M., 1995, Mice doubly deficient in the genes for P0 and myelin basic protein show that both proteins contribute to the formation of the major dense line in peripheral nerve myelin. *J Neurosci* **15**:4488–95.
Mauch, D. H., Nagler, K., Schumacher, S., Goritz, C., Muller, E. C., Otto, A., and Pfrieger, F. W., 2001, CNS synaptogenesis promoted by glia-derived cholesterol. *Science* **294**:1354–7.
Michailov, G. V., Sereda, M. W., Brinkmann, B. G., Fischer, T. M., Haug, B., Birchmeier, C., Role, L., Lai, C., Schwab, M. H., and Nave, K. A., 2004, Axonal neuregulin-1 regulates myelin sheath thickness. *Science* **304**:700–3.
Miller, R. H., 2002, Regulation of oligodendrocyte development in the vertebrate CNS. *Prog Neurobiol* **67**:451–67.
Miron, V. E., Zehntner, S. P., Kuhlmann, T., Ludwin, S. K., Owens, T., Kennedy, T. E., Bedell, B. J., and Antel, J. P., 2009, Statin therapy inhibits remyelination in the central nervous system. *Am J Pathol* **17**:1880–90.
Mirsky, R., Winter, J., Abney, E. R., Pruss, R. M., Gavrilovic, J., and Raff, M. C., 1980, Myelin-specific proteins and glycolipids in rat Schwann cells and oligodendrocytes in culture. *J Cell Biol* **84**:483–94.
Morell, P., and Jurevics, H., 1996, Origin of cholesterol in myelin. *Neurochem Res* **21**:463–70.
Morell, P., and Norton, W. T., 1980, Myelin. *Sci Am* **242**:88–90, 92, 96 passim.
Muse, E. D., Jurevics, H., Toews, A. D., Matsushima, G. K., and Morell, P., 2001, Parameters related to lipid metabolism as markers of myelination in mouse brain. *J Neurochem* **76**:77–86.
Musse, A. A., Gao, W., Homchaudhuri, L., Boggs, J. M., and Harauz, G., 2008, Myelin basic protein as a "PI(4,5)P2-modulin": a new biological function for a major central nervous system protein. *Biochemistry* **47**:10372–82.
Nagarajan, R., Le, N., Mahoney, H., Araki, T., and Milbrandt, J., 2002, Deciphering peripheral nerve myelination by using Schwann cell expression profiling. *Proc Natl Acad Sci U S A* **99**:8998–9003.
Nave, K. A., and Salzer, J. L., 2006, Axonal regulation of myelination by neuregulin 1. *Curr Opin Neurobiol* **16**:492–500.
Nawaz, S., Kippert, A., Saab, A. S., Werner, H. B., Lang, T., Nave, K. A., and Simons, M., 2009, Phosphatidylinositol 4,5-bisphosphate-dependent interaction of myelin basic protein with the plasma membrane in oligodendroglial cells and its rapid perturbation by elevated calcium. *J Neurosci* **29**:4794–807.
Noble, M., Murray, K., Stroobant, P., Waterfield, M. D., and Riddle, P., 1988, Platelet-derived growth factor promotes division and motility and inhibits premature differentiation of the oligodendrocyte/type-2 astrocyte progenitor cell. *Nature* **333**:560–2.

Osono, Y., Woollett, L. A., Herz, J., and Dietschy, J. M., 1995, Role of the low density lipoprotein receptor in the flux of cholesterol through the plasma and across the tissues of the mouse. *J Clin Invest* **95**:1124–32.

Peles, E., and Salzer, J. L., 2000, Molecular domains of myelinated axons. *Curr Opin Neurobiol* **10**:558–65.

Pertusa, M., Morenilla-Palao, C., Carteron, C., Viana, F., and Cabedo, H., 2007, Transcriptional control of cholesterol biosynthesis in Schwann cells by axonal neuregulin 1. *J Biol Chem* **282**:28768–78.

Quarles, R. H., 2002, Myelin sheaths: glycoproteins involved in their formation, maintenance and degeneration. *Cell Mol Life Sci* **59**:1851–71.

Raff, M. C., Lillien, L. E., Richardson, W. D., Burne, J. F., and Noble, M. D., 1988, Platelet-derived growth factor from astrocytes drives the clock that times oligodendrocyte development in culture. *Nature* **333**:562–5.

Rechthand, E., and Rapoport, S. I., 1987, Regulation of the microenvironment of peripheral nerve: role of the blood-nerve barrier. *Prog Neurobiol* **28**:303–43.

Richardson, W. D., Pringle, N., Mosley, M. J., Westermark, B., and Dubois-Dalcq, M., 1988, A role for platelet-derived growth factor in normal gliogenesis in the central nervous system. *Cell* **53**:309–19.

Rowitch, D. H., 2004, Glial specification in the vertebrate neural tube. *Nat Rev Neurosci* **5**: 409–19.

Rubin, L. L., and Staddon, J. M., 1999, The cell biology of the blood-brain barrier. *Annu Rev Neurosci* **22**:11–28.

Saher, G., Brugger, B., Lappe-Siefke, C., Mobius, W., Tozawa, R., Wehr, M. C., Wieland, F., Ishibashi, S., and Nave, K. A., 2005, High cholesterol level is essential for myelin membrane growth. *Nat Neurosci* **8**:468–75.

Saher, G., Quintes, S., Mobius, W., Wehr, M. C., Kramer-Albers, E. M., Brugger, B., and Nave, K. A., 2009, Cholesterol regulates the endoplasmic reticulum exit of the major membrane protein P0 required for peripheral myelin compaction. *J Neurosci* **29**:6094–104.

Salzer, J. L., 2003, Polarized domains of myelinated axons. *Neuron* **40**:297–318.

Sato, R., 2009, SREBPs: protein interaction and SREBPs. *Febs J* **276**:622–7.

Shanmugaratnam, J., Berg, E., Kimerer, L., Johnson, R. J., Amaratunga, A., Schreiber, B. M., and Fine, R. E., 1997, Retinal Muller glia secrete apolipoproteins E and J which are efficiently assembled into lipoprotein particles. *Brain Res Mol Brain Res* **50**:113–20.

Sherman, D. L., and Brophy, P. J., 2005, Mechanisms of axon ensheathment and myelin growth. *Nat Rev Neurosci* **6**:683–90.

Simons, M., Kramer, E. M., Macchi, P., Rathke-Hartlieb, S., Trotter, J., Nave, K. A., and Schulz, J. B., 2002, Overexpression of the myelin proteolipid protein leads to accumulation of cholesterol and proteolipid protein in endosomes/lysosomes: implications for Pelizaeus-Merzbacher disease. *J Cell Biol* **157**:327–36.

Simons, M., Kramer, E. M., Thiele, C., Stoffel, W., and Trotter, J., 2000, Assembly of myelin by association of proteolipid protein with cholesterol- and galactosylceramide-rich membrane domains. *J Cell Biol* **151**:143–54.

Simons, M., and Raposo, G., 2009, Exosomes - vesicular carriers for intercellular communication. *Curr Opin Cell Biol.* **21**:575–81

Simons, M., and Trajkovic, K., 2006, Neuron-glia communication in the control of oligodendrocyte function and myelin biogenesis. *J Cell Sci* **119**:4381–9.

Simons, M., and Trotter, J., 2007, Wrapping it up: the cell biology of myelination. *Curr Opin Neurobiol* **17**:533–40.

Smith, M. E., 1968, The turnover of myelin in the adult rat. *Biochim Biophys Acta* **164**:285–93.

Snipes, G. J., and Suter, U., 1997, Cholesterol and myelin. *Subcell Biochem* **28**:173–204.

Spady, D. K., Huettinger, M., Bilheimer, D. W., and Dietschy, J. M., 1987, Role of receptor-independent low density lipoprotein transport in the maintenance of tissue cholesterol balance in the normal and WHHL rabbit. *J Lipid Res* **28**:32–41.

Taveggia, C., Zanazzi, G., Petrylak, A., Yano, H., Rosenbluth, J., Einheber, S., Xu, X., Esper, R. M., Loeb, J. A., Shrager, P., Chao, M. V., Falls, D. L., Role, L., and Salzer, J. L., 2005, Neuregulin-1 type III determines the ensheathment fate of axons. *Neuron* **47:**681–94.

Temple, S., and Raff, M. C., 1986, Clonal analysis of oligodendrocyte development in culture: evidence for a developmental clock that counts cell divisions. *Cell* **44:**773–9.

Thery, C., Ostrowski, M., and Segura, E., 2009, Membrane vesicles as conveyors of immune responses, *Nat Rev Immunol*. **9**(8): 581–93.

Trajkovic, K., Dhaunchak, A. S., Goncalves, J. T., Wenzel, D., Schneider, A., Bunt, G., Nave, K. A., and Simons, M., 2006, Neuron to glia signaling triggers myelin membrane exocytosis from endosomal storage sites. *J Cell Biol* **172:**937–48.

Trajkovic, K., Hsu, C., Chiantia, S., Rajendran, L., Wenzel, D., Wieland, F., Schwille, P., Brugger, B., and Simons, M., 2008, Ceramide triggers budding of exosome vesicles into multivesicular endosomes. *Science* **319:**1244–7.

Verheijen, M. H., Chrast, R., Burrola, P., and Lemke, G., 2003, Local regulation of fat metabolism in peripheral nerves. *Genes Dev* **17:**2450–64.

Verheijen, M. H., Camargo, N., Verdier, V., Nadra, K., de Preux Charles, A. S., Medard, J. J., Luoma, A., Crowther, M., Inouye, H., Shimano, H., Chen, S., Brouwers, J. F., Helms, J. B., Feltri, M. L., Wrabetz, L., Kirschner, D., Chrast, R., and Smit, A. B., 2009, SCAP is required for timely and proper myelin membrane synthesis. *Proc Natl Acad Sci U S A* **106:**21383–8.

Webster, H. D., 1971, The geometry of peripheral myelin sheaths during their formation and growth in rat sciatic nerves. *J Cell Biol* **48:**348–67.

Weimbs, T., and Stoffel, W., 1992, Proteolipid protein (PLP) of CNS myelin: positions of free, disulfide-bonded, and fatty acid thioester-linked cysteine residues and implications for the membrane topology of PLP. *Biochemistry* **31:**12289–96.

Woodhoo, A., and Sommer, L., 2008, Development of the Schwann cell lineage: from the neural crest to the myelinated nerve. *Glia* **56:**1481–90.

Chapter 19
Cholesterol and Ion Channels

Irena Levitan, Yun Fang, Avia Rosenhouse-Dantsker, and Victor Romanenko

Abstract A variety of ion channels, including members of all major ion channel families, have been shown to be regulated by changes in the level of membrane cholesterol and partition into cholesterol-rich membrane domains. In general, several types of cholesterol effects have been described. The most common effect is suppression of channel activity by an increase in membrane cholesterol, an effect that was described for several types of inwardly-rectifying K^+ channels, voltage-gated K^+ channels, Ca^{+2} sensitive K^+ channels, voltage-gated Na^+ channels, N-type voltage-gated Ca^{+2} channels and volume-regulated anion channels. In contrast, several types of ion channels, such as epithelial amiloride-sensitive Na^+ channels and Transient Receptor Potential channels, as well as some of the types of inwardly-rectifying and voltage-gated K^+ channels were shown to be inhibited by cholesterol depletion. Cholesterol was also shown to alter the kinetic properties and current-voltage dependence of several voltage-gated channels. Finally, maintaining membrane cholesterol level is required for coupling ion channels to signalling cascades. In terms of the mechanisms, three general mechanisms have been proposed: (i) specific interactions between cholesterol and the channel protein, (ii) changes in the physical properties of the membrane bilayer and (iii) maintaining the scaffolds for protein-protein interactions. The goal of this review is to describe systematically the role of cholesterol in regulation of the major types of ion channels and to discuss these effects in the context of the models proposed.

Keywords Ion channels · Cholesterol · Lipid rafts

19.1 Introduction

During the last decade, a growing number of studies have demonstrated that the level of membrane cholesterol is a major regulator of ion channel function (*reviewed by*

I. Levitan (✉)
Department of Medicine, Pulmonary Section, University of Illinois at Chicago, 840 S. Wood Str, 60612, Chicago, IL, USA
e-mail: levitan@uic.edu

Maguy et al., 2006; Martens et al., 2004). It is also becoming increasingly clear that the impact of cholesterol on different types of ion channels is highly heterogeneous. The most common effect is cholesterol-induced decrease in channel activity that may include decrease in the open probability, unitary conductance and/or the number of active channels on the membrane. This effect was observed in several types of K^+ channels, voltage-gated Na^+ and Ca^{+2} channels, as well as in volume-regulated anion channels. However, there are also several types of ion channels, such as epithelial Na^+ channels (eNaC) and transient receptor potential (Trp) channels that are inhibited by the removal of membrane cholesterol. Finally, in some cases changes in membrane cholesterol affect biophysical properties of the channel such as the voltage dependence of channel activation or inactivation. Clearly, therefore, more than one mechanism has to be involved in cholesterol-induced regulation of different ion channels.

Two general mechanisms have been proposed for cholesterol regulation of ion channels. One possibility is that cholesterol may interact directly and specifically with the transmembrane domains of the channels protein. Direct interaction between channels and cholesterol as a boundary lipid was first proposed in a "lipid belt" model by Marsh and Barrantes (1978) suggesting that cholesterol may be a part of a lipid belt or a "shell" constituting the immediate perimeter of the channel protein (Barrantes, 2004; Criado et al., 1982; Marsh and Barrantes, 1978). Figure 19.1 schematically shows the dynamic exchange between the lipid shell of an acetylcholine receptor protein and the bulk of the membrane (Barrantes, 2004). The role of cholesterol in the regulation of acetylcholine receptor is described in detail in Chapter 17 of this book. More recently, our studies demonstrated that inwardly-rectifying K^+ channel are sensitive to the chiral nature of the sterol analogue providing further support for the hypothesis that sensitivity of these channels to cholesterol can be due to specific sterol-protein interactions (Romanenko et al., 2002). An alternative mechanism proposed by Lundbaek and colleagues (Lundbaek and Andersen, 1999; Lundbaek et al., 1996) suggested that cholesterol may regulate ion channels by hydrophobic mismatch between the transmembrane domains

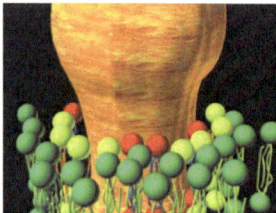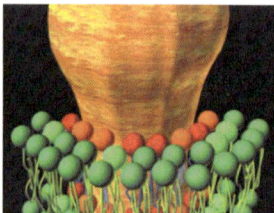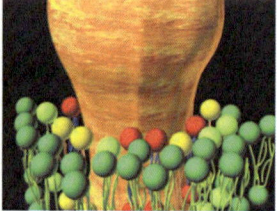

Fig. 19.1 Regulation of an ion channel by annular lipids (from Barrantes (2004)). The diagram schematically shows a channel protein surrounded by specific lipid molecules that constitute the annular "belt" around the channel. The three panels illustrate the exchange process between the annular lipids and the bulk lipids of the membrane. A cholesterol molecule is proposed to be part of the lipid belt surrounding the channel. © Barrantes (2004). Originally published in *Brain Research Reviews* 47:71–95

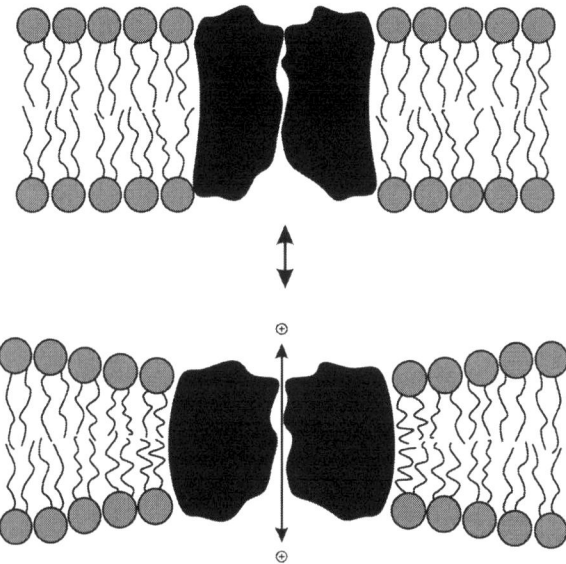

Fig. 19.2 Hydrophobic coupling between channel conformational changes and lipid bilayer deformations. The diagram schematically shows a transition between the closed and the open states of an ion channel that is accompanied with a deformation of the lipid bilayer in the vicinity of the membrane. Membrane deformation involves compression and bending of the membrane leaflets, which is suggested to contribute the energetic cost of the channel opening. In this model, an increase in stiffness of the lipid bilayer is expected to increase the cost of the transition resulting in the inhibition of channel activity. © Lundbaek et al. (2004). Originally published in the *Journal of General Physiology* 123: 599–621 (Reproduced with permission)

and the lipid bilayer. More specifically, it was proposed that when a channel goes through a change in conformation state within the viscous medium of the lipid membrane it may induce deformation of the lipid bilayer surrounding the channel. If this is the case, then a stiffer less deformable membrane will increase the energy that is required for the transition, as described schematically in Fig. 19.2. It is important to note that the two mechanisms are not mutually exclusive. A lipid shell surrounding a channel may also affect the hydrophobic interactions between the channels and the lipids and increase the deformation energy required for the transitions between closed and open states. Finally, obviously, cholesterol may also affect the channels indirectly through interactions with different signalling cascades.

Another important factor in understanding the mechanisms of cholesterol regulation of ion channels is the association of the channels with cholesterol-rich membrane domains, typically called membrane or lipid rafts. While the exact nature and composition of these domains remains controversial, they are generally defined as "small (10–200 nm), heterogeneous, highly dynamic, sterol- and sphingolipid-enriched domains that compartmentalize cellular processes" (Pike, 2006). Indeed, a variety of ion channels have been shown to be associated with these domains. It is important to note, however, that within lipid rafts, channels may be regulated by any of the mechanisms described above: (i) by direct binding of cholesterol, which is abundant in the rafts; (ii) by an increase in membrane stiffness within the raft domains, also known as "ordered domains" due to high order of lipid packing and, of course (iii) by the interactions with multiple signalling molecules that are segregated within the raft domains. Thus, association with lipid rafts provides an additional level of complexity to how ion channels or any other membrane proteins can be regulated by cholesterol rather than a specific mechanism. In this chapter, we will systematically describe what is known about the effects of cholesterol on all major types of ion channels and discuss these effects in context of the three mechanisms described above.

We will also discuss the evidence for and the implications of cholesterol regulation of ion channels under hypercholesterolemic conditions *in vivo*. Indeed, the presence of high levels of cholesterol in the blood, termed hypercholesterolemia, contributes significantly to the development of many human cardiovascular diseases such as coronary heart disease and stroke. Human hypercholesterolemia is typically associated with diet and genetic factors, or can be direct result of other disorders such as diabetes mellitus and an underactive thyroid. Furthermore, hypercholesterolemia was shown to cause dysfunction of many cell types including endothelial cells, endothelial progenitor cells, smooth muscle cells, monocytes, macrophage, T lymphocytes, platelets and cardiomyocytes. In this chapter, we will therefore also discuss the current understanding of ion channel modulation by *in vivo* hypercholesterolemia that may underlie key mechanistic events of development of human diseases.

19.2 Cholesterol Regulation of K^+ Channels

K^+ channels are a highly heterogeneous group constituting the largest and the most diverse group of ion channels that includes several structurally different classes, such as two-transmembrane domains inwardly-rectifying K^+ channels (Kir), four-transmembrane domains two pore K^+ (2PK) channels, and six-transmembrane domains voltage-gated (Kv) and Ca^{+2}-activated K^+ (K_{Ca}) channels. Changes in the level of membrane cholesterol were shown to regulate multiple types of K^+ channels belonging to Kir, Kv and K_{Ca} families. However, specific cholesterol effects vary significantly between different families or even between the members of the same family. The following section summarizes the similarities and the differences in cholesterol sensitivity of different types of K^+ channels.

19.2.1 Inwardly Rectifying K^+ (Kir) Channels

Kir channels constitute one of the major classes of K^+ channels that are responsible for the maintenance of membrane potential and K^+ homeostasis in a variety of cell types, including heart, brain, vascular cells and pancreas (Bichet et al.,

2003; Kubo et al., 2005; Nichols and Lopatin, 1997). Kir channels open at resting membrane potential and their main physiological roles is regulating membrane excitability, heart rate and vascular tone (Bichet et al., 2003; Kubo et al., 2005; Nichols and Lopatin, 1997). Structurally, Kir channels are tetramers with a basic subunit consisting of two transmembrane domains, a pore loop, and an N-terminus and a C-terminus cytoplasmic domain. Kir channels are classified into seven sub-families (Kir1–7) identified by distinct biophysical properties, such as degree of current rectification and unitary conductance, and by their sensitivities to different mediators (*reviewed by* Kubo et al., 2005; Nichols and Lopatin, 1997). Three of these sub-families, Kir2, Kir4 and Kir6 channels, have been shown to be sensitive to cholesterol, but the ways cholesterol affects different types of Kir channels are significantly different. Cholesterol sensitivity of other Kir channels has not yet been established.

19.2.2 Kir2 Channels

Kir2 channels are ubiquitously expressed in a variety of cell types, including heart, neurons, vascular smooth muscle and endothelial cells (Kubo et al., 2005). Todate, four members of the Kir2 family have been identified (Kir2.1–2.4). All four members of the Kir2 family are suppressed by the elevation of membrane cholesterol and enhanced by cholesterol depletion, but there are some differences in their cholesterol sensitivity, with Kir2.1 and Kir2.2 being most sensitive and Kir2.3 being least sensitive (Romanenko et al., 2004a; Romanenko et al., 2002). Surprisingly, in spite of 2–3 fold decrease in whole cell Kir2 currents, no or little effect was observed in the single channel properties of the channels: unitary conductance was not affected at all and the open probability was decreased less than 10% (Romanenko et al., 2004a; Romanenko et al., 2002). Moreover, changes in membrane cholesterol have no effect either on Kir2.1 expression, as estimated by Western blot analysis and by immunostaining, or on its plasma membrane level, as estimated by tagging the extracellular domains of the channels (Romanenko et al., 2004a), as was described earlier (Zerangue et al., 1999). Taken together, these observations led us to the hypothesis that Kir2 channels exist in the plasma membrane in two modes: "active channels" that flicker between the closed and the open states with high open probability and "silent channels" that are stabilized in their closed state.

The first clue about the mechanism came from comparing the effects of cholesterol and of its optical isomer, epicholesterol (Romanenko et al., 2002). The two stereoisomers, native cholesterol (3β-hydroxy-5-cholestene) and epicholesterol (3α-hydroxy-5-cholestene) differ only in the rotational angle of the hydroxyl group at position 3 and are known to have similar effects on membrane ordering and lipid packing (Demel et al., 1972; Xu and London, 2000). However, our observations showed that the effects of the two sterols on Kir2 channels are completely different: while, as described above, cholesterol suppresses Kir2 channels, partial substitution of native cholesterol by epicholesterol resulted in significant increase in Kir2 current suggesting that it is a specific lipid-channel interaction rather than changes in

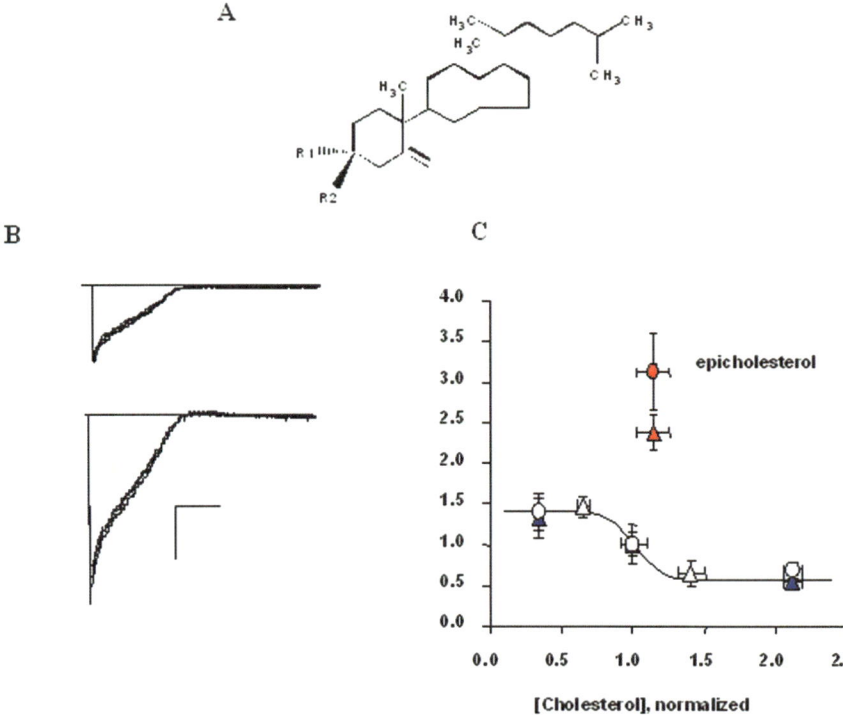

Fig. 19.3 Chiral analogues of cholesterol have opposite effects on endothelial Kir currents. (**A**) Structure of cholesterol and epicholesterol. Cholesterol: R1=H, R2=OH; epicholesterol: R1=OH, R2=H. (**B**) Typical current traces recorded from a cell exposed to MßCD-epicholesterol and from a control cell. (**C**) Functional dependence of Kir current density on cholesterol and epicholesterol. Adapted from Romanenko et al. (2002)

the physical properties of the membrane that is responsible for cholesterol-induced suppression of Kir2 channels (Fig. 19.3).

More recently, we have identified a specific region of the Kir2.1 channels that plays a critical role in the sensitivity of these channels to cholesterol (Fig. 19.4) (Epshtein et al., 2009). Surprisingly, the region critical for the sensitivity of Kir2 channels to cholesterol was identified not in the transmembrane domain of the channels, as expected, but in the C terminus cytosolic domain. More specifically, cholesterol sensitivity of Kir2 channels depends on the CD loop, a specific region of the C-terminus of the cytosolic domain of the channel. Within this loop, the L222I mutation in Kir2.1 abrogates the sensitivity of the channel to cholesterol, whereas a reverse mutation in the corresponding position in Kir2.3, I214L, has the opposite effect, increasing cholesterol sensitivity. Furthermore, the L222I mutation has a dominant negative effect on cholesterol sensitivity of Kir2.1 WT. Mutations of two additional residues in the CD loop in Kir2.1, N216D and K219Q, partially affect the sensitivity of the channel to cholesterol. We proposed, therefore, that the residues of the CD-loop are involved in "docking" of the C-terminus of Kir2.1 to the inner

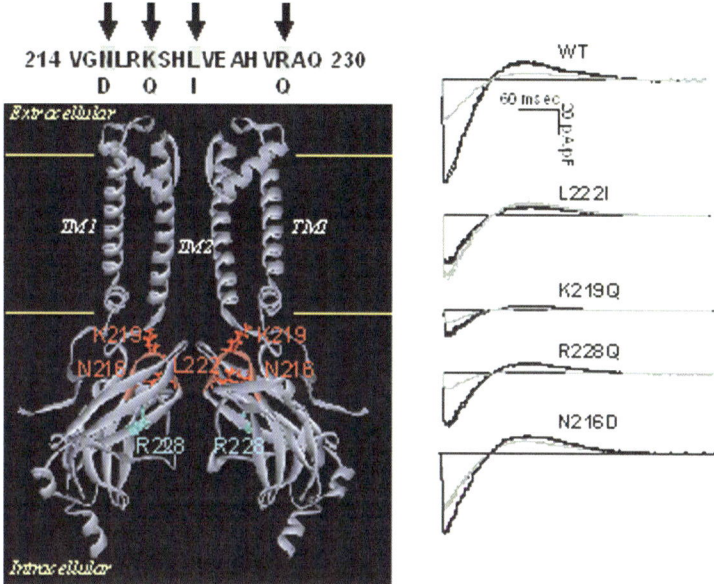

Fig. 19.4 Identification of a cytoplasmic domain critical for the sensitivity of Kir2.1 channels to cholesterol. (**A**) Sequence of Kir WT with marked PIP_2-sensitive mutations analyzed for sensitivity to cholesterol and the homology model showing two opposite facing subunits of the channel with the positions of these residues. (**B**) Typical current traces of Kir2.1-WT, Kir2.1-R228Q, Kir2.1-K219Q and Kir2.1-N216D in control cells (*grey*) and in cells depleted of cholesterol (*black*) *From* Epshtein et al. (2009)

leaflet of the membrane and facilitating its interaction with membrane cholesterol. In this model, when a channel is in the "docking" configuration it may interact with cholesterol, which in turn is proposed to stabilize the channel in a closed state. It is important to note, however, that in the framework of this hypothesis, the critical residues of the CD loop do not necessarily interact with cholesterol directly. Alternatively, it is possible that their role is to maintain the channels in a "docking" conformation state that allows cholesterol to bind to another part of the channel. Finally, it is also possible that residues in the CD-loop facilitate the hydrophobic interaction between the TM domains of the channel and the lipid core of the plasma membrane.

19.2.3 Kir3 Channels

Kir3 channels are a family of G-protein gated channels activated by G protein βγ subunits and play a major role in the inhibitory regulation of neuronal excitability (Kir3.2 and Kir 3.1/3.2 heteromers) and in the regulation of the heart rate (Kir3.4, Kir3.1/3.4 heteromers) (Kubo et al., 2005). Loss of Kir3 channels leads to hyperexcitability in the brain, hyperactivity and seizures, as well as cardiac abnormalities. Little is known, however, about cholesterol sensitivity of these channels. Specifically, it was shown that regulation of Kir3.1/3.2 channels by Neural Cell Adhesion Molecule (NCAM) was compromised by lovastatin, a drug that lowers cholesterol (Delling et al., 2002). It was also shown that deletion of a number of genes that have been shown to alter the lipid composition of yeast membrane significantly affect functional expression of Kir3.2 channels in the yeast membranes, suggesting that membrane lipids may play an important role in the regulation of Kir3.2 trafficking and/or function (Haass et al., 2007). Clearly, more studies are needed to investigate how cholesterol affects Kir3 channels.

19.2.4 Kir4 Channels

Kir4 channels are also expressed in multiple cell types, including glial and kidney epithelial cells. Two Kir4 channels have been identified: Kir4.1 implicated in the control of K^+ buffering and homeostasis, and Kir4.2 whose physiological functions are not well established yet (Kubo et al., 2005). Recently, (Hibino and Kurachi, 2007) have shown that Kir4.1 channels are cholesterol sensitive, but in contrast to Kir2 channels, Kir4.1 channels were shown to be inhibited by cholesterol depletion. Furthermore, the authors suggested that the loss of Kir4.1 activity may be due to the dissociation of the channels from a regulatory phospholipid PI(4,5)P2 that is known to be required for the activation of multiple ion channels including Kir channels (Hilgemann et al., 2001; Logothetis et al., 2007), that also resides in cholesterol-rich membrane domains (Pike and Casey, 1996). This interpretation, however, does not explain the dramatic difference between Kir4.1 and Kir2 channels in terms of their sensitivity to cholesterol because Kir2 also require PI(4,5)P2 for their function. The source of this difference is not yet clear.

19.2.5 Kir6 (K_{ATP}) Channels

Kir6 (K_{ATP}) channels are heteromultimeric Kir channels with a pore formed by a tetramer of Kir6 subunits with each Kir6 subunit associated with one ATP-binding cassette (ABC) sulfonylurea receptor (SUR) protein (*reviewed by* Nichols, 2006; Zingman et al., 2007). Two members of the Kir6 channel family have been identified: Kir6.1 channels are expressed in vascular smooth muscle cells and Kir6.2 channels in pancreatic β-cells, heart, and brain (Kubo et al., 2005; Nichols, 2006; Zingman et al., 2007). It is known that K_{ATP} channels are under the control of protein kinases A (PKA) and C (PKC), with PKA activating the channels and PKC having an inhibitory effect (Edwards and Weston, 1993). Sampson and colleagues (Sampson et al., 2007; Sampson et al., 2004) showed that cholesterol depletion affects Kir6.1 by interfering with the coupling of the channels to PKA, thus inhibiting PKA-dependent activation of the channels.

19.3 Association of Kir Channels with Cholesterol-Rich Membrane Domains (Lipid Rafts)

As discussed briefly above, multiple studies have shown that different types of ion channels are associated with lipid rafts. Indeed, members of almost all of the Kir families were shown to partition into cholesterol-rich membrane domains: Kir2 s (Kir2.1 and Kir2.3) (Tikku et al., 2007), Kir3.1/3.4 (Delling et al., 2002), Kir 4 (Hibino and Kurachi, 2007) and Kir6.1 (Sampson et al., 2004). However, partitioning of different Kir channels into lipid rafts may be associated with completely different or even opposite functional effects: for example, Kir2 channels are suppressed by cholesterol whereas Kir4 channels are suppressed by cholesterol depletion. The variability of these effects may be due to specific raft environments or due to structural differences between the channels. Furthermore, we have shown recently that Kir2.1-L222I mutant that is not sensitive to cholesterol also partitions into lipid rafts (Epshtein et al., 2009), demonstrating that partitioning into rafts does not necessarily indicate that the channels are functionally regulated by changes in membrane cholesterol. Thus, while partitioning into cholesterol-rich membrane domains is frequently associated with functional dependence of ion channels on membrane cholesterol, it is not always the case. Furthermore, within the rafts, channels may be regulated by completely different mechanisms.

19.3.1 Regulation of Kir Channels by Plasma Hypercholesterolemia In Vivo

<u>Kir2:</u> Our group has demonstrated that diet-induced hypercholesterolemia in a porcine model results in significant suppression of Kir current in freshly-isolated aortic endothelial cells and in bone-marrow derived progenitor cells (Fang et al., 2006; Mohler Iii et al., 2007). Removing cholesterol surplus *ex vivo* resulted in full recovery of the current. We have also shown previously that endothelial Kir current is underlain by Kir2.1 and Kir2.2 channels (Fang et al., 2005). In addition, hypercholesterolemia resulted in significant membrane depolarization of endothelial cells and notable loss of flow-mediated vasodilatation of the femoral artery estimated *in vivo* in the same model (Fang et al., 2006). We proposed that since endothelial Kir channels are sensitive to shear stress and constitute one of the putative flow sensors (Davies, 1995; Olesen et al., 1988), impairment of flow-induced vasodilatation may be due to the suppression of endothelial Kir channels (Fang et al., 2006). These observations suggest that hypercholesterolemia-mediated suppression of Kir channels may be an important factor in dysfunction of mature endothelium and endothelial progenitor cells and contribute to the initiation and development of atherosclerosis.

Kir6 (K_{ATP}) channels are also regulated by hypercholesterolemia *in vivo,* but the mode of the regulation is still controversial. In a porcine model, (Mathew and Lerman, 2001) showed that pinacidil, a K_{ATP} opener, and glibenclamide, a

K_{ATP} blocker, have stronger effects on coronary blood flow under hypercholesterolemic conditions than under control conditions, suggesting functional activation of Kir6 under hypercholesterolemia. In contrast, Genda et al. (2002) demonstrated that diet-induced hypercholesterolemia suppressed K_{ATP} opening in microvasculature, which led to a delay of infarct healing under no-reflow phenomenon after acute myocardial infarction in a rabbit model. Lee et al. (2004) also reported that hypercholesterolemia resulted in impairment of myocardial K_{ATP} channels, which contributed to left ventricular hypertrophy in a rabbit model. One possibility to account for these differences in the effects of hypercholesterolemia on K_{ATP} activity may be the difference in the lipid profiles between hypercholesterolemic rabbits and pigs. Pongo et al. (2001) reported dysfunction of protein kinase C– K_{ATP} channel coupling in rabbit coronary arteries under hypercholesterolemia, providing a mechanistic explanation of hypercholesterolemia-induced suppression of K_{ATP} activity. Hypercholesterolemia may also regulate Kir channels on the level of mRNA expression. Ren et al. (2001) demonstrated that hypercholesterolemia enhanced the mRNA of Kir6.2, but down-regulated Kir3.1 mRNA in smooth muscle cells while Kir2.1 and Kir6.1 transcripts were unchanged in a rat model.

In summary, sensitivity to cholesterol and association with cholesterol-rich membrane domains appears to be a common feature for multiple types of Kir channels but there is a strong diversity in the effects of cholesterol on channel function, and while some Kir channels, specifically Kir2s, are suppressed by cholesterol, other Kirs, specifically Kir4.1, require cholesterol for their function. Our recent studies provided the first insights into the structural requirements for cholesterol sensitivity of Kir2 channels, establishing the basis for further investigations and detailed understanding of how Kir channels interact with cholesterol. Furthermore, we propose that suppression of Kir channels in hypercholesterolemic conditions may play a major role in the development of cardiovascular disease.

19.4 Voltage-Gated K^+ (Kv) Channels

Voltage-gated K^+ channels are the largest and the most diverse family of K^+ channels that include about 40 members, classified into 12 subfamilies (Kv1–12 sub-families) (Gutman et al., 2005). Kv channels open in response to membrane depolarization and their main physiological role is termination of action potentials and returning of the membrane potential back to its resting state. A large diversity of Kv channels with different activation and inactivation properties underlie the plethora of firing patterns in different excitable cells. Kv channels are also expressed in a variety of non-excitable cells, including different types of immune cells where they are known to play important roles in the immune response. Structurally, Kv channels are tetramers formed either of identical subunits or between different Kv subunits with each subunit consisting of 6 transmembrane helices (S1–S6) linked by extracellular and cytosolic loops. The fifth and the sixth helices (S5–S6) and the connecting pore region form the conducting pore and are homologous to the two transmembrane helices of Kir channels (Gutman et al., 2005). Several types of Kv channels belonging to different sub-families have been identified in lipid rafts and shown to be sensitive to changes in the level of membrane cholesterol, but molecular mechanisms underlying cholesterol sensitivity of Kv channels are mostly unknown.

19.4.1 Kv1 Channels

Kv1 channels constitute the *Shaker*-related family of delayed rectifiers consisting of eight members (Kv1.1–1.8) expressed in brain, heart, skeletal muscle, pancreas and blood cells (Gutman et al., 2005). The role of cholesterol in regulation of Kv1 channels was investigated mainly for two members of the family, Kv1.3 and Kv1.5, but specific effects observed in different studies are highly controversial. *Kv1.3* is expressed in a variety of cell types and plays a critical role in the regulation of membrane potential and calcium signalling and apoptosis of T lymphocytes (Bock et al., 2003; Gutman et al., 2005; Hajdú et al., 2003). (Bock et al., 2003) showed that Kv1.3 partition almost exclusively to cholesterol-rich membrane domains in a T-lymphocytes cell line and that cholesterol depletion significantly decreases Kv1.3 activity. However, Hajdú et al., (2003) observed an opposite effect in primary lymphocytes. Specifically, Hajdú et al., (2003) showed that elevation of membrane cholesterol in primary lymphocytes resulted in a significant decrease in Kv1.3 current density, slowing down both activation and inactivation kinetics and resulting in the right-shift in the voltage dependence of activation. These changes are consistent with a decrease in the open probability of the channels upon cholesterol loading. In terms of partitioning into lipid rafts, Kv1.3 was demonstrated to cluster with a T cell antigen receptor complex (Panyi et al., 2003) and localize to the immunological synapse where it also colocalizes with a lipid raft marker G_{M1}(Panyi et al., 2004). Kv1.3 also partitions into lipid rafts when expressed heterologously in HEK cells (Vicente et al., 2008) but its membrane distribution is uniform rather than clustered (O'Connell et al., 2004), which may be consistent with partitioning into small but highly abundant lipid rafts. Kv1.4 channels were also shown to partition to lipid rafts in neurons (Wong and Schlichter, 2004), but not in pancreatic β-cells (Xia et al., 2004). *Kv1.5*, a channel that underlies an ultrarapid-activating K^+ current in heart muscle (Gutman et al., 2005) is also significantly affected by cellular cholesterol. First, Martens et al. (2001) showed that cholesterol depletion resulted in a hyperpolarizing shift in the voltage dependence of both activation and inactivation of the current, which would be expected to have a significant impact on the duration of action potentials that are controlled by Kv1.5 channels. More recently, Abi-Char et al. (2007) showed that cholesterol depletion results in a slow progressive increase of the Kv1.5-based component of the ultrarapid delayed rectifier current (I_{Kur}) in atrial cardiomyocytes. Kv1.5 was also shown to partition into lipid rafts but the relative distribution between rafts and non-rafts, as well as association of the channels with caveolae are still controversial and may vary in different cells and in different experimental conditions (Abi-Char et al., 2007; Eldstrom et al., 2006; Martens et al., 2001; McEwen et al., 2008; Vicente et al., 2008).

19.4.2 Kv2 Channels

Kv2 channels are the *Shab*-related family of delayed rectifiers consisting of two members Kv2.1 and Kv2.2, both playing key roles in the regulation of neuron excitability (Gutman et al., 2005). Kv2.1 was the first member of the Kv family to be found in lipid rafts and identified as sensitive to membrane cholesterol (Martens et al., 2000; Xia et al., 2004). More specifically, Kv2.1 partitions into non-caveolae lipid rafts and its membrane distribution is significantly different from that of Kv1 channels: Kv2.1 channels appear as clusters and have very limited lateral mobility whereas Kv1.3 and Kv1.4 were distributed more homogenously and were significantly more mobile (O'Connell et al., 2004). More recently, distinct cellular distributions of Kv2.1 and Kv1.4 were also demonstrated in freshly-isolated cardiomyocytes (O'Connell et al., 2008). Clustering of Kv2.1 channels was also significantly different in atrial and ventricular myocytes providing additional evidence that sub-cellular organization of Kv channels is cell-specific, which again may underlie the variability of cholesterol effects observed in different cell types (O'Connell et al., 2008). Interestingly, cholesterol depletion resulted in a significant increase in Kv2.1 cluster size (O'Connell et al., 2004), an effect quite unexpected because cholesterol depletion is typically believed to disrupt protein complexes in raft domains. The effect of cholesterol depletion on Kv2.1 mobility was also significantly different from its effect on Kv1.3 and Kv1.4. These observations suggest that all three channels partition into distinct types of lipid rafts, which may also explain the differences in their functional dependence on cholesterol depletion. Functionally, cholesterol depletion resulted in a significant hyperpolarizing shift of Kv2.1 inactivation from $V_{1/2}$ of ~ -16 mV to ~-52 mV (Martens et al., 2000). However, Xia et al. (2004) showed that in pancreatic β-cells, the Kv2.1 current is strongly suppressed by MβCD-induced cholesterol depletion. While the difference between the two observations is not clear, it was suggested to be due to possible differences in the microenvironment between rafts in insulin-secreting cells and in mouse L cells. Additionally, cholesterol depletion was shown to increase voltage-gated K^+ current in Drosophila neurons (Kenyon cells), the current that is putatively underlain by the *Shab* gene, a Drosophila homolog of Kv2.1 (Gasque et al., 2005). The reasons for these differences are not clear and may reflect the diversity of specific lipid environments and/or association of the channels with different regulatory subunits.

19.4.3 Other Kv Channels

Kv3 channels constitute a *Shaw*-related family that consists of four members (Kv3.1–3.4) (Gutman et al., 2005). Kv3.2 and Kv3.3 were shown to partition into non raft fractions in pancreatic α-cells (Xia et al., 2007). *Kv4* are a *Shal*-related family of Kv channels that includes three members (Kv4.1–4.3) (Gutman et al., 2005). All three members of the Kv4 family were shown to partition into lipid rafts in some (Wong and Schlichter, 2004; Xia et al., 2007), but not all studies (Martens et al.,

2000). Thus, functional dependence of Kv4 channels on membrane cholesterol is not yet established. Two more types of Kv channels, Kv7.1 (KCNQ) and Kv11.1 (hERG1) were also found to partition into rafts (Balijepalli et al., 2006). Cholesterol depletion results in a positive shift in the voltage dependence of activation and acceleration of the deactivation rate of Kv11.1 channels, both when the channels are expressed in HEK cells and in mouse myocytes (Balijepalli et al., 2006).

19.4.4 Regulation of Kv Channels by Plasma Hypercholesterolemia

Several studies have shown that plasma hypercholesterolemia has a major impact on the function of Kv channels. First, Jiang et al. (1999) reported that opening of Kv channels in aortic smooth muscle cells is significantly reduced in a hypercholesterolemic mouse model that lacks apolipoprotein E (Apo E) and low density lipoprotein receptor genes (LDLR (Apo E$^-$/LDLR$^-$ mice). Consistently, Ghanam et al. (2000) demonstrated that 4-aminopyridine, a blocker of voltage-dependent K$^+$ channels, strongly inhibited the acetylcholine-induced relaxation in normal rabbit cerebral arteries but not in cholesterol-fed rabbits, indicating impairment of Kv channel-mediated vasodilation under hypercholesterolemia. Finally, Heaps et al. (2005) showed that Kv channel-mediated, adenosine-induced vasodilatation was impaired in coronary arterioles of hypercholesterolemic pigs and that Kv currents in coronary arteriolar smooth muscle cells were significantly reduced. This, hypercholesterolemia impairs Kv channel activity in vascular smooth muscle cells and the consequent Kv channel-mediated cellular function in the cardiovascular system. The impact of hypercholesterolemia on other types of Kv channels has not yet been established.

In Summary, multiple types of Kv channels have been found to partition into lipid rafts but, similar to Kir channels, the functional impact of cholesterol on the activity and biophysical properties of different Kv channels appear to be highly variable. One of the important insights of these studies is the discovery that different types of Kv channels may partition into distinct types of membrane domains in the same cells (Fig. 19.5) (Martens et al., 2000; Martens et al., 2001; O'Connell and Tamkun, 2005). It is also clear that cholesterol-dependent changes in the voltage characteristics of Kv channels are expected to have major impacts on the termination of action potentials and firing patterns of a variety of excitable cells.

19.5 Ca^{2+}-Activated K$^+$ Channels

Three families of Ca^{2+}-activated K$^+$ channels are recognized on the basis of their genetic, biophysical, and pharmacologic properties: large (BK, K$_{Ca}$1.1, maxi-K, or Slo), intermediate (IK, K$_{Ca}$3.1) and small (SK, K$_{Ca}$2) conductance channels. In addition to the well known Ca^{2+}-activated K$_{Ca}$1.1, the large conductance K$^+$ channel family also includes the homologous Na$^+$ activated K$_{Ca}$4.1 and K$_{Ca}$4.2 channels

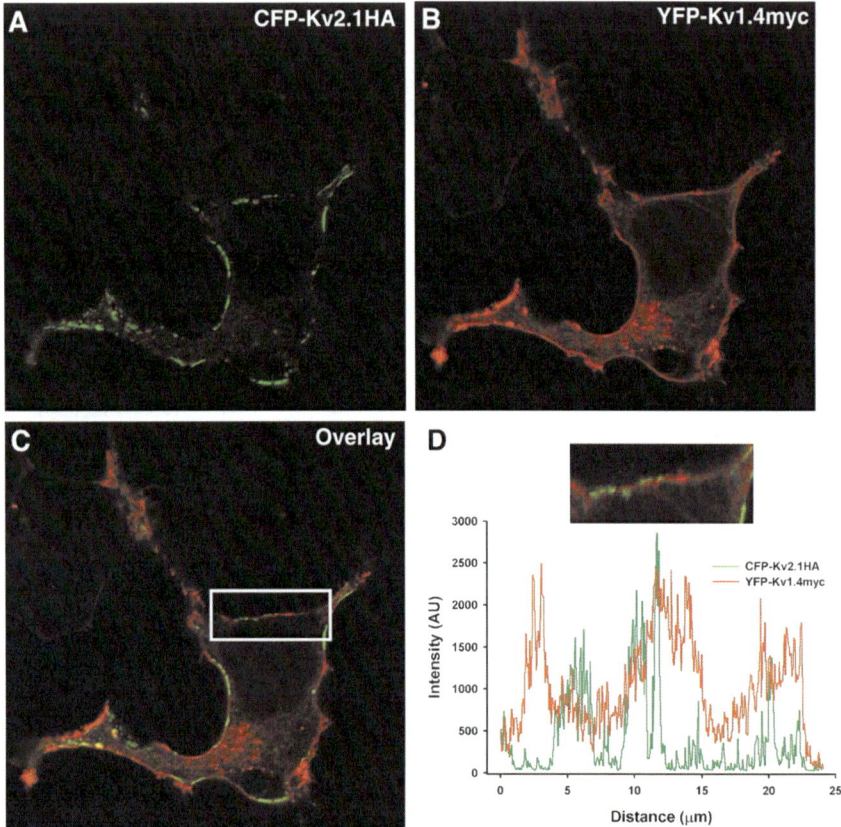

Fig. 19.5 Partitioning of different of Kv channels into distinct membrane domains. Lack of colocalization between Kv2.1 and Kv1.4 channels co-expressed in the same cells. (**A**) Kv2.1-CFP, (**B**) Kv1.4-YFP, (**C**) the overlay between Kv2.1 and Kv1.4 with Kv2.1-CFP pseudocolored *green* and Kv1.4-YFP pseudocolored *red*. (**D**) Fluorescence intensity profiles of Kv2.1-CFP and Kv1.4-YFP showing no or little correlation. *From* O'Connell and Tamkun (2005), published in the *Journal of Cell Science* 118: 2155–2166. Reproduced with permission)

(Slack and Slick, respectively) and the pH-sensitive $K_{Ca}5.1$ channel (Salkoff et al., 2006). The ubiquitous $K_{Ca}1.1$ channel controls a diverse array of physiological functions including the regulation of smooth muscle tone (Khan et al., 2001; Sprossmann et al., 2009), the excitability of neurons (Sah and Faber, 2002) and electrolyte secretion in the salivary glands and colon (Perry and Sandle, 2009; Romanenko et al., 2007). $K_{Ca}4$ channels are found in the brain and in several types of myocytes (Bhattacharjee and Kaczmarek, 2005), and expression of $K_{Ca}5$ is restricted to the testis (Santi et al., 2009; Schreiber et al., 1998). The intermediate-conductance $K_{Ca}3.1$ channels are primarily expressed in non-excitable cells and are involved in many physiological functions ranging from red blood cell volume regulation (Begenisich et al., 2004; Vandorpe et al., 1998) to mitogen activation of

T-lymphocytes (Ghanshani et al., 2000; Logsdon et al., 1997). SK channels play an important role in neuronal excitability and function in the CNS and periphery (Bond et al., 1999; Bond et al., 2005). Additionally, several Ca^{2+}-activated K^+ channels are expressed in vascular smooth muscles and endothelium and are important for regulation of vascular functions (Feletou, 2009; Ledoux et al., 2006).

19.5.1 BK

Among the large-conductance K^+ channels, only $K_{Ca}1.1$ has been reported to be sensitive to the membrane cholesterol level. Cholesterol-dependent regulation of $K_{Ca}1.1$ current varies dramatically, from strong suppression to upregulation, depending on the cell or tissue type studied. The inhibitory effect of cholesterol on $K_{Ca}1.1$ currents or, alternatively, current upregulation by cholesterol level reduction has been demonstrated in endothelial (Wang et al., 2005), smooth muscle (Bolotina et al., 1989; Brainard et al., 2005), neuronal (Lin et al., 2006) cells and colonic epithelia (Lam et al., 2004). This cholesterol-dependent inhibition is likely to be due to changes in the open probability and not in the unitary conductance of the channel (Bolotina et al., 1989; Chang et al., 1995; Crowley et al., 2003; Lin et al., 2006; Yuan et al., 2004). Conversely, downregulation of the channel by lowering the cholesterol level has been demonstrated in rat uterine myocytes and in glioma cell (Shmygol et al., 2007; Weaver et al., 2007). Finally, lack of any effect of cholesterol has been reported for heterologous expression of the α subunit of the channel cloned from rat brain (King et al., 2006) and for channels in parotid acinar cells (Romanenko et al., 2009). The latter study demonstrated that while the amplitude and the biophysical properties of $K_{Ca}1.1$ and $K_{Ca}3.1$ currents were not affected by cholesterol depletion, the functional interaction between the two channels was dependent on the membrane cholesterol.

Currently, no single mechanism rationalizing the diversity of cholesterol effects on $K_{Ca}1.1$ channel has been proposed. Several studies suggested that the channel modulation is mainly due to changes in the biophysical properties of the lipid bilayer (Bolotina et al., 1989; Chang et al., 1995). The hypothesis was supported by cholesterol-dependent modulation of BK sensitivity to temperature changes, which is consistent with the known effects of cholesterol on lateral elastic stress within the lipid bilayer as a function of temperature (Chang et al., 1995). Crowley et al. (2003) found that cholesterol modulated the activity of the channels reconstituted in bilayers composed of POPE/POPS mixture, but failed to do so in pure POPE bilayers. It has been hypothesized that cholesterol modulation of the bilayer stress energy was masked in pure POPE because of high initial degree of curvature stress. Interestingly, the same human $K_{Ca}1.1$ channel α-subunit expressed in HEK293 cells alone or in combination with any of the four known β subunits was cholesterol-insensitive (King et al., 2006).

In several studies cholesterol-dependent modulation of the channel activity in native cellular membranes has been attributed to its localization to lipid rafts. Membrane fractionation methods and co-localization with the lipid raft marker caveolin has been used to demonstrate the association of the channels with the raft membrane microdomains. It was also demonstrated that cholesterol depletion or interfering with caveolin scaffolding or expression decreases raft association of the channels (Babiychuk et al., 2004; Brainard et al., 2005; Bravo-Zehnder et al., 2000; Shmygol et al., 2007; Wang et al., 2005; Weaver et al., 2007). While in most of these studies, destabilization of lipid rafts upregulated BK currents, Weaver et al (2007) showed that cholesterol depletion inhibits an intricate functional association of BK channels and IP$_3$ receptors resulting in strong reduction of K$^+$ currents. Nevertheless, Lam et al. (2004) found that the cholesterol-sensitive BK channel is not present in low-density membrane fractions. Finally, Romanenko et al (2009) demonstrated that the interaction between K$_{Ca}$1.1 and K$_{Ca}$3.1 channels is cholesterol-dependent but does not require caveolin-1. This suggests that the participation of the channels in at least one type of lipid rafts, caveolae, is not required for the channel interaction. It is possible however, that other membrane domains are involved (*e.g.* flotillin-enriched microdomains (Langhorst et al., 2005).

BK channels were also shown to be sensitive to plasma hypercholesterolemia. First, earlier studies have shown that BK channels play a larger role in endothelium-dependent vasodilatation of carotid artery in hypercholesterolemic rabbits than in control animals (Najibi and Cohen, 1995; Najibi et al., 1994). It was suggested that an enhanced role of vascular BK channels under hypercholesterolemic conditions may be due to lower basal channel activity, in which case the availability of BK channels to acetylcholine stimulus may be increased (Sobey, 2001). In contrast, Jeremy and McCarron (2000) have shown that diet-induced hypercholesterolemia in rabbits significantly suppressed K$_{Ca}$-mediated vasodilatation in response to acetylcholine and bradikinin, as estimated by measuring hind limb vascular conductance. The effect was blocked by carybdotoxin, a blocker of BK channels. In addition, it was also shown that hypercholesterolemia decreases the expression of the BK β1 subunit that sensitizes the BK channel to intracellular Ca^{2+} (Du et al., 2006). Interestingly, Wiecha et al. (1997) reported a significantly higher BK activity in human smooth muscle cells obtained from atherosclerotic plaques, compared to those isolated from control media segments of human coronary arteries. Thus, several lines of evidence suggest that BK activity is modulated by hypercholesterolemia in a tissue-specific way.

19.5.2 SK and IK

As mentioned above, Ca^{2+}-activated K$^+$ channels are involved in regulation of a variety of vascular functions. Specifically, opening of SK and IK channels is a key step in endothelium-dependent hyperpolarization and relaxation of the vascular wall (Ledoux et al., 2006). Absi et al. (2007) demonstrated that cholesterol depletion of a vascular wall with intact epithelium resulted in SK-dependent vasorelaxation while IK-dependent hyperpolarization was unaffected. Moreover, the arterial SK but not IK channel protein was in low density membrane fraction and co-localized with caveolin-1 (Absi et al., 2007). Conversely, impairment of endothelium-dependent vasodilatation mediated by Ca^{2+}-activated K^+ channels was observed in hind limb vasculature of hypercholesterolemic rabbits (Jeremy and McCarron, 2000). At the same time, Toyama et al. (2008) recently demonstrated upregulation of $K_{Ca}3.1$ expression in coronary vessels of patients with coronary artery disease. Furthermore, $K_{Ca}3.1$ expression was also upregulated in smooth muscle cells, macrophages, and T lymphocytes found in atherosclerotic lesions of ApoE$^{-/-}$ mice, one of the main genetic models of atherosclerosis (Toyama et al., 2008). These authors further demonstrated that increased IK activity plays a key role in regulating the proliferation, migration, ROS, and cytokine production in vascular cells, consequently contributing to atherogenesis. Application of IK blockers in *ApoE*$^{-/-}$ mice markedly reduced the development of atherosclerosis, indicating IK as a novel therapeutic target for atherosclerosis under hypercholesterolemia. In contrast, expression of SK channels in aortas of hypercholesterolemic Apo E$^{-/-}$ mice was significantly reduced compared with wild-type controls (Zhou et al., 2007).

In Summary, similarly to the two groups of K^+ channels described above, while the effects of cholesterol on Ca^{2+}-activated K^+ channels are complex and controversial, multiple lines of evidence demonstrate that cholesterol sensitivity of these channels may play an important role in the development of cardiovascular disease.

19.6 Na$^+$ Channels

19.6.1 Voltage-Gated Na$^+$ (Na$_v$) Channels

Voltage-gated Na$^+$ channels are members of the same superfamily of ion channels as the voltage-gated K^+ channels described above (Catterall et al., 2005a). The pore forming α-subunits of Na$_v$ channels consist of four domains, homologous to the basic subunit of the Kv channels with six transmembrane helices (S1–S6) (Catterall et al., 2005a). However, in contrast to the highly diverse Kv channels, Na$_v$ channels are relatively similar and contain only one family of channels consisting of nine members (Na$_v$1.1–1.9). The main physiological role of Na$_v$ channels is the initiation and propagation of action potentials in a variety of excitable cells including neurons, different types of muscle cells and neuroendocrine cells (Catterall et al., 2005a). Two general mechanisms have been described for cholesterol sensitivity of Na$_v$ channels: partitioning to caveolae (cardiac Na$_v$1.5 channels) (Yarbrough et al., 2002) and regulation by the elastic properties of the membrane lipid bilayer (skeletal Na$_v$1.4 channels) (Lundbaek et al., 2004). It was also shown that plasma hypercholesterolemia resulting in a ∼2-fold increase in the level membrane cholesterol in rabbit ventricular myocytes is associated with a significant decrease in the current density of voltage-gated Na$^+$ currents (Wu et al., 1995). Specifically, plasma hypercholesterolemia resulted in decreased current density, slower recovery from inactivation, and more negative potential for inactivation of the sodium inward currents in hyperlipidemic myocytes (Wu et al., 1995).

One of the first studies to test the role of cholesterol in the regulation of voltage-gated Na^+ channels showed that cholesterol dramatically alters the sensitivity of Na_V channels to pentobarbital (Rehberg et al., 1995). In this study, Na_V channels were isolated from human cortical brain tissues and incorporated into PE/PC lipid bilayers with cholesterol added up to 50% weight/weight. An increase in cholesterol alone had no effect on the single channel properties of the channels, but it significantly inhibited the pentobarbital-induced block of the channels. Furthermore, competitive inhibition of the pentobarbital block was observed at very low concentrations of cholesterol with an EC_{50} of less than 2%, which is below cholesterol levels found in neuronal cells (Rehberg et al., 1995). In terms of the mechanism, it was proposed that the competitive effect could be either due to changes in the physical properties of the lipid bilayer or due to a direct interaction with the channel or with the anesthetic. The first mechanism was investigated in more details for the skeletal $Na_V 1.4$ channels as described below.

$Na_V 1.4$ channels responsible for the initiation of transmission of action potentials in skeletal muscle (Catterall et al., 2005a) were also shown to be significantly dependent on membrane cholesterol when expressed in HEK cells (Lundbaek et al., 2004). Specifically, depletion of membrane cholesterol resulted in a significant hyperpolarizing shift in current inactivation, an effect that was partially reversible by cholesterol depletion. Cholesterol enrichment, on the other hand, had no effect on the inactivation of the current. Surprisingly, however, cholesterol enrichment did have a significant effect on the voltage parameters of current activation whereas cholesterol depletion had no effect. Cholesterol depletion also decreased the peak amplitude of the current. In this case, it was proposed that cholesterol sensitivity of voltage-gated Na^+ channels can be attributed to the sensitivity of the channels rather than to the physical properties of the membrane bilayer because the effect of cholesterol depletion was similar to the effect induced by exposing the cells to 10 µM Triton X-100 or 2.5 mM β-octyl-glucoside (βOG), two micelle-forming amphiphilic compounds which are known to alter the physical properties of the membrane lipid bilayer (Lundbaek et al., 2004). More specifically, Lundbaek et al., (2004) proposed that $Na_V 1.4$ channels are regulated by the elastic properties of the membrane.

$Na_V 1.5$ channels are responsible for the initial upstroke of the action potential in atrial and ventricular myocytes, as well as contribute to the propagation of the action potential in the heart (Balijepalli and Kamp, 2008). $Na_V 1.5$ channels partition into cholesterol-rich membrane fractions and directly interact with caveolin-3 (Yarbrough et al., 2002). Furthermore, blocking the interaction with caveolin-3 using anti-caveolin antibodies abrogates direct $G_s\alpha$-mediated regulation of $Na_V 1.5$ current by β-adrenergic stimulation (Palygin et al., 2008; Yarbrough et al., 2002). It was proposed that activation of β-adrenergic activation of $G_s\alpha$ results in opening of the necks of the caveolae and thus recruiting caveolae $Na_V 1.5$ channels to the sarcolemma (Yarbrough et al., 2002). It was also proposed that caveolin-3 may directly stabilize the interaction between the channels and $G_s\alpha$ subunits (Palygin et al., 2008). While the role of cholesterol was not tested directly in these studies, multiple studies have shown that changes in the level of membrane cholesterol both

in vitro and *in vivo* have profound effects on the integrity of caveolae, suggesting that β-adrenergic stimulation of cardiac $Na_v1.5$ channels is cholesterol dependent.

19.6.2 Epithelial Na^+ Channels (eNaC)

Amiloride-sensitive Na^+ channels constitute another family of Na^+ channels expressed mostly in different types of epithelial cells (hence the name epithelial Na^+ channels, eNaC), which are responsible for Na^+ absorption regulating Na^+ and fluid homeostasis, which in turn play important roles in the control of blood pressure (de la Rosa et al., 2000). eNaC channels are heteromultimeric proteins formed by pore-forming α subunits plus two regulatory subunits (β and γ) (de la Rosa et al., 2000). Several studies have shown that eNaC subunits partition into lipid rafts (Hill et al., 2002; Hill et al., 2007; Prince and Welsh, 1999), but in other studies eNaC was found not to associate with rafts (Hanwell et al., 2002). Hill et al. (2002) also showed that MβCD-induced cholesterol depletion results in partial redistribution of β and γ subunits to higher density fractions, corresponding to a shift in the distribution of caveolin. It is noteworthy that a similar shift was observed in cells exposed to aldosterone, a hormone known to be involved in sodium channel regulation. More recently, it was also shown that eNaC directly interacts with caveolin-1 (Lee et al., 2009).

Functionally, the predominant effect of cholesterol on eNaC is inhibition of channel activity by cholesterol depletion. First, Shlyonsky et al. (2003) showed that functional eNaC channels can be found only in low-density Triton-insoluble fractions (rafts) whereas channels that partition into Triton-soluble fractions appear to be non-functional, as determined by incorporation of reconstituted proteoliposomes into lipid bilayers. Surprisingly, in this study no effect of cholesterol depletion was detected on transepithelial Na^+ current or single channel eNaC activity. To explain this controversy Shlyonsky et al. (2003) suggested that microdomains surrounding eNaC channels in A6 renal epithelial cells may be resistant to cholesterol depletion. More recently, however, several studies have shown that cholesterol depletion results in a decrease in the basal and/or hormone-stimulated Na^+ transport (Balut et al., 2006; Hill et al., 2007; West and Blazer-Yost, 2005), and in a decrease in both the open probability (Balut et al., 2006) and the surface expression of eNaC channels (Hill et al., 2007). It is also noteworthy that cholesterol depletion of apical and basal membranes has significantly different effects (Balut et al., 2006; West and Blazer-Yost, 2005). Interestingly, Wei et al. (2007) showed that the effect of cholesterol depletion was significantly facilitated by membrane stretch.

In general, loss of channel activity after cholesterol depletion suggests that the scaffold of a lipid raft and interactions with other proteins are required for eNaC normal function. However, recently it was shown that caveolin-1 is actually a negative regulator of eNaC (Lee et al., 2009), suggesting that it is an interaction with other regulatory molecules that could be responsible for the inhibition of the channels by cholesterol depletion.

19.7 Ca⁺ Channels

19.7.1 Voltage-Gated Ca^{+2} (Ca_v) Channels

Voltage-gated Ca^{+2} channels are also members of the superfamily of ion channels that include voltage-gated K^+ and Na^+ channels and they are responsible for calcium influx in response membrane depolarization controlling muscle contraction, secretion, neurotransmission and gene expression in a variety of cell types (Catterall et al., 2005b). The pore forming α-subunit of Ca_v channels is homologous to that of Na_v channels (Catterall et al., 2005b). The family of Ca_v channels includes 10 members that belong to one of three subfamilies: the Ca_v1 subfamily that underlies L-type Ca^{+2} currents, the Ca_v2 subfamily that includes P/Q, N and R-type Ca^{+2} currents and the Ca_v3 subfamily that underlies T-type Ca^{+2} currents (Catterall et al., 2005b). Changes in membrane cholesterol were shown to regulate members of all three subfamilies Ca_v channels: L-type Ca^{2+} channels (Balijepalli et al., 2006; Bowles et al., 2004; Cox and Tulenko, 1995; Pouvreau et al., 2004), N-type Ca^{2+} channels (Lundbaek et al., 1996; Toselli et al., 2005), and P/Q-type Ca^{2+} channels (Taverna et al., 2004). However, specific effects are highly diverse and vary not only between the channel types but also between different cell types and experimental conditions, even for the same types of channels.

19.7.2 L-Type Ca^{2+} Channels

One of the first reports of cholesterol sensitivity of L-type Ca^{2+} channels demonstrated that L-type Ca^{2+} currents are augmented in freshly dispersed myocytes from a rabbit portal vein under dietary hypercholesterolemia (Cox and Tulenko, 1995). Consistent with these observations, (Pouvreau et al., 2004) showed that depleting cellular cholesterol in mice skeletal muscle cells resulted in a significant reduction of L-type Ca^{2+} channels and ~15 mV positive shift in activation. Kinetics of both activation and inactivation were slowed down. Cholesterol depletion also resulted in disorganization of the T-tubule system, a decrease in the number of caveolae connected to the plasma membrane, and a decrease in cell capacitance which could partially explain the loss of Ca^{2+} current. However, the decrease in current appeared to be stronger than predicted on the basis of the loss of electrical capacitance suggesting that removal of cholesterol also affects the properties of the channels. Unexpectedly, cholesterol depletion facilitated activation of the channels by the saturating concentrations of Bay K 8644, suggesting that while the basal activity of the channels is inhibited by cholesterol depletion, their sensitivity to stimulation is enhanced. More recently, it was shown that $Ca_v1.2$ channels colocalize with caveolin-3 in cardiac myocytes and that MβCD-induced cholesterol depletion completely abolishes β2-adrenergic stimulation of $Ca_v1.2$ channels in these cells (Balijepalli et al., 2006; Tsujikawa et al., 2008). It is important to note that this effect is consistent with the loss of β-adrenergic stimulation of the $Na_v1.5$ current by anti-caveolin antibodies that was described earlier (Yarbrough et al., 2002).

In contrast, Jennings et al. (1999) showed that L-type Ca^{2+} channels in gall bladder cells are strongly inhibited by enriching the cells with cholesterol, using the cholesterol-saturated cyclodextrin complex. A similar effect was also demonstrated in coronary smooth muscle cells (Bowles et al., 2004). Consistent with these observations, L-type Ca^{2+} channels were also shown to be inhibited by hypercholesterolemic serum when the channels were recorded in myocytes isolated from conduit coronary arteries. Suppression of the channels was fully reversed by MβCD-induced cholesterol depletion (Bowles et al., 2004). However, hypercholesterolemia had no effect on L-type Ca^{2+} channels in myocytes isolated from arterioles of the same pigs. Furthermore, cholesterol enrichment had no effect on L-type Ca^{2+} channels in neuroblastoma-glioma cells (Toselli et al., 2005).

19.7.3 N-Type Ca^{2+} Channels

N-type Ca^{2+} channels have also been shown to have different sensitivities to membrane cholesterol in different cell types. An increase in membrane cholesterol was shown to alter the inactivation properties but not to affect the activation of N-types Ca^{2+} channels in neuroblastoma IMR32 cells (Lundbaek et al., 1996), whereas in neuroblastoma-glioma hybrid cells an increase in membrane cholesterol strongly suppressed N-type Ca^{2+} channels, without having a significant effect on the inactivation of the current (Toselli et al., 2005). The mechanism underlying cholesterol sensitivity of N-type Ca^{2+} channels in neuroblastoma-glioma hybrid cells seems to be strikingly similar to that of inwardly-rectifying K^+ channels (Romanenko et al., 2004a). Specifically, an increase in membrane cholesterol had no effect on the unitary conductance of the channels and a small effect on the open probability of the channels. The most prominent effect of cholesterol was an increase in the number of sweeps that had no channel activity, suggesting that the channels are stabilized in the closed state (Toselli et al., 2005). Interestingly, activity of N-type Ca^{2+} channels was also significantly reduced by transfecting the cells with caveolin-1 (Toselli et al., 2005), suggesting that N-type Ca^{2+} channels may also be suppressed by the interaction with caveolin. However, Toselli et al. (2005) have also shown that caveolin-induced suppression of the current was not due to the interaction of the channels with the caveolin inhibitory scaffolding domain. Therefore, Toselli et al. suggested that since caveolin is a cholesterol-binding protein that facilitates cholesterol transport to the plasma membrane, the negative effect of caveolin on N-type Ca^{2+} channels can be attributed to an increase in membrane cholesterol. Thus, it is interesting to take into account that cholesterol may affect not only the interactions of the channels with caveolin but also that caveolin may affect the interactions of the channels with cholesterol.

19.8 Transient Receptor Potential (TRP) Channels

Transient receptor potential or TRP channels are relatively non-selective ion channels, which are activated and regulated by a wide variety of stimuli. For example, TRP channels respond to temperature, touch, pain, osmolarity, pheromones, and taste (Clapham, 2003). They are ubiquitously expressed in many cell types and tissues. The mammalian TRP channels superfamily consists of six related protein families but with sequence identity as low as 20% (Clapham, 203). Among these, the three major families are the vanilloid (TRPV), the canonical (TRPC) and the melastatin (TRPM) TRP channels (Clapham, 203; Pedersen et al., 2005). Members of each of these three families have been shown to be sensitive to cholesterol and/or localize in cholesterol-rich lipid rafts.

19.8.1 TRPV Channels

Among the TRPV channels that include five members (TRPV1,2,4-6), the activity of TRPV1 has been shown to be sensitive to membrane cholesterol content, suggesting the raft association is pivotal for the physiological role of the channel (Liu et al., 2006). Specifically, it has been demonstrated that there is a close link between cholesterol levels, the amount of TRPV1 in the plasma membrane and TRPV1-mediated responses in primary sensory neurons. Cholesterol depletion following treatment with MβCD reduced significantly whole cell inward currents thus reducing the level of immunoreactivity of TRPV1 on the surface of the dorsal root ganglion neurons. Thus, it was suggested that TRPV1 is localized in cholesterol-rich microdomains, whose integrity determines the function and membrane expression of TRPV1 (Liu et al., 2006).

19.8.2 TRPC Channels

Functional expression of TRPC channels (seven members: TRPC1–7) has been shown to enhance store-mediated Ca^{2+} entry in a variety of mammalian cells (Brownlee et al., 2004; Grimaldi et al., 1999; Hartwig, 1992). Several TRPC channels have been shown to segregate into lipid rafts (Brownlow and Sage (2005), and there is growing evidence that some members of the TRPC channels family assemble in cholesterol-rich caveolae domains in order to participate in Ca^{2+} influx pathways. Specifically, TRPC1, TRPC4 and TRPC5 were found to be predominantly associated with detergent-resistant, insoluble platelet fractions, from which they were partially released following cholesterol depletion following MβCD treatment (Brownlow and Sage (2005). Furthermore, it has been shown that when cholesterol depletion of the plasma membrane disrupts lipid raft domains, it reduces both thapsigargin- and thrombin-evoked Ca^{2+} entry in human platelets (Brownlow et al., 2004). Together, these results suggest that TRPC1, TRPC4, and TRPC5 are associated with lipid raft domains in human platelets (Brownlow and Sage (2005).

A similar observation was reported for the distribution of TRPC1 in sympathetic neurons, in which TRPC1 was found to be localized in caveolae where it is associated with signalling proteins (Beech, 2005; Delmas, 2004). These data are also consistent with the partition of TRPC1 into detergent insoluble membrane fractions that was observed in several additional cell types, including HSG cells, THP-1 monocytic cells, neutrophils and skeletal myoblasts (Berthier et al., 2004; Brazer et al., 2003; Formigli et al., 2009; Kindzelskii et al., 2004; Lockwich et al., 2000). Further investigation elucidated the molecular determinants of the localization of TRPC1 in caveolae. It was determined that TRPC1 segregates into lipid rafts by binding to the raft-associated protein caveolin-1 (Lockwich et al., 2000), whose scaffolding domain is necessary for anchoring the channel to caveolae and for its regulation (Brazer et al., 2003; Kwiatek et al., 2006). Specifically, the N-terminus of TRPC1 contains a cav-1 binding motif and its C-terminus contains a caveolin scaffolding consensus binding domain that allows for its physical and functional interaction with caveolin-1 in the caveolae of human pulmonary artery endothelial cells (Kwiatek et al., 2006; Remillard and Yuan, 2006). The critical role that lipid raft domains play in the activation of TRPC1 channels has also been confirmed using caveolin knock-out mice (Murata et al., 2007).

Furthermore, excess of cholesterol (Kannan et al., 2007) or 7-ketocholesterol (Berthier et al., 2004) has been shown to induce TRPC1 redistribution to raft fractions. Colocalization of TRPC1 with caveolin-1 was reduced following cholesterol depletion by MβCD, and similarly to the case in human platelets (Brownlow and Sage (2005), Ca^{2+} inflow was reduced in MβCD-treated caudal arteries (Bergdahl et al., 2003). Since 7-ketocholesterol is a major component of oxLDL, the effect of oxLDL on TRPC1 function was also investigated (Ingueneau et al., 2008). It was found that oxLDL-induced TRPC1 translocation depended on actin cytoskeleton and was associated with a significant increase in the concentration of 7-ketocholesterol (a major oxysterol in oxLDL) into caveolar membranes. Since the cells were treated mildly with oxLDL, in these experiments, the caveolar content of cholesterol was unchanged (Ingueneau et al., 2008).

A different point of view on the relevance of lipid rafts to the function of TRPC1 was obtained by investigating the relationship between TRPC1 and STIM1, an ER Ca^{2+} protein that was suggested to be involved in coupling of ER Ca^{2+} entry channels (Lewis, 2007; Liou et al., 2005; Roos et al., 2005). In this context it has been shown for HSG and HEK cells that TRPC1 and STIM1 associated with each other within the lipid raft domains, and this association was dynamically regulated by the status of the ER Ca^{2+} store. This indicates that lipid raft domains facilitate the store-dependent interaction of STIM1 with TRPC1 (Pani et al., 2008) and suggests that TRPC1 functions as a component of store-operated channels only when linked to STIM1 in lipid rafts (Beech et al., 2009).

For TRPC3, the data is inconclusive. It was first reported that in HEK cells, TRPC3 was assembled in a multimolecular complex with key Ca^{2+} signalling proteins. It was thus suggested that caveolar localization of TRPC3 determines the activation and regulation of TRPC3 (Lockwich et al., 2001). Later it was reported that TRPC3 was evenly distributed between insoluble and soluble platelet fractions, and that its distribution was unaffected by MβCD (Beech, 2005). It was thus suggested that TRPC3 did not associate with lipid raft domains in human platelets (Beech, 2005). A year later it was reported that in HEK cells its activity was sensitive to membrane cholesterol content, and cholesterol-loading of cells had a positive effect on signals related to TRPC3 (Graziani et al., 2006). Furthermore, increased surface expression of TRPC3 was identified as a prominent event associated with cholesterol-induced TRPC3 activation (Graziani et al., 2006). Further experiments would be required to elucidate the conditions that determine the distribution of TRPC3 in lipid rafts.

As noted above, similarly to TRPC1, TRPC4 has also been associated with lipid raft domains in human platelets (Brownlow and Sage (2005). This observation is in agreement with data obtained from interstitial cells of Cajal, which are considered to be the pacemaker cells in gastrointestinal tracts. Specifically, TRPC4 was found to be located mostly in caveolae and colocalized with caveolin-1 (Torihashi et al., 2002).

TRPC6, similarly to TRPC3, showed a similar distribution between insoluble and soluble platelet fractions, but its distribution was unaffected by MβCD, suggesting that it is not associated with lipid raft domains in human platelets, TRPC6 (Brownlow and Sage (2005). A similar observation was reported for the distribution of TRPC6 in sympathetic neurons, where TRPC6 was evenly distributed throughout the plasma membrane (Beech, 2005; Delmas, 2004). On the other hand, it was demonstrated in HEK cells that cholesterol can have an indirect effect on channel function (Huber et al., 2006). Podocin, a cholesterol binding protein was shown to interact and regulate the activity of TRPC6 in a cholesterol-dependent manner. Cholesterol depletion using MβCD inhibited the effect of podocin on the channel (Beech et al., 2009; Huber et al., 2006).

19.9 TRPM Channels

Amongst the TRPM channels (eight members: TRPM1–8), two channels have been investigated in this context, TRPM7 and TRPM8. TRPM7 accumulates at the cell membrane in response to bradykinin-stimulation (Langeslag et al., 2007). Cell fractionation by sucrose gradient and differential centrifugation demonstrated that in bradykinin-stimulated cells, TRPM7 localized in fractions corresponding to caveolae whereas in basal conditions, TRPM7 was almost undetectable in cholesterol-rich fractions (Yogi et al., 2009). This suggests that TRPM7/caveolae/lipid raft association may facilitate TRPM7 scaffolding to cell membrane receptors. In sensory neurons, TRPM8 was found to be localized in cholesterol-rich lipid rafts both *in-vivo* and in heterologous expression systems (Morenilla-Palao et al., 2009). It has been shown that lipid-raft segregation of TRPM8 is favored by glycosylation at the Asn934, and mutating this asparagine to a glutamine that prevents glycosylation at this position, reduces the amount of TRPM8 channels that are associated with lipid-rafts by approximately 50%, without affecting the total amount of expressed protein or protein cell surface trafficking (Morenilla-Palao et al., 2009). This suggests that lipid raft association modulates TRPM8 activity. Specifically, both menthol- and cold- mediated responses are enhanced when the channel is located outside the lipid raft, and lipid raft disruption shifts the threshold for TRPM8 activation to warmer temperatures (Morenilla-Palao et al., 2009).

19.10 Cl⁻ Channels

In general, Cl⁻ channels can be loosely classified based on their gating mode to voltage-gated, ligand-gated, Ca^{2+}-activated channel superfamilies and channels regulated by cell volume (VRAC) and by cyclic nucleotide-dependent kinases (Cystic Fibrosis Transmembrane Conductance Regulator, CFTR) (Jentsch et al., 2002; Nilius and Droogmans, 2003; Suzuki et al., 2006). Multiple channels from all the classes have been found to be regulated by cholesterol, except Ca^{2+}-activated Cl⁻ channels - currently there is no direct evidence for cholesterol sensitivity of these channels.

19.10.1 Voltage-Gated Cl⁻ Channels

The families of ClC channels and the mitochondrial porins (voltage-dependent anion channels, VDAC) can be assigned to this group of channels. There are nine known mammalian isoforms of ClC channels with different tissue distribution and function (Jentsch, 2008). Among these channels only ClC-2 has been demonstrated to be regulated by cholesterol. Though ClC-2 is broadly expressed, its biological significance is still vague (Jentsch, 2008; Thiemann et al., 1992). When expressed in HEK293 cells, CLC-2 channel protein was found to be concentrated in cholesterol-rich microdomains and this association was disrupted by cholesterol depletion. Membrane cholesterol was shown to regulate the activation kinetics of the channel resulting in a shift in the activation curve, as well as to regulate channel trafficking (Hinzpeter et al., 2007). Identification of ClC-2 and ClC-3 channels as a part of the neuronal fusion pore, a process that requires cholesterol, suggests that cholesterol-sensitivity of these channels may affect pore formation (Cho et al., 2007; Jena, 2008). However, a specific functional link between the channels and cholesterol has not been established yet.

The voltage-dependent anion channel (VDAC) in the mitochondrial outer membrane is a major metabolite pathway across the membrane and plays an essential role in intracellular signalling and apoptosis (Rostovtseva and Bezrukov, 2008; Shoshan-Barmatz et al., 2006). Sterols such as cholesterol and ergosterol have been found to form a complex with purified VDAC protein (De Pinto et al., 1989; Freitag et al., 1982). Further, Popp et al. (1995) demonstrated that cholesterol is necessary for functional reconstitution of VDAC and that voltage dependence of the channel depends on the sterol used for the reconstitution. Several studies suggested that mitochondrial proteins, including VDAC, are present in lipid rafts of mitochondrial membranes (Foster and Chan, 2007). However, Zheng et al. (2009) showed that raft localization of none of the mitochondrial proteins was affected by cholesterol depletion, which suggested that previous findings could be an artifact of the raft preparation methods. Nevertheless, similarly to the plasma membrane, distinct glycosphingolipid enriched microdomains are present in the mitochondrial membrane with VDAC preferentially partitioning into these domains. Moreover, activation of death receptors recruited other components of the permeability transition pore, including pro-apoptotic Bax and tBid, to the same microdomains (Garofalo et al., 2005). Electrophysiological studies have demonstrated that VDAC voltage gating can be regulated by tBid in planar membranes (Rostovtseva et al., 2004; Rostovtseva and Bezrukov, 2008). Finally, cholesterol depletion abrogated the death-inducing pore formation and prevented mitochondrial depolarization (Christenson et al., 2008; Garofalo et al., 2005; Lucken-Ardjomande et al., 2008; Martinez-Abundis et al., 2007).

19.10.2 CFTR

CFTR is a Cl⁻ channel regulated by cyclic AMP and GMP and is critically involved in water and salt transport in multiple types of epithelial tissues. Genetic defects in the CFTR gene and the consequent malfunction of the channel lead to cystic fibrosis (CF), a fatal disease that affects the lungs, liver, pancreas, and the gastrointestinal tract (Rowntree and Harris, 2003). In addition, CFTR has been found to regulate various proteins, including transporters such as eNaC and K^+ and Cl⁻ channels (Berdiev et al., 2009; Schwiebert et al., 1995; Yoo et al., 2004).

Cholesterol-dependent regulation of CFTR currents have been described for its most common mutant deltaF508, which is defective in folding, membrane targeting, and stability but retains some of the channel's functionality (Cheng et al., 1990). Cholesterol depletion partially rescued the CFTR-mediated anion efflux in 3T3 fibroblasts expressing the mutant channel, which was attributed to enhanced plasma membrane targeting and retention of the channel (Lim et al., 2007). Similarly, reduced temperature, chemical chaperons or proteasome inhibitors were previously suggested to partially reverse the folding defect of the deltaF508 mutant (Jiang et al., 1998; Kopito, 1999). In multiple native and expressed cell systems, the CFTR protein was found to be associated with lipid rafts (Dudez et al., 2008; Grassme et al., 2003; Kowalski and Pier, 2004; Vij et al., 2009), but changes in membrane cholesterol were shown to have no effect on the activity of the channels (Lam et al., 2004; Singh et al., 2000; Wang et al., 2008). On the other hand, it has been shown that cholesterol may modulate CFTR-dependent regulation of various pathways, such as NFκB mediated IL-8 signalling (Vij et al., 2009), gap junction modulation by TNF-a (Dudez et al., 2008), and apoptosis induced by *P. aeruginosa* infection (Kowalski and Pier, 2004). In these studies disruption of lipid rafts and dissociation of CFTR from signalling complexes has been proposed as the mechanism of the cholesterol-dependent modulation.

19.11 Volume-Regulated Anion Channel (VRAC)

Maintenance of cell volume and osmotic homeostasis is the most prominent function of VRAC channels (coupled with swelling-activated conductive K^+ transporters) in most animal cells (Hoffmann et al., 2009). Additionally, VRAC channels are also involved in cell cycle progression, proliferation, and apoptosis (Nilius and Droogmans, 2001; Stutzin and Hoffmann, 2006). The properties and regulatory pathways of VRAC have been extensively studied; however the molecular identity of the channel is still unknown. Moreover, it is also debated as to VRAC, which can conduct amino acids such as taurine, is the main conducting pathway for swelling-induced release of organic osmolytes (Shennan, 2008).

Our studies have shown that membrane cholesterol has a negative effect on VRAC, as is manifested by an increase in amplitude and/or rate of swelling-induced current development in cholesterol-depleted cells and by the suppression of the current by cholesterol enrichment (Byfield et al., 2006; Levitan et al., 2000; Romanenko et al., 2004b). Similar observations were also reported by other investigators (Klausen et al., 2006). Likewise, swelling-induced efflux of anionic osmolytes in several cell types was enhanced by cholesterol depletion and suppressed by cholesterol enrichment (Cheema and Fisher, 2008; Lambert, 2004; Lim et al., 2006; Ortenblad et al., 2003). In terms of the mechanism, cholesterol-dependent regulation of VRAC and swelling-induced osmolyte release was ascribed to changes in the physical properties of the membrane rather than to specific sterol-protein interactions, because substitution of endogenous cholesterol by its analogues, such as epicholesterol, produced no effect on VRAC (Fig. 19.6) (Byfield et al., 2006; Cheema and Fisher, 2008; Romanenko et al., 2004b). Remarkably, cholesterol depletion upregulated VRAC activated by a supra-maximal concentration of GTPγS (Rho GTPase agonist) and, on the contrary, cholesterol sensitivity was lost when VRAC was activated by "strong" hypotonic shock. Moreover, at high cholesterol levels only swelling-activated VRAC was cholesterol-sensitive but not GTPγS-activated VRAC. It was therefore suggested that cholesterol may affect VRAC development by more than one mechanism (Levitan et al., 2000; Romanenko et al., 2004b).

Association with lipid rafts has also been hypothesized as a mechanism for cholesterol sensitivity of VRAC. Indeed, several studies demonstrate correlation of caveolin-1 expression with VRAC activation and swelling-induced taurine release. Paradoxically, ablation of caveolin-1 expression, which destabilized lipid rafts as did cholesterol depletion, inhibited VRAC (Trouet et al., 2001; Trouet et al., 1999; Ullrich et al., 2006). Caveolae/lipid rafts were shown to serve as scaffolds for multiple proteins that are involved in VRAC regulation, including Rho GTPases, as well as anchoring sites for actin filaments (Allen et al., 2007; Levitan and Gooch, 2007). Specifically, in ELA cells Klausen et al. (2006) showed that F-actin and Rho-Rho kinase modulate VRAC magnitude and activation rate, respectively, and that cholesterol depletion potentiates VRAC at least in part by preventing the hypotonicity-induced decrease in Rho activity and eliciting actin polymerization. A complex interplay of raft-dependent factors may underlie the controversial effects of cholesterol and caveolin.

19.12 Mechanosensitive Channels

Finally, we will discuss cholesterol sensitivity of mechanically sensitive ion channels that are expressed ubiquitously in a variety of tissues and are activated by membrane stretch (Hamill and McBride, 1996; Sachs and Morris, 1998). It

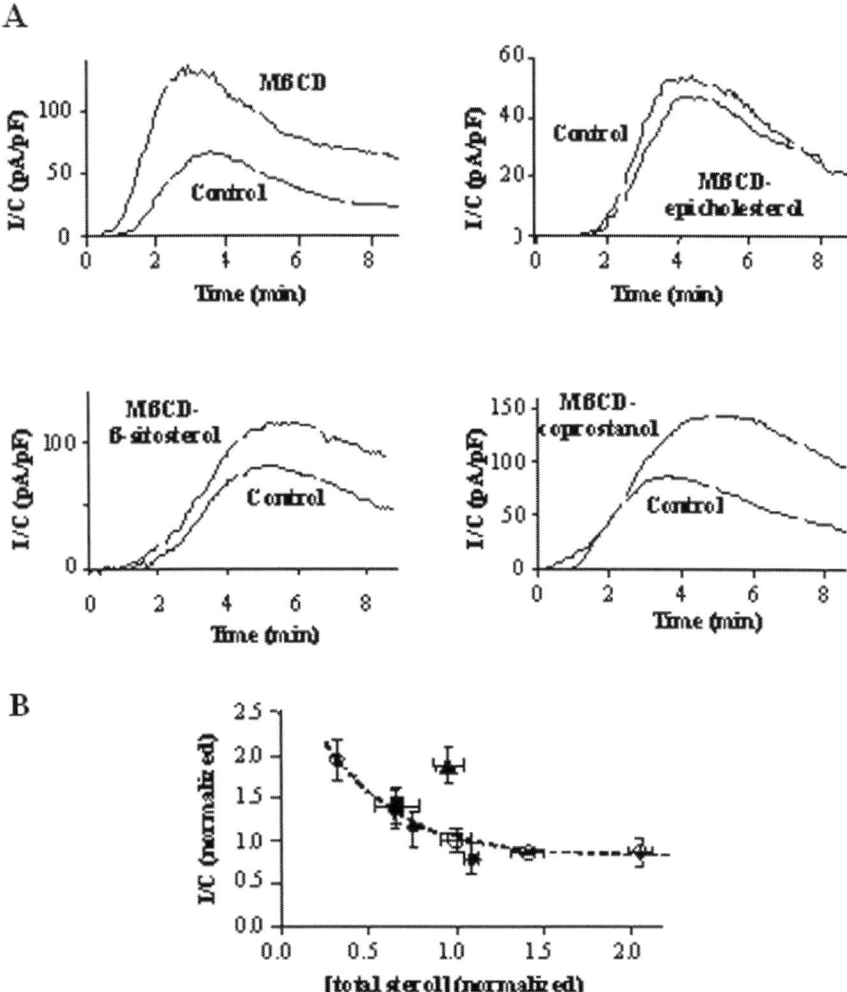

Fig. 19.6 Differential regulation of GTPγS-activated VRAC by cholesterol depletion and substitution with its analogues. (**A**) The time-courses of VRAC current densities recorded in cells treated as indicated. (**B**) Normalized VRAC currents plotted as a function of the total sterol level in cells either depleted of or enriched with cholesterol (*open circles*) and in the cells, in which endogenous cholesterol was substituted with epicholesterol (*diamonds*), sitosterol (*square*), or coprostanol (*triangle*). In contrast to other two analogues, coprostanol could not substitute for cholesterol in regulation of VRAC, which is consistent with lack of strong effect of coprostanol on the physical properties of the membrane (Adapted from Romanenko et al., 2004b)

has been suggested that lipid bilayer and submembranous cytoskeleton structures are involved in the control of mechanosensitive channels (Hamill and Martinac, 2001). Very little, however, is known about cholesterol sensitivity of endogenous stretch-activated channels in eukaryotic cells.

Within human Leukemia K562 cells cholesterol-depletion with MβCD resulted in suppression of the activity of mechanosensitive channels (Morachevskaya et al., 2007). It was shown that while cholesterol depletion did not affect the unitary conductance, the open probability of the channel was significantly decreased. At the same time, F-actin revealed reorganization of cortical cytoskeleton in these leukemia cells after cholesterol depletion. The integrity of F-actin is critical for the stiffening of cellular membranes, which has been shown to increase following cholesterol depletion (Byfield et al., 2004). It was thus suggested that F-actin rearrangement underlies changes in the mechanical properties of the cell surface, presumably induced by lipid raft destruction and thus mediates the modulation of mechanosensitive channel activity due to cholesterol depletion (Morachevskaya et al., 2007). In B-lymphocytes, the principal cellular mediators of specific humoral immune response to infection, mechanical stimulation reversibly activates LK_{bg} large conductance background-type K^+ channels (Nam et al., 2007). It has been suggested that mechanosensitive activation of LK_{bg} channels is mediated by PLC-dependent hydrolysis of PIP_2. Inhibition of LK_{bg} channels by PIP_2 was found to be partially reversible. Interestingly, cholesterol depletion achieved by application of MβCD, induced full recovery of LK_{bg} activity from inhibition by PIP_2, and facilitated its stretch activation (Nam et al., 2007).

19.13 Concluding Remarks and Future Directions

The main conclusion of this chapter is that cholesterol clearly is a major regulator of ion channel function. As described in detail above, changes in the level of membrane cholesterol regulate the activity and the biophysical properties of numerous ion channels including members of all major classes of channels. It is also clear from the variability of cholesterol effects on different types of ion channels that more than one mechanism may underlie the sensitivity of the channels to cholesterol.

However, much less is known about the specific mechanisms responsible for these effects. Furthermore, even the most basic question of whether cholesterol regulates the channels directly or by altering other signalling pathways, which in turn regulate channel activity, is still open in most cases. The most straightforward approach to discriminate between these possibilities is to test whether cholesterol regulates the channels in non-cellular environments, such as lipid planar bilayers or liposomes. Indeed, several types of channels, such as large-conductance Ca^{+2}-sensitive K^+ (BK) channels, were shown to be suppressed by cholesterol when incorporated into the bilayers, but for the majority of the channels this information is still missing. Moreover, to completely exclude the possibility that cholesterol effects ion channels are indirect, it would be necessary to test these effects on purified channels rather than on channels isolated together with the complex environment of the surrounding plasma membrane. The major constraint in performing these experiments is the difficulty of purifying mammalian channels and it remains as a challenge for future studies.

Another basic question is whether cholesterol regulates ion channels by altering the physical properties of the lipid bilayer or by specific sterol-protein interactions. In several studies, this question was addressed by testing whether there is a correlation between the effect of cholesterol on membrane fluidity and on channel function. Clearly, however, it is not enough to discriminate between these possibilities, because since it is well known that cholesterol alters membrane fluidity, other cholesterol effects will correlate with the changes in fluidity regardless of whether fluidity is indeed responsible for these effects. A better approach is to compare the effects of multiple sterols that are similar to cholesterol in terms of their effects on membrane fluidity and other physical properties of the membrane. Using this approach, we showed that Kir2 channels are sensitive to the chiral nature of cholesterol, whereas VRAC channels are not, suggesting that Kir2 channels are regulated by specific sterol-protein interactions whereas VRAC channels are regulated by changes in the physical properties of the membrane. Applying this approach to purified channels will unambiguously determine whether cholesterol specifically interacts with ion channels.

Finally, the fundamental question remains as to what are the structural determinants of the sensitivity of ion channels to cholesterol. Our recent studies provided the first clues about the structural requirements of cholesterol sensitivity of an ion channel for Kir2 channels, but no information exists to-date on other ion channels. It is possible, however, that since Kir channels are homologous with the basic subunits of voltage-gated K^+, Na^+ and Ca^{+2} channels, with the two-transmembrane helices of Kir corresponding to the fifth and sixth transmembrane helices of the voltage-gated channels, identifying the structural basis for cholesterol sensitivity of Kir channels should provide the clues into how cholesterol interact with other channels.

Acknowledgements We thank Drs. Francisco Barrantes, Olaf Andersen and Michael Tamkun for allowing us to include illustrations from their work. I also thank Dr. Barrantes and Dr. Andersen for many discussions of these topics that we had over recent years. This work was supported by NIH grants HL073965 and HL083298 for IL

References

Abi-Char, J.l., Maguy, A., Coulombe, A., Balse, E., Ratajczak, P., Samuel, J.-L., Nattel, S., and Hatem, S.p.N., 2007, Membrane cholesterol modulates Kv1.5 potassium channel distribution and function in rat cardiomyocytes. *The Journal of Physiology* **582**: 1205.

Absi, M., Burnham, M.P., Weston, A.H., Harno, E., Rogers, M., and Edwards, G., 2007, Effects of methyl beta-cyclodextrin on EDHF responses in pig and rat arteries; association between SK(Ca) channels and caveolin-rich domains. *Br J Pharmacol.* **151**: 332–40.

Allen, J.A., Halverson-Tamboli, R.A., and Rasenick, M.M., 2007, Lipid raft microdomains and neurotransmitter signalling. *Nat Rev Neurosci* **8**: 128–140.

Babiychuk, E.B., Smith, R.D., Burdyga, T., Babiychuk, V.S., Wray, S., and Draeger, A., 2004, Membrane cholesterol regulates smooth muscle phasic contraction. *J Membr Biol* **198**: 95–101.

Balijepalli, R.C., Foell, J.D., Hall, D.D., Hell, J.W., and Kamp, T.J., 2006, Localization of cardiac L-type Ca2+ channels to a caveolar macromolecular signalling complex is required for beta2-adrenergic regulation. *PNAS* **103**: 7500–7505.

Balijepalli, R.C., and Kamp, T.J., 2008, Caveolae, ion channels and cardiac arrhythmias. *Progress in Biophysics and Molecular Biology* **98:** 149.

Balut, C., Steels, P., Radu, M., Ameloot, M., Driessche, W.V., and Jans, D., 2006, Membrane cholesterol extraction decreases Na+ transport in A6 renal epithelia. *Am J Physiol Cell Physiol* **290:** C87–94.

Barrantes, F.J., 2004, Structural basis for lipid modulatioon of nicotinic acetylcholine receptor function. *Brain Research Reviews* **47:** 71–95.

Beech, D.J., 2005, TRPC1: store-operated channel and more. *Pflugers Arch.* **451:** 53–60.

Beech, D.J, Bahnasi, Y.M., Dedman, A.M., and Al-Shawaf, E., 2009, TRPC channel lipid specificity and mechanisms of lipid regulation. *Cell Calcium* **45:** 583–8.

Begenisich, T., Nakamoto, T., Ovitt, C.E., Nehrke, K., Brugnara, C., Alper, S.L., and Melvin, J.E., 2004, Physiological roles of the intermediate conductance, Ca2+-activated potassium channel Kcnn4. *J Biol Chem* **279:** 47681–47687.

Berdiev, B.K., Qadri, Y.J., and Benos, D.J., 2009, Assessment of the CFTR and ENaC association. *Mol Biosyst* **5:** 123–127.

Bergdahl, A., Gomez, M.F., Dreja, K., Xu, S.Z., Adner, M., Beech, D.J., Broman, J., Hellstrand, P., and Swärd, K., 2003, Cholesterol depletion impairs vascular reactivity to endothelin-1 by reducing store-operated Ca^{2+} entry dependent on TRPC1. *Circ. Res.* **93:** 839–47.

Berthier, A., Lemaire-Ewing, S., Prunet, C., Monier, S., Athias, A., Bessède, G., Pais de Barros, J.P., Laubriet, A., Gambert, P., Lizard, G., and Néel, D., 2004, Involvement of a calcium-dependent dephosphorylation of BAD associated with the localization of Trpc-1 within lipid rafts in 7-ketocholesterol-induced THP-1 cell apoptosis. *Cell Death Differ.* **11:** 897–905.

Bhattacharjee, A., and Kaczmarek, L.K., 2005, For K+ channels, Na+ is the new Ca2+. *Trends Neurosci* **28:** 422–428.

Bichet, D., Haass, F.A., and Jan, L.Y., 2003, Merging functional studies with structures of inward-rectifier K(+) channels. *Nat Rev Neurosci.* **4:** 957–967.

Bock, J., Szabo, I., Gamper, N., Adams, C., and Gulbins, E., 2003, Ceramide inhibits the potassium channel Kv1.3 by the formation of membrane platforms. *Biochemical and Biophysical Research Communications* **305:** 890.

Bolotina, V., Omelyanenko, V., Heyes, B., Ryan, U., and Bregestovski, P., 1989, Variations of membrane cholesterol alter the kinetics of Ca2+-dependent K+ channels and membrane fluidity in vascular smooth muscle cells. *Pflugers Archives* **415:** 262–268.

Bond, C.T., Maylie, J., and Adelman, J.P., 1999, Small-conductance calcium-activated potassium channels. *Ann N Y Acad Sci* **868:** 370–378.

Bond, C.T., Maylie, J., and Adelman, J.P., 2005, SK channels in excitability, pacemaking and synaptic integration. *Curr Opin Neurobiol.* **15:** 305–311.

Bowles, D.K., Heaps, C.L., Turk, J.R., Maddali, K.K., and Price, E.M., 2004, Hypercholesterolemia inhibits L-type calcium current in coronary macro-, not microcirculation. *J. Appl. Physiol.* **96:** 2240–2248.

Brainard, A.M., Miller, A.J., Martens, J.R., and England, S.K., 2005, Maxi-K channels localize to caveolae in human myometrium: a role for an actin-channel-caveolin complex in the regulation of myometrial smooth muscle K^+ current. *Am. J. Physiol. Cell Physiol.* **289:** C49–57.

Bravo-Zehnder, M., Orio, P., Norambuena, A., Wallner, M., Meera, P., Toro, L., Latorre, R., and Gonzalez, A., 2000, Apical sorting of a voltage- and Ca^{2+}-activated K^+ channel alpha -subunit in Madin-Darby canine kidney cells is independent of N-glycosylation. *Proc. Natl. Acad. Sci. U.S.A.* **97:** 13114–13119.

Brazer, S.C., Singh, B.B., Liu, X., Swaim, W., and Ambudkar, I.S., 2003, Caveolin-1 contributes to assembly of store-operated Ca^{2+} influx channels by regulating plasma membrane localization of TRPC1. *J. Biol. Chem.* **278:** 27208–15.

Brownlow, S.L., Harper, A.G., Harper, M.T., and Sage, S.O., 2004, A role for hTRPC1 and lipid raft domains in store-mediated calcium entry in human platelets. *Cell Calcium* **35:** 107–13.

Brownlow, S.L., and Sage, S.O., 2005, Transient receptor potential protein subunit assembly and membrane distribution in human platelets. *Thromb. Haemost.* **94:** 839–45.

Byfield, F.J., Aranda-Espinoza, H., Romanenko, V.G., Rothblat, G.H., and Levitan, I., 2004, Cholesterol depletion increases membrane stiffness of aortic endothelial cells. *Biophys. J.* **87**: 3336–3343.

Byfield, F.J., Hoffman, B.D., Romanenko, V.G., Fang, Y., Crocker, J.C., and Levitan, I., 2006, Evidence for the role of cell stiffness in modulation of volume-regulated anion channels. *Acta Physiol.* **187**: 285–294.

Catterall, W.A., Goldin, A.L., and Waxman, S.G., 2005a, International Union of Pharmacology. XLVII. Nomenclature and Structure-Function Relationships of Voltage-Gated Sodium Channels. *Pharmacol. Rev.* **57**: 397–409.

Catterall, W.A., Perez-Reyes, E., Snutch, T.P., and Striessnig, J., 2005b, International Union of Pharmacology. XLVIII. Nomenclature and Structure-Function Relationships of Voltage-Gated Calcium Channels. *Pharmacol. Rev.* **57**: 411–425.

Chang, H.M., Reitstetter, R., Mason, R.P., and Gruener, R., 1995, Attenuation of channel kinetics and conductance by cholesterol: an interpretation using structural stress as a unifying concept. *J. Membrane Biol.* **143**: 51–63.

Cheema, T.A., and Fisher, S.K., 2008, Cholesterol regulates volume-sensitive osmolyte efflux from human SH-SY5Y neuroblastoma cells following receptor activation. *J. Pharmacol. Exp. Ther.* **324**: 648–657.

Cheng, S.H., Gregory, R.J., Marshall, J., Paul, S., Souza, D.W., White, G.A., O'Riordan, C.R., and Smith, A.E., 1990, Defective intracellular transport and processing of CFTR is the molecular basis of most cystic fibrosis. *Cell* **63**: 827–834.

Cho, W.J., Jeremic, A., Jin, H., Ren, G., and Jena, B.P., 2007, Neuronal fusion pore assembly requires membrane cholesterol. *Cell Biol. Int.* **31**: 1301–1308.

Christenson, E., Merlin, S., Saito, M., and Schlesinger, P., 2008, Cholesterol effects on BAX pore activation. *J. Mol. Biol.* **381**: 1168–1183.

Clapham, D.E., 2003, TRP channels as cellular sensors. *Nature* **426**: 517–24.

Cox, R.H., and Tulenko, T.N., 1995, Altered contractile and ion channel function in rabbit portal vein with dietary atherosclerosis. *Am. J. Physiol.* **268**: H2522–2530.

Criado, M., Eibl, H., and Barrantes, F.J., 1982, Effects of lipids on acetylcholine receptor. Essential need of cholesterol for maintenance of agonist-induced state transitions in lipid vesicles. *Biochemistry* **21**: 3622–3629.

Crowley, J.J., Treistman, S.N., and Dopico, A.M., 2003, Cholesterol antagonizes ethanol potentiation of human brain BKCa channels reconstituted into phospholipid bilayers. *Mol. Pharmacol.* **64**: 365–372.

Davies, P.F., 1995, Flow-mediated endothelial mechanotransduction. *Physiol. Rev.* **75**: 519–560.

de la Rosa, D.A., Canessa, C.M., Fyfe, G.K., and Zhang, P., 2000, Structure and Regulation of Amiloride-Sensitive Sodium Channels. *Ann. Rev. Physiol.* **62**: 573.

De Pinto, V., Benz, R., Caggese, C., and Palmieri, F., 1989, Characterization of the mitochondrial porin from Drosophila melanogaster. *Biochim. Biophys. Acta* **987**: 1–7.

Delling, M., Wischmeyer, E., Dityatev, A., Sytnyk, V., Veh, R.W., Karschin, A., and Schachner, M., 2002, The Neural Cell Adhesion Molecule Regulates Cell-Surface Delivery of G-Protein-Activated Inwardly Rectifying Potassium Channels Via Lipid Rafts. *J. Neurosci.* **22**: 7154–7164.

Delmas, P., 2004, Assembly and gating of TRPC channels in signalling microdomains. *Novartis Found. Symp.* **258**: 75–89; discussion 89–102, 263–6.

Demel, R.A., Bruckdorfer, K.R., and van Deenen, L.L.M., 1972, Structural requirements of sterols for the interaction with lecithin at the air-water interface. *Biochem. Biophys. Acta* **255**: 311–320.

Du, P., Cui, G.B., Wang, Y.R., Zhang, X.Y., Ma, K.J., and Wei, J.G., 2006, Down regulated expression of the beta1 subunit of the big-conductance Ca^{2+} sensitive K^+ channel in sphincter of Oddi cells from rabbits fed with a high cholesterol diet. *Acta Biochim. Biophys. Sin. (Shanghai).* **38**: 893–899.

Dudez, T., Borot, F., Huang, S., Kwak, B.R., Bacchetta, M., Ollero, M., Stanton, B.A., and Chanson, M., 2008, CFTR in a lipid raft-TNFR1 complex modulates gap junctional intercellular communication and IL-8 secretion. *Biochim. Biophys. Acta* **1783:** 779–788.

Edwards, G., and Weston, A.H., 1993, The Pharmacology of ATP-Sensitive Potassium Channels. *Ann. Rev. Pharmacol. Toxicol.* **33:** 597–637.

Eldstrom, J., Van Wagoner, D.R., Moore, E.D., and Fedida, D., 2006, Localization of Kv1.5 channels in rat and canine myocyte sarcolemma. *FEBS Lett.* **580:** 6039.

Epshtein, Y., Chopra, A., Rosenhouse-Dantsker, A., Kowalsky, G., D.E., L., and Levitan, I., 2009, Identification of a C-terminus domain critical for the sensitivity of Kir2.1 channels to cholesterol. *Proc. Natl. Acad. Sci. U.S.A.* **106:** 8055–8060.

Fang, Y., Mohler, E.R., III, Hsieh, E., Osman, H., Hashemi, S.M., Davies, P.F., Rothblat, G.H., Wilensky, R.L., and Levitan, I., 2006, Hypercholesterolemia suppresses inwardly rectifying k^+ channels in aortic endothelium in vitro and in vivo. *Circ. Res.* **98:** 1064–1071.

Fang, Y., Schram, G., Romanenko, V.G., Shi, C., Conti, L., Vandenberg, C.A., Davies, P.F., Nattel, S., and Levitan, I., 2005, Functional expression of Kir2.x in human aortic endothelial cells: the dominant role of Kir2.2. *Am. J. Physiol. Cell Physiol.* **289:** C1134–1144.

Feletou, M., 2009, Calcium-activated potassium channels and endothelial dysfunction: therapeutic options? *Br. J. Pharmacol.* **156:** 545–562.

Formigli, L., Sassoli, C., Squecco, R., Bini, F., Martinesi, M., Chellini, F., Luciani, G., Sbrana, F., Zecchi-Orlandini, S., Francini, F., and Meacci, E., 2009, Regulation of transient receptor potential canonical channel 1 (TRPC1) by sphingosine 1-phosphate in C2C12 myoblasts and its relevance for a role of mechanotransduction in skeletal muscle differentiation. *J. Cell Sci.* **122:** 1322–33.

Foster, L.J., and Chan, Q.W., 2007, Lipid raft proteomics: more than just detergent-resistant membranes. *Subcell Biochem.* **43:** 35–47.

Freitag, H., Genchi, G., Benz, R., Palmieri, F., and Neupert, W., 1982, Isolation of mitochondrial porin from Neurospora crassa. *FEBS Lett.* **145:** 72–76.

Garofalo, T., Giammarioli, A.M., Misasi, R., Tinari, A., Manganelli, V., Gambardella, L., Pavan, A., Malorni, W., and Sorice, M., 2005, Lipid microdomains contribute to apoptosis-associated modifications of mitochondria in T cells. *Cell Death Differ.* **12:** 1378–1389.

Gasque, Labarca, and Darszon. 2005, Cholesterol-depleting compounds modulate K^+-currents in Drosophila Kenyon cells. *FEBS Lett.* **579:** 5129.

Genda, S., Miura, T., Miki, T., Ichikawa, Y., and Shimamoto K., 2002, K(ATP) channel opening is an endogenous mechanism of protection against the no-reflow phenomenon but its function is compromised by hypercholesterolemia. *J Am Coll Cardiol.* **40:** 1339–1346.

Ghanam, K., Javellaud, J., Ea-Kim, L., and Oudart, N., 2000, Effects of treatment with 17beta-estradiol on the hypercholesterolemic rabbit middle cerebral artery. *Maturitas* **34:** 249–260.

Ghanshani, S., Wulff, H., Miller, M.J., Rohm, H., Neben, A., Gutman, G.A., Cahalan, M.D., and Chandy, K.G., 2000, Up-regulation of the IKCa1 potassium channel during T-cell activation. Molecular mechanism and functional consequences. *J. Biol. Chem.* **275:** 37137–37149.

Grassme, H., Jendrossek, V., Riehle, A., von Kurthy, G., Berger, J., Schwarz, H., Weller, M., Kolesnick, R., and Gulbins, E., 2003, Host defense against Pseudomonas aeruginosa requires ceramide-rich membrane rafts. *Nat. Med.* **9:** 322–330.

Graziani, A., Rosker, C., Kohlwein, S.D., Zhu, M.X., Romanin, C., Sattler, W., Groschner, K., and Poteser, M., 2006, Cellular cholesterol controls TRPC3 function: evidence from a novel dominant-negative knockdown strategy. *Biochem. J.* **396:** 147–55.

Grimaldi, M., Favit, A., Alkon, D.L., 1999, cAMP-induced cytoskeleton rearrangement increases calcium transients through the enhancement of capacitative calcium entry. *J. Biol. Chem.* **274:** 33557–64.

Gutman, G.A., Chandy, K.G., Grissmer, S., Lazdunski, M., McKinnon, D., Pardo, L.A., Robertson, G.A., Rudy, B., Sanguinetti, M.C., Stuhmer, W., and Wang, X., 2005, International Union of Pharmacology. LIII. Nomenclature and Molecular Relationships of Voltage-Gated Potassium Channels. *Pharmacol. Rev.* **57:** 473–508.

Haass, F.A., Jonikas, M., Walter, P., Weissman, J.S., Jan, Y.-N., Jan, L.Y., and Schuldiner, M., 2007, Identification of yeast proteins necessary for cell-surface function of a potassium channel. *Proc. Natl. Acad. Sci. U.S.A.* **104:** 18079–18084.

Hajdú, P., Varga, Z., Pieri, C., Panyi, G., and Gáspár, R.J., 2003, Cholesterol modifies the gating of Kv1.3 in human T lymphocytes. *Pflugers Arch.* **445:** 674–682.

Hamill, O.P., and Martinac, B., 2001, Molecular basis of mechanotransduction in living cells. *Physiol. Rev.* **81:** 685–740.

Hamill, O.P., and McBride, J.D.W., 1996, The pharmacology of mechanogated membrane ion channels. *Pharmacol. Rev.* **48:** 231–252.

Hanwell, D., Ishikawa, T., Saleki, R., and Rotin, D., 2002, Trafficking and Cell Surface Stability of the Epithelial Na+ Channel Expressed in Epithelial Madin-Darby Canine Kidney Cells. *J. Biol. Chem.* **277:** 9772–9779.

Hartwig, J.H., 1992, Mechanisms of actin rearrangements mediating platelet activation. *J. Cell Biol.* **118:** 1421–42.

Heaps, C.L., Tharp, D.L., and Bowles, D.K., 2005, Hypercholesterolemia abolishes voltage-dependent K^+ channel contribution to adenosine-mediated relaxation in porcine coronary arterioles. *Am. J. Physiol. Heart Circ. Physiol.* **288:** H568–576.

Hibino, H., and Kurachi, Y., 2007, Distinct detergent-resistant membrane microdomains (lipid rafts) respectively harvest K^+ and water transport systems in brain astroglia. *Euro. J. Neurosci.* **26:** 2539–2555.

Hilgemann, D.W., Feng, S., and Nasuhoglu, C., 2001, The complex and intriguing lives of PIP2 with ion channels and transporters. *Sci. STKE.* **111:** RE19.

Hill, W.G., An, B., and Johnson, J.P., 2002, Endogenously expressed epithelial sodium channel is present in lipid rafts in A6 cells. *J. Biol. Chem.* **277:** 33541–33544.

Hill, W.G., Butterworth, M.B., Wang, H., Edinger, R.S., Lebowitz, J., Peters, K.W., Frizzell, R.A., and Johnson, J.P., 2007, The Epithelial Sodium Channel (ENaC) Traffics to Apical Membrane in Lipid Rafts in Mouse Cortical Collecting Duct Cells. *J. Biol. Chem.* **282:** 37402–37411.

Hinzpeter, A., Fritsch, J., Borot, F., Trudel, S., Vieu, D.L., Brouillard, F., Baudouin-Legros, M., Clain, J., Edelman, A., and Ollero, M., 2007, Membrane cholesterol content modulates ClC-2 gating and sensitivity to oxidative stress. *J. Biol. Chem.* **282:** 2423–2432.

Hoffmann, E.K., Lambert, I.H., and Pedersen, S.F., 2009, Physiology of cell volume regulation in vertebrates. *Physiol. Rev.* **89:** 193–277.

Huber, T.B., Schermer, B., Muller, R.U., Hohne, M., Bartram, M., Calixto, A., Hagmann, H., Reinhardt, C., Koos, F., Kunzelmann, K., Shirokova, E., Krautwurst, D., Harteneck, C., Simons, M., Pavenstädt, H., Kerjaschki, D., Thiele, C., Walz, G., Chalfie, M., and Benzing, T., 2006, Podocin and MEC-2 bind cholesterol to regulate the activity of associated ion channels. *Proc. Natl. Acad. Sci. U.S.A.* **103:** 17079–86.

Ingueneau, C., Huynh-Do, U., Marcheix, B., Athias, A., Gambert. P., Nègre-Salvayre, A., Salvayre, R., and Vindis, C., 2008, TRPC1 is regulated by caveolin-1 and is involved in oxidized LDL-induced apoptosis of vascular smooth muscle cells. *J. Cell Mol. Med.* (Epub ahead of print).

Jena, B.P., 2008, Porosome: the universal molecular machinery for cell secretion. *Mol. Cells* **26:** 517–529.

Jennings, L.J., Xu, Q.-W., Firth, T.A., Nelson, M.T., and Mawe, G.M., 1999, Cholesterol inhibits spontaneous action potentials and calcium currents in guinea pig gallbladder smooth muscle. *Am. J. Physiol.* **277:** G1017–1026.

Jentsch, T.J., 2008, CLC chloride channels and transporters: from genes to protein structure, pathology and physiology. *Crit. Rev. Biochem. Mol. Biol.* **43:** 3–36.

Jentsch, T.J., Stein, V., Weinreich, F., and Zdebik, A.A., 2002, Molecular Structure and Physiological Function of Chloride Channels. *Physiol. Rev.* **82:** 503–568.

Jeremy, R.W., and McCarron, H., 2000, Effect of hypercholesterolemia on Ca2+-dependent K+ channel-mediated vasodilatation in vivo. *Am. J. Physiol.* **279:** H1600–1608.

Jiang, C., Fang, S.L., Xiao, Y.F., O'Connor, S.P., Nadler, S.G., Lee, D.W., Jefferson, D.M., Kaplan, J.M., Smith, A.E., and Cheng, S.H., 1998, Partial restoration of cAMP-stimulated

CFTR chloride channel activity in DeltaF508 cells by deoxyspergualin. *Am. J. Physiol.* **275:** C171–178.

Jiang, J., Thorén, P., Caligiuri, G., Hansson, G.K., and Pernow, J., 1999, Enhanced phenylephrine-induced rhythmic activity in the atherosclerotic mouse aorta via an increase in opening of KCa channels: relation to Kv channels and nitric oxide. *Br. J. Pharmacol.* **128:** 637–646.

Kannan, K.B., Barlos, D., and Hauser, C.J., 2007, Free cholesterol alters lipid raft structure and function regulating neutrophil Ca^{2+} entry and respiratory burst: correlations with calcium channel raft trafficking. *J. Immunol.* **178:** 5253–61.

Khan, R.N., Matharoo-Ball, B., Arulkumaran, S., and Ashford, M.L., 2001, Potassium channels in the human myometrium. *Exp. Physiol.* **86:** 255–264.

Kindzelskii, A.L., Sitrin, R.G., and Petty, H.R., 2004, Cutting edge: optical microspectrophotometry supports the existence of gel phase lipid rafts at the lamellipodium of neutrophils: apparent role in calcium signalling. *J. Immunol.* **172:** 4681–5.

King, J.T., Lovell, P.V., Rishniw, M., Kotlikoff, M.I., Zeeman, M.L., and McCobb, D.P., 2006, Beta2 and beta4 subunits of BK channels confer differential sensitivity to acute modulation by steroid hormones. *J. Neurophysiol.* **95:** 2878–2888.

Klausen, T.K., Hougaard, C., Hoffmann, E.K., and Pedersen, S.F., 2006, Cholesterol modulates the volume-regulated anion current in Ehrlich-Lettre ascites cells via effects on Rho and F-actin. *Am. J. Physiol. Cell. Physiol.* **291:** C757–771.

Kopito, R.R., 1999, Biosynthesis and degradation of CFTR. *Physiol. Rev.* **79:** S167–173.

Kowalski, M.P., and Pier, G.B., 2004, Localization of Cystic Fibrosis Transmembrane Conductance Regulator to Lipid Rafts of Epithelial Cells Is Required for Pseudomonas aeruginosa-Induced Cellular Activation. *J. Immunol.* **172:** 418–425.

Kubo, Y., Adelman, J.P., Clapham, D.E., Jan, L.Y., Karschin, A., Kurachi, Y., Lazdunski, M., Nichols, C.G., Seino, S., and Vandenberg, C.A., 2005, International Union of Pharmacology. LIV. Nomenclature and Molecular Relationships of Inwardly Rectifying Potassium Channels. *Pharmacol. Rev.* **57:** 509–526.

Kwiatek, A.M., Minshall, R.D., Cool, D.R., Skidgel, R.A., Malik, A.B., and Tiruppathi, C., 2006, Caveolin-1 regulates store-operated Ca^{2+} influx by binding of its scaffolding domain to transient receptor potential channel-1 in endothelial cells. *Mol. Pharmacol.* **70:** 1174–83.

Lam, R.S., Shaw, A.R., and Duszyk, M., 2004, Membrane cholesterol content modulates activation of BK channels in colonic epithelia. *Biochim. Biophys. Acta* **1667:** 241–248.

Lambert, I.H., 2004, Regulation of the cellular content of the organic osmolyte taurine in mammalian cells. *Neurochem. Res.* **29:** 27–63.

Langeslag, M., Clark, K., Moolenaar, W.H., van Leeuwen, F.N., and Jalink, K., 2007, Activation of TRPM7 channels by phospholipase C-coupled receptor agonists. *J. Biol. Chem.* **282:** 232–9.

Langhorst, M.F., Reuter, A., and Stuermer, C.A., 2005, Scaffolding microdomains and beyond: the function of reggie/flotillin proteins. *Cell Mol. Life Sci.* **62:** 2228–2240.

Ledoux, J., Werner, M.E., Brayden, J.E., and Nelson, M.T., 2006, Calcium-activated potassium channels and the regulation of vascular tone. *Physiol. (Bethesda)* **21:** 69–78.

Lee, I.-H., Campbell, C.R., Song, S.-H., Day, M.L., Kumar, S., Cook, D.I., and Dinudom, A., 2009, The Activity of the Epithelial Sodium Channels Is Regulated by Caveolin-1 via a Nedd4-2-dependent Mechanism. *J. Biol. Chem.* **284:** 12663–12669.

Lee, T.-M., Lin, M.-S., Chou, T.-F., Tsai, C.-H., and Chang, N.-C., 2004, Effect of pravastatin on left ventricular mass by activation of myocardial KATP channels in hypercholesterolemic rabbits. *Atherosclerosis* **176:** 273.

Levitan, I., Christian, A.E., Tulenko, T.N., and Rothblat, G.H., 2000, Membrane cholesterol content modulates activation of volume-regulated anion current (VRAC) in bovine endothelial cells. *J. Gen. Physiol.* **115:** 405–416.

Levitan, I., and Gooch, K.J., 2007, Lipid Rafts in Membrane-Cytoskeleton Interactions and Control of Cellular Biomechanics: Actions of oxLDL. *Antioxidants & Redox Signalling* **9:** 1519–1534.

Lewis, R.S., 2007, The molecular choreography of a store-operated calcium channel. *Nature* **446**: 284–7.

Lim, C.H., Bijvelds, M.J., Nigg, A., Schoonderwoerd, K., Houtsmuller, A.B., de Jonge, H.R., and Tilly, B.C., 2007, Cholesterol depletion and genistein as tools to promote F508delCFTR retention at the plasma membrane. *Cell Physiol. Biochem.* **20**: 473–482.

Lim, C.H., Schoonderwoerd, K., Kleijer, W.J., de Jonge, H.R., and Tilly, B.C., 2006, Regulation of the cell swelling-activated chloride conductance by cholesterol-rich membrane domains. *Acta Physiol.* **187**: 295–303.

Lin, M.W., Wu, A.Z., Ting, W.H., Li, C.L., Cheng, K.S., and Wu, S.N., 2006, Changes in membrane cholesterol of pituitary tumor (GH3) cells regulate the activity of large-conductance Ca^{2+}-activated K^+ channels. *Chin. J. Physiol.* **49**: 1–13.

Liou, J., Kim, M.L., Heo, W.D., Jones, J.T., Myers, J.W., Ferrell, J.E., Jr., and Meyer, T., 2005, STIM is a Ca^{2+} sensor essential for Ca^{2+}-store-depletion-triggered Ca^{2+} influx. *Curr Biol* **15**: 1235–41.

Liu, M., Huang, W., Wu, D., Priestley, J.V., 2006, TRPV1, but not P2X, requires cholesterol for its function and membrane expression in rat nociceptors. *Eur. J. Neurosci.* **24**: 1–6.

Lockwich, T., Singh, B.B., Liu, X., and Ambudkar, I.S., 2001, Stabilization of cortical actin induces internalization of transient receptor potential 3 (Trp3)-associated caveolar Ca^{2+} signalling complex and loss of Ca^{2+} influx without disruption of Trp3-inositol trisphosphate receptor association. *J. Biol. Chem.* **276**: 42401–8.

Lockwich, T.P., Liu, X., Singh, B.B., Jadlowiec, J., Weiland, S., and Ambudkar, I.S., 2000, Assembly of Trp1 in a signalling complex associated with caveolin-scaffolding lipid raft domains. *J. Biol. Chem.* **275**: 11934–42.

Logothetis, D.E., Jin, T., Lupyan, D., and Rosenhouse-Dantsker, A., 2007, Phosphoinositide-mediated gating of inwardly rectifying K^+ channels. *Pflugers Arch.* **455**: 83–95.

Logsdon, N.J., Kang, J., Togo, J.A., Christian, E.P., and Aiyar, J., 1997, A novel gene, hKCa4, encodes the calcium-activated potassium channel in human T lymphocytes. *J. Biol. Chem.* **272**: 32723–32726.

Lucken-Ardjomande, S., Montessuit, S., and Martinou, J.C., 2008, Bax activation and stress-induced apoptosis delayed by the accumulation of cholesterol in mitochondrial membranes. *Cell Death Differ.* **15**: 484–493.

Lundbaek, J.A., and Andersen, O.S., 1999, Spring constants for channel-induced lipid bilayer deformations estimates using gramicidin channels. *Biophys. J.* **76**: 889–895.

Lundbaek, J.A., Birn, P., Hansen, A.J., and Andersen, O.S., 1996, Membrane stiffness and channel function. *Biochemistry* **35**: 3825–3830.

Lundbaek, J.A., Birn, P., Hansen, A.J., Sogaard, R., Nielsen, C., Girshman, J., Bruno, M.J., Tape, S.E., Egebjerg, J., Greathouse, D.V., Mattice, G.L., Koeppe, R.E., II, and Andersen, O.S., 2004, Regulation of Sodium Channel Function by Bilayer Elasticity: The Importance of Hydrophobic Coupling. Effects of Micelle-forming Amphiphiles and Cholesterol. *J. Gen. Physiol.* **123**: 599–621.

Maguy, A., Hebert, T.E., and Nattel, S., 2006, Involvement of lipid rafts and caveolae in cardiac ion channel function. *Cardiovasc. Res.* **69**: 798.

Marsh, D., and Barrantes, F.J., 1978, Immobilized lipid in acetylcholine receptor-rich membranes from Torpedo marmorata. *Proc. Natl. Acad. Sci. U.S.A.* **75**: 4329–4333.

Martens, J.R., Navarro-Polanco, R., Coppock, E.A., Nishiyama, A., Parshley, L., Grobaski, T.D., and Tamkun, M.M., 2000, Differential Targeting of Shaker-like Potassium Channels to Lipid Rafts. *J. Biol. Chem.* **275**: 7443–7446.

Martens, J.R., O'Connell, K., and Tamkun, M., 2004, Targeting of ion channels to membrane microdomains: localization of KV channels to lipid rafts. *Trends Pharmacol. Sci.* **25**: 16–21.

Martens, J.R., Sakamoto, N., Sullivan, S.A., Grobaski, T.D., and Tamkun, M.M., 2001, Isoform-specific Localization of Voltage-gated K+ Channels to Distinct Lipid Raft Populations. Targeting of Kv1.5 to caveolae. *J. Biol. Chem.* **276**: 8409–8414.

Martinez-Abundis, E., Garcia, N., Correa, F., Franco, M., and Zazueta, C., 2007, Changes in specific lipids regulate BAX-induced mitochondrial permeability transition. *FEBS J.* **274**: 6500–6510.

Mathew, V., and Lerman, A., 2001, Altered effects of potassium channel modulation in the coronary circulation in experimental hypercholesterolemia. *Atherosclerosis* **154**: 329–335.

McEwen, D.P., Li, Q., Jackson, S., Jenkins, P.M., and Martens, J.R., 2008, Caveolin Regulates Kv1.5 Trafficking to Cholesterol-Rich Membrane Microdomains. *Mol. Pharmacol.* **73**: 678–685.

Mohler Iii, E.R., Fang, Y., Gusic Shaffer, R., Moore, J., Wilensky, R.L., Parmacek, M., and Levitan, I., 2007, Hypercholesterolemia suppresses Kir channels in porcine bone marrow progenitor cells in vivo. *Biochem. Biophys. Res. Comm.* **358**: 317–324.

Morachevskaya, E., Sudarikova, A., and Negulyaev, Y., 2007, Mechanosensitive channel activity and F-actin organization in cholesterol-depleted human leukaemia cells. *Cell Biol. Int.* **31**: 374–381.

Morenilla-Palao, C., Pertusa, M., Meseguer, V., Cabedo, H., and Viana, F., 2009, Lipid raft segregation modulates TRPM8 channel activity. *J. Biol. Chem.* **284**: 9215–24.

Murata, T., Lin, M.I., Stan, R.V., Bauer, P.M., Yu, J., and Sessa, W.C., 2007, Genetic evidence supporting caveolae microdomain regulation of calcium entry in endothelial cells. *J. Biol. Chem.* **282**: 16631–43.

Najibi, S., and Cohen, R.A., 1995, Enhanced role of K^+ channels in relaxations of hypercholesterolemic rabbit carotid artery to NO. *Am. J. Physiol. Heart Circ. Physiol.* **269**: H805–811.

Najibi, S., Cowan, C.L., Palacino, J.J., and Cohen, R.A., 1994, Enhanced role of potassium channels in relaxations to acetylcholine in hypercholesterolemic rabbit carotid artery. *Am J. Physiol. Heart Circ. Physiol.* **266**: H2061–2067.

Nam, J.H., Lee, H.-S., Nguyen, Y.H., Kang, T.M., Lee, S.W., Kim, H.-Y., Kim, S.J., Earm, Y.E., and Kim, S.J., 2007, Mechanosensitive activation of K^+ channel via phospholipase C-induced depletion of phosphatidylinositol 4,5-bisphosphate in B lymphocytes. *J. Physiol.* **582**: 977.

Nichols, C., and Lopatin, A., 1997, Inward rectifier potassium channels. *Annu. Rev. Physiol.* **59**.

Nichols, C.G., 2006, KATP channels as molecular sensors of cellular metabolism. *Nature* **440**: 470.

Nilius, B., and Droogmans, G., 2001, Ion channels and their functional role in vascular endothelium. *Physiol. Rev.* **81**: 1415–1459.

Nilius, B., and Droogmans, G., 2003, Amazing chloride channels: an overview. *Acta Physiol. Scand.* **177**: 119–147.

O'Connell, K.M.S., Martens, J.R., and Tamkun, M.M., 2004, Localization of Ion Channels to Lipid Raft Domains within the Cardiovascular System. *Trends Cardiovasc. Med.* **14**: 37.

O'Connell, K.M.S., and Tamkun, M.M., 2005, Targeting of voltage-gated potassium channel isoforms to distinct cell surface microdomains. *J. Cell Sci.* **118**: 2155–2166.

O'Connell, K.M.S., Whitesell, J.D., and Tamkun, M.M., 2008, Localization and mobility of the delayed-rectifer K+ channel Kv2.1 in adult cardiomyocytes. *Am. J. Physiol. Heart Circ. Physiol.* **294**: H229–237.

Olesen, S.-P., Clapham, D.E., and Davies, P.F., 1988, Hemodynamic shear stress activates a K+ current in vascular endothelial cells. *Nature* **331**: 168–170.

Ortenblad, N., Young, J.F., Oksbjerg, N., Nielsen, J.H., and Lambert, I.H., 2003, Reactive oxygen species are important mediators of taurine release from skeletal muscle cells. *Am J. Physiol. Cell Physiol.* **284**: C1362–1373.

Palygin, O.A., Pettus, J.M., and Shibata, E.F., 2008, Regulation of caveolar cardiac sodium current by a single Gs{alpha} histidine residue. *Am. J. Physiol. Heart Circ. Physiol.* **294**: H1693–1699.

Pani B, Ong HL, Liu X, Rauser K, Ambudkar IS, Singh BB., 2008, Lipid rafts determine clustering of STIM1 in endoplasmic reticulum-plasma membrane junctions and regulation of store-operated Ca^{2+} entry (SOCE). *J. Biol. Chem.* **283**: 17333–40.

Panyi, G., Bagdany, M., Bodnar, A., Vamosi, G., Szentesi, G., Jenei, A., Matyus, L., Varga, S., Waldmann, T.A., Gaspar, R., and Damjanovich, S., 2003, Colocalization and nonrandom distribution of Kv1.3 potassium channels and CD3 molecules in the plasma membrane of human T lymphocytes. *Proc. Natl. Acad. Sci. U.S.A.* **100:** 2592–2597.

Panyi, G., Vamosi, G., Bacso, Z., Bagdany, M., Bodnar, A., Varga, Z., Gaspar, R., Matyus, L., and Damjanovich, S., 2004, Kv1.3 potassium channels are localized in the immunological synapse formed between cytotoxic and target cells. *Proc. Natl. Acad. Sci. U.S.A.* **101:** 1285–1290.

Pedersen, S.F., Owsianik, G., and Nilius, B., 2005, TRP channels: an overview. *Cell Calcium* **38:** 233–52.

Perry, M.D., and Sandle, G.I., 2009, Regulation of colonic apical potassium (BK) channels by cAMP and somatostatin. *Am. J. Physiol. Gastrointest. Liver Physiol.* **297:** G159–167.

Pike, L., and Casey, L., 1996, Localization and turnover of phosphatidylinositol 4,5-bisphospate in caveolin-enriched membrane domains. *J. Biol. Chem.* **271:** 26453–26456.

Pike, L.J., 2006, Rafts defined: a report on the Keystone symposium on lipid rafts and cell function. *J. Lipid Res.* **47:** 1597–1598.

Pongo, E., Balla, Z., Mubagwa, K., Flameng, W., Edes, I., Szilvassy, Z., and Ferdinandy, P., 2001, Deterioration of the protein kinase C-KATP channel pathway in regulation of coronary flow in hypercholesterolaemic rabbits. *Euro. J. Pharmacol.* **418:** 217.

Popp B, Schmid A, and Benz R., 1995, Role of sterols in the functional reconstitution of water-soluble mitochondrial porins from different organisms. *Biochemistry.* **34:** 3352–3361.

Pouvreau, S., Berthier, C., Blaineau, S., Amsellem, J., Coronado, R., and Strube, C., 2004, Membrane cholesterol modulates dihydropyridine receptor function in mice fetal skeletal muscle cells. *J. Physiol. (Lond.)* **555:** 365–381.

Prince, L.S., and Welsh, M.J., 1999, Effect of subunit composition and Liddle's syndrome mutations on biosynthesis of ENaC. *Am. J. Physiol. Cell Physiol.* **276:** C1346–1351.

Rehberg, B., Urban, B.W., and Duch, D.S., 1995, The membrane lipid cholesterol modulates anesthetic actions on a human brain ion channel. *Anesthesiology* **82:** 749–758.

Remillard, C.V., and Yuan, J.X., 2006, Transient receptor potential channels and caveolin-1: good friends in tight spaces. *Mol. Pharmacol.* **70:** 1151–4.

Ren, Y.J., Xu, X.H., Zhong, C.B., Feng, N., and Wang, X.L., 2001, Hypercholesterolemia alters vascular functions and gene expression of potassium channels in rat aortic smooth muscle cells. *Acta Pharmacol. Sin.* **22:** 274–278.

Romanenko, V.G., Fang, Y., Byfield, F., Travis, A.J., Vandenberg, C.A., Rothblat, G.H., and Levitan, I., 2004a, Cholesterol sensitivity and lipid raft targeting of Kir2.1 channels. *Biophys. J.* **87:** 3850–3861.

Romanenko, V.G., Nakamoto, T., Srivastava, A., Begenisich, T., and Melvin, J.E., 2007, Regulation of membrane potential and fluid secretion by Ca^{2+}-activated K^+ channels in mouse submandibular glands. *J. Physiol.* **581:** 801–817.

Romanenko, V.G., Roser, K.S., Melvin, J.E., and Begenisich, T., 2009, The role of cell cholesterol and the cytoskeleton in the interaction between IK1 and maxi-K channels. *Am J. Physiol. Cell Physiol.* **296:** C878–888.

Romanenko, V.G., Rothblat, G.H., and Levitan, I., 2002, Modulation of endothelial inward rectifier K^+ current by optical isomers of cholesterol. *Biophys. J.* **83:** 3211–3222.

Romanenko, V.G., Rothblat, G.H., and Levitan, I., 2004b, Sensitivity of volume-regulated anion current to cholesterol structural analogues. *J. Gen. Physiol.* **123:** 77–88.

Roos, J., DiGregorio, P.J., Yeromin, A.V., Ohlsen, K., Lioudyno, M., Zhang, S., Safrina, O., Kozak, J.A., Wagner, S.L., Cahalan, M.D., Veliçelebi, G., and Stauderman, K.A., 2005, STIM1, an essential and conserved component of store-operated Ca^{2+} channel function. *J. Cell. Biol.* **169:** 435–45.

Rostovtseva, T.K., Antonsson, B., Suzuki, M., Youle, R.J., Colombini, M., and Bezrukov, S.M., 2004, Bid, but not Bax, regulates VDAC channels. *J. Biol. Chem.* **279:** 13575–13583.

Rostovtseva, T.K., and Bezrukov, S.M., 2008, VDAC regulation: role of cytosolic proteins and mitochondrial lipids. *J. Bioenerg. Biomembr.* **40:** 163–170.

Rowntree, R.K., and Harris, A., 2003, The phenotypic consequences of CFTR mutations. *Ann. Hum. Genet.* **67**: 471–485.

Sachs, F., and Morris, C., 1998, Mechanosensitive ion channels in nonspecialized cells. *Rev. Physiol. Biochem. Pharmacol.* **132**: 1–78.

Sah, P., and Faber, E.S., 2002, Channels underlying neuronal calcium-activated potassium currents. *Prog. Neurobiol.* **66**: 345–353.

Salkoff, L., Butler, A., Ferreira, G., Santi, C., and Wei, A., 2006, High-conductance potassium channels of the SLO family. *Nat. Rev. Neurosci.* **7**: 921–931.

Sampson, L.J., Davies, L.M., Barrett-Jolley, R., Standen, N.B., and Dart, C., 2007, Angiotensin II-activated protein kinase C targets caveolae to inhibit aortic ATP-sensitive potassium channels. *Cardiovasc. Res.* **76**: 61–70.

Sampson, L.J., Hayabuchi, Y., Standen, N.B., and Dart, C., 2004, Caveolae localize protein kinase A signalling to arterial ATP-sensitive potassium channels. *Circ. Res.* **95**: 1012–1018.

Santi, C.M.D., Butler, A., Kuhn, J., Wei, A.D., and Salkoff, L.D., 2009, Bovine and mouse SLO3 K^+ channels:evolutionary divergence points to a RCK1 region of critical function. *J. Biol. Chem.***284**: 21589–98.

Schreiber, M., Wei, A., Yuan, A., Gaut, J., Saito, M., and Salkoff, L., 1998, Slo3, a novel pH-sensitive K^+ channel from mammalian spermatocytes. *J. Biol. Chem.* **273**: 3509–3516.

Schwiebert, E.M., Egan, M.E., Hwang, T.H., Fulmer, S.B., Allen, S.S., Cutting, G.R., and Guggino, W.B., 1995, CFTR regulates outwardly rectifying chloride channels through an autocrine mechanism involving ATP. *Cell* **81**: 1063–1073.

Shennan, D.B., 2008, Swelling-induced taurine transport: relationship with chloride channels, anion-exchangers and other swelling-activated transport pathways. *Cell Physiol. Biochem.* **21**: 15–28.

Shlyonsky, V.G., Mies, F., and Sariban-Sohraby, S., 2003, Epithelial sodium channel activity in detergent-resistant membrane microdomains. *Am. J. Physiol. Renal Physiol.* **284**: F182–188.

Shmygol, A., Noble, K., and Wray, S., 2007, Depletion of membrane cholesterol eliminates the Ca^{2+}-activated component of outward potassium current and decreases membrane capacitance in rat uterine myocytes. *J. Physiol.* **581**: 445–456.

Shoshan-Barmatz, V., Israelson, A., Brdiczka, D., and Sheu, S.S., 2006, The voltage-dependent anion channel (VDAC): function in intracellular signalling, cell life and cell death. *Curr. Pharm. Des.* **12**: 2249–2270.

Singh, A.K., Schultz, B.D., Katzenellenbogen, J.A., Price, E.M., Bridges, R.J., and Bradbury, N.A., 2000, Estrogen inhibition of cystic fibrosis transmembrane conductance regulator-mediated chloride secretion. *J. Pharmacol. Exp. Ther.* **295**: 195–204.

Sobey, C.G., 2001, Potassium Channel Function in Vascular Disease. *Arterioscler Thromb. Vasc. Biol.* **21**: 28–38.

Sprossmann, F., Pankert, P., Sausbier, U., Wirth, A., Zhou, X.B., Madlung, J., Zhao, H., Bucurenciu, I., Jakob, A., Lamkemeyer, T., Neuhuber, W., Offermanns, S., Shipston, M.J., Korth, M., Nordheim, A., Ruth, P., and Sausbier, M., 2009, Inducible knockout mutagenesis reveals compensatory mechanisms elicited by constitutive BK channel deficiency in overactive murine bladder. *FEBS J.* **276**: 1680–1697.

Stutzin, A., and Hoffmann, E.K., 2006, Swelling-activated ion channels: functional regulation in cell-swelling, proliferation and apoptosis. *Acta Physiol. (Oxf)* **187**: 27–42.

Suzuki, M., Morita, T., and Iwamoto, T., 2006, Diversity of Cl$^-$ channels. *Cell Mol. Life Sci.* **63**: 12–24.

Taverna, E., Saba, E., Rowe, J., Francolini, M., Clementi, F., and Rosa, P., 2004, Role of Lipid Microdomains in P/Q-type Calcium Channel (Cav2.1) Clustering and Function in Presynaptic Membranes. *J. Biol. Chem.* **279**: 5127–5134.

Thiemann, A., Grunder, S., Pusch, M., and Jentsch, T.J., 1992, A chloride channel widely expressed in epithelial and non-epithelial cells. *Nature* **356**: 57–60.

Tikku, S., Epshtein, Y., Collins, H., Travis, A.J., Rothblat, G.H., and Levitan, I., 2007, Relationship between Kir2.1/Kir2.3 activity and their distribution between cholesterol-rich and cholesterol-poor membrane domains. *Am. J. Physiol. Cell Physiol.* **293:** C440–450.

Torihashi, S., Fujimoto, T., Trost, C., and Nakayama, S., 2002, Calcium oscillation linked to pacemaking of interstitial cells of Cajal: requirement of calcium influx and localization of TRP4 in caveolae. *J. Biol. Chem.* **277:** 19191–7.

Toselli, M., Biella, G., Taglietti, V., Cazzaniga, E., and Parenti, M., 2005, Caveolin-1 Expression and Membrane Cholesterol Content Modulate N-Type Calcium Channel Activity in NG108-15 Cells. *Biophys. J.* **89:** 2443–2457.

Toyama, K., Wulff, H., Chandy, K.G., Azam, P., Raman, G., Saito, T., Fujiwara, Y., Mattson, D.L., Das, S., Melvin, J.E., Pratt, P.F., Hatoum, O.A., Gutterman, D.D., Harder, D.R., and Miura, H., 2008, The intermediate-conductance calcium-activated potassium channel KCa3.1 contributes to atherogenesis in mice and humans. *J. Clin. Invest.* **118:** 3025.

Trouet, D., Hermans, D., Droogmans, G., Nilius, B., and Eggermont, J., 2001, Inhibition of Volume-Regulated Anion Channels by Dominant-Negative Caveolin-1. *Biochem. Biophys. Res. Comm.* **284:** 461.

Trouet, D., Nilius, B., Jacobs, A., Remacle, C., Droogmans, G., and Eggermont, J., 1999, Caveolin-1 modulates the activity of the volume-regulated chloride channel. *J. Physiol.* **520 Pt 1:** 113–119.

Tsujikawa, H., Song, Y., Watanabe, M., Masumiya, H., Gupte, S.A., Ochi, R., and Okada, T., 2008, Cholesterol depletion modulates basal L-type Ca2+ current and abolishes its - adrenergic enhancement in ventricular myocytes. *Am. J. Physiol. Heart Circ. Physiol.* **294:** H285–292.

Ullrich, N., Caplanusi, A., Brone, B., Hermans, D., Lariviere, E., Nilius, B., Van Driessche, W., and Eggermont, J., 2006, Stimulation by caveolin-1 of the hypotonicity-induced release of taurine and ATP at basolateral, but not apical, membrane of Caco-2 cells. *Am. J. Physiol. Cell Physiol.* **290:** C1287–1296.

Vandorpe, D.H., Shmukler, B.E., Jiang, L., Lim, B., Maylie, J., Adelman, J.P., de Franceschi, L., Cappellini, M.D., Brugnara, C., and Alper, S.L., 1998, cDNA cloning and functional characterization of the mouse Ca^{2+}-gated K^+ channel, mIK1. Roles in regulatory volume decrease and erythroid differentiation. *J. Biol. Chem* **273:** 21542–21553.

Vicente, R., Villalonga, N., Calvo, M., Escalada, A., Solsona, C., Soler, C., Tamkun, M.M., and Felipe, A., 2008, Kv1.5 Association Modifies Kv1.3 Traffic and Membrane Localization. *J. Biol. Chem.* **283:** 8756–8764.

Vij, N., Mazur, S., and Zeitlin, P.L., 2009, CFTR is a negative regulator of NFkappaB mediated innate immune response. *PLoS ONE* **4:** e4664.

Wang, D., Wang, W., Duan, Y., Sun, Y., Wang, Y., and Huang, P., 2008, Functional coupling of Gs and CFTR is independent of their association with lipid rafts in epithelial cells. *Pflugers Arch.* **456:** 929–938.

Wang, X.L., Ye, D., Peterson, T.E., Cao, S., Shah, V.H., Katusic, Z.S., Sieck, G.C., and Lee, H.C., 2005, Caveolae targeting and regulation of large conductance Ca^{2+}-activated K^+ channels in vascular endothelial cells. *J. Biol. Chem.* **280:** 11656–11664.

Weaver, A.K., Olsen, M.L., McFerrin, M.B., and Sontheimer, H., 2007, BK channels are linked to inositol 1,4,5-triphosphate receptors via lipid rafts: a novel mechanism for coupling $[Ca^{2+}](i)$ to ion channel activation. *J. Biol. Chem.* **282:** 31558–31568.

Wei, S.-P., Li, X.-Q., Chou, C.-F., Liang, Y.-Y., Peng, J.-B., Warnock, D., and Ma, H.-P., 2007, Membrane Tension Modulates the Effects of Apical Cholesterol on the Renal Epithelial Sodium Channel. *J. Membrane Biol.* **220:** 21.

West, A., and Blazer-Yost, B., 2005, Modulation of basal and peptide hormone-stimulated Na transport by membrane cholesterol content in the A6 epithelial cell line. *Cell Physiol. Biochem.* **16:** 263–270.

Wiecha, J., Schläger, B., Voisard, R., Hannekum, A., Mattfeldt, T., and Hombach, V., 1997, Ca^{2+}-activated K^+ channels in human smooth muscle cells of coronary atherosclerotic plaques and coronary media segments. *Basic Res. Cardiol.* **92:** 233–239.

Wong, W., and Schlichter, L.C., 2004, Differential Recruitment of Kv1.4 and Kv4.2 to Lipid Rafts by PSD-95. *J. Biol. Chem.* **279:** 444–452.

Wu, C.C., Su, M.J., Chi, J.F., Chen, W.J., Hsu, H.C., and Lee, Y.T., 1995, The effect of hypercholesterolemia on the sodium inward currents in cardiac myocyte. *J Mol. Cell Cardiol.* **27:** 1263–1269.

Xia, F., Gao, X., Kwan, E., Lam, P.P.L., Chan, L., Sy, K., Sheu, L., Wheeler, M.B., Gaisano, H.Y., and Tsushima, R.G., 2004, Disruption of Pancreatic {beta}-Cell Lipid Rafts Modifies Kv2.1 Channel Gating and Insulin Exocytosis. *J. Biol. Chem.* **279:** 24685–24691.

Xia, F., Leung, Y.M., Gaisano, G., Gao, X., Chen, Y., Manning Fox, J.E., Bhattacharjee, A., Wheeler, M.B., Gaisano, H.Y., and Tsushima, R.G., 2007, Targeting of Voltage-Gated K^+ and Ca^{2+} Channels and Soluble N-Ethylmaleimide-Sensitive Factor Attachment Protein Receptor Proteins to Cholesterol-Rich Lipid Rafts in Pancreatic {alpha}-Cells: Effects on Glucagon Stimulus-Secretion Coupling. *Endocrinology* **148:** 2157–2167.

Xu, X., and London, E., 2000, The effect of sterol structure on membrane lipid domains reveals how cholesterol can induce lipid domain formation. *Biochemistry* **39:** 843–849.

Yarbrough, T.L., Lu, T., Lee, H.-C., and Shibata, E.F., 2002, Localization of Cardiac Sodium Channels in Caveolin-Rich Membrane Domains: Regulation of Sodium Current Amplitude. *Circ. Res.* **90:** 443–449.

Yogi, A., Callera, G.E., Tostes, R., and Touyz, R.M., 2009, Bradykinin regulates calpain and proinflammatory signalling through TRPM7-sensitive pathways in vascular smooth muscle cells. *Am. J. Physiol. Regul. Integr. Comp. Physiol.* **296:** R201–7.

Yoo, D., Flagg, T.P., Olsen, O., Raghuram, V., Foskett, J.K., and Welling, P.A., 2004, Assembly and trafficking of a multiprotein ROMK (Kir 1.1) channel complex by PDZ interactions. *J. Biol. Chem.* **279:** 6863–6873.

Yuan, C., O'Connell, R.J., Feinberg-Zadek, P.L., Johnston, L.J., and Treistman, S.N., 2004, Bilayer thickness modulates the conductance of the BK channel in model membranes. *Biophys. J.* **86:** 3620–3633.

Zerangue, N., Schwappach, B., Jan, Y.N., and Jan, L.Y., 1999, A new ER trafficking signal regulates the subunit stoichiometry of plasma membrane K(ATP) channels. *Neuron* **22:** 537–548.

Zheng, Y.Z., Berg, K.B., and Foster, L.J., 2009, Mitochondria do not contain lipid rafts, and lipid rafts do not contain mitochondrial proteins. *J Lipid Res.* **50:** 988–998.

Zhou, Z., Jiang, D.J., Jia, S.J., Xiao, H.B., Xiao, B., and Li, Y.J., 2007, Down-regulation of endogenous nitric oxide synthase inhibitors on endothelial SK3 expression. *Vascul. Pharmacol.* **47:** 265–271.

Zingman, L.V., Alekseev, A.E., Hodgson-Zingman, D.M., and Terzic, A., 2007, ATP-sensitive potassium channels: metabolic sensing and cardioprotection. *J. Appl. Physiol.* **103:** 1888–1893.

Chapter 20
The Cholesterol-Dependent Cytolysin Family of Gram-Positive Bacterial Toxins

Alejandro P. Heuck, Paul C. Moe, and Benjamin B. Johnson

Abstract The cholesterol-dependent cytolysins (CDCs) are a family of β-barrel pore-forming toxins secreted by Gram-positive bacteria. These toxins are produced as water-soluble monomeric proteins that after binding to the target cell oligomerize on the membrane surface forming a ring-like pre-pore complex, and finally insert a large β-barrel into the membrane (about 250 Å in diameter). Formation of such a large transmembrane structure requires multiple and coordinated conformational changes. The presence of cholesterol in the target membrane is absolutely required for pore-formation, and therefore it was long thought that cholesterol was the cellular receptor for these toxins. However, not all the CDCs require cholesterol for binding. Intermedilysin, secreted by *Streptoccocus intermedius* only binds to membranes containing a protein receptor, but forms pores only if the membrane contains sufficient cholesterol. In contrast, perfringolysin O, secreted by *Clostridium perfringens*, only binds to membranes containing substantial amounts of cholesterol. The mechanisms by which cholesterol regulates the cytolytic activity of the CDCs are not understood at the molecular level. The C-terminus of perfringolysin O is involved in cholesterol recognition, and changes in the conformation of the loops located at the distal tip of this domain affect the toxin-membrane interactions. At the same time, the distribution of cholesterol in the membrane can modulate toxin binding. Recent studies support the concept that there is a dynamic interplay between the cholesterol-binding domain of the CDCs and the excess of cholesterol molecules in the target membrane.

Keywords Cholesterol · Membranes · Pore-forming toxins · Cholesterol-dependent cytolysins · Membrane structure · Cholesterol activity · Transmembrane beta-barrel · Transmembrane pore · Fluorescence spectroscopy · Perfringolysin · Lipid cluster

A.P. Heuck (✉)
Department of Biochemistry and Molecular Biology, University of Massachusetts, Amherst, MA 01003, USA
e-mail: heuck@biochem.umass.edu

Abbreviations

CDCs	cholesterol-dependent cytolysins
PFO	perfringolysin O
ILY	intermedilysin
PLY	pneumolysin
SLO	streptolysin O
ALO	anthrolysin
TMH/s	transmembrane β-hairpin/s
D4	domain 4
L1, L2, and L3	loop 1, loop 2 and loop 3

20.1 Introduction

The cholesterol-dependent cytolysins (CDCs) are a growing group of β-barrel pore-forming toxins secreted by Gram-positive bacteria (Farrand et al., 2008; Gelber et al., 2008; Heuck et al., 2001; Jefferies et al., 2007; Mosser and Rest, 2006), and the first members were discovered more than a century ago (*see* Alouf et al., 2006 for a historical background on the CDCs). To date, there are complete amino acid sequences for 28 species distributed among the phyla of *Firmicutes* (genera of *Bacillus*, *Paenibacillus*, *Lysinibacillus*, *Listeria*, *Streptococcus*, and *Clostridium*), and of *Actinobacteria* (genera of *Arcanobacterium* and *Gardenella*) (Table 20.1). Most of the CDCs have a cleavable signal sequence and are therefore secreted to the extracellular medium via the general secretion system (*see* Harwood and Cranenburgh, 2008). A few exceptions are species of the genus *Streptoccocus* (*S. pneumoniae, S. mitis,* and *S. pseudoneumoniae*) that produce CDC without a signal sequence. The secretion mechanism for these CDCs is unclear (Jefferies et al., 2007; Marriott et al., 2008). After secretion to the extracellular medium, the CDCs fold into water-soluble monomeric proteins, travel and bind to the target membrane, and oligomerize on the membrane surface forming characteristic arcs and ring-like structures which are responsible for cytolysis. Several reviews have been published describing the recent advances in the structural and mechanistic studies of the CDCs (Alouf et al., 2006; Giddings et al., 2006; Gilbert, 2005; Rossjohn et al., 2007; Tweten, 2005). Here, we will focus on the role played by cholesterol during the transformation of the CDC from a water-soluble monomer to a membrane-inserted oligomeric complex. Although the cholesterol-dependent inhibition of the activity for these toxins was one of the first biochemical properties attributed to the family (Arrhenius, 1907), the molecular mechanism of the cholesterol-toxin interaction remains as one of the least understood aspects in the study of the CDC family.

20.2 Mechanism of Pore Formation

The 28 CDC family members listed in Table 20.1 show a significant degree of amino acid identity (from 28.1 to 99.6%) and similarity (greater than 45.7%), with amino acid sequences ranging from 471 to 665 amino acids in length. A comparison of

20 The Cholesterol-Dependent Cytolysin Family of Toxins

Table 20.1 Homologs in Gram-positive species compose the CDC family. Twenty-eight CDC family members from divergent phyla have been identified by amino acid sequence. The protein three letter code for each homolog (as defined in Fig. 20.1) is followed by its phylogenetic relationship to the PFO standard. Because many of the CDC family are expressed with variable N-terminus, PFO relationship is expressed in bold for the conserved core only (corresponding to amino acids 38–500 of PFO) and in parentheses for the full length form. The lengths of the respective polypeptides are presented. Percentages of identity and similarity were calculated as indicated in Fig. 20.1 legend, * subsp. equisimilis

		%Identity	%Similarity	Length	ID
PHYLUM					
Firmicutes					
CLASS					
Bacilli					
ORDER					
Bacillales					
FAMILY					
Bacillaceae					
GENUS					
Bacillus					
SPECIES					
B. anthracis	ALO	**72** (68)	**88** (83)	**462** (512)	ZP_03017964.1
B. thurigiensis	TLO	**74** (69)	**88** (83)	**462** (512)	YP_037419
B. cereus	CLO	**74** (69)	**88** (84)	**462** (512)	YP_002369889.1
B. weihenstephanensis	WLO	**74** (69)	**87** (83)	**462** (512)	ABY46062
Listeriaceae					
Listeria					
L. monocytogenes	LLO	**43** (40)	**66** (62)	**469** (529)	DQ838568.1
L. seeligeri	LSO	**45** (41)	**67** (63)	**469** (530)	P31830.1
L. ivanovii	ILO	**46** (43)	**66** (62)	**469** (528)	AAR97343.1
Planococcaceae					
Lysinibacillus					
L. sphaericus	SPH	**76** (72)	**90** (87)	**463** (506)	YP_001699692.1
Paenibacillaceae					
Paenibacillus					
P. alvei	ALV	**75** (71)	**87** (84)	**462** (501)	P23564
Brevibacillus					
B. brevis	BVL	**73** (69)	**88** (84)	**464** (511)	YP_002770211.1
Lactobacillales					
Streptococcaceae					
Streptococcus					
*S. dysgalactiae**	SLOe	**67** (56)	**83** (70)	**463** (571)	BAD77791
S. pyogenes	SLO	**67** (56)	**83** (70)	**463** (571)	NP_268546.1
S. canis	SLOc	**66** (55)	**82** (69)	**463** (574)	Q53957
S. pseudonemoniae	PSY	**46** (43)	**67** (63)	**466** (471)	ACJ76900
S. pneumoniae	PLY	**46** (43)	**67** (64)	**466** (471)	ABO21366.1
S. mitis	MLY	**46** (43)	**67** (63)	**466** (471)	ABK58695
S. suis	SLY	**41** (40)	**65** (63)	**465** (497)	ABE66337.1
S. intermedius	ILY	**41** (37)	**65** (59)	**469** (532)	B212797.1
S. mitis (Lectinolysin)	LLY	**39** (29)	**62** (47)	**463** (665)	BAE72438.1

Table 20.1 (continued)

Clostridia						
Clostridiales						
Clostridiaceae						
Clostridium						
C. perfringens	PFO			463 (500)	NP_561079
C. butyricum	BRY	69 (65)	85 (82)	462 (513)	ZP_02950902.1
C. tetani	TLY	60 (55)	78 (72)	464 (527)	NP_782466.1
C. botulinumB	BLYb	60 (49)	78 (63)	464 (602)	YP_001886995.1
C. botulinum E3	BLYe	60 (48)	77 (60)	464 (602)	YP_001921918.1
C. botulinumC	BLYc	60 (56)	79 (74)	463 (518)	ZP_02620972.1
C. novyi	NVL	58 (54)	78 (73)	463 (514)	YP_878174.1
Actinobacteria						
Actinobacteria						
Bifidobacteriales						
Bifidobacteriaceae						
Gardenella						
G. vaginallis	VLY	40 (39)	65 (60)	466 (516)	EU522488.1
Actinomycetales						
Actinomycetaceae						
Arcanobacterium						
A. pyogenes	PLO	41 (38)	60 (56)	469 (534)	U84782.2

the primary structure of these proteins shows that they share a very low degree of similarity at their N-terminus, in part because different species employ distinct signal sequences for secretion, but also because some of the CDC members possess additional domains located in this region (e.g., Farrand et al., 2008). If we consider just the conserved core shared by all CDCs and required for pore-formation activity [amino acids 38–500 in perfringolysin O (PFO)], the amino acid identity and similarity among different members becomes higher than 36.7 and 58%, respectively (sequence length of analyzed sequences range from 462 to 469, Fig. 20.1). Therefore, from the analysis of the primary structure of these toxins we can anticipate that all the CDCs will exhibit similar activities and three-dimensional structures.

The first crystal structure for a CDC was solved for PFO by Rossjohn and colleagues (1997). The crystal structure for two other CDCs, intermedylisin (ILY) and anthrolysin (ALO), have been solved so far, and all of them share similar secondary and tertiary structure (Bourdeau et al., 2009; Polekhina et al., 2005). They have a high β-strand content and their structures have been divided into four domains, with the C-terminal domain (domain 4 or D4) being the only independent and continuous domain (Fig. 20.2A) (Polekhina et al., 2006).

PFO secreted by the pathogen *Clostridium perfringens* is a prototypical CDC (Tweten et al., 2001). To describe the general mechanism of pore-formation for the CDC we will depict the current knowledge of the PFO cytolytic mechanism which starts with the binding of the toxin to the target membrane and concludes with the insertion of a large transmembrane β-barrel (Fig. 20.2A).

20 The Cholesterol-Dependent Cytolysin Family of Toxins

	PFO	SPH	ALO	CLO	TLO	WLO	ALV	BVL	BRY	TLY	BLYb	BLYe	BLYc	NVL	SLO	SLOc	SLOe	LLY	PLY	MLY	PSY	SLY	ILY	ILO	LSO	LLO	VLY	PLO
Perfringolysin O		76	72	74	74	74	74	75	73	69	60	60	60	58	67	66	67	39	46	46	46	41	41	46	45	43	40	41
Sphaericolysin	90		80	82	82	82	75	77	80	69	58	57	56	54	66	65	65	37	42	42	42	41	37	45	45	43	37	38
Anthrolysin O	88	93		97	97	97	73	76	76	65	56	56	55	55	62	61	62	37	41	41	42	40	39	44	43	41	41	39
Cereolysin O	88	93	99		98	98	73	76	76	67	57	58	58	54	63	62	63	37	41	41	41	40	39	44	43	42	40	39
Thuringiensilysin O	88	93	99	99		98	75	77	77	66	57	57	57	54	63	62	63	37	41	41	41	40	39	44	43	42	40	39
Weihenstephanensilysin	87	93	99	99	99		75	77	77	66	57	57	57	54	63	62	63	37	41	41	41	40	39	43	43	42	40	39
Alveolysin	87	89	88	88	88	88		72		65	58	58	57	53	63	63	63	37	41	41	41	40	39	43	43	42	39	39
Brevilysin	88	93	91	91	92	92	87		84	68	57	58	57	57	64	63	64	39	42	42	43	40	40	44	44	42	39	40
Butyriculysin	85	86	84	84	84	84	81	84		75	55	55	57	56	63	63	63	37	43	43	45	40	39	45	44	42	40	39
Tetanolysin O	78	79	77	78	77	77	76	75	75		82	81	81	77	55	55	55	42	45	45	45	45	42	50	50	48	43	44
Botulinolysin B	77	75	75	75	75	75	75	74	76	93		95	81	73	55	55	55	43	46	46	46	45	43	51	49	48	43	44
Botulinolysin E3	77	77	75	75	75	75	75	74	75	93	98		74	72	54	54	54	42	44	45	45	45	42	51	49	49	43	43
Botulinolysin C	79	79	78	78	78	77	77	77	74	93	90	90		83	54	54	54	42	46	46	46	46	46	49	49	48	45	43
Novyilysin	77	78	79	79	79	78	76	77	77	90	88	87	93		52	52	52	42	44	44	45	45	44	50	50	47	44	44
Streptolysin O	83	81	79	80	79	80	79	80	78	74	73	74	74	75		100	99	39	42	42	42	41	38	43	43	41	39	39
Streptolysin O c	82	80	79	79	79	79	79	79	81	74	73	73	74	74	99		99	39	41	41	41	40	37	44	43	42	39	39
Streptolysin O e	83	81	79	80	79	79	79	80	81	74	73	73	74	75	100	99		39	41	41	42	41	37	44	43	42	39	39
Lectinolysin	62	61	61	61	61	61	62	62	62	63	65	68	64	64	64	63	64		51	51	51	46	54	40	41	40	59	41
Pneumolysin	67	66	65	66	66	66	66	66	66	66	68	69	66	67	65	64	65	73		99	99	51	53	43	43	43	53	41
Mitilysin	67	66	66	66	66	66	65	66	67	68	69	69	67	67	65	64	65	73	99		100	50	53	44	43	43	53	41
Pseudopneumolysin	67	66	66	66	66	66	66	66	66	68	69	69	67	69	65	64	65	73	100	100		50	53	44	43	43	53	40
Suilysin	65	65	66	66	66	66	65	66	64	65	66	68	66	67	65	64	65	68	73	74	74		47	48	45	46	52	45
Intermedilysin	64	67	64	65	64	65	64	63	64	66	70	71	68	67	65	61	62	77	74	74	74	68		42	42	42	60	42
Ivanolysin	66	66	65	66	66	66	65	65	65	70	71	70	71	71	65	65	65	64	67	67	67	71	66		79	81	42	43
Seeligeriolysin O	67	66	65	66	66	66	66	66	67	71	70	71	71	71	66	66	66	64	66	66	66	69	64	91		85	41	43
Listeriolysin O	66	65	63	64	64	64	64	63	65	72	71	71	70	70	66	66	66	65	65	67	67	69	66	94	95		40	43
Vaginolysin	63	62	62	64	64	64	63	62	62	62	64	64	64	66	62	61	62	77	73	73	73	69	79	63	62	62		60
Pyolysin	60	58	58	59	59	59	60	61	61	61	60	60	60	63	60	60	60	62	61	61	61	64	63	63	62	62	60	

Fig. 20.1 Analysis of the primary structure for the CDCs reveals a high degree of identity and similarity among them. Only the sequence for the conserved core of the CDCs was used for the analysis (corresponding to PFO amino acids 38–500). If more than one sequence was available for individual species, only one was used in the analysis. The databank access numbers are provided in Table 20.1. Sequence relationships were calculated using the MatGat 2.02 alignment program using the BLOSUM 62 matrix and open and extension gap penalties of 12 and 1, respectively (Campanella et al., 2003). The identity scores occupy the upper triangle (in bold) with scores higher than 70% shaded in dark gray, and those at 50–70% in light gray. Similarity scores in the lower triangle where shaded in dark gray if higher than 80% and in light gray if between 70 and 80%

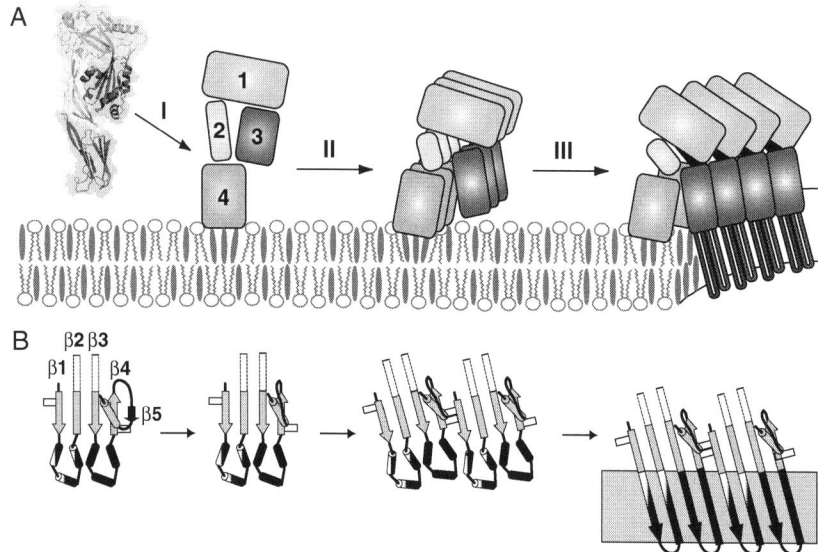

Fig. 20.2 Pore formation mechanism for the CDCs. Secreted as water-soluble monomeric proteins, the toxins bind to the target membrane and oligomerize into a ring-like structure called the pre-pore complex. A poorly understood conformational change then leads to the insertion of the TMHs into the bilayer to form the aqueous pore. (**A**) Stages of PFO pore formation. The defined PFO structural domains are numbered. The membrane bilayer is depicted with cholesterol molecules (ovals) intercalated between the phospholipid constituents. Membrane binding is accomplished as D4 interacts with membrane regions having free cholesterol molecules available. Subsequent allosteric rearrangements within the monomer promote oligomerization and pore-formation. (**B**) Conformational changes in domain 3 of PFO are required for monomer–monomer association and β-barrel pore formation. Each stage corresponds to the stage shown above in (**A**). The TMH1 is shown as bicolor and the TMH2 in black. The small β5 strand is shown as a black loop. The aromatic residues involved in the alignment of the β-strands are shown as open rectangles. Adapted from Ramachandran et al. (2004), with permission

Upon encountering a cholesterol-containing membrane, PFO oligomerizes and spontaneously inserts into the bilayer to form a large transmembrane pore (∼35–50 monomers per oligomer; approximately 250 Å in diameter, Fig. 20.2), (Czajkowsky et al., 2004; Dang et al., 2005; Mitsui et al., 1979; Olofsson et al., 1993). The C-terminus of PFO (D4) encounters the membrane first (Fig. 20.2A, I, Heuck et al., 2000; Nakamura et al., 1995; Ramachandran et al., 2002). The binding of D4 triggers the structural rearrangements required to initiate the oligomerization of PFO monomers (Ramachandran et al., 2004; Soltani et al., 2007a) and formation of a pre-pore complex on the membrane surface (Fig. 20.2A, II, Heuck et al., 2003; Shepard et al., 2000; Tilley et al., 2005). Pore formation commences when two amphipathic β-hairpins from each PFO molecule insert and span the membrane (Fig. 20.2A, III, Hotze et al., 2002; Shatursky et al., 1999; Shepard et al., 1998). The concerted insertion of two transmembrane β-hairpins (TMHs) from ∼35 PFO monomers then creates the large transmembrane β-barrel that penetrates the membrane (Dang et al.,

2005; Tilley et al., 2005). This general mechanism of pore-formation is followed by most CDCs, however, some variations have been observed for specific members and they will be described in the following sections.

20.2.1 Localizing the Target Membrane

The first step in the CDC cytolytic cascade is the recognition of the target cell (Fig. 20.2A, I). The CDC binds to the target membrane by recognizing a specific membrane lipid, cholesterol, or by recognizing a membrane-anchored protein in the case of ILY (Giddings et al., 2004). Cholesterol-recognition provides specificity towards eukaryotic cells in general, and the glycosylphosphatidylinositol-anchored protein CD59 provides specificity for human cells. While it has been shown that ILY interacts with the CD59 receptor forming a 1:1 complex (Lachapelle et al., 2009), the interaction of other CDCs with cholesterol is less well understood. Independently of the recognition mechanism, it appears that all CDCs bind to the target membranes via D4 (Nagamune et al., 2004; Soltani et al., 2007a).

20.2.2 Grouping Forces on the Membrane Surface: Pre-pore Formation

After successful recognition of the target membrane, the CDC oligomerize in the membrane surface to form a membrane-bound pre-pore complex (Fig. 20.2, II). Formation of a pre-pore complex seems to be a common feature of the β-barrel pore-forming toxins (Heuck et al., 2001; Miller et al., 1999; Shepard et al., 2000; Walker et al., 1992). In general, the secreted monomeric proteins do not oligomerize spontaneously in solution, and it has been shown that the binding of the toxins to the target membrane is required to trigger the monomer-monomer association (Abdel Ghani et al., 1999; Lachapelle et al., 2009; Ramachandran et al., 2004). Although oligomerization has been observed in the absence of membranes for certain CDCs (e.g., pneumolysin, (PLY) Gilbert et al., 1998; Solovyova et al., 2004), it only occurs when the toxin concentration is relatively high (in the micromolar range or higher), compared to the low concentration needed for efficient oligomerization when incubated with natural membranes. The difference in efficiency between oligomerization in solution and at the surface of a cell membrane suggests that the cells in some way promote the association of toxin monomers. In general, oligomerization of β-barrel pore-forming toxins requires the exposure of hidden polypeptide regions involved in the monomer-monomer interaction (Heuck and Johnson, 2005; Heuck et al., 2001). In the CDC, this process is triggered by conformational changes induced by protein-lipid interactions (e.g., PFO, Ramachandran et al., 2004) or by conformational changes induced by protein-protein interactions (e.g., ILY, Lachapelle et al., 2009).

Ramachandran et al. (2004) have shown that in the water-soluble form of the toxin, oligomerization is prevented by blocking access to one edge of a core β-sheet in the monomer (Fig. 20.2B). This blockage prevents its association with the edge of the core β-sheet in the neighboring monomer, thus impeding formation of an extended β-sheet. Specifically and importantly, premature association of PFO molecules (before they bind to the appropriate membrane surface) is prevented by the presence of β5, a short polypeptide loop that hydrogen bonds to β4 in the monomer, and thereby prevents its interaction with the β1 strand in the adjacent monomer. This feature is conserved in all crystal structures so far reported for the CDCs (i.e., PFO, ILY, and ALO).

The structural changes associated with converting a CDC from a water-soluble monomer to a membrane-inserted oligomer extend through much of the molecule. The binding of D4 to the membrane surface immediately elicits a conformational change in domain 3, more than 70 Å above the membrane (Abdel Ghani et al., 1999; Heuck et al., 2000; Ramachandran et al., 2002, 2004, 2005). This conformational change rotates β5 away from β4 and thereby exposes β4 to the aqueous medium where it can associate with the always-exposed β1 strand of another PFO molecule, to initiate and promote oligomerization (Fig. 20.2B).

Such an extensive network of structural linkages within a CDC can be advantageous because it reduces the chance of prematurely entering a structural transition that exposes a TMH. By allosterically linking different domains or regions of the protein, the system can couple separate interactions (e.g., binding to the membrane and binding to another subunit) and thereby ensure that pore formation proceeds only when the necessary criteria are met. Given the important allosteric communication between the membrane binding domain and the pore-forming domain, it is not surprising that the most conserved regions on these proteins are located among inter-domain segments, forming an almost continuous path with its origin at the tip of D4 and terminus at the segments that form the amphipathic TMHs (Fig. 20.3). Interestingly, while most of the surface exposed residues of the CDCs are not very conserved, the residues at the surface of the D4 tip, involve in membrane interaction, are highly conserved.

Establishment of an oligomeric complex in the membrane surface facilitates the formation of a transmembrane pore because the insertion of a single amphipathic β-hairpin into a membrane is not energetically favored. In a hydrophobic environment that lacks hydrogen bond donors or acceptors, isolated β-hairpins cannot achieve the hydrogen-bond formation necessary to lower the thermodynamic cost of transferring the polar atoms of the polypeptide backbone into the hydrocarbon interior (White and Wimley, 1999). However, this energy barrier is circumvented if the β-strands are inserted as β-sheets and form closed structures such as a β-barrel. For monomeric β-barrel membrane proteins such as OmpA, a concerted folding mechanism has been observed in vitro, in which the hydrogen bonds formed between adjacent β-chains presumably favor the insertion of the β-barrel into the membrane (Kleinschmidt, 2006; Tamm et al., 2004). Similarly, the formation of a pre-pore complex may be required to allow the concerted, and perhaps simultaneous, insertion of the β-hairpins from individual monomers, thereby overcoming the energetic barrier of inserting non-hydrogen-bonded β-strands into the membrane bilayer. Whereas it is clear that the formation of a complete ring (or pre-pore complex) on the membrane surface will minimize the energetic requirements for

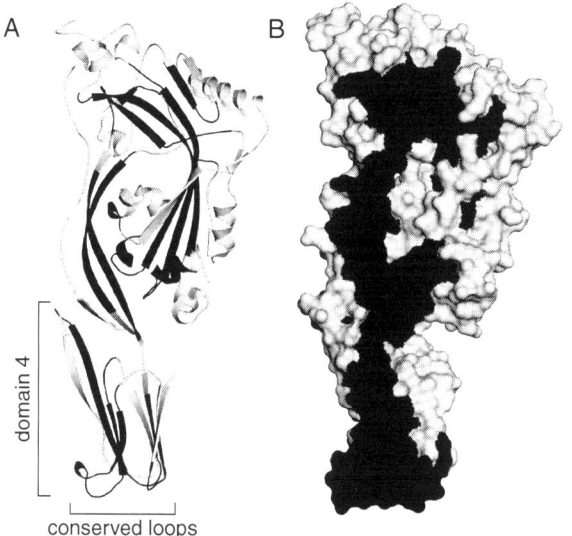

Fig. 20.3 Comparison of PFO homologs reveals a conserved core backbone. Alignment and comparison of the composite members of the CDC family reveals conserved regions that extend from the tip of the membrane recognition domain, D4, through the regions involved in oligomerization and membrane insertion. (**A**) Cartoon representation of PFO with the conserved residues shown in black. (**B**) Surface representation of PFO the conserved core highlighted in black. It is postulated that this conserved backbone is especially adapted to allosterically communicate successful, cholesterol-dependent membrane binding, and thus permit subsequent conformational adaptations that favor oligomerization and pore formation. Alignment of the 28 CDC sequences was effected using the PRALINE multiple sequence alignment tool using a BLOSUM62 matrix with open and extension gap penalties set at 12 and 1, respectively, a PSI-BLAST pre-profile processing with iterations set at 3, e-value cut off set at 0.01, non-redundant data bases, and a DSSP-defined secondary structure search using PSIPRED (Simossis et al., 2005). PFO structure representation was rendered using PyMol (DeLano Scientific LLC)

inserting a β-barrel into the membrane, it is likely that the insertion of incomplete rings can also occur if monomer recruitment into the oligomer slows down. In the absence of additional monomers, the incomplete pre-pore complexes observed in vitro (or metastable arc structures) will be trapped, and they may have enough time to insert into the membrane and form a pore (Gilbert, 2005). Insertion of an arc may well form a transmembrane pore by itself, or in association with other arcs (double arc structures, *see* Palmer et al., 1998). A minimal number of monomers must be required to overcome the energetic barrier of inserting an arc-like β-sheet into the membrane. It has been shown that independently of the toxin/lipid ratio, the pores formed by PFO and streptolysin O (SLO) are at least large enough to allow the passage of proteins with an approximate diameter of 100 Å (Heuck et al., 2003).

In summary, a coordinated train of events regulates the proper assembly of the CDC oligomeric complex at the surface of the target membrane. Formation of these oligomeric structures facilitates the insertion of numerous TMHs, which are required to form the large transmembrane β-barrel.

20.2.3 Perforating the Membrane: Insertion of a Large β-Barrel

A characteristic of the CDC that distinguishes them from most other β-barrel pore-forming toxins is the use of two amphipathic β-hairpins *per* monomer to form the large transmembrane barrel (Heuck and Johnson, 2005; Heuck et al., 2001; Shatursky et al., 1999). In the water-soluble monomeric configuration of the CDC these TMHs are folded as short α-helices, presumably to minimize the exposure of the hydrophobic surfaces (Heuck and Johnson, 2005). These helices, located at either side of the central β-sheet in domain 3, extend and insert into the membrane bilayer (Shatursky et al., 1999; Shepard et al., 1998). The conversion of short α-helices to amphipathic β-hairpins constituted a new paradigm for how pore-forming toxins transform from a water-soluble to membrane-inserted conformation. This structural transformation has been recently found in eukaryotic pore-forming proteins, as revealed by the structure of the membrane attack complex/perforin superfamily members (Hadders et al., 2007; Rosado et al., 2007). After insertion, the hydrophobic surfaces of the TMHs are exposed to the non-polar lipid core of the membrane and the hydrophilic surfaces face the aqueous pore. A concerted mechanism of insertion ensures that the hydrophilic surfaces of the hairpins remain exposed to the aqueous medium, and not to the hydrophobic core of the membrane. Such a coordinated insertion requires the displacement of membrane bilayer lipids as the aqueous pore is formed in the membrane.

The creation of a circular hole, having a radius of nearly 150 Å, in a liposomal membrane requires the displacement of about 1000 phospholipid molecules in each monolayer (or about 800 phospholipids plus 800 cholesterol molecules, because the average surface area occupied by one phospholipid molecule plus one cholesterol molecule is ∼90 Å^2 in a 1:1 phospholipid/cholesterol mixture) (Heuck et al., 2001; Lecuyer and Dervichian, 1969). Analysis of the release of markers encapsulated in liposomes when using limiting concentrations of PFO or SLO have shown that both the small markers and the large markers are released at the same rate. Therefore, it appears that all of these lipid molecules leave or are displaced from the pore formed by these CDCs at the same time (Heuck et al., 2003), though not all agree (Palmer et al., 1998).

A direct comparison of the cytolytic mechanism of PFO and ILY showed that whereas ILY does not require cholesterol for binding, pore-formation is subsequently entirely dependent on the presence of cholesterol in the target membrane (Giddings et al., 2003). Employing a series of ILY mutants that block pore formation at different stages, Hotze and colleagues have shown that ILY remains engaged with its receptor (human CD59) throughout the assembly of the pre-pore complex, but it is released from CD59 upon the transition to the membrane-inserted oligomer (Lachapelle et al., 2009). Upon release from the receptor, ILY is anchored to the membrane via D4 suggesting that this domain still conserves the cholesterol-binding properties of other CDC members (note that insertion of the ILY β-barrel does not occur if cholesterol is depleted from the membrane).

After pre-pore formation, the insertion of the PFO TMHs requires the appropriate intermonomer β-strand alignment. Ramachandran et al. (2004) suggested that the π-stacking interaction between Y181 and F318 guides the alignment of the TMHs of adjacent monomers (Fig. 20.2B). Interestingly, while Y181 is completely conserved in the 28 members of the CDC family, F318 is not. Instead of phenylalanine, this position is occupied by valine in lectinolysin, vaginolysin, and PLY, by isoleucine in ILY, and alanine in pyolysin. It will be interesting to determine if a mutation of the conserved PFO-Y181-equivalent in ILY results in a pre-pore blocked derivative, as observed in PFO.

20.3 The Role of Cholesterol in Membrane Binding

Among all the different lipids that shape the vast diversity of cell membranes, the presence of cholesterol is a distinguishing feature of mammalian cells. The CDCs have evolved to take advantage of this feature of mammalian membranes, and their ability to perforate the target membrane is totally dependent on the presence of cholesterol (Giddings et al., 2003; Palmer, 2004).

In liposomal membranes containing only phosphatidylcholine and cholesterol, more than 30 mole % cholesterol is required for CDCs such as tetanolysin (Alving et al., 1979), SLO (Rosenqvist et al., 1980), and PFO (Heuck et al., 2000; Ohno-Iwashita et al., 1992), to bind and create a pore in the bilayer. For PFO, no binding at all is detected when the cholesterol concentration in the liposomal membrane is less than ~30 mole% of the total lipids (Flanagan et al., 2009; Heuck et al., 2000; Nelson et al., 2008). Thus, if cholesterol acts solely as a receptor, and hence as a PFO binding ligand, reducing the cholesterol concentration in the bilayer should only affect the kinetics of the cytolytic process. In other words, lowering the amount of cholesterol in the membrane should result in a longer time required for PFO to form a transmembrane pore. However, the sharp transition observed in the binding isotherm of PFO suggests that the basis of this recognition is more complex than a simple encounter frequency between PFO and individual cholesterol molecules (Heuck et al., 2000).

20.3.1 Domain 4 and Membrane Recognition

The initial members of the CDC family were characterized by their sensitivity to oxygen and cholesterol (Alouf et al., 2006). Toxins isolated from culture supernatants were inactivated by exposure to oxygen present in the air or when pre-incubated with cholesterol. While the oxygen-dependent inactivation of the toxins could be reversed by incubation with thiol-based reducing agents, inactivation by pre-incubation with cholesterol was not reversible. A direct consequence of these findings was that the discovery of new CDC members was strongly influenced by the search for these two distinguishing features in the newly encountered hemolytic toxins: i.e. inhibition by oxygen and cholesterol. Therefore, it is not surprising that the first sequences obtained for CDCs revealed that all of them contained a conserved undecapeptide which was critical for cholesterol recognition, and a unique cysteine in this segment that was sensitive to aerobic oxidation. This correlation led researchers to postulate that the conserved undecapeptide, and attendant cysteine constituted the cholesterol-binding site for the CDC. However, advancements in recombinant DNA technology soon allowed researchers to show that this unique cysteine was not essential for cholesterol recognition. First, the replacement of this cysteine with alanine rendered a protein that remained hemolytic (Michel et al., 1990; Pinkney et al., 1989; Saunders et al., 1989; Shepard et al., 1998). Second, the sequence of newly discovered CDC members showed that this cysteine was indeed replaced by alanine during the evolution of different Gram-positive species (Billington et al., 2001; Nagamune et al., 2000).

New protein homologues of the CDCs are being revealed as new genomes are sequenced, and these new family members show greater variability in the amino acid sequence of this segment. The multi-sequence alignment for the 28 CDC sequences shows that 20% of the CDCs contain amino acid substitutions in the conserved undecapeptide. Based on this newly accumulated evidence, the original view of the conserved undecapeptide as the cholesterol binding site is being replaced by alternative models for membrane-binding. It has been shown that one of the CDCs, intermedilysin (ILY) recognize the target membrane by the specific binding to a human protein receptor CD59, and it is therefore possible that other members may also bind to the target membrane by as yet unidentified protein receptors (Bourdeau et al., 2009). In addition to the undecapeptide, other well conserved peptide loops located at the tip of D4 may contribute to the cholesterol recognition motif (Ramachandran et al., 2002; Soltani et al., 2007a; Soltani et al., 2007b).

20.3.1.1 The Conserved Loops

PFO D4 has a 4 stranded β-sandwich structure that interacts with the membrane surface only at one end, via the distal loops that interconnect the eight β-strands that form the domain (Fig. 20.4A, Ramachandran et al., 2002; Rossjohn et al., 1997; Soltani et al., 2007a). Superimposition of the D4 α-carbons for PFO, ALO, and ILY reveals that the global structure of D4 is well conserved among these members. The main differences arise in the conformation of the undecapeptide, involved in toxin-membrane interaction, and in the loops that are close to the domain 2-D4 interface (Fig. 20.4A).

Three of the four loops located at the distal tip of D4 are highly conserved among the CDC members: the conserved undecapeptide (also known as the Trp-rich loop), L1, and L2 (Fig. 20.4B). The L3 loop is less conserved and is located farther away from the unique cysteine residue. Recent data obtained by Tweten and colleagues suggest that in addition to the undecapeptide, the other D4 loops

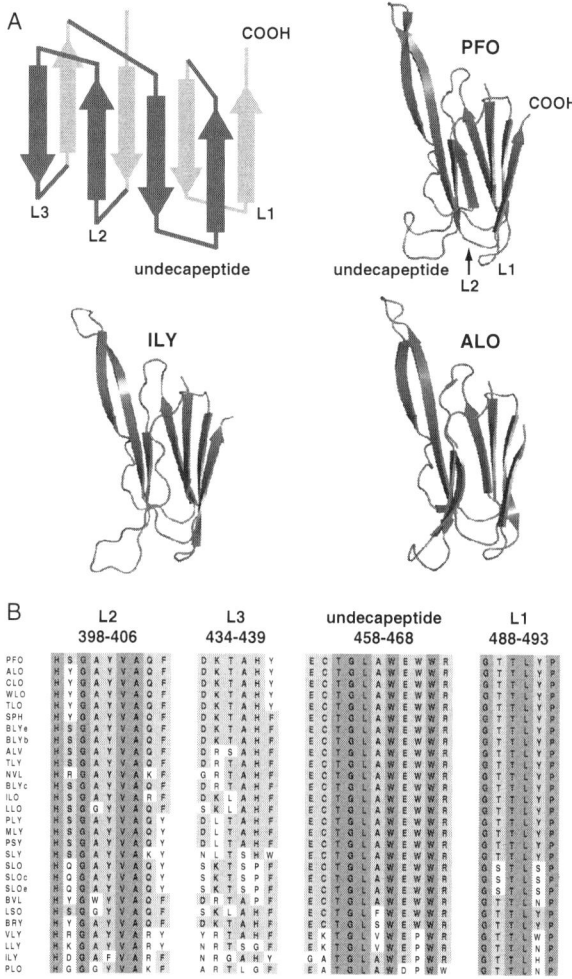

Fig. 20.4 The three dimensional structure of D4 is highly conserved in the CDC family. (**A**) Comparison of D4 from three CDC homologs highlights the conserved architecture of this C-terminal domain. A cartoon, upper left, clarifies the threading of 2 β-sheets and loops in the β-sandwich and indicates the spatial organization of the undecapeptide, L1, and L2. The α-backbone for the D4 domains of PFO, ILY, and ALO were superimposed using PyMol (DeLano Scientific LLC; available at www.pymol.org). (**B**) Alignment of the sequence for the 28 CDC family members reveals substantial conservation in loops L1, L2 and the undecapeptide. While integrity of the undecapeptide was long recognized for being critical to the cholesterol-dependent activity of these toxins, other loops are also important. Residues conserved in all sequences are shaded in black, and highly conserved residues are shaded in gray. Protein names are as in Fig. 20.2. Residue numbers correspond to the PFO sequence. Multiple sequence alignment was effected as indicated in Fig. 20.3

(L1–L3) may also play a role in the cholesterol-dependent recognition of the CDC (Soltani et al., 2007b). Single amino acid modifications in these loops prevented the binding of PFO to cholesterol-rich liposomes, and abolished the pre-pore to pore transition for ILY in a cholesterol-dependent manner. Both of these events involve the association of the D4 with the cholesterol-containing membrane. It has become clear that the three-dimensional arrangement of the undecapeptide and the L1–L3 loops is important for the association of the CDC with the cholesterol-containing membrane (Giddings et al., 2003; Polekhina et al., 2005; Soltani et al., 2007a, b).

Interestingly, changes in the pH of the medium which affect the conformation of D4 also influence the cholesterol-toxin interaction. A reduction of the pH from 7.5 to 6.0 induces a conformational change in PFO causing the tryptophan residues to be more exposed to the aqueous solvent, and also alters the threshold for the minimal cholesterol concentration required to trigger binding of PFO to liposomal membranes (Nelson et al., 2008). Since no major changes are expected to occur in the structure of the membrane in between pH 7.5 and 6.0, one can assume that protonation of certain amino acids in PFO may alter the D4 conformation, and as a consequence, its ability to recognize cholesterol in the target membrane. A related effect has been observed for listeriolysin O (LLO), a CDC recognized for having an optimum acidic pH for activity (Bavdek et al., 2007). However, the loss of activity of LLO at neutral pH can be rescued by increasing the concentration of cholesterol in the membrane.

Given that conformational changes in D4 can alter the cholesterol-dependent properties of the CDC, one can speculate that the conformational change triggered by the binding of ILY to the CD59 receptor (Soltani et al., 2007a), may modulate the cholesterol-dependent association with the membrane required for pore-formation.

Unfortunately, despite the various high-resolution structures available for the CDCs, and the multiple functional data obtained by modification of amino acids located at the D4 loops, it is still unclear how cholesterol modulates the conformational changes required to anchor the toxin to the membrane and to insert a large transmembrane β-barrel. Furthermore, is not clear if the binding of PFO (and related CDCs) is triggered by the binding of a single cholesterol molecule (Geoffroy and Alouf, 1983; Nollmann et al., 2004; Polekhina et al., 2005), or by the recognition of a more complex cholesterol-arrangement in the bilayer structure (Bavdek et al., 2007; Flanagan et al., 2009; Heuck and Johnson, 2005; Heuck et al., 2007; Nelson et al., 2008).

20.3.2 Searching for Cholesterol in the Membrane

The binding of a protein domain to a membrane surface is in general, a two-step process that involves the initial formation of non-specific collisional complex, followed by the formation of a tightly bound complex. The first step is diffusional and may involve electrostatic interactions, and the second step stabilizes the initial interaction by membrane penetration of non-polar amino acids and/or specific interactions between the protein and the membrane lipids (Cho and Stahelin, 2005). The initial membrane association locates non-polar amino acids close to the interfacial region of the bilayer, facilitating their exposure to the hydrophobic core. Non-polar amino acids are not usually exposed to the protein surface, and therefore conformational changes are required to expose them to the membrane.

Exposure of the aromatic residues located in the undecapeptide occurs upon membrane binding, though they do not penetrate deeply into the bilayer core (Heuck et al., 2003; Nakamura et al., 1998; Sekino-Suzuki et al., 1996). The sensitivity of the undecapeptide to amino acid changes suggests that the exposure of aromatic amino acids and membrane binding requires precise conformational changes and/or a particular three-dimensional conformation. A conformational change in the undecapeptide that modulates cholesterol binding and membrane anchoring has been suggested for PFO (Rossjohn et al., 1997), however the binding site for cholesterol, if any, remains elusive.

It has become apparent that in addition to the three dimensional structure of the binding-domain, the arrangement of the cholesterol molecules in the bilayer is also critical for successful binding. In a membrane, the cholesterol molecules are mobile in the non-polar core of the bilayer with an orientation nearly parallel to the acyl chains of the phospholipids. The non-polar hydrocarbon tail of the molecule orients towards the center of the bilayer, and the 3-β-OH group locates close to the ester bonds formed by the fatty acid chains and the glycerol backbone of the phospholipids near the membrane-water interface. Compared to the phospholipid head groups, the polar group of the cholesterol molecule is not highly exposed at the membrane surface. Therefore, it is not strange that at such relatively low concentrations, few cholesterol molecules should be available to interact with water-soluble molecules (e.g., cholesterol oxidase, cyclodextrins or CDCs) (Lange et al., 1980).

20.3.2.1 Cholesterol Availability in Membrane Bilayers

In multi-component membranes, the availability of cholesterol at the membrane surface is regulated by the interactions between cholesterol and other the components of the membrane (phospholipids, glycolipids and proteins). The more the cholesterol interacts with the othere membrane components, the less available it will be to interact with extra-membranous molecules. Factors that affect the interaction of cholesterol with phospholipids are the length of the acyl chains, the presence of double bonds in these chains, the size of the polar head-groups, and the ability of the phospholipid to form hydrogen bonds with the hydroxyl group of cholesterol (Ohvo-Rekilä et al., 2002).

When cholesterol is added to a membrane containing a single phospholipid species, two phases appear in a concentration-dependent manner (Mouritsen and Zuckermann, 2004; Sankaram and Thompson, 1991). This suggests that instead of randomly distributing among the membrane phospholipids, cholesterol associates with the phospholipids, presumably forming stoichiometric complexes (Radhakrishnan and Mcconnell, 1999). When the phospholipids are in excess, most of the cholesterol molecules form complexes with phospholipids. These complexes are immiscible in the pure phospholipid phase and therefore a two-phase mixture appears in the membrane. Increasing the cholesterol concentration will increase the population of the complexes until they form a single phase containing the complexes with a minor presence of uncomplexed phospholipids and cholesterol molecules. Beyond this point, the added cholesterol molecules (free cholesterol) will mix with the complexes until they reach the solubility limit and precipitate out of the membrane (Mason et al., 2003). Cholesterol molecules do not form stable single bilayers in aqueous solution, so when present in excess they cannot form a new stable and extended phase. The free cholesterol molecules in excess are likely to have a tendency to "fly" away from the membrane, and outside the membrane due to their low solubility they will be prone to associate to form multi-bilayer crystals in aqueous solution (Harris, 1988).

The formation of phospholipid-cholesterol complexes can explain the low interaction detected between cyclodextrins and cholesterol when the membrane sterol is present in low amounts (Mcconnell and Radhakrishnan, 2003). An alternative model to account for this behavior was proposed by Huang and Feigenson (1999). These authors propose that the hydrophobic effect positions the phospholipid head groups toward the membrane surface to protect the hydrophobic molecule of cholesterol from the unfavorable contact with water. When the concentration of cholesterol in the membrane achieves and exceeds the protective capacity of the head-groups, the tendency for the sterol molecules to exit the membrane will increase.

Both models provide a reasonable explanation for the increased accessibility of cholesterol at high sterol/phospholipid ratios, and the consensus is that they are not mutually exclusive (Lange and Steck, 2008; Mesmin and Maxfield, 2009). Binding (and/or pore-formation) of the CDCs occurs at high cholesterol concentration where free cholesterol becomes available, and therefore any of these models can be used to explain the experimental observations.

In more complex lipids mixtures, when more than one phospholipid is present in the membrane, the total cholesterol content will distribute unevenly between any formed phases (Goñi et al., 2008; Veatch and Keller, 2002). How much cholesterol is present in each phase will be governed by the interaction between cholesterol and the components (lipids and proteins) present in the phases (Epand, 2006).

20.3.2.2 The Role of Other Lipids

The pioneering work of Ohno-Iwashita and colleagues on the binding of PFO to membranes showed that the phospholipid composition affects the arrangement of cholesterol in the membrane (*see also* Chapter 22). Using a protease-nicked derivate of PFO they showed that the binding of the toxin was not only influenced by the total amount of cholesterol present in the membrane, but also by the phospholipid composition. They found that this PFO derivative preferentially binds to cholesterol-rich membranes composed of phospholipids with 18-carbon acyl chains (Ohno-Iwashita et al., 1992, 1991). An effect on cholesterol state in the membrane by ceramides and glycerolipids was also suggested by Zitzer et al. (2003), based on their studies of SLO pore-formation in liposomal membranes prepared with different phospholipids. Lipids having a conical molecular shape appear to effect a change in the energetic state of membrane cholesterol that in turn augments the interaction of the sterol with the cholesterol-specific cytolysin. Interestingly, these authors also showed that SLO was active when membranes were prepared solely with the enantiomeric cholesterol, suggesting that the effect associated with the presence of cholesterol may be other than a site specific binding event (Zitzer et al., 2003).

A more systematic analysis of the interaction of PFO D4 with membranes prepared with different phospholipds and sterols revealed that PFO binding to the bilayer and the initiation of the sequence of events that culminate in the formation of a transmembrane pore depend on the availability of free cholesterol at the membrane surface (Flanagan et al., 2002; Flanagan et al., 2009; Nelson et al., 2008). These studies also showed that changes in the acyl chain packing of the phospholipids and cholesterol in the membrane core do not correlate with PFO binding. Taken together, all these studies suggest than the binding of PFO (and SLO) to the

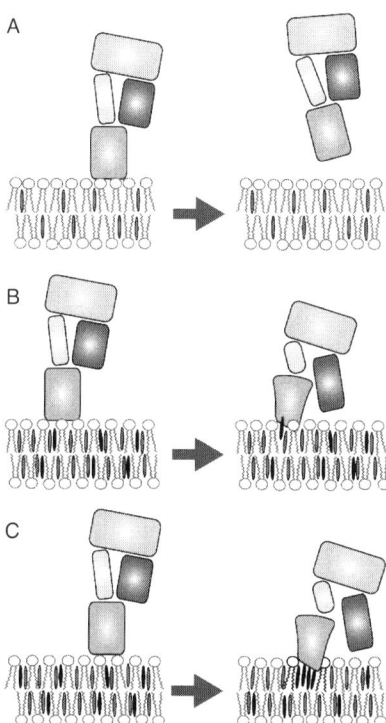

Fig. 20.5 PFO only binds to membranes containing free cholesterol molecules. Examples of mechanisms for cholesterol-dependent anchoring of PFO to the membrane surface: (**A**) PFO cannot stably bind to the bilayer if there are no free cholesterol molecules available in the membrane surface. (**B**) At high cholesterol concentrations free cholesterol molecules become available (*black ovals*), and D4 can anchor to the bilayer. In this example, a single cholesterol molecule binds to D4 and induces the conformational changes required to expose the D4 loops to the bilayer core. (**C**) Alternatively, the interplay between D4 and the membrane result in the redistribution of the lipids at the surface, clustering the free cholesterol molecules underneath the tip of D4. Anchoring may be accomplished by the interaction of multiple hydroxyl groups located in the cholesterol-rich cluster and the conserved amino acids of the loops

membrane is triggered when the concentration of cholesterol exceeds the association capacity of the phospholipids, and this cholesterol excess is then free to associate with the toxin (Fig. 20.5).

The requirement of such high cholesterol content in membranes was initially associated with the binding of PFO to cholesterol-rich domains (or membrane rafts) (Ohno-Iwashita et al., 2004; Waheed et al., 2001). However, recent results indicate that this assertion may require further analysis and consideration. It was found that the incorporation of sphingomyelin, a necessary component for the formation of membrane rafts, inhibited rather than promoted the binding of PFO to membranes (Flanagan et al., 2009). No correlation was found between PFO binding, and the amount of the detergent-resistant fraction in membranes, a fraction usually associated with membrane rafts (Flanagan et al., 2009). Incorporation of sterols that promote the formation of ordered membrane domains was not critical to promoting the PFO-membrane interaction (Nelson et al., 2008). Therefore, one needs to be cautious when employing PFO as a probe to reveal the presence of membrane rafts in cellular membranes. Rather than recognizing a particular membrane "raft", PFO seems to bind to membranes containing free cholesterol (or where cholesterol has a high chemical activity).

20.3.2.3 Cholesterol Alone Is Enough

It was long known that incubation of SLO (Duncan and Schlegel, 1975; Johnson et al., 1980), PFO (Mitsui et al., 1979), cereolysin (Cowell and Bernheimer, 1978), alveolysin (Johnson et al., 1980), PLY (Johnson et al., 1980), and LLO (Vazquez-Boland et al., 1989) with cholesterol dispersed in aqueous solution produced the typical aggregated sterol-toxin complexes. For PFO and SLO, typical ring- and arc-like structures were observed after incubation with cholesterol at concentrations above its solubility limit (i.e., higher than 5 μM Duncan and Schlegel, 1975, Haberland and Reynolds, 1973, Harris et al., 1998, Mitsui et al., 1979).

To clarify the role of cholesterol in PFO cytolysis, the extent to which the different steps of the cytolytic mechanism could be elicited solely by the presence of cholesterol was analyzed (Heuck et al., 2007). Using site-directed fluorescence labelling of PFO in combination with multiple independent fluorescence techniques (Heuck and Johnson, 2002; Johnson, 2005), it was revealed that a selective interaction between the undecapeptide and the D4 loops with cholesterol dispersed in aqueous solution is indistinguishable from the interaction of PFO with cholesterol-containing membranes. Binding solely to cholesterol aggregates in aqueous solution is sufficient to initiate the coupled conformational changes that extend throughout the toxin molecule from the tip of D4 to the TMHs. Moreover, it was found that the topology of D4 bound to cholesterol aggregates was identical to the one observed in liposomal membranes, and that the binding of PFO to cholesterol aggregates was sufficient to trigger the conformational change in domain 3 that has been associated with oligomerization (Heuck et al., 2007; Ramachandran et al., 2004). As previously observed for SLO in cholesterol micro-crystals (Harris et al., 1998), oligomerization and formation of typical arc and ring structures were observed in the presence of cholesterol microcrystals. Surprisingly, none of these changes were produced by epicholesterol, a sterol that differs from cholesterol only in that the hydroxyl group is directed axially instead of equatorial (Heuck et al., 2007).

Taking advantage of the inability of PFO to recognize epicholesterol, competition experiments were done to examine how cholesterol packing in the bilayer affects the interactions with the membrane. More than 48 mole % cholesterol is required for PFO to bind to POPC-cholesterol liposomes (Flanagan et al., 2009). However, when the epicholesterol was mixed with cholesterol to maintain the concentration of total sterols constant at 48 mole %, and to reduce the net amount of cholesterol in the membrane, it was shown that in this case considerable binding of PFO was found with as little as 19 mole % cholesterol. Epicholesterol apparently intercalates in the bilayer and competes with cholesterol for association with phospholipids, as reported for other membrane intercalating agents (Lange et al., 2005). These data therefore confirmed that there are at least two distinctive states of cholesterol in a typical membrane bilayer: one in which cholesterol is readily accessible for binding to proteins such as PFO (free cholesterol), and one in which the sterol is associated with surrounding membrane components that reduce its exposure to the surface (e.g., phospholipid headgroups may obscure access to sterols associated with phospholipid acyl chains).

The selective binding of PFO to cholesterol aggregates and not to epicholesterol aggregates, suggests that the failure to bind epicholesterol when incorporated in membrane bilayers is not related to the packing or association of this sterol with the phospholipids. This failure is rather caused by the inappropriate orientation of the hydroxyl group (Murari et al., 1986), which it may be required for the specific docking of the sterol molecule to a binding pocket located in D4 (Fig. 20.5B, Rossjohn et al., 2007). Alternatively, the hydroxyl group may need to be properly exposed at the surface of a lipid cluster, that may then act as a platform for the anchoring of the D4 loops (Fig. 20.5C). Such a cluster may be preformed on the membrane before binding, or formed as a result of the interaction of D4 with the bilayer surface. Redistribution of lipids after protein-binding has been observed for LLO (Gekara et al., 2005), and other proteins (e.g., Heimburg et al., 1999).

The PFO and SLO specific binding to cholesterol aggregates and microcrystals (Harris et al., 1998; Heuck et al., 2007), together with the need for more than 30 mole% cholesterol in membranes to trigger binding (Flanagan et al., 2009; Heuck et al., 2000; Nelson et al., 2008), suggest that the role of cholesterol in the cytolytic mechanism of the CDC may be more complex than solely binding to a specific binding site. An alternative explanation would be the need of a cluster of cholesterol molecules at the membrane surface to provide a docking platform for the D4 loops (Gekara et al., 2005, Heimburg et al., 1999, Heuck and Johnson, 2005). Interestingly, the binding of pore-forming toxins to lipid clusters have been reported for *Staphylococcus aureus* α-hemolysin (Valeva et al., 2006), and the need for small cholesterol clusters have been recently suggested for the binding of LLO to membranes (Bavdek et al., 2007). Further work is needed to unambiguously determine the mechanism by which cholesterol specifically anchors the CDC to the target membrane.

20.4 Conclusions and Future Perspectives

Recent studies support the concept that there is a complex interplay between the structural arrangement of the CDC D4 loops and the distribution of cholesterol in the target membrane (Bavdek et al., 2007; Flanagan et al., 2009; Giddings et al., 2003; Heuck and Johnson, 2005; Nelson et al., 2008; Polekhina et al., 2005; Ramachandran et al., 2002; Soltani et al., 2007a; Soltani et al., 2007b). Modifications in the lipid composition alter the cholesterol arrangement in the membrane, and as a consequence, the binding of the CDC (Flanagan et al., 2009; Nelson et al., 2008). At the same time, modifications to the structure of the CDC due to mutations, changes in the pH of the medium or other factors, apparently modifies the threshold for the amount of cholesterol required to trigger binding (Bavdek et al., 2007; Nelson et al., 2008; Moe & Heuck, unpublished).

The presence of free cholesterol molecules at the membrane surface seems to be critical to trigger the binding of most CDCs. A direct inference from these findings is that the exposure of cholesterol at the membrane surface may be facilitated by the action of other membrane-damaging toxins or enzymes secreted by these pathogens like, for example phospholipase C. Such toxins cleave the head-groups of phospholipids, and consequently increase the exposure of cholesterol molecules (or availability of free cholesterol) to the membrane surface. Cooperation between the CDC and different phospholipase C molecules contribute to the pathogenesis of at least two organisms. A synergic effect has been reported for the action of PFO and α-toxin in clostridial myonecrosis (Awad et al., 2001), and both phospholipase C and LLO have been identified as key factors for the vacuolar dissolution and cell-to-cell spreading mechanism of *Listeria monocytogenes* (Alberti-Segui et al., 2007).

Complete understanding of the mechanism of pore formation for the CDCs at the molecular level will require high-resolution structures of the initial (water-soluble monomer), the final (membrane-inserted pore/oligomer), and any intermediate pre-pore state involved in the cytolytic process (including complexes with receptors or lipids). Great progress has been achieved to this end, but there is much more to be accomplished. A few crystal structures for monomeric CDCs are currently available (PFO, ILY, ALO, Bourdeau et al., 2009 ; Polekhina et al., 2005; Rossjohn et al., 1997), and the low resolution structure for the pre-pore complex and the membrane-inserted oligomer of PLY have been obtained by cryo-electron microscopy (Tilley et al., 2005).

It has become clear that the analysis of complex biological systems, in particular those involving membranes, benefits from the combination of high-resolution structural techniques (e.g., X-ray crystallography, nuclear magnetic resonance and electron microscopy) and spectroscopic analysis of probes incorporated at specific positions in the proteins (e.g., electron paramagnetic resonance, fluorescence spectroscopy) (Cowieson et al., 2008; Heuck and Johnson, 2002; Hubbell et al., 2000). In addition to providing structural information, by monitoring the spectral signal of these probes as a function of time, one can determine the kinetics of the discrete steps of the pore-formation mechanism (Heuck et al., 2000, , 2003) and the dynamics of the structural transformations (Columbus and Hubbell, 2002).

Understanding the CDC function in the establishment of the diseases caused by various Gram-positive pathogens is far from complete (Marriott et al., 2008; Schnupf and Portnoy, 2007). The actual role of CDCs in bacterial pathogenesis may be more complex than merely forming a transmembrane pore. For example, it has been proposed that SLO is involved in protein translocation during *Streptococcus pyogenes* infection (Madden et al., 2001; Meehl and Caparon, 2004).

The involvement of protein receptors in the mechanism of certain CDC is another area that requires further investigation. The discovery of the ILY receptor illuminated two distinct roles for cholesterol in the cytolytic mechanism of this CDC (Giddings et al., 2003). ALO's strong preference for targeting the apical side of gut epithelial cells suggests that a receptor (other than cholesterol) may be present in these cells (Bourdeau et al., 2009). Clearly, there is much to be discovered concerning the complex and fascinating roles played by the CDC in bacterial pathogenesis.

Acknowledgments Work in the authors' laboratory was supported by a Scientist Development Grant from the American Heart Association to A.P.H

References

Abdel Ghani, E. M., Weis, S., Walev, I., Kehoe, M., Bhakdi, S. and Palmer, M., 1999, Streptolysin O: inhibition of the conformational change during membrane binding of the monomer prevents oligomerization and pore formation. *Biochemistry* **38:** 15204–15211.

Alberti-Segui, C., Goeden, K. R. and Higgins, D. E., 2007, Differential function of *Listeria monocytogenes* listeriolysin O and phospholipases C in vacuolar dissolution following cell-to-cell spread. *Cell. Microbiol.* **9:** 179–195.

Alouf, J. E., Billington, S. J. and Jost, B. H., 2006, Repertoire and general features of the family of cholesterol-dependent cytolysins. In Alouf, J. E. and Popoff, M. R. (Eds.) *The Comprehensive Sourcebook of Bacterial Protein Toxins.* 3rd ed., pp. 643–658, Oxford, England, Academic Press.

Alving, C. R., Habig, W. H., Urban, K. A. and Hardegree, M. C., 1979, Cholesterol-dependent tetanolysin damage to liposomes. *Biochim. Biophys. Acta* **551:** 224–228.

Arrhenius, S., 1907. *Immunochemistry. The application of the principles of physical chemistry to the study of the biological antibodies.* New York, The Macmillian Company.

Awad, M. M., Ellemor, D. M., Boyd, R. L., Emmins, J. J. and Rood, J. I., 2001, Synergistic effects of alpha-toxin and perfringolysin O in *Clostridium perfringens*-mediated gas gangrene. *Infect. Immun.* **69:** 7904–7910.

Bavdek, A., Gekara, N. O., Priselac, D., Gutierrez Aguirre, I., Darji, A., Chakraborty, T., Macìœek, P., Lakey, J. H., Weiss, S. and Anderluh, G., 2007, Sterol and pH interdependence in the binding, oligomerization, and pore formation of listeriolysin O. *Biochemistry* **46:** 4425–4437.

Billington, S. J., Songer, J. G. and Jost, B. H., 2001, Molecular characterization of the pore-forming toxin, pyolysin, a major virulence determinant of *Arcanobacterium pyogenes*. *Vet. Microbiol.* **82:** 261–274.

Bourdeau, R. W., Malito, E., Chenal, A., Bishop, B. L., Musch, M. W., Villereal, M. L., Chang, E. B., Mosser, E. M., Rest, R. F. and Tang, W.-J., 2009, Cellular functions and x-ray structure of anthrolysin O, a cholesterol-dependent cytolysin secreted by *Bacillus anthracis*. *J. Biol. Chem.* **284:** 14645–14656.

Campanella, J., Bitincka, L. and Smalley, J., 2003, MatGAT: An application that generates similarity/identity matrices using protein or DNA sequences. *BMC Bioinf.* **4:** 29.

Cho, W. and Stahelin, R. V., 2005, Membrane-protein interactions in cell signaling and membrane trafficking. *Annu. Rev. Biophys. Biomol. Struct.* **34:** 119–151.

Columbus, L. and Hubbell, W. L., 2002, A new spin on protein dynamics. *Trends Biochem. Sci.* **27:** 288–295.

Cowell, J. L. and Bernheimer, A. W., 1978, Role of cholesterol in the action of cereolysin on membranes. *Arch. Biochem. Biophys.* **190:** 603–610.

Cowieson, N. P., Kobe, B. and Martin, J. L., 2008, United we stand: combining structural methods. *Curr. Opin. Struct. Biol.* **18:** 617–622.

Czajkowsky, D. M., Hotze, E. M., Shao, Z. and Tweten, R. K., 2004, Vertical collapse of a cytolysin prepore moves its transmembrane beta-hairpins to the membrane. *EMBO J.* **23:** 3206–3215.

Dang, T. X., Hotze, E. M., Rouiller, I., Tweten, R. K. and Wilson-Kubalek, E. M., 2005, Prepore to pore transition of a cholesterol-dependent cytolysin visualized by electron microscopy. *J. Struct. Biol.* **150:** 100–108.

Duncan, J. L. and Schlegel, R., 1975, Effect of streptolysin O on erythrocyte membranes, liposomes, and lipid dispersions. A protein-cholesterol interaction. *J. Cell Biol.* **67:** 160–174.

Epand, R. M., 2006, Cholesterol and the interaction of proteins with membrane domains. *Prog. Lipid Res.* **45:** 279–294.

Farrand, S., Hotze, E., Friese, P., Hollingshead, S. K., Smith, D. F., Cummings, R. D., Dale, G. L. and Tweten, R. K., 2008, Characterization of a streptococcal cholesterol-dependent cytolysin with a Lewis y and b Specific Lectin Domain. *Biochemistry* **47:** 7097–7107.

Flanagan, J. J., Heuck, A. P. and Johnson, A. E. (2002) Cholesterol-phospholipid interactions play an important role in perfringolysin O binding to membrane. *FASEB J.,* **16,** A929.

Flanagan, J. J., Tweten, R. K., Johnson, A. E. and Heuck, A. P., 2009, Cholesterol exposure at the membrane surface is necessary and sufficient to trigger perfringolysin O binding. *Biochemistry* **48:** 3977–3987.

Gekara, N. O., Jacobs, T., Chakraborty, T. and Weiss, S., 2005, The cholesterol-dependent cytolysin listeriolysin O aggregates rafts via oligomerization. *Cell Microbiol.* **7:** 1345–1356.

Gelber, S. E., Aguilar, J. L., Lewis, K. L. T. and Ratner, A. J., 2008, Functional and phylogenetic characterization of vaginolysin, the human-specific cytolysin from *Gardnerella vaginalis. J. Bacteriol.* **190:** 3896–3903.

Geoffroy, C. and Alouf, J. E., 1983, Selective purification by thiol-disulfide interchange chromatography of alveolysin, a sulfhydryl-activated toxin of *Bacillus alvei*. Toxin properties and interaction with cholesterol and liposomes. *J. Biol. Chem.* **258:** 9968–9972.

Giddings, K. S., Johnson, A. E. and Tweten, R. K., 2003, Redefining cholesterol's role in the mechanism of the cholesterol-dependent cytolysins. *Proc. Natl. Acad. Sci. USA* **100:** 11315–11320.

Giddings, K. S., Johnson, A. E. and Tweten, R. K., 2006, Perfringolysin O and Intermedilysin: Mechanisms of Pore Formation by the Cholesterol-Dependent Cytolysins. *In* Alouf, J. E. and Popoff, M. R. (Eds.) *The Comprehensive Sourcebook of Bacterial Protein Toxins.* 3rd ed., pp. 671–679, Oxford, England, Academic Press.

Giddings, K. S., Zhao, J., Sims, P. J. and Tweten, R. K., 2004, Human CD59 is a receptor for the cholesterol-dependent cytolysin intermedilysin. *Nat. Struct. Mol. Biol.* **11:** 1173–1178.

Gilbert, R. J., 2005, Inactivation and activity of cholesterol-dependent cytolysins: what structural studies tell us. *Structure (Camb.)* **13:** 1097–1106.

Gilbert, R. J. C., Rossjohn, J., Parker, M. W., Tweten, R. K., Morgan, P. J., Mitchell, T. J., Errington, N., Rowe, A. J., Andrew, P. W. and Byron, O., 1998, Self-interaction of pneumolysin, the pore-forming protein toxin of Streptococcus pneumoniae. *J. Mol. Biol.* **284:** 1223–1237.

Goñi, F. M., Alonso, A., Bagatolli, L. A., Brown, R. E., Marsh, D., Prieto, M. and Thewalt, J. L., 2008, Phase diagrams of lipid mixtures relevant to the study of membrane rafts. *Biochim. Biophys. Acta, Mol. Cell. Biol. Lipids* **1781:** 665–684.

Haberland, M. E. and Reynolds, J. A., 1973, Self-association of cholesterol in aqueous solution. *Proc. Natl. Acad. Sci. USA* **70:** 2313–2316.

Hadders, M. A., Beringer, D. X. and Gros, P., 2007, Structure of C8 α-MACPF reveals mechanism of membrane attack in complement immune defense. *Science* **317:** 1552–1554.

Harris, J. R., 1988, Electron microscopy of cholesterol. *Micron Microsc. Acta* **19,** 19–31.

Harris, J. R., Adrian, M., Bhakdi, S. and Palmer, M., 1998, Cholesterol-streptolysin O interaction: An EM study of wild-type and mutant streptolysin O. *J. Struct. Biol.* **121:** 343–355.

Harwood, C. R. and Cranenburgh, R., 2008, Bacillus protein secretion: an unfolding story. *Trends Microbiol.,* **16,** 73–79.

Heimburg, T., Angerstein, B. and Marsh, D., 1999, Binding of peripheral proteins to mixed lipid membranes: Effect of lipid demixing upon binding. *Biophys. J.* **76:** 2575–2586.

Heuck, A. P., Hotze, E. M., Tweten, R. K. and Johnson, A. E., 2000, Mechanism of membrane insertion of a multimeric β-barrel protein: perfringolysin O creates a pore using ordered and coupled conformational changes. *Mol. Cell* **6:** 1233–1242.

Heuck, A. P. and Johnson, A. E., 2002, Pore-forming protein structure analysis in membranes using multiple independent fluorescence techniques. *Cell Biochem. Biophys.* **36:** 89–101.

Heuck, A. P. and Johnson, A. E., 2005, Membrane recognition and pore formation by bacterial pore-forming toxins. *In* Tamm, L. K. (Ed.) *Protein-Lipid Interactions. From Membrane Domains to Cellular Networks*, pp. 165–188, Weinheim, Wiley-VCH.

Heuck, A. P., Savva, C. G., Holzenburg, A. and Johnson, A. E., 2007, Conformational changes that effect oligomerization and initiate pore formation are triggered throughout perfringolysin O upon binding to cholesterol. *J. Biol. Chem.* **282:** 22629–22637.

Heuck, A. P., Tweten, R. K. and Johnson, A. E., 2001, beta-Barrel pore-forming toxins: intriguing dimorphic proteins. *Biochemistry* **40:** 9065–9073.

Heuck, A. P., Tweten, R. K. and Johnson, A. E., 2003, Assembly and topography of the prepore complex in cholesterol-dependent cytolysins. *J. Biol. Chem.* **278:** 31218–31225.

Hotze, E. M., Heuck, A. P., Czajkowsky, D. M., Shao, Z., Johnson, A. E. and Tweten, R. K., 2002, Monomer-monomer interactions drive the prepore to pore conversion of a beta -barrel-forming cholesterol-dependent cytolysin. *J. Biol. Chem.* **277:** 11597–11605.

Huang, J. and Feigenson, G. W., 1999, A microscopic interaction model of maximum solubility of cholesterol in lipid bilayers. *Biophys. J.* **76:** 2142–2157.

Hubbell, W. L., Cafiso, D. S. and Altenbach, C., 2000, Identifying conformational changes with site-directed spin labeling. *Nat. Struct. Mol. Biol.* **7:** 735–739.

Jefferies, J., Nieminen, L., Kirkham, L.-A., Johnston, C., Smith, A. and Mitchell, T. J., 2007, Identification of a secreted cholesterol-dependent cytolysin (Mitilysin) from *Streptococcus mitis*. *J. Bacteriol.* **189:** 627–632.

Johnson, A. E., 2005, Fluorescence approaches for determining protein conformations, interactions and mechanisms at membranes. *Traffic* **6:** 1078–1092.

Johnson, M. K., Geoffroy, C. & Alouf, J. E. (1980) Binding of cholesterol by sulfhydryl-activated cytolysins. *Infect. Immun.,* **27,** 97–101.

Kleinschmidt, J. H. (2006) Folding kinetics of the outer membrane proteins OmpA and FomA into phospholipid bilayers. *Chem. Phys. Lipids,* **141,** 30–47.

Lachapelle, S., Tweten, R. K. and Hotze, E. M., 2009, Intermedilysin-receptor interactions during assembly of the pore complex: assembly intermediates increase host cell susceptibility to complement-mediated lysis. *J. Biol. Chem.* **284:** 12719–12726.

Lange, Y., Cutler, H. B. and Steck, T. L., 1980, The effect of cholesterol and other intercalated amphipaths on the contour and stability of the isolated red cell membrane. *J. Biol. Chem.* **255:** 9331–9337.

Lange, Y. and Steck, T. L., 2008, Cholesterol homeostasis and the escape tendency (activity) of plasma membrane cholesterol. *Prog. Lipid Res.* **47:** 319–332.

Lange, Y., Ye, J. and Steck, T. L., 2005, Activation of membrane cholesterol by displacement from phospholipids. *J. Biol. Chem.* **280:** 36126–36131.

Lecuyer, H. and Dervichian, D. G., 1969, Structure of aqueous mixtures of lecithin and cholesterol. *J. Mol. Biol.* **45:** 39–57.

Madden, J. C., Ruiz, N. and Caparon, M., 2001, Cytolysin-mediated translocation (CMT): a functional equivalent of type III secretion in gram-positive bacteria. *Cell* **104:** 143–152.

Marriott, H. M., Mitchell, T. J. and Dockrell, D. H., 2008, Pneumolysin: a double-edged sword during the host-pathogen interaction. *Curr. Mol. Med.* **8:** 497–509.

Mason, P. R., Tulenko, T. N. and Jacob, R. F., 2003, Direct evidence for cholesterol crystalline domains in biological membranes: role in human pathobiology. *Biochim. Biophys. Acta* **1610:** 198–207.

Mcconnell, H. M. and Radhakrishnan, A., 2003, Condensed complexes of cholesterol and phospholipids. *Biochim. Biophys. Acta* **1610:** 159–73.

Meehl, M. A. and Caparon, M. G., 2004, Specificity of streptolysin O in cytolysin-mediated translocation. *Mol. Microbiol.* **52:** 1665–1676.

Mesmin, B. and Maxfield, F. R., 2009, Intracellular sterol dynamics. *Biochim. Biophys. Acta, Mol. Cell. Biol. Lipids* **1791:** 636–645.

Michel, E., Reich, K. A., Favier, R., Berche, P. and Cossart, P., 1990, Attenuated mutants of the intracellular bacterium *Listeria monocytogenes* obtained by single amino acid substitutions in listeriolysin O. *Mol. Microbiol.* **4:** 2167–2178.

Miller, C. J., Elliott, J. L. and Collier, R. J., 1999, Anthrax protective antigen: prepore-to-pore conversion. *Biochemistry* **38:** 10432–10441.

Mitsui, K., Sekiya, T., Okamura, S., Nozawa, Y. and Hase, J., 1979, Ring formation of perfringolysin O as revealed by negative stain electron microscopy. *Biochim. Biophys. Acta* **558:** 307–313.

Mosser, E. and Rest, R., 2006, The *Bacillus anthracis* cholesterol-dependent cytolysin, Anthrolysin O, kills human neutrophils, monocytes and macrophages. *BMC Microbiol.* **6:** 56.

Mouritsen, O. G. and Zuckermann, M. J., 2004, What's so special about cholesterol? *Lipids* **39:** 1101–1113.

Murari, R., Murari, M. P. and Baumann, W. J., 1986, Sterol orientations in phosphatidylcholine liposomes as determined by deuterium NMR. *Biochemistry* **25:** 1062–1067.

Nagamune, H., Ohkura, K., Sukeno, A., Cowan, G., Mitchell, T. J., Ito, W., Ohnishi, O., Hattori, K., Yamato, M., Hirota, K., Miyake, Y., Maeda, T. and Kourai, H., 2004, The human-specific action of intermedilysin, a homolog of streptolysin O, is dictated by domain 4 of the protein. *Mol. Microbiol.* **48:** 677–692.

Nagamune, H., Whiley, R. A., Goto, T., Inai, Y., Maeda, T., Hardie, J. M. and Kourai, H., 2000, Distribution of the intermedilysin gene among the anginosus group streptococci and correlation between intermedilysin production and deep-seated infection with *Streptococcus intermedius*. *J. Clin. Microbiol.* **38:** 220–226.

Nakamura, M., Sekino, N., Iwamoto, M. and Ohno-Iwashita, Y., 1995, Interaction of .theta.-toxin (perfringolysin O), a cholesterol-binding cytolysin, with liposomal membranes: change in the aromatic side chains upon binding and insertion. *Biochemistry* **34:** 6513–6520.

Nakamura, M., Sekino-Suzuki, N., Mitsui, K.-I. and Ohno-Iwashita, Y., 1998, Contribution of tryptophan residues to the structural changes in perfringolysin O during interaction with liposomal membranes. *J. Biochem.* **123:** 1145–1155.

Nelson, L. D., Johnson, A. E. and London, E., 2008, How interaction of perfringolysin O with membranes is controlled by sterol structure, lipid structure, and physiological low pH: insights into the origin of perfringolysin O-lipid raft interaction *J. Biol. Chem.* **283:** 4632–4642.

Nollmann, M., Gilbert, R., Mitchell, T., Sferrazza, M. and Byron, O., 2004, The role of cholesterol in the activity of pneumolysin, a bacterial protein toxin. *Biophys. J.* **86:** 3141–3151.

Ohno-Iwashita, Y., Iwamoto, M., Ando, S. and Iwashita, S., 1992, Effect of lipidic factors on membrane cholesterol topology - mode of binding of θ-toxin to cholesterol in liposomes. *Biochimica et Biophysica Acta* **1109:** 81–90.

Ohno-Iwashita, Y., Iwamoto, M., Mitsui, K.-I., Ando, S. and Iwashita, S., 1991, A cytolysin, θ-toxin, preferentially binds to membrane cholesterol surrounded by phospholipids with 18-carbon hydrocarbon chains in cholesterol-rich region. *J. Biochem.* **110:** 369–375.

Ohno-Iwashita, Y., Shimada, Y., Waheed, A., Hayashi, M., Inomata, M., Nakamura, M., Maruya, M. and Iwashita, M., 2004, Perfringolysin O, a cholesterol-binding cytolysin, as a probe for lipid rafts. *Anaerobe* **10:** 125–134.

Ohvo-Rekilä, H., Ramstedt, B., Leppimäki, P. and Peter Slotte, J., 2002, Cholesterol interactions with phospholipids in membranes. *Prog. Lipid Res.* **41:** 66–97.

Olofsson, A., Hebert, H. and Thelestam, M., 1993, The projection structure of Perfringolysin O (*Clostridium perfringens* θ-toxin). *FEBS Lett.* **319:** 125–127.

Palmer, M., 2004, Cholesterol and the activity of bacterial toxins. *FEMS Microbiol. Lett.* **238:** 281–289.

Palmer, M., Harris, R., Freytag, C., Kehoe, M., Tranum-Jensen, J. and Bhakdi, S., 1998, Assembly mechanism of the oligomeric streptolysin O pore: the early membrane lesion is lined by a free edge of the lipid membrane and is extended gradually during oligomerization. *EMBO J.* **17:** 1598–1605.

Pinkney, M., Beachey, E. and Kehoe, M., 1989, The thiol-activated toxin streptolysin O does not require a thiol group for cytolytic activity. *Infect. Immun.* **57:** 2553–2558.

Polekhina, G., Feil, S. C., Tang, J., Rossjohn, J., Giddings, K. S., Tweten, R. K. and Parker, M. W., 2006, Comparative three-dimensional structure of cholesterol-dependent cytolysins. *In* Alouf, J. E. and Popoff, M. R. (Eds.) *The Comprehensive Sourcebook of Bacterial Protein Toxins.* Third ed., pp. 659–670, Oxford, England, Academic Press.

Polekhina, G., Giddings, K. S., Tweten, R. K. and Parker, M. W., 2005, Insights into the action of the superfamily of cholesterol-dependent cytolysins from studies of intermedilysin. *Proc. Natl. Acad. Sci. USA* **102:** 600–605.

Radhakrishnan, A. and Mcconnell, H. M., 1999, Condensed complexes of cholesterol and phospholipids. *Biophys. J.* **77:** 1507–1517.

Ramachandran, R., Heuck, A. P., Tweten, R. K. and Johnson, A. E., 2002, Structural insights into the membrane-anchoring mechanism of a cholesterol-dependent cytolysin. *Nat. Struct. Mol. Biol.* **9:** 823–827.

Ramachandran, R., Tweten, R. K. and Johnson, A. E., 2004, Membrane-dependent conformational changes initiate cholesterol-dependent cytolysin oligomerization and intersubunit beta-strand alignment. *Nat. Struct. Mol. Biol.* **11:** 697–705.

Ramachandran, R., Tweten, R. K. and Johnson, A. E., 2005, The domains of a cholesterol-dependent cytolysin undergo a major FRET-detected rearrangement during pore formation. *Proc. Natl. Acad. Sci. USA* **102:** 7139–7144.

Rosado, C. J., Buckle, A. M., Law, R. H. P., Butcher, R. E., Kan, W.-T., Bird, C. H., Ung, K., Browne, K. A., Baran, K., Bashtannyk-Puhalovich, T. A., Faux, N. G., Wong, W., Porter, C. J., Pike, R. N., Ellisdon, A. M., Pearce, M. C., Bottomley, S. P., Emsley, J., Smith, A. I., Rossjohn, J., Hartland, E. L., Voskoboinik, I., Trapani, J. A., Bird, P. I., Dunstone, M. A. and Whisstock, J. C., 2007, A common fold mediates vertebrate defense and bacterial attack. *Science* **317:** 1548–1551.

Rosenqvist, E., Michaelsen, T. E. and Vistnes, A. I., 1980, Effect of streptolysin O and digitonin on egg lecithin/cholesterol vesicles. *Biochim. Biophys. Acta* **600:** 91–102.

Rossjohn, J., Feil, S. C., Mckinstry, W. J., Tweten, R. K. and Parker, M. W., 1997, Structure of a cholesterol-binding, thiol-activated cytolysin and a model of its membrane form. *Cell* **89:** 685–692.

Rossjohn, J., Polekhina, G., Feil, S. C., Morton, C. J., Tweten, R. K. and Parker, M. W., 2007, Structures of perfringolysin O suggest a pathway for activation of cholesterol-dependent cytolysins. *J. Mol. Biol.* **367:** 1227–1236.

Sankaram, M. B. and Thompson, T. E., 1991, Cholesterol-induced fluid-phase immiscibility in membranes. *Proc. Natl. Acad. Sci. USA* **88:** 8686–8690.

Saunders, F. K., Mitchell, T. J., Walker, J. A., Andrew, P. W. and Boulnois, G. J., 1989, Pneumolysin, the thiol-activated toxin of *Streptococcus pneumoniae*, does not require a thiol group for in vitro activity. *Infect. Immun.* **57:** 2547–2552.

Schnupf, P. and Portnoy, D. A., 2007, Listeriolysin O: a phagosome-specific lysin. *Microbes Infect.* **9:** 1176–1187.

Sekino-Suzuki, N., Nakamura, M., Mitsui, K.-I. and Ohno-Iwashita, Y., 1996, Contribution of individual tryptophan residues to the structure and activity of θ-toxin (perfringolysin O), a cholesterol-binding cytolysin. *Eur. J. Biochem.* **241:** 941–947.

Shatursky, O., Heuck, A. P., Shepard, L. A., Rossjohn, J., Parker, M. W., Johnson, A. E. and Tweten, R. K., 1999, The mechanism of membrane insertion for a cholesterol-dependent cytolysin: A novel paradigm for pore-forming toxins. *Cell* **99:** 293–299.

Shepard, L. A., Heuck, A. P., Hamman, B. D., Rossjohn, J., Parker, M. W., Ryan, K. R., Johnson, A. E. and Tweten, R. K., 1998, Identification of a membrane-spanning domain of the thiol-activated pore-forming toxin *Clostridium perfringens* perfringolysin O: an alpha-helical to beta-sheet transition identified by fluorescence spectroscopy. *Biochemistry* **37:** 14563–14574.

Shepard, L. A., Shatursky, O., Johnson, A. E. and Tweten, R. K., 2000, The mechanism of pore assembly for a cholesterol-dependent cytolysin: formation of a large prepore complex precedes the insertion of the transmembrane beta-hairpins. *Biochemistry* **39:** 10284–10293.

Simossis, V. A., Kleinjung, J. and Heringa, J., 2005, Homology-extended sequence alignment. *Nucl. Acids Res.* **33:** 816–824.

Solovyova, A. S., Nollmann, M., Mitchell, T. J. and Byron, O., 2004, The solution structure and oligomerization behavior of two bacterial toxins: pneumolysin and perfringolysin O. *Biophys. J.* **87:** 540–552.

Soltani, C. E., Hotze, E. M., Johnson, A. E. and Tweten, R. K., 2007a, Specific protein-membrane contacts are required for prepore and pore assembly by a cholesterol-dependent cytolysin. *J. Biol. Chem.* **282:** 15709–15716.

Soltani, C. E., Hotze, E. M., Johnson, A. E. and Tweten, R. K., 2007b, Structural elements of the cholesterol-dependent cytolysins that are responsible for their cholesterol-sensitive membrane interactions. *Proc. Natl. Acad. Sci. USA* **104:** 20226–20231.

Tamm, L. K., Hong, H. and Liang, B., 2004, Folding and assembly of beta-barrel membrane proteins. *Biochim. Biophys. Acta* **1666:** 250–263.

Tilley, S. J., Orlova, E. V., Gilbert, R. J., Andrew, P. W. and Saibil, H. R. (2005) Structural basis of pore formation by the bacterial toxin pneumolysin. *Cell* **121:** 247–256.

Tweten, R. K., 2005, Cholesterol-dependent cytolysins, a family of versatile pore-forming toxins. *Infect. Immun.* **73:** 6199–6209.

Tweten, R. K., Parker, M. W. and Johnson, A. E., 2001, The cholesterol-dependent cytolysins. *Curr. Top. Microbiol. Immunol.* **257:** 15–33.

Valeva, A., Hellmann, N., Walev, I., Strand, D., Plate, M., Boukhallouk, F., Brack, A., Hanada, K., Decker, H. and Bhakdi, S., 2006, Evidence that clustered phosphocholine head groups serve as sites for binding and assembly of an oligomeric protein pore. *J. Biol. Chem.* **281:** 26014–26021.

Vazquez-Boland, J. A., Dominguez, L., Rodriguez-Ferri, E. F., Fernandez-Garayzabal, J. F. and Suarez, G., 1989, Preliminary evidence that different domains are involved in cytolytic activity and receptor (cholesterol) binding in listeriolysin O, the *Listeria monocytogenes* thiol-activated toxin. *FEMS Microbiol. Lett.* **53:** 95–99.

Veatch, S. L. and Keller, S. L., 2002, Organization in lipid membranes containing cholesterol. *Phys. Rev. Lett.* **89:** 268101.

Waheed, A., Shimada, Y., Heijnen, H. F. G., Nakamura, M., Inomata, M., Hayashi, M., Iwashita, S., Slot, J. W. and Ohno-Iwashita, Y., 2001, Selective binding of perfringolysin O derivative to cholesterol-rich membrane microdomains (rafts). *Proc. Natl. Acad. Sci. USA* **98:** 4926–4931.

Walker, B., Krishnasastry, M., Zorn, L. and Bayley, H., 1992, Assembly of the oligomeric membrane pore formed by Staphylococcal alpha-hemolysin examined by truncation mutagenesis. *J. Biol. Chem.* **267**: 21782–21786.

White, S. H. and Wimley, W. C., 1999, Membrane protein folding and stability: physical principles. *Annu. Rev. Biophys. Biomol. Struct.* **28**: 319–365.

Zitzer, A., Westover, E. J., Covey, D. F. and Palmer, M., 2003, Differential interaction of the two cholesterol-dependent, membrane-damaging toxins, streptolysin O and Vibrio cholerae cytolysin, with enantiomeric cholesterol. *FEBS Lett.* **553**: 229–231.

Chapter 21
Cholesterol Specificity of Some Heptameric β-Barrel Pore-Forming Bacterial Toxins: Structural and Functional Aspects

J. Robin Harris and Michael Palmer

Abstract Apart from the thiol-specific/cholesterol-dependent cytolysin family of toxins (*see* Chapter 20) there are a number of other unrelated bacterial toxins that also have an affinity for plasma membrane cholesterol. Emphasis is given here on the *Vibrio cholerae* cytolysin (VCC) and the cytolysins from related *Vibrio* species. The inhibition of the cytolytic activity of these toxins by prior incubation with extracellular cholesterol or low density lipoprotein emerges as a unifying feature, as does plasma membrane cholesterol depletion. Incubation of VCC with cholesterol produces a heptameric oligomer, which is not equivalent to the pre-pore since it is unable to penetrate the plasma membrane. In structural terms, the precise sequence of VCC monomer binding to membrane, oligomer formation and pore insertion through the bilayer has yet to be fully defined. Several other bacterial toxins have a dependency for cholesterol, although the available data is limited in most cases.

Keywords *Vibrio cholerae* cytolysin · VCC · Hemolysin · Cholesterol · β-Barrel · Pore-forming toxin

21.1 Introduction

Although much has been discovered concerning the cholesterol-dependency for oligomerization and pore formation by the streptolysin/perfringolysin/pneumolysin family of gram-positive bacterial thiol-specific/cholesterol-depedent cytolysins (CDCs) (*see* Chapter 20), cholesterol is also required for the oligomerization and membrane pore-formation for several other *non-thiol-specific* β-barrel-forming bacterial cytolysins. In this Chapter, the properties of some of these cytolysins will be presented, with some emphasis upon the most extensively studied, the *Vibrio cholerae* cytolysin (VCC), and other members of the *Vibrio* family. In all cases, it is the interaction of the cytolysin monomer cholesterol-binding domain that is

---

J.R. Harris (✉)
Institute of Zoology, University of Mainz, Mainz, D-55099, Germany
e-mail: rharris@uni-mainz.de

of significance. However, the binding motif of VCC with membrane cholesterol (and/or phospholipids) that enables the necessary attachment to occur, leading to a conformational change and increased hydrophobicity of the membrane-penetrating polypeptide sequence remains unknown. Protein conformational change leads to oligomer formation and, similar to the CDCs, bilayer penetration by a different domain to that responsible for the initial attachment, then creating the transmembrane ion channel/pore. The preformed VCC oligomer, in solution or attached to LDL, is unable to penetrate the plasma membrane of erythrocytes or other cells, exemplified for numerous *non-thiol-specific* (and the thiol-specific CDCs), by the inhibition of cytolytic activity following the prior addition of extracellular cholesterol or low density lipoprotein (LDL). The stability of the toxin oligomer and pore in SDS is a characteristic feature of these proteins, clearly related to the inherent stability of the β-barrel pore structure. Detailed comparisons of the properties of the two main classes of CDCs have been presented (Palmer, 2004; Zitzer et al., 2001, 2003). The significance of the partial protein unfolding/conformational change that occurs when bacterial toxins interact with lipids to form a membrane-penetrating β-barrel-containing pore is now widely accepted as a fundamental structural feature of these proteins (Galdiero et al., 2007; Geny and Popoff, 2006).

21.2 *Vibrio cholerae* Cytolysin (VCC)

Vibrio cholerae is the pathogenic gram-negative bacterium responsible for the water-borne gastrointestinal disease, cholera. The causative endotoxin responsible for cholera from the O1/O139 *V. cholerae* strains is termed the cholera enterotoxin (CT). Other *V. cholerae* strains, in particular the non-O1/non-O139 strains and the *V. cholerae* O1 biotype *eltor* strain do not produce CT, but release a cytolysin/hemolysin (Honda and Finkelstein, 1979; Zitzer et al., 1993), currently generally termed the *V. cholerae* cytolysin (VCC), although the term El Tor hemolysin is still used by some authors (Shinoda and Miyoshi, 2006).

VCC is a water-soluble protein with a molecular mass of $\sim$ 63 kDa, encoded by the *hlyA* gene of *V. cholerae*. A precursor protein of 82 kDa is cleaved proteolytically to release the cytolytically active N-terminal 63 kDa fragment (Yamamoto et al., 1990; Nagamune et al., 1996). More specifically, this VCC precursor has been shown to be cleaved and activated by a cellular metalloproteinase (Valeva et al., 2004). That VCC/El Tor hemolysin produces an ion channel in plasma membranes and planar bilayer membranes containing cholesterol was established long before the oligomerization and structure of the membrane-bound cytolysin pore was determined (Ikigai et al., 1996, 1997; Yuldasheva et al., 2001). Nevertheless, interest in the conductance and anion selectivity profile of the VCC pore continues, in relation to the cholesterol requirement for pore formation and the molecular structure of the pore (Pantano and Montecucco, 2006; Krasilnikov et al., 2007). In addition, the role of this VCC anion channel during the induction of cellular vacuolation and apoptosis remains of prime interest (Zitzer et al., 1997a; Moshioni et al., 2002; Saka et al.,

2008), as do the mechanisms the body may have established to protect itself against attack by VCC (Gutierrez et al., 2007; Valeva et al., 2008).

As mentioned above, and in accord with the characteristic property of toxins of the CDC family, addition of cholesterol prior to interaction with living cells blocks the cytolytic activity of VCC. This indicates that the initial monomer attachment, oligomerization, pre-pore and pore formation must occur on and in the surface of cholesterol-containing membranes, and that pre-formed cholesterol-induced VCC oligomers, even if free in solution, are unable to penetrate cell membranes.

21.2.1 Structure of the VCC Oligomer

Although initial biochemical experiments relating to VCC pore formation did not provide an absolute value for the number of subunits within the VCC (a pentamer was initially suggested), they showed the temperature- and cholesterol-dependence of pore formation (Zitzer et al., 1995, 1997b, 2000). Importantly, these studies also showed the value of transmission electron microscopy for the assessment of the VCC pore attachment to membranes and liposomes (Figs. 21.1 and 21.2), and the solubilization of the membrane-bound pore by the addition of the surfactant deoxycholate. Subsequently, further emphasis was placed upon the role of cholesterol interaction with the VCC monomer (Zitzer et al., 2001; Ikigai et al., 2006) and from fluorimetric analysis definition of the polypeptide domain of the monomer (an extended anti-parallel β-strand hairpin), that collectively upon oligomerization form the membrane-penetrating β-barrel of the VCC pore (Valeva et al., 2005). That the transformation of the VCC monomer into the SDS-stable pore, apparently

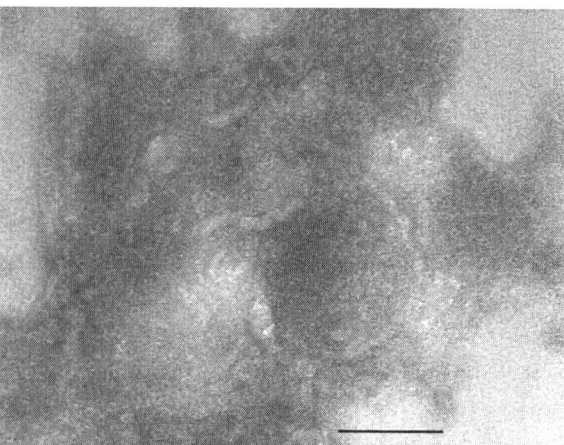

Fig. 21.1 Sheep erythrocyte membrane fragments treated with VCC monomer for 30 min at room temperature (∼ 22°C), negatively stained with 2% uranyl acetate (pH 4.5). Note the dense coating of the membrane surface with VCC pores. The scale bar indicates 100 nm

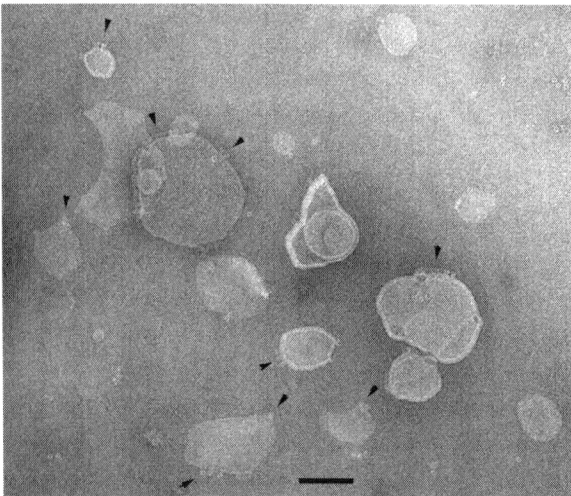

Fig. 21.2 Liposomes containing sphingomyelin-phosphatidyl choline-cholesterol following interaction with VCC monomer. VCC pores in side-on profile can be seen bound to the liposome edges (*arrows*) and on the flattened liposome surfaces. Negatively stained with 5% ammonium molybdate, 1.0% trehalose, pH 6.9. The scale bar indicates 100 nm

induced spontaneously by a predominantly hydrophobic interaction with membrane cholesterol, is due to a localized unfolding of the protein has been supported by the biochemical study of Chattopadhyay and Banerjee (2003). Interestingly, these workers utilized urea to produce controlled partial VCC (HlyA) unfolding; in 1.75 M urea they induced oligomer formation and at higher urea concentration (8 M) the more complete protein subunit unfolding could be reversed, showing that renaturation then also yielded the VCC oligomer.

In a comparative TEM study Harris et al. (2002) assessed the interaction of VCC monomer with microcrystalline cholesterol, some cholesterol esters and cholesterol derivatives. Negative stain TEM data was obtained using the holey-carbon spreading technique, whereby cholesterol-VCC samples were spread and supported in a thin film of negative stain alone (Harris and Scheffler, 2002). In a time-dependent manner, VCC oligomers were initially detected at the edges of cholesterol microcrystals, where the stacked cholesterol bilayers will expose a more hydrophobic sterol ring surface than the planar surface where the 3-β-OH groups will be exposed. Nevertheless, with increasing time the planar cholesterol surface also becomes coated with VCC oligomers (see Fig. 21.3) and oligomers tend to be released into solution. Image processing of individual VCC oligomers, which have a strong tendency to be orientated as rings at the fluid/air interface, yielded strong evidence for the presence of a seven-fold rotational symmetry (Fig. 21.4). 19-Hydroxycholestrerol, 7β-hydroxycholesterol, cholesteryl acetate and β-estradiol all efficiently induced VCC oligomer formation, but cholesterol stearate and oleate, and the C22 (2-trifluoroacetyl) naphthyl analogue of cholesterol all

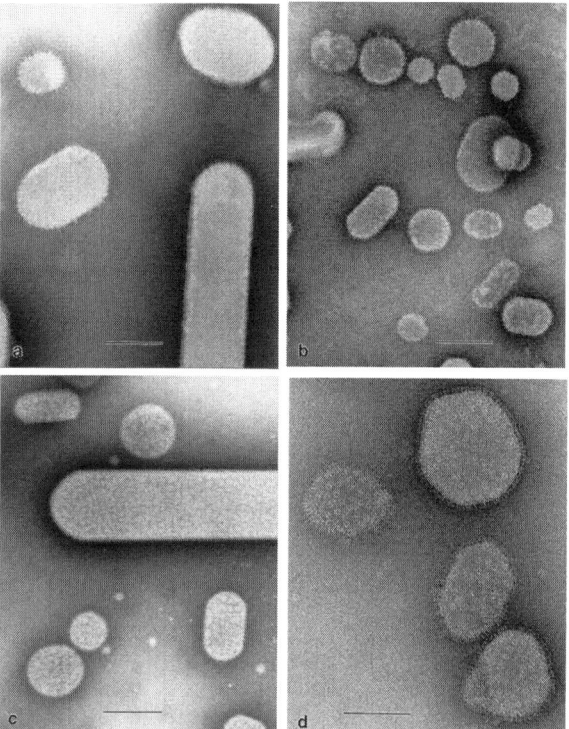

Fig. 21.3 Cholesterol microcrystals following: (**a**) interaction with VCC monomer (0.1 mg/ml) for 15 min at room temperature ($\sim$ 22°C); (**b**) interaction with VCC for 30 min at room temperature; (**c**) and (**d**) interaction with VCC at room temperature for 1 and 24 h, respectively. Note that at the short incubation times (**a**) and (**b**), the VCC has oligomerized predominantly at the edge of the cholesterol microcrystals and that with increasing time the planar surface also becomes coated with oligomers. Samples were negatively stained with 2% ammonium molybdate (pH 7.0). The scale bars indicate 200 nm (**a-c**) and 100 nm (**d**). From Harris et al. (2002), with permission from Elsevier

failed to induce oligomerization (Harris et al., 2002). Stigmasterol showed a slight tendency to produce VCC oligomers and ergosterol even less. Overall, this sterol survey indicated that the 3β-OH group of cholesterol is not an essential requirement for VCC oligomerization, indicating that a more generalized hydrophobic interaction may be involved. This may represent a difference between VCC and the CDC family of cytolysins, which usually do not interact with sterols that that have a blocked or modified 3βOH group. Solubilization of the VCC oligomers formed on cholesterol using the neutral surfactant octyl-glucoside yielded a suspension of individual oligomers, which unfortunately showed a considerable tendency to cluster (Fig. 21.5). In a further attempt to define the stereo-specificity of cholesterol interaction VCC, together with streptolysin O (SLO), Zitzer et al. (2003) used enantiomeric cholesterol-containing liposomes and found that VCC showed very little interaction

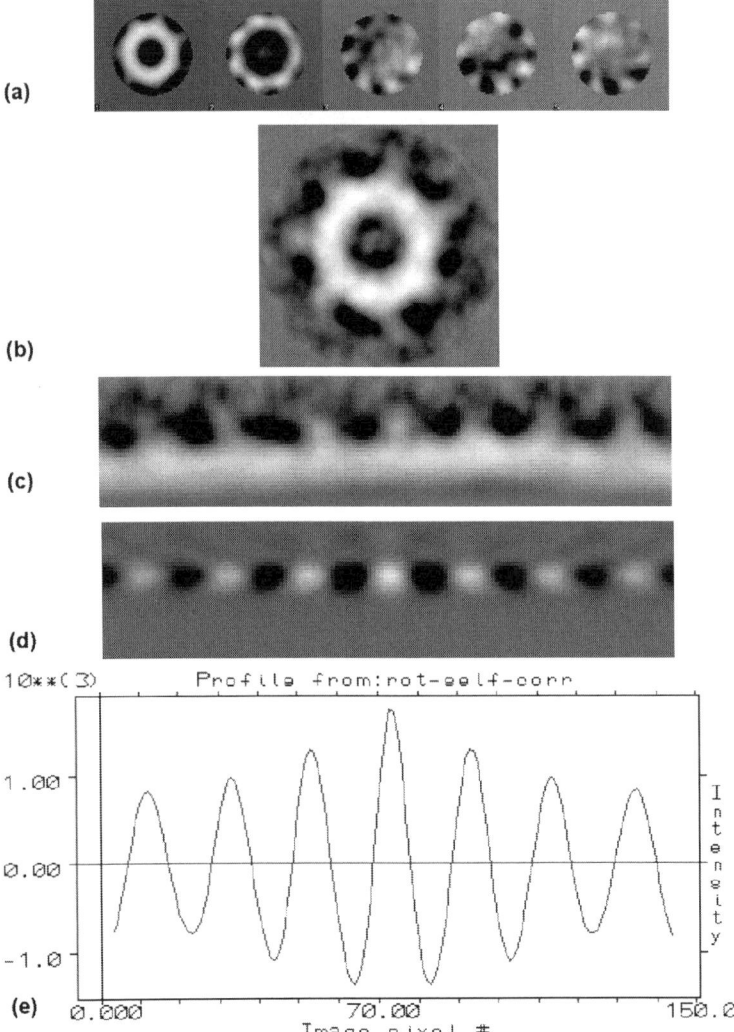

Fig. 21.4 Sevenfold rotational symmetry of the VCC oligomer: (**a**) The first five eigenimages created within IMAGIC-5 during the initial mulitivariate statistical analysis (MSA) of the VCC oligomer, showing sevenfold rotational symmetry. (**b**) A single heptameric VCC oligomer class average, used for the second MSA. (**c**) When cut open to display cylindrical coordinates this image shows seven vertical projections, which following rotational self-correlation are shown as seven transmission maxima (**d**). When expressed graphically, the profile from the rotational self-correlation shows seven peaks of intensity (**e**). From Harris et al. (2002) with permission from Elsevier

(expressed by calcein release), whereas streptolysin O was only slightly less active than when cholesterol was present in the liposomes. With cholesterol microcrystals the difference between the action of these two toxins was less clearly defined (JRH, previously unpublished data: Fig. 21.6). With VCC, oligmers form predominantly at

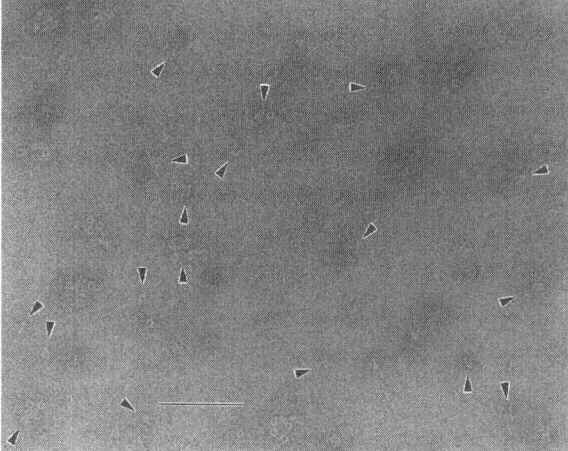

Fig. 21.5 VCC oligomers released from cholesterol by solubilization with 100 mM octylglucoside. Oligomer clusters are present, together with individual oligomers, orientated end-on and side-on. The side-on images (arrowheads) show the characteristic channel/pore structure of the oligomer; this correlates well with the VCC oligmer images visible at the edge of cholesterol microcrystals (see Fig. 21.2). Negatively stained with 2% ammonium molybdate. The scale bar indicates 100 nm. From Harris et al. (2002), with permission from Elsevier

the edges of the enantiomeric cholesterol microcrystals, essentially the same as with cholesterol, whereas SLO formed characteristic but disorganized oligomers and arcs on the cholesterol planar surface. In sum, there are considerable differences in the stereospecificity of cholesterol interaction for VCC vs the CDCs.

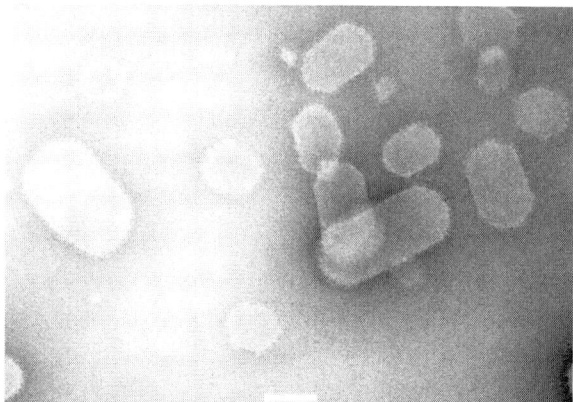

Fig. 21.6 The interaction of VCC with enantiomeric cholesterol microcrystals (15 min at room temperature). As with cholesterol, oligomer formation has occurred predominantly at the edge of the microcrystals. Negatively stained with 2% ammonium molybdate. The scale bar indicates 200 nm

Olson and Gouaux (2003), in a structural comparison of the VCC oligomer (then thought to be pentameric) and the *S. aureus* α-toxin, showed that VCC contains a core component that is closely related to that of *S. aureus* α-toxin and related hemolysins. These authors extended their study by producing the 2.8 Å crystal structure of the ~ 80 kDa pro-toxin monomer of VCC, together with crystals of the VCC oligomer at a lower resolution (3.5 Å), from which a clear non-crystallographic seven-fold symmetry was determined, thereby confirming the TEM data of Harris et al. (2002). For both the *S. aureus* α-toxin and *V. cholerae* cytolysin, a protein unfolding event has to take place, following a hydrophobic interaction with phospholipid and cholesterol-containing biological membranes, liposomes or pure cholesterol, to generate the 14-stranded (7-hairpin) β-barrel pore that penetrates the lipid bilayer.

Olson and Gouaux (2005) were unable to explain the lipid requirement for VCC assembly from their pro-toxin monomer structure and advanced only a tenuous model for the interaction of VCC with biological membranes, with comparison to the *S. aureus* α-toxin. The role of a putative lectin-like sequence in the VCC monomer remains unclear; unlike the anthrax PA63 monomer, it is not involved in cellular internalisation. Clearly, a higher resolution X-ray structure of the soluble surfactant-induced VCC oligomer and also of the in situ membrane-bound VCC pore is still awaited, and from which it should be possible to define the residues in the 65 kDa VCC monomer that interact hydrophobically with membrane cholesterol (possibly also involving sphingomyelin and other phospholipids) to partially unfold the polypeptide chain and generate the multiple-β-sheet hairpin pre-pore leading to membrane-penetration of the heptameric β-barrel. Recently, He and Olson (2010) have produced a low resolution (2nm) 3D reconstruction of the VCC pore from cryoEM images. The VCC pore was solubilized from cholesterol-containing liposomes with 40 mM hexaethylene glycol monodecyl ether. Docking of the crystal structure of the VCC protoxin (Olson and Gouaux, 2005) assisted the interpretation of the location of the cytolytic and lectin domains of the VCC pore. Furthermore, the "spikes" visible around the toxin core (see Fig 21.4b; also Fig. 21.2 of He and Olson, 2010) were advanced as the putative carbohydrate-binding sites.

The numerous oligomeric toxin β-barrels, of varying diameter and strand number, are all essentially structurally equivalent to a sequence of circular/closed crossed β-sheets.

The existence of a membrane-bound VCC pre-pore has been claimed by Löhner et al. (2009), using disulphide cross-linked genetically engineered mutant VCC molecules. In the non-reduced state, these mutant VCC oligomers bound to erythrocytes and erythrocyte membranes but did not penetrate the membrane bilayer. Subsequent reduction enabled the β-barrel-forming sequence to unfold, create a membrane-penetrating pore, thus leading to haemolysis. Mixed VCC wildtype-mutant hybrid oligomers apparently created pores with a reduced diameter permeability channel. This perhaps suggests a somewhat loosely coupled membrane insertion of the subunits of the oligomer. Such a transient membrane-bound pre-pore has to be clearly distinguished from the VCC oligomers formed in the presence of cholesterol and released into solution, where the further attachment to and penetration of a membrane bilayer is blocked, in all probability by oligomers containing β-barrels that already have bound cholesterol molecules which then block membrane penetration.

21.2.2 Fibril Formation by VCC and Other Toxins In Vitro

That a partial unfolding of toxin proteins is closely involved with their self-assembly into membrane-penetrating ring-like pores has emerged as a fundamental feature of these molecules. It has also been found that the thermostable direct hemolysin (TDH) from *Vibrio parahaemolyticus* has the property of transforming at ~60 to 70°C from a toxic dimer of a ~21 kDa subunit into β-sheet-rich non-toxic fibrils (Fukui et al., 2005). Remarkably, heating above 80°C yields an unfolded protein, which upon rapid cooling produces refolding and a toxic pore-forming protein. Apoprotein B, in sodium deoxycholate mixed micelles also undergoes also undergoes temperature-dependent protein unfolding (Walsh and Atkinson, 1986). Deoxycholate (DOC), at low concentrations has been used to induce the formation of *S. aureus* α-toxin oligomers (Bhakdi et al., 1981), which can be solubilized by higher DOC concentrations. DOC solublizes VCC oligomers from liposomes (Zitzer et al., 1997b), but without further change in protein conformation. Also for VCC, there was early evidence for temperature-induced inactivation and reactivation.

In our hands, the VCC monomer has been found to form pre-pore oligomers on the surface of sodium DOC micellar aggregates, when the surfactant is present at a low concentration (e.g. 1 mM, and over a period of time, these bound oligomers are released into solution, similar to the situation with octylglucoside (Fig. 21.5). However, at a higher DOC concentration (e.g. 25 to 50 mM) it has been found that the VCC forms stable fibrillar structures (Fig. 21.7).

In view of the increasing interest in oligomer and membrane-penetrating pore formation by several cytotoxic amyloid peptides and also the formation of crossed β-sheet amyloid fibrils (*see also* Chapter 2), combined with the hypothesis that these events may mimic the partial protein unfolding established for bacterial pore-forming toxins (Lashuael and Lansbury, 2006; Yoshiike et al., 2007), the data shown in Fig. 21.7 for VCC and the reversible heat-dependent fibril formation by *Vibrio parahaemolyticus* TDH mentioned above, may represent complimentary views of this structural hypothesis. Clearly, further biophysical investigations need to be performed to firmly establish this β-sheet/β-barrel structural parallel. Nevertheless, support for this concept has come from the study of Srisailam et al. (2002) who investigated the transformation of the all β-barrel acidic fibroblast growth factor from *Notopthalmus viridescens* and found partially-structured intermediates leading to fibril formation. Furthermore, evidence for a correlation between human amyloid fibrils and bacterial amyloids has been advanced and reviewed by Epstein and Chapman (2008). Indeed, the broad concept that the repeated β-sheet hairpin found in both the smaller bacterial toxin pores discussed above, as well as

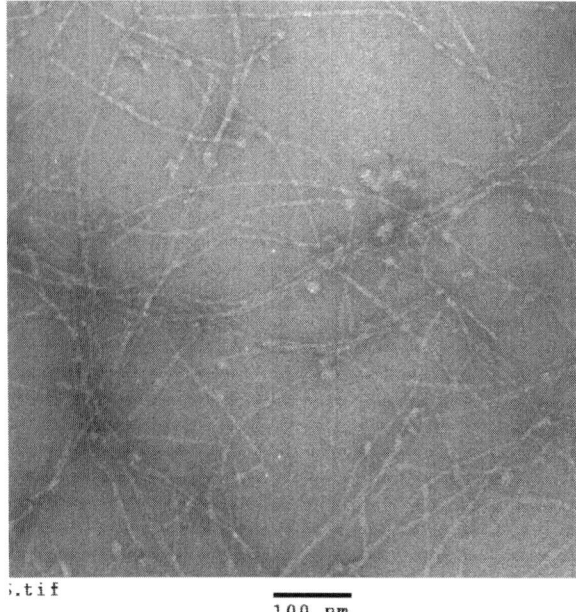

Fig. 21.7 VCC treated with 50 mM sodium deoxycholate. Instead of forming oligomers, the cytolysin has transformed into fibrils, structurally similar to amyloid fibrils. Negatively stained with 2% uranyl acetate. The scale bar indicates 200 nm

the larger pores produced by the CDC perfringolysin/streptolysin family of thiol-specific/cholesterol-dependent cytolysins (*see* Chapter 20), can be extended to include the membrane-penetrating pore-like amyloid oligomers and *infinite* crossed-beta sheets of highly polymorphic fibrillar and tubular structures, created by partial unfolding of the extensive range of amyloid-forming peptides and proteins appears to be steadily gaining acceptance (Zamotin et al., 2006).

21.3 Hemolysins/Cytolysins from Other *Vibrio* Species

Vibrio vulnificus is classified as a slightly halophilic estuarine bacterium, which causes wound infections and septicemia. Inhibition of the haemolytic activity of the *V. vulmificus* hemolysin by prior incubation with cholesterol was shown 25 years ago by Shinoda et al. (1985). The close genetic homology of the hemolysins from both *V. vulnificus* and *V. mimicus* to the El Tor hemolysin (VCC) was presented by Yamamoto et al. (1990) and Kim et al. (1997), firmly establishing this family of *Vibrio* hemolysins/cytolysins (reviewed by Miyoshi et al., 2004; Shinoda and Miyoshi, 2006).

Strong supportive evidence for the involvement of cholesterol in the cytolytic activity of *V. vulnificus* hemolysin/cytolysin (VVC) has come from J.-S. Kim and his co-workers who have actively pursued their investigations in recent years. The cytotoxicity of VVC on cultured pulmonary endothelial cells was shown to be inhibited by cholesterol (Kim, 1997) and using the same cell line Choi et al. (2004) claimed that this cytolysin formed anion-selective transmembrane pores. In an investigation of several cholesterol detivatives, including 7-dehydrocholesterol, cholesterol esters, deoxycholate and cholestane, Kim and Kim (2002) showed that only cholesterol and 7-dehydrocholesterol induced oligomerization and inhibition of the cytolytic activity of VVC. Subsequently it was shown that low density lipoprotein also has the ability to inactivate VVC by inducing oligomerization (Park et al., 2005). Reduction of membrane cholesterol by methyl-β-cyclodextrin also inhibited VVC (Yu et al., 2007), firmly established the parallel between VVC and VCC with respect to their requirement for cholesterol and the conclusion that membrane cholesterol is the likely receptor for the monomer of these cytolysins, prior to oligomerization, pre-pore formation and membrane penetration.

21.4 Cholesterol Dependency of Heptameric and Other β-Barrel-Forming Hemolysins/Toxins from Non-*Vibrio* Species

Interestingly, early studies on the gram positive *Staphylococcus aureus* α-hemolysin indicated that sodium deoxycholate would initiate oligomer formation, then thought to be a hexamer (Bhakdi et al., 1981; Tobjes et al. 1985), but later shown to be a heptamer (Galdiero and Gouaux, 2004; Gouaux, 1998). Although Forti and Menestrina (1989) claimed that cholesterol was involved in *S. aureus* α-hemolysin oligomerization, no firm evidence has subsequently emerged to involve membrane cholesterol and the *S. aureus* α-hemolysin is now believed to have a requirement for membrane phospholipid headgroups (Galdiero and Gouaux, 2004). It should be remembered, however, that the *Staphylococcus* α-hemolysin family of toxins is only distantly related to the gram-negative heptameric β-barrel toxins. Nevertheless, it can reasonably be considered that similar shape changes to the toxin monomer, resulting in monomer-monomer pre-pore formation might occur within the membrane hydrophobic environment, as defined by Olson and Gouaux (2003) for VCC.

The gram-positive bacterium *Bacillus anthracis* produces a toxin complex, comprising the anthrax protective antigen (PA83) carrier, along with the lethal factor (LF) and oedema factor (EF). Following endocytosis of the cell surface PA83/LF/EF-transmembrane receptor complex, endomembrane insertion of the proteolytically cleaved PA63 pre-pore is thought to occur within the acidic environment of the endosome (Milne et al., 1994; Gao and Schulten, 2006; Puhar and Montecucco, 2007). To-date, no unequivocal evidence has been presented to indicate which lipid component of the endosomal membrane, if any, might participate in the further unfolding and integration of the pre-formed PA63 heptamer as a transmembrane pore, through which the *B. anthracis* oedema and lethal factors are apparently able to enter the cytosol. However, Abrami et al. (2003) have suggested that oligomerization of PA63 triggers its association with cell surface cholesterol-containing lipid raft domains, promoting endocytosis of the complete toxin complex.

Recent in vitro experiments (JRH, unpublished observations) using a cholesterol microcrystal suspension have indicated that PA63 pre-pore heptamers do cluster on the cholesterol microcrystals at pH 5.0 (Fig. 21.8), but earlier permeability studies showed that in the absence of cholesterol, phospholipid bilayers and liposomes could integrate the PA63 pore (Blaustein et al., 1989; Koehler and Collier, 1991). Addition of PA63 to cholesterol-sphingomyelin-phosphatidylcholine liposomes at pH 5 was found to lead to integration of the PA63 pre-pore, but also showing evidence for bilayer/liposome disruption (JRH, unpublished data). X-ray crystallographic studies have shown the structure of the PA63 and the PA63 –receptor complex (Petosa et al., 1997; Santelli et al., 2004; Lacy et al. 2004), and Ren et al. (2004) showed by cryo-electron microscopy that structural changes occur on binding the lethal factor to PA63.

There are several other examples of toxins that bind to cholesterol, or cholesterol-containing plasma membrane lipid raft domains, which will now be briefly mentioned. The *Clostridium perfringens* ε-toxin has been shown to target detergent-insoluble raft domains of Madin-Darby canine kidney cells, with reduced cytotoxicity following cholesterol depletion of the cells with methyl-β-cyclodextrin (Miyata et al., 2002). The *C. perfringens* ε-toxin toxin is know to form a heptameric pore,

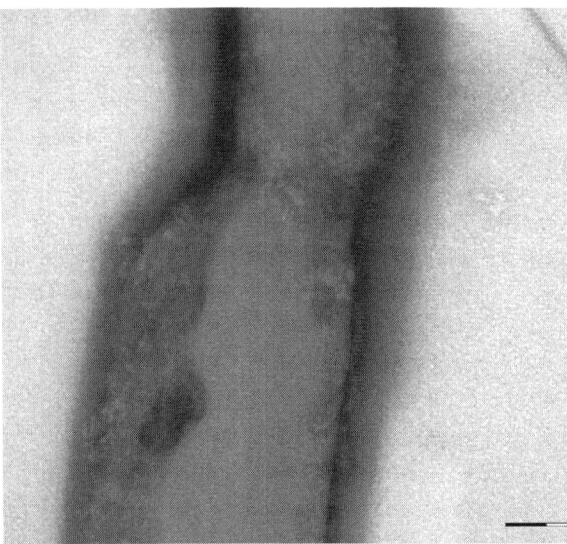

Fig. 21.8 Cholesterol microcrystals following interaction with anthrax PA63 prepores at pH 5. The toxin has clustered onto the edges of the cholesterol microcrystals, indicating an affinity with the hydrophobic edges of the stacked bilayers. Negatively stained with 2% uranyl acetate. The scale bar indicates 100 nm

which apparently has a more clearly defined association with cholesterol than does the toxin aerolysin from *Aeromonas hydrophila*, although Ferguson et al. (1997) showed that another related *Aeromonas hydrophila* cytotoxic enterotoxin (Act) does have a requirement for cholesterol. A similar situation exists for the cellular cholesterol-dependent pore formation by *C. difficile* toxin A, which was also shown to induce ion-permeable pores in bilayers at low pH, only in the present of cholesterol (Giesmann et al., 2006). Cholera toxin cell surface internalization in caveolae (Orlandi and Fishman, 1998) and transport from endosomes to the Golgi apparatus (Shogomori and Futerman, 2001) both require the presence of cholesterol, as demonstrated by methyl-β-cyclodextrin cholesterol depletion and competition by filipin-treatment of cells.

Using an in vitro cholesterol-binding assay Hayward et al. (2005) demonstrated that *Salmonella* (SipB) and *Shigella* (IpaB) type III secretion system translocation components bind to cholesterol with a high affinity, facilitating delivery of the bacteria into mammalian cells. Similarly, Sato et al. (2006) showed by cholesterol depletion that the *Yershina enterocolitica* entry into macrophages was greatly reduced. An *E. coli* enterohemolysin has also been found to have cholesterol-binding properties, apparently clearly distinguishable from the thiol-activated CDCs (Figueirêdo et al., 2003).

Recently, Mosser and Res (2006) described a new toxin from *Bacillus anthracis*, termed anthralysin 0 (ALO or AnlO) which is a cholesterol-dependent cytolysin. Its close homologue, anthrolysin B has sphingomyelinase activity (Popova et al., 2006). The oligomeric structure of the AnlO pore has yet to be shown; however, AnlO may be a member of the thiol-activated cholesterol-dependent toxin family (*see* Chapter 20). The non-bacterial toxin osterolysin, from the edible oyster mushroom, has been shown to have cholesterol dependency (Rebolj et al., 2006). However, it must be acknowledged that other pore-forming toxins, such as the octameric *E. coli* hemolysin E (Wallace et al., 2000; Tzokov et al., 2005) and the *P. aeruginosa* cytolysin, *S. aureus* α- toxin, as well as anthrax PA63 may possess a broader requirement for phospholipid (or surfactant in vitro), rather than for cholesterol alone.

21.5 Conclusions

This survey of the cholesterol requirement of the heptameric β-barrel forming cytolysins from the *Vibrio* and other bacterial species leads to the conclusion that the naturally occurring cell membrane lipid, cholesterol, has been successfully targeted by numerous cytolytic bacteria. Despite that fact that high resolution structural analysis is not yet available for the membrane-integrated heptameric pore, much of fundamental and clinical interest has emerged through recent years. It can be predicted that X-ray crystallography will continue to provide additional important data in the foreseeable future, relating in particular to the initial cholesterol-binding sites and the lipid bilayer contact sites of the membrane-penetrating β-barrel of the pore.

References

Abrami, L., Liu, S., Cosson, P., Leppla, S.H., Van Der Goot F.G., 2003, Anthrax toxin triggers endocytosis of its receptor via a lipid raft-mediated clathrin-dependent process. J. Cell Biol. **160**:321–328.

Bhakdi, S., Füssle, R., Tranum-Jensen, J., 1981, Staphylococal α-toxin: Oligomerization of hydrophilic monomers to form amphiphilic hexamers induced through contact with deoxycholate micelles. Proc. Natl. Acad. Sci. USA **78**:5475–5479.

Blaustein, R.O., Koehler, T.M., Collier, R.J., Finkelstein, A., 1989, Anthrax toxin: Channel-forming activity of protective antigen in planar phospholipid bilayers. Proc. Natl. Acad. Sci. USA **86**:2209–2213.

Chattopadhyay, K., Banerjee, K.K., 2003, Unfolding of Vibrio cholerae hemolysin induced oligomerization of the toxin monomer. J. Biol. Chem. **278**:38470–38475.

Choi, B.-H., Park, B.-H, Kwak, Y.-G., 2004, Vibrio vulnificus cytolysin forms anion-selective pores on the CPAE cells, a pulmonary endothelial cell line. Korean J. Phjysiol. Pharmacol. **8**: 259–264.

Epstein, E.A., and Chapman, M.R., 2008, Polymerizing the fibre between bacteria and host cells: the biogenesis of functional amyloid fibres. Cell. Microbiol. **10**:1413–1420.

Ferguson, M.R., Xu, X.J., Houston, C.W., Petterson, J.W., Coppenhaver, D.H., Popov, V.L., Chopra, A.K., 1997, Hyperproduction, purification, and mechanism of action of the cytotoxic enterotoxin produced by Aeromonas hydrophilia. Infect. Immun. **65**:4299–4308.

Figueirêdo, P.M., Catani,C.F., Yano, T., 2003, Thiol-independent activity of a cholesterol-binding enterohemolysin produced by enteropathogenic Escherichia coli. Braz. J. Med. Biol.Res. **36**:1495–1499.

Forti, S., Menestrina,G., 1989, Staphylococcal alpha-toxin increases the permeability of lipid vesicles by cholesterol- and pH-dependent assembly of oligomeric channels. Eir. J. Biochem. **181**:767–773.

Fukui, T., Shiraki, K., Hamada, D., Hara, K., Miyata, T., Fujiwara, S., Mayanagi, K., Tanaagihara, K., Iida, T., Fukusaki, E., Imanaka, T., Honda, T., Tanagihara, I., 2005, Thermostable direct hemolysin of Vibrio parahaemolyticus is a bacterial reversible amyloid toxin. Biochemistry **44**:9825–9832.

Galdiero, S., Galdiero, M., Pedone, C., 2007, β-Barrel membrane bacterial proteins: Structure function, assembly and interaction with lipids. Curr. Protein Peptide Sci. **8**:63–82.

Galdiero, S., Gouaux, E., 2004, High resolution crystallographic studies of alpha-hemolysin-phospholipid complexes define heptamer-lipid head group interactions: implication for understanding protein-lipid interactions. Protein Sci. **13**:1503–1511.

Gao, M., Schulten, K., 2006, Onset of anthrax toxin pore formation. Biophys. J. **90**:3267–3279.

Geny, B., Popoff, M.R., 2006, Bacterial protein toxins and lipids: pore formation or toxin entry into cells. Biol. Cell **98**:667–678.

Giesmann, T. Jank, T., Gerhard, R., Maier, E., Jut, I., Benz, R., Aktories, K., 2006, Cholesterol-dependent pore formation of Clostridium difficile toxin A. J. Biol. Chem. **281**:10808–10815.

Gouaux, E., 1998, α-Hemolysin from Staphylococcus aureus: An archetype of β-barrel, channel-forming toxins. J. Struct. Biol. **121**:110–122.

Gutierrez, M.G., Saka, H.A., Chinen, I., Zoppino, F.C.M., Yoshimori, T., Bocco, J.L., Colombo, M.I., 2007, Protective role of autophagy against Vibrio cholerae cytolysin, a pore-foprming toxin from V. cholerae. Proc. Natl. Acad. Sci. USA **104**:1829–1834.

Harris, J.R., Bhakdi, S., Meissner, U., Scheffler, D., Bittman, R., Li, G., Zitzer, A., Palmer, M., 2002, Interaction of the Vibrio cholerae cytolysin (VCC) with cholesterol, some cholesterol esters, and cholesterol derivatives: a TEM study. J. Struct. Biol. **139**:122–135.

Harris, J.R., Scheffler, D., 2002, Routine preparation of air-dried negatively stained and unstained specimens on holey carbon support films: A review of applications. Micron **33**: 461–480.

Hayward, AR.D., Cain, R.J., McGhie,E.J., Phillips, N., Garner, M.J., Koronakis, V., 2005, Cholesterol binding by the bacterial type III translocon is essential for virulence effector delivery into mammalian cells. *Molec. Microbiol.* **56**:590–603.

He, Y., Olson, R., 2010, Three-dimensional structure of the detergent-solubilized Vibrio cholerae cytolysin (VCC) heptamer by electron cryomicroscopy. *J. Struct. Biol.* **169**:6–13.

Honda, T., Finkelstein, R.A., 1979, Purification and characterization of a hemolysin produced by *Vibrio cholera* biotype El Tor: another toxic substance produced by cholera vibrios. *Infect. Immun.* **26**:1020–1027.

Ikigai, H., Akatsuka, A., Tsujiyama, H., Nakae, T., Shimamura, T., 1996, Mechanism of membrane damage by El Tor hemolysin of *Vibrio choleerae* O1. *Infect. Immun.* **64**:2968–2973.

Ikigai, H., Ono, T., Iwata, M., Nakae, T., Shimamura, T., 1997. El Tor hemolysin of *Vibrio cholerae* O1 forms channels in planar lipid membranes. *FEMS Miorobiol. Lett.* **150**:249–254.

Ikigai, H., Otsuru, H., Yamamoto, K., Shimamura, T., 2006, Structural requirements of cholesterol for binding to *Vibrio cholerae* hemolysin. *Microbiol. Immunol.* **50**:751–757.

Kim, B.-S., Kim, J.-S., 2002, Cholesterol induce oligomerization of *Vibrio vulnificus* cytolysin specifically. *Exp. Molec. Med.* **34**:239–242.

Kim, G.-T., Lee, J.-Y., Huh, S.-H., Yu, J.-H., Kong, I.-S., 1997, Nucleotide sequence of the *vmbH* gene encoding hemolysin from *Vibrio mimicus. Biochim. Biophys. Acta* **1360**:102–104.

Kim, J.-S., 1997, Cytotoxicity of Vitrio vulnificus cytolysin on pulmonary endothelial cells. *Exp. Molec. Med.* **29**:117–121.

Koehler, T.M., Collier, R.J., 1991, Anthrax toxin protective antigen: low-pH-induced hydrophobicity and channel formation in liposomes. *Mol. Microbiol.* **5**:1501–1506.

Krasilnikov, O.V., Merzlyak, P.G., Lima, V.L.M., Zitzer, A.O., Valeva, A., Yuldasheva, L.N., 2007, Pore formation by *Vibrio cholerae* cytolysin requires cholesterol in both monolayers of the target membrane. *Biochimie* **89**:271–277.

Lacy, D.B., Wiglesworth, D.J., Melnyk, R.A., Harrison, S.C., Collier, R.J., 2004, Structure of heptameric protective antigen bound to an anthrax receptor: A role for receptor in pH-dependent pore formation. *Proc. Natl. Acad. Sci. USA* **101**:13147–13151.

Lashuael, H.A., Lansbury P.T., 2006, Are amyloid diseases cased by protein aggregates that mimic bacterial pore-forming proteins? *Quart. Rev. Biophys.* **39**:167–201.

Löhner, S., Walev, I., Boukhallouk, F., Palmer, M., Shakdi, S., Valeva, A., 2009, Pores formation by *Vibrio cholerae* cytolysin follows the same archetypical mode as β-barrel toxins from gram-positive organisms. *The FASEB J.* doi: 10.1096/fj.08-127688.

Milne, J.C., Furlong, D., Hanna, P.C., Wall, J.S., Collier, R.J., 1994, Anthrax protective antigen forms oligomers during intoxication of mammalian cells. *J. Biol. Chem.* **269**: 20607–20612.

Miyata. S., Minami, J., Tamai, E., Matsushita, O., Shimamoto,S., Okabe, A., 2002, *Clostridium perfringens* ε-toxin forms a heptameric pore within the detergent-insoluble microdomains of Madin-Darby canine kidney cells and rat synaptosomes. *J. Biol.Chem.* **277**:39463–39468.

Miyoshi, S.-I., Morita, A., Teranishi, T., Tomochika, K.-I., Yamamoto, S., Shinoda, S., 2004, An exocellular cytolysin produced by *Vibrio vulnificus* CDC B3547, a clinical isolate in Biotype 2 (Serovar E). *J. Toxicol. Toxin Rev.* **23**:111–121.

Moshioni, M., Tombola, F., de Bernard, M., Coelho, A., Zitzer, A., Zoratti, M., Montecucco, C., 2002, The Vibrio cholerae hemolysin anion channel is required for cell vacuolation and death. *Cell Microbiol.* **4**:397–409.

Mosser, E.M., Res, R.F., 2006, The *Bacillus anthracis* cholesterol-dependent cytolysin, Anthrolysin O, kills human neutrophils, monocytes and macrophages. *BMC Microbiol.* **6**:56 Doi: 10.1186/1417-2180-6-56.

Nagamune, K., Yamamoto, K., Naka, A., Matsuyama, J., Miwatani, T., Honda, T., 1996, In vitro proteolytic processing and activation of the recombinant precursor of El Tor cytolysin/hemolysin (pro-HlyA) of Vibrion cholerae by soluble hemagglutinin/protease of V. cholearae, trypsin, and other proteases. *Infect. Immune.* **64**:4655–4658.

Olson, R., Gouaux, E., 2003, *Vibrio cholerase cytolysin* is composed of an α-hemolysin-like core. *Protein Sci.* **12**:379–383.

Olson, R., Gouaux, E., 2005, Crystal structure of the *Vibrio cholerae* cytolysin (VCC) protoxin andits assembly into a heptameric transmembrane pore. *J. Mol. Biol.* **350**:997–1016.

Orlandi, P.A., Fishman, P.H., 1998, Filipin-dependent inhibition of cholera toxin: Evidence for toxin internalisation and activation through caveolae-like domains. *J. Cell Biol.* **141**: 905–915.

Palmer, M., 2004, Cholesterol and the activity of bacterial toxins. *FEMS Microbiol. Lett.* **238**: 281–289.

Pantano, S., Montecucco, C., 2006, A molecular model of the *Vibrio cholerae* cytolysin transmembrane pore. *Toxicon* **47**:35–40.

Park, K-H., Yang, H.-B., Kim, H.-G., Lee, Y.-R., Hur, H., Kim, J.-S., Koo, B.-S., Han, M.-K., Kim, J.-H., Jeong, Y.-J., Kim, J.-S., 2005, Low density lipoprotein inactivates *Vibrio vulnificus* cytolysin through the oligomerization of toxin monomer. *Med. Microbiol. Immunol.* **194**: 137–141.

Petosa, C., Collier, R.J., Klimpel, K.R., Leppla, S.H., Liddington, R.C., 1997, Crystal structure of the anthrax toxin protective antigen. *Nature* **385**:833–838.

Popova, T.C., Millis, B., Bradburne, C., Nazarenko, S., Bailey, C., Chandhoke, V., Papov, S.C., 2006, Acceleration of epithelial cell syndecan-I sheddingby anthrax haemolytic virulence factors. *BMC Microbiol.* 6: 8 doi:10.1186/1471-2180-6-8.

Puhar, A., Montecucco, C., 2007, Where and how do anthrax toxins exit endosomes and intoxicate host cells? *Trends Microbiol.* **15**:477–482.

Rebolj, K., Ulrih, N.P., Maček, P., Sepčić, 2006, Steroid structural requirements for interaction of ostreolysin, a lipid-raft binding cytolysin, with lipid monolayers and bilayers. *Biochim. Biophys. Acta* **1758**:1662–1670.

Ren, G., Quispe, J., Leppla, S.H., Mitra, A.K., 2004, Large-scale structural changes accompany binding of lethal factor to anthrax protective antigen: A cryo-electron microscopic study. *Structure* **12**:2059–2066.

Saka, H.A., Bidinost, C., Sola, C., Carranza, P., Collino, C., Oritz, S., Echenique, J.R., Bocco, J.L., 2008, Vibrio cholerae cytolysin is essential for high enterotoxicity and apoptosis induction produced by a cholera toxin gene-negative V. cholerae non-O1, non-O139 strain. *Microb. Pathog.* **44**:118–128.

Santelli, E., Bankston, L.A., Leppla, S.H., Liddington, R.C., 2004, Crystal structure of a complex between anthrax toxin and its host cell receptor. *Nature* **430**:905–908.

Sato, Y., Kaneko, K., Sasahara, T., Inoue, M., 2006, Novel pathogenic mechanism in a clinical isolate of Yersina enterocolitica KU14. *J. Microbiol.* **44**:98–105.

Shinoda, S., Miyoshi, S.-I., 2006, Hemolysin of *vibrio cholerae* and other *vibrio* species. In: The Comprehensive Sourcebook of Bacterial Protein Toxins, (Eds) Alouf, X and Popoff, X., Elsevier Ltd, pp. 748–762.

Shinoda, S., Miyoshi, S.-I., Yamanaka, H., Miyoshi-Nakahara, N., 1985, Some properties of *Vibrio vulnificus* hemolysin. *J. Microbiol. Immunol.* **29**:583–590.

Shogomori, H., Futerman, A.H., 2001, Cholesterol depletion by methyl-β-cyclodextrin blocks cholera toxin transport from endosomes to the Golgi apparatus ain hippocampal neurons. *J. Neurochem.* **78**:991–999.

Srisailam, S., Wang, H.-M., Kumart, T.K,S., Rajalingam, D., Sivaraja, V., Sheu, H.-S., Chang, Y.-C., Yu, C., 2002, Amyloid-like fibril formation in an all β-barrel protein involves the formation of partially structured intermediate(s). *J. Biol. Chem.* **277**:19027–19036.

Tobjes, N., Wallace, B.A.., Bayley, H., 1985, Secondary structure and assembly mechanism of an oligomeric channel protein. *Biochemistry* **24**:1915–1920.

Tzkov, S.B., Wyborn, N.R., Stillman, T.J., Jamieson, S., Czudnochowski, N., Artymiuk, P.J., Green, J., Bullough, P.A., 2005, Structure of the hemolysin E (HlyE, ClyA, and SheA) channel in its membrane-bound form. *J. Biol. Chem.* **281**:233042–23049.

Valeva, A., Walev, I., Boukhjallouk, F., Wassenaar, T.M., Heinz, N., Hedderich, J., Lautwein, S., Möcking, M., Weis, S., Zitzer, A., Bhakdi, S., 2005, Identification of the membrane penetrating domain of Vibrio cholerae cytolysin as a beta-barrel structure. *Mol. Microbiol.* **57**:124–131.

Valeva, A., Walev, I., Weis, S., Boukhallouk, F., Wassenaar, T.M., Bhakdi, S., 2008, Pro-inflammatory feedback activation cycle evoked by attack of Vibrio cholerae cytolysin on human neutrophil granulocytes. *Med. Microbiol. Immunol.* **197:**285–293.

Valeva, A., Walev, I., Weis, S., Boukhallouk, F., Wassenaar, T.M., Endres, K., Fahrenholz, F., Bhakdi, S., Zitzer, A., 2004, A cellular metalloproteinase activates Vibrio cholerae procytolysin. *J. Biol. Chem.* **279:**25143–25148.

Wallace, A.J., Stillman, T.J., Atkins, A., Jamieson, S.J., Bulloch, P.A., Green, J., Artymuik, P.J., 2000, *E. coli* hemolysin E (HlyE, ClyA, SheA): X-ray crystal structure of the toxin and observation of membrane pores by electron microscopy. *Cell* **100:**265–276.

Walsh, M.T., Atkinson, D., 1986, Physical properties of apoprotein B in mixed micelles with sodium deoxycholate and in a vesicle with dimyristoyl phosphatidylcholine. *J. Lipid Res.* **27:**316–325.

Yamamoto, K., Ichinose, Y., Shinagawa, H., Makino, K., Nakata, A., Iwanaga, M., Honda, T., Miwatani, T., 1990, Two-step processing for activation of the cytolysin/hemolysin of Vibrio cholerae O1 biotype El Tor: nucleotide sequence of the structural gene (hlyA) and characterization of the processed products. *Infect. Immun.* **58:**4106–4116.

Yamamoto, K., Wright, A.C., Kaper, J.B., Morris, J.G., 1990, The cytolysin gene of *Vibrio vulnificus*: Sequence and relationship to the *Vibrio cholerae* El Tor hemolysin gene. *Infect. Immun.* **58:**2706–2709.

Yoshiike, Y., Kayed, R., Milton, S.C., Takashima, A., Glabe, C.G., 2007, Pore-forming proteins share structural and functional homology with amyloid oligomers. *Neuromol. Med.* **9:**270–275.

Yu, H.N., Lee, Y.R., Park, K.H., Rah, S.Y., Noh, E.M., Song, E.K., Han, M.K., Kim, B.S., Lee, S.H., Kim, J.S., 2007, Membrane cholesterol is required for activity of Vibrio vulnificus cytolysin. *Arch. Microbiol.* **187:**467–473.

Yuldasheva, L.N., Merzlyak, P.G., Zitzer, A.O., Rodrigues, C.G., Bhakdi, S., Krasilnikov, O.V., 2001, Lumen geometry of ion channels formed by Vibrio cholerae EL Tor cytolysin elucidated by nonelectrolyte exclusion. *Biochim. Biophys. Acta* **1512:**53–63.

Zamotin, V., Gharibyan, A., Gibanova, N.V., Lavrikova, M.A., Dolgikh, D.A., Kirpichnikov, M.P., Kostanyan, I.A., Morozova-Roche, L.A., 2006, Cytotoxicity of ablebetin oligomers depends on cross-β-sheet formation. *FEBS Lett.* **580:**2451–2457.

Zitzer, A., Bittman, R., Verbicky, C.A., Erukulla, R.K., Bhadki, S., Weis, S., Valeva, A., Palmer, M., 2001, Coupling of cholesterol and cone-shaped lipids in bilayers augments membrane permeabilization by the cholesterol-specific toxins streptolysin O and *Vibrio cholerae* cytolysin. *J. Biol. Chem.* **276:**14628–14633.

Zitzer, A., Harris, J.R., Kemminer, S.E., Zitzer, O., Bhakdi, S., Meuthing, J., Palmer, M., 2000, *Vibrio cholerae* cytolysin: assembly and membrane insertion of the oligomeric pore are tightly linked and are not detectably restricted by membrane fluidity. *Biochim. Biophys. Acta* **1509:**264–274.

Zitzer, A., Palmer, M., Weller, U., Wassenaar, T., Biermann, C., Tranum-Jensen, J., Bhakdi, S., 1997b, Mode of primary binding to target membranes and pore formation induced by Vibrio cholerae cytolysin (hemolysin). *Eur. J. Biochem.* **247:**209–216.

Zitzer, A., Walev, I., Palmer, M., Bhakdi, S., 1995, Characterization of Vibrio cholerae El Tor cytolysin as an oligomerizing pore-forming toxin. *Med. Microbiol. Immunol.* **184:**37–44.

Zitzer, A., Wassenaar, T.M., Walev, I., Bhakdi,S., 1997a, Potent membrane-permeabilizing and cytocidal action of Vibrio cholerae cytolysin on human intestinal cells. *Infect. Immun.* **65:** 1293–1298.

Zitzer, A., Westover, E.J., Covey, D.F., Palmer, M., 2003, Differential interaction of the two cholesterol-dependent membrane-damaging toxins, streptolysin O and *Vibrio cholerae* cytolysin with enantiomeric cholesterol. *FEBS Lett.* **553:**229–231.

Zitzer, A., Westover, E.J., Covey, D.F., Palmer, M., 2003, Differential interaction of the two cholesterol-dependent, membrane-damaging toxins, streptolysin O and *Vibrio cholerae* cytolysin, with enantiomeric cholesterol. *FEBS Lett.* **553**:229–231.

Zitzer, A.O., Nakisbekov, N.O., Li, A.V., Semiotrochev, V.L., Kiseliov, Yu.L., Muratkhodjaev, J.N., Krasilnikov, O.V., Ezepchuk, Yu.V., 1993, Entero-cytolysin (EC) from Vibro cholerae non-O1 (some properties and pore-forming activity). *Zentralbl. Bakteriol.* **279**:494–504.

Chapter 22
Cholesterol-Binding Toxins and Anti-cholesterol Antibodies as Structural Probes for Cholesterol Localization

Yoshiko Ohno-Iwashita, Yukiko Shimada, Masami Hayashi, Machiko Iwamoto, Shintaro Iwashita, and Mitsushi Inomata

Abstract Cholesterol is one of the major constituents of mammalian cell membranes. It plays an indispensable role in regulating the structure and function of cell membranes and affects the pathology of various diseases. In recent decades much attention has been paid to the existence of membrane microdomains, generally termed lipid "rafts", and cholesterol, along with sphingolipids, is thought to play a critical role in raft structural organization and function. Cholesterol-binding probes are likely to provide useful tools for analyzing the distribution and dynamics of membrane cholesterol, as a structural element of raft microdomains, and elsewhere within the cell. Among the probes, non-toxic derivatives of perfringolysin O, a cholesterol-binding cytolysin, bind cholesterol in a concentration-dependent fashion with a strict threshold. They selectively recognize cholesterol in cholesterol-enriched membranes, and have been used in many studies to detect microdomains in plasma and intracellular membranes. Anti-cholesterol antibodies that recognize cholesterol in domain structures have been developed in recent years. In this chapter, we describe the characteristics of these cholesterol-binding proteins and their applications to studies on membrane cholesterol localization.

Keywords Cholesterol-binding toxin · Perfringolysin O · Anti-cholesterol antibody · Lipid raft · Microdomain · Plasma membrane

Abbreviations

PFO,	perfringolysin O;
EGFP,	enhanced green fluorescent protein;
DRMs,	detergent-resistant membranes;
βCD,	β-cyclodextrin;
SFKs,	Src-family protein kinases;

Y. Ohno-Iwashita (✉)
Faculty of Pharmacy, Iwaki Meisei University, 5-5-1 Chuodai Iino, Iwaki City, Fukushima 970-8551, Japan
e-mail: yiwast@iwakimu.ac.jp

TD,   Tangier disease;
NPC,   Niemann-Pick disease type C

22.1 Introduction

Cholesterol is one of the major constituents of mammalian cell membranes. It affects the structure and function of biological membranes by determining the physico-chemical nature of membranes. Cholesterol also influences the pathology of atherosclerosis, Alzheimer's disease and cholesterol-storage disorders such as Niemann-Pick disease type C (NPC). The lateral and transbilayer distributions of cholesterol in cell membranes are not uniform, suggesting that cholesterol is involved in the formation of functional membrane domains. One such membrane domain, generally termed lipid rafts, has attracted attention in recent decades, and the dynamic clustering of sphingolipids and cholesterol in cell membranes plays a crucial role in the formation of these domains. Studies on the distribution and dynamics of cholesterol in membranes have been hindered by the lack of appropriate cholesterol probes. For a long time filipin has been used to detect membrane cholesterol, despite it's cytolytic properties. The development of non-toxic cholesterol-binding probes has provided much useful information on the distribution and dynamics of cholesterol in membranes. Non-toxic derivatives of perfringolysin O (PFO, also known as θ-toxin), a cholesterol-binding cytolysin have been used in many studies to detect cholesterol in plasma and intracellular membranes. In recent years anti-cholesterol antibodies have also been developed for this purpose. In this chapter, we describe the characteristics of these cholesterol-binding proteins and their application to studies on membrane cholesterol.

22.2 Cholesterol-Binding Toxins

The toxins under consideration belong to a family of so-called "thiol-activated cytolysins" (Billington et al., 2000; Palmer, 2001), also termed cholesterol-binding cytolysins (Alouf, 2000; Gilbert, 2002) or cholesterol-dependent cytolysins (CDCs) (Tweten et al., 2001). These toxins are expected to be good experimental tools for detecting cholesterol in membranes, due to their strict specificity for binding to cholesterol. More than 20 cytolysins produced by gram-positive bacteria belong to this toxin family (Alouf, 2000), including PFO produced by *Clostridium perfringens*, streptolysin O by *Streptococcus pyogenes,* and pneumolysin by *Streptococcus pneumoniae*. Members of this toxin family share a high degree of homology in their amino acid sequences and are believed to share common biological properties (Billington et al., 2000; Alouf, 2000).

Among the cytolysin family members, PFO has been well characterized in terms of its structure-function relationship. The action of PFO can be divided into the following three steps: i) binding to cholesterol in the membrane, ii) self-assembly on the membrane to form prepore oligomers, and iii) formation of transmembrane pores leading to cell lysis. Crystallographic studies have shown the PFO molecule to comprise four domains that are rich in β-sheet structure (Rossjohn et al., 1997). It has been shown that domain 4, the C-terminal domain, possesses membrane cholesterol-binding activity (Shimada et al., 2002), and thus participates in the first step. The relationship between the structure and mode of action of the CDCs is described in detail elsewhere (*see* Chapter 20).

PFO binds specifically and with high affinity to cholesterol in artificial membranes (Ohno-Iwashita et al., 1991; 1992; Nakamura et al., 1995) and intact cells (Ohno-Iwashita et al., 1988; 1990). Taking advantage of this property, non-cytolytic derivatives of PFO were produced as probes for detecting membrane cholesterol (Ohno-Iwashita et al., 1988; 1990; Iwamoto et al., 1997; Shimada et al., 2002). Characterization of the toxin derivatives revealed them to be useful for detecting cholesterol-rich membrane microdomains, the so-called lipid rafts (Waheed et al., 2001; Ohno-Iwashita et al., 2004; Shimada et al., 2005). In this section we describe methods for preparing these toxin derivatives, their binding characteristics and their application to the detection of membrane cholesterol.

22.2.1 Preparation of Non-cytolytic Derivatives of Perfringolysin O

PFO derivatives that are non-cytolytic but retain their cholesterol-binding properties were prepared by two methods (Fig. 22.1). One method involves a two-step modification of PFO that blocks toxin oligomerization, the 2nd step of toxin action as described above. The toxin is first subjected to enzymic digestion and nicked between the 144th and 145th amino acids, by proteolysis with subtilisin Carlsberg (Ohno-Iwashita et al., 1986; 1988). The product, Cθ (Carlsberg protease-nicked PFO/θ-toxin), is a complex of two fragments (molecular sizes ∼38 kDa and ∼15 kDa) (Ohno-Iwashita et al., 1986). Cθ causes cytolysis at 37°C but not at temperatures below 20°C, even though it binds to membrane cholesterol at both temperatures. Cθ forms oligomeric structures on erythrocytes only at high temperatures (Ohno-Iwashita et al., 1986; Iwamoto et al., 1993). In the second modification

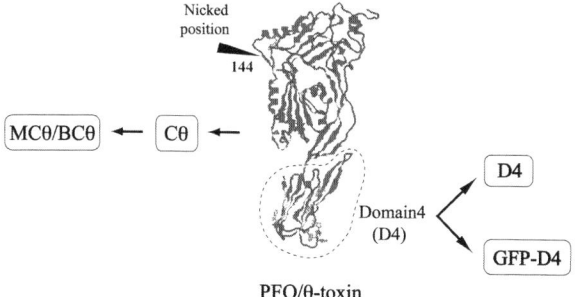

Fig. 22.1 Preparation procedures for non-cytolytic derivatives of PFO/θ-toxin

step, Cθ is either methylated or biotinylated at the ε-amino groups of lysine residues and the α-amino group at the N terminus. The final products, MCθ (methylated Cθ) and BCθ (biotinylated Cθ), bind to cell membranes, but cause no membrane damage even at 37°C (Ohno-Iwashita et al., 1990; Iwamoto et al., 1997).

Another non-cytolytic PFO derivative is prepared by isolating the cholesterol-binding domain of the toxin, a C-terminal fragment of 110 amino acids corresponding to domain 4 (residues 363–472) (Shimada et al., 2002) (Fig. 22.1). PFO comprises four β-sheet-rich domains, and only domain 4 is structurally autonomous (Rossjohn et al., 1997). Several lines of evidence suggest that a cholesterol-binding site is located within domain 4 (Iwamoto et al., 1990; Sekino-Suzuki et al., 1996; Nakamura et al., 1998; Jacobs et al., 1999). A C-terminal fragment obtained by trypsin digestion (T2; residues 277–472) (*see* Fig. 22.2) binds to cholesterol and to cholesterol-containing membranes (Iwamoto et al., 1990). In addition, various toxins mutated in the tryptophan-rich motif (residues 430–440; Fig. 22.2) in domain 4 have significantly reduced membrane-binding activities (Sekino-Suzuki et al., 1996; Nakamura et al., 1998). It has also been shown that the C-terminal amino acid residues are required for maintaining the overall structure and cholesterol-binding properties of the toxin (Shimada et al., 1999; 2002).

D4, the isolated domain 4, and several N-terminal and C-terminal truncated fragments of D4, were prepared and characterized to determine the shortest cholesterol-binding unit (Fig. 22.2). D4 has cholesterol-binding activity comparable to that of the full-size toxin (Shimada et al., 2002). Circular dichroism measurements have revealed that D4 has a β-sheet-rich secondary structure, in good agreement with that predicted from the crystal structure (Rossjohn et al., 1997). On the other hand, ΔN-D4, which is truncated by eight-amino acids from the N-terminus of D4, and ΔC-D4, in which only two amino acids are deleted from the C-terminus of D4, are

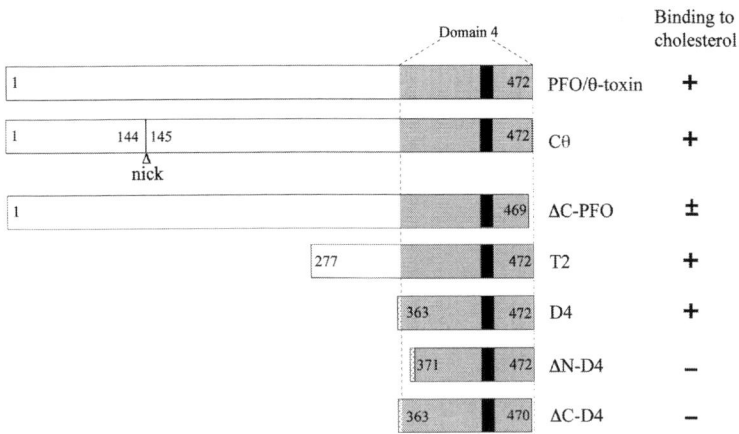

Fig. 22.2 PFO derivatives and their binding activities with cholesterol. D4, ΔN-D4 and ΔC-D4 contain an extra six amino acids (hatched box) at their N termini, which are derived from the His-tag used for purification (Shimada et al., 2002). The black rectangles represent the tryptophan-rich motif in domain 4

unstable and degraded during purification (Shimada et al., 2002). These observations indicate that D4 is the minimal functional and structurally stable unit with the same cholesterol-binding activity as the full-size protein toxin. For the purpose of monitoring membrane cholesterol in living cells, a fused protein (EGFP-D4) comprising enhanced green-fluorescent protein (EGFP) and D4 was also constructed (Shimada et al., 2002).

22.2.2 Binding Properties of Perfringolysin O Derivatives

22.2.2.1 Specific Binding to Cholesterol

All toxin derivatives, Cθ, MCθ, BCθ, and D4, bind specifically to cholesterol (Shimada et al., 2002; Ohno-Iwashita et al., 1988; 1990; Iwamoto et al., 1997). When lipids were separated on TLC plates, incubated with PFO or its derivatives, and immunostained to detect binding with an anti-PFO antibody, the toxin derivatives were shown to bind only to cholesterol, and not to phospholipids, as in the case of PFO (Fig. 22.3) (Shimada et al., 2002; Ohno-Iwashita et al., 1988; 1990; Iwamoto et al., 1997). PFO and its derivatives also bind to some cholesterol analogues with variable intensities (Fig. 22.4A, B) (Ohno-Iwashita et al., 1988). The 3β-OH group on the steroid nucleus of cholesterol is strictly required for toxin binding; the toxin derivatives never bind to cholesterol analogues lacking the 3β-OH

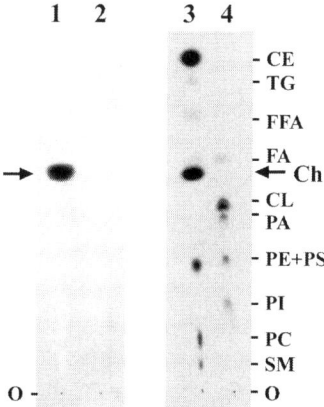

Fig. 22.3 Detection of lipid components that bind MCθ. Lipids were applied to a TLC plate and developed in two solvent systems. Half of the plate was then incubated with MCθ and lipid components that bind MCθ were detected by immunostaining with anti-θ-toxin antibody (*lanes* 1 and 2). The other half was used for the detection of lipids (*lanes* 3 and 4). The migration position of cholesterol (Ch) is shown by arrows. The migration positions of other lipids are shown as follows: O, origin; SM, sphingomyelin; PC, phosphatidylcholine; PI, phosphatidylinositol; PE, phosphatidylethanolamine; PS, phosphatidylserine; PA, phosphatidic acid; CL, cardiolipin; FA, fatty alcohol; FFA, free fatty acid; TG, triglyceride; CE, esterified cholesterol. Modified from Ohno-Iwashita et al. (1990) with permission from Elsevier

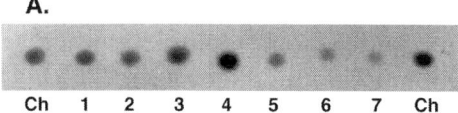
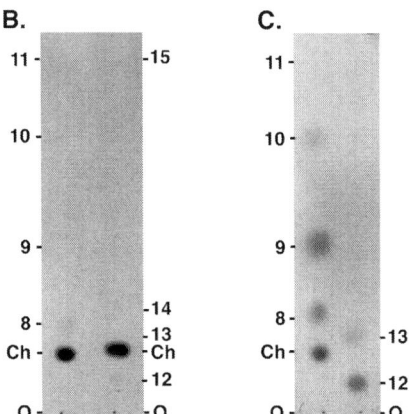

Fig. 22.4 Binding of various cholesterol analogues to PFO and digitonin. Cholesterol analogues were applied to TLC plates and developed in a solvent system of hexane/diisopropylether/acetic acid (65/35/2). Two plates (**A and B**) were then incubated with PFO (10 μg/ml), and sterols that bind PFO were detected by immunostaining with anti-θ-toxin antibody. The other plate (**C**) was incubated with 0.5 mM digitonin, and sterols that bind digitonin were detected by the orcinol reaction. Ch, cholesterol; 1, β-cholestanol; 2, 7-dehydrocholesterol; 3, desmosterol; 4, lathosterol; 5, ergosterol; 6, β-sitosterol; 7, stigmasterol; 8, lanosterol; 9, 5α-cholestan-3-one; 10, cholesterol acetate; 11, 5α-cholestane; 12, 20-α-hydroxycholesterol; 13, epicoprostanol; 14, 4-cholesten-3-one; 15, 5β-cholestane; O, origin

(5α-cholestane and 5β-cholestane), or in which the 3β-OH is replaced with 3α-OH (epicoprostanol) or =O (5α-cholestan-3-one and 4-cholesten-3-one), or is esterified (cholesterol acetate) (Fig. 22.4B). The structural variety of the aliphatic side chains could also affect the binding intensity of the toxin (ergosterol, β-sitosterol, stigmasterol, lanosterol and 20-α-hydroxycholesterol in Fig. 22.4A, B). Changes in the ring structure are less effective (β-cholestanol, 7-dehydrocholesterol, and lathosterol in Fig. 22.4A). The structural requirement for toxin binding is distinct from that for the binding of digitonin, a small cholesterol-binding molecule (Fig. 22.4C). For example, digitonin, but not PFO, binds to 5α-cholestan-3-one (compare Fig. 22.4B, C); thus 3β-OH is not strictly required for digitonin binding.

PFO derivatives exhibit high affinity ($K_d \sim 10^{-7} \sim {}^{-9}$M) for cholesterol in artificial membranes (Ohno-Iwashita et al., 1991; 1992) and intact cell membranes, including erythrocytes (Ohno-Iwashita et al., 1988; 1990) and lymphoma B cells (Ohno-Iwashita et al., 1990). A study using surface plasmon resonance revealed that the association and dissociation rate constants (k_{on} and k_{off}, respectively) and the dissociation constant (K_d) of D4 to cholesterol-containing liposomes are comparable to those of a non-cytolytic full-size toxin (Shimada et al., 2002). This indicates that the large deletion of the N-terminal region corresponding to domains 1–3 from the toxin does not influence its binding kinetics.

22.2.2.2 Selective Binding to Cholesterol-Enriched Membranes

To understand the characteristics of toxin binding to cholesterol more precisely, the binding properties were examined using artificial membranes with various lipid compositions (Ohno-Iwashita et al., 1991; 1992). Figure 22.5 shows the relationship between cholesterol mol% and total toxin-binding sites in liposomes. PFO derivatives bind only to cholesterol-enriched liposomes, while their binding is negligible when the amount of cholesterol is below 25 mol% (Fig. 22.5; Ohno-Iwashita et al., 1992), indicating that toxin binding is highly dependent on cholesterol content and that there is a threshold concentration of membrane cholesterol for toxin binding.

Toxin binding to cultured cells was visualized using BCθ (biotinylated Cθ) and an avidin-fluorescent dye. Figure 22.6A shows BCθ binding to cell surface cholesterol in human epidermoid A431 cells. BCθ binding completely disappears (Fig. 22.6B) when cell cholesterol is depleted by 30% with β-cyclodextrin (βCD), a reagent that removes cholesterol from the cell surface (Christian et al., 1997; Ilangumaran and Hoessli, 1998). This is in remarkable contrast to filipin staining (Fig. 22.6C, D). Filipin, a polyene antibiotic, is a cholesterol-binding reagent used to stain cholesterol (Miller, 1984; Severs, 1997). Since filipin causes membrane damage, it can be used only to stain fixed cells. Filipin staining is significantly retained after cholesterol is depleted by 30% (Fig. 22.6D), implying that the mode of BCθ binding to cell cholesterol differs from that of filipin. It is likely that BCθ binds to a specific population of cholesterol, while filipin binds indiscriminately to cell cholesterol.

A similar relationship has been observed between toxin binding and the cholesterol content of intact cells other than A431 cells. When the cholesterol content of human erythrocytes (Ohno-Iwashita et al., 2004), platelets (Waheed et al., 2001),

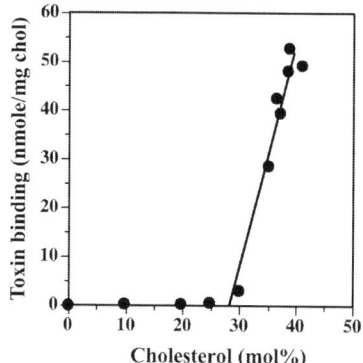

Fig. 22.5 Binding of a PFO derivative to liposomes with various cholesterol contents. Cholesterol contents in liposomes are expressed as mol percentage. Total toxin-binding sites were determined by Scatchard analysis (Ohno-Iwashita et al., 1992). From Ohno-Iwashita et al. (2004) with permission from Elsevier

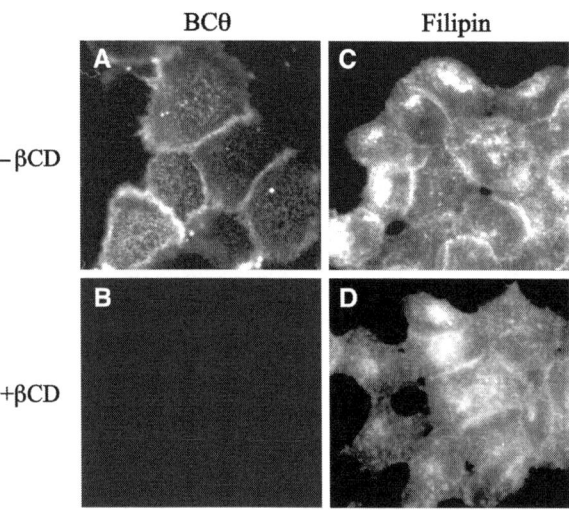

Fig. 22.6 Staining of A431 cells with BCθ or filipin. A431 cells were treated with (**B and D**) or without (**A and C**) βCD, fixed and then incubated with either BCθ (**A and B**) or filipin (**C and D**). BCθ-treated cells were then incubated with cy3-avidin for fluorescence visualization

human diploid fibroblasts (Nakamura et al., 2003) or MOLT-4 cells (Shimada et al., 2002) is reduced by one-third by βCD treatment, BCθ binding is almost completely abolished.

The above results strongly suggest the possibility that PFO derivatives bind selectively to microdomains called lipid rafts. Lipid rafts are membrane microdomains that are enriched in cholesterol and sphingolipids (Fig. 22.7) (Simons and Ikonen, 1997; Simons and Toomre, 2000; Brown and London, 2000; Edidin, 2001). It has been suggested that signalling molecules, such as Src-family protein kinases (SFKs), assemble in lipid rafts, where they play a role in signal transduction and many other cellular events. Cholesterol has been suggested to play an essential role in both the structural maintenance and function of lipid rafts (Hooper, 1999; Ostermeyer et al., 1999; Simons and Ikonen, 2000; Simons and Ehehalt, 2002; Fielding and Fielding, 2003). Lipid rafts have often been prepared as detergent-resistant membranes (DRMs) (Schuck et al., 2003), although recent reports have pointed out that DRMs are not necessarily equivalent to lipid rafts (London and Brown 2000; Shogomori and Brown, 2003). As a first step to examine whether PFO derivatives bind to lipid rafts, DRMs were prepared and PFO binding was examined (Waheed et al., 2001; Shimada et al., 2005; Inomata et al., 2006). BCθ-bound cells were treated with TritonX-100, homogenized and fractionated on sucrose-density gradients. Specific marker molecules of lipid rafts/caveolae, such as ganglioside GM1, flotillin and caveolin-1, are concentrated in the low-density fractions (DRMs/raft fractions, #3–5 in Fig. 22.8). Cell-bound BCθ is predominantly recovered in the DRMs/raft fractions. Cholesterol gives two peaks, one in DRMs/raft fractions and the other in high-density non-raft fractions (Fig. 22.8). BCθ

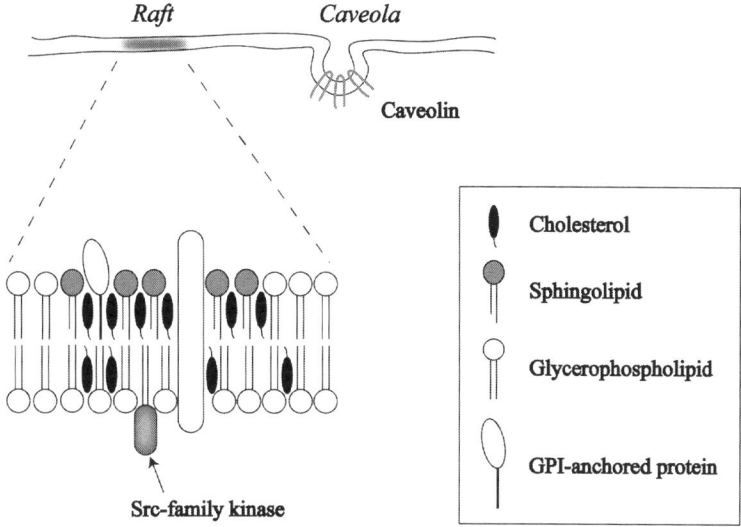

Fig. 22.7 Schematic illustration of lipid rafts and caveolae. Cell surface invaginations called caveolae comprise a specialized subpopulation of lipid rafts. See reviews for details (Simons and Ikonen, 1997; Simons and Toomre, 2000; Brown and London, 2000; Edidin, 2001)

is localized with cholesterol in DRMs/raft fractions but not in the non-raft fractions. This observation is consistent with the previous result that BCθ binds to a particular population of cholesterol (Fig. 22.6). D4 also binds to cholesterol in DRMs/raft fractions (Shimada et al., 2002).

Interestingly, an electron microscopic analysis of membrane vesicles in DRMs/raft fractions revealed that some, but not all, of the vesicles in the DRMs/raft fractions associate with BCθ (Waheed et al., 2001; Shimada et al., 2005). Next, the BCθ-bound vesicles were further purified from DRMs/raft fractions of Jurkat cells by affinity chromatography using avidin-magnet beads. Lipid analysis showed that the BCθ-bound membrane vesicles are highly enriched in cholesterol, while unbound vesicles are poor in cholesterol despite the fact that they are recovered in DRMs/raft fractions (Shimada et al., 2005). This indicates that the BCθ-bound vesicles are a cholesterol-rich membrane subpopulation in DRMs/raft fractions. Furthermore, ganglioside GM1, flotillin and SFKs, molecules assumed to be lipid raft-associated, are also predominantly recovered in the BCθ-bound, cholesterol-enriched membrane subpopulation (Shimada et al., 2005). These findings indicate that BCθ binds to membranes that fulfil the biochemical criteria of lipid rafts, and therefore suggest that BCθ recognizes a specific kind of lipid raft (hereafter called cholesterol-rich microdomains). In addition to raft markers, the BCθ-bound vesicles from DRMs of Jurkat cells contain a variety of signalling molecules (Shimada et al., 2005), indicating that the above method for isolating particular cholesterol-rich membrane vesicles provides a useful tool for analyzing functional membrane domains concentrated with signalling molecules (*see below*). It is assumed that several types of lipid rafts with differing lipid and protein compositions perform

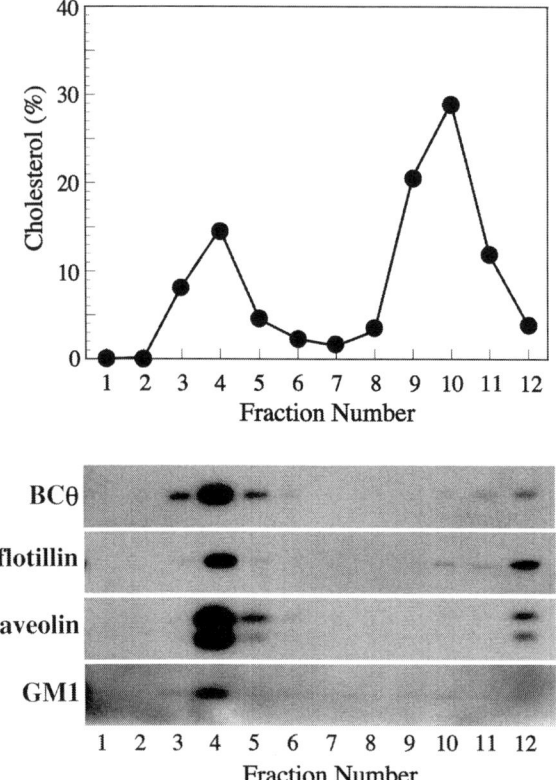

Fig. 22.8 Cell-bound BCθ is predominantly recovered in detergent-insoluble, low-density membrane fractions (DRMs/raft fractions). BCθ-bound human diploid fibroblasts (TIG-1) were treated with Triton X-100 and subjected to sucrose-density gradient fractionation. Distribution patterns of BCθ, cholesterol, and raft markers (GM1 ganglioside, flotillin and caveolin-1) were determined

different functions (Madore et al., 1999; Roper et al., 2000; Gomez-Mouton et al., 2001). It has even been suggested that lipid rafts can be generated in the absence of high concentrations of cholesterol (Pike, 2006). Therefore, it would be interesting to examine whether the cholesterol-poor membrane fraction in DRMs is another subpopulation of lipid rafts.

22.2.3 Application of Perfringolysin O Derivatives to the Detection of Cholesterol-Rich Membranes

Various tools have been used to identify lipid rafts and to clarify their physiological significance. Bacterial toxins that target components in lipid rafts are often used as raft markers (Parton, 1994; Harder et al., 1998; Yamaji et al., 1998; Fivaz et al.,

1999). For example, lysenin, a toxin secreted by earthworms, and cholera toxin have been employed to detect sphingomyelin and ganglioside GM1, respectively, both of which are enriched in lipid rafts. However, no tool has been reported for the selective detection of cholesterol in lipid rafts. Although filipin is a cholesterol-binding reagent (Miller, 1984; Severs, 1997), it disrupts membrane integrity and thus cannot be used for studies on living cells. As described in the previous section, PFO derivatives specifically bind to membrane cholesterol without disrupting membrane integrity. Furthermore, toxin binding requires high cholesterol concentrations with a threshold in targeted membranes. Taking advantage of these properties, PFO derivatives, especially BCθ and EGFP-D4, have been used to visualize cholesterol in membranes and found to provide a powerful tool for cytochemical and biochemical studies of cholesterol-rich membrane microdomains/lipid rafts in cell physiology and diseases. In this section we describe some typical studies.

22.2.3.1 BCθ as a Marker of Cholesterol-Rich Microdomains on the Cell Surface

Since PFO derivatives are not membrane-penetrating, they have been used to detect cholesterol-rich microdomains on the cell surface (Waheed et al., 2001; Shimada et al., 2002; Terashita et al., 2002; Heijnen et al., 2003; Kokubo et al., 2003; Nakamura et al., 2003; Aoki et al., 2003; Nagafuku et al., 2003; Ohno-Iwashita et al., 2004; Tashiro et al., 2004; Ohsaki et al., 2004; Tani-ichi et al., 2005; Ishii et al., 2005; van Lier et al., 2005; Sato et al., 2005; Koseki et al., 2007). Many molecules have been suggested to localize in cholesterol-rich microdomains in plasma membranes, due to their co-localization with cell-bound PFO derivatives in fluorescence and electron microscopic studies. Such molecules include signalling molecules such as NAP-22 (neuronal acidic protein of 22 kDa; also known as CAP-23 and BAPS 1; Terashita et al., 2002), SFKs (Heijnen et al., 2003; Sato et al., 2005), a linker for the activation of T cells (LAT) (Nagafuku et al., 2003), raft markers such as GM1 (Sato et al., 2005), flotillin (Kokubo et al., 2003) and caveolin (Fujimoto et al., 1997; Aoki et al., 2003). However, co-localization is not definitive due to the limitations of optical resolution in the case of fluorescence microscopic studies (Pike, 2006).

NAP-22, a major component of brain-derived rafts (Maekawa et al., 1993; 2003), is a cortical cytoskeleton-associated and calmodulin-binding protein, and participates in the regulation of actin dynamics (Frey et al., 2000). Double-staining analysis has shown the protein to co-localize with BCθ, suggesting its localization in cholesterol-rich microdomains (Terashita et al., 2002). Cholesterol depletion with βCD solubilizes NAP-22 from brain rafts, indicating its cholesterol-dependent localization in microdomains (Maekawa et al., 1999).

The localization of SFKs upon differentiation of mouse F9 embryonal carcinoma cells was analyzed using BCθ (Sato et al., 2005). SFKs do not co-localize with BCθ in undifferentiated F9 cells, but do co-localize when the cells differentiate into primitive endoderm following retinoic acid treatment. The disruption of cholesterol-enriched microdomains by βCD delays differentiation, suggesting that localization of SFKs in the microdomains is important for differentiation.

The dynamic behavior of cholesterol-rich microdomains in platelets has been visualized using BCθ combined with confocal- and immunoelectron-microscopy (Waheed et al., 2001; Heijnen et al., 2003). In resting platelets, cholesterol-rich membrane microdomains are uniformly distributed on the cell surface (Heijnen et al., 2003). Upon platelet activation cholesterol-enriched microdomains accumulate at the extended tips of the filopodia and at the leading edge of spreading cells (Heijnen et al., 2003). This accumulation process is accompanied by simultaneous enrichment in c-Src, CD63, and β_3-integrin in the microdomain clusters. Under cholesterol-depleted conditions c-Src disappears from the filopodia, and thrombin-induced platelet aggregation is impaired (Waheed et al., 2001; Heijnen et al., 2003). These observations demonstrate that cholesterol-rich microdomains in platelets are dynamic entities in the membrane that co-cluster with c-Src and CD63 in specialized domains on the cell surface, providing a possible mechanism to act as signalling centers.

Flotillin has often been used as a marker of lipid rafts, especially in cells that lack caveolin (Stuermer et al., 2001). The co-localization of flotillin and BCθ-labelled cholesterol was examined in rat brain tissue by pre-embedding immuno-electron microscopy (Kokubo et al., 2003). Flotillin-1- and BCθ-labelled areas appear patchy and mostly merged on a part of plasma membranes with small processes and secondary lysosome membranes (Kokubo et al., 2003). Approximately 80% of BCθ-positive micropatches are flotillin-1-positive. Thus flotillin-1 and BCθ-labelled cholesterol could act as ultra-structural raft markers in neural tissues.

There are two types of cholesterol-rich microdomains in plasma membranes: caveolar and non-caveolar microdomains. Caveolae are uncoated round invaginations in the plasma membrane and are generally considered to be specialized microdomains (Anderson, 1998). When cultured cells are visualized with BCθ coupled with either fluorescein-avidin D or colloidal gold-streptavidin, fine dot labelling is distributed over the entire cell surface and does not coincide with the labelling pattern of caveolin, a protein that forms the framework of caveolae, as long as the cells are kept on ice (Fujimoto et al., 1997). This observation suggests that most of BCθ labels non-caveolar microdomains under these conditions. In contrast, when cell-bound BCθ is cross-linked with avidin and the temperature is raised to 37°C, the probe forms discrete large patches and becomes sequestered in caveolae (Fujimoto et al., 1997). Interestingly, other markers of microdomains, such as glycolipids and sphingomyelin, are also sequestered in caveolae when cross-linked with antibodies or other multivalent ligands (Mayor et al., 1994; Fujimoto 1996). This phenomenon may reflect a close functional relationship between caveolar and non-caveolar microdomains (Abrami et al., 2001).

22.2.3.2 Detection of Cholesterol-Rich Microdomains in the Inner Leaflet of the Plasma Membrane

It is commonly accepted that microdomains enriched in cholesterol and sphingolipids are present in the outer leaflet of the plasma membrane (Simons and Ikonen, 1997; Brown and London, 1997). Recent studies have suggested that an equivalent domain organization could be also present in the inner (cytoplasmic) leaflet of the plasma membrane (Mayor and Rao, 2004; Kusumi et al., 2004). The coupling of these domains on the outer and inner leaflets might be important for signal transduction *via* lipid rafts (Simons and Toomre, 2000; Kusumi et al., 2004). However, little is known about the properties of inner membrane leaflet microdomains. To address the question of whether cholesterol-rich microdomains exist in the inner leaflet, we developed a new probe by using domain 4 (D4) of PFO (Hayashi et al., 2006). It has been reported that when D4 interacts with membranes, it does not span the entire membrane bilayer, but only its tip is shallowly embedded in the non-polar interior of the bilayer (Ramachandran et al., 2002). This indicates that D4, rather than recognizing cholesterol in both leaflets of the bilayer, instead primarily recognizes cholesterol in just the one leaflet facing the toxin. Thus, it was expected that if D4 is expressed inside cells, it could capture cholesterol in the inner leaflet of the plasma membrane. We expressed D4 as a fusion protein with EGFP in cultured cells. Biochemical analysis of stable cell clones of mouse embryonic fibroblasts that express EGFP-D4 inside cells showed that more than half of the expressed EGFP-D4 was bound to DRMs/raft fractions in which cholesterol and other raft markers are enriched (Hayashi et al., 2006). Depletion of membrane cholesterol with βCD reduced the amount of EGFP-D4 localized in DRMs/raft fractions, indicating that EGFP-D4 binds to cholesterol-rich microdomains. Subcellular fractionation experiments showed that most of the EGFP-D4 recovered in DRMs/raft fractions was bound to the plasma membrane rather than to intracellular membranes (Hayashi et al., 2006). The above results strongly suggest the existence of cholesterol-rich microdomains in the inner leaflet of the plasma membrane. This expression system provides a potential tool for visualizing inner leaflet cholesterol-rich microdomains in living cells. Along with this expression system, we prepared fluorescent dye-conjugated PFO derivatives, such as Alexa dye-labelled BCθ and D4, as tools for detecting cell-surface (outer leaflet) cholesterol-rich microdomains in living cells (Shimada et al., 2002). Simultaneous visualizaion of the inner- and outer-leaflet microdomains by combining these tools might contribute to understanding their dynamic behavior in cell signalling.

22.2.3.3 Electron Microscopic Analysis of Microdomains in Intracellular Membranes

Since cholesterol-rich microdomains are hypothesized to play an important role in cholesterol distribution and transport (Simons and Ikonen, 2000; Fielding and Fielding, 2003), a selective marker for cholesterol in microdomains may provide a good tool for the study of intracellular cholesterol trafficking. BCθ has been used to study the *in situ* distribution of cholesterol-rich membranes in cryosections of cells examined by immuno-electron microscopy (Möbius et al., 2002). In cryosections of EBV-transformed human B-cell line RN, strong labelling was observed in internal vesicles of multivesicular bodies and exosomes, in addition to the plasma membranes (Möbius et al., 2002). When this method was applied to examining the distribution of cholesterol in the endocytic pathway of human B lymphocytes, BCθ labelling was detected in recycling tubulovesicles and in two types of multivesicular bodies, while it was mostly absent from lysosomes (Möbius et al., 2003). This observation is consistent with a previous report that purified lysosomes contain only trace amounts of cholesterol (Schoer et al., 2000). BCθ-labelled cholesterol distributes differently to lysobisphosphatidic acid, which localizes predominantly in lysosomes and late endosomes (Kobayashi et al., 1998). These findings support the view that plasma membrane cholesterol recycles constitutively *via* recycling endosomes (Hao et al., 2002) and that cholesterol is efficiently removed from late endosomes (Blanchette-Mackie, 2000).

22.2.3.4 Microdomains in T-cell Receptor Signalling

Based on the observation that DRMs, membranes assumed to be lipid rafts, are enriched in molecules involved in T-cell receptor (TCR) signalling, such as SFKs (Lck and Fyn) (Montixi et al., 1998; Xavier et al., 1998) and LAT (Zhang et al., 1998), it has been suggested that lipid rafts act as platforms for signal transduction in T cells (Alonso and Millan, 2001). However, experiments in which βCD is used to remove cholesterol have provided conflicting results (Marwali et al., 2003; Tu et al., 2004), and thus the role of cholesterol in TCR signalling remains unclear. As described earlier (22.2.2.2), BCθ-labelling experiments have shown that DRMs contain both cholesterol-enriched and cholesterol-poor membrane subpopulations. To address a role of the cholesterol-enriched membrane subpopulation in TCR signalling, DRMs prepared from unstimulated and antibody-stimulated Jurkat T cells were separated into BCθ-labelled (cholesterol-enriched) and -unlabelled (cholesterol-poor) membrane fractions using avidin-magnet beads (Shimada et al., 2005). Lck, Fyn, and LAT, in addition to the raft markers flotillin and GM1, were predominantly recovered in the cholesterol-enriched membrane fraction in both unstimulated and stimulated cells. On the other hand, CD3ϵ, CD3ζ and Zap70, other components required for TCR signalling, were localized in the cholesterol-poor membranes under unstimulated conditions. However, when TCR was stimulated, some of these molecules were recruited to the cholesterol-enriched membranes. These results indicate that the segregation from and recruitment of these molecules to cholesterol-enriched membranes could contribute to the on/off switching of TCR signalling. In addition to positive regulation (stimulation) of TCR signalling, lipid rafts might also participate in its negative regulation. It has been suggested that phosphorylated PAG (phosphoprotein associated with glycosphingolipid-enriched membrane microdomains), an adaptor molecule located in lipid rafts, recruits Csk (C-terminal Src kinase) to lipid rafts, where Csk negatively regulates SFKs and maintains the "off" state of signalling under unstimulated conditions (Brdicka et al., 2000; Davidson et al., 2003). Experiments separating BCθ-labelled and -unlabelled membranes showed that Csk and PAG are also localized in cholesterol-enriched membranes, suggesting that cholesterol-rich microdomains are involved in not only stimulation but also negative regulation of TCR signalling (Shimada et al., 2005).

T cell functions such as antigen-induced proliferation and IL-2 production decline in elderly humans and aged mice, and it has been suggested that alterations in lipid raft-mediated TCR signalling are associated with this decline (Miller, 2000; Pawelec et al., 2001; Larbi et al., 2004). Age-associated alterations in proteins involved in the positive regulation of TCR signalling have been reported from several laboratories (Tamura et al., 2000; Tamir et al., 2000; Larbi et al., 2004). Recently, impairments in the negative regulatory system in aged mouse T cells have been reported (Inomata et al., 2007). These alterations might cause an imbalance between the positive and negative regulatory systems and lead to a dysregulation of TCR signalling. In addition to changes in protein components, alterations in the composition/organization of lipids in lipid rafts might cause their functional dysregulation. We recently found that the amount of BCθ bound to $CD4^+$ T cells from aged mice is significantly higher than that bound to $CD4^+$ T cells from young mice (M. Inomata and Y. Ohno-Iwashita, unpublished results), implying age-related changes in the lipid composition and/or membrane organization of mouse T cells. Interestingly, we also found that the cholesterol contents of both whole cells and DRMs are nearly the same in T cells from young and aged mice. In most cases so far examined, an increase in BCθ binding accompanies an increase in the cholesterol content of the targeted membranes; however, in this case, BCθ binding increases without a change in cholesterol content. One possible explanation for this observation is a change in the distribution of cholesterol between the outer and inner leaflets with age, without significant alteration in the cholesterol content of the entire membrane. In support of this view, there is an interesting report that transmembrane cholesterol distribution changes with age in the mouse synaptic plasma membrane: the amount of cholesterol in the outer leaflet increases while that of the inner leaflet decreases (Igbavboa et al., 1996). However, the transmembrane distribution of cholesterol remains a matter of debate due to the limitations of the techniques used for its study. In this and other reports, transmembrane cholesterol distribution was examined by incubating cells with fluorescent cholesterol analogues or a cholesterol-modifying enzyme; therefore, the possibility that the transmembrane distribution of cholesterol changes during incubation cannot be excluded. Recently, Murate et al. (2008) succeeded in distinguishing outer and inner leaflet cholesterol in red blood cells by an SDS-digested freeze-fracture replica method (SDS-FRL method). It is expected that quick freezing in SDS-FRL results in minimal disturbance of the distribution of lipids, and thus this method provides a promising approach to the direct examination of transmembrane lipid asymmetry in biomembranes by electron microscopy (Fujimoto et al., 2006). Studies using this method to examine whether the transmembrane distribution of cholesterol changes with aging are currently under way.

22.2.3.5 Microdomains in Cholesterol Storage/Transport Disorders

Tangier Disease

Cells from patients with Tangier disease (TD) have mutation(s) in the ATP-binding cassette transporter-A1 (ABCA1) gene, resulting in a deficiency in apoA-I-mediated cholesterol efflux and a subsequent accumulation of intracellular lipids as lipid droplets. The defect is closely related to the development of atherosclerosis in TD patients (Brooks-Wilson et al., 1999). Recently, it was revealed that cholesterol-enriched microdomains in the plasma membrane are dramatically elevated in macrophages from ABCA1-knockout mice and TD patients as detected by BCθ (Koseki et al., 2007). This observation is consistent with another report showing that ABCA1 over-expression disrupts raft microdomains (Landry et al., 2006). Interestingly, the increase in microdomains in TD macrophages, which may be caused by the deposition of free cholesterol in lipid rafts, is accompanied by an accelerated lipopolysaccharide-induced TNFα release (Koseki et al., 2007). The close relationship between TNFα release and the increased level of microdomains was demonstrated using the cholesterol-modulating agents βCD and nystatin (Koseki et al., 2007). These observations suggest that the progress of premature atherosclerosis in TD patients may be accelerated by enhanced inflammation through an abnormality in the microdomains.

Niemann-Pick Disease Type C (NPC)

The endocytic pathway plays an important role in cholesterol homeostasis since it is the entrance point and delivery site for LDL-derived cholesterol. The endocytic pathway is affected in cholesterol-associated diseases, such as NPC (Blanchette-Mackie, 2000) and hypercholesterolemia (Brown and Goldstein, 1986). A defect in NPC1, the gene product responsible for Niemann-Pick disease type C (*see* Chapter 11), causes an accumulation of cholesterol in lysosomes (Blanchette-Mackie, 2000). Notably, BCθ labelled cholesterol accumulates in intracellular organelles in NPC1-deficient cells as demonstrated by fluorescent microscopic studies on fixed specimens (Sugii et al., 2003). In contrast to the intracellular accumulation of cholesterol, the level of cell surface cholesterol-rich microdomains seems to be decreased in NPC1-deficient cells. It has been reported that cell surface cholesterol labelling with BCθ disappears in living hippocampal neurons when the cells are treated with class 2 amphiphiles to mimic NPC1-deficient conditions (Tashiro et al., 2004). Such a decrease in cell surface cholesterol-rich domains might be related to the functional loss in neurons and neurodegeneration in NPC brains.

Recently, studies using BCθ revealed an abnormal cholesterol distribution on the surface of NPC1-deficient cells (Ohsaki et al., 2004). Multiple large speckles were stained on the plasma membrane of NPC1-deficient Chinese hamster ovary cells, but not on wild type cells. Such large speckles were also detected on the surface of cultured skin fibroblasts from an NPC patient. Analyses by immuno-electron microscopy showed these speckles to be cholesterol-rich vesicles, suggesting that the active shedding of cholesterol-rich vesicles from the plasma membrane may take place under NPC1-deficient conditions, at least in these two types of cells.

BCθ has also been used to show cholesterol accumulation in an NPC1-deficient mouse brain model (Reid et al., 2004; 2008). Unusual cholesterol accumulation was clearly demonstrated as early as postnatal day 9 in various regions of the NPC1-deficient mouse brain (Reid et al., 2004). Cholesterol accumulation at this early stage is not detected by conventional filipin staining. Although it is not clear that this BCθ staining corresponds to cholesterol-rich microdomains as mentioned above, membrane regions with higher cholesterol concentrations than some threshold were induced even at this early stage. It will be interesting to characterize these cholesterol-rich domains in detail, and BCθ staining will provide a very sensitive tool to detect this kind of cholesterol accumulation.

22.3 Anti-cholesterol Antibodies

22.3.1 Characterization and Binding Properties

Cholesterol had long been assumed to be a non-immunogenic molecule because it is so widely distributed and plays important biological roles in mammals. However, it was subsequently shown that cholesterol is an excellent immunogen. It was reported that the sera of most healthy individuals contain varying levels of natural autoantibodies against cholesterol (Alving et al., 1989). In addition, several laboratories have produced polyclonal and monoclonal antibodies against cholesterol by injection of cholesterol-loaded liposomes or cholesterol crystals (Swartz et al., 1988; Perl-Treves et al., 1996; Kessler et al., 1996; Dijkstra et al., 1996; Bíró et al., 2007). Monoclonal antibodies against cholesterol were first isolated by Swartz et al. (1988). They immunized mice with liposomes containing 71 mol% cholesterol together with lipid A as adjuvant and obtained IgM-type monoclonal antibodies. The antibodies cause complement-dependent immune damage in liposomes containing 71 mol% cholesterol, but not liposomes containing 43 mol% cholesterol. They also recognize cholesterol in the crystalline form. Perl-Treves et al. (1996), and Kessler et al. (1996) also generated IgM-type monoclonal antibodies against cholesterol monohydrate crystals. They immunized mice with either cholesterol crystals or dinitrobenzene crystals. Interestingly, among the antibodies thus obtained, one antibody (58B1) from mice immunized with 1,4-dinitrobenzene crystals displayed the highest degree of specificity to cholesterol crystals. Since it recognized cholesterol in the crystalline form but not the isolated cholesterol molecule, they speculate that the antibody does not recognize the cholesterol molecule itself, but a repetitive motif of molecular moieties exposed on the surface of the cholesterol crystal (Addadi et al., 2003).

Dijkstra et al. (1996) generated polyclonal IgM-type antibodies against cholesterol by immunizing mice with cholesterol-rich liposomes and lipid A. Immunization with liposomes containing more than 56 mol% cholesterol raised the serum antibody titer. They compared the binding specificity of the polyclonal antibodies with a monoclonal antibody (2C5-6) reported by Swartz et al. (1988), and showed that the antibodies are similar to each other (Dijkstra et al., 1996) in that they both recognize cholesterol and structurally related sterols containing a 3β-hydroxyl group. The binding activity was significantly diminished if the 3β-hydroxyl group was altered by epimerization, oxidation, or esterification. In addition, both antibodies reacted with intact human very-low-density lipoproteins and low-density lipoproteins (LDL), but not with high-density lipoproteins (HDL).

Recently, Bíró et al. (2007) obtained IgG-type monoclonal antibodies against cholesterol. They immunized mice with cholesterol-rich liposomes similar to those used by Dijkstra et al. (1996) and obtained two IgG3-isotype antibodies, AC1 and AC8. There were some differences in the binding specificity of these antibodies from that of previously reported IgM-type antibodies. The IgM-type antibodies bound to 25-hydroxy-cholesterol (Dijkstra et al., 1996), while the IgG-type antibodies do not (Bíró et al., 2007), despite the fact that this sterol contains a free 3β-hydroxyl group. In addition, the IgG-type antibodies recognize cholesterol in intact HDL (Bíró et al., 2007), whereas the IgM-type antibodies do not (Dijkstra et al., 1996). In the latter case, the inaccessibility of IgM-type antibodies to HDL cholesterol may be partly due to their larger molecular size, since the high surface density of proteins in HDL is likely to shield a portion of the cholesterol epitopes.

22.3.2 Application to Cellular Cholesterol Detection

Several anti-cholesterol antibodies have been used to detect cellular cholesterol. Kruth et al. (2001) targetted plasma membrane cholesterol with the IgM-type monoclonal antibody 58B1 that specifically recognizes ordered cholesterol arrays. The antibody detected cell surface cholesterol on fibroblasts and macrophages only after artificial cholesterol enrichment, by pre-culture with Acyl-CoA:cholesterol acyltransferase (ACAT) inhibitor to block the esterification of excess cellular cholesterol together with LDL or acetylated LDL (Kruth et al., 2001). The results suggest that the monoclonal antibody recognizes unique cholesterol microdomains that are induced in the plasma membrane of cholesterol-enriched cells. Induction of the plasma membrane microdomains could be blocked by agents that inhibit cholesterol trafficking to the plasma membrane, and by cholesterol acceptors that remove cholesterol from the plasma membrane (Kruth et al., 2001). However, the microdomains detected by Mab 58B1 do not seem to be equivalent to lipid rafts, since they were sensitive to extraction with ice-cold 1% Triton X-100 (Kruth et al., 2001).

An IgG-type antibody is more convenient for immunological analyses than an IgM-type antibody, for practical reasons involving molecular size, valency and available techniques for handling. Bíró et al. (2007) stained human and murine lymphocytes and monocyte-macrophage cell lines with the IgG-type monoclonal antibodies AC1 and AC8. The antibodies bound to live cells with low avidity, but their binding was enhanced by mild digestion of the cell surface with papain. The expression of some protruding extracellular proteins, such as CD44 and CD45R, was decreased by the papain treatment. These large protruding membrane proteins may shield cell surface cholesterol from antibody access. These binding characteristics are in contrast to the cell binding of PFO derivatives: PFO derivatives bind to cells without any protease pretreatment. The different molecular sizes of IgG (~170 kDa) and PFO derivatives (53 kDa) might explain this difference in binding. The antibodies bound to papain-treated cells were highly colocalized with cholera toxin B, anti-thy1 and anti-caveolin-1 antibodies, indicating that lipid rafts or caveolae can be considered preferential binding sites for anti-cholesterol antibodies in the plasma membrane (Bíró et al., 2007; Gombos et al., 2008). Such co-localization was also detected in activated T cells. While the IgM-type antibodies bind to cells only after artificial cholesterol enrichment as described above (Kruth et al., 2001), IgG type antibodies do not require such pretreatment for binding (Bíró et al., 2007). Thus, the binding sites of these two classes of antibody on the plasma membrane appear to be different.

22.4 Conclusions

Useful cholesterol-binding probes have been developed to study the localization and organization of cholesterol in membranes and cells. Non-toxic derivatives of PFO detect cholesterol in cholesterol-rich membrane microdomains that are suggested to be a kind of lipid raft. They have been used to detect these microdomains in plasma and intracellular membranes of various cell types. Fluorescent PFO derivatives make it possible to visualize the outer and inner leaflet microdomains simultaneously. This method will contribute to analyzing dynamic changes that take place in the microdomains upon trans-membrane signalling through the plasma membrane, and help the understanding of their roles in cell physiology and diseases. In recent years, single particle tracking of resident raft molecules has been used to analyze the dynamic behavior of lipid rafts (Kusumi et al., 2004). Single particle tracking of cholesterol using fluorescent PFO derivatives may provide useful information concerning the dynamic behavior of cholesterol-rich microdomains in cells. Several anti-cholesterol antibodies can also detect cholesterol in membrane domains. Among the available antibodies, those of the IgG type are considered to bind preferentially to lipid rafts/caveolae in the plasma membrane, while IgM-type antibodies seem to bind to membrane domains other than lipid rafts. Further characterization of their binding specificities and the use of Fc antibody fragments should provide useful structural probes for heterogeneously distributed cholesterol in membranes.

Acknowledgements Work in the authors' laboratory is supported in part by Grants-in-Aid for Science Research from the Japan Society for the Promotion of Science (to Y. O.-I. and to M. I.). We thank Dr. M.M. Dooley-Ohto for reading the manuscript.

References

Abrami, L., Fivaz, M., Kobayashi, T., Kinoshita, T., Parton, R.G., van der Goot, F.G., 2001, Cross-talk between caveolae and glycosylphosphatidylinositol-rich domains. *J. Biol. Chem.* **276**: 30729–30736.

Addadi, L., Geva, M., Kruth, H.S., 2003, Structural information about organized cholesterol domains from specific antibody recognition. Biochim. *Biophys.* Acta **1610**: 208–216.

Alonso, M.A., Millan, J., 2001, The role of lipid rafts in signalling and membrane trafficking in T lymphocytes. *J. Cell Sci.* **114**: 3957–3965.

Alouf, J.E., 2000, Cholesterol-binding cytolytic protein toxins. *Int. J. Med. Microbiol.* **290**: 351–356.

Alving, C.R., Swarz, G.M., Jr., Wassef, N.M., 1989, Naturally occurring autoantibodies to cholesterol in humans. *Biochem. Soc. Trans.* **17**: 637–639.

Anderson, R.G.W., 1998, The caveolae membrane system. *Annu. Rev. Biochem.* **67**: 199–225.

Aoki, T., Kogure, S., Kogo, H., Hayashi, M., Ohno-Iwashita, Y., Fujimoto, T., 2003, Sequestration of cross-linked membrane molecules to caveolae in two different pathways. *Acta Histochem. Cytochem.* **36**: 165–171.

Billington, S.J., Jost, B.H., Songer, J.G., 2000, Thiol-activated cytolysins: structure, function and role in pathogenesis. *FEMS Microbiol. Lett.* **182**: 197–205.

Bíró, A., Cervenak, L., Balogh, A., Lorincz, A., Uray, K., Horváth, A., Romics, L., Matkó, J., Füst, G., László, G., 2007, Novel anti-cholesterol monoclonal immunoglobulin G antibodies as probes and potential modulators of membrane raft-dependent immune functions. *J. Lipid Res.* **48**: 19–29.

Blanchette-Mackie, E.J., 2000, Intracellular cholesterol trafficking: role of the NPC1 protein. *Biochim. Biophys. Acta* **1486**: 171–183.

Brdicka, T., Pavlistova, D., Leo, A., Bruyns, E., Korinek, V., Angelisova, P., Scherer, J., Shevchenko, A., Hilgert, I., Cerny, J., Drbal, K., Kuramitsu, Y., Kornacker, B., Horejsi, V., Schraven, B., 2000, Phosphoprotein associated with glycosphingolipid-enriched microdomains (PAG), a novel ubiquitously expressed transmembrane adaptor protein, binds the protein tyrosine kinase csk and is involved in regulation of T cell activation. *J. Exp. Med.* **191**: 1591–1604.

Brooks-Wilson, A., Marcil, M., Clee, S.M., Zhang, L.H., Roomp, K., van Dam, M., Yu, L., Brewer, C., Collins, J.A., Molhuizen, H.O., Loubser, O., Ouelette, B.F., Fichter, K., Ashbourne-Excoffon, K.J., Sensen, C.W., Scherer, S., Mott, S., Denis, M., Martindale, D., Frohlich, J., Morgan, K., Koop, B., Pimstone, S., Kastelein, J.J., Genest, J. Jr, Hayden, M.R., 1999, Mutations in ABC1 in Tangier disease and familial high-density lipoprotein deficiency. *Nat. Genet.* **22**: 336–345.

Brown, D.A., London, E., 1997, Structure of detergent-resistant membrane domains: does phase separation occur in biological membranes? *Biochem. Biophys. Res. Commun.* **240**:1–7.

Brown, D.A., London, E., 2000, Structure and function of sphingolipid- and cholesterol-rich membrane rafts. *J. Biol. Chem.* **275**: 17221–17224.

Brown, M.S., Goldstein, J.L., 1986, A receptor-mediated pathway for cholesterol homeostasis. *Science* **232**: 34–47.

Christian, A.E., Haynes, M.P., Phillips, M.C., Rothblat, G.H., 1997, Use of cyclodextrins for manipulating cellular cholesterol content. *J. Lipid Res.* **38**: 2264–2272.

Davidson, D., Bakinowski, M., Thomas, M.L., Horejsi, V., Veillette, A., 2003, Phosphorylation-dependent regulation of T-cell activation by PAG/Cbp, a lipid raft-associated transmembrane adaptor. *Mol. Cell. Biol.* **23**: 2017–2028.

Dijkstra, J., Swartz, G.M., Jr., Raney, J.J., Aniagolu, J., Toro, L., Nacy, C.A., Green, S.J., 1996, Interaction of anti-cholesterol antibodies with human lipoproteins. *J. Immunol.* **157**: 2006–2013.

Edidin, M., 2001, Shrinking patches and slippery rafts: scales of domains in the plasma membrane. *Trends Cell Biol.* **11**: 492–496.

Fielding, C.J., Fielding, P.E., 2003, Relationship between cholesterol trafficking and signaling in rafts and caveolae. *Biochim. Biophys. Acta* **1610**: 219–228.

Fivaz, M., Abrami, L., van der Goot, F.G., 1999, Landing on lipid rafts. *Trends. Cell Biol.* **9**: 212–213.

Frey, D., Laux, T., Xu, L., Schneider, C., Caroni, P., 2000, Shared and unique roles of CAP23 and GAP43 in actin regulation, neurite outgrowth, and anatomical plasticity. *J. Cell Biol.* **149**: 1443–1454.

Fujimoto, K., Umeda M., Fujimoto, T., 1996, Transmembrane phospholipid distribution revealed by freeze-fracture replica labeling. *J. Cell Sci.* **109**: 2453–2460.

Fujimoto, T., 1996, GPI-anchored proteins, glycosphingolipids, and sphingomyelin are sequestered to caveolae only after crosslinking. *J. Histochem. Cytochem.* **44**: 929–941.

Fujimoto, T., Hayashi, M., Iwamoto, M., Ohno-Iwashita, Y., 1997, Crosslinked plasmalemmal cholesterol is sequestered to caveolae: analysis with a new cytochemical probe. *J. Histochem. Cytochem.* **45**: 1197–1205.

Gilbert, R.J., 2002, Pore-forming toxins. *Cell. Mol. Life Sci.* **59**: 832–844.

Gombos, I., Steinbach, G., Pomozi, I., Balogh, A., Vámosi, G., Gansen, A., László, G., Garab, G., Matkó, J., 2008, Some new faces of membrane microdomains: a complex confocal fluorescence, differential polarization, and FCS imaging study on live immune cells. *Cytometry A* **73**: 220–229.

Gomez-Mouton, C., Abad, J.L., Mira, E., Lacalle, R.A., Gallardo, E., Jimenez-Baranda, S., Illa, I., Bernad, A., Manes, S., Martinez-A., C., 2001, Segregation of leading-edge and uropod components into specific lipid rafts during T cell polarization. *Proc. Natl. Acad. Sci. USA* **98**: 9642–9647.

Hao, M., Lin, S.X., Karylowski, O.J., Wustner, D., McGraw, T.E., Maxfield, F.R., 2002, Vesicular and non-vesicular sterol transport in living cells. The endocytic recycling compartment is a major sterol storage organelle. *J. Biol. Chem.* **277**: 609–617.

Harder, T., Scheiffele, P., Verkade, P., Simons, K., 1998, Lipid domain structure of the plasma membrane revealed by patching of membrane components. *J. Cell Biol.* **141**: 929–942.

Hayashi, M., Shimada, Y., Inomata, M., Ohno-Iwashita, Y., 2006, Detection of cholesterol-rich microdomains in the inner leaflet of the plasma membrane. *Biochem. Biophys. Res. Commun.* **351**: 713–718.

Heijnen, H.F.G., Van Lier, M., Waaijenborg, S., Ohno-Iwashita, Y., Waheed, A.A., Inomata, M., Gorter, G., Möebius, W., Akkerman, J.W.N., Slot, J.W., 2003, Concentration of rafts in platelet filopodia correlates with recruitment of c-Src and CD63 to these domains. *J. Thromb. Haemost.* **1**:1161–1173.

Hooper, N.M., 1999, Detergent-insoluble glycosphingolipid/cholesterol-rich membrane domains, lipid rafts and caveolae. *Mol. Membr. Biol.* **16**: 145–156.

Igbavboa, U., Avdulov, N. A., Schroeder, F., Wood, W.G., 1996, Increasing age alters transbilayer fluidity and cholesterol asymmetry in synaptic plasma membranes of mice. *J. Neurochem.*, **66**: 1717–1725.

Ilangumaran, S., Hoessli, D.C., 1998, Effects of cholesterol depletion by cyclodextrin on the sphingolipid microdomains of the plasma membrane. *Biochem. J.* **335**: 433–440.

Inomata, M., Shimada, Y., Hayashi, M., Kondo, H., Ohno-Iwashita, Y., 2006, Detachment-associated changes in lipid rafts of senescent human fibroblasts. *Biochem. Biophys. Res. Commun.* **343**: 489–495.

Inomata, M., Shimada, Y., Hayashi, M., Shimizu, J., Ohno-Iwashita, Y., 2007, Impairment in a negative regulatory system for TCR signaling in CD4$^+$ T cells from old mice. *FEBS Lett.* **581**: 3039–3043.

Ishii, H., Mori, T., Shiratsuchi, A., Nakai, Y., Shimada, Y., Ohno-Iwashita, Y., Nakanishi, Y., 2005, Distinct localization of lipid rafts and externalized phosphatidylserine at the surface of apoptotic cells. *Biochem. Biophys. Res. Commun.* **327**: 94–99.

Iwamoto, M., Ohno-Iwashita, Y., Ando S., 1990, Effect of isolated C-terminal fragment of theta-toxin (perfringolysin O) on toxin assembly and membrane lysis. *Eur. J. Biochem.* **194**: 25–31.

Iwamoto, M., Nakamura, M., Mitsui, K., Ando, S., Ohno-Iwashita, Y., 1993, Membrane disorganization induced by perfringolysin O (theta-toxin) of *Clostridium perfringens* – effect of toxin binding and self-assembly on liposomes. *Biochim. Biophys. Acta* **1153**: 89–96.

Iwamoto, M., Morita, I., Fukuda, M., Murota, S., Ando, S., Ohno-Iwashita, Y., 1997, A biotinylated perfringolysin O derivative: a new probe for detection of cell surface cholesterol. *Biochim. Biophys. Acta* **1327**: 222–230.

Jacobs, T., Cima-Cabal, M.D., Darji, A., Mendez, F.J., Vazquez, F., Jacobs, A.A.C., Shimada, Y., Ohno-Iwashita, Y., Weiss, S., de los Toyos, J.R., 1999, The conserved undecapeptide shared by thiol-activated cytolysins is involved in membrane binding. *FEBS Lett.* **459**: 463–466.

Kessler, N., Perl-Treves, D., Addadi, L., 1996, Monoclonal antibodies that specifically recognize crystals of dinitrobenzene. *FASEB J.* **10**: 1435–1442.

Kobayashi, T., Stang, E., Fang, K.S., de Moerloose, P., Parton, R.G., Gruenberg, J., 1998, A lipid associated with the antiphospholipid syndrome regulates endosome structure and function. *Nature* **392**: 193–197.

Kokubo, H., Helms, J.B., Ohno-Iwashita, Y., Shimada, Y., Horikoshi, Y., Yamaguchi, H., 2003, Ultrastructural localization of flotillin-1 to cholesterol-rich membrane microdomains, rafts, in rat brain tissue. *Brain Res.* **965**: 83–90.

Koseki, M., Hirano, K., Masuda, D., Ikegami, C., Tanaka, M., Ota, A., Sandoval, J.C., Nakagawa-Toyama, Y., Sato, S.B., Kobayashi, T., Shimada Y., Ohno-Iwashita Y., Matsuura, F., Shimomura, I., Yamashita, S., 2007, Increased lipid rafts and accelerated lipopolysaccharide-induced tumor necrosis factor-alpha secretion in Abca1-deficient macrophages. *J. Lipid Res.* **48**: 299–306.

Kruth, H.S., Ifrim, I., Chang, J., Addadi, L., Perl-Treves, D., Zhang, W.Y., 2001, Monoclonal antibody detection of plasma membrane cholesterol microdomains responsive to cholesterol trafficking. *J. Lipid Res.* **42**: 1492–1500.

Kusumi, A, Koyama-Honda, I, Suzuki, K., 2004, Molecular dynamics and interactions for creation of stimulation-induced stabilized rafts rom small unstable steady-state rafts. *Traffic* **5**: 213–230.

Landry, Y.D., Denis, M., Nandi, S., Bell, S., Vaughan, A.M., Zha, X., 2006, ATP-binding cassette transporter A1 expression disrupts raft membrane microdomains through its ATPase-related functions. *J. Biol. Chem.* **281**: 36091–36101.

Larbi, A., Douziech, N., Dupuis, G., Khalil, A., Pelletier, H., Guerard, K.P., Fülöp, T. Jr., 2004, Age-associated alterations in the recruitment of signal-transduction proteins to lipid rafts in human T lymphocytes. *J. Leukoc. Biol.* **75**: 373–381.

London, E., Brown, D.A., 2000, Insolubility of lipids in triton X-100: physical origin and relationship to sphingolipid/cholesterol membrane domains (rafts). *Biochim. Biophys. Acta* **1508**:182–195.

Madore, N., Smith, K.L., Graham, C.H., Jen, A., Brady, K., Hall, S., Morris, R., 1999, Functionally different GPI proteins are organized in different domains on the neuronal surface. *EMBO J.* **18**: 6917–6926.

Maekawa, S., Maekawa, M., Hattori, S., Nakamura, S., 1993, Purification and molecular cloning of a novel acidic calmodulin binding protein from rat brain. *J. Biol. Chem.* **268**: 13703–13709.

Maekawa, S., Sato, C., Kitajima, K., Funatsu, N., Kumanogoh, H., Sokawa, Y., 1999, Cholesterol-dependent localization of NAP-22 on a neuronal membrane microdomain (raft). *J. Biol. Chem.* **274**: 21369–21374.

Maekawa, S., Iino, S., Miyata, S., 2003, Molecular characterization of the detergent-insoluble cholesterol-rich membrane microdomain (raft) of the central nervous system. *Biochim. Biophys. Acta* **1610**: 261–270.

Marwali, M.R., Rey-ladino, J., Dreolini, L., Shaw, D., Takei, F., 2003, Membrane cholesterol regulates LFA-1 function and lipid raft heterogeneity. *Blood* **102**: 215–222.

Mayor, S, Rao, M., 2004, Rafts: scale-dependent, active lipid organization at the cell surface. *Traffic* **5**:231–240.

Mayor, S., Rothberg, K.G., Maxfield, F.R., 1994, Sequestration of GPI-anchored proteins in caveolae triggered by cross-linking. *Science* **264**: 1948–1951.

Miller, R.A., 2000, Effect of aging on T lymphocyte activation. *Vaccine* **18**: 1654–1660.

Miller, R.G., 1984, The use and abuse of filipin to localize cholesterol in membranes. *Cell Biol. Int. Rep.* **8**: 519–535.

Möbius, W., Ohno-Iwashita, Y., van Donselaar, E.G., Oorschot, V.M.J., Shimada, Y., Fujimoto, T., Heijnen, H.F.G., Geuze, H.J., Slot, J.W., 2002, Immunoelectron microscopic localization of cholesterol using biotinylated and non-cytolytic perfringolysin O. *J. Histochem. Cytochem.* **50**: 43–55.

Möbius, W., van Donselaar, E., Ohno-Iwashita, Y., Shimada, Y., Heijnen, H.F.G., Slot, J.W., Geuze, H.J., 2003, Recycling compartments and the internal vesicles of multivesicular bodies harbor most of the cholesterol found in the endocytic pathway. *Traffic* **4**: 222–231.

Montixi, C., Langlet, C., Bernard, A.M., Thimonier, J., Dubois, C., Wurbel, M.A., Chauvin, J.P., Pierres, M., He, H.T., 1998, Engagement of T cell receptor triggers its recruitment to low-density detergent-insoluble membrane domains. *EMBO J.* **17**: 5334–5348.

Murate, M., Shimada, Y., Ohno-Iwashita, Y., Umeda, M., Kobayashi, T., 2008, Revisiting lipid asymmetry in red blood cells. *Flippases 2008 Abstract Book*: 15.

Nagafuku, M., Kabayama, K., Oka, D., Kato, A., Tani-ichi, S., Shimada, Y., Ohno-Iwashita, Y., Yamasaki, S., Saito, T., Iwabuchi, K., Hamaoka, T., Inokuchi, J., Kosugi, A., 2003, Reduction of glycosphingolipid levels in lipid rafts affects the expression state and function of glycosylphosphatidylinositol-anchored proteins but does not impair signal transduction via the T cell receptor. *J. Biol. Chem.* **278**: 51920–51927.

Nakamura, M., Sekino, N., Iwamoto, M., Ohno-Iwashita, Y., 1995, Interaction of theta-toxin (perfringolysin O), a cholesterol-binding cytolysin, with liposomal membranes: change in the aromatic side chains upon binding and insertion. *Biochemistry* **34**: 6513–6520.

Nakamura, M., Sekino-Suzuki, N., Mitsui, K., Ohno-Iwashita, Y., 1998, Contribution of tryptophan residues to the structural changes in perfringolysin O during interaction with liposomal membranes. *J. Biochem. (Tokyo)* **123**: 1145–1155.

Nakamura, M., Kondo, H., Shimada, Y., Waheed, A.A., Ohno-Iwashita, Y., 2003, Cellular aging-dependent decrease in cholesterol in membrane microdomains of human diploid fibroblasts. *Exp. Cell Res.* **290**: 381–390.

Ohno-Iwashita, Y., Iwamoto, M., Mitsui, K., Kawasaki, H., Ando, S., 1986, Cold-labile hemolysin produced by limited proteolysis of theta-toxin from *Clostridium perfringens*. *Biochemistry* **25**: 6048–6053.

Ohno-Iwashita, Y., Iwamoto, M., Mitsui, K., Ando, S., Nagai, Y., 1988, Protease-nicked θ-toxin of *Clostridium perfringens*, a new membrane probe with no cytolytic effect, reveals two classes of cholesterol as toxin-binding sites on sheep erythrocytes. *Eur. J. Biochem.* **176**: 95–101.

Ohno-Iwashita, Y., Iwamoto, M., Ando, S., Mitsui, K., Iwashita, S., 1990, A modified theta-toxin produced by limited proteolysis and methylation: a probe for the functional study of membrane cholesterol. *Biochim. Biophys. Acta* **1023**: 441–448.

Ohno-Iwashita, Y., Iwamoto, M., Mitsui, K., Ando, S., Iwashita, S., 1991, A cytolysin, θ-toxin, preferentially binds to membrane cholesterol surrounded by phospholipids with 18-carbon hydrocarbon chains in cholesterol-rich region. *J. Biochem. (Tokyo)* **110**: 369–375.

Ohno-Iwashita, Y., Iwamoto, M., Ando, S., Iwashita, S., 1992, Effect of lipidic factors on membrane cholesterol topology – mode of binding of theta-toxin to cholesterol in liposomes. *Biochim. Biophys. Acta* **1109**: 81–90.

Ohno-Iwashita, Y., Shimada, Y., Waheed, A.A., Hayashi, M., Inomata, M., Nakamura, M., Maruya, M., Iwashita, S., 2004, Perfringolysin O, a cholesterol-binding cytolysin, as a probe for lipid rafts. *Anaerobe* **10**: 125–134.

Ohsaki, Y., Sugimoto, Y., Suzuki, M., Kaidoh, T., Shimada, Y., Ohno-Iwashita, Y., Davies, J.P., Ioannou, Y.A., Ohno, K., Ninomiya, H., 2004, Reduced sensitivity of Niemann-Pick C1-deficient cells to θ-toxin (perfringolysin O): sequestration of toxin to raft-enriched membrane vesicles. *Histochem. Cell Biol.* **121**: 263–272.

Ostermeyer, A.G., Beckrich, B.T., Ivarson, K.A., Grove, K.E., Brown, D.A., 1999, Glycosphingolipids are not essential for formation of detergent-resistant membrane rafts in melanoma cells. methyl-beta-cyclodextrin does not affect cell surface transport of a GPI-anchored protein. *J. Biol. Chem.* **274**: 34459–34466.

Palmer, M., 2001, The family of thiol-activated, cholesterol-binding cytolysins. *Toxicon* **39**: 1681–1689.

Parton, R.G., 1994, Ultrastructural localization of gangliosides; GM1 is concentrated in caveolae. *J. Histochem. Cytochem.* **42**: 155–166.

Pawelec, G., Hirokawa, K., Fülöp, T. Jr., 2001, Altered T cell signalling in ageing. *Mech. Ageing Dev.* **122**: 1613–1637.

Perl-Treves, D., Kessler N., Izhaky, D., Addadi, L., 1996, Monoclonal antibody recognition of cholesterol monohydrate crystal faces. *Chem. Biol.* **3**: 567–577.

Pike, L.J., 2006, Rafts defined: a report on the Keystone Symposium on Lipid Rafts and Cell Function. *J. Lipid Res.* **47**:1597–1598.

Ramachandran, R., Heuck, A.P., Tweten, R.K., Johnson, A.E., 2002, Structural insights into the membrane-anchoring mechanism of a cholesterol-dependent cytolysin. *Nat. Struct. Biol.* **9**: 823–827.

Reid, P.C., Sakashita, N., Sugii, S., Ohno-Iwashita, Y., Shimada, Y., Hickey, W.F., Chang, T.Y., 2004, A novel cholesterol stain reveals early neuronal cholesterol accumulation in the Niemann-Pick type C1 mouse brain. *J. Lipid Res.* **45**: 582–591.

Reid, P.C., Lin, S., Vanier, M.T., Ohno-Iwashita, Y., Harwood, H.J., Jr., Hickey, W.F., Chang, C.C.Y., Chang, T.Y., 2008, Partial blockage of sterol biosynthesis with a squalene synthase inhibitor in early postnatal Niemann-Pick type C npc^{nih} null mice brains reduces neuronal cholesterol accumulation, abrogates astrogliosis, but may inhibit myelin maturation. *J. Neurosci. Methods* **168**: 15–25.

Roper, K., Corbeil, D., Huttner, W.B., 2000, Retention of prominin in microvilli reveals distinct cholesterol-based lipid micro-domains in the apical plasma membrane. *Nat. Cell Biol.* **2**: 582–592.

Rossjohn, J., Feil, S.C., McKinstry, W.J., Tweten, R.K., Parker, M.W., 1997, Structure of a cholesterol-binding, thiol-activated cytolysin and a model of its membrane form. *Cell* **89**: 685–692.

Sato, T., Zakaria, A.M., Uemura, S., Ishii, A., Ohno-Iwashita, Y., Igarashi, Y., Inokuchi. J., 2005, Role for up-regulated ganglioside biosynthesis and association of Src family kinases with microdomains in retinoic acid-induced differentiation of F9 embryonal carcinoma cells. *Glycobiology* **15**: 687–699.

Schoer, J.K., Gallegos, A.M., McIntosh, A.L., Starodub, O., Kier, A.B., Billheimer, J.T., Schroeder, F., 2000, Lysosomal membrane cholesterol dynamics. *Biochemistry* **39**: 7662–7677.

Schuck, S., Honsho, M., Ekroos, K., Shevchenko, A., Simons, K., 2003, Resistance of cell membranes to different detergents. Proc. Natl. Acad. Sci. USA **100**: 5795–5800.

Sekino-Suzuki, N., Nakamura, M., Mitsui, K., Ohno-Iwashita, Y., 1996, Contribution of individual tryptophan residues to the structure and activity of theta-toxin (perfringolysin O), a cholesterol-binding cytolysin. *Eur. J. Biochem.* **241**: 941–947.

Severs, N.J., 1997, Cholesterol cytochemistry in cell biology and disease. *Subcell. Biochem.* **28**: 477–505.

Shimada, Y., Nakamura, M., Naito, Y., Nomura, K., Ohno-Iwashita, Y., 1999, C-terminal amino acid residues are required for the folding and cholesterol binding property of perfringolysin O, a pore-forming cytolysin. *J. Biol. Chem.* **274**: 18536–18542.

Shimada, Y., Maruya, M., Iwashita, S., Ohno-Iwashita, Y., 2002, The C-terminal domain of perfringolysin O is an essential cholestereol-binding unit targeting to cholesterol-rich microdomains. *Eur. J. Biochem.* **269**: 6195–6203.

Shimada, Y., Inomata, M., Suzuki, H., Hayashi, M., Waheed, A.A., Ohno-Iwashita, Y., 2005, Separation of a cholesterol-enriched microdomain involved in T-cell signal transduction. *FEBS J.* **272**: 5454–5463.

Shogomori, H., Brown, D.A., 2003, Use of detergents to study membrane rafts: the good, the bad, and the ugly. *Biol. Chem.* **384**: 1259–1263.

Simons, K., Ikonen, E., 1997, Functional rafts in cell membranes. *Nature* **387**: 569–572.

Simons, K., Ikonen, E., 2000, How cells handle cholesterol. *Science* **290**: 1721–1726.

Simons, K., Toomre, D., 2000, Lipid rafts and signal transduction. *Nat. Rev. Mol. Cell. Biol.* **1**: 31–39.

Simons, K., Ehehalt, R., 2002, Cholesterol, lipid rafts, and disease. *J. Clin. Invest.* **110**: 597–603.

Stuermer, C.A., Lang, D.M., Kirsch, F., Wiechers, M., Deininger, S.O., Plattner, H., 2001, Glycosylphosphatidyl inositol-anchored proteins and fyn kinase assemble in noncaveolar plasma membrane microdomains defined by reggie-1 and -2. *Mol. Biol. Cell* **12**: 3031–3045.

Sugii, S., Reid, P.C., Ohgami, N., Shimada, Y., Maue, R.A., Ninomiya, H., Ohno-Iwashita, Y., Chang, T.Y., 2003, Biotinylated theta-toxin derivative as a probe to examine intracellular cholesterol-rich domains in normal and Niemann-Pick type C1 cells. *J. Lipid Res.* **44**: 1033–1041.

Swartz, G.M., Jr., Gentry, M.K., Amende, L.M., Blanchette-Mackie, E.J., Alving, C.R., 1988, Antibodies to cholesterol. *Proc. Natl. Acad. Sci. USA* **85:** 1902–1906.

Tamir, A., Eisenbraun, M.D., Garcia, G.G., Miller, R.A., 2000, Age-dependent alterations in the assembly of signal transduction complexes at the site of T cell/APC interaction. *J. Immunol.* **165**:1243–1251.

Tamura, T., Kunimatsu, T., Yee, S.T, Igarashi, O., Utsuyama, M., Tanaka, S., Miyazaki, S., Hirokawa, K., Nariuchi, H., 2000, Molecular mechanism of the impairment in activation signal transduction in CD4(+) T cells from old mice. *Int. Immunol.* **12**:1205–1215.

Tani-ichi, S., Maruyama, K., Kondo, N., Nagafuku, M.,Kabayama, K., Inokuchi, J., Shimada, Y., Ohno-Iwashita, Y., Yagita, H., Kawano, S., Kosugi, A., 2005, Structure and function of lipid rafts in human activated T cells. *Int. Immunol.* **17**: 749–758.

Tashiro, Y., Yamazaki, T., Shimada, Y., Ohno-Iwashita, Y., Okamoto, K., 2004, Axon-dominant localization of cell-surface cholesterol in cultured hippocampal neurons and its disappearance in Niemann-Pick type C model cells. *Eur. J. Neurosci.* **20**: 2015–2021.

Terashita, A., Funatsu, N., Umeda, M., Shimada, Y., Ohno-Iwashita, Y., Epand, R.M., Maekawa, S., 2002, Lipid binding activity of a neuron-specific protein NAP-22 studied in vivo and in vitro. *J. Neurosci. Res.* **70**: 172–179.

Tu, X., Huang, A., Bae, D., Slaughter, N., Whitelegge, J., Crother, T., Bickel, P.E., Nel, A., 2004, Proteome analysis of lipid rafts in Jurkat cells characterizes a raft subset that is involved in NF-kappaB activation. *J. Proteome Res.* **3**: 445–454.

Tweten, R.K., Parker, M.W., Johnson, A.E., 2001, The cholesterol-dependent cytolysins. *Curr. Top. Microbiol. Immunol.* **257**: 15–33.

Van Lier, M., Lee, F., Farndale, R.W., Gorter, G., Verhoef, S., Ohno-Iwashita , Y., Akkerman, J.W.N., Heijnen, H.F.G., 2005, Adhesive surface determines raft composition in platelets adhered under flow. *J. Thromb. Haemost.* **3**: 2514–2525.

Waheed, A.A., Shimada, Y., Heijnen, H.F.G., Nakamura, M., Inomata, M., Hayashi, M., Iwashita, S., Slot, J.W., Ohno-Iwashita, Y., 2001, Selective binding of perfringolysin O derivative to cholesterol-rich membrane microdomains (rafts*). Proc. Natl. Acad. Sci. USA* **98**: 4926–4931.

Xavier, R., Brennan, T., Li, Q., McCormack, C., Seed, B., 1998, Membrane compartmentation is required for efficient T cell activation. *Immunity* **8**: 723–732.

Yamaji, A., Sekizawa, Y., Emoto, K., Sakuraba, H., Inoue, K., Kobayashi, H., Umeda, M., 1998, Lysenin, a novel sphingomyelin-specific binding protein. *J. Biol. Chem.* **273**: 5300–5306.

Zhang, W., Trible, R.P., Samelson, L.E., 1998, LAT palmitoylation: its essential role in membrane microdomain targeting and tyrosine phosphorylation during T cell activation. *Immunity* **9**: 239–246.

Index

A
Acetylcholine receptor (AChR)
    cholesterol interactions, 469–470, 481–482
    endocytosis, 467, 473–477
    molecular dynamics simulations of, 482
    nanoclusters of, 467–468, 479–481
    stability of, 467, 469, 471
    trafficking, 467–469, 472
Actin, 96, 239, 241–243, 257, 400, 405, 410, 413, 474–477, 481, 531, 535, 537, 607
Acyl-CoA:cholesterol acyltransferase (ACAT), 1, 3, 16, 24, 28–29, 159, 161, 167, 213, 281, 289, 339, 346, 362, 496, 614
β2-adrenergic receptor, 19, 79, 82, 448–453, 456–457
Affinity labelling, 3, 30–32, 84–85, 407, 409
Agrin, 471, 475–476, 481
Allopregnanolone, 324, 326–328
Alphavirus
    E1 protein, 98
    fusion proteins, 98–99
Alzheimer's disease (AD), 47–65, 187, 211–212, 320, 325, 328, 387, 598
Amiloride-sensitive Na+ channels, 509, 527
Amylin, 57, 64–65
Amyloidogenic disorder, 47–65
Amyloid plaque, 59–60, 63
Amyloid precursor protein (APP), 48, 84, 160, 169, 257, 320
Amyloid-β (Aβ), 48, 53, 211–212
    Aβ fibrillogenesis, 47, 51–55
    Aβ oligomerization, 51, 53, 65
    Aβ pore formation, 47, 52–53
Ankyrin motifs, 166, 170
Anti-cholesterol antibodies, 597–615
Antidepressant, 391, 393
Antioxidant, 48, 51, 60, 234
Apical plasma membrane
    building units of, 407–409
    cholesterol-based microdomains, 401, 404–411, 415
    dynamics of, 401, 411–415
    ganglioside association with, 409–410
Apolipoprotein (apo)
    ABCA-1 interaction, 205–206
    apoA-1, 183, 185, 197–202, 205–213, 359–360
    apoB-100, 184, 233–234, 359
    apoE4 allele, 57, 60, 211
    apoE, 54, 57–60, 63, 120, 183, 185–195, 201–203, 205, 210–214, 234, 236–238, 240, 242, 320, 346, 354, 359, 497, 525
    discoidal particles, 194–195, 198, 201, 204–205, 211
    interaction with lipids, 191–197
    lipid solubilizing propterties of, 197–202
    primary and secondary structures of, 185–187
    quaternary structure of, 190–191
    spherical particles, 96, 194
    tertiary structure of, 187–190, 192, 196–197, 213, 430, 432–435, 498, 554
Astrocyte, 60, 62–64, 211, 323–324, 327–328, 393, 401, 495, 497
Astrocytosis, 50
Atherogenesis
    zebrafish model of, 243–244
Atherosclerosis, 47, 62–63, 120, 187, 203, 212, 230–233, 236–238, 240, 242–244, 284, 354–359, 517, 525, 598, 612
ATP-binding cassette
    ABCA1, 203
    transporters G5 and G8, 338, 350–356
Autophagosome, 326–327
Autophagy, 166, 325–328
Axon, 16, 320, 323, 325, 489–495, 499–503

B

Bacillus anthracis toxin complex
    PA63 pre-pore, 589–590
Bacteria
    and cholesterol oxidase, 137–139, 141, 150
    gram-positive CDCs, 15, 551–571, 579, 589
Bacterial toxin, 270, 551–571, 579–591, 606
β-barrel, 52–53, 159, 164, 167, 552, 554, 556–561, 564, 579–591
Bile acid
    -protein interactions, 115–120
    synthesis/biosynthesis of, 109–130, 287
Biliary cholesterol, 286, 301–302, 343, 354, 357
Biotinylated Cθ (BCθ), 1, 15–16, 279, 291, 296–297, 600–601, 603–613
Bodipy-cholesterol, 22–24, 27–28, 291
Brain
    cholesterol in, 49, 61, 63, 211, 323–324, 496–497
Brevibacterium sterolicum, 144
Brush border enzymes, 407
Brush border membrane (BBM), 337–363, 399–400, 403
β-sheet, 4, 15, 52–54, 64, 167, 557–560, 563, 586–587, 599–600
α-bungarotoxin (α-BTX), 468, 471, 473–476

C

Cancer, 172, 303, 387, 393, 401, 412, 415
Caveolae, 6–9, 15–16, 140, 255–256, 282, 286–287, 290–291, 298, 348, 382, 519–520, 524–528, 530–532, 535, 591, 604–605, 608, 615
Caveolin, 3, 6–9, 31–32, 63, 162, 255, 279–306, 348, 382, 388, 474, 498, 524, 526–529, 531–532, 535, 604, 606–608, 614
Cell surface, 14, 16, 18, 53, 86–87, 92, 119, 185, 204, 231, 243, 266, 297, 344, 348, 402, 407, 411, 467–468, 470, 472–477, 479–480, 497, 532, 537, 589–591, 603, 605, 607–609, 612, 614
Central nervous system (CNS), 83, 211, 322, 324, 392, 415, 473, 490–491, 494, 496–497, 499–503, 523
Ceramide, 37, 61, 168, 428
Ceramide transfer protein (CERT), 160, 166, 168–169, 428
Cerebrospinal fluid (CSF), 58, 60, 211, 412
Cholera toxin B, 409, 471, 475, 614
Cholestatrienol, 22–26, 449

Cholesterol
    accumulation, 16, 49, 60, 62–64, 129, 173, 229, 231, 238, 242, 244, 301–305, 322–323, 325–328, 612–613
    analogues, 3, 12–13, 22–33, 407, 446, 447, 449–450, 452, 601–602, 611
    binding to Aβ, 51–57
    binding motif, 1, 3, 19, 32, 79, 85, 270, 361, 389, 457
    binding protein, 18–21, 25, 27, 30–32, 77, 80–81, 280, 286, 301–302, 306, 357, 361, 414–415, 497–499, 529, 532, 598
    binding site, 3–5, 15–16, 19, 21, 31–33, 79–82, 84–85, 99, 341, 429–430, 439–440, 445, 449, 451, 453, 456–457, 482, 562, 591, 600
    binding studies, 99, 361
    -binding toxins, 597–615
    -binding virus proteins, 99
    biosynthesis, 128, 268–270, 319, 338–339, 341, 384, 451, 454, 456–457, 472, 493–494, 496–497, 499–500, 504
    chemical structure, 258, 442
    chiral analogues of, 514
    depletion, 10, 12, 51, 93, 97, 261, 327, 344, 407, 413, 446–447, 455–456, 467, 471–472, 475–477, 479–481, 499–503, 509, 513, 516–517, 519–521, 523–524, 526–537, 579, 590–591, 607
    diet, 50, 244, 303–304, 339, 343, 347, 413–414
    dynamics, 284–287, 298, 301
    efflux, 9, 60, 110, 120, 161, 202, 207, 211, 213, 237, 283–284, 286, 288, 321, 343, 359, 611–612
    fluorescent analogues, 22–29, 32–33, 295, 449, 611
    homeostasis, 49, 60–61, 63, 110, 120, 123, 125, 160, 172–173, 211–212, 268, 319–324, 339, 343, 361, 500, 612
    interaction with proteins, 253–271, 285
    intestinal absorption of, 351–352
    and ion channels, 509–538
    lowering, 51–52, 63, 268, 327, 339, 355, 496
    metabolism, 19, 48–51, 60, 62–63, 130, 169, 230, 288, 302, 306, 320, 323, 342, 355–356, 360, 493
    microcrystal, 54–58, 569, 582–585, 590
    and myelin biogenesis, 489–504
    in NPC disease, 320, 322–325, 327
    3β-OH group, 15, 55, 601
    oxidases

Index 625

    applications of, 140–142
    catalytic/enzyme mechanism of,
        146–150
    redox properties of, 142–143
    structure of, 143–146
  oxidation, 48, 60–62, 64, 140–141
  -PEG 600, 54–55
  photoreactive, 3, 30–32
  -poor microdomains, 281–285, 290–293,
      297–299
  probe/marker
      BCθ, 291–292, 607–608
  reduction, 413
  reporter molecules, 1–33
  reversible binding of, 425–435
  and sigma-1 receptor, 21, 381–394
  spin-labelled, 29–30
  START complexes, 2, 4, 11, 19–20, 160,
      162, 425–435
  structural probe for, 597–615
  synthesis, 50, 61, 63, 81, 111, 120,
      123, 127, 129, 160, 162, 167,
      175, 269, 305, 322–324, 327, 355,
      499–500, 502
  transport
      soluble binding proteins, 162
Cholesterol 24-hydroxylase (CYP 46A1), 3,
      122, 497
Cholesterol 7α-hydroxylase (CYP7A1),
      110–111, 113–115, 120, 129
Cholesterol consensus motif (CCM), 1, 4, 19,
      78, 82, 440, 450–451, 498
Cholesterol-dependent cytolysins (CDCs)
  conformational change, 551, 556–558, 580
  domain 4 (D4) of
      conserved loops, 559, 562–564
  homologues of, 562
  non-cytolytic derivatives of, 599
  oligomeric complex, 15, 552, 558–559
  pore formation, 4, 47, 52, 65, 87, 533–534,
      552–557, 564, 566, 570, 579–581, 587,
      589, 591
  pre-pore formation, 557–559, 561, 589
  primary sequence of, 559, 562
Cholesterol recognition amino acid consensus
      motif (CRAC), 2, 4, 19, 21, 77–78,
      81–91, 94–95, 99, 167, 260–264,
      266–268, 270–271, 361, 451, 498
  segment/consensus sequence/motif,
      260–261, 264, 268
Cholesterol reverse transport (RCT) proteins,
      110–111, 183–214, 230, 232, 240, 243,
      280, 285–286, 353, 359

Cholesterol-rich microdomains
  detection of
      in inner membrane leaflet, 609
Cholesteryl ester, 184, 209, 229–230, 232, 235,
      238, 241, 243, 261, 301, 357–358, 496
Cilium, 400, 402–405, 408, 410, 415
Clathrin, 6, 28, 344, 348–349, 467, 474–476
Clostridium perfringens ε-toxin, 590
Confocal imaging, 291–293, 297
Confocal laser scanning microscopy, 407, 409
Connexin, 491
Consensual model for cholesterol binding,
      434–435
Copper
  LDL oxidation by, 233–234
Coronary artery disease (CAD), 203, 229,
      231, 525
Cortical cytoskeleton, 481, 537, 607
Crystal structure, 3–4, 19, 61, 82, 118, 120,
      122–123, 143, 147, 149, 188, 196,
      425–428, 430, 433, 439, 448–451, 453,
      456–457, 554, 558, 570, 586, 600
Cyclodextrins (CDs)
  methyl-β-cyclodextrin (mβCD/M-β-
      CDx/βCD), 2, 9–12, 20, 24–25, 78,
      80, 98, 280, 292–294, 298–300, 352,
      407, 413–414, 440, 446–447, 454–455,
      468, 471, 476, 514, 520, 527–532, 537,
      589–591
Cytochrome P450, 16, 61, 81, 110, 113, 384
Cytolysin, 3–4, 98, 256, 296, 551–571,
      579–580, 583, 586, 588–589, 591, 598
Cytoskeleton, 161, 169, 171, 242, 257, 349,
      400, 405–406, 413, 477, 481, 531, 537,
      607

D

Dansyl-cholestanol, 23, 28–29, 280, 291–292
7-Dehydrocholesterol (7-DHC), 12, 440,
      589, 602
  reductase, 268, 497
Dehydroergosterol, 3, 12, 21–29, 280, 290–291
Dementia, 47–48, 62, 320
Demyelination, 63, 496
Detergent
  -based methods, 6–8, 286
  -free methods, 7–8
Detergent resistant membranes (DRM), 2, 7,
      78, 92, 99, 254, 260, 263, 280, 282,
      285, 400, 470, 471–472, 497–499, 597,
      604–606, 609–611
Diet
  cholesterol, 50, 244, 303–304, 339, 343,
      347, 413–414

Dimyristoyl PC (DMPC), 184, 198–201, 207
Domain 4 (D4)
  tryptophan-rich motif, 600
Down syndrome (DS), 60
Drug targets, 349–350, 444, 454, 458, 461
Dynamin, 6, 474, 476

E

Electrocyte, 472, 479
Electron microscopy
  analysis of microdomains, 609–610
  See also Transmission electron microscopy
Electron spin resonance (ESR) spectroscopy, 29, 448, 469–470, 482
Endoplasmic reticulum (ER)
  lipid rafts in, 382, 388
  and sigma-1 receptor, 386
Endosome, 14, 25, 28, 53, 161, 166, 168, 170, 172–173, 211, 289, 320–324, 427, 473, 476, 497–498, 589, 591, 610
Enhanced green fluorescent protein (EGFP), 15, 597, 601, 607, 609
Epithelial cell, 6, 27, 50, 352, 399–415, 516, 527, 571
Extracellular membrane vesicles, 400–401, 411–412
Ezetimibe, 289, 338–345, 347–349, 354–355, 357–358, 362
Ezrin-radixin-moesin (ERM) proteins, 404

F

Farnesoid X receptor (FXR), 110, 116–120, 122, 129–130
Fibrillogenesis, 51–55, 58, 62
Fibril(s)
  formation, 53–54, 56–57, 61, 65, 587–588
Filamentous virus particles, 96
Filipin, 13–16, 28, 33, 49, 64, 84, 173, 326–327, 387, 447, 591, 598, 603–604, 607, 612
Flavin adenine dinucleotide (FAD), 137–139, 142–145, 148, 150–152
Flotillin, 7–8, 383, 524, 604–608, 610
Fluorescence
  images, 295–296, 299–300, 479
  spectroscopy, 4, 570
Fluorescence lifetime imaging microscopy (FLIM), 256–257
Fluorescent sterols, 3, 22, 25–27, 33, 282, 290–291
Foam cell, 62, 238, 240, 242–244

Forster resonance energy transfer (FRET), 33, 117, 257, 280, 291–297, 440, 446, 449–450, 475
FPEG-cholesterol, 23, 28

G

Galanin receptor, 13–14, 447
Gangliosides
  GM1, 53, 92, 97, 256, 292–293, 388, 391–392, 408–410, 471, 481, 519, 604–607, 610
Gap junction, 491, 534
Glial cells, 58, 64, 211, 401, 492
Glioblastoma, 412
Glycosphingolipids, 263, 383, 391, 472, 491
Golgi apparatus (Golgi), 27, 165–171, 173, 210, 260, 326, 351, 382, 391, 467, 591
G-protein coupled receptors (GCPRs)
  and membrane cholesterol, 439–459
Gram-positive bacteria
  CDC homologues of, 562

H

α-helices, 59, 81–82, 86, 91, 185–190, 192, 194, 196, 198–199, 205, 213, 453, 560
α/β helix grip fold, 426, 428, 434–435
Hemagglutinin (HA), 78, 80, 90, 92–93, 95, 97, 259
Heme
  LDL oxidation by, 234–235
Hemolysin, 141, 569, 580, 586–589, 591
Hepatic cholesterol, 288, 301–305, 348
High density lipoprotein (HDL)
  and inflammation, 212–213
  remodeling of particles, 195–197
  and reverse cholesterol transport (RCT), 202–212
  See also Apolipoproteins
Hippocampal membrane, 455–456
HIV
  envelope glycoprotein, 85
  fusion protein gp41, 85–86
  gag protein, 81
  gp160, 81
  peptide studies, 88–90
  polypeptide constructs, 88
  pre-TM, 86–88
Huntington disease, 325
Hydrophobic cavity, 3–4, 9–10
Hydrophobic domain, 50, 57
Hydroxyl group, 10, 21, 28, 113–115, 117–118, 120, 122, 126, 128–129, 137, 139, 141, 145–146, 164, 196, 441–443, 451, 489, 513, 565, 567, 569, 613–614

Index 627

3-Hydroxy-3-methylglutaryl coenzyme A (HMG-CoA) reductase, 18, 51, 110, 112, 124–125, 127–129, 268–269, 321, 493–494, 496
  inhibitor, 51
Hypercholesterolemia, 51, 62, 130, 140, 230–232, 339, 355, 512, 517–518, 521, 524–525, 528–529, 612
Hypomyelination, 500–501

I
Immune and complement complexes, 242
Immuno-electron microscopy, 608–609, 612
Immunolabelling, 410
Inclusion body myositis (IBM), 50
Inflammation, 50, 63, 212, 235, 320, 328, 612
Influenza M2 protein
  cholesterol binding of, 82–85
  ion channel function of, 93
  M2 tetramer, 93
  maturation of, 92–93
  membrane fission, 97–98
  post-TM sequence, 94–95
Influenza virus
  filamentous particles, 96
  morphogenesis and budding, 96–97
Insig proteins, 110, 125, 127–129
Integral membrane proteins, 404–405, 444, 447–448, 475
Intermedilysin (ILY), 552, 555, 562
Intestinal absorption of cholesterol, 352
Intracellular cholesterol-binding proteins, 280, 286
Intracellular cholesterol trafficking, 279–306, 325, 327, 609
Ion channels
  association with lipid rafts
    and cholesterol, 517–518
  Ca+ channels
    N-type Ca2+ channels, 529
    voltage-gated Ca+2 (Cav) channels, 528
  Cl⁻ channels, 533–534
  K⁺ channels
    Ca2⁺-activated K+ channels, 521–523
    Inwardly rectifying K+ (Kir) channels, 512–513
    voltage-gated K+ (Kv) channels, 518–519
  mechanosensitive channels, 535–537
  Na⁺ channels
    epithelial Na+ channels (eNaC), 527
    voltage-gated Na+ (Nav) channels, 525–527

  transient receptor potential or TRP channels, 529–532
  volume-regulated anion channel (VRAC), 534–535
Iron, 51, 60, 234, 237, 407–408

L
Lecithin-cholesterol acyltransferase (LCAT), 184, 187, 195–197, 202–203, 210
Ligand-binding, 117–118, 120–121, 385–386, 390–391
Ligand-binding domain (LBD), 110, 116–118, 120–123, 165
Lipid
  belt, 469, 480, 510
  bilayer, 17, 52, 95, 126, 139, 199, 201, 205, 389, 413–415, 442, 450, 511, 523, 525–526, 536–538, 586
  cluster, 569
  domains, 22, 25, 27–28, 89, 256, 470–472, 475, 480
  monolayer, 27, 52, 61, 63
  phases, 17, 448, 469, 565
    perturbation by proteins, 258
  rafts
    reconstitution of, 391–392
Lipoprotein, 2, 25, 54–55, 58, 63, 110–111, 123, 160–161, 183–214, 229–244, 257, 280, 305, 319, 322, 337, 339, 352, 357–359, 447, 496–497, 502, 521, 580, 589
Liposomal membrane, 560–561, 564, 566, 568
Liposome, 3, 15, 29, 31, 89, 93, 98, 167, 173, 261, 321, 354, 358–359, 470, 527, 537, 560, 564, 569, 581–584, 587, 590, 603, 613–614
12/15-Lipoxygenase, 230, 235
Liver fatty acid-binding protein (L-FABP), 280–281, 287–289, 301–305
Liver X-activated receptor (LXR), 61–62, 110, 120–123, 129, 160, 166, 169–170, 172, 204, 230, 242, 288, 337, 342, 345–346, 353, 355, 360
Low density lipoprotein (LDL)
  aggregated, 231, 242–243
  in cholesterol transport, 238
  modified, 231–232
  non-oxidative enzymatic modifications, 238
  oxidation, 232–235, 242
Low density lipoprotein receptor (LDLR), 123, 184, 202, 230–234, 237–239, 243, 257, 325, 496, 502, 521

Lubrol WX, 7, 405–409, 413, 415
Lysosome, 14, 161, 289–290, 320–323, 325–326, 427, 498, 608, 610, 612

M

Macrophage
  lipoprotein uptake by, 232, 243
  uptake of modified LDL, 238–243
Macropinocytosis, 232, 238–239, 241–243
MDCK cells, 291, 402, 404–405, 407–411
Mechanosensitive channels, 535–537
Membrane
  -associated proteins, 114
  bilayer, 12, 16, 64, 85, 89, 201, 205, 207, 254, 400, 441, 443, 449, 453–454, 469, 489, 491, 526, 556, 558, 560, 565, 569, 586–587, 609
  cholesterol
    availability, 565–566
  curvature, 382, 403, 410, 413, 415
  domains, 2, 7, 60, 79, 85, 160, 205, 253–271, 411, 443, 511, 516–522, 524, 568, 598, 605, 615
  fluidity, 2, 11–13, 52, 54, 97, 475, 490, 499, 538
  pearling, 413
  perforation, 560–561
  pore formation, 52–53, 587
  protrusions, 399–401, 403–405, 407–411, 413, 415
  rafts, 6, 79–80, 92, 95, 97–99, 568
  vesicle, 340, 358, 361, 400–401, 411–415, 605
Metal ions, 48, 51–52, 60, 340
Mevalonate pathway, 111, 123
Mevinolin, 346, 372
Microdomain, 3, 255, 280, 284, 400, 405, 408–411, 413–415, 608
Microglia, 62–63, 211, 323–324, 328
Microvillus, 289, 349, 400, 408, 410, 413–414
Mitochondria, 19, 27, 140, 288–289, 326, 381–385, 387, 390, 426, 430
Mitochondria-associated ER membrane (MAM), 381, 384–392
Molten globule, 189, 429–430, 433–435
Monoclonal antibody, 78, 88, 96, 401, 411, 613–614
Mutagenesis, 114–115, 146–147, 187, 351, 452
Myasthenia gravis, 474
Myelin
  biogenesis of, 489–504
  cholesterol-binding proteins in, 497–499
  composition, 490–491
  lipids of, 491, 493
  P0 protein, 497–498, 501–503
  sheath, 489, 491, 498, 500, 503
  source of cholesterol, 495–497
  structure, 490–491
Myelin associated glycoprotein (MAG), 261, 491, 498–499, 503
Myelination, 328, 387, 492–496, 499, 501, 503
Myelin basic protein (MBP), 78, 88, 491, 502
Myeloperoxidase (MPO), 230, 235–237

N

NADPH Oxidases, 235, 237
NBD-cholesterol, 22–24, 26–27, 29, 280, 287, 302
Nerve terminals, 50
Neuraminidase (NA), 78, 80, 83, 90, 92, 94, 97, 283, 510, 521, 525–527
Neuregulin 1 (NRG1), 493
Neurofibrillary tangle, 64, 320, 323
Neuromuscular junction (NMJ), 468, 470–471, 473, 475, 477, 481
Neuron, 53, 489, 495, 520
Neuronal acidic protein of 22kDa (NAP-22), 83, 94, 607
Neuronal death, 325
Neuronal plasticity, 392
Neuropsychiatric disorders, 392–393, 454
Neurotoxicity, 50–52, 62
Nicotinic acetylcholine receptor, 5, 30–31, 447, 451–452, 467–482
  See also Acetylcholine receptor
Niemann-Pick C1-Like 1 (NPC1L1) protein, 321, 338–351, 354–355
Niemann-Pick C (NPC) disease
  autophagy in, 325–327
  cholesterol accumulation in, 322–323
  impairment of cholesterol transport in, 324–325
  suppression of brain cholesterol synthesis in, 323–324
  treatment for, 327–328
Nitric oxide synthases, 235, 237–238
Nodes of Ranvier, 490–491, 494
Non-annular cholesterol-binding sites, 445, 449–452, 456–457
Non-annular lipid, 447–449
Non-caveolar microdomains, 608
Nonidet P-40, 21
Nuclear receptors, 49, 61, 110–111, 116–118, 120–122, 129, 342, 345

O

Octyl glucoside, 526, 583
Oligodendrocyte
  cholesterol depletion in, 499–503
Oligodendrocyte precursor cells (OPCs), 494–495
Oligomerization, 51, 53, 64, 87, 190, 386, 449, 556–559, 568, 579–581, 583, 589–590, 599
Oxidation, 48, 50, 60–62, 64, 138–141, 146–150, 152, 231–237, 287–288, 302, 305, 454–455, 562, 613
Oxygen channel, 150–153
Oxysterol, 21, 61, 113, 118, 122, 126, 159–175, 204, 288–289, 328, 531
Oxysterol-binding proteins (OSBP)
  family members, 164–167
  ligand-binding domain of, 165
  organelle-specific targeting of, 165–166
  OSBP homology domain (OHD), 160, 164
  OSBP-related proteins (ORPs), 162
  phylogenetic distribution of, 162–164
  in sterol transport and signalling, 166–174
Oxysterol related proteins (ORP), 160, 163–166, 174–175, 288–289
Oxytocin and cholecystokinin (CCK) receptors, 440, 445–447

P

P0 protein, 497–498, 501–503
Palmitoylation, 91, 94–95, 410, 452–454
Parkinson's disease, 64, 325
Patocytosis, 242–243
Pattern recognition receptors (PRRs), 230, 239–242
PDZ-binding domain, 405
Pepstatin A, 57
Perfringolysin O (PFO), 1, 4, 15–16, 33, 256, 552–570, 597–604, 607, 609, 614
  non-cytolytic derivatives
    binding properties of, 601–606
Peripheral nervous system (PNS), 490–492, 498–499, 501–503
Phagocytosis
  of aggregated LDL, 242–243
Phase separation, 89, 414–415
Phosphatidylcholine (PC), 17, 19, 26–27, 31, 54, 78–79, 81, 164, 184, 193, 195–196, 198, 201, 209, 233, 254, 262, 265, 281, 286, 288, 346, 359, 426, 428, 502, 526, 561, 590, 601
Phosphatidylethanolamine (PE), 426, 526, 601

Phosphatidylserine (PS), 59, 164–165, 204, 230, 240, 285, 339, 382, 384, 601
Phospholipid (PL), 18, 25, 27, 54, 141, 171, 173, 184, 190, 196, 202, 213, 230, 262, 269, 283–285, 287, 343, 346, 441–443, 448, 469, 499, 516, 556, 560, 565–566, 569, 586, 589–591
Photoaffinity labelling, 85, 407, 409
Photoreactive sterols/cholesterol analogues, 3, 30–32
Photoreceptors, 26, 401, 403–404, 445
Placental alkaline phosphatase, 407–408
Plasma membrane (PM)
  of cultured cells, 298–300
  detection of cholesterol in inner leaflet, 608–609
Polyenes, 3, 13–14, 32–33, 141, 454, 603
Polyethylene glycol (PEG), 23, 28, 54–55, 57–58, 402, 410
Pore-forming toxins, 52, 557, 569, 591
Postsynaptic membrane, 468–470, 475
Potassium (K^+) channels
  inwardly rectifying K+ (Kir) channels, 512–513
  voltage-gated K+ (Kv) channels, 518–519
Pregnenolone, 61, 390, 427, 433
Probes for cholesterol, 3, 16, 18, 22, 24–25, 28–29, 31–33, 598
Prohibitin domain-containing (PHB) proteins, 381, 383
Prominin-1
  extracellular membrane vesicles, 400, 411–413
  membrane topology of, 341, 349, 384, 401–402
  and photoreceptors, 403–404
  and plasma membrane protrusions, 400–405, 407–415
Protein(s)
  colocalization with cholesterol, 259–270
  lipidated, 259–260
  toxin, 15, 601
Proteolipid proteins (PLP), 31, 491, 493, 498
Psychostimulant, 392
Purkinje cell, 323, 326–327

R

Radiolabelled cholesterol, 20–21, 358
Radioligand, 4, 5, 12, 19–21, 32
Rafts
  detection of, 256
  properties of, 256–257
Real-time multiphoton imaging, 290–291

Reverse cholesterol transport (RCT) proteins, 110, 183–213, 230, 232, 240, 243, 280, 285–286, 353, 359
Rhodopsin, 24–25, 444–446, 449–451, 453

S

SCAP protein (SREBP cleavage activating protein), 2, 110, 123–125, 160, 268, 271, 321, 341, 500
Scavenger receptor
    CD36, 239–240, 337, 356, 362
    SR-BI, 202, 203, 207–210, 213, 230, 240, 338–339, 356–362
Schwann cell
    cholesterol depletion in, 499–503
Schwann cell precursors (SCPs), 492–493
Secretase, 48–51, 54, 170
Seladin-1, 50
Sendai virus F-protein, 99
Senile plaque, 49–51
Serotonin$_{1A}$ (5-HT$_{1A}$) receptor
    cholesterol-binding motif, 457
Serum amyloid A (SAA), 184, 212–213
24S-hydroxycholesterol, 61
Sigma-1 receptor chaperone
    and cancer, 393
    ligand-binding profile of, 390–391
    localization of, 384–386
    molecular function of, 389–390
    structure of, 384–386
Signalling, 79, 116, 129, 162, 164–165, 169–171, 174, 241, 244, 327, 341, 353, 382, 386, 388–389, 391, 426, 470, 495, 511–512, 519, 530–531, 533–534, 604–605, 607–611
Signal transduction, 6, 140, 255–258, 382–388, 391–393, 400, 443, 604, 609–610
Smith-Lemli-Opitz syndrome (SLOS), 440, 454, 456, 496
Soluble cholesterol, 9, 19, 32, 54–55, 57, 286
Sonication, 7–8, 198
Sphingolipid, 6, 93, 161, 281, 322, 386, 388, 428, 472
Sphingomyelinase, 17, 141, 161, 238, 591
Sphingomyelin (SM), 13, 79–81, 89, 98, 160–161, 166–169, 173, 175, 209, 254–255, 257, 284–285, 455, 568, 582, 586, 590, 601, 607–608
Spin-label, 29–30, 469
SREBP cleavage activating protein (SCAP), 2, 110, 123–125, 160, 268, 271, 321, 341, 500–502
Staphylococcus aureus α-toxin, 569, 589

StAR-related lipid transfer (START) domain
    specific binding of cholesterol, 434–435
Statin, 51, 231, 338, 355–356
Steroid, 2, 18–19, 84, 113, 115, 117, 121, 137–140, 143–147, 149–151, 356–357, 359, 382, 441–442, 446, 601
Steroid acute regulatory protein (StAR), 2, 4, 19, 162, 288, 360, 425–435
Steroidogenesis, 19, 426–427, 433
Steroidogenic acute regulatory transport (START) proteins, 2, 4, 19–20, 160, 162–163, 288–289, 303, 350, 361, 425–435, 493–494
Sterol
    degradation of, 128
    domains of, 126, 128
    -sensing proteins, 123–128
    and sigma-1 receptor, 386–387
    transport, 166–174, 337–363
Sterol carrier protein-2 (SCP-2)
    overexpression and gene ablation, 301–302
Sterol regulatory element binding protein (SREBP), 2, 18, 110, 123–127, 129, 160–167, 169, 268–269, 288, 321, 338, 341, 346, 355, 360, 381–382, 494, 500–502
Stimulated emission depletion microscopy (STED), 257, 467–468, 477, 479
Streptococcus pyogenes, 15, 571, 598
Streptolysin O (SLO), 15–16, 235, 521, 552–553, 555, 559–561, 563, 566–569, 583–585, 598
Streptomyces sp., 144, 148
Structural studies, 53–57, 96, 147, 151
Sucrose density gradient, 406, 604, 606
α-Synuclein, 64

T

Tangier disease (TD), 203, 598, 611–612
Tau fibrillization, 63–65
T-cell receptor signalling, 610–611
Toll-like receptors (TLRs), 241
Trafficking, 6, 8, 22, 28–29, 33, 49, 63, 79, 86, 109–110, 161–162, 165, 171, 173–174, 204, 211–212, 279–306, 320, 322, 324–325, 327, 340–341, 344, 351–352, 358, 362–363, 393, 400–401, 411, 443, 468–469, 472, 486, 498–499, 501, 503, 516, 532–533, 609, 614
Transbilayer coupling, 257–258
Transferrin-like iron-binding protein, 407–408
Transgenic mice, 50, 59, 63, 236, 346, 493
Transgenic model, 59, 122

Index 631

Transmembrane beta-barrel, 554, 556–557, 560, 564
Transmembrane pore, 556, 558–559, 561, 567, 571, 589, 599
Transmembrane segment/domain, 86, 96, 126, 128, 165, 172, 204, 208, 259, 261–262, 320–321, 340, 351, 356, 361, 384–386, 401–402, 405, 410–411, 415, 441–442, 444, 451–453, 457, 482, 497–498, 510, 512–514
Transmembrane β-hairpins (TMHs), 556, 558, 560–561, 568
Transmission electron microscope (TEM), 54–55, 582, 586
Triton X-100, 6–8, 19, 79, 255, 388, 400, 405–409, 413, 415, 472, 479, 526, 606, 614
Trophic factor, 382, 388–389, 391–394, 495
Tryptophan-rich motif, 600
Tubular membranes, 413
Two-state model, 430–434

V

Very low density lipoprotein (VLDL), 58–59, 184, 489, 197, 202, 208, 210, 230, 239, 243, 357, 613
Vesicle, 13, 24, 26, 28, 31, 49, 64, 78, 89, 93, 97, 123–126, 160–161, 165, 170, 184, 193, 198–201, 208, 269, 340–341, 354, 358, 361, 382, 384, 400–401, 411–415, 470, 472, 497, 502, 605, 609
Vibrio cholerae cytolysin (VCC)
  fibril formation by, 587–588
  oligomer of, 581–587
Vibrio vulnificus hemolysin/cytolysin (VVC), 588–589
Villin, 405
Viruses
  budding, 77, 80–81, 87, 92, 95–99
  fission, 79, 95, 97–98
  fusion, 79–80, 82–83, 85–90, 98–99
  lipidomics, 80–81
Voltage-gated K+ (Kv) channels, 518–519

X

X-ray crystallography, 79, 82, 84, 117, 193, 570

Z

Zinc, 51, 116, 360